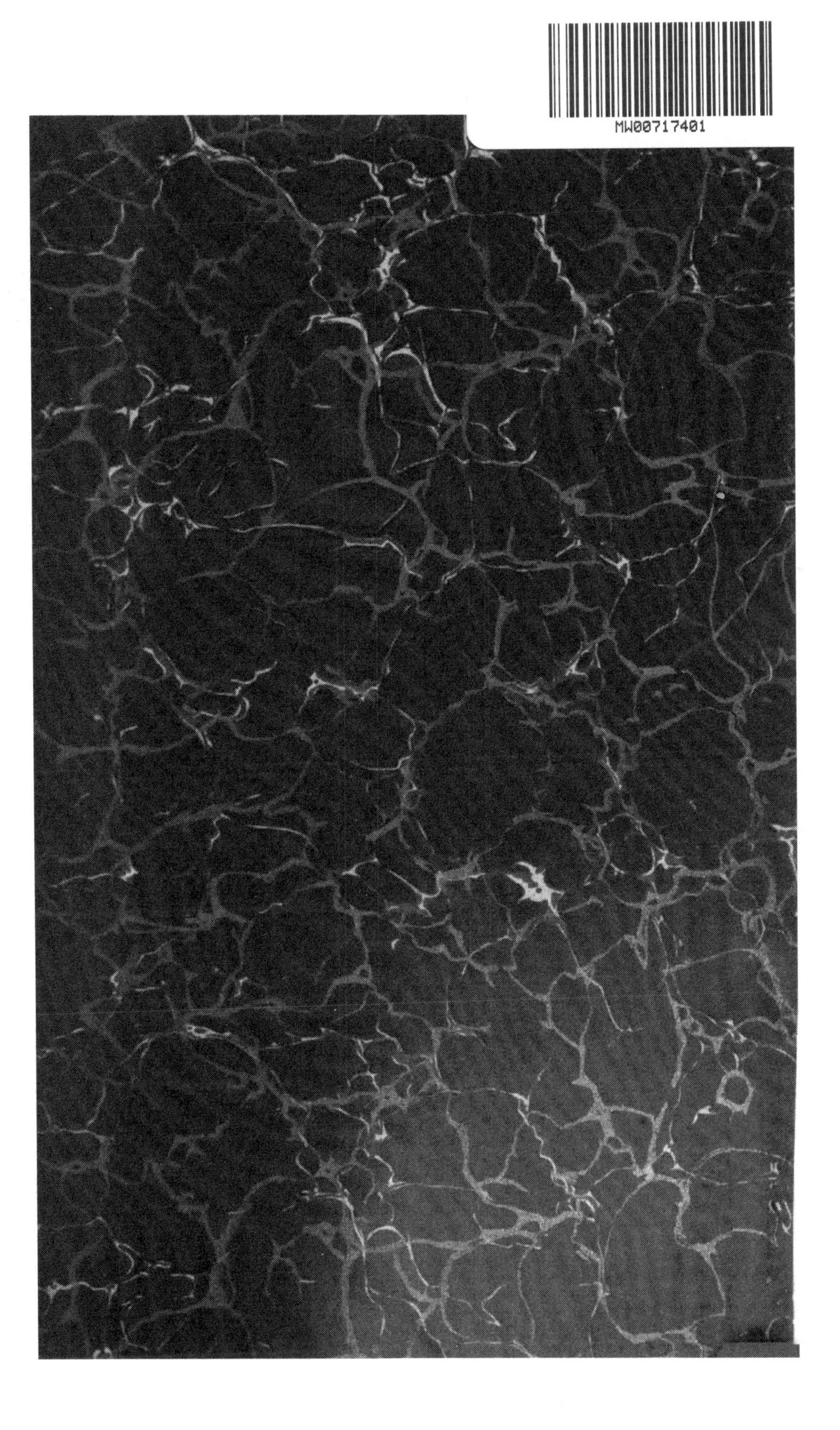

INTERNATIONAL LIBRARY OF TECHNOLOGY

A SERIES OF TEXTBOOKS FOR PERSONS ENGAGED IN THE ENGINEERING PROFESSIONS AND TRADES OR FOR THOSE WHO DESIRE INFORMATION CONCERNING THEM, FULLY ILLUSTRATED AND CONTAINING NUMEROUS PRACTICAL EXAMPLES AND THEIR SOLUTIONS

LOADS IN STRUCTURES
PROPERTIES OF SECTIONS
MATERIALS OF STRUCTURAL ENGINEERING
BEAMS AND GIRDERS
COLUMNS AND STRUTS
DETAILS OF CONSTRUCTION
GRAPHICAL ANALYSIS OF STRESSES

SCRANTON
INTERNATIONAL TEXTBOOK COMPANY
51

PRINTED IN THE UNITED STATES.

PREFACE

The International Library of Technology is the outgrowth of a large and increasing demand that has arisen for the Reference Libraries of the International Correspondence Schools on the part of those who are not students of the Schools. As the volumes composing this Library are all printed from the same plates used in printing the Reference Libraries above mentioned, a few words are necessary regarding the scope and purpose of the instruction imparted to the students of—and the class of students taught by—these Schools, in order to afford a clear understanding of their salient and unique features.

The only requirement for admission to any of the courses offered by the International Correspondence Schools, is that the applicant shall be able to read the English language and to write it sufficiently well to make his written answers to the questions asked him intelligible. Each course is complete in itself, and no textbooks are required other than those prepared by the Schools for the particular course selected. The students themselves are from every class, trade, and profession and from every country; they are, almost without exception, busily engaged in some vocation, and can spare but little time for study, and that usually outside of their regular working hours. The information desired is such as can be immediately applied in practice, so that the student may be enabled to exchange his present vocation for a more congenial one, or to rise to a higher level in the one he now pursues. Furthermore, he wishes to obtain a good working knowledge of the subjects treated in the shortest time and in the most direct manner possible.

In meeting these requirements, we have produced a set of books that in many respects, and particularly in the general plan followed, are absolutely unique. In the majority of subjects treated the knowledge of mathematics required is limited to the simplest principles of arithmetic and mensuration, and in no case is any greater knowledge of mathematics needed than the simplest elementary principles of algebra, geometry, and trigonometry, with a thorough, practical acquaintance with the use of the logarithmic table. To effect this result, derivations of rules and formulas are omitted, but thorough and complete instructions are given regarding how, when, and under what circumstances any particular rule, formula, or process should be applied; and whenever possible one or more examples, such as would be likely to arise in actual practice—together with their solutions—are given to illustrate and explain its application.

In preparing these textbooks, it has been our constant endeavor to view the matter from the student's standpoint, and to try and anticipate everything that would cause him trouble. The utmost pains have been taken to avoid and correct any and all ambiguous expressions—both those due to faulty rhetoric and those due to insufficiency of statement or explanation. As the best way to make a statement, explanation, or description clear, is to give a picture or a diagram in connection with it, illustrations have been used almost without limit. The illustrations have in all cases been adapted to the requirements of the text, and projections and sections or outline, partially shaded, or full-shaded perspectives, have been used, according to which will best produce the desired results. Half-tones have been used rather sparingly, except in those cases where the general effect is desired rather than the actual details.

It is obvious that books prepared along the lines mentioned must not only be clear and concise beyond anything heretofore attempted, but they must also possess unequaled value for reference purposes. They not only give the maximum of information in a minimum space, but this information is so ingeniously arranged and correlated, and the indexes are so full and complete, that it can at once be made available to the reader. The numerous examples and explanatory remarks, together with the absence of long demonstrations and abstruse mathematical calculations, are of great assistance in helping one to select the proper formula, method, or process and in teaching him how and when it should be used.

Two of the volumes composing this library, of which this is the first, are devoted to structural engineering. It has been our aim to present the fundamental laws that serve as a basis for structural engineering, in as clear and concise a manner as possible, thus assuring a thorough understanding of the succeeding and more difficult papers in which these laws are applied. The present volume deals mainly with the laws of stresses, that is, with the methods for ascertaining their magnitude, direction, and points of application, both the analytical and the graphical methods being used. Other papers, perhaps of equal importance, are those which treat of the means for distributing the stresses; they consider not only the material most suitable for the purpose, but also the most efficient and economical shape in which to arrange the material. It may be added that of the subjects here mentioned, some are perhaps treated in a more thorough and simple manner than is to be found in any other treatise; we refer in particular to the papers entitled Beams and Girders and Columns and Struts.

The method of numbering the pages, cuts, articles, etc. is such that each subject or part, when the subject is divided into two or more parts, is complete in itself; hence, in order to make the index intelligible, it was necessary to give each subject or part a number. This number is placed at the top of each page, on the headline, opposite the page number; and to distinguish it from the page number it is preceded by the printer's section mark (§). Consequently, a reference such as § 16, page 26, will be readily found by looking along the inside edges of the headlines until § 16 is found, and then through § 16 until page 26 is found.

INTERNATIONAL TEXTBOOK COMPANY.

CONTENTS

LOADS IN STRUCTURES

FLOOR, ROOF, AND WIND LOADS

DEAD LOAD

1. The weight of the material used in the permanent structure of a building produces loads on the floor systems, the columns, and the foundations. These loads are called the **dead loads** and include the weight of the structural framework, walls, floors, partitions, and roofs. In fact, the weight of every piece of material used in the construction of the building is included in the dead load.

Before the dead load can be computed, the weight of various materials must be known, and those in common use are given in the following tables. The units in which these weights are expressed are those most often employed in making estimates of loads in engineering calculations. Thus, Table I gives the weight, per cubic foot, of the materials usually measured by that unit, together with the weight, per cubic inch, of a few often measured in inches; while Table II gives the weights of such materials as are used in the construction of floors, roofs, ceilings, etc., where the quantities are generally expressed in square feet.

2. Weight of Fireproof Floors.—If fireproof floors are of standard construction, their weight may be determined from the weights given by the manufacturers of the particular type to be used; they may be found for most systems in the tables in *Fireproofing*. Where the fireproof-floor system is of special construction, that is, different from the standard commercial construction, a careful estimate of the dead load

TABLE I

WEIGHT OF MATERIALS

Name of Material	Average Weight	
	Pounds per Cu. In.	Pounds per Cu. Ft.
Aluminum	.096	166
Asphalt pavement composition		130
Bluestone		160
Brass	.302	523
Brickwork, in lime mortar		120
Brickwork, in cement mortar		130
Bronze	.319	552
Cement, Portland		80 to 100
Cement, Rosendale		56 to 60
Concrete, in cement		140
Copper, cast	.319	550
Earth, dry and loose		72 to 80
Earth, dry and moderately rammed		90 to 100
Gneiss, common		168
Gneiss, in loose piles		96
Granite		165 to 170
Gravel		117 to 125
Iron, cast	.260	450
Iron, wrought	.277	480
Lead, commercial cast	.412	712
Limestone		170
Marble		164
Masonry, granite or limestone		165
Masonry, granite or limestone rubble		150
Masonry, granite or limestone dry rubble		138
Masonry, sandstone		145
Mortar, hardened		90 to 100
Quartz, common pure		165
Sand, pure quartz, dry		90 to 106
Sandstone, building, dry		144 to 151
Slate		160 to 180
Snow, fresh fallen		5 to 12
Steel, structural	.283	490
Terra cotta		110
Terra-cotta masonry work		112
Tile		110 to 120

NOTE.—While it is not necessary for the student to memorize all of this table, it is well to keep in mind the weights of the materials printed in *Italics*.

TABLE II

WEIGHT OF MATERIALS

Name of Material	Average Weight per Square Foot Pounds
Corrugated galvanized iron No. 20, unboarded	2¼
Copper, 16-ounce, standing seam	1¼
Felt and asphalt, without sheathing	2
Glass, ⅛ inch thick	1¾
Hemlock sheathing, 1 inch thick	2½
Lead, about ⅛ inch thick	6 to 8
Lath-and-plaster ceiling (ordinary)	6 to 8
Mackite, 1 inch thick, with plaster	10
Neponset roofing felt, 2 layers	½
Spruce sheathing, 1 inch thick	2
Slate, 3/16 inch thick, 3 inches double lap	6¾
Slate, ⅛ inch thick, 3 inches double lap	4½
Shingles, 6 inches by 18 inches, one-third to weather	2
Skylight of glass, 3/16 inch to ¼ inch, including frame	4 to 10
Slag roof, 4-ply	4
Tin, IX	¾
Tiles, 10½ inches by 6¼ inches by ⅝ inch; 5¼ inches to weather (plain)	18
Tiles, 14½ inches by 10½ inches; 7¼ inches to weather (Spanish)	8½
White-pine sheathing, 1 inch thick	2½
Yellow-pine sheathing, 1 inch thick	4

per square foot of floor surface should be made. The volume of all materials that are measured by the cubic inch or cubic foot should be obtained by the rules and methods in *Geometry and Mensuration*, and the load obtained by multiplying by the unit weights of the materials found in Table I. The area covered by materials, such as flooring, sheathing, roof covering, etc., that are measured by the square foot, should be computed and multiplied by the weight, per square foot, given in Table II, to obtain the load.

In making calculations for the dead load of floors where the section through the floor shows irregularities in thickness and consequently in volume and weight, it is necessary to consider the section through a **panel**, which is the space between two floor beams. The section of the floor is taken

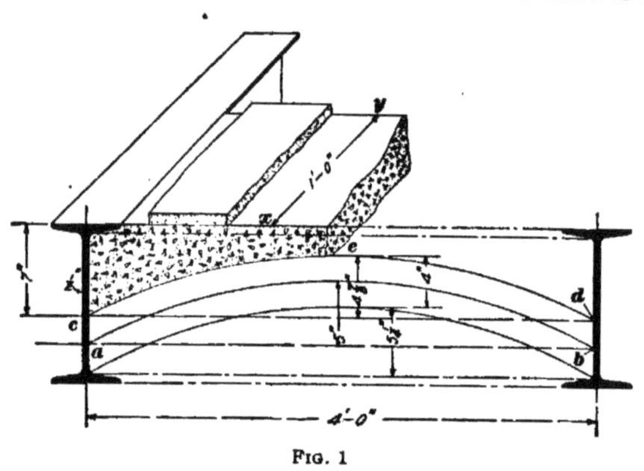

FIG. 1

1 foot in depth, as designated at xy, Fig. 1, so that when the entire weight of the section has been obtained the average weight per square foot can be found by dividing by the panel distance, or the distance between the floor beams.

EXAMPLE.—What is the amount of dead load per square foot of floor surface on the floor system shown in Fig. 1?

SOLUTION.—The sectional area of the brick arch is practically equal to the product of the length of the arc on the center line ab by the thickness of the arch, which in this instance is 4 in. The length of the chord of the arc ab is 47½ in., while the rise is 5 in. From these dimensions the length of the arc on the center line ab may be found by substituting in the formula $l = \frac{4\sqrt{c^2 + 4h^2} - c}{3}$, given in *Geometry and Mensuration*, in which c equals the chord and h the rise of the arc, the value of l is found to equal

$$\frac{4\sqrt{(47.5 \times 47.5) + (4 \times 5 \times 5)} - 47.5}{3} = 48.8883 \text{ in.}$$

Then the sectional area of the brick arch equals $\frac{48.8883 \times 4}{144} =$ 1.358 sq. ft. Since the calculation is for a portion of the floor system 1 ft. in depth or length, the area of the section of the arch also equals, numerically, the cubical contents, so that the weight of the brick arch 1 ft. wide is equal to 1.358 multiplied by 120, the weight per cubic foot of brickwork laid in lime mortar, obtained from Table I, or 162.96 lb.

The area of the section of the concrete is equal to the area of a rectangle, in this case 7 in. × 47½ in., from which must be deducted the area of the segment of the circle included between the arc $c\,e\,d$ and the chord $c\,d$. In order to obtain the area of this segment calculate the radius of the arc $c\,e\,d$ by applying the formula $r = \frac{c^2 + 4\,h^2}{8\,h}$, given in *Geometry and Mensuration*. The quantities c and h represent, as before, the chord and the rise and are equal, respectively, to 47.5 in. and 4.875 in. Substituting these values in the formula,

$$r = \frac{(47.5 \times 47.5) + (4 \times 4.875 \times 4.875)}{8 \times 4.875} = 60.29 \text{ in.}$$

The area of the segment is equal to the area of the sector minus the area of the triangle formed by the chord and the radii, or, as designated in Fig. 2, the area of the shaded portion is equal to the area $c\,e\,d\,o$ minus the area of the triangle $c\,d\,o$. The area of the sector may be found by the formula $a = \frac{l\,r}{2}$, given in *Geometry and Mensuration*, in which l equals the length of the arc and r the radius. The arc $c\,d$ has a smaller rise than that of arc $a\,b$ and will therefore be shorter. Its length, found by the formula just given, is 48.82 in. Inserting this value and that of the radius in the formula,

$$a = \frac{48.82 \times 60.29}{2} = 1{,}471.68 \text{ sq. in.}$$

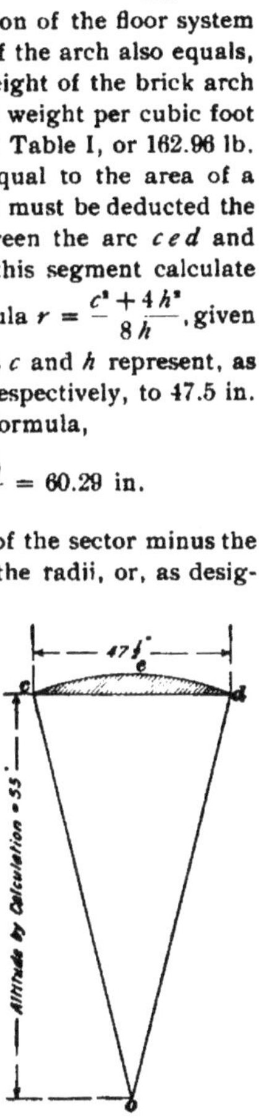

Fig. 2

The area of the triangle to be deducted from the sector is equal to one-half the product of the base and the altitude. From Figs. 1 and 2, the base equals 47.5 in. and the altitude equals the radius minus the rise of the arc, or $60.29 - 4.875 = 55.415$ in., and consequently the area of the triangle $c\,d\,o$, as designated in Fig. 2, is $\frac{47.5 \times 55.415}{2} = 1{,}316.1063$ sq. in. Since the area of the sector $c\,e\,d\,o$ equals 1,471.68 sq. in. and the triangle $c\,d\,o$ has an area of 1,316.1063 sq. in., the area of the segment $c\,e\,d$ equals the difference between these quantities, or 155.57 sq. in.; hence, the area of concrete is $(7 \times 47\frac{1}{4})$ − 155.57 = 176.93 sq. in. The weight, from Table I, is 140 lb. per cu. ft.; since the length of the concrete section is 1 ft., its weight equals $(176.93 \div 144) \times 140 = 172$ lb. The steel beam shown in Fig. 1 weighs 40 lb. per lineal ft. From these calculations, the entire weight of a panel section of the floor system for 1 ft. in length or per lineal foot may be itemized as follows:

Weight of brick arch	162.96 lb.
Weight of concrete	172.00 lb.
Weight of steel beam	40.00 lb.
Total	374.96 lb.

This amount is the dead load on 4 sq. ft., and hence the dead load per square foot is $374.96 \div 4 = 93.74$ lb. Ans.

3. Where the floor construction is of uniform weight and thickness throughout, as in mill construction, designated in Fig. 3, the calculations for the dead load can be made directly for 1 square foot of floor surface. The size of the girders or floor beams is seldom known before the dead load has been determined, so that it is necessary to assume their size

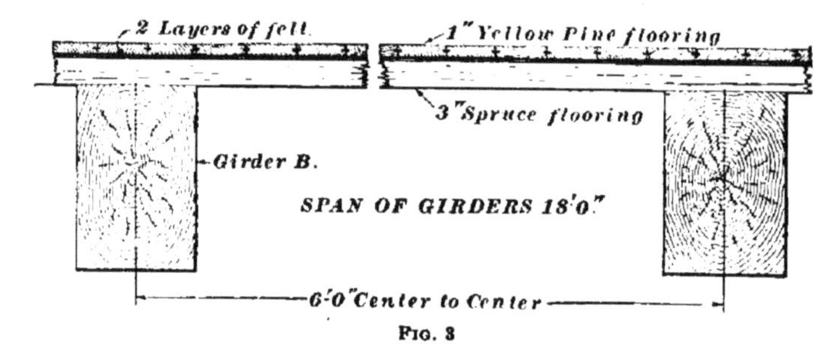

FIG. 3

and to add the weight of the assumed girders or beams in calculating the dead load. When considering the amount of dead weight supported by a beam or girder, it is customary to consider the same as made up of one-half the panel situated on either side of the beam. After the dead load has been found and the size of the girder accurately determined, the assumed weight can be checked by the actual weight.

EXAMPLE.—In Fig. 3, what is the total dead load on the girder *B*?

SOLUTION.—The weight of the materials per square foot may be obtained from Table II and be tabulated as follows:

Yellow-pine flooring, 1 in. thick	4 lb. per sq. ft.
Two layers of felt	½ lb. per sq. ft.
Rough spruce flooring, 3 in. thick	6 lb. per sq. ft.
Assume the weight of the girder	8 lb. per sq. ft.
Total dead load of floor surface	18½ lb. per sq. ft.

The area of the floor carried by the girder is 6 × 18 = 108 sq. ft. Then, 108 × 18½ = 1,998 lb., the entire dead load on the girder *B*. Ans.

4. Dead Load on Roof Trusses.—This includes the weight of the roof covering, sheathing, and the weight of the roof trusses. The weight of the roof covering and the sheathing may be calculated from the unit weights given in Table II. The weight of the roof trusses, or *principals*, as they are termed, is not known until they have been designed and must be assumed in the original calculation. The weight of roof trusses depends on the material of which they are constructed, the span, and the distance they are placed apart and also on the rise and the type of construction, though these two latter factors are neglected in the usual empirical formulas.

The approximate weight of wooden and iron or steel roof trusses may be determined by the following formula:

$$W = a D L \left(1 + \frac{L}{10}\right) \qquad (1)$$

in which W = approximate weight of truss, in pounds;
a = constant, for wood .50, for iron or steel .75;
D = distance, in feet, from center to center of trusses;
L = span of the truss, in feet.

This formula may be expressed as follows:

Rule.—*Multiply the constant for the material of which the truss is composed by the distance, in feet, from center to center of trusses by the span of the truss, in feet; the product of this result and 1 plus one-tenth of the span of the truss, in feet, is the approximate weight of the truss, in pounds.*

5. After the weight of the principals, or roof trusses, has been determined by formula **1,** it is usual, in order to find the panel loads, or weight created at the connections of the truss, to determine what weight per square foot of roof surface it is necessary to add to the weight of the covering in order to provide for the weight of the principals or trusses. This weight may be found by dividing W, as determined from formula **1,** by the actual area on the slope of the roof supported by one truss, or, it may be determined directly from the following formula, which is evolved from formula **1** by dividing by $D L$ sec x; this expression represents twice the length of the slope multiplied by the distance from center to center of trusses. The value w, or the approximate weight of the truss in pounds per square foot of roof surface, can be obtained from the following formula:

$$w = \frac{a\,(10 + L)}{10 \sec x} \qquad \textbf{(2)}$$

in which a and L have the same values as above and x equals the angle of rafter member with horizontal.

This may be stated in the form of a rule as follows:

Rule.—*Multiply the constant by 10 plus the span of the truss, in feet; divide this product by 10 times the secant of the angle that the rafter member makes with the horizontal, which gives the approximate weight of the truss, in pounds per square foot of roof surface.*

Example.—Determine the weight, per square foot, that it is necessary to add to the weight of the roof covering to provide for the weight of the principals, when the steel trusses have a span of 72 feet and a rise of 18 feet.

Solution.—The roof slope has a pitch of 6 in. for every foot horizontal and the angle x of the slope with the horizontal is found from Table VII to be 26° 33′. The sec x = hypotenuse ÷ adjacent side or sec $x = \frac{\sqrt{36^2 + 18^2}}{36} = 1.118$. The hypotenuse in this case is identical with the roof slope, the length of which is found by the formula: hypotenuse$^2 = \left(\frac{\text{span}}{2}\right)^2 +$ rise$^2 = 36^2 + 18^2$. The hypotenuse is therefore $\sqrt{36^2 + 18^2}$, which value has been inserted in the formula.

The secant x may also be found by means of the formula: $\sec x = \frac{1}{\cos x}$; $\cos 26° 33'$ being .8945, $\sec x = \frac{1}{.8945} = 1.118$. Substituting the values of a, L, and $\sec x$ in formula **2**.

$$w = \frac{.75(10 + 72)}{10 \times 1.118} = 5.5 \text{ lb. Ans.}$$

From formula **2**, the following table has been calculated. It gives the weight that it is necessary to add to a square

TABLE III

WEIGHT OF ROOF TRUSSES

$w = \frac{a(10 + L)}{10 \sec x}$ = Form. 2	Span Feet	Pounds per Square Foot of Roof Surface			
Character of Truss		½ Pitch	⅓ Pitch	¼ Pitch	⅙ Pitch
Wood	30	1.417	1.63	1.79	1.90
	35	1.588	1.87	2.01	2.13
	40	1.764	2.08	2.24	2.37
	45	1.941	2.29	2.46	2.61
	50	2.115	2.49	2.68	2.85
	55	2.293	2.70	2.91	3.08
	60	2.470	2.91	3.13	3.32
	65	2.646	3.12	3.35	3.56
	70	2.823	3.33	3.58	3.80
	75	2.999	3.54	3.80	4.03
	80	3.176	3.75	4.03	4.27
Iron or steel	30	2.117	2.45	2.69	2.85
	35	2.382	2.81	3.02	3.20
	40	2.647	3.12	3.35	3.56
	45	2.911	3.44	3.69	3.92
	50	3.176	3.74	4.02	4.28
	55	3.440	4.05	4.37	4.62
	60	3.705	4.37	4.70	4.98
	65	3.965	4.68	5.03	5.34
	70	4.235	5.00	5.37	5.70
	75	4.499	5.31	5.40	6.05
	80	4.764	5.63	6.05	6.41

foot of roof covering in order to provide, in the amount of the unit dead load, for the weight of the principals or trusses.

The term pitch used in Table III may have more than one interpretation; it may be the quotient $\frac{\text{rise}}{\text{span}}$ or $\frac{\text{rise}}{\frac{\text{span}}{2}}$.

Throughout this Course it will mean: $\frac{\text{rise}}{\text{span}}$. For instance, $\frac{1}{4}$ pitch means a rise one-fourth of the total span; $\frac{1}{6}$ pitch a rise one-sixth of the total span, etc.

EXAMPLES FOR PRACTICE

1. A 2″ × 3″ wrought-iron bar is 36⅛ inches long. What is its weight? Ans. 60.66 lb.

2. The outside diameter of a cast-iron column is 10 inches, and the thickness of the material composing the column is ¾ inch. What is its weight per foot of length? Ans. 68 lb.

3. The wall of a brick building, laid in cement mortar, is 24 inches thick, 36 feet high, and 100 feet long; in it are located 20 window openings, 2 feet 6 inches wide by 6 feet high. What is the weight of this wall? Ans. 858,000 lb.

4. What is the weight of a structural steel angle 6″ × 6″ × ½″ × 20′ long? Ans. 390.54 lb.

5. The roof of a building is made of No. 20 corrugated galvanized iron, laid on 1-inch spruce boarding. What is the weight of the roof covering per square foot? Ans. 4¼ lb.

6. What will be the difference in weight per square foot between a 4-ply slag roof, laid on 3-inch tongued-and-grooved yellow-pine planking, and a $\frac{3}{16}$-inch slate roof laid on 2-inch hemlock sheathing, covered with Neponset roofing felt, two layers thick? Ans. 3¾ lb.

7. The span of a steel roof truss is 40 feet, and its rise 10 feet. Referring to Table III, what weight per square foot of roof surface should be assumed so as to allow for the weight of the principal or roof truss? Ans. 3.35 lb.

LIVE LOAD

6. Besides the dead load, which includes the weight of all the material used in the structure itself, there is a load due to the weight of people, machinery, and merchandise; this load is called the **live load.** The live load comprises people in the building, furniture, movable stocks of goods, small safes, and varying weights of any character. Large safes and extremely heavy machinery require some special provision, usually embodied in the construction. Table IV gives the live loads per square foot, recommended as good practice in conservative building construction.

TABLE IV

Character of Building	Pounds
Dwellings	70
Offices	70
Hotels and apartment houses	70
Theaters	120
Churches	120
Ballrooms and drill halls	120
Factories	from 150 up
Warehouses	from 150 to 250 up

The load of 70 pounds will probably never be realized in dwellings; but inasmuch as a city house may, at times, be used for some purpose other than that of a dwelling, it is not generally advisable to use a lighter load. In the case of a country house, a hotel, or a building of like character, where economy demands it, and its actual use for a long time, for some fixed purpose, is almost certain, a live load of 40 pounds per square foot of floor surface is ample for all rooms not used for public assembly.

For rooms thus used, a live load of 80 pounds will be sufficient, experience having demonstrated that a floor cannot be crowded to more. If the desks and chairs are fixed, as in a schoolroom or church, a live load of more than 40 to 50 pounds will never be attained. Retail stores should have floors proportioned for a live load of 100 pounds and upwards. Wholesale stores, machine shops, etc., should have the floors proportioned for a live load of not less than 150 pounds per square foot. The floors of printing houses and binderies, especially where the accumulation of heavy stock, such as bound volumes and calendered paper, is likely to occur, should be proportioned for a live load of at least 250 pounds per square foot. Special provision should be made in floor systems for heavy presses, trimmers, and cutters and the beams should be proportioned for twice the static load likely to occur from such machines.

The static load in factories seldom exceeds 40 to 50 pounds per square foot of floor surface, and, therefore, in the majority of cases, a live load of 100 pounds, including the effects of vibrations due to moving machinery, is ample. The conservative rule is, in general, to assume loads not less than the above, and to proportion the beams so as to avoid excessive deflection. Stiffness is as important a factor as strength.

7. Warehouse Floors.—In the design of warehouse floors, the character of the material to be stored should always be considered and the data should be obtained regarding the manner of storing, the bulk of the packages, and the weight of the load per square foot. With the view of furnishing reliable data to manufacturers, architects, and engineers, the Boston Manufacturers' Mutual Fire Insurance Company has prepared from its extensive experience the following table, which gives the greatest possible loads that can be placed on warehouse floors, with the usual system of loading, and the space that the merchandise occupies. Where the floor space and the cubical contents of the load are given in the table, the height of the load above the finished floor may be obtained by dividing the volume of the load by the floor area covered. For instance, the floor space occupied by white linen rags in a bale is 8.5 square feet, and the cubical contents are 39.5 cubic feet; then the height of the loading is $39.5 \div 8.5 = 4.65$ feet. It is unusual and hardly possible, in the absence of hoists, to place such materials on the floor more than one bale in thickness, and the same thing applies to merchandise in barrels on the side and end. Where no data other than the weight per cubic foot is given in the table, it signifies that the possible height of the load is only limited by the height of the room, or the *headroom*, as it is called. With a live load of such merchandise, the floor system must be designed for the maximum load, which consists of the weight of the merchandise covering the entire floor area to a depth of not less than 6 feet.

The building ordinances of the principal American cities are particularly emphatic with reference to warehouse floors. For instance, the building laws of Greater New York stipulate that, in all warehouses, storehouses, factories, workshops, and stores where heavy materials are kept and stored, or machinery introduced, the weight that each floor will safely sustain upon each superficial foot, or upon each varying part of such floor, shall be estimated by the owner or occupant, or by a competent person employed by the owner or occupant.

Such estimate shall be reduced to writing or printed forms, stating the material, size, distance apart, and span of beams and girders, posts or columns to support floors, and its correctness shall be sworn to by the person making the same; it further being required that this estimate shall be filed in the office of the Department of Buildings.

But if the commissioners of buildings shall have cause to doubt the correctness of said estimate, they are empowered to revise and correct the same, and for the purpose of such revision the officers and employes of the Department of Buildings may enter any building and remove so much of any floor or other portion thereof as may be required to make necessary measurements and examination.

When the correct estimate of the weight that the floors in any such buildings will safely sustain has been ascertained as herein provided, the Department of Buildings shall approve the same. Thereupon the owner or occupant of said building, or any portion thereof, shall post a copy of such approved estimate in a conspicuous place on each story or varying parts of each story of the building to which it relates. No person shall place on any floor of any building any greater load than the safe load as correctly estimated and ascertained.

TABLE V

WEIGHTS OF MERCHANDISE FOR CALCULATING LIVE LOADS

Materials		Measurements		Approximate Weights		
		Floor Area Sq. Ft.	Contents Cu. Ft.	Total Pounds	Pounds per Sq. Ft.	Pounds per Cu. Ft.
Cotton, etc.	Bale	8.1	44.2	515	64	12
	Bale of compressed	4.1	21.6	550	134	25
	Bale of American Cotton Co.	4.0	11.0	263	66	24
	Bale of Planters Compress Co.	2.3	7.2	254	110	35
	Bale of jute	2.4	9.9	300	125	30
	Bale of jute lashings	2.6	10.5	450	172	43
	Bale of manila	3.2	10.9	280	88	26
	Bale of hemp	8.7	34.7	700	81	20
	Bale of sisal	5.3	17.0	400	75	24
Cotton goods	Bale of unbleached jeans	4.0	12.5	300	72	24
	Piece of duck	1.1	2.3	75	68	33
	Bale of brown sheetings	3.6	10.1	235	65	23
	Case of bleached sheetings	4.8	11.4	330	69	30
	Case of quilts	7.2	19.0	295	41	16
	Bale of print cloths	4.0	9.3	175	44	19
	Case of prints	4.5	13.4	420	93	31
	Bale of tickings	3.3	8.8	325	99	37
	Skein of cotton yarns					11
	Burlaps			130		30
	Jute bagging	1.4	5.3	100	70	24
Grain	Wheat in bags	4.2	4.2	165	39	39
	Wheat in bulk					44

Grain	Wheat in bulk					39
	Wheat in bulk, mean					41
	Flour in barrels on side	4.10	5.40	218	53	40
	Flour in barrels on end	3.10	7.10	218	70	31
	Corn in bags	3.60	3.60	112	31	31
	Corn meal in barrels	3.70	5.90	218	59	37
	Oats in bags	3.30	3.60	96	29	27
	Bale of hay	5.00	20.00	284	57	14
	Hay, Derrick compressed	1.75	5.25	125	72	24
	Straw	1.75	5.25	100	57	19
	Tow	1.75	5.25	150	87	29
	Excelsior	1.75	5.25	100	57	19
Paper	Calendered book					50
	Supercalendered book					69
	News paper					38
	Straw board					33
	Leather board					59
	Writing					64
	Wrapping					10
	Manila					37
Rags in bales	White linen	8.50	39.50	910	107	23
	White cotton	9.20	40.00	715	78	18
	Brown cotton	7.60	30.00	442	59	15
	Paper shavings	7.50	34.00	507	68	15
	Sacking	16.00	65.00	450	28	7
	Woolen	7.50	30.00	600	80	20
	Jute butts	2.80	11.00	400	143	36

TABLE V—(*Continued*)

Materials		Measurements		Approximate Weights		
		Floor Area Sq. Ft.	Contents Cu. Ft.	Total Pounds	Pounds per Sq. Ft.	Pounds per Cu. Ft.
Wool	Bale, East India	3.0	12	340	113	28
	Bale, Australian	5.8	26	385	66	15
	Bale, South American	7.0	34	1,000	143	29
	Bale, Oregon	6.9	33	482	70	15
	Bale, California	7.5	33	550	73	17
	Bag of wool	5.0	30	200	40	7
	Sack of scoured wool					5
Woolen goods	Case of flannels	5.5	12.7	220	40	17
	Case of flannels, heavy	7.1	15.2	330	46	22
	Case of dress goods	5.5	22	460	84	21
	Case of cassimeres	10.5	28	550	52	20
	Case of underwear	7.3	21	350	48	16
	Case of blankets	10.3	35	450	44	13
	Case of horse blankets	4.0	14	250	63	18
Miscellaneous	Box of tin	2.7	.5	139	99	278
	Box of glass					60
	Crate of crockery	9.9	39.6	1,600	162	40
	Cask of crockery	13.4	42.5	600	52	14
	Coal, anthracite, broken					54
	Coal, anthracite, moderately shaken					58
	Coal, anthracite, solid					93
	Coal, anthracite, heaped bushel					80
	Coal, bituminous, solid					84

Miscellaneous . .	Coal, bituminous, broken, loose . . .					54
	Coal, bituminous, heaped bushel . . .					74
	Coke, loose					30 to 50
	Bale of leather	7.3	12.2	190	26	16
	Bale of goat skin	11.2	16.7	300	27	18
	Bale of raw hides	6.0	30.0	400	67	13
	Bale of raw hides, compressed	6.0	30.0	700	117	23
	Bale of sole leather	12.6	8.9	200	22	16
	Pile of sole leather					17
	Barrel of granulated sugar	3.0	7.5	317	106	42
	Barrel of brown sugar	3.0	7.5	240	113	45
	Cheese					30
	Hogshead of bleaching powder	11.8	39.2	1,200	102	31
	Hogshead of soda ash	10.8	29.2	1,800	167	62
	Box of indigo	3.0	9.0	385	128	43
	Box of cutch	4.0	3.3	150	38	45
	Box of sumac	1.6	4.1	160	100	39
	Caustic soda in iron drum	4.3	6.8	600	140	88
	Barrel of starch	3.0	10.5	250	83	23
	Barrel of pearl alum	3.0	10.5	350	117	33
	Box of extract logwood	1.1	.8	55	52	70
	Barrel of lime	3.6	4.5	225	63	50
	Barrel of American cement	3.8	5.5	325	86	59
	Barrel of English cement	3.8	5.5	400	105	73
	Barrel of plaster	3.7	6.1	325	88	53
	Barrel of rosin	3.0	9.0	430	143	48
	Barrel of lard oil	4.3	12.3	422	98	34
	Rope					42

EXAMPLE.—What will be the entire live load coming on a large girder supporting a portion of a church floor, if the floor area to be supported is 600 square feet?

SOLUTION.—From the list given in Table IV, 120 lb. is usually considered safe for a live load in a church. Therefore, 600 × 120 = 72,000 lb., the total live load on the girder. Ans.

EXAMPLES FOR PRACTICE

1. What will be the entire live load on the floor of a church 50 feet by 120 feet? Ans. 720,000 lb.

2. What live load will a joist in a city dwelling be required to bear, the distance between centers being 14 inches, and the span of the joist 20 feet? Ans. 1,633 lb.

3. A steel beam, supporting a portion of the floor in an office building, sustains an area of 80 square feet. What will be the live load coming on the beam? Ans. 5,600 lb.

4. A warehouse used for the storage of South American wool is 40 feet wide and 80 feet long inside. The girders extend across the building and divide it lengthwise into 5 bays; provided the floor construction and the girders weigh 20 pounds per square foot of surface, what is the total dead and live load on each girder? Ans. 104,320 lb.

---

8. In proportioning the live loads on floors, the architectural engineer cannot always exercise his own judgment, for if the building is to be erected in a large city, the live load must comply with the building laws. As these are not uniform in the several cities the following table is given to show the stipulated live loads in the four largest cities in the United States:

TABLE VI

THE ALLOWABLE LIVE LOADS ON FLOORS IN DIFFERENT CITIES

Character of Building	Pounds per Square Foot			
	New York	Philadelphia	Chicago	Boston
Buildings for public assembly	90	100	120	150
Buildings for ordinary stores, light manufacturing, and light storage	120	100	120	
Dwellings, apartment houses, hotels, tenement houses, or lodging houses	60	40	70	50
Office buildings, first floor	150	100	100	100
Office buildings, above first floor	75	100	100	100
Public buildings, except schools				150
Roofs, pitch less than 20°	50	25	30	25*
Roofs, pitch more than 20°	30	25	30	25*
Schools or places of instruction	75			80
Stables or carriage houses less than 500 square feet in area	75	40		
Stables or carriage houses more than 500 square feet in area	75	100		
Stores for heavy materials, warehouses, and factories	150		150	250
Sidewalks	300			

NOTE.—In this table the values given for roofs are not live loads, for a snow load can hardly be classified with live loads. The roof loads, in the last column, marked with the asterisk (*), do not include the wind load, and the building laws of Boston require that a proper allowance for the wind load exerting a pressure of 30 pounds per square foot of vertical surface shall be made in designing roofs.

SNOW AND WIND LOADS

9. In calculating the weight on roofs, there are two other loads always to be considered when obtaining the stresses on the various members of the truss; these are *snow* and *wind loads*. When the roof is comparatively flat, that is, when the rise of the roof is under 12 inches per foot of horizontal distance, the **snow load** is estimated at 12 pounds per square foot; for roofs that have a steep slope, or a rise of more than 12 inches per foot of horizontal distance, it is good practice to assume the snow load to be 8 pounds per square foot. In northern climates, such as Canada, snow loads 50 per cent. greater than the above should be assumed.

10. Wind Pressure.—The wind pressure depends on the velocity with which the air is moving. United States Government tests have determined that the pressure per square foot on a vertical surface is approximately represented by the formula

$$p = .004\ V^2 \qquad (3)$$

in which p = pressure, in pounds per square foot, of vertical surface;

V = velocity of wind, in miles per hour.

This formula may be expressed in a rule thus:

Rule.—*The wind pressure, in pounds per square foot of vertical surface, is obtained by multiplying the square of the velocity of the wind, in miles per hour, by .004.*

The velocity of the wind varies from a pleasant breeze of 2 or 3 miles per hour to a violent hurricane or tornado of 100 or more miles per hour. Careful records, extending over a period of years, show that the velocity of the wind seldom attains 100 miles per hour; probably not more than once in the lifetime of the structure. In cyclonic storms, the velocity of the wind greatly exceeds 100 miles per hour, and structures cannot be built that will withstand their fury.

By applying formula **3**, it will be found that a wind having a velocity of 100 miles per hour will exert a pressure of 40 pounds per square foot of vertical surface and this is the pressure usually assumed by conservative engineers in providing for the resistance to wind pressure on structures. However, where the surface is of great area, as the side of a large office building, a pressure of 30 pounds per square foot is considered ample; for, the average unit pressure on a large surface is never so great as the maximum unit pressure on a small surface.

11. Curved surfaces, such as would be presented by circular towers and stacks, and flat surfaces not in a vertical plane, as roofs, are subjected to less pressure than flat vertical surfaces. The pressure on a cylindrical surface is about one-half the pressure on a flat surface having the same width as the diameter of the cylinder and the same height.

12. On roofs the wind pressure is always assumed as acting normal (that is, perpendicular) to the slope. In Fig. 4, the outline *a b c* of a roof is shown; the force *d* is normal to the slope *a b*, and represents the assumed pressure of the wind on the roof. The wind generally acts in a horizontal direction, as shown by the full arrow *e*. The reason for assuming the force *d* instead of *e* as the active wind pressure is for convenience in determining the stresses in the members of a roof frame or truss, which is explained in *Graphical Statics*. At present it may suffice to say that in order to estimate the effect of the force *e*, it is supposed to consist of two

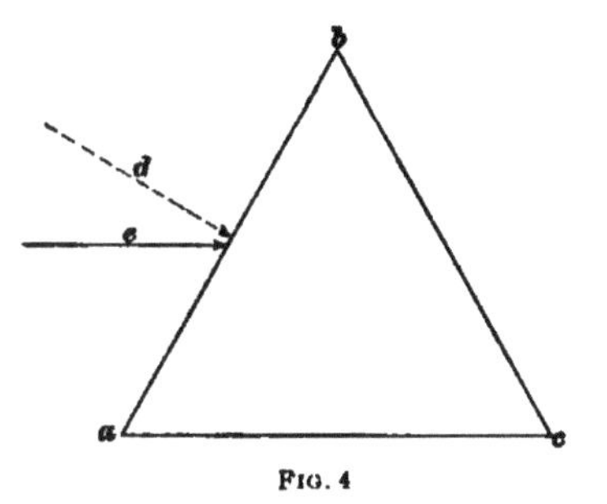

Fig. 4

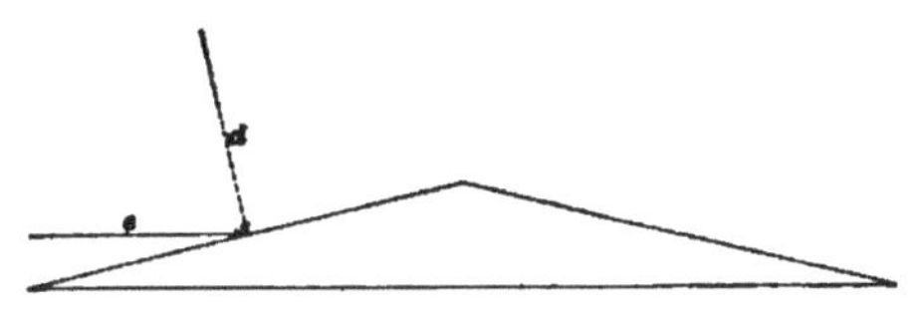

Fig. 5

components, one of which is the force *d* and the other a force acting upwards and in a direction parallel with the slope. The latter is not taken into consideration. The wind, blowing with a horizontal pressure of 40 pounds, strikes the roof at an angle; consequently, the pressure *d*, normal to the slope, is considerably less than 40 pounds, unless the slope of the roof is very steep. Referring to Figs. 4 and 5, it is clear that the horizontal force e of the wind on the slope of the roof, shown in Fig. 4, is almost as intense as on a vertical surface; on the extremely flat roof in Fig. 5, however, the wind exerts hardly any force normal to the slope, because it strikes the slope at such an acute angle, and therefore has a tendency to slide along its surface. The more acute the angle between the lines e and d, the greater the pressure normal to the slope; whereas, the greater the angle, the less the pressure normal to the slope, until they approximate a right angle with each other, when the pressure on the roof may be disregarded. In the design of roof trusses a horizontal wind pressure of 40 pounds per square foot is usually assumed.

13. All necessary data for calculating the wind pressure on a roof are given in Table VII, but it may be interesting to the student to know how the results given therein are obtained, and therefore the following explanation is offered. The pressure normal to the slope is generally determined by what is known as *Hutton's formula*. This formula is trigonometric, and may be expressed as follows:

$$p = p' \sin x^{1.84 \cos x - 1} \qquad (4)$$

in which p = pressure, in pounds per square foot, normal to slope of roof;

p' = wind pressure, in pounds per square foot, of vertical surface;

x = internal angle of roof with the horizontal.

This formula may be expressed as follows:

Rule.—*The pressure, in pounds per square foot, normal to the slope of the roof is equal to the product of the wind pressure, in pounds per square foot of vertical surface, and the sine of the angle that the roof slope makes with the horizontal, having an exponent of 1.84 times the cosine of the same angle, minus 1.*

For further explanation and application of the formula, assume a roof slope of 30° and the usual horizontal wind pressure of 40 pounds per square foot. The formula becomes, on substitution, $p = 40 \sin 30^{\circ\, 1.84 \cos 30^\circ - 1}$, and since sin 30° equals .5000 and cos 30° equals .86603, by further substitution $p = 40 = .5000^{1.84 \times .86603 - 1}$, or $p = 40 \times .5000^{.5935}$. Referring to any logarithmic table, the calculation for this last expression is as follows:

$$\text{Log } .5000 = \bar{1}.69897 \text{ and } \bar{1}.69897 \times .5935 = -.1787$$

Then,

$$\begin{array}{r} \log 40 = 1.60206 \\ -.17870 \\ \hline 1.42336 \end{array}$$

The number corresponding to this logarithm is 26.5+, the value of p or the pressure in pounds per square foot normal to the slope of the roof.

The following table, which gives the normal wind pressure on roofs of different slopes, has been calculated for a horizontal wind pressure of 40 pounds per square foot for the benefit of those who are not familiar with logarithms:

TABLE VII

WIND PRESSURE NORMAL TO THE SLOPE OF ROOF

Rise	Angle of Slope With Horizontal	Pitch, Proportion of Rise to Span	Wind Pressure Normal to Slope Pounds per Sq. Ft.
4 inches per foot horizontal. .	18° 25′	$\frac{1}{6}$	16.8
6 inches per foot horizontal. .	26° 33′	$\frac{1}{4}$	23.7
8 inches per foot horizontal. .	33° 42′	$\frac{1}{3}$	29.1
12 inches per foot horizontal. .	45° 0′	$\frac{1}{2}$	36.1
16 inches per foot horizontal. .	53° 7′	$\frac{2}{3}$	38.7
18 inches per foot horizontal. .	56° 20′	$\frac{3}{4}$	39.3
24 inches per foot horizontal. .	63° 27′	1	40.0

14. A diagram of the formula for obtaining the wind pressure normal to the slope, such as shown in Fig. 6, is interesting and provides a convenient means of determining the amount of the pressure for any slope. This may be found by projecting the point which represents the upper end of the slope on the arc horizontally, until it intersects the curve of normal pressure; the multiplier directly under this point multiplied by the assumed wind pressure on a vertical surface will give the normal pressure for the given

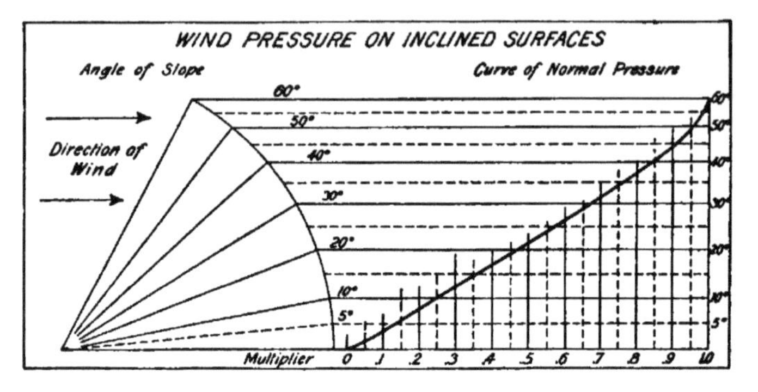

Fig. 6

slope. For instance, assume that it is desired to obtain the normal pressure on a roof whose slope forms an angle of 35°

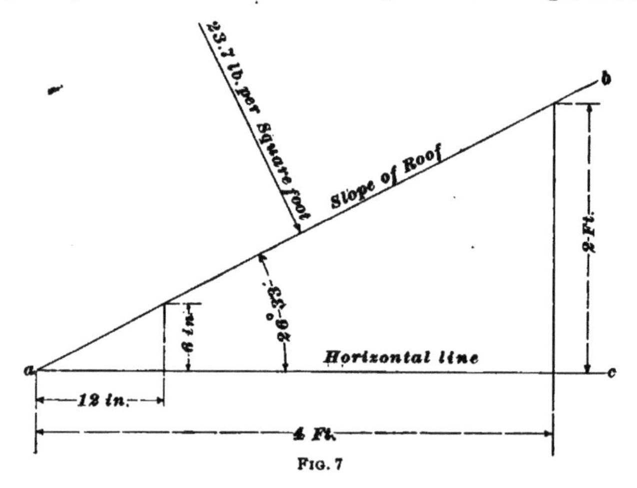

Fig. 7

with the horizontal; mark off 35° on the arc and project this point horizontally until it intersects the curve of normal pressure, then note the value of the multiplier directly under the point of intersection, which in this case is .75. Assuming a wind pressure of 40 pounds per square foot on a vertical surface, the normal pressure is found to be $40 \times .75 = 30$ pounds per square foot.

15. In order to more fully explain Table VII assume the conditions shown in Fig. 7. The rise in the slope *a b* is 6 inches for every 12 inches on the horizontal line *a c*; for instance, at 4 feet from *a* on the horizontal line *a c*, the rise is four times 6 inches, or 2 feet, the angle included between the line of slope *a b* and the horizontal base line *a c* is 26° 33′, and the pressure normal to the slope, according to Table VII, is 23.7 pounds per square foot. Since the rise at the center is equal to one-half the length of one-half the span, the total rise is one-quarter of the span. Under these conditions, the pitch of the roof, that is, the ratio of the rise to the span, is $\frac{1}{4}$, and the roof is said to be $\frac{1}{4}$ *pitch*.

EXAMPLE.—(*a*) What will be the dead load per square foot of roof surface, on a roof with a 12-inch rise per foot horizontal, the span of the trusses being 50 feet, the roof covering 1-inch white-pine sheathing, 2 layers of Neponset roofing felt, and $\frac{1}{8}$-inch slate 3-inch lap? (*b*) What will be the wind pressure per square foot normal to the slope? (*c*) If the roof trusses are placed 12 feet apart, what will be the entire dead load on one truss? Fig. 8 shows a plan with elevation and detail section of the roof.

SOLUTION.—(*a*) It is first necessary to obtain the length of the line of slope *a b*; this is done by calculating the hypotenuse of the triangle, or by laying the figure out to scale and measuring. In this case it is found that *a b* measures about 35 ft. 4 in., equal to 35.33 ft. The area of the roof supported by one truss is $2 \times 35.33 \times 12 = 847.92$ sq. ft. By referring to Table III, it is seen that the approximate weight of a roof truss of $\frac{1}{2}$ pitch and with a span of 50 ft. is 3.176 lb. per sq. ft. of roof surface. Using the approximate value of 3.2 lb., the dead load per square foot of roof surface is, then, as follows:

Weight of supporting truss	3.2 lb. per sq. ft.
Weight of white-pine sheathing, 1 inch thick	2.5 lb. per sq. ft.
Weight of 2 layers of Neponset roofing felt .	.5 lb. per sq. ft.
Weight of slate ($\frac{1}{8}$ inch thick)	4.5 lb. per sq. ft.
Total	10.7 lb. per sq. ft.

The weight of the purlins supporting the sheathing has not been estimated in the above, it being safe in this case to assume that the weight used for the principals, or trusses, is sufficient to cover this item. A snow and accidental load of 12 lb. per sq. ft. of roof surface should also be added to the dead load to get the entire vertical load on the roof.

(*b*) The wind pressure normal to the slope of this roof, according to Table VII for a ½-pitch roof is 36.1 lb., say 36 lb. per sq. ft. Ans.

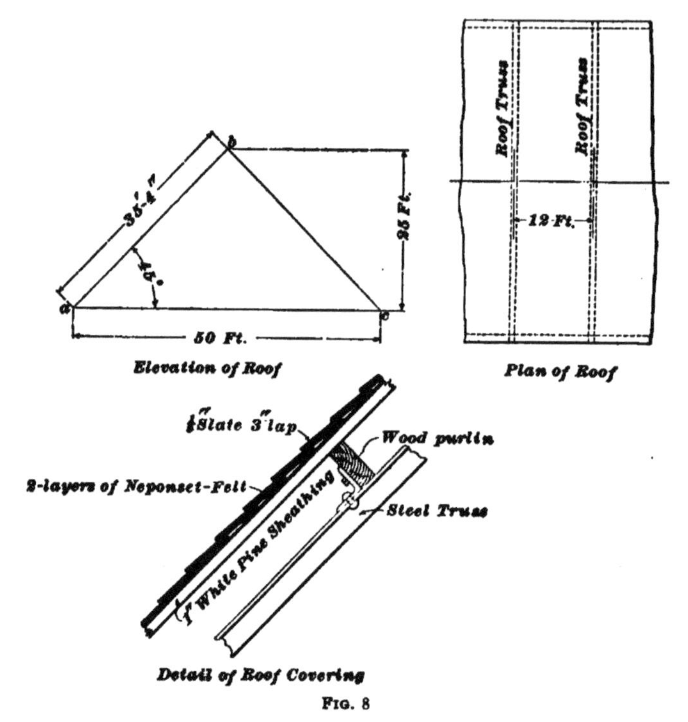

FIG. 8

(*c*) The area of the roof supported by one truss is, as previously found, 847.92 sq. ft., and the dead load, 10.7 lb. per sq. ft. Then, $847.92 \times 10.7 = 9{,}072.74$ lb. to be supported by one truss, not including the snow load. Ans.

16. Engineering, it must be remembered, is not an exact science, the results obtained depending more or less on the judgment and experience of the designer. When, for instance, the wind is blowing a hurricane, snow never lodges on a roof, the slates, shingles, and sheathing being themselves exposed to sudden removal. If, therefore, the full wind pressure be assumed, the snow load may, in most cases, be neglected, especially if it is desired to build an economical roof. However, it is not well for the student to make such assumptions until his experience and judgment are sufficiently developed to enable him to make true deductions.

EXAMPLES FOR PRACTICE

1. With the wind blowing at a velocity of 36 miles per hour, what is the pressure, in pounds per square foot of vertical surface?
Ans. 5.18 lb.

2. The area of one slope of a ½-pitch roof is 800 square feet. What is the entire pressure on the slope of the roof, provided the maximum horizontal wind pressure is taken at 40 pounds per square foot?
Ans. 28,880 lb.

3. In a ¼-pitch roof the trusses are 20 feet apart, and the length of the roof slope is 40 feet. What wind load is there on each roof truss, if the horizontal pressure is 40 pounds per square foot? Ans. 18,960 lb.

4. The purlins supporting a ¾-pitch roof are placed 6 feet apart, and the trusses are 12 feet from center to center. What is the maximum load due to the wind on each purlin, providing the greatest horizontal pressure is 40 pounds per square foot? Ans. 2,830 lb.

5. The angle that the slope of a roof makes with the horizontal is 40°. Provided the wind exerts a pressure of 30 pounds per square foot of vertical surface, what is the pressure normal to the slope?
Ans. 25.50 lb.

ACCIDENTAL AND SUDDENLY APPLIED LOADS

17. Careful designers sometimes make allowance for the **accidental load** caused by a heavy body falling on the floor, or by a mass of snow dropping from one roof to another. The latter may usually be ignored, because it is taken care of in the factor of safety, within the limit of which every member in the structure is designed.

Members subjected to suddenly applied loads are seldom encountered in building construction, and still less frequently are members required to resist the effect of impact, or the blow imparted by a falling load. The beams supporting the mechanism at the heads of elevator shafts are at times subjected not only to suddenly applied loads but also to falling loads; therefore, they should always be proportioned to withstand at least a suddenly applied load.

18. Where a load is placed suddenly on a beam, the stress produced is twice as great as if the same load had been at rest; that is, a beam to sustain a suddenly applied load should have twice the transverse strength required to sustain the same load at rest.

A falling load produces a much greater stress on a beam than a load suddenly applied, owing to the impact produced by the momentum of the falling body. It is usual in considering the effect of a falling load on a beam to determine the amount of a statical load that would produce the same results; the formula used to accomplish this is as follows:

$$W_1 = W\left(1 + \sqrt{\frac{2ah}{d} + 1}\right) \qquad (5)$$

in which W_1 = static load that would produce same stress in beam as falling load;
W = falling load;
h = height load falls, in inches;
d = deflection of beam, in inches;
a = constant.

The value of d may be determined as described in *Beams and Girders*, Part 1, while the value of a must be determined by the formula

$$a = \frac{1}{1 + .489\frac{W_2}{W}} \qquad (6)$$

in which W_2 = combined weight of beam and dead load that it supports;
W = amount of falling load, as before.

This may be expressed by a rule as follows:

Rule.—*The value of the constant is equal to 1 divided by the sum of 1 and .489 times the quotient obtained by dividing the combined weight of the beam and the dead load that it supports, by the falling load.*

19. Formula **5** may be stated as follows:

Rule.—*To the quotient of 2 times the constant times the distance the load falls, in inches, divided by the deflection of the beam in inches, add 1; the square root of this result is added to 1 and their sum multiplied by the falling load gives the static load that will produce the same strain in a beam as the falling load.*

EXAMPLE.—The drop test of a fireproof-floor system is to be made by letting a weight of 300 pounds fall through a distance of 3 feet. Provided the deflection of the floor beams under the falling load is $\frac{1}{8}$ inch and the weight of the beam with the dead load is 3,000 pounds, what static concentrated load is it necessary to figure on, that it may equal the falling load in its effect?

SOLUTION.—From formula **6**, the value of a equals $\dfrac{1}{1 + .489\,\dfrac{W_2}{W}}$, or by substitution, $a = \dfrac{1}{1 + .489 \times \frac{3000}{300}} = .1698.$

The value of a having been found, the equivalent static load required by the example may be obtained from formula **5** where $W_1 = W\left(1 + \sqrt{\dfrac{2\,a\,h}{d} + 1}\right)$, or by substitution,

$$W_1 = 300\left(1 + \sqrt{\frac{2 \times .1698 \times 36}{.125} + 1}\right) = 3{,}282 \text{ lb. Ans.}$$

EXAMPLES FOR PRACTICE

1. The static load from an elevator car on the steel beams at the head of the shaft is 4,000 pounds; what load should be figured on to compensate for the weight being suddenly applied in starting the car?
Ans. 8,000 lb.

2. The beams supporting a loading platform have a span of 20 feet and are frequently subjected to a load of 500 pounds falling a distance of 2 feet. Provided the deflection of the beam is $\frac{3}{16}$ inch, what static load will be equivalent to the effect of the falling load if the beam and the dead load on it weigh 1,000 pounds? Ans. 6,210 lb.

THE DISPOSITION OF LOADS

20. In warehouses built especially for the storage of heavy merchandise, where the floors are likely at any time to be fully loaded, the beams, girders, columns, and foundations are always proportioned for the entire live and dead load on all floors. However, where the building exceeds four or five stories in height and is used for any other purpose than for storage, as, for instance, a modern office building, it is customary to assume that certain members, while proportioned for the entire dead load, carry only a certain percentage of the live loads.

In an office building, or similar structure, it is highly improbable that all the floors or all parts of the same floor will be fully loaded at the same time, and in view of this fact it is considered good practice, while proportioning the floor beams for the full live load, to calculate only 90 per cent. of the live load on the girders. It is usual to proportion the columns supporting the roof and the top floor for the full live load. The live loads on the columns, in each successive tier, from the floor above is reduced 5 per cent. until 50 per cent. of the live load is reached, when such reduced loads are used for all remaining floors to the basement. The economy obtained by this disposition of the live load is best observed from Table VIII, which gives the distribution of the assumed live loads on the columns in the several tiers of an eighteen-story office building.

While this system of graduating the live loads on the columns from floor to floor is generally practiced, the amount of reduction at each floor is a matter that depends on the judgment of the designer. The percentage of reduction is often fixed by the building laws of the city, with which the designer must comply. The reduction of 5 per cent. at each floor, the economy of which is shown in the table, is conservative and in most cases will be found to be in accordance with the building departments of the principal American cities.

TABLE VIII

SHOWING REDUCTIONS OF LIVE LOADS FROM FLOOR TO FLOOR

Floors	Live Load in Lb. per Sq. Ft. on Each Floor	Live Load in Lb. per Sq. Ft. of Floor on Columns From Floor Above at 5% Reduction	Live Load in Lb. per Sq. Ft. of Floor on Columns From All Floors Above If No Reduction Were Made	Live Load in Lb. per Sq. Ft. of Floor on Columns From All Floors Above Increment of Reduction	Theoretical Percentage of Saving Instituted by the Reduction of 5% at Each Floor
	a	$a_1 = a - .05a$	Σa	Σa_1	$\frac{\Sigma a - \Sigma a_1}{\Sigma a}$
Roof	20	20.00	20	20.00	
18	60	60.00	80	80.00	
17	60	57.00	140	137.00	2.1
16	60	54.15	200	191.15	4.4
15	60	51.44	260	242.59	6.7
14	60	48.87	320	291.46	8.9
13	60	46.43	380	337.89	11.1
12	60	44.11	440	382.00	13.2
11	60	41.90	500	423.90	15.2
10	60	39.80	560	463.70	17.2
9	60	37.81	620	501.51	19.1
8	60	35.92	680	537.43	21.0
7	60	34.12	740	571.55	22.8
6	60	32.41	800	603.96	24.5
5	60	30.79	860	634.75	26.2
4	60	30.00	920	664.75	27.7
3	60	30.00	980	694.75	29.1
2	60	30.00	1,040	724.75	30.3
1	60	30.00	1,100	754.75	31.4

21. In the design of the type of building known as **skeleton construction**, that is, one in which all floors and walls are supported on beams and girders that transmit the loads to columns and, in turn, are supported on ample foundation footings, it is necessary to fix on the general arrangement, disposition, and approximate dimensions of the component parts before the dead load can be computed. After the calculations are made and the structural details are designed, the actual dead load should be checked to see whether it approximates the assumed load. If any considerable variation is found it can be provided for by increasing or diminishing the weight or thickness of the rolled steel shapes making up the structural members, the sizes of which have already been determined.

Where permanent partitions exist, they should always be figured in the dead load; and where they are directly above a beam or a girder, the member should be proportioned to sustain the additional weight without appreciable deflection. Where movable partitions occur or where there is a probability of the location of permanent partitions being changed, it is usual to add 20 pounds per square foot of floor surface to the dead load to take care of such contingencies.

The foundations of an office building should be proportioned for the entire dead load and none of the live load, the latter being provided for by making the unit pressure on the footings and piers well within the safe unit bearing value of the soil. In this way unequal settlement is prevented, as explained in *Foundations*.

PROPERTIES OF SECTIONS

CENTER OF GRAVITY

1. Introduction.—The calculations involved in the design of such built-up members of a building as steel columns and plate girders—members that are formed by the combination of several of the simple sections produced by the rolling mills—require a knowledge of certain mathematical properties of the simpler sections, together with the methods by which these properties may be calculated. In many cases, the exact determination of the required properties is based on complicated mathematical principles; there are, however, numerous formulas and practical methods by means of which the values for all sections used in ordinary practice may be determined, either exactly or with a degree of approximation sufficiently close for all practical purposes.

2. The **center of gravity** of a body, or of a system of bodies, is that point from which, if the body or system were suspended, it would be in equilibrium. If the body or system were suspended from any other point than the center of gravity, and in such a manner as to be free to turn about the point of suspension, the body would rotate until the center of gravity reached a position directly under the point of suspension.

3. Center of Gravity of Plane Figures.—If a plane figure has an *axis of symmetry*, this axis passes through its center of gravity. If the figure has two axes of symmetry, its center of gravity is at their point of intersection.

The *center of gravity of a triangle* lies on a line drawn from a vertex to the middle point of the opposite side, and at a distance from that side equal to one-third of the length of the line; or it is at the intersection of lines drawn from the vertexes to the middle of the opposite sides.

The perpendicular distance of the center of gravity of a triangle from the base is equal to one-third of the altitude.

The *center of gravity of a parallelogram* is at the intersection of its two diagonals; consequently, it is midway between its sides.

The *center of gravity of an irregular four-sided figure* may be found by first dividing it, by a diagonal, into two triangles and joining the centers of gravity of the triangles by a straight line; then, by means of the other diagonal, divide the figure into two other triangles, and join their centers of gravity by another straight line; the center of gravity of the figure is at the intersection of the lines joining the centers of gravity of the two sets of triangles.

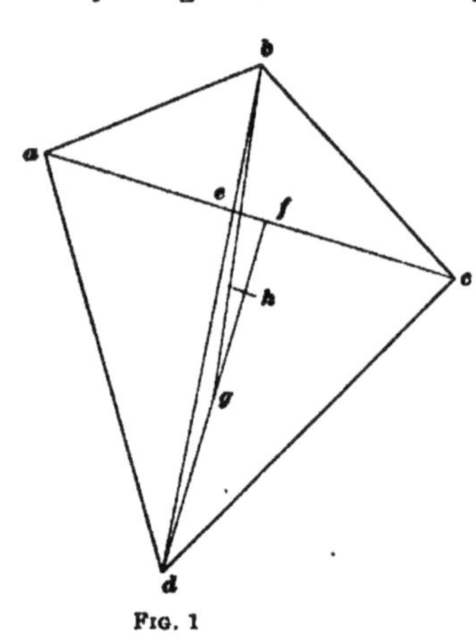

Fig. 1

Another method by which to locate the center of gravity of an irregular four-sided figure is illustrated in Fig. 1. Draw the diagonals ac and bd, and from their intersection e, measure the distance to any vertex, as ae. From the opposite vertex, lay off this distance, as at cf. Then from f, draw a line to one of the other vertexes, as fd, and bisect this line as at g. Connect g and b and lay off one-third of its length from g at the point h. This point is the center of gravity of the figure.

The distance of the *center of gravity of the surface of a half circle* from the center is equal to the product of the radius multiplied by .424.

THE NEUTRAL AXIS

4. Explanation of Neutral Axis.—When a beam is subjected to a bending stress, the fibers on the concave side of the beam are shortened while those on the convex side are lengthened; hence, there is a compressive stress on the concave side of the beam and a tensile stress on the convex side. Therefore, there will be a plane near the center of the beam in which the fibers are neither lengthened nor shortened and in which there is no stress; this plane is termed the **neutral plane**, or **neutral surface**. A **cantilever** beam, or beam supported at one end, is shown in Fig. 2, having a load W applied at the end. In this case the upper side of the beam is lengthened and the lower side is shortened, while in a simple beam, or one having a support at each end, the reverse is the case. A line dc representing the intersection of the neutral plane $a\,b\,c\,d$ with a cross-section of the beam is termed the **neutral axis**. This axis always passes through the center of gravity of the cross-section of the beam until the elastic limit is reached.

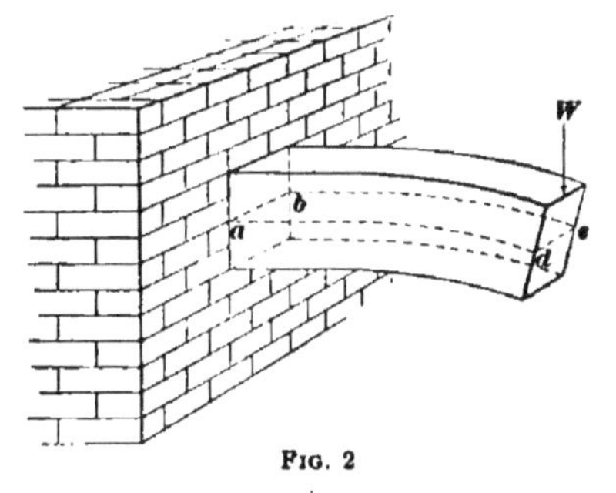

Fig. 2

LOCATION OF NEUTRAL AXIS

5. Mechanical Method.—It is evident, from Fig. 2, that the neutral axis is perpendicular to the direction in which the load acts on the beam; therefore, to find the neutral axis of a section with reference to a set of loads applied in a given direction, it is only necessary to pass a line through the center of gravity perpendicular to the direction of the load.

A simple approximate method of locating the center of gravity and neutral axis of a section is shown in Fig. 3. Draw the outline of the section, either full size or to some

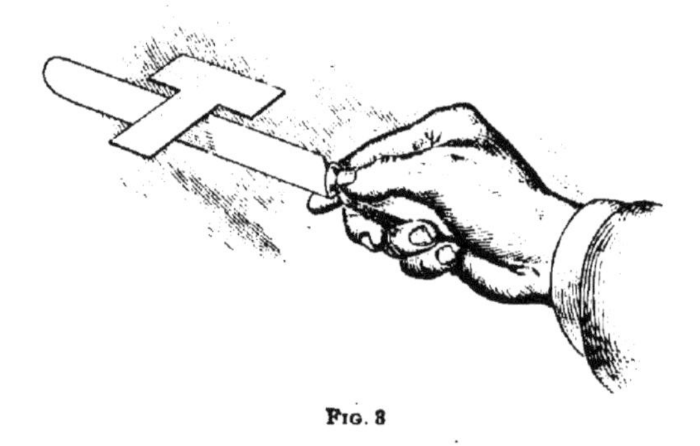

FIG. 3

convenient scale, on a piece of heavy cardboard. Cut out the section and balance it carefully over a knife edge, as shown in the figure; the line along which it rests on the edge of the knife is a line passing through the center of gravity, and by locating two such lines in different directions, the center of gravity will be found at their point of intersection.

6. Locating the Neutral Axis by Means of the Principle of Moments.—A convenient method of locating the neutral axis is based on the principle that the moment of any figure, with respect to a given line as an axis or origin of moments, is equal to the sum of the

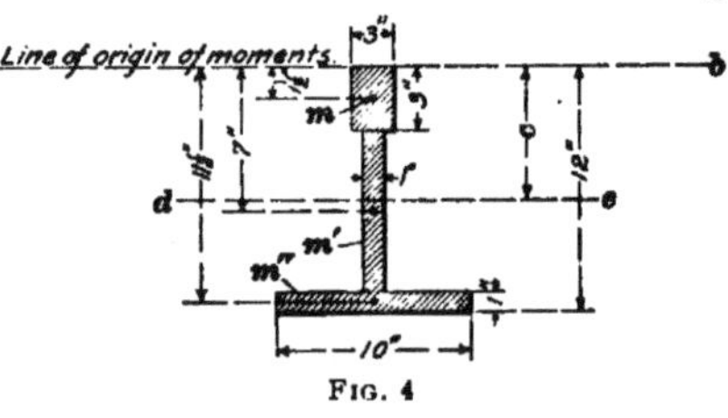

FIG. 4

moments of its separate parts with respect to the same axis. The moment of a figure about a given axis is the product of its area by the perpendicular distance from the center of gravity of the figure to the given axis. Thus, in Fig. 4, take the line $a\,b$ as the line of origin of moments. Divide the figure into the three rectangles m, m', m''. In accordance with the principles stated in Art. **3**, the center of gravity of each of these rectangles is midway between its edges; the distances of the respective centers from the axis are, therefore, $1\frac{1}{2}$, 7, and $11\frac{1}{2}$ inches. The areas of the figures are, respectively, 3×3 or 9 square inches, 8×1 or 8 square inches, and 10×1 or 10 square inches. The moments of these areas about the axis $a\,b$ are:

Section m	$9 \times 1\frac{1}{2}$	=	13.5
Section m'	8×7	=	56.0
Section m''	$10 \times 11\frac{1}{2}$	=	115.0
	Total,		184.5

The area of the whole section is equal to $9 + 8 + 10 = 27$ square inches, the sum of the areas of the rectangles; the distance c of its neutral axis from the line of the origin of moments is, therefore, $184.5 \div 27 = 6.83$ inches, or nearly $6\frac{27}{32}$ inches.

It is not necessary that the line of the origin of moments should coincide with an edge of the figure, as in Fig. 4, since any other line parallel with the direction of the required neutral axis gives the same results; in most cases, however, it will be found more convenient to take the axis about which the moments are calculated on one of the extreme edges of the section.

Since the section shown in Fig. 4 is symmetrical with respect to an axis perpendicular to the neutral axis, it is evident that its center of gravity is on their intersection. If, however, there were no axis of symmetry, the center of gravity could have been located by taking a second line perpendicular to $a\,b$ as an origin of moments and finding the neutral axis parallel to it. The intersection of this neutral axis with the one first found is the center of gravity of the section. In accordance with the principles illustrated in this example, we have the following rule:

Rule.—*To find the neutral axis of any section, first divide it into a number of simple parts, each of whose areas and centers* *of gravity can be readily found; then find the sum of the moments of the areas of each of these parts with respect to an axis parallel to the required neutral axis; finally, divide this sum by the sum of the areas of the parts of the section. The result will be the perpendicular distance from the axis of the origin of moments to the required neutral axis.*

7. Application of the Rule to a Built-Up Section. Fig. 5 shows a section of the rafter member of a large roof truss formed of a $\frac{3}{8}''\times 16''$ web-plate and a $\frac{3}{8}''\times 12''$ flange plate, the two joined by two $4''\times 4''\times\frac{1}{2}''$ angles. What is the distance from the neutral axis of the section to the top edge of

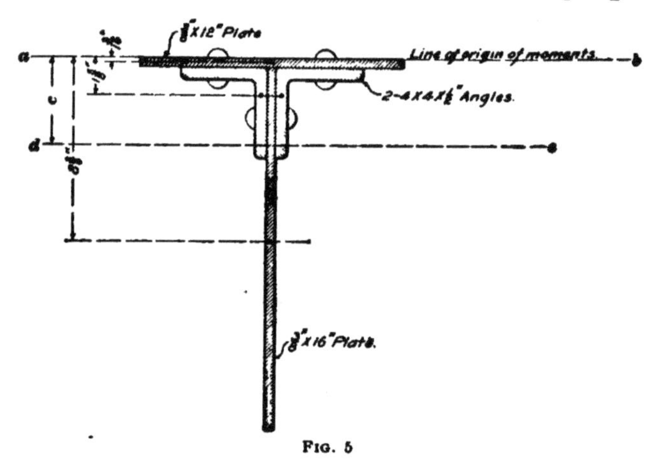

FIG. 5

the flange plate? By means of the principles given, the centers of gravity of the two rectangular plates are easily located as shown. The centers of gravity of the angles might also be located by applying the rule given in the preceding article; this, however, is unnecessary, since the center of gravity can be obtained directly by reference to the tables of the properties of rolled sections. Referring to Table V, the center of gravity of a $4''\times 4''\times\frac{1}{2}''$ angle is found to be 1.18 inches from the back of a flange, thus giving us the distance $1.18 + .375 = 1.555$, or about $1\frac{14}{25}$ inches from the top edge of the flange plate to the axis through the centers of gravity of the angles. From the same table it is also found that the area of the section of a $4'' \times 4'' \times \frac{1}{2}''$ angle is 3.75 square inches.

The area of the section of the flange plate is $\frac{3}{8} \times 12 = 4.5$ square inches, and of the web-plate, $\frac{3}{8} \times 16 = 6$ square inches; the area of the whole section is, therefore, $2 \times 3.75 + 4.5 + 6 = 18$ square inches.

The moments of the areas of the separate sections, with respect to the line *a b*, are as follows:

Flange plate,	$4.5 \times \frac{3}{16} =$	.84
Two angles,	$2 \times 3.75 \times 1\frac{9}{16} =$	11.70
Web-plate,	$6 \times 8\frac{3}{8} =$	50.25
	Total,	62.79

The distance *c* from the top edge of the flange plate to the neutral axis *d e* of the section is, therefore, $62.79 \div 18 = 3.48$ inches.

8. Graphical Method of Locating the Neutral Axis. Let it be required to determine the position of the neutral axis of the cast-iron beam section shown in Fig. 6. First, divide the depth of the section into any number of parts—as has been done in this case by the dotted lines *w′ x*, *y z*, etc.—whose areas and centers of gravity can readily be found. Then compute the area and locate the center of gravity of each part. In Fig. 6, the area of the top slice is 12 square inches, the area of each of the slices in the web of the section is 3 square inches, and the area of the bottom flange is 28 square inches. Assume some scale whose unit of length represents 1 square inch; for example, in this case, let $\frac{1}{16}$ inch represent 1 square inch. Then, commencing at some point *a*, lay off on the line *a l* lengths *a b*, *b c*, *c f*, etc., which represent the respective areas of the successive parts into which the section of the beam has been divided, beginning at the top part. Thus, with the scale of $\frac{1}{16}$ inch = 1 square inch, the line *a b*, which represents an area of 12 square inches, is $\frac{12}{16} = \frac{3}{4}$ inch long; each of the lines *b c*, *c f*, etc. is $\frac{3}{16}$ inch long; and the line *k l* is $\frac{28}{16} = 1\frac{3}{4}$ inches long.

From the points *a* and *l*, draw lines at any convenient angle to *a l* intersecting at the point *m*. Then from the points *b*, *c*, *f*, etc., draw the lines *b m*, *c m*, *f m*, etc. Through the center

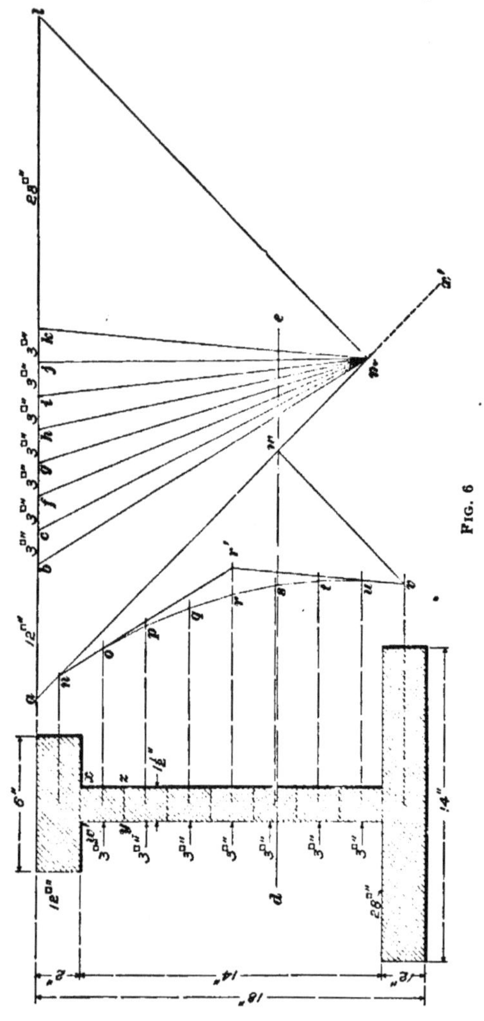

Fig. 6

of gravity of each of the parts of the section, draw indefinite lines parallel to *a l*.

From the point *n*, where the line through the center of gravity of the top section intersects the line *a m*, draw the line *n o* parallel to the line *b m* until it intersects the line passing through the center of gravity of the second slice in the point *o*; draw the line *o p* parallel to *c m*; *p q* parallel to *f m*; *q r* parallel to *g m*; *r s* parallel to *h m*; *s t* parallel to *i m*; *t u* parallel to *j m*; and *u v* parallel to *k m*.

From the point *v*, which is the point at the intersection of the line *u v* with the line drawn through the center of gravity of the last elementary section, draw the line *v w* parallel to *m l*; then its intersection *w* with the line *a m* is a point on the required neutral axis. If the line *a m* is so short that the line *v w* fails to cut it, it may be extended indefinitely, as shown at *m x'*, so as to make it intersect with the line *v w*. Having found the point *w*, draw the horizontal line *d e* through it. This line is the required neutral axis of the figure, and passes through its center of gravity.

This method of determining the position of the neutral axis and center of gravity may be applied to any irregular-shaped section, and in many cases may be found more convenient than the mathematical method.

When, as in Fig. 6, the section is made up of several regular parts whose centers of gravity can be readily located, it is not necessary to subdivide any one of these parts. Thus, the center of gravity of the web of the beam is located on the horizontal line through *r*, and its area is represented by the distance *b k*. We can therefore draw from *n* a line parallel to *b m* until it intersects the horizontal line through the center of gravity of the web member; then, from this point, draw a line parallel to *k m* until it intersects the horizontal line through the lower section at the point *v*. The point *v*, as thus located, is identical with the point previously found when the web section was divided into the small parts, and the line drawn from *v* parallel to *l m* until it intersects *a m*, locates the point *w* on the neutral axis, as before. If, however, the center of gravity of the web cannot be readily located, it is better to divide it into small parts, as in the first method.

THE MOMENT OF INERTIA

9. The term **moment of inertia** is a mathematical expression that depends on the distribution of either the material of a body or the area of a surface with respect to a given axis. In other words, the moment of inertia is a term that expresses the relative value of all of the infinitely small portions that make up the entire section of a beam or column. The portions of the area farthest from the neutral axis of a section are much more valuable in resisting stress than those portions adjacent to the axis, so that some value must be obtained that will express the efficiency of the entire section to resist transverse stress when compared with any other section. This value is called the moment of inertia. As applied to the area of a plane figure, the moment of inertia, with respect to an axis lying in the same plane, is numerically equal to the sum of the products obtained by multiplying each of the elementary areas of which the figure is composed by the square of its distance from the given axis.

By **elementary area** is meant an area smaller than any with which we are accustomed to deal in ordinary calculations; it is, therefore, impossible to find an exact expression for the moment of inertia of a figure by the methods of calculation in ordinary use. By means of calculus, however, exact formulas have been deduced, by which the moments of inertia of many of the simple geometrical forms, with respect to axes through their centers of gravity, have been found. There are also a number of approximate methods by which the moment of inertia of an irregular section, to which these formulas do not apply, may be found. Further, the moment of inertia of any section that can be divided into parts, the moments of each of which, with respect to an axis through its center of gravity, can be found, is easily calculated by one of the following principles.

COMPUTING THE MOMENT OF INERTIA

10. Analytical Method.—To illustrate a simple approximate method of computing the moment of inertia of a figure, consider the relation of the small I section shown in Fig. 7 to the axis *de* through its center of gravity. Divide the section

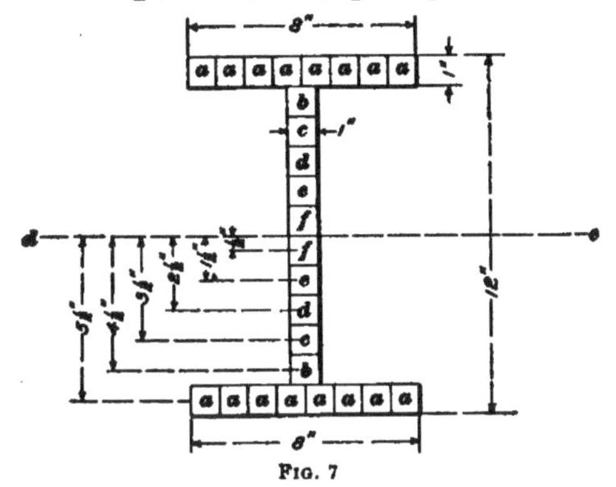

FIG. 7

into a number of little squares (in this case, each with an area of 1 square inch) and consider the distance of each square from the axis to be the distance from the axis to its center of gravity. The products of the area of each square, multiplied by the square of its distance from the axis, are as follows:

$$\begin{aligned}
&\text{Squares } a,\ 1\times(5\tfrac{1}{2})^2 = \tfrac{121}{4}\\
&\text{Squares } b,\ 1\times(4\tfrac{1}{2})^2 = \tfrac{81}{4}\\
&\text{Squares } c,\ 1\times(3\tfrac{1}{2})^2 = \tfrac{49}{4}\\
&\text{Squares } d,\ 1\times(2\tfrac{1}{2})^2 = \tfrac{25}{4}\\
&\text{Squares } e,\ 1\times(1\tfrac{1}{2})^2 = \tfrac{9}{4}\\
&\text{Squares } f,\ 1\times(\tfrac{1}{2})^2 = \tfrac{1}{4}
\end{aligned}$$

Adding these products for all the squares, we have:

$$\begin{aligned}
&16 \text{ squares } a,\ \tfrac{121}{4}\times 16 = 484\\
&2 \text{ squares } b,\ \tfrac{81}{4}\times 2 = 40\tfrac{1}{2}\\
&2 \text{ squares } c,\ \tfrac{49}{4}\times 2 = 24\tfrac{1}{2}\\
&2 \text{ squares } d,\ \tfrac{25}{4}\times 2 = 12\tfrac{1}{2}\\
&2 \text{ squares } e,\ \tfrac{9}{4}\times 2 = 4\tfrac{1}{2}\\
&2 \text{ squares } f,\ \tfrac{1}{4}\times 2 = \tfrac{1}{2}\\
&\qquad\text{Total, } 566\tfrac{1}{2}
\end{aligned}$$

which is the sum of the products of each of the small areas multiplied by the square of its distance from the axis. This result, however, is only a rough approximation to the moment of inertia, owing to the fact that the assumed areas are so large. The actual value of the moment of inertia of the section, as will be shown subsequently, is $568\frac{2}{3}$.

11. Rules and Formulas for Moments of Inertia. In Table I exact formulas are given for computing the moment of inertia of such regular figures as are most often used in structural design; it also gives approximate formulas for computing this factor for common rolled sections. The tables of properties of rolled sections published by the rolling mills give accurate values of the moment of inertia of all the principal sections used in the construction of buildings, so that it is not generally necessary to make the calculations for these sections; the approximate formulas in Table I are, however, sometimes useful in making calculations when the tables published by the rolling mills are not at hand.

Rule.—*To find the moment of inertia, with respect to any axis of any figure whose moment of inertia, with respect to a parallel axis through its center of gravity, is known, add its moment of inertia, with respect to the axis through its center of gravity, to the product of its area multiplied by the square of the distance from its center of gravity to the required axis.*

This rule may be expressed by the formula

$$I' = I + a x^2 \qquad (1)$$

in which I' = required moment of inertia;

I = moment of inertia of the section, with respect to the axis through its center of gravity and parallel to the given axis;

a = area of the figure;

x = distance from its center of gravity to the required axis.

The moment of inertia, with respect to an axis through its center of gravity, of any section that can be divided into a number of parts, the moments of inertia of each of which, with respect to an axis through its center of gravity parallel to the given axis, is known, is equal to the sum of the moments of inertia of its parts with respect to the given axis. Since the moment of inertia of any figure, with respect to any axis, is expressed by the formula $I' = I + a\,x^2$, if we denote the sum of the moments of the separate figures making up a section, with respect to an axis through the center of gravity of that section, by $\Sigma\, I'$ (in which the Greek letter Σ, read *sigma*, means *sum of*), we have

$$I_s = \Sigma\, I' = \Sigma(I + a\,x^2) \qquad (2)$$

which is a general formula often used to denote the moment of inertia I_s of any built-up section.

12. Graphical Methods of Computing Moments of Inertia.—There are several graphical methods of computing the moment of inertia, one of which is an extension of the graphical method of locating the center of gravity and neutral axis that was described in Art. 8 and illustrated by Fig. 6. Thus, let it be required to determine the moment of inertia, with respect to the axis *d e*, of the beam section shown in Fig. 6. Using the same scale as that to which the section was drawn, compute or measure the area of the figure enclosed by the lines *n o p q . . . v w n*; multiply this area by the area of the section—shown graphically by the length of the line *a l*—and the product will be the moment of inertia of the section, with respect to the axis *d e*, through its center of gravity. For example, suppose that the section shown in Fig. 6 has been drawn to a scale of ¼ inch = 1 inch. Using this scale, and computing the area of the figure enclosed by the lines *n o p q . . . v w n*, we find it to be 43.36 square inches. The area of the section is 61 square inches; therefore, according to the rule, its moment, with respect to the axis *d e*, is 43.36 × 61 = 2,644.96.

For finding the moment of inertia, it is necessary to divide the section into a number of parts, for it is evident that the area of the figure enclosed by the lines *n r' v w n* is considerably less than that of the figure *n o p q . . . v w n*, obtained by dividing the web into the small sections and drawing the lines of the diagram for each.

The area of the figure *nopq . . . vwn* may be computed by extending the horizontal lines through *o*, *p*, *q*, etc., so as to divide it into a series of triangles and trapezoids. The dimensions of these can be readily measured, and their areas can be calculated by means of the principles of mensuration.

This method of computing the moment of inertia will be found convenient in the case of very irregular sections, to which the methods previously given can be applied only with considerable difficulty. The accuracy of the result will, in general, be greater when the section is drawn to a large scale and divided into a comparatively large number of parts.

EXAMPLE 1.—What is the moment of inertia of the section of a 10″ × 16″ beam about an axis through its center of gravity parallel to its shorter side?

SOLUTION.—From Table I, the formula for the moment of inertia of a solid rectangle is $I = \frac{bd^3}{12}$. Substituting the given dimensions, we have

$$I = \frac{10 \times 16^3}{12} = 3{,}413\tfrac{1}{3}. \text{ Ans.}$$

EXAMPLE 2.—Using the formula given in Table I, compute the least moment of inertia of the section of a cast-iron column shown in Fig. 8.

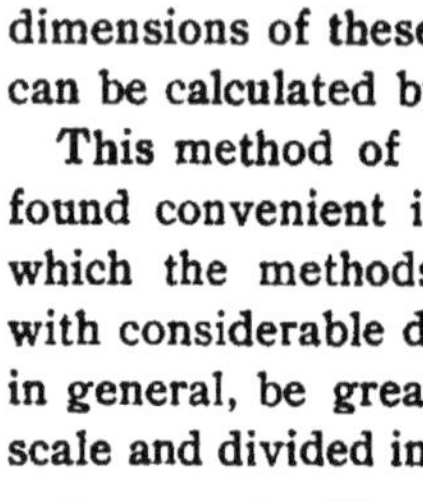

SOLUTION.—It can readily be seen that the least moment of inertia will be with reference to an axis parallel with the web. Hence, substituting in the formula, we have

$$I = \frac{2sb^3 + ht^3}{12}$$

$$= \frac{2 \times 1 \times 8^3 + (8 \times 1^3)}{12} = 86. \text{ Ans.}$$

FIG. 8

EXAMPLE 3.—Referring to the rules and formulas in Art. **11**, compute the moment of inertia, with respect to the axis *de* through its center of gravity, of the section shown in Fig. 7.

SOLUTION.—This section is made up of three rectangles, the moments of inertia of which, with respect to the given axis, can be found by means of formula **1**. The moment of inertia of one of the flanges, with respect to an axis through its center of gravity parallel to *de* is $I = \frac{8 \times 1^3}{12} = \frac{2}{3}$. The area of this figure is $8 \times 1 = 8$ sq. in., and the distance of its center of gravity from de is $5\frac{1}{2}$ in.; therefore, its moment of inertia, with respect to de, is $I' = \frac{2}{3} + 8 \times (5\frac{1}{2})^2 = 242\frac{2}{3}$. The axis through the center of gravity of the web section coincides with the axis de; hence, the moment of inertia of this section, with respect to de, is

$$I' = \frac{1 \times 10^3}{12} = 83\frac{1}{3}$$

Then, the moment of inertia I_s of the whole section $= \Sigma I' = 242\frac{2}{3} + 242\frac{2}{3} + 83\frac{1}{3} = 568\frac{2}{3}$. Ans.

EXAMPLE 4.—What is the moment of inertia, with respect to the axis de, of the column section shown in Fig. 9?

SOLUTION.—The moment of inertia of one of the flange plates, with respect to an axis through its center of gravity, parallel to the axis de, is $\frac{12 \times (\frac{3}{8})^3}{12} = .05$; the area of the plate is $12 \times \frac{3}{8} = 4.5$ sq. in., and the distance of its center of gravity from the axis is $6\frac{3}{16}$ in. Therefore, the moment of inertia of the plate, with respect to the axis de, is $.05 + 4.5 \times (6\frac{3}{16})^2 = 172.33$. From Table V, the area of a $4'' \times 4'' \times \frac{1}{2}''$ angle is 3.75 sq. in., and the distance of its center of gravity from the back of a flange is 1.18 in. The distance of the center of gravity of the angle from the axis de, in accordance with the dimensions given in the figure, is $6 - 1.18 = 4.82$ in. In accordance with the formula $\frac{t(a-c)^3 + a c^3 - (a-t)(c-t)^3}{3}$ given in Table II, for finding the moment of inertia of an angle with equal legs, the moment of inertia of the $4'' \times 4'' \times \frac{1}{2}''$ angle, with respect to the axis through its center of gravity, is

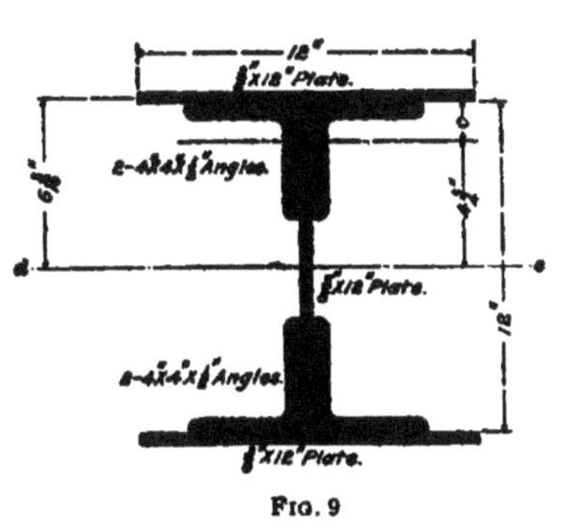

FIG. 9

$$\frac{.5(4 - 1.18)^3 + 4 \times 1.18^3 - (4 - .5)(1.18 - .5)^3}{3} = 5.54$$

The moment of inertia of the angle, with respect to de, is, therefore, $5.54 + 3.75 \times 4.82^2 = 92.65$. The center of gravity of the web-plate lies on the axis de; therefore, the moment of inertia of the plate, with respect to de, is $\frac{\frac{3}{8} \times 12^3}{12} = 54$. The moment of inertia of the whole section, with respect to de, is, therefore, $2 \times 172.33 + 4 \times 92.65 + 54 = 769.26$. Ans.

EXAMPLE 5.—What is the moment of inertia, with respect to the axis de, of the column section shown in Fig. 10?

SOLUTION.—The moment of inertia of one of the cover-plates, with respect to an axis through its center of gravity, parallel to de, is $\frac{12 \times (\frac{1}{2})^3}{12} = .125$; the area of the plate is $12 \times \frac{1}{2} = 6$ sq. in., and the distance of its center of gravity from de is $5\frac{1}{4}$ in.; therefore, its moment of inertia, with respect to de, is $I' = .125 + 6 \times (5\frac{1}{4})^2 = 165.5$. From Table IV, the area of a 10-in. 15-lb. channel is 4.46 sq. in., and its moment of inertia, with respect to an axis through its center of gravity, corresponding in this case with the axis de, is 66.9; therefore, the moment of inertia of the whole section, with respect to de, is $2 \times 165.5 + 2 \times 66.9 = 464.8$. Ans.

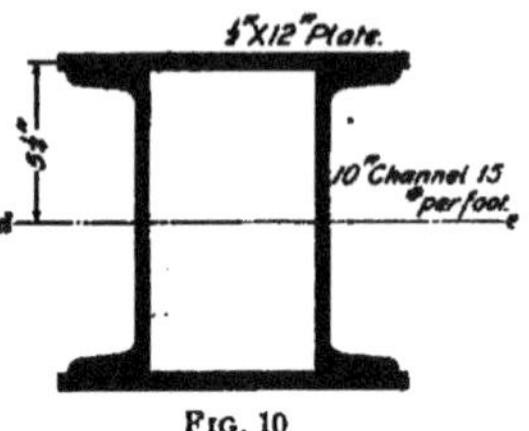

FIG. 10

EXAMPLES FOR PRACTICE

1. What is the moment of inertia of a hollow, square column section 12 inches outside and 10 inches inside? Ans. 894⅔

2. Find the moment of inertia of a $4'' \times 6'' \times \frac{1}{2}''$ angle, with respect to an axis parallel to its long leg, using the formula given in Table II. Ans. 6.27

3. What is the moment of inertia, with respect to an axis parallel to the web, of a **Z** bar 5 inches in depth, ½ inch thick, and having legs 3¼ inches in length? Use the formula given in Table II. Ans. 9.05

SECTION MODULUS

13. The **section modulus** of a section under **transverse stress,** or the stress to which a beam is subjected under a load which tends to bend it, is the moment about the neutral axis of such a portion of the section that if its area is multiplied by the **extreme fiber stress,** the **resisting moment** of this portion will equal the resisting moment of the variable stress in the actual section, from the neutral axis to the extreme fibers. By the term resisting moment is meant the value which expresses the resistance of the beam to the bending moment produced upon it by the loads which it supports. The resisting moment of any beam section is always equal to the section modulus multiplied by the extreme fiber stress, which is the strength value of

ELEMENTS OF

Sections	Area of Section A	Distance From Neutral Axis to Extremities of Section c and c_1
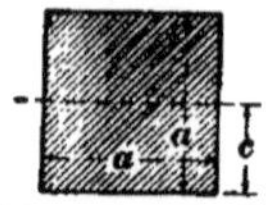	a^2	$c = \frac{a}{2}$
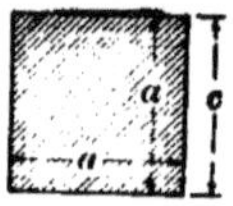	a^2	$c = a$
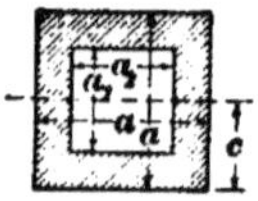	$a^2 - a_1^2$	$c = \frac{a}{2}$
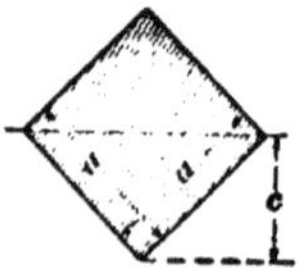	a^2	$c = \frac{a}{\sqrt{2}} = .707\,a$
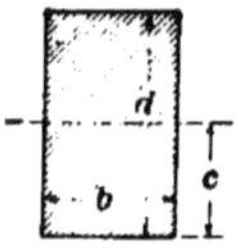	$b\,d$	$c = \frac{d}{2}$
	$b\,d$	$c = d$
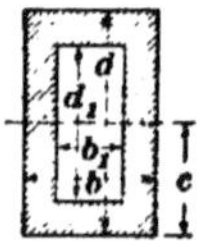	$b\,d - b_1\,d_1$	$c = \frac{d}{2}$
	$b\,d$	$c = \frac{b\,d}{\sqrt{b^2 + d^2}}$
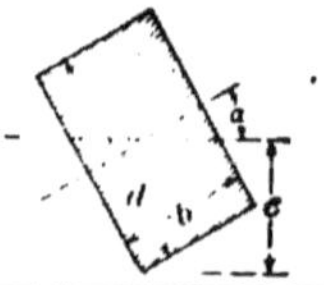	$b\,d$	$c = \frac{d \cos a + b \sin a}{2}$

E I

UAL SECTIONS

Moment of Inertia I	Section Modulus $S = \frac{I}{c}$	Radius of Gyration $r = \sqrt{\frac{I}{A}}$
$\frac{a^4}{12}$	$\frac{a^3}{6}$	$\frac{a}{\sqrt{12}} = .289\,a$
$\frac{a^4}{3}$	$\frac{a^3}{3}$	$\frac{a}{\sqrt{3}} = .577\,a$
$\frac{a^4 - a_1^4}{12}$	$\frac{a^4 - a_1^4}{6\,a}$	$\sqrt{\frac{a^2 + a_1^2}{12}}$
$\frac{a^4}{12}$	$\frac{a^3}{6\sqrt{2}} = .118\,a^3$	$\frac{a}{\sqrt{12}} = .289\,a$
$\frac{b\,d^3}{12}$	$\frac{b\,d^2}{6}$	$\frac{d}{\sqrt{12}} = .289\,d$
$\frac{b\,d^3}{3}$	$\frac{b\,d^2}{3}$	$\frac{d}{\sqrt{3}} = .577\,d$
$\frac{b\,d^3 - b_1\,d_1^3}{12}$	$\frac{b\,d^3 - b_1\,d_1^3}{6\,d}$	$\sqrt{\frac{b\,d^3 - b_1\,d_1^3}{12\,(b\,d - b_1\,d_1)}}$
$\frac{b^3\,d^3}{6\,(b^2 + d^2)}$	$\frac{b^2\,d^2}{6\sqrt{b^2 + d^2}}$	$\frac{b\,d}{\sqrt{6\,(b^2 + d^2)}}$
$\frac{b\,d}{12}(d^2 \cos^2 a + b^2 \sin^2 a)$	$\frac{b\,d}{6}\left(\frac{d^2 \cos^2 a + b^2 \sin^2 a}{d \cos a + b \sin a}\right)$	$\sqrt{\frac{d^2 \cos^2 a + b^2 \sin^2 a}{12}}$

Sections	Area of Section A	Distance From Neutral Axis to Extremities of Section c and c_1
	$\frac{\pi d^2}{4} = .785\, d^2$	$c = \frac{d}{2}$
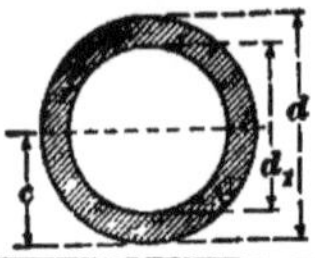	$\frac{\pi (d^2 - d_1^2)}{4} = .785\,(d^2 - d_1^2)$	$c = \frac{d}{2}$
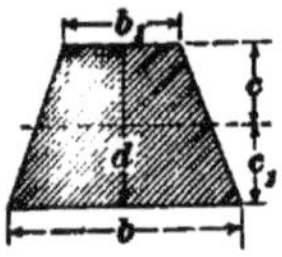	$\frac{b + b_1}{2} \cdot d$	$c_1 = \frac{b + 2\,b_1}{b + b_1} \cdot \frac{d}{3}$ $c = \frac{b_1 + 2\,b}{b + b_1} \cdot \frac{d}{3}$
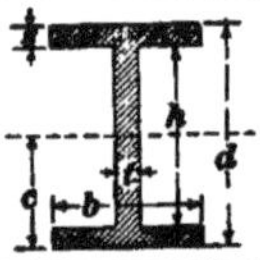	$b\,d - h\,(b - t)$	$c = \frac{d}{2}$
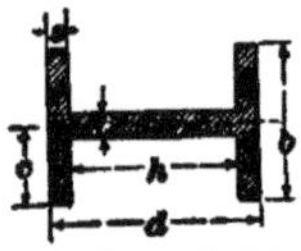	$b\,d - h\,(b - t)$	$c = \frac{b}{2}$
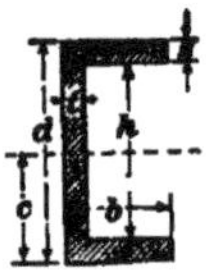	$b\,d - h\,(b - t)$	$c = \frac{d}{2}$
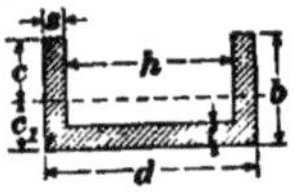	$b\,d - h\,(b - t)$	$c_1 = \frac{2\,b^2 s + h\,t^2}{2\,A}$ $c = b - c_1$
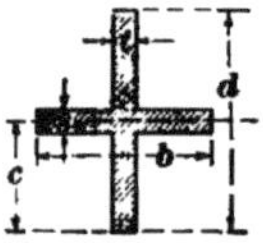	$t\,d + s\,(b - t)$	$c = \frac{d}{2}$
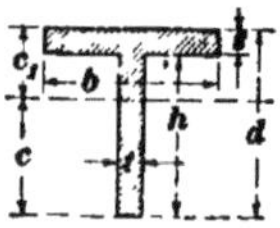	$b\,s + h\,t$	$c_1 = \frac{d^2 t + s^2 (b - t)}{2\,A}$ $c = d - c_1$
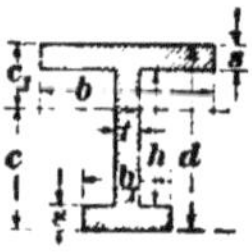	$b\,s + h\,t + b_1\,s$	$c_1 = \frac{t\,d^2 + s^2 (b - t) + s\,(b_1 - t)\,(2\,d - s}{2\,A}$ $c = d - c_1$

Continued)

Moment of Inertia I	Section Modulus $S = \frac{I}{c}$	Radius of Gyration $r = \sqrt{\frac{I}{A}}$
$\frac{\pi d^4}{64} = .049\, d^4$	$\frac{\pi d^3}{32} = .098\, d^3$	$\frac{d}{4}$
$\frac{\pi (d^4 - d_1^4)}{64} = .049\,(d^4 - d_1^4)$	$\frac{\pi}{32}\left(\frac{d^4 - d_1^4}{d}\right) = .098\left(\frac{d^4 - d_1^4}{d}\right)$	$\frac{\sqrt{d^2 + d_1^2}}{4}$
$\frac{b^2 + 4bb_1 + b_1^2}{36(b + b_1)} \cdot d^3$	$\frac{b^2 + 4bb_1 + b_1^2}{12(b_1 + 2b)} \cdot d^2$	$\frac{d}{6(b + b_1)}\sqrt{2(b^2 + 4bb_1 + b_1^2)}$
$\frac{bd^3 - h^3(b - t)}{12}$	$\frac{bd^3 - h^3(b - t)}{6d}$	$\sqrt{\frac{bd^3 - h^3(b - t)}{12[bd - h(b - t)]}}$
$\frac{2sb^3 + ht^3}{12}$	$\frac{2sb^3 + ht^3}{6b}$	$\sqrt{\frac{2sb^3 + ht^3}{12[bd - h(b - t)]}}$
$\frac{bd^3 - h^3(b - t)}{12}$	$\frac{bd^3 - h^3(b - t)}{6d}$	$\sqrt{\frac{bd^3 - h^3(b - t)}{12[bd - h(b - t)]}}$
$\frac{2sb^3 + ht^3}{3} - Ac_1^2$	$\frac{I}{b - c_1}$	$\sqrt{\frac{I}{A}}$
$\frac{td^3 + s^3(b - t)}{12}$	$\frac{td^3 + s^3(b - t)}{6d}$	$\sqrt{\frac{td^3 + s^3(b - t)}{12[td + s(b - t)]}}$
$\frac{tc^3 + bc_1^3 - (b - t)(c_1 - s)^3}{3}$	$\frac{I}{d - c_1}$	$\sqrt{\frac{tc^3 + bc_1^3 - (b - t)(c_1 - s)^3}{3(bs + ht)}}$
$\frac{bc_1^3 + b_1c^3 - (b - t)(c_1 - s)^3}{3} - \frac{(b_1 - t)(c - s)^3}{3}$	$\frac{I}{d - c_1}$	$\left[\frac{bc_1^3 + b_1c^3 - (b - t)(c_1 - s)^3}{3(bs + ht + b_1s)} - \frac{(b_1 - t)(c - s)^3}{3(bs + ht + b_1s)}\right]^{\frac{1}{2}}$

VALUES FOR STAN

	Moment of Inertia, Axis $y-y$ I
	$\frac{t(a-c)^3 + a c^3 - (a-t)}{3}$
	$\frac{t(a-c)^3 + b c^3 - (b-t)}{3}$
	$\frac{b d^3}{12} - \frac{h^4 - t^4}{16}$
	$e\left[\frac{k^2}{16} + \left(d - \frac{2s+k}{2}\right)^2\right] + \quad + \frac{b_1 s_1^3 + 2 b s^3}{6} - A (c$
	$\frac{b d^3}{12} - \frac{h^4 - t^4}{8}$
$t)$	$\frac{l^3(3t + t_1) + 4 b n_1^3 - 2}{12} - A (c_1 - n_1)^2$
	$\frac{a b^3 - a_1 (b - 2t)}{12}$

$t)^2$	$2c^4 - 2(c\cdot$
$t)^2$	$t(b - c_2)^2 +$
	$\frac{1}{3}\left(2sb^2 +\right.$
$+s_1)^2$ 3	$\frac{ek^2}{16}$ $+\frac{s_1b_1}{}$
	$\frac{b^2}{6}$
	$\frac{sb^2 + s_1t_1^2 + lt^2}{12}$ $+\frac{l(t_1 - t)}{}$
	$b(a + a_1)$

the material of which the section is composed, per square inch of area; and the material is only subjected to this extreme stress at the outside edge of the section.

The value designated as the section modulus is more clearly explained by referring to Fig. 11, in which $ABCD$ shows a section of a rectangular beam. EF represents the neutral axis, which passes through the center of gravity, and GH is a line situated at the edge of the section and represents the position of the extreme fibers. Then $\frac{d}{2}$ is the distance from the extreme fibers to the neutral axis. The stress in the fibers on the neutral axis is zero, but it gradually increases until the edge of the section is reached, where it is maximum. Therefore, while the area is uniform, the stress varies with the distance of the fiber from the neutral axis.

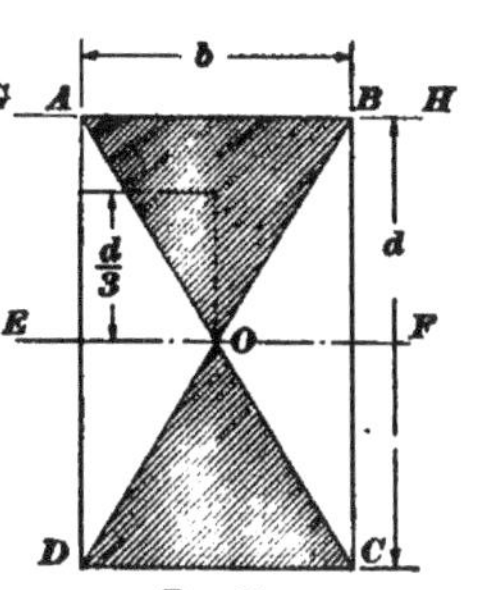

FIG. 11

Draw the diagonals AC and BD; the shaded portions ABO and DCO represent the area that, if considered as having a uniform stress equal to the stress in the extreme fibers of the section, offers the same resistance as the entire section having the variable stress. Then the moment of this shaded portion about the neutral axis gives the section modulus of the beam. The combined area of the shaded portions is equal to $\frac{bd}{2}$, and the lever arm, which is the distance from the center of gravity of each section to the neutral axis, is $\frac{d}{3}$. The moment, therefore, is $\frac{bd}{2} \times \frac{d}{3} = \frac{bd^2}{6}$, and the section modulus for a rectangular beam may be expressed by the formula

$$S = \frac{bd^2}{6} \qquad (3)$$

in which S = section modulus;
d = depth of beam, in inches;
b = breadth, or width, of beam, in inches.

14. In Fig. 12 is shown the graphical method of obtaining the section modulus for a **T** bar. The neutral axis is located at AB and the greatest distance from AB to the extreme fiber is c; from a and b draw lines to the center of gravity of the section o. Below the neutral axis lay off the line CD parallel to the neutral axis AB and at a distance from it equal to c. Project the points d and e to d' and e' and from these points draw lines to o; also project f and g to f' and g' and from these points draw lines to o. Then the shaded portions $h\,i\,o\,j\,k\,l\,m$ and $a\,o\,b$ represent areas that, if considered as having a uniform stress equal to that in the extreme fiber, will produce a stress equal to the total resisting stress, which is not uniform in the entire section. The moments of these areas about the neutral axis will give the section modulus of the section.

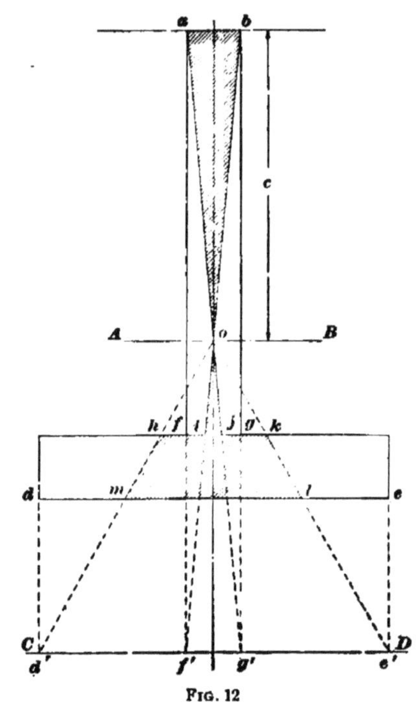

Fig. 12

The same method may be applied to an irregular section, as illustrated in Fig. 13. In this case the top and bottom edges of the section are equally distant from the neutral axis. Several points are projected from the surface of the section to the lines CD and EF and from these points lines are drawn to the point O on the neutral axis. Where the lines thus drawn intersect the horizontal lines drawn through the points in question, will be located points on the surface of the shaded portion.

The outline of the figure may then be drawn through the points thus located and the section determined. The moment of this section about the neutral axis gives the section modulus. In considering an irregular section, it is very necessary to apply the graphical method for determining the section modulus, as it is extremely difficult to evolve a convenient formula for obtaining it.

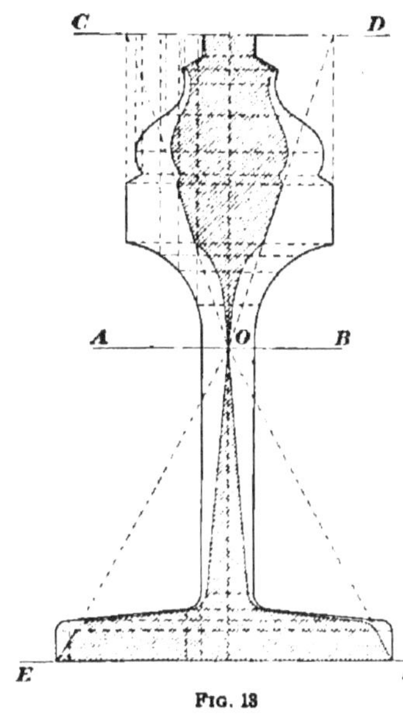

FIG. 13

15. The section modulus for any section whose moment of inertia is known, is found by dividing the moment of inertia by the greatest distance of the neutral axis from the outside fibers of the section. This may be expressed by the formula

$$S = \frac{I}{c} \qquad (4)$$

in which I = moment of inertia, with respect to the neutral axis;

S = section modulus;

c = distance from neutral axis to farthest edge of section.

EXAMPLE 1.—What will be the section modulus of a yellow-pine beam, 10 inches wide by 12 inches in depth?

SOLUTION.— d = 12 in., b = 10 in. Therefore, by applying formula **3**, we have

$$S = \frac{b\,d^2}{6} = \frac{10 \times 12^2}{6} = \frac{1{,}440}{6} = 240. \text{ Ans.}$$

EXAMPLE 2.—What is the section modulus of the cast-iron lintel shown in Fig. 14?

SOLUTION.—From Table I, we have the formula $S = \frac{I}{d - c_1}$. As it is necessary to have the area in order to determine the value of c_1, in the above formula, this will be determined by the formula in the first column, or $A = bs + ht$. Substituting the values, we have $A = 8 \times \frac{3}{4} + 9\frac{1}{4} \times \frac{3}{4} = \frac{207}{16}$. Then substituting in the formula

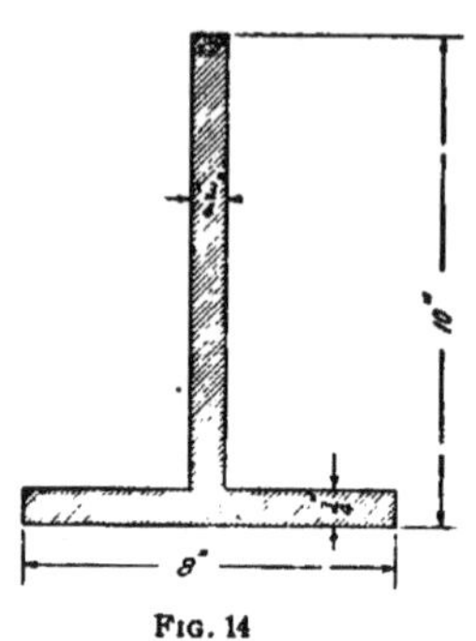

FIG. 14

$$c_1 = \frac{d^2 t + s^2 (b - t)}{2A}, \text{ gives}$$

$$c_1 = \frac{10^2 \times \frac{3}{4} + (\frac{3}{4})^2 (8 - \frac{3}{4})}{2 \times \frac{207}{16}} = 3.056 \text{ in.}$$

$$c = d - c_1 = 10 - 3.056 = 6.944 \text{ in.}$$

The moment of inertia is obtained from the formula

$$I = \frac{tc^3 + bc_1^3 - (b - t)(c_1 - s)^3}{3}$$

which gives

$$I = \frac{\frac{3}{4} \times 6.944^3 + 8 \times 3.056^3 - (8 - \frac{3}{4})(3.056 - \frac{3}{4})^3}{3} = 130.181$$

Then, $$S = \frac{I}{d - c_1} = \frac{130.181}{6.944} = 18.747. \text{ Ans.}$$

EXAMPLES FOR PRACTICE

1. What is the section modulus of a trapezoidal section whose parallel sides are 6 and 9 inches and the distance between them is 6 inches? Ans. $249\frac{3}{4}$

2. Find the section modulus of a hollow, rectangular, cast-iron lintel that measures 10 inches by 12 inches outside and is $\frac{5}{8}$ inch thick. Ans. 89.026

3. The structural member of a skylight has the section of a cross, the vertical web of which is 5 inches in depth; the cross-bar extends $1\frac{1}{4}$ inches each side of the center line of the vertical bar and the center line of the horizontal bar bisects that of the vertical bar. Provided the thickness of the metal is $\frac{3}{8}$ inch, what will be the section modulus? Ans. 1.5662

RADIUS OF GYRATION

16. In computing the strength of columns, frequent use is made of a property of a section that is numerically equal to the square root of the quotient of its moment of inertia, with respect to an axis through its center of gravity, divided by its area. This property is called the **radius of gyration** of the section, with respect to the given axis. It is usually expressed by the letter r, and its value, with respect to a given axis, for any section whose area A and moment of inertia I, with respect to the same axis, are known, is given by the formula

$$r = \sqrt{\frac{I}{A}} \quad (5)$$

The form in which the radius of gyration appears in most formulas for calculating the strength of columns is its square; hence, it is convenient to express the above relation by the formula

$$r^2 = \frac{I}{A} \quad (6)$$

which gives directly the value to be substituted in the column formulas. Formulas for computing the least radius of gyration and its square for the sections most often used in the design of structures, are given in Table I. The tables of the properties of rolled sections also give accurate values of r and r^2 for the sections used in the examples given in the following pages.

EXAMPLE.—Compute the square of the radius of gyration, with respect to the axis de, of the column section shown in Fig. 9.

SOLUTION.—By referring to example 4, Art. **12**, the moment of inertia of the section is found to be $I = 737.06$, and the area of the section is $3 \times 4.5 + 4 \times 3.75 = 28.50$ sq. in. Substituting these values in formula **6**, we have

$$r^2 = \frac{737.06}{28.50} = 25.86. \quad \text{Ans.}$$

EXAMPLES FOR PRACTICE

1. Find the radius of gyration of a round wooden column having a diameter of 12 inches. Ans. 3

2. What is the square of the least radius of gyration of the column section shown in Fig. 8? Ans. 3.583

3. Compute the radius of gyration of the section of a hollow cast-iron column, 8 inches by 12 inches outside and ¾ inch thick. Ans. 4.349

---

17. The diagram shown in Fig. 15 gives the width of flange, weight in pounds, per foot, and thickness of

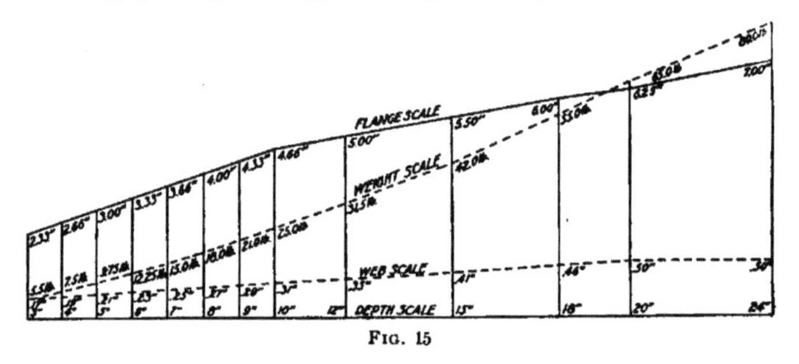

Fig. 15

web for beams of the usual depths, these values being for standard beams of minimum weight.

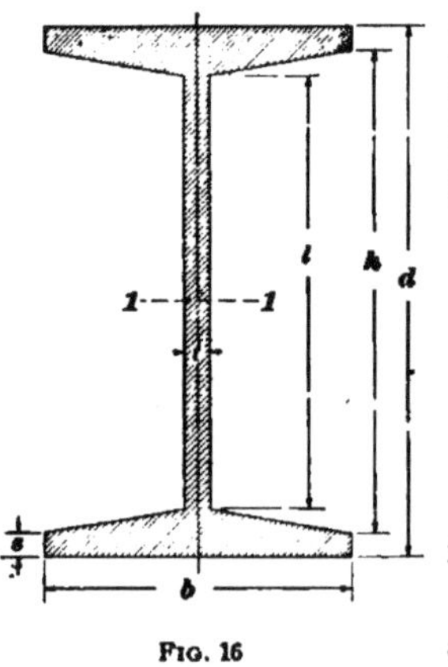

Fig. 16

The following formulas, adopted by the Association of American Steel Manufacturers, are of value in determining the properties of I beams. (See Fig. 16.)

Weight per foot = area × 3.4.

$$\text{Area} = td + 2s(b-t) + \frac{(b-t)^2}{12}.$$

$$\text{Section modulus} = S = \frac{2I}{d}.$$

$$\text{Slope of flange} = g = \frac{h-l}{b-t} = \frac{1}{6}$$

for standard beams.

I = moment of inertia, neutral axis (*1–1*) parallel to flange.

$I = \frac{1}{12}\left[b\,d^3 - \frac{1}{4g}(h^4 - l^4)\right]$, or $\frac{b\,d^3}{12} - \frac{h^4 - l^4}{8}$ for standard beams.

18. Fig. 17 shows the width of flange, thickness of web, and weight per foot for channel beams of the usual depths

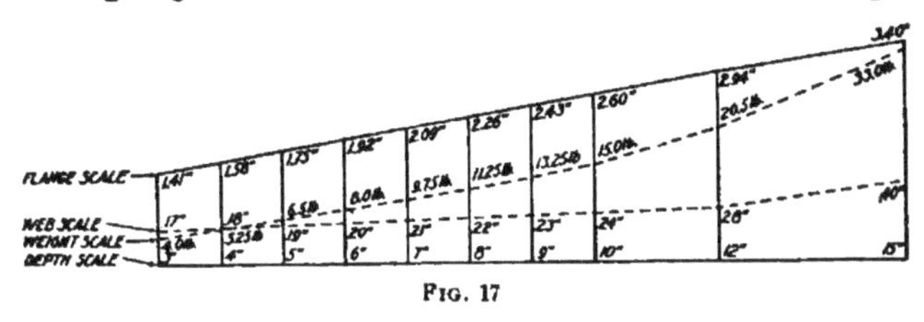

Fig. 17

and, as in the diagram for **I** beams, these values are for channels of minimum weight.

The following formulas are given in connection with Fig. 18.

Weight per foot = area × 3.4.

Area $= t\,d + 2\,s\,(b - t) + \frac{(b - t)^2}{6}$.

Section modulus $= S = \frac{2\,I}{d}$.

Slope of flange $= g = \frac{h - l}{2\,(b - t)}$, or $\frac{1}{6}$ for standard channels.

I = moment of inertia, neutral axis (1–1) parallel to flange.

$I = \frac{1}{12}\left[b\,d^3 - \frac{1}{8g}(h^4 - l^4)\right]$, or $\frac{b\,d^3}{12} - \frac{h^4 - l^4}{16}$ for standard channels.

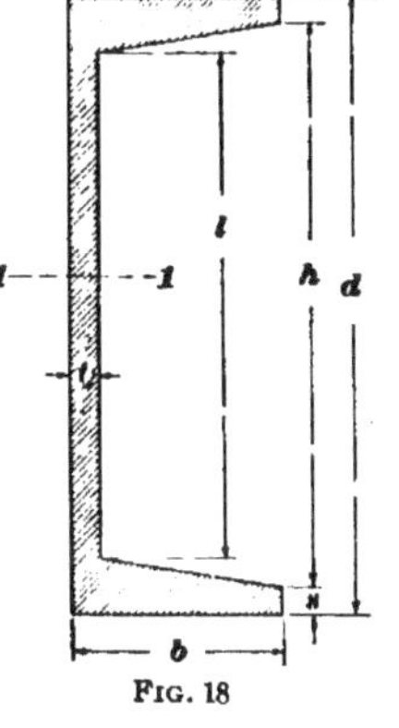

Fig. 18

19. Table I gives formulas for computing the area, moment of inertia, section modulus, and radius of gyration of usual sections, also the distance from the neutral axis to the extremities of the section. These formulas are useful in obtaining the properties of sections that are not given in the table of properties for the several standard rolled sections.

20. Table II gives formulas for the moment of inertia of the standard rolled sections with respect to their different axes, also the distance from the neutral axis to the back of the section; where a third axis is considered it gives the angle between the third axis and the vertical one.

The slope of the flanges of the I beams and channels is $\frac{1}{6}$, that is, $\frac{h-l}{b-t} = \frac{1}{6}$. In the formulas relating to the deck beam, e represents the area of the head.

21. Tables III to IX give the dimensions and properties of structural-steel shapes that conform to the American standards, adopted January, 1896. Such tables are found in most of the handbooks published by the different steel manufacturers. The first five columns give the dimensions, area, and weight of the section considered, and as all these sections are of structural steel, there is a fixed relation between the weight and area and one may be obtained from the other. The weight of 1 square inch of structural steel 1 foot long is 3.4 pounds; hence, the weight of 1 foot of any section is equal to its sectional area multiplied by 3.4, as stated in Arts. **17** and **18,** in connection with the weight of I beams and channels.

The values for the section modulus, radius of gyration, and moment of inertia are required in finding the strength of the section under different conditions, and are used in calculating the strength of beams, girders, columns, and posts. The moment of inertia is particularly valuable because the section modulus and radius of gyration may be obtained from it, as explained in Arts. **15** and **16.**

The uses of the coefficients of strength and coefficients of deflection are illustrated in the examples given in Art. **22.**

The values in the columns headed F and F' are the safe uniformly distributed loads, in pounds, including the weight of the beam, for a beam 1 foot long; they have been computed for fiber stresses of 16,000 and 12,500 pounds per square inch, respectively. Hence, the safe load for any span is equal to the coefficient F or F' divided by the span, in feet.

The formulas from which these coefficients were obtained are $F = \frac{8}{3} \times 16{,}000 \times S$ and $F' = \frac{8}{3} \times 12{,}500 \times S$, in which S is the section modulus.

The values for N and N', the coefficients of deflection for uniform and center loads, respectively, were obtained from the formulas $N = \frac{W l^3}{76.8 E I}$ and $N' = \frac{W l^3}{48 E I}$, in which W equals 1,000 pounds, l equals 12 inches, E equals 29,000,000, and I equals the moment of inertia about the axis *1–1*. Therefore, these coefficients represent the deflection, in inches, of a beam 1 foot long having a load of 1,000 pounds. Multiplying the proper coefficient by the cube of the span, in feet, and by the number of 1,000-pound units in the given load, will give the deflection of a beam for any load and span.

Tables XI, XII, and XIII are useful in calculating the strength of these built-up sections when they are used as posts or struts in structural work. The angles given in these tables are not all standard, but the standard sections may be selected by referring to Tables V and VI.

22. The following examples illustrate the application of the tables giving the values for the properties of the several rolled sections.

In making the selection of the beams from the tables, it is customary to select the deepest beam having the least weight and the greatest section modulus. For instance, if a beam is required to have a section modulus of 30, and from the table it is observed that a 10-inch 40-pound beam has a section modulus of 31.7 and that a 12-inch 31.5-pound beam has a section modulus of 36, it is evident that it would be more economical to select the deeper, and consequently, the safer beam, thus saving a weight of 5 pounds per foot. As steel beams are sold usually by the pound price, such a saving would be great in a large order. This economical practice should always be employed when the several additional inches in the depth of the beam will not interfere with the required headroom in the architectural design.

EXAMPLE 1.—What is the deflection of a 20-inch 65-pound I beam that carries a center load of 28,000 pounds and has a span of 20 feet?

SOLUTION.—The amount of deflection is obtained by multiplying the coefficient of deflection for beams with center loads (column 15,

Table III) by the cube of the span, in feet, and the number of 1,000-lb. units in the load. Hence, the deflection equals

$$.00000106 \times 20^3 \times \frac{28,000}{1,000} = .237 \text{ in. Ans.}$$

EXAMPLE 2.—What size of I beam is required to support a uniform load of 800 pounds per foot over a span of 25 feet, considering a safe unit fiber stress of 16,000 pounds?

SOLUTION.—According to Art. **21**, the safe uniformly distributed load for any span is equal to the coefficient F or F' divided by the span, in feet. By transposing the terms of the equation it is found that F or F' = load × span. The coefficient required for the load in this case is 800 × 25 × 25 = 500,000. From column 12, Table III, the coefficient of strength for a 15-in. 42-lb. beam is found to be 628,270. The weight of the beam is 42 × 25 = 1,050 lb., which requires a coefficient of 1,050 × 25 = 26,250; then 628,270 − 26,250 gives a net coefficient of 602,020. Therefore, a 15-in. 42-lb. beam is the size required. If the load on the beam is concentrated at the center, the coefficient of strength required is twice as great. Ans.

EXAMPLE 3.—Two channels, placed back to back with webs vertical, form the support of an overhead mechanism for an elevator. The load is concentrated at the center of the channels and is equal to 10,000 pounds; provided the span of the channels is 8 feet and they are 7 inches in depth and weigh 14.75 pounds per foot, what will be the deflection?

SOLUTION.—The deflection is equal to the coefficient multiplied by the cube of the span and the number of 1,000-lb. units in the load. From column N', Table IV, the coefficient of deflection for a 7-in. 14.75-lb. channel is .0000457. As there are two channels, each may be considered as supporting one-half of the load or 5,000 lb. Then the deflection will be equal to

$$.0000457 \times 8^3 \times \frac{5,000}{1,000} = .117 \text{ in. Ans.}$$

EXAMPLE 4.—What is the section modulus of a rectangular bar 3 inches by ¾ inch, using the moment of inertia obtained from Table X?

SOLUTION.—The section modulus is determined by the formula $S = \frac{I}{c}$, given in Art. **15**, and from the table the moment of inertia is found to be 1.69. The neutral axis passes through the center of the rectangle and consequently the distance c is equal to 1½ in. Substituting these values in the formula,

$$S = \frac{1.69}{1.5} = 1.127. \text{ Ans.}$$

EXAMPLES FOR PRACTICE

1. Two channels, placed back to back, are to be used to support a uniform load of 1,000 pounds per lineal foot over a span of 20 feet. Using a safe unit fiber stress of 16,000 pounds, what size channels will be the most economical for the purpose?

Ans. Two 12-in. 20½-lb. channels

2. What would be the deflection in the previous example?

Ans. .488 in.

3. What size of I beam should be used to support a center load of 10,000 pounds over a span of 12 feet, using a safe unit fiber stress of 16,000 pounds?

Ans. 10-in. 25-lb. I beam

4. An I beam to support a given load is required to have a section modulus of 30; what would be the size and weight of the most economical section?

Ans. 12-in. 31½-lb. I beam

5. From the fact that the available headroom is limited, it is necessary to use several 8-inch I beams placed side by side in order to support a uniformly distributed load of 20,000 pounds. The span of the beams is 20 feet. How many beams weighing 25.25 pounds are required to support this load, provided the allowable deflection is 1 inch, and a safe unit fiber stress of 16,000 pounds is used?

Ans. Three beams

TABLE III

PROPERTIES OF STANDARD I BEAMS

1	2	3	4	5	6	7	8	9	10	11	12	13	14	15
Depth of Beam Inches	Weight per Foot Pounds	Area of Section Square Inches	Thickness of Web Inches	Width of Flange Inches	Moment of Inertia, Axis 1-1 Inches	Section Modulus, Axis 1-1 Inches	Radius of Gyration, Axis 1-1 Inches	Moment of Inertia, Axis 2-2 Inches	Radius of Gyration, Axis 2-2 Inches	Increase of Thickness of Web for Each Pound Increase in Weight. Inches	Coefficient of Strength		Coefficient of Deflection	
											For Fiber Stress of 16,000 Pounds per Square Inch for Buildings	For Fiber Stress of 12,500 Pounds per Square Inch for Bridges	Uniform Load	Center Load
d		A	t	b	I	s	r	I'	r'	f	F	F'	N	N'
3	5.50	1.63	.17	2.33	2.5	1.7	1.23	.46	.53	.098	17,650	13,790	.00031253	.00050006
3	6.50	1.91	.26	2.42	2.7	1.8	1.19	.53	.52		19,140	14,950	.00028827	.00046124
3	7.50	2.21	.36	2.52	2.9	1.9	1.15	.60	.52		20,710	16,180	.00026644	.00042630
4	7.50	2.21	.19	2.66	6.0	3.0	1.64	.77	.59	.074	31,810	24,850	.00013009	.00020815
4	8.50	2.50	.26	2.73	6.4	3.2	1.59	.85	.58		33,890	26,480	.00012209	.00019535
4	9.50	2.79	.34	2.81	6.7	3.4	1.54	.93	.58		35,980	28,110	.00011500	.00018400
4	10.50	3.09	.41	2.88	7.1	3.6	1.52	1.01	.57		38,070	29,750	.00010868	.00017389
5	9.75	2.87	.21	3.00	12.1	4.8	2.05	1.23	.65	.059	51,590	40,300	.00006417	.00010267
5	12.25	3.60	.36	3.15	13.6	5.4	1.94	1.45	.63		58,100	45,390	.00005698	.00009117
5	14.75	4.34	.50	3.29	15.1	6.1	1.87	1.70	.63		64,630	50,490	.00005122	.00008195
6	12.25	3.61	.23	3.33	21.8	7.3	2.46	1.85	.72	.049	77,460	60,520	.00003561	.00005698
6	14.75	4.34	.35	3.45	24.0	8.0	2.35	2.09	.69		85,270	66,610	.00003235	.00005177
6	17.25	5.07	.47	3.57	26.2	8.7	2.27	2.36	.68		93,110	72,740	.00002963	.00004741

7	15.00	4.42	.25	3.66	36.2	10.4	2.86	2.67	.78	.042	110,410	86,260	.00002142	.00003427
7	17.50	5.15	.35	3.76	39.2	11.2	2.76	2.94	.76		119,400	93,290	.00001980	.00003168
7	20.00	5.88	.46	3.87	42.2	12.1	2.68	3.24	.74		128,560	100,430	.00001839	.00002943
8	18.00	5.33	.27	4.00	56.9	14.2	3.27	3.78	.84	.037	151,660	118,490	.00001364	.00002183
8	20.25	5.96	.35	4.08	60.2	15.0	3.18	4.04	.82		160,510	125,400	.00001289	.00002062
8	22.75	6.69	.44	4.17	64.1	16.0	3.10	4.36	.81		170,970	133,570	.00001210	.00001936
8	25.25	7.43	.53	4.26	68.0	17.0	3.03	4.71	.80		181,430	141,740	.00001140	.00001825
9	21.00	6.31	.29	4.33	84.9	18.9	3.67	5.16	.90	.033	201,300	157,260	.00000914	.00001462
9	25.00	7.35	.41	4.45	91.9	20.4	3.54	5.65	.88		217,930	170,260	.00000844	.00001350
9	30.00	8.82	.57	4.61	101.9	22.6	3.40	6.42	.85		241,460	188,640	.00000762	.00001219
9	35.00	10.29	.73	4.77	111.8	24.8	3.30	7.31	.84		264,990	207,020	.00000694	.00001110
10	25.00	7.37	.31	4.66	122.1	24.4	4.07	6.89	.97	.029	260,470	203,500	.00000635	.00001017
10	30.00	8.82	.45	4.80	134.2	26.8	3.90	7.65	.93		286 250	223,630	.00000578	.00000925
10	35.00	10.29	.60	4.95	146.4	29.3	3.77	8.52	.91		312,390	244,050	.00000530	.00000848
10	40.00	11.76	.75	5.10	158.7	31.7	3.67	9.50	.90		338,530	264,480	.00000489	.00000782
12	31.50	9.26	.35	5.00	215.8	36.0	4.83	9.50	1.01	.025	383,670	299,740	.00000360	.00000575
12	35.00	10.29	.44	5.09	228.3	38.0	4.71	10.07	.99		405,800	317,030	.00000340	.00000544
12	40.00	11.76	.56	5.21	245.9	41.0	4.57	10.95	.96		437,170	341,540	.00000316	.00000505
15	42.00	12.48	.41	5.50	441.8	58.9	5.95	14.62	1.08	.020	628,270	490,840	.00000176	.00000281
15	45.00	13.24	.46	5.55	455.8	60.8	5.87	15.09	1.07		648,310	506,490	.00000170	.00000272
15	50.00	14.71	.56	5.65	483.4	64.5	5.73	16.04	1.04		687,530	537,130	.00000161	.00000257
15	55.00	16.18	.66	5.75	511.0	68.1	5.62	17.06	1.03		726,740	567,770	.00000152	.00000243
15	60.00	17.65	.76	5.84	538.6	71.8	5.52	18.17	1.01		765,960	598,410	.00000144	.00000231
18	55.00	15.93	.46	6.00	795.6	88.4	7.07	21.19	1.15	.016	942,880	736,620	.00000098	.00000156
18	60.00	17.65	.56	6.10	841.8	93.5	6.91	22.38	1.13		997,680	779,440	.00000092	.00000148
18	65.00	19.12	.64	6.18	881.5	97.9	6.79	23.47	1.11		1,044,740	816,200	.00000088	.00000141
18	70.00	20.59	.72	6.26	921.2	102.4	6.69	24.62	1.09		1,091,800	852,970	.00000084	.00000135
20	65.00	19.08	.50	6.25	1,169.5	117.0	7.83	27.86	1.21	.015	1,247,490	974,600	.00000066	.00000106
20	70.00	20.59	.58	6.33	1,219.8	122.0	7.70	29.04	1.19		1,301,110	1,016,490	.00000064	.00000102
20	75.00	22.06	.65	6.40	1,268.8	126.9	7.58	30.25	1.17		1,353,400	1,057,340	.00000061	.00000098
24	80.00	23.32	.50	7.00	2,087.2	173.9	9.46	42.86	1.36	.0123	1,855,310	1,449,460	.00000037	.00000060
24	85.00	25.00	.57	7.07	2,167.8	180.7	9.31	44.35	1.33		1,926,950	1,505,430	.00000036	.00000057
24	90.00	26.47	.63	7.13	2,238.4	186.5	9.20	45.70	1.31		1,989,700	1,554,450	.00000035	.00000056
24	95.00	27.94	.69	7.19	2,309.0	192.4	9.09	47.10	1.30		2,052,440	1,603,470	.00000034	.00000054
24	100.00	29.41	.75	7.25	2,379.6	198.3	8.99	48.55	1.28		2,115,190	1,652,490	.00000033	.00000052

TABLE IV

PROPERTIES OF STANDARD CHANNELS

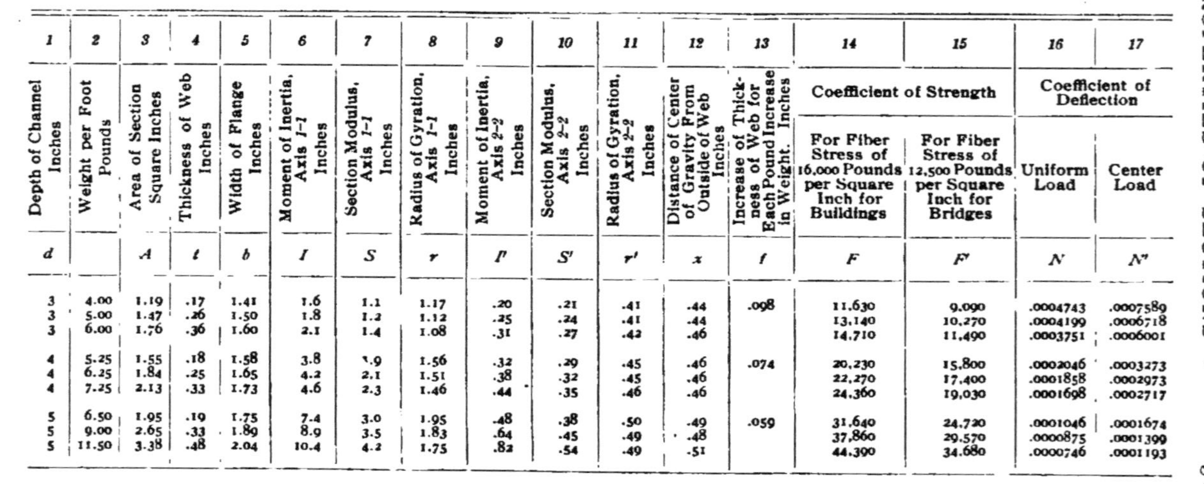

1	2	3	4	5	6	7	8	9	10	11	12	13	14	15	16	17
Depth of Channel Inches	Weight per Foot Pounds	Area of Section Square Inches	Thickness of Web Inches	Width of Flange Inches	Moment of Inertia, Axis 1-1 Inches	Section Modulus, Axis 1-1 Inches	Radius of Gyration, Axis 1-1 Inches	Moment of Inertia, Axis 2-2 Inches	Section Modulus, Axis 2-2 Inches	Radius of Gyration, Axis 2-2 Inches	Distance of Center of Gravity From Outside of Web Inches	Increase of Thickness of Web for Each Pound Increase in Weight. Inches	Coefficient of Strength		Coefficient of Deflection	
													For Fiber Stress of 16,000 Pounds per Square Inch for Buildings	For Fiber Stress of 12,500 Pounds per Square Inch for Bridges	Uniform Load	Center Load
d		A	t	b	I	S	r	I'	S'	r'	x	f	F	F'	N	N'
3	4.00	1.19	.17	1.41	1.6	1.1	1.17	.20	.21	.41	.44	.098	11,630	9,090	.0004743	.0007589
3	5.00	1.47	.26	1.50	1.8	1.2	1.12	.25	.24	.41	.44		13,140	10,270	.0004199	.0006718
3	6.00	1.76	.36	1.60	2.1	1.4	1.08	.31	.27	.42	.46		14,710	11,490	.0003751	.0006001
4	5.25	1.55	.18	1.58	3.8	1.9	1.56	.32	.29	.45	.46	.074	20,230	15,800	.0002046	.0003273
4	6.25	1.84	.25	1.65	4.2	2.1	1.51	.38	.32	.45	.46		22,270	17,400	.0001858	.0002973
4	7.25	2.13	.33	1.73	4.6	2.3	1.46	.44	.35	.46	.46		24,360	19,030	.0001698	.0002717
5	6.50	1.95	.19	1.75	7.4	3.0	1.95	.48	.38	.50	.49	.059	31,640	24,720	.0001046	.0001674
5	9.00	2.65	.33	1.89	8.9	3.5	1.83	.64	.45	.49	.48		37,860	29,570	.0000875	.0001399
5	11.50	3.38	.48	2.04	10.4	4.2	1.75	.82	.54	.49	.51		44,390	34,680	.0000746	.0001193

6	8.00	2.38	.20	1.92	13.0	4.3	2.34	.70	.50	.54	.52	.049	46,210	36,100	.0000597	.0000855
6	10.50	3.09	.32	2.04	15.1	5.0	2.21	.88	.57	.53	.50		53,750	42,000	.0000513	.0000821
6	13.00	3.82	.44	2.16	17.3	5.8	2.13	1.07	.65	.53	.52		61,600	48,120	.0000448	.0000717
6	15.50	4.56	.56	2.28	19.5	6.5	2.07	1.28	.74	.53	.55		69,440	54,250	.0000397	.0000636
7	9.75	2.85	.21	2.09	21.1	6.0	2.72	.98	.63	.59	.55	.042	64,270	50,210	.0000368	.0000588
7	12.25	3.60	.32	2.20	24.2	6.9	2.59	1.19	.71	.57	.53		73,650	57,540	.0000321	.0000514
7	14.75	4.34	.42	2.30	27.2	7.8	2.50	1.40	.79	.57	.53		82,740	64,690	.0000286	.0000457
7	17.25	5.07	.53	2.41	30.2	8.6	2.44	1.62	.87	.56	.55		91,950	71,840	.0000257	.0000411
7	19.75	5.81	.63	2.51	33.2	9.5	2.39	1.85	.96	.56	.58		101,100	78,990	.0000234	.0000374
8	11.25	3.35	.22	2.26	32.3	8.1	3.10	1.33	.79	.63	.58	.037	86,140	67,300	.0000240	.0000384
8	13.75	4.04	.31	2.35	36.0	9.0	2.98	1.55	.87	.62	.56		95,990	75,000	.0000216	.0000345
8	16.25	4.78	.40	2.44	39.9	10.0	2.89	1.78	.95	.61	.56		106,450	83,170	.0000194	.0000311
8	18.75	5.51	.49	2.53	43.8	11.0	2.82	2.01	1.02	.60	.57		116,910	91,340	.0000177	.0000283
8	21.25	6.25	.58	2.62	47.8	11.9	2.76	2.25	1.11	.60	.59		127,370	99,510	.0000162	.0000260
9	13.25	3.89	.23	2.43	47.3	10.5	3.49	1.77	.97	.67	.61	.033	112,170	87,630	.0000164	.0000262
9	15.00	4.41	.29	2.49	50.9	11.3	3.40	1.95	1.03	.66	.59		120,540	94,170	.0000153	.0000244
9	20.00	5.88	.45	2.65	60.8	13.5	3.21	2.45	1.19	.65	.58		144,070	112,550	.0000128	.0000204
9	25.00	7.35	.61	2.81	70.7	15.7	3.10	2.98	1.36	.64	.62		167,590	130,930	.0000110	.0000176
10	15.00	4.46	.24	2.60	66.9	13.4	3.87	2.30	1.17	.72	.64	.029	142,680	111,470	.0000116	.0000186
10	20.00	5.88	.38	2.74	78.7	15.7	3.66	2.85	1.34	.70	.61		167,940	131,210	.0000099	.0000158
10	25.00	7.35	.53	2.89	91.0	18.2	3.52	3.40	1.50	.68	.62		194,090	151,630	.0000085	.0000136
10	30.00	8.82	.68	3.04	103.2	20.6	3.42	3.99	1.67	.67	.65		220,230	172,060	.0000075	.0000120
10	35.00	10.29	.82	3.18	115.5	23.1	3.35	4.66	1.87	.67	.69		246,380	192,480	.0000067	.0000107
12	20.50	6.03	.28	2.94	128.1	21.4	4.61	3.91	1.75	.81	.70	.025	227,750	177,930	.0000061	.0000097
12	25.00	7.35	.39	3.05	144.0	24.0	4.43	4.53	1.91	.78	.68		256,000	200,000	.0000054	.0000086
12	30.00	8.82	.51	3.17	161.6	26.9	4.28	5.21	2.09	.77	.68		287,370	224,510	.0000048	.0000077
12	35.00	10.29	.64	3.30	179.3	29.9	4.17	5.90	2.27	.76	.69		318,750	249,020	.0000043	.0000069
12	40.00	11.76	.76	3.42	196.9	32.8	4.09	6.63	2.46	.75	.72		350,120	273,530	.0000039	.0000063
15	33.00	9.90	.40	3.40	312.6	41.7	5.62	8.23	3.16	.91	.79	.020	444,520	347,280	.0000025	.0000040
15	35.00	10.29	.43	3.43	319.9	42.7	5.57	8.48	3.22	.91	.79		455,030	355,500	.0000024	.0000039
15	40.00	11.76	.52	3.52	347.5	46.3	5.44	9.39	3.43	.89	.78		494,250	386,130	.0000022	.0000036
15	45.00	13.24	.62	3.62	375.1	50.0	5.32	10.29	3.63	.88	.79		533,470	416,770	.0000021	.0000033
15	50.00	14.71	.72	3.72	402.7	53.7	5.23	11.22	3.85	.87	.80		572,680	447,410	.0000019	.0000031
15	55.00	16.18	.82	3.82	430.2	57.4	5.16	12.19	4.07	.87	.82		611,900	478,050	.0000018	.0000029

TABLE V

PROPERTIES OF STANDARD ANGLES HAVING EQUAL LEGS

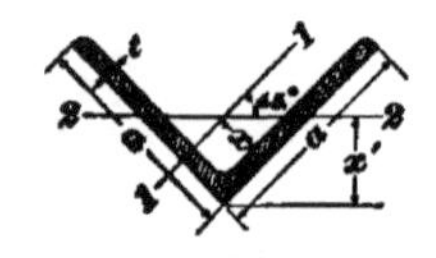

1	2	3	4	5	6	7	8	9	10	11	12
Dimensions Inches	Thickness Inches	Weight per Foot Pounds	Area of Section Square Inches	Distance of Center of Gravity From Back of Flange. Inches	Moment of Inertia, Axis 1-1 Inches	Section Modulus, Axis 1-1 Inches	Radius of Gyration, Axis 1-1 Inches	Distance of Center of Gravity From External Apex. Inches	Least Moment of Gyration, Axis 2-2 Inches	Section Modulus, Axis 2-2 Inches	Least Radius of Gyration, Axis 2-2 Inches
$a \times a$	t		A	x	I	S	r	x'	I'	S'	r'
$\frac{3}{4} \times \frac{3}{4}$	$\frac{1}{8}$	.58	.17	.23	.009	.017	.22	.33	.004	.011	.14
$\frac{3}{4} \times \frac{3}{4}$	$\frac{3}{16}$	.84	.25	.25	.012	.024	.22	.36	.005	.014	.14
1×1	$\frac{1}{8}$	.80	.23	.30	.022	.031	.30	.42	.009	.021	.19
1×1	$\frac{3}{16}$	1.16	.34	.32	.030	.044	.30	.45	.013	.028	.19
1×1	$\frac{1}{4}$	1.49	.44	.34	.037	.056	.29	.48	.016	.034	.19
$1\frac{1}{4} \times 1\frac{1}{4}$	$\frac{1}{8}$	1.02	.30	.36	.044	.049	.38	.51	.018	.035	.24
$1\frac{1}{4} \times 1\frac{1}{4}$	$\frac{3}{16}$	1.47	.43	.38	.061	.071	.38	.54	.025	.047	.24
$1\frac{1}{4} \times 1\frac{1}{4}$	$\frac{1}{4}$	1.91	.56	.40	.077	.091	.37	.57	.033	.057	.24
$1\frac{1}{4} \times 1\frac{1}{4}$	$\frac{5}{16}$	2.32	.68	.42	.090	.109	.36	.60	.040	.066	.24
$1\frac{1}{2} \times 1\frac{1}{2}$	$\frac{1}{8}$	1.23	.36	.42	.08	.072	.47	.60	.031	.053	.30
$1\frac{1}{2} \times 1\frac{1}{2}$	$\frac{3}{16}$	1.79	.53	.44	.11	.104	.46	.63	.045	.072	.29
$1\frac{1}{2} \times 1\frac{1}{2}$	$\frac{1}{4}$	2.34	.69	.47	.14	.134	.45	.66	.058	.088	.29
$1\frac{1}{2} \times 1\frac{1}{2}$	$\frac{5}{16}$	2.86	.84	.49	.16	.162	.44	.69	.070	.101	.29
$1\frac{1}{2} \times 1\frac{1}{2}$	$\frac{3}{8}$	3.35	.98	.51	.19	.188	.44	.72	.082	.114	.29
$1\frac{1}{2} \times 1\frac{1}{2}$	$\frac{7}{16}$	3.81	1.12	.53	.21	.214	.43	.75	.094	.126	.29
$1\frac{3}{4} \times 1\frac{3}{4}$	$\frac{3}{16}$	2.11	.62	.51	.18	.14	.54	.72	.073	.10	.34
$1\frac{3}{4} \times 1\frac{3}{4}$	$\frac{1}{4}$	2.76	.81	.53	.23	.19	.53	.75	.094	.13	.34
$1\frac{3}{4} \times 1\frac{3}{4}$	$\frac{5}{16}$	3.39	1.00	.55	.27	.23	.52	.78	.113	.15	.34
$1\frac{3}{4} \times 1\frac{3}{4}$	$\frac{3}{8}$	3.98	1.17	.57	.31	.26	.51	.81	.133	.16	.34
$1\frac{3}{4} \times 1\frac{3}{4}$	$\frac{7}{16}$	4.56	1.34	.59	.35	.30	.51	.84	.152	.18	.34
$1\frac{3}{4} \times 1\frac{3}{4}$	$\frac{1}{2}$	5.10	1.50	.61	.38	.33	.50	.87	.171	.20	.34
2×2	$\frac{3}{16}$	2.43	.71	.57	.27	.19	.62	.80	.11	.14	.39
2×2	$\frac{1}{4}$	3.19	.94	.59	.35	.25	.61	.84	.14	.17	.39
2×2	$\frac{5}{16}$	3.92	1.15	.61	.42	.30	.60	.87	.17	.20	.39
2×2	$\frac{3}{8}$	4.62	1.36	.64	.48	.35	.59	.90	.20	.22	.39
2×2	$\frac{7}{16}$	5.30	1.56	.66	.54	.40	.59	.93	.23	.25	.38
2×2	$\frac{1}{2}$	5.95	1.75	.68	.59	.45	.58	.96	.26	.27	.38
$2\frac{1}{2} \times 2\frac{1}{2}$	$\frac{3}{16}$	3.1	.90	.69	.55	.20	.78	.98	.22	.22	.49
$2\frac{1}{2} \times 2\frac{1}{2}$	$\frac{1}{4}$	4.0	1.19	.72	.70	.39	.77	1.01	.29	.28	.49
$2\frac{1}{2} \times 2\frac{1}{2}$	$\frac{5}{16}$	5.0	1.46	.74	.80	.48	.76	1.05	.35	.33	.49
$2\frac{1}{2} \times 2\frac{1}{2}$	$\frac{3}{8}$	5.9	1.73	.76	.98	.57	.75	1.08	.41	.38	.48
$2\frac{1}{2} \times 2\frac{1}{2}$	$\frac{7}{16}$	6.8	2.00	.78	1.11	.65	.75	1.11	.46	.42	.48
$2\frac{1}{2} \times 2\frac{1}{2}$	$\frac{1}{2}$	7.7	2.25	.81	1.23	.72	.74	1.14	.52	.46	.48
$2\frac{1}{2} \times 2\frac{1}{2}$	$\frac{9}{16}$	8.5	2.50	.83	1.34	.80	.73	1.17	.58	.49	.48

TABLE V—(Continued)

PROPERTIES OF STANDARD ANGLES HAVING EQUAL LEGS

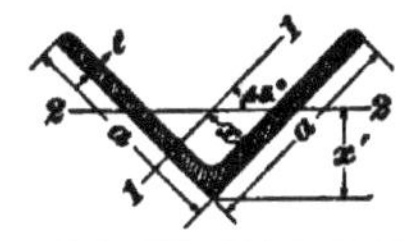

1	2	3	4	5	6	7	8	9	10	11	12
Dimensions Inches	Thickness Inches	Weight per Foot Pounds	Area of Section Square Inches	Distance of Center of Gravity From Back of Flange. Inches	Moment of Inertia, Axis 1-1 Inches	Section Modulus, Axis 1-1 Inches	Radius of Gyration, Axis 1-1 Inches	Distance of Center of Gravity From External Apex. Inches	Least Moment of Gyration, Axis 2-2 Inches	Section Modulus, Axis 2-2 Inches	Least Radius of Gyration, Axis 2-2 Inches
$a \times a$	t		A	x	I	S	r	x'	I'	S'	r'
3 × 3	1/4	4.9	1.44	.84	1.24	.58	.93	1.19	.50	.42	.59
3 × 3	5/16	6.0	1.78	.87	1.51	.71	.92	1.22	.61	.50	.59
3 × 3	3/8	7.2	2.11	.89	1.76	.83	.91	1.26	.72	.57	.58
3 × 3	7/16	8.3	2.43	.91	1.99	.95	.91	1.29	.82	.64	.58
3 × 3	1/2	9.4	2.75	.93	2.22	1.07	.90	1.32	.92	.70	.58
3 × 3	9/16	10.4	3.06	.95	2.43	1.19	.89	1.35	1.02	.76	.58
3 × 3	5/8	11.4	3.36	.98	2.62	1.30	.88	1.38	1.12	.81	.58
3 × 3	11/16	12.4	3.65	1.00	2.81	1.40	.88	1.41	1.22	.86	.58
3½ × 3½	5/16	7.1	2.09	.99	2.45	.98	1.08	1.40	.99	.71	.69
3½ × 3½	3/8	8.4	2.48	1.01	2.87	1.15	1.07	1.43	1.16	.81	.68
3½ × 3½	7/16	9.8	2.87	1.04	3.26	1.32	1.07	1.46	1.33	.91	.68
3½ × 3½	1/2	11.1	3.25	1.06	3.64	1.49	1.06	1.50	1.50	1.00	.68
3½ × 3½	9/16	12.3	3.62	1.08	3.99	1.65	1.05	1.53	1.66	1.09	.68
3½ × 3½	5/8	13.5	3.98	1.10	4.33	1.81	1.04	1.56	1.82	1.17	.68
3½ × 3½	11/16	14.8	4.34	1.12	4.65	1.96	1.04	1.59	1.97	1.24	.67
3½ × 3½	3/4	15.9	4.69	1.15	4.96	2.11	1.03	1.62	2.13	1.31	.67
3½ × 3½	13/16	17.1	5.03	1.17	5.25	2.25	1.02	1.65	2.28	1.38	.67
3½ × 3½	7/8	18.3	5.36	1.19	5.53	2.39	1.02	1.68	2.43	1.45	.67
4 × 4	5/16	8.2	2.40	1.12	3.71	1.29	1.24	1.58	1.50	.95	.79
4 × 4	3/8	9.7	2.86	1.14	4.36	1.52	1.23	1.61	1.77	1.10	.79
4 × 4	7/16	11.3	3.31	1.16	4.97	1.75	1.23	1.64	2.02	1.23	.78
4 × 4	1/2	12.8	3.75	1.18	5.56	1.97	1.22	1.67	2.28	1.36	.78
4 × 4	9/16	14.2	4.18	1.21	6.12	2.19	1.21	1.71	2.52	1.48	.78
4 × 4	5/8	15.7	4.61	1.23	6.66	2.40	1.20	1.74	2.76	1.59	.77
4 × 4	11/16	17.1	5.03	1.25	7.17	2.61	1.19	1.77	3.00	1.70	.77
4 × 4	3/4	18.5	5.44	1.27	7.66	2.81	1.19	1.80	3.23	1.80	.77
4 × 4	13/16	19.9	5.84	1.29	8.14	3.01	1.18	1.83	3.46	1.89	.77
4 × 4	7/8	21.2	6.23	1.31	8.59	3.20	1.17	1.86	3.69	1.99	.77
6 × 6	3/8	14.8	4.36	1.64	15.39	3.53	1.88	2.32	6.19	2.67	1.19
6 × 6	7/16	17.2	5.06	1.66	17.68	4.07	1.87	2.34	7.13	3.04	1.19
6 × 6	1/2	19.6	5.75	1.68	19.91	4.61	1.86	2.38	8.04	3.37	1.18
6 × 6	9/16	21.9	6.43	1.71	22.07	5.14	1.85	2.41	8.94	3.70	1.18
6 × 6	5/8	24.2	7.11	1.73	24.16	5.66	1.84	2.45	9.81	4.01	1.17
6 × 6	11/16	26.4	7.78	1.75	26.19	6.17	1.83	2.48	10.67	4.31	1.17
6 × 6	3/4	28.7	8.44	1.78	28.15	6.66	1.83	2.51	11.52	4.59	1.17
6 × 6	13/16	30.9	9.09	1.80	30.06	7.15	1.82	2.54	12.35	4.86	1.17
6 × 6	7/8	33.1	9.73	1.82	31.92	7.63	1.81	2.57	13.17	5.12	1.16
6 × 6	15/16	35.3	10.37	1.84	33.72	8.11	1.80	2.60	13.98	5.37	1.16
6 × 6	1	37.4	11.00	1.86	35.46	8.57	1.80	2.64	14.78	5.61	1.16

TABLE VI

PROPERTIES OF STANDARD ANGLES HAVING UNEQUAL LEGS

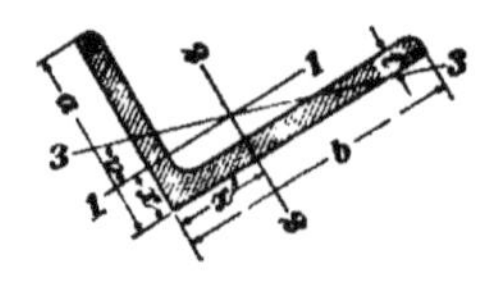

1	2	3	4	5	6	7	8	9	10	11	12	13	14
Dimensions Inches	Thickness Inches	Weight per Foot Pounds	Area of Section Square Inches	Distance of Center of Gravity From Back of Longer Flange Inches	Moment of Inertia, Axis 1-1 Inches	Section Modulus, Axis 1-1 Inches	Radius of Gyration, Axis 1-1 Inches	Distance of Center of Gravity From Back of Shorter Flange. Inches	Moment of Inertia, Axis 2-2 Inches	Section Modulus, Axis 2-2 Inches	Radius of Gyration, Axis 2-2 Inches	Tangent of Angle	Least Radius of Gyration, Axis 2-2 Inches
$b \times a$	t		A	x	I	S	r	x'	I'	S'	r'	a	r''
2½ × 2	3/16	2.7	.81	.51	.29	.20	.60	.76	.51	.29	.79	.632	.43
2½ × 2	1/4	3.6	1.06	.54	.37	.25	.59	.79	.65	.38	.78	.626	.42
2½ × 2	5/16	4.5	1.31	.56	.45	.31	.58	.81	.79	.47	.78	.620	.42
2½ × 2	3/8	5.3	1.55	.58	.51	.36	.58	.83	.91	.55	.77	.614	.42
2½ × 2	7/16	6.1	1.78	.60	.58	.41	.57	.85	1.03	.62	.76	.607	.42
2½ × 2	1/2	6.8	2.00	.63	.64	.46	.56	.88	1.14	.70	.75	.600	.42
2½ × 2	9/16	7.6	2.21	.65	.69	.51	.56	.90	1.24	.77	.75	.592	.42
3 × 2½	1/4	4.5	1.31	.66	.74	.40	.75	.91	1.17	.56	.95	.684	.53
3 × 2½	5/16	5.5	1.62	.68	.90	.49	.74	.93	1.42	.69	.94	.680	.53
3 × 2½	3/8	6.5	1.92	.71	1.04	.58	.74	.96	1.66	.81	.93	.676	.52
3 × 2½	7/16	7.5	2.21	.73	1.18	.66	.73	.98	1.88	.93	.92	.672	.52
3 × 2½	1/2	8.5	2.50	.75	1.30	.74	.72	1.00	2.08	1.04	.91	.666	.52
3 × 2½	9/16	9.4	2.78	.77	1.42	.82	.72	1.02	2.28	1.15	.91	.661	.52
3 × 2½	5/8	10.4	3.05	.79	1.53	.90	.71	1.04	2.46	1.26	.90	.655	.52
3½ × 2½	1/4	4.9	1.44	.61	.78	.41	.74	1.11	1.80	.75	1.12	.506	.54
3½ × 2½	5/16	6.0	1.78	.64	.94	.50	.73	1.14	2.19	.93	1.11	.501	.54
3½ × 2½	3/8	7.2	2.11	.66	1.09	.59	.72	1.16	2.56	1.09	1.10	.496	.54
3½ × 2½	7/16	8.3	2.43	.68	1.23	.68	.71	1.18	2.91	1.26	1.09	.491	.54
3½ × 2½	1/2	9.4	2.75	.70	1.36	.76	.70	1.20	3.24	1.41	1.09	.486	.53
3½ × 2½	9/16	10.4	3.06	.73	1.49	.84	.70	1.23	3.55	1.56	1.08	.480	.53
3½ × 2½	5/8	11.4	3.36	.75	1.61	.92	.69	1.25	3.85	1.71	1.07	.472	.53
3½ × 2½	11/16	12.4	3.65	.77	1.72	.99	.69	1.27	4.13	1.85	1.06	.463	.53
3½ × 2½	3/4	13.4	3.94	.79	1.83	1.07	.68	1.29	4.40	1.99	1.06	.461	.54
3½ × 3	5/16	6.6	1.93	.81	1.58	.72	.90	1.06	2.33	.95	1.10	.724	.63
3½ × 3	3/8	7.8	2.30	.83	1.85	.85	.90	1.08	2.72	1.13	1.09	.721	.62
3½ × 3	7/16	9.0	2.65	.85	2.09	.98	.89	1.10	3.10	1.29	1.08	.718	.62
3½ × 3	1/2	10.2	3.00	.88	2.33	1.10	.88	1.13	3.45	1.45	1.07	.714	.62
3½ × 3	9/16	11.4	3.34	.90	2.55	1.21	.87	1.15	3.79	1.61	1.07	.711	.62
3½ × 3	5/8	12.5	3.67	.92	2.76	1.33	.87	1.17	4.11	1.76	1.06	.707	.62
3½ × 3	11/16	13.6	4.00	.94	2.96	1.44	.86	1.19	4.41	1.91	1.05	.703	.62
3½ × 3	3/4	14.7	4.31	.96	3.15	1.54	.85	1.21	4.70	2.05	1.04	.698	.62
3½ × 3	13/16	15.7	4.62	.98	3.33	1.65	.85	1.23	4.98	2.20	1.04	.694	.62
3½ × 3	7/8	16.8	4.92	1.00	3.50	1.75	.84	1.25	5.24	2.33	1.03	.689	.63
4 × 3	5/16	7.1	2.09	.76	1.65	.73	.89	1.26	3.38	1.23	1.27	.554	.65
4 × 3	3/8	8.4	2.48	.78	1.92	.87	.88	1.28	3.96	1.46	1.26	.551	.64
4 × 3	7/16	9.8	2.87	.80	2.18	.99	.87	1.30	4.52	1.68	1.25	.547	.64
4 × 3	1/2	11.1	3.25	.83	2.42	1.12	.86	1.33	5.05	1.89	1.25	.543	.64
4 × 3	9/16	12.3	3.62	.85	2.66	1.23	.86	1.35	5.55	2.09	1.24	.538	.64
4 × 3	5/8	13.5	3.98	.87	2.87	1.35	.85	1.37	6.03	2.30	1.23	.534	.64
4 × 3	11/16	14.8	4.34	.89	3.08	1.46	.84	1.39	6.49	2.49	1.22	.529	.64
4 × 3	3/4	15.9	4.69	.92	3.28	1.57	.84	1.42	6.93	2.68	1.22	.524	.64
4 × 3	13/16	17.1	5.03	.94	3.47	1.68	.83	1.44	7.35	2.87	1.21	.518	.64
4 × 3	7/8	18.3	5.36	.96	3.66	1.79	.83	1.46	7.75	3.05	1.20	.512	.64

TABLE VI—(Continued)

PROPERTIES OF STANDARD ANGLES HAVING UNEQUAL LEGS

1	2	3	4	5	6	7	8	9	10	11	12	13	14
Dimensions Inches	Thickness Inches	Weight per Foot Pounds	Area of Section Square Inches	Distance of Center of Gravity From Back of Longer Flange Inches	Moment of Inertia, Axis 1-1 Inches	Section Modulus, Axis 1-1 Inches	Radius of Gyration, Axis 1-1 Inches	Distance of Center of Gravity From Back of Shorter Flange. Inches	Moment of Inertia, Axis 2-2 Inches	Section Modulus, Axis 2-2 Inches	Radius of Gyration, Axis 2-2 Inches	Tangent of Angle	Least Radius of Gyration, Axis 2-2 Inches
$b \times a$	t		A	x	I	S	r	x'	I'	S'	r'	a	r''
5 × 3	5/16	8.2	2.40	.68	1.75	.75	.85	1.68	6.26	1.89	1.61	.368	.66
5 × 3	3/8	9.7	2.86	.70	2.04	.89	.84	1.70	7.37	2.24	1.61	.364	.65
5 × 3	7/16	11.3	3.31	.73	2.32	1.02	.84	1.73	8.43	2.58	1.60	.361	.65
5 × 3	1/2	12.8	3.75	.75	2.58	1.15	.83	1.75	9.45	2.91	1.59	.357	.65
5 × 3	9/16	14.2	4.18	.77	2.83	1.27	.82	1.77	10.43	3.23	1.58	.353	.65
5 × 3	5/8	15.7	4.61	.80	3.06	1.39	.82	1.80	11.37	3.55	1.57	.349	.64
5 × 3	11/16	17.1	5.03	.82	3.29	1.51	.81	1.82	12.28	3.86	1.56	.345	.64
5 × 3	3/4	18.5	5.44	.84	3.51	1.62	.80	1.84	13.15	4.16	1.55	.340	.64
5 × 3	13/16	19.9	5.84	.86	3.71	1.74	.80	1.86	13.98	4.46	1.55	.336	.64
5 × 3	7/8	21.2	6.23	.88	3.91	1.85	.79	1.88	14.78	4.75	1.54	.331	.64
5 × 3½	5/16	8.7	2.56	.84	2.72	1.02	1.03	1.59	6.60	1.94	1.61	.489	.77
5 × 3½	3/8	10.4	3.05	.86	3.18	1.21	1.02	1.61	7.78	2.29	1.60	.485	.76
5 × 3½	7/16	12.0	3.53	.88	3.63	1.39	1.01	1.63	8.90	2.64	1.59	.482	.76
5 × 3½	1/2	13.6	4.00	.91	4.05	1.56	1.01	1.66	9.99	2.99	1.58	.479	.75
5 × 3½	9/16	15.2	4.46	.93	4.45	1.73	1.00	1.68	11.03	3.32	1.57	.476	.75
5 × 3½	5/8	16.7	4.92	.95	4.83	1.90	.99	1.70	12.03	3.65	1.56	.472	.75
5 × 3½	11/16	18.3	5.37	.97	5.20	2.06	.98	1.72	12.99	3.97	1.56	.468	.75
5 × 3½	3/4	19.8	5.81	1.00	5.55	2.22	.98	1.75	13.92	4.28	1.55	.464	.75
5 × 3½	13/16	21.2	6.25	1.02	5.89	2.37	.97	1.77	14.81	4.58	1.54	.460	.75
5 × 3½	7/8	22.7	6.67	1.04	6.21	2.52	.96	1.79	15.67	4.88	1.53	.455	.75
5 × 3½	15/16	24.1	7.09	1.06	6.52	2.67	.96	1.81	16.49	5.17	1.53	.451	.75
6 × 3½	3/8	11.6	3.42	.79	3.34	1.23	.99	2.04	12.86	3.24	1.94	.350	.77
6 × 3½	7/16	13.5	3.96	.81	3.81	1.41	.98	2.06	14.76	3.75	1.93	.347	.76
6 × 3½	1/2	15.3	4.50	.83	4.25	1.59	.97	2.08	16.59	4.24	1.92	.344	.76
6 × 3½	9/16	17.1	5.03	.86	4.67	1.77	.96	2.11	18.37	4.72	1.91	.341	.75
6 × 3½	5/8	18.9	5.55	.88	5.08	1.94	.96	2.13	20.08	5.19	1.90	.338	.75
6 × 3½	11/16	20.6	6.06	.90	5.47	2.11	.95	2.15	21.74	5.65	1.89	.334	.75
6 × 3½	3/4	22.3	6.56	.93	5.84	2.27	.94	2.18	23.34	6.10	1.89	.331	.75
6 × 3½	13/16	24.0	7.06	.95	6.20	2.43	.94	2.20	24.89	6.55	1.88	.327	.75
6 × 3½	7/8	25.7	7.55	.97	6.55	2.59	.93	2.22	26.39	6.98	1.87	.323	.75
6 × 3½	15/16	27.3	8.03	.99	6.88	2.74	.93	2.24	27.84	7.41	1.86	.320	.75
6 × 3½	1	28.9	8.50	1.01	7.21	2.90	.92	2.26	29.15	7.80	1.85	.317	.75
6 × 4	3/8	12.3	3.61	.94	4.90	1.60	1.17	1.94	13.47	3.32	1.93	.446	.88
6 × 4	7/16	14.2	4.18	.96	5.60	1.85	1.16	1.96	15.46	3.83	1.92	.443	.87
6 × 4	1/2	16.2	4.75	.99	6.27	2.08	1.15	1.99	17.40	4.33	1.91	.440	.87
6 × 4	9/16	18.1	5.31	1.01	6.91	2.31	1.14	2.01	19.26	4.83	1.90	.438	.87
6 × 4	5/8	19.9	5.86	1.03	7.52	2.54	1.13	2.03	21.07	5.31	1.90	.434	.86
6 × 4	11/16	21.8	6.40	1.06	8.11	2.76	1.13	2.06	22.82	5.78	1.89	.431	.86
6 × 4	3/4	23.6	6.94	1.08	8.68	2.97	1.12	2.08	24.51	6.25	1.88	.428	.86
6 × 4	13/16	25.4	7.46	1.10	9.23	3.18	1.11	2.10	26.15	6.70	1.87	.425	.86
6 × 4	7/8	27.2	7.98	1.12	9.75	3.39	1.11	2.12	27.73	7.15	1.86	.421	.86
6 × 4	15/16	28.9	8.50	1.14	10.26	3.59	1.10	2.14	29.26	7.59	1.86	.418	.86
6 × 4	1	30.6	9.00	1.17	10.75	3.79	1.09	2.17	30.75	8.02	1.85	.414	.86

TABLE VII

PROPERTIES OF Z BARS

1	2	3	4	5	6	7	8	9	10	11	12	13	14	15	16	17
													Coefficient of Strength		Coefficient of Deflection	
Depth of Bar Inches	Length of Legs Inches	Thickness of Web and Legs Inches	Weight per Foot Pounds	Area of Section Square Inches	Moment of Inertia, Axis 1-1 Inches	Section Modulus, Axis 1-1 Inches	Radius of Gyration, Axis 1-1 Inches	Moment of Inertia, Axis 2-2 Inches	Section Modulus, Axis 2-2 Inches	Radius of Gyration, Axis 2-2 Inches	Tangent of Angle α	Least Radius of Gyration, Axis 3-3 Inches	For Fiber Stress of 16,000 Pounds per Square Inch	For Fiber Stress of 12,500 Pounds per Square Inch	Uniform Load	Center Load
b	a	t		A	I	S	r	I'	S'	r'		r''	F	F'	N	N'
3	2 11/16	1/4	6.7	1.97	2.87	1.92	1.21	2.81	1.10	1.19	.986	.55	20,400	16,000	.000270	.000432
3 1/16	2 3/4	5/16	8.4	2.48	3.64	2.38	1.21	3.64	1.40	1.21	1.000	.55	25,400	19,800	.000213	.000341
3	2 11/16	3/8	9.7	2.86	3.85	2.57	1.16	3.92	1.57	1.17	.990	.54	27,400	21,400	.000201	.000322
3 1/16	2 3/4	7/16	11.4	3.36	4.57	2.98	1.17	4.75	1.88	1.19	.975	.55	31,800	24,800	.000170	.000272
3	2 11/16	1/2	12.5	3.69	4.59	3.06	1.12	4.85	1.99	1.15	.965	.53	32,600	25,500	.000169	.000271
3 1/16	2 3/4	9/16	14.2	4.18	5.26	3.43	1.12	5.68	2.30	1.17	.951	.54	36,600	28,600	.000148	.000236

4	$3\frac{1}{16}$	$\frac{1}{4}$	8.2	2.41	6.28	3.14	1.62	4.23	1.44	1.33	.778	.67	33,500	26,200	.000123	.000198
$4\frac{1}{16}$	$3\frac{1}{8}$	$\frac{5}{16}$	10.3	3.03	7.94	3.91	1.62	5.46	1.84	1.34	.788	.68	41,700	32,600	.000098	.000156
$4\frac{1}{8}$	$3\frac{3}{16}$	$\frac{3}{8}$	12.4	3.66	9.63	4.67	1.62	6.77	2.26	1.36	.798	.69	49,800	38,900	.000081	.000129
4	$3\frac{1}{16}$	$\frac{7}{16}$	13.8	4.05	9.66	4.83	1.54	6.73	2.37	1.29	.794	.66	51,500	40,200	.000080	.000129
$4\frac{1}{16}$	$3\frac{1}{8}$	$\frac{1}{2}$	15.8	4.66	11.18	5.50	1.55	7.96	2.77	1.31	.804	.67	58,700	45,900	.000069	.000111
$4\frac{1}{8}$	$3\frac{3}{16}$	$\frac{9}{16}$	17.9	5.27	12.74	6.18	1.55	9.26	3.19	1.32	.814	.68	65,900	51,500	.000061	.000097
4	$3\frac{1}{16}$	$\frac{5}{8}$	18.9	5.55	12.11	6.05	1.48	8.73	3.18	1.25	.808	.65	64,600	50,500	.000064	.000103
$4\frac{1}{16}$	$3\frac{1}{8}$	$\frac{11}{16}$	20.9	6.14	13.52	6.65	1.48	9.95	3.58	1.27	.818	.67	71,000	55,500	.000057	.000092
$4\frac{1}{8}$	$3\frac{3}{16}$	$\frac{3}{4}$	23.0	6.75	14.97	7.26	1.49	11.24	4.00	1.29	.828	.68	77,400	60,500	.000052	.000083
5	$3\frac{1}{4}$	$\frac{5}{16}$	11.6	3.40	13.36	5.34	1.98	6.18	2.00	1.35	.611	.75	57,000	44,500	.000058	.000093
$5\frac{1}{16}$	$3\frac{5}{16}$	$\frac{3}{8}$	13.9	4.10	16.18	6.39	1.99	7.65	2.45	1.37	.619	.76	68,200	53,300	.000048	.000077
$5\frac{1}{8}$	$3\frac{3}{8}$	$\frac{7}{16}$	16.4	4.81	19.07	7.44	1.99	9.20	2.92	1.38	.628	.76	79,400	62,000	.000041	.000065
5	$3\frac{1}{4}$	$\frac{1}{2}$	17.9	5.25	19.19	7.68	1.91	9.05	3.02	1.31	.616	.74	81,900	64,000	.000040	.000065
$5\frac{1}{16}$	$3\frac{5}{16}$	$\frac{9}{16}$	20.2	5.94	21.83	8.62	1.92	10.51	3.47	1.33	.623	.75	92,000	71,900	.000036	.000057
$5\frac{1}{8}$	$3\frac{3}{8}$	$\frac{5}{8}$	22.6	6.64	24.53	9.57	1.92	12.06	3.94	1.35	.631	.76	102,100	79,800	.000032	.000051
5	$3\frac{1}{4}$	$\frac{11}{16}$	23.7	6.96	23.68	9.47	1.84	11.37	3.91	1.28	.619	.73	101,000	78,900	.000033	.000052
$5\frac{1}{16}$	$3\frac{5}{16}$	$\frac{3}{4}$	26.0	7.64	26.16	10.34	1.85	12.83	4.37	1.30	.626	.74	110,200	86,100	.000030	.000047
$5\frac{1}{8}$	$3\frac{3}{8}$	$\frac{13}{16}$	28.3	8.33	28.70	11.20	1.86	14.37	4.84	1.31	.633	.76	119,500	93,300	.000027	.000043
6	$3\frac{1}{2}$	$\frac{3}{8}$	15.6	4.59	25.32	8.44	2.35	9.11	2.75	1.41	.519	.83	90,000	70,300	.000031	.000049
$6\frac{1}{16}$	$3\frac{9}{16}$	$\frac{7}{16}$	18.3	5.39	29.80	9.83	2.35	10.94	3.27	1.43	.526	.83	104,900	81,900	.000026	.000042
$6\frac{1}{8}$	$3\frac{5}{8}$	$\frac{1}{2}$	21.0	6.19	34.36	11.22	2.36	12.87	3.81	1.44	.532	.84	119,700	93,500	.000023	.000036
6	$3\frac{1}{2}$	$\frac{9}{16}$	22.7	6.68	34.64	11.55	2.28	12.59	3.91	1.37	.520	.81	123,200	96,200	.000022	.000036
$6\frac{1}{16}$	$3\frac{9}{16}$	$\frac{5}{8}$	25.4	7.46	38.87	12.82	2.28	14.41	4.44	1.39	.526	.82	136,800	106,800	.000020	.000032
$6\frac{1}{8}$	$3\frac{5}{8}$	$\frac{11}{16}$	28.1	8.25	43.18	14.10	2.29	16.34	4.98	1.41	.532	.84	150,400	117,500	.000018	.000029
6	$3\frac{1}{2}$	$\frac{3}{4}$	29.3	8.63	42.12	14.04	2.21	15.44	4.94	1.34	.519	.81	149,800	117,000	.000018	.000029
$6\frac{1}{16}$	$3\frac{9}{16}$	$\frac{13}{16}$	31.9	9.39	46.13	15.22	2.22	17.27	5.47	1.36	.525	.82	162,300	126,800	.000017	.000027
$6\frac{1}{8}$	$3\frac{5}{8}$	$\frac{7}{8}$	34.6	10.17	50.22	16.40	2.22	19.18	6.02	1.37	.530	.83	174,900	136,700	.000015	.000025
$7\frac{1}{2}$	3	$\frac{3}{8}$	16.3	4.78	38.19	10.18	2.83	5.59	1.99	1.08	.29	.72	108,600	84,800	.000020	.000033
8	3	$\frac{3}{8}$	16.9	4.97	44.64	11.16	3.00	5.60	1.99	1.06	.27	.72	119,000	93,000	.000017	.000028
10	3	$\frac{3}{8}$	19.4	5.72	76.87	15.37	3.67	5.60	1.99	.99	.19	.71	163,900	128,100	.000010	.000016

TABLE VIII—PROPERTIES OF T BARS

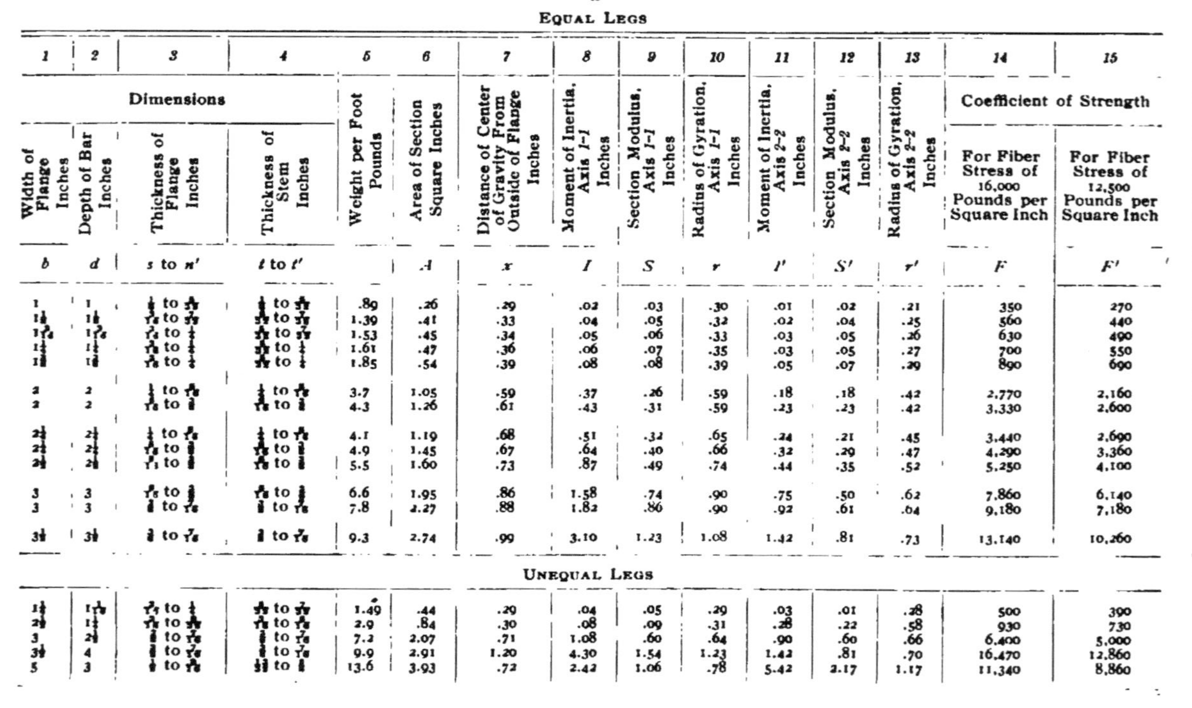

EQUAL LEGS

1	2	3	4	5	6	7	8	9	10	11	12	13	14	15
Dimensions													Coefficient of Strength	
Width of Flange Inches	Depth of Bar Inches	Thickness of Flange Inches	Thickness of Stem Inches	Weight per Foot Pounds	Area of Section Square Inches	Distance of Center of Gravity From Outside of Flange Inches	Moment of Inertia, Axis 1-1 Inches	Section Modulus, Axis 1-1 Inches	Radius of Gyration, Axis 1-1 Inches	Moment of Inertia, Axis 2-2 Inches	Section Modulus, Axis 2-2 Inches	Radius of Gyration, Axis 2-2 Inches	For Fiber Stress of 16,000 Pounds per Square Inch	For Fiber Stress of 12,500 Pounds per Square Inch
b	d	s to n'	t to t'		A	x	I	S	r	I'	S'	r'	F	F'
1	1	1/8 to 5/32	1/8 to 5/32	.89	.26	.29	.02	.03	.30	.01	.02	.21	350	270
1 1/4	1 1/4	3/16 to 7/32	5/32 to 7/32	1.39	.41	.33	.04	.05	.32	.02	.04	.25	560	440
1 7/16	1 7/16	3/16 to 1/4	5/32 to 7/32	1.53	.45	.34	.05	.06	.33	.03	.05	.26	630	490
1 1/2	1 1/2	3/16 to 1/4	5/32 to 1/4	1.61	.47	.36	.06	.07	.35	.03	.05	.27	700	550
1 3/4	1 3/4	3/16 to 1/4	5/32 to 1/4	1.85	.54	.39	.08	.08	.39	.05	.07	.29	890	690
2	2	1/4 to 5/16	1/4 to 5/16	3.7	1.05	.59	.37	.26	.59	.18	.18	.42	2,770	2,160
2	2	5/16 to 3/8	5/16 to 3/8	4.3	1.26	.61	.43	.31	.59	.23	.23	.42	3,330	2,600
2 1/4	2 1/4	1/4 to 5/16	1/4 to 5/16	4.1	1.19	.68	.51	.32	.65	.24	.21	.45	3,440	2,690
2 1/2	2 1/2	5/16 to 3/8	5/16 to 3/8	4.9	1.45	.67	.64	.40	.66	.32	.29	.47	4,290	3,360
2 1/2	2 1/2	5/16 to 3/8	5/16 to 3/8	5.5	1.60	.73	.87	.49	.74	.44	.35	.52	5,250	4,100
3	3	5/16 to 3/8	5/16 to 3/8	6.6	1.95	.86	1.58	.74	.90	.75	.50	.62	7,860	6,140
3	3	3/8 to 7/16	3/8 to 7/16	7.8	2.27	.88	1.82	.86	.90	.92	.61	.64	9,180	7,180
3 1/2	3 1/2	3/8 to 7/16	3/8 to 7/16	9.3	2.74	.99	3.10	1.23	1.08	1.42	.81	.73	13,140	10,260

UNEQUAL LEGS

b	d	s to n'	t to t'	Weight	A	x	I	S	r	I'	S'	r'	F	F'
1 1/2	1 1/16	7/32 to 1/4	5/32 to 7/32	1.49	.44	.29	.04	.05	.29	.03	.01	.28	500	390
2 1/2	1 1/2	3/16 to 5/32	3/16 to 5/16	2.9	.84	.30	.08	.09	.31	.28	.22	.58	930	730
3	2 1/2	3/8 to 7/16	3/8 to 7/16	7.2	2.07	.71	1.08	.60	.64	.90	.60	.66	6,400	5,000
3 1/2	4	3/8 to 7/16	3/8 to 7/16	9.9	2.91	1.20	4.30	1.54	1.23	1.42	.81	.70	16,470	12,860
5	3	1/2 to 9/16	17/32 to 5/8	13.6	3.93	.72	2.42	1.06	.78	5.42	2.17	1.17	11,340	8,860

TABLE IX

PROPERTIES AND PRINCIPAL DIMENSIONS OF STANDARD T RAILS

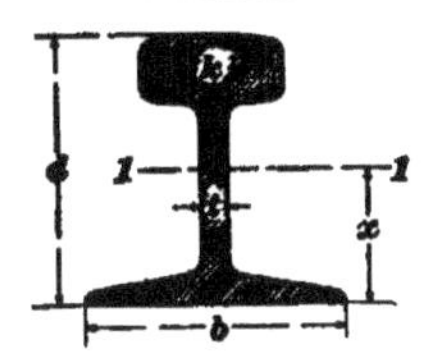

1	2	3	4	5	6	7	8	9
							Axis 1-1	
Weight per Yard Pounds	Area Square Inches	b Inches	d Inches	k Inches	t Inches	x Inches	Moment of Inertia I	Section Modulus S
8	.78	1 1/2	1 1/2	13/16	5/32	.75	.23	.31
12	1.18	1 7/8	1 7/8	1 1/16	3/16	.92	.55	.58
16	1.57	2 1/4	2 1/4	1 3/16	7/32	1.10	1.13	.99
20	2.00	2 3/8	2 3/8	1 3/8	1/4	1.2	1.5	1.2
25	2.5	2 3/4	2 3/4	1 1/2	19/64	1.4	2.4	1.7
30	2.9	3 5/16	3	1 11/16	11/32	1.5	3.7	2.4
35	3.4	3 3/8	3 1/16	1 27/32	13/32	1.5	4.3	2.8
40	3.9	3 1/2	3 1/2	1 27/32	23/64	1.7	6.0	3.4
45	4.4	3 3/4	3 3/4	1 59/64	3/8	1.8	7.6	3.9
50	4.9	3 7/8	3 7/8	2 1/8	7/16	1.9	10.1	5.1
55	5.4	4 1/16	4 1/16	2 1/4	15/32	2.0	12.2	5.9
60	5.9	4 1/4	4 1/4	2 3/8	31/64	2.1	14.7	6.7
65	6.4	4 7/16	4 7/16	2 13/32	1/2	2.1	17.0	7.4
70	6.9	4 5/8	4 5/8	2 7/16	33/64	2.2	20.0	8.4
75	7.4	4 13/16	4 13/16	2 15/32	17/32	2.3	23.0	9.1
80	7.8	5	5	2 1/2	35/64	2.4	26.7	10.1
85	8.3	5 3/16	5 3/16	2 9/16	9/16	2.5	30.5	11.2
90	8.8	5 3/8	5 3/8	2 5/8	9/16	2.6	35.2	12.6
100	9.8	5 3/4	5 3/4	2 3/4	9/16	2.8	44.4	15.0
150	14.7	6	6	4 1/4	1	3.0	69.3	22.9

TABLE X

MOMENT OF INERTIA OF RECTANGULAR SECTIONS

Depth in Inches	Width of Rectangle in Inches						
	1/4	5/16	3/8	7/16	1/2	9/16	5/8
2	.17	.21	.25	.29	.33	.38	.42
3	.56	.70	.84	.98	1.13	1.27	1.41
4	1.33	1.67	2.00	2.33	2.67	3.00	3.33
5	2.60	3.26	3.91	4.56	5.21	5.86	6.51
6	4.50	5.63	6.75	7.88	9.00	10.13	11.25
7	7.15	8.93	10.72	12.51	14.29	16.08	17.86
8	10.67	13.33	16.00	18.67	21.33	24.00	26.67
9	15.19	18.98	22.78	26.58	30.38	34.17	37.97
10	20.83	26.04	31.25	36.46	41.67	46.87	52.08
11	27.73	34.66	41.59	48.53	55.46	62.39	69.32
12	36.00	45.00	54.00	63.00	72.00	81.00	90.00
13	45.77	57.21	68.66	80.10	91.54	102.98	114.43
14	57.17	71.46	85.75	100.04	114.33	128.63	142.92
15	70.31	87.89	105.47	123.05	140.63	158.20	175.78
16	85.33	106.67	128.00	149.33	170.67	192.00	213.33
17	102.35	127.94	153.53	179.12	204.71	230.30	255.89
18	121.50	151.88	182.25	212.63	243.00	273.38	303.75
19	142.90	178.62	214.34	250.07	285.79	321.52	357.24
20	166.67	208.33	250.00	291.67	333.33	375.00	416.67
21	192.94	241.17	289.41	337.64	385.88	434.11	482.34
22	221.83	277.29	332.75	388.21	443.67	499.13	554.58
23	253.48	316.85	380.22	443.59	506.96	570.33	633.70
24	288.00	360.00	432.00	504.00	576.00	648.00	720.00
25	325.52	406.90	488.28	569.66	651.04	732.42	813.80
26	366.17	457.71	549.25	640.79	732.33	823.88	915.42
27	410.06	512.58	615.09	717.61	820.13	922.64	1,025.16
28	457.33	571.67	686.00	800.33	914.67	1,029.00	1,143.33
29	508.10	635.13	762.16	889.18	1,016.21	1,143.23	1,270.26
30	562.50	703.13	843.75	984.38	1,125.00	1,265.63	1,406.25
32	682.67	853.33	1,024.00	1,194.67	1,365.33	1,536.00	1,706.67
34	818.83	1,023.54	1,228.25	1,432.96	1,637.67	1,842.38	2,047.08
36	972.00	1,215.00	1,458.00	1,701.00	1,944.00	2,187.00	2,430.00
38	1,143.17	1,428.96	1,714.75	2,000.54	2,286.33	2,572.13	2,857.92
40	1,333.33	1,666.67	2,000.00	2,333.33	2,666.67	3,000.00	3,333.33
42	1,543.50	1,929.38	2,315.25	2,701.13	3,087.00	3,472.88	3,858.75
44	1,774.67	2,218.33	2,662.00	3,105.67	3,549.33	3,993.00	4,436.67
46	2,027.83	2,534.79	3,041.75	3,548.71	4,055.67	4,562.63	5,069.58
48	2,304.00	2,880.00	3,456.00	4,032.00	4,608.00	5,184.00	5,760.00
50	2,604.17	3,255.21	3,906.25	4,557.29	5,208.33	5,859.38	6,510.42
52	2,929.33	3,661.67	4,394.00	5,126.33	5,858.67	6,591.00	7,323.33
54	3,280.50	4,100.63	4,920.75	5,740.88	6,561.00	7,381.13	8,201.25
56	3,658.67	4,573.33	5,488.00	6,402.67	7,317.33	8,232.00	9,146.67
58	4,064.83	5,081.04	6,097.25	7,113.46	8,129.67	9,145.87	10,162.08
60	4,500.00	5,625.00	6,750.00	7,875.00	9,000.00	10,125.00	11,250.00

TABLE X—(Continued)

MOMENT OF INERTIA OF RECTANGULAR SECTIONS

Depth in Inches	Width of Rectangle in Inches					
	$\frac{11}{16}$	$\frac{3}{4}$	$\frac{13}{16}$	$\frac{7}{8}$	$\frac{15}{16}$	1
2	.46	.50	.54	.58	.63	.67
3	1.55	1.69	1.83	1.97	2.11	2.25
4	3.67	4.00	4.33	4.67	5.00	5.33
5	7.16	7.81	8.46	9.11	9.77	10.42
6	12.38	13.50	14.63	15.75	16.88	18.00
7	19.65	21.44	23.22	25.01	26.80	28.58
8	29.33	32.00	34.67	37.33	40.00	42.67
9	41.77	45.56	49.36	53.16	56.95	60.75
10	57.29	62.50	67.71	72.92	78.13	83.33
11	76.26	83.19	90.12	97.05	103.98	110.92
12	99.00	108.00	117.00	126.00	135.00	144.00
13	125.87	137.31	148.75	160.20	171.64	183.08
14	157.21	171.50	185.79	200.08	214.38	228.67
15	193.36	210.94	228.52	246.09	263.67	281.25
16	234.67	256.00	277.33	298.67	320.00	341.33
17	281.47	307.06	332.65	358.24	383.83	409.42
18	334.13	364.50	394.88	425.25	455.63	486.00
19	392.96	428.69	464.41	500.14	535.86	571.58
20	458.33	500.00	541.67	583.33	625.00	666.67
21	530.58	578.81	627.05	675.28	723.52	771.75
22	610.04	665.50	720.96	776.42	831.87	887.33
23	697.07	760.44	823.81	887.18	950.55	1,013.92
24	792.00	864.00	936.00	1,008.00	1,080.00	1,152.00
25	895.18	976.56	1,057.94	1,139.32	1,220.70	1,302.08
26	1,006.96	1,098.50	1,190.04	1,281.58	1,373.13	1,464.67
27	1,127.67	1,230.19	1,332.70	1,435.22	1,537.73	1,640 25
28	1,257.67	1,372.00	1,486.33	1,600.67	1,715.00	1,829.33
29	1,397.29	1,524.31	1,651.34	1,778.36	1,905.39	2,032.42
30	1,546.88	1,687.50	1,828.13	1,968.75	2,109.38	2,250.00
32	1,877.33	2,048.00	2,218.67	2,389.33	2,560.00	2,730.67
34	2,251.79	2,456.50	2,661.21	2,865.92	3,070.63	3,275.33
36	2,673.00	2,916.00	3,159.00	3,402.00	3,645.00	3,888.00
38	3,143.71	3,429.50	3,715.29	4,001.08	4,286.88	4,572.67
40	3,666.67	4,000.00	4,333.33	4,666.67	5,000.00	5,333.33
42	4,244.63	4,630.50	5,016.38	5,402.25	5,788.13	6,174.00
44	4,880.33	5,324.00	5,767.67	6,211.33	6,655.00	7,098.67
46	5,576.54	6,083.50	6,590.46	7,097.42	7,604.38	8,111.33
48	6,336.00	6,912.00	7,488.00	8,064.00	8,640.00	9,216.00
50	7,161.46	7,812.50	8,463.54	9,114.58	9,765.63	10,416.67
52	8,055.67	8,788.00	9,520.33	10,252.67	10,985.00	11,717.33
54	9,021.38	9,841.50	10,661.63	11,481.75	12,301.88	13,122.00
56	10,061.33	10,976.00	11,890.67	12,805.33	13,720.00	14,634.67
58	11,178.29	12,194.50	13,210.71	14,226.92	15,243.12	16,259.33
60	12,375.00	13,500.00	14,625.00	15,750.00	16,875.00	18,000.00

TABLE XI

RADII OF GYRATION FOR TWO ANGLES, HAVING EQUAL LEGS, PLACED BACK TO BACK

Dimensions Inches	Thickness Inches	Area of Two Angles Square Inches	Radii of Gyration					
			r_0	r_1	r_2	r_3	r_4	r_5
3/4 × 3/4	1/8	.34	.22	.32	.42	.48	.53	.65
3/4 × 3/4	3/16	.49	.22	.33	.44	.49	.55	.67
1 × 1	1/8	.47	.30	.42	.52	.57	.62	.74
1 × 1	1/4	.88	.29	.45	.55	.60	.66	.77
1 1/4 × 1 1/4	1/8	.60	.38	.52	.62	.67	.72	.83
1 1/4 × 1 1/4	5/16	1.37	.36	.56	.66	.71	.77	.88
1 1/2 × 1 1/2	3/16	1.05	.46	.64	.73	.78	.83	.94
1 1/2 × 1 1/2	3/8	1.97	.44	.67	.77	.82	.88	.99
1 3/4 × 1 3/4	3/16	1.24	.54	.74	.83	.88	.93	1.03
1 3/4 × 1 3/4	7/16	2.68	.51	.78	.88	.93	.98	1.09
2 × 2	3/16	1.43	.62	.84	.93	.98	1.03	1.13
2 × 2	5/16	2.30	.60	.86	.95	1.00	1.05	1.16
2 × 2	7/16	3.12	.59	.88	.98	1.03	1.08	1.19
2 1/4 × 2 1/4	3/16	1.62	.70	.94	1.03	1.08	1.12	1.22
2 1/4 × 2 1/4	3/8	3.09	.67	.97	1.06	1.11	1.16	1.27
2 1/2 × 2 1/2	1/4	2.38	.77	1.05	1.14	1.19	1.24	1.34
2 1/2 × 2 1/2	3/8	3.47	.75	1.07	1.16	1.21	1.26	1.36
2 1/2 × 2 1/2	1/2	4.50	.74	1.09	1.19	1.24	1.29	1.39
2 3/4 × 2 3/4	3/16	1.99	.86	1.14	1.23	1.28	1.32	1.42
2 3/4 × 2 3/4	5/16	3.24	.84	1.16	1.25	1.30	1.35	1.45
2 3/4 × 2 3/4	7/16	4.43	.83	1.18	1.28	1.32	1.37	1.47
3 × 3	1/4	2.88	.93	1.26	1.34	1.39	1.43	1.53
3 × 3	7/16	4.87	.91	1.28	1.37	1.42	1.47	1.57
3 × 3	5/8	6.72	.88	1.32	1.41	1.46	1.51	1.61
3 1/2 × 3 1/2	3/8	4.97	1.07	1.48	1.56	1.61	1.66	1.75
3 1/2 × 3 1/2	5/8	7.97	1.04	1.52	1.61	1.66	1.71	1.81
3 1/2 × 3 1/2	13/16	10.05	1.02	1.55	1.65	1.70	1.75	1.85
4 × 4	5/16	4.80	1.24	1.67	1.76	1.80	1.85	1.94
4 × 4	9/16	8.37	1.21	1.71	1.80	1.85	1.89	1.99
4 × 4	13/16	11.68	1.18	1.75	1.85	1.89	1.94	2.04
4 1/2 × 4 1/2	5/16	5.43	1.40	1.87	1.96	2.00	2.05	2.14
4 1/2 × 4 1/2	1/2	8.50	1.38	1.90	1.99	2.04	2.08	2.18
4 1/2 × 4 1/2	5/8	10.47	1.36	1.92	2.01	2.06	2.10	2.20
5 × 5	3/8	7.22	1.56	2.09	2.17	2.22	2.26	2.35
5 × 5	1/2	9.50	1.54	2.10	2.19	2.24	2.28	2.38
5 × 5	5/8	11.72	1.52	2.12	2.21	2.26	2.30	2.40
6 × 6	7/16	10.12	1.87	2.50	2.58	2.63	2.67	2.76
6 × 6	5/8	14.22	1.84	2.53	2.62	2.66	2.71	2.80
6 × 6	7/8	19.47	1.81	2.57	2.66	2.70	2.75	2.85

TABLE XII

RADII OF GYRATION FOR TWO ANGLES, HAVING UNEQUAL LEGS, PLACED BACK TO BACK

Dimensions Inches	Thickness Inches	Area of Two Angles Square Inches	Radii of Gyration					
			r_0	r_1	r_2	r_3	r_4	r_5
2 × 1⅜	3/16	1.20	.63	.54	.62	.67	.72	.83
2 × 1⅜	⅜	2.25	.61	.56	.66	.71	.76	.88
2 × 1½	3/16	1.24	.63	.59	.68	.73	.78	.88
2 × 1½	⅜	2.34	.61	.62	.72	.77	.82	.93
2½ × 1¼	3/16	1.34	.80	.44	.52	.58	.63	.74
2½ × 1¼	⅜	2.53	.78	.47	.57	.62	.68	.79
2½ × 1½	3/16	1.43	.80	.55	.64	.69	.74	.84
2½ × 1½	⅜	2.72	.78	.58	.68	.73	.78	.89
2½ × 1¾	3/16	1.52	.80	.67	.75	.80	.85	.95
2½ × 1¾	¼	2.00	.79	.68	.77	.81	.86	.97
2½ × 2	3/16	1.62	.79	.79	.88	.92	.97	1.07
2½ × 2	⅜	3.09	.77	.82	.91	.96	1.01	1.12
2½ × 2	½	4.00	.75	.84	.94	.99	1.04	1.15
2¾ × 1½	3/16	1.52	.89	.53	.62	.67	.72	.82
2¾ × 1½	5/16	2.46	.87	.55	.65	.70	.75	.86
2¾ × 1½	7/16	3.34	.85	.58	.68	.73	.78	.89
3 × 2	3/16	1.80	.97	.75	.83	.88	.93	1.03
3 × 2	5/16	2.93	.95	.76	.85	.90	.95	1.05
3 × 2	7/16	3.99	.93	.79	.88	.93	.98	1.09
3 × 2½	¼	2.63	.95	1.00	1.09	1.13	1.18	1.28
3 × 2½	⅜	3.84	.93	1.02	1.11	1.16	1.21	1.31
3 × 2½	9/16	5.55	.91	1.05	1.15	1.20	1.25	1.35
3½ × 2½	¼	2.88	1.12	.96	1.04	1.09	1.13	1.23
3½ × 2½	½	5.50	1.09	1.00	1.09	1.14	1.19	1.29
3½ × 2½	11/16	7.30	1.06	1.03	1.13	1.18	1.23	1.33
3½ × 3	5/16	3.87	1.10	1.21	1.30	1.35	1.39	1.49
3½ × 3	9/16	6.68	1.07	1.25	1.34	1.39	1.44	1.54
3½ × 3	13/16	9.24	1.04	1.30	1.40	1.45	1.50	1.60

TABLE XII—(Continued)

RADII OF GYRATION FOR TWO ANGLES, HAVING UNEQUAL LEGS, PLACED BACK TO BACK

Dimensions Inches	Thickness Inches	Area of Two Angles Square Inches	Radii of Gyration					
			r_0	r_1	r_2	r_3	r_4	r_5
4 × 3	5/16	4.18	1.27	1.17	1.25	1.30	1.34	1.44
4 × 3	9/16	7.24	1.24	1.21	1.30	1.34	1.39	1.49
4 × 3	13/16	10.05	1.21	1.25	1.35	1.40	1.45	1.55
4 × 3½	5/16	4.49	1.26	1.42	1.50	1.55	1.59	1.69
4 × 3½	½	7.00	1.23	1.44	1.53	1.58	1.63	1.72
4 × 3½	⅝	8.59	1.22	1.46	1.55	1.60	1.65	1.75
4½ × 3	⅜	5.34	1.44	1.14	1.22	1.27	1.31	1.41
4½ × 3	½	7.00	1.42	1.15	1.24	1.29	1.34	1.44
4½ × 3	⅝	8.59	1.40	1.18	1.27	1.31	1.36	1.46
5 × 3	5/16	4.80	1.61	1.09	1.17	1.22	1.26	1.36
5 × 3	9/16	8.37	1.58	1.13	1.22	1.26	1.31	1.41
5 × 3	13/16	11.68	1.55	1.17	1.27	1.32	1.37	1.47
5 × 3½	⅜	6.09	1.60	1.34	1.42	1.46	1.51	1.60
5 × 3½	⅝	9.84	1.56	1.37	1.46	1.51	1.56	1.66
5 × 3½	⅞	13.34	1.53	1.42	1.51	1.56	1.61	1.71
5 × 4	⅜	6.47	1.59	1.58	1.66	1.71	1.75	1.85
5 × 4	½	8.50	1.57	1.60	1.68	1.73	1.78	1.87
5 × 4	⅝	10.47	1.55	1.62	1.71	1.75	1.80	1.90
6 × 3½	⅜	6.84	1.94	1.26	1.34	1.39	1.43	1.53
6 × 3½	⅝	11.09	1.90	1.30	1.39	1.43	1.48	1.58
6 × 3½	⅞	15.09	1.87	1.34	1.44	1.49	1.53	1.64
6 × 4	⅜	7.22	1.93	1.50	1.58	1.62	1.67	1.76
6 × 4	⅝	11.72	1.90	1.53	1.62	1.67	1.71	1.81
6 × 4	⅞	15.97	1.86	1.58	1.67	1.71	1.76	1.86
7 × 3½	7/16	8.80	2.26	1.16	1.29	1.33	1.38	1.47
7 × 3½	½	10.00	2.25	1.22	1.30	1.35	1.39	1.48
7 × 3½	⅝	12.34	2.24	1.24	1.32	1.37	1.42	1.51
7 × 3½	13/16	15.74	2.21	1.27	1.36	1.41	1.46	1.56
7 × 3½	1	19.00	2.19	1.31	1.40	1.45	1.50	1.60

TABLE XIII

RADII OF GYRATION FOR TWO ANGLES, HAVING UNEQUAL LEGS, PLACED BACK TO BACK

Dimensions Inches	Thickness Inches	Area of Two Angles Square Inches	Radii of Gyration					
			r_0	r_1	r_2	r_3	r_4	r_5
2 × 1⅜	3/16	1.20	.41	.92	1.01	1.06	1.11	1.22
2 × 1⅜	⅜	2.25	.38	.95	1.05	1.10	1.15	1.26
2 × 1½	3/16	1.24	.44	.90	.99	1.05	1.09	1.20
2 × 1½	⅜	2.34	.42	.93	1.09	1.14	1.19	1.29
2½ × 1¼	3/16	1.34	.33	1.21	1.31	1.36	1.41	1.51
2½ × 1¼	⅜	2.53	.32	1.25	1.35	1.40	1.45	1.56
2½ × 1½	3/16	1.43	.42	1.17	1.26	1.31	1.36	1.47
2½ × 1½	⅜	2.72	.40	1.20	1.30	1.35	1.40	1.51
2½ × 1¾	3/16	1.52	.51	1.13	1.23	1.27	1.32	1.43
2½ × 1¾	¼	2.00	.50	1.14	1.24	1.29	1.34	1.44
2½ × 2	3/16	1.62	.60	1.10	1.19	1.24	1.29	1.39
2½ × 2	⅜	3.09	.58	1.13	1.23	1.28	1.33	1.43
2½ × 2	½	4.00	.56	1.15	1.25	1.30	1.35	1.46
2¾ × 1½	3/16	1.52	.41	1.31	1.40	1.45	1.50	1.60
2¾ × 1½	5/16	2.46	.40	1.33	1.43	1.48	1.53	1.63
2¾ × 1½	7/16	3.34	.39	1.36	1.45	1.51	1.56	1.66
3 × 2	3/16	1.80	.58	1.37	1.46	1.51	1.56	1.66
3 × 2	5/16	2.93	.57	1.39	1.48	1.53	1.58	1.68
3 × 2	7/16	3.99	.55	1.41	1.51	1.56	1.61	1.71
3 × 2½	¼	2.63	.75	1.31	1.40	1.45	1.50	1.60
3 × 2½	⅜	3.84	.74	1.33	1.42	1.47	1.52	1.63
3 × 2½	9/16	5.55	.72	1.37	1.46	1.51	1.56	1.66
3½ × 2½	¼	2.88	.74	1.58	1.67	1.72	1.76	1.86
3½ × 2½	½	5.50	.70	1.62	1.72	1.77	1.81	1.92
3½ × 2½	11/16	7.30	.69	1.66	1.75	1.80	1.86	1.96
3½ × 3	5/16	3.87	.90	1.52	1.61	1.66	1.71	1.80
3½ × 3	9/16	6.68	.87	1.57	1.66	1.71	1.76	1.86
3½ × 3	13/16	9.24	.85	1.61	1.71	1.76	1.81	1.91

TABLE XIII—(Continued)

RADII OF GYRATION FOR TWO ANGLES, HAVING UNEQUAL LEGS, PLACED BACK TO BACK

Dimensions Inches	Thickness Inches	Area of Two Angles Square Inches	Radii of Gyration					
			r_0	r_1	r_2	r_3	r_4	r_5
4 × 3	5/16	4.18	.89	1.79	1.88	1.93	1.97	2.07
4 × 3	9/16	7.24	.86	1.83	1.93	1.97	2.02	2.12
4 × 3	13/16	10.05	.83	1.88	1.97	2.02	2.08	2.18
4 × 3½	5/16	4.49	1.07	1.73	1.81	1.86	1.91	2.00
4 × 3½	½	7.00	1.04	1.76	1.85	1.89	1.94	2.04
4 × 3½	⅝	8.59	1.02	1.78	1.87	1.92	1.97	2.07
4½ × 3	⅜	5.34	.86	2.07	2.16	2.21	2.26	2.35
4½ × 3	½	7.00	.85	2.09	2.18	2.23	2.28	2.38
4½ × 3	⅝	8.59	.83	2.11	2.21	2.26	2.31	2.41
5 × 3	5/16	4.80	.85	2.33	2.42	2.47	2.52	2.61
5 × 3	9/16	8.37	.82	2.37	2.47	2.52	2.57	2.67
5 × 3	13/16	11.68	.80	2.42	2.52	2.57	2.62	2.72
5 × 3½	⅜	6.09	1.02	2.27	2.36	2.41	2.45	2.55
5 × 3½	⅝	9.84	.99	2.31	2.40	2.45	2.50	2.60
5 × 3½	⅞	13.34	.96	2.36	2.45	2.50	2.55	2.65
5 × 4	⅜	6.47	1.20	2.20	2.29	2.34	2.38	2.48
5 × 4	½	8.50	1.18	2.22	2.31	2.36	2.41	2.50
5 × 4	⅝	10.47	1.17	2.24	2.33	2.38	2.43	2.53
6 × 3½	⅜	6.84	.99	2.81	2.90	2.95	3.00	3.09
6 × 3½	⅝	11.09	.96	2.86	2.95	3.00	3.05	3.15
6 × 3½	⅞	15.09	.93	2.90	3.00	3.05	3.10	3.20
6 × 4	⅜	7.22	1.17	2.74	2.83	2.87	2.92	3.02
6 × 4	⅝	11.72	1.13	2.78	2.87	2.92	2.97	3.06
6 × 4	⅞	15.97	1.11	2.82	2.92	2.97	3.02	3.12
7 × 3½	7/16	8.80	.95	3.37	3.47	3.52	3.56	3.66
7 × 3½	½	10.00	.94	3.39	3.48	3.53	3.58	3.67
7 × 3½	⅝	12.34	.93	3.40	3.50	3.55	3.60	3.70
7 × 3½	13/16	15.64	.91	3.45	3.54	3.59	3.64	3.74
7 × 3½	1	19.00	.89	3.48	3.58	3.63	3.68	3.78

MATERIALS OF STRUCTURAL ENGINEERING

(PART 1)

FOUNDATION SOILS

EXAMINATION OF SOILS

1. Methods of Investigation.—Before any permanent structure is founded upon the earth, an accurate knowledge of the character of the soil at and below the foundation line should be obtained by examination. If the building is to be placed in a well-settled locality, this information may possibly be obtained from the history of the adjacent excavations; although, unless the geological formation of the ground is regular and uniform, this should not be relied on. If the building is to be placed in an unsettled locality, or if, in a settled locality, it is to be of greater weight than the adjoining buildings, maps of the geological formation, if available, should be consulted and may furnish valuable information as to the probable bearing capacity of the soils. If the building is an important one and doubt exists as to the capacity of the soil to sustain its weight, borings should be made for some distance below the bottom of the footings in order to determine the character of the underlying strata. These borings can be made by methods similar to those used in the driving of an ordinary pipe well, from which samples of the material passed through can be obtained as the pipe is driven down.

Tests may also be made by driving a gas pipe, say 2 inches or more in diameter, with a maul or hammer, and withdrawing the pipe from time to time to obtain samples of the material passed through. If the pipe passes through gravel or hard material, difficulty will be experienced in extracting the material after the pipe is withdrawn. This is at best a crude method and should not be used if accurate results are desired.

A better method of making this test is to sink a pipe and use an auger to bring up the material. A common wood auger with levers 2 or 3 feet long, turned by two men, will bring up samples that may be sufficient to determine the nature of the soil, but such samples cannot be taken as indicating the compactness of the soil, as the driving of the pipe compresses the material inside of it and the auger subsequently disturbs it. The same thing is true of the samples taken from a driven well, as the material inside the pipe is first compressed and then taken out with an auger, or, if the material is hard, broken up with a bit or chisel, and then removed with a sand pump.

In making tests with pipes, too great care cannot be exercised in arriving at conclusions regarding the character of the underlying material.

The soil may also be tested by digging an ordinary well 4 to 6 feet in diameter, curbing it with wood as the excavation proceeds. This method will permit accurate conclusions to be formed as to the character of the material in its natural position. If the ground is wet, it will be necessary to make the curbing water-tight and to remove the water.

If the site of the proposed structure is in a locality of apparently recent geological formation, such as the bottom lands in the valleys, more care must be exercised in obtaining the required information than is necessary in localities having an older geological history. In land formed of alluvial deposits a wide variation may be found between two contiguous sites. The difference may have been caused by changes in the river currents and channels due to floods or other natural causes, which may cause gravel to be deposited in considerable layers in one spot and soft material in an adjoining spot. In alluvial soils, deep borings should always be made, as a shallow excavation or test well may disclose gravel that is of variable thickness and may be underlaid with softer and yielding material. If the gravel is thick enough, it may be advisable not to go below it, but in any event the actual knowledge of the conditions should be ascertained before decision as to the character of the foundations is made.

2. Determination of Bearing Capacity.—After a full knowledge of the nature of the soil has been obtained, if the material is compressible and spread foundations are to be used, a test of the actual bearing capacity of the soil should be made by loading a platform of certain area and measuring accurately the settlements under increasing loads.

In testing the bearing capacity of the soil on which the New York State Capitol at Albany was erected, a measured load was applied to a square foot and also to a square yard. For the first test a timber mast 12 inches square, held in a vertical position by guys, was fitted with a cross-frame to hold the weights. A hole 3 feet deep was dug in the blue clay at the bottom of the foundation. The hole was 18 inches square at the top and 14 inches square at the bottom. Small stakes were driven in the ground on lines radiating from the center of the hole. The tops of the stakes were brought exactly to the same level so that any change in the surface of the ground adjacent to the hole could readily be detected by means of a straightedge. The foot of the mast was placed in the hole and the weights applied. No change in the surface of the adjacent ground was observed until the load reached 5.9 tons per square foot, when an uplift of the surrounding earth was observed in the form of a ring with an irregular rounded surface. Similar experiments were made by applying the load to a square yard, with essentially the same result. The loads were allowed to remain some time and the settlements observed.

Before building the Congressional Library at Washington, similar experiments were made with a frame, which rested upon four corner posts, each a foot square. The frame could therefore be moved from place to place upon wheels and a number of tests were made on different parts of the site.

3. Experimental Precautions.—In making experimental tests, it is necessary to take extraordinary precautions to avoid obtaining erroneous results, and more particularly is extra care necessary if the bearing area tested is a small one, as slight errors either way may vitiate the correctness of the finding.

In placing posts on the ground, such as 12″ × 12″ timbers, great care is necessary to see that they are evenly and gently placed upon the surface to be tested; and if only a square foot is tested, allowance must be made for the greater resistance to displacement which the surrounding ground offers to a small area, as compared with a larger area. It is likewise well to consider that the employment of a square testing area gives four cutting edges, a condition likely to produce a greater and more rapid settlement than the parallel edges of a footing course.

CHARACTERISTICS OF FOUNDATION SOILS

4. The capacity of the earth to support superimposed loads without lateral displacement or crushing depends on the character of the soil, which may vary from solid rock through all intermediate stages to a soft or semifluid condition, as mud, silt, or marshy matter. Rock furnishes the most stable foundation. Gravel, which is composed of waterworn pieces of rock, is the next best. Hard dry sand, wet sand if confined, dry clay, wet clay, silt, and marshy matter decrease in bearing capacity approximately in the order named.

5. Rock.—The ultimate crushing strength of **rock** can only be determined from tests made on cubes of it in the testing machine, or, if the variety is well known, from the history of the stone. The ultimate strength of the rock varies from about 15 tons, the crushing strength of cemented gravel in the case of decayed rock and some of the softest varieties, up to about 1,800 tons per square foot for the hardest rocks. In making tests of cubes of rock for the ultimate strength, in order to secure accurate results, the beds must be truly parallel and reduced to a true plane.

The safe bearing value of rock is considered to be not less than about one-tenth the ultimate crushing strength of the cubes. Under ordinary conditions it may safely be said that all rock, except that of the softest and most friable character, will safely sustain any load that may be placed upon it by a building of ordinary character, the area of the foundations being determined by the safe stresses that may be placed on the material forming the foundation courses of the superstructure. In ordinary practice, the safe bearing value of sound rock is taken from 14 to 20 tons per square foot.

6. Gravel.—Broken rock from which the earthy substance has been worn away by the action of water, or *detritus of rock*, as it is called, composes **gravel**. If gravel is found in a compact mass and in layers or strata of considerable thickness, it forms a most excellent foundation, compression within the safe limits of loading being so small that it can be neglected as a factor in design. In founding a building upon gravel, care must be taken to see that the stratum is of sufficient thickness and that the material underneath it is not such that it would be displaced by the load coming upon the strata above it. If a bed of gravel overlays a dry material, such as clay or sandy clay, it can be regarded as an excellent foundation; but if the clay is wet, a test of the ultimate bearing capacity should be made. The safe bearing loads upon gravel are usually taken from 6 to 10 tons per square foot on hard cemented gravel and from 4 to 6 tons per square foot on loose and sandy gravel, although the lower limit is considered the safer one to use under ordinary circumstances.

Strata, or sometimes merely pockets, of gravel are frequently found between layers of alluvial mud in land that has been formed through the action of water. In building upon such soil, care must be taken that the gravel is not merely a local deposit, and soundings should be made over the whole site before the foundation is designed.

7. Sand.—Sandy soils vary from fine sand to coarse gravel. Dry **sand** is incompressible, and if beyond the danger of scour or of being carried away by running water is one of the best foundations obtainable. Where foundations are built upon sand, it is necessary to make sure provision that water will not remove it from underneath the footing courses. Dry sand when confined may be considered as having the same safe load as hard cemented gravel; but no rule can be laid down for the safe load of wet sand.

8. Clay.—Soils that are composed of **clay** vary greatly in hardness and bearing capacity, and range from slate or shale to soft, wet, or semiliquid clay. The bearing capacity of clays may roughly be said to vary directly in proportion to the water contained in them. It is therefore evident that their bearing capacity may be improved by draining them and thereby removing the water. Hard clay, such as slate or shale, will support the same load as rock of equal hardness, while soft, wet clay will squeeze out laterally when any pressure, such as ordinarily occurs in building, is placed upon it. Wet clay is at best a dangerous material for foundations, because it has a tendency to squeeze out around the edges of the footings, or, if it is not soft enough for this, settlement of the building takes place by the load pressing the water out of the clay and thereby reducing its volume.

In making an excavation below the water-line, clay may be found quite hard when first uncovered, but as it has a great capacity for absorbing water, it may swell up and increase in volume and become unfit for a foundation that would permit of no settlement.

If coarse sand or gravel is found mixed with the clay its bearing capacity is increased in the same proportion as these materials are mixed with it. It must be remembered, however, that clay is an unctuous material, that is, of a fatty or greasy nature, and acts as a lubricant between the coarser particles, and therefore sand and gravel mixed with a very slight proportion of clay, if the material is wet, has a much smaller bearing capacity than the sand and gravel without the clay, but if the material is dry the bearing capacity is very much greater than if it is wet.

The city of Chicago is underlaid very largely with a stratum of clay of variable thickness, and the Building Laws there make the following restrictions: If the soil is a layer of pure clay at least 15 feet thick, without a mixture of any foreign substance, excepting gravel, it shall not be loaded more than at the rate of 3,500 pounds per square foot. If the soil is dry and thoroughly compressed, it may be loaded not to exceed 4,500 pounds per square foot. If the soil is a mixture of clay and sand, it shall not be loaded more than at the rate of 3,000 pounds per square foot. With the loads authorized by the Building Department of Chicago, the settlement of a building, due to the compression of the soil, amounts to from 3 inches to 5 inches.

From the experimental tests made on the soil at the State Capitol at Albany, it was found that the ultimate bearing capacity of the clay soil was less than 6 tons per square foot, and the building was designed to impose a load of 2 tons per square foot. This soil contained from 60 to 90 per cent. of alumina clay, the remainder being fine silicious sand. It was also found to contain from 27 to 43 per cent., usually about 40 per cent., of water, and samples of it weighed from 81 to 101 pounds per cubic foot.

In the soil underlying the Congressional Library at Washington, the ultimate bearing capacity of the soil, which was yellow clay mixed with sand, was 13½ tons per square foot, and the building was designed to impose a pressure of 2½ tons per square foot on this material.

Hard, stiff clay, when dry, may be estimated to have a safe bearing capacity of from 4 to 6 tons, but if it is allowed to become wet, its resistance is greatly decreased and it is probably not good for more than 1½ to 2 tons; even then the compression and consequent settlement may be such as will make it necessary to design all the footings so that the settlement on each will be the same, and a consideration of the safe capacity of such soil to properly bear the weight of the structure put upon it leads directly to the question of the design of foundations for which settlement has to be provided.

Where subsequent settlement of the foundations might cause cracks in the superstructure, the footing of a heavier part of the building, such as a tower, for instance, is temporarily loaded, the load being removed as the superstructure is built. In the Chicago Auditorium, a large tower rises 94 feet above the main building, and as a precaution against the heavy concentrated weight of the tower cracking the adjoining wall, a direct load, approximately equal to the weight of the finished tower, was placed upon the tower footings, and gradually removed as the walls were carried up.

In Chicago, when piles are not resorted to and the building is founded upon the compressible soil, the wall piers are built detached and are carefully calculated for the dead load that comes upon them. The live load is either omitted entirely or a very small proportion of it is used.

The great difficulty in designing a building to meet such conditions of supporting ground is that of accurately proportioning the foundations where the interior columns carry but little dead load, as, for instance, the interior columns in a store or other building having an open interior. In the case of office buildings, the dead load being a considerable part of the total load, the problem is easy of solution, but in store buildings it is a problem requiring great judgment to reach a satisfactory conclusion.

In designing foundations for which a settlement must be allowed, great care must be taken to have the center of resistance under the center of gravity of the load, as, with a yielding material, the overloading of one side or one edge may cause an undue compression of that side or edge and thereby tilt the footings with disastrous consequences to the superstructure. For compressible soil, it is necessary to spread the foundation over a large area, so as to have a small load per square foot of surface. This is usually accomplished by means of I beams bearing on or embedded in a concrete slab or plate, but in buildings of no great importance this spreading of the footing is often done by means of timbers. When timbers are used, however, it is absolutely necessary that they be below the permanent water-line, as otherwise they will decay; it is also necessary that the bending stresses on the timbers be kept within very small limits, as otherwise the deflection of the timber under load will cause settlement.

9. Mud, Silt, Marsh, or Semiliquid Soils.—With soils of this character, it is necessary either to drive piles to a hard substratum, if it can be found, or to sink caissons to the hard material below, the method to be pursued usually depending on the importance of the building and the value of the land. With ordinary buildings, the pile foundation is undoubtedly the simplest and the cheapest method, but in New York city, in the lower part of the island, it is found necessary to go down to the bed rock, and in the most recent important structures there pneumatic caissons have been sunk to bed rock and afterwards filled with concrete, so arranged as to form a continuous retaining wall around the building. The material inside of this retaining wall is excavated and the space utilized for various purposes, such as basements, subbasements, and office or store purposes, the light, heat, and air being furnished by mechanical means.

Rankin gives the following formula for ascertaining the unit bearing capacity P_u of soft soil:

$$P_u = w\,h\left(\frac{1 + \sin a}{1 - \sin a}\right)$$

in which w = weight of soil, in pounds per cubic foot;
h = depth of marsh, in feet;
a = angle of repose of the soil.

For example, if a equals 5°, the supporting power of the soil is 1.19 $w\,h$ per unit of area; if a equals 10°, it is 1.42 $w\,h$; and if a equals 15°, it is 1.69 $w\,h$. The weight of mud, silt, and quicksand varies from about 100 to 130 pounds per cubic foot.

The city of New Orleans is underlaid with alluvial soil, that is, a soil deposited by water, and experiments there made on quicksand would indicate that with a load of ½ ton to 1 ton per square foot, the settlement would not be excessive.

In designing the foundation for any building, the particular character of the building, its cost, and permanence must determine the character of the footings. If the foundation is of soft material, the possibility of adjacent excavations wrecking the building must be carefully considered. If it is necessary to establish the foundation of a building on soft, yielding material, either too soft for a grillage or spread foundation, piles must usually be driven. If, also, the character of the building is such that it cannot be designed to insure equal settlement of all its pieces or parts, and a firm bearing soil cannot be found within convenient distance, piles long enough to sustain the load coming upon them should be driven into the ground. The piles can be used only in soils that are permanently saturated with water, as, if exposed to alternate wetting and drying, or if all the piling is not thoroughly immersed, fungus soon destroys the wood, and the building resting on the piles must fail.

10. Quicksand.—The term **quicksand** is applied to any silicious material so saturated with water that it will flow more or less easily when its natural condition of equilibrium is destroyed by excavating pits, trenches, shafts, or tunnels. Quicksand may be found on the surface, underlaid by firm material, or it may be found in a stratum of greater or less thickness, confined by firm strata both above and below. If the quicksand is on the surface, and underlaid by a hard material, it is no more difficult to deal with than water, but when found between strata of harder material at considerable depth, great difficulty will be ordinarily experienced in working in it, for the reason that as the depth below the surface increases, the pressure increases and the flow of the liquid or semiliquid materials allows the stratum on top of it to settle, bringing great pressure on the sides of the caissons, sheet piling, or cribbing, either crushing it or throwing it out of line. The material falling on top of the quicksand increases the amount of material to be excavated and usually adds very greatly to the cost of the foundation of the structure.

Owing to the fact that quicksand flows and runs so readily, the greatest precaution must be exercised in working in this soil adjacent to the foundations of a building or structure. Instances have been known where the undue removal of sand and water, while boring a hole for a direct plunger elevator, imperiled the adjacent foundations by removing some of the stratum from beneath the footings with the detritus from the hole. It is likewise a necessary precaution to have all sewers and drain lines running near footings and on a lower level, and that are built in soils partaking of the nature of quicksand, so carefully constructed and inspected that there can be no possibility of bleeding the foundation stratum.

11. If quicksand is encountered in building construction, the only successful method of dealing with it is to use sheet piling in the excavation, keeping it in position by horizontal bars and bracing. The piling must be of sufficient thickness to withstand the bending strains due to the pressure of the quicksand. These piles should be tongued and grooved or connected with loose slip tongues. If the excavation can be carried on at the same time with the driving, it is possible to drive the piling with a heavy maul, or, if the material is soft enough, it may be driven with a maul before the excavation is made. If the material is of a clayey nature, and the excavation can not be carried along as the piling is driven, it may be impossible to drive it any distance with the maul. In such cases, a portable steam pile driver, which fits on the head of the pile, built on the same lines as a steam rock drill, may be used to advantage. By means of this driver, the piles may be driven to a considerable depth before the excavation is made, and as the material is thrown out of the trench or pit, shores can be put in place and the trenches carried down for some distance without any displacement of the adjacent material unless it is so thin that it will slip down under the piles and come up into the trench, in which case, if the amount of material coming into the trench is considerable, the whole temporary structure must collapse because of the displacement of the material behind it. In such cases, several methods of founding may be followed. One is the freezing process, in which the earth adjacent to the excavation is frozen hard by means of brine from a portable refrigerating plant, which is circulated through pipes that are sunk into the ground around the sides of the proposed excavation. After the ground is frozen, the excavation can be carried on as with other hard material.

Another method is by sinking either pneumatic or open caissons through the material. If the material is very soft and fluid, the pneumatic caisson must be employed. Open caissons of hollow brick or concrete, or iron cylinders, or timber-lined shafts can be sunk to great depths through soft material and ultimately filled with concrete and masonry. In sinking these open caissons, the material is excavated from the inside and weights are built or placed upon the caissons, causing them to sink. If brick or concrete cylinders are used, the bottom must rest on a timber or iron curb with a cutting edge.

Pneumatic caissons usually consist of a strong grillage of timber, laid diagonally and crosswise, strongly bolted together, beneath which a strong cutting edge of timber, sheathed with iron, is provided. The height of this cutting edge is sufficient to permit the work to be carried on beneath the under side of the caisson. An air pressure sufficient to prevent the material from running in around the cutting edge is maintained in this open space. As the material is excavated from beneath it, the caisson sinks. The excavated material is taken out in buckets through the air lock. As the caisson sinks, the masonry is built upon it until the cutting edge finally reaches the hard material, after which the space between the cutting edge and the under side of the timber platform is filled with concrete.

Another method of dealing with quicksand is to solidify it by liquid cement by means of pipes sunk into the ground. Liquid hydraulic cement is forced down through one pipe and suction is maintained in the other; by this means the cement is drawn from the supply pipe over to the exhaust pipe and the intervening material is saturated with it. By changing the position of the pipes often enough, a stratum or area of quicksand saturated with cement can be produced, which, upon hardening, forms an artificial stone.

12. Culm.—It is frequently necessary in the anthracite regions to build at least temporary structures upon culm or refuse from mines. This material is composed of fine particles of coal and slate, being exceedingly springy or elastic when in beds of considerable thickness and is a dangerous material upon which to build. This is especially true when it is not thoroughly confined or the building is so situated that its foundations are not well within the slope of repose, or slope which the material naturally assumes when dumped or piled loosely. Another great danger in building upon culm and similar materials exists in the liability of the material to be washed away by heavy rains, fire-streams, or leaky mains. Under no condition should culm be subjected to a greater pressure than $\frac{1}{2}$ to $\frac{3}{4}$ ton per square foot.

13. Foundations Over Mines, Etc.—Large buildings have been destroyed by the failure of the principal piers built over long-disused cisterns or vaults, which in the course of time have been forgotten and filled over. A principal pier in one instance did not break through the covering of the cistern until the building was far along in its construction, so that the failure of the pier precipitated the destruction of almost the entire building.

In the hard-coal section of this country, cities of importance are frequently undermined by galleries or veins which have been worked, and these are often near the surface. When heavy buildings are erected in such localities, the position of the building with reference to these mine workings must be investigated; then, if the roof of the working does not seem to offer adequate strength, some means of support must be provided. The working, if abandoned, may be filled, or a better method would be to extend piers through to the solid floor.

One great danger exists in such localities from the fact that surface water is likely to run along the hard-pan or shaly roof of the working, and thence down through fissures in some abandoned working. In such an instance a tunnel will frequently be formed by the action of the subterranean stream, and where a heavy foundation is placed above, damaging settlement is likely to occur which may possibly cause the failure of the building.

BEARING VALUE OF FOUNDATION SOILS

14. There is some difference of opinion regarding the safe bearing value of foundation soils, due probably to the difficulty of arriving at any experimental results that will have a general application. Conservative engineering practice, however, dictates that the greatest unit pressure on the different foundation soils shall not exceed the values given in the following table:

TABLE I

SAFE BEARING VALUES OF DIFFERENT FOUNDATION SOILS

Materials	Tons per Square Foot
Granite rock formation	30
Limestone, compact beds	25
Sandstone, compact beds	20
Shale formation, or soft friable rock	8 to 10
Gravel and sand, compact	6 to 10
Gravel, dry and coarse, packed and confined .	6
Gravel and sand, mixed with dry clay . . .	4 to 6
Clay, absolutely dry and in thick beds . . .	4
Clay, moderately dry and in thick beds . . .	3
Clay, soft (similar to Chicago clay)	1 to 1½
Sand, compact, well-cemented, and confined .	4
Sand, clean and dry, in natural beds and confined	2
Earth, solid, dry, and in natural beds . . .	4

BUILDING LAWS REGARDING FOUNDATION SOILS

15. The observance of the revised building laws of the several cities is considered good engineering practice, for they are usually the results of careful investigations and records of long experience. The following, quoted from the New York Building Law, is interesting and gives bearing values that are well within the safe limits.

Where no test of the sustaining power of the soil is made, different soils, excluding mud, at the bottom of the footings shall be deemed to safely sustain the following loads to the superficial foot, namely:

Soft clay, 1 ton per square foot.

Ordinary clay and sand together, in layers, wet and springy, 2 tons per square foot.

Loam, clay, or fine sand, firm and dry, 3 tons per square foot.

Very firm, coarse sand, stiff gravel, or hard clay, 4 tons per square foot, or as otherwise determined by the Commissioner of Buildings having jurisdiction.

Where a test is made of the sustaining power of the soil, the Commissioner of Buildings shall be notified so that he may be present in person or by representative. The record of the test shall be filed in the Department of Buildings.

When a doubt arises as to the safe sustaining power of the earth upon which a building is to be erected, the Department of Buildings may order borings to be made, or direct the sustaining power of the soil to be tested by and at the expense of the owner of the proposed building.

MATERIALS OF STRUCTURAL ENGINEERING

(PART 2)

MASONRY

INTRODUCTION

1. The materials of masonry construction include stone, brick, terra cotta, and the cementing materials—cement, lime, and sand—used in the manufacture of the mortars. That an intelligent working knowledge of these materials may be had, it is necessary to know something of the elements that compose them.

Though an extensive knowledge of chemistry and geology would be valuable to the structural engineer, his practice involves only an elementary knowledge of these subjects. This acquaintance is necessary, however, in order that he may know the chemical and physical nature of the materials with which he builds. Therefore, the first portion of this Section is devoted to a brief description of the principal elements of nature that combine to constitute stones and minerals, and following this is given a condensed geological description that will furnish an insight into the physical formation of the rocks from which building stones are obtained.

2. Elements and Compounds.—Every substance or mass of matter is either an *element*, a *compound*, or a *mixture*.

Any substance that can be decomposed or divided into separate substances is called a **compound**. For example, if an electric current is passed through water, the water slowly disappears and two gases are formed. These gases are entirely unlike, and neither resembles the water from which it was produced. Likewise, lime can be divided into two other substances, calcium and oxygen.

There are substances, however, like iron, gold, sulphur, and arsenic, that have never been decomposed into other substances. These are called **elements.**

In referring to an element, it is customary to simply use the *symbol,* which is usually the first letter of the name. Thus, *H* stands for hydrogen, *C* for carbon, etc.

CHEMICAL ELEMENTS

3. About seventy elements have been found in the earth, but of these, fourteen in various combinations form nearly the entire crust of the earth. These elements, named in the approximate order of their importance, are oxygen, silicon, aluminum, iron, calcium, magnesium, potassium, sodium, carbon, hydrogen, phosphorus, sulphur, chlorine, and manganese, to which may be added nitrogen, for though it does not enter into the rocks to a great extent, it forms about 79 per cent. of the air.

As all these elements are ready to form combinations with others, they are seldom found uncombined and there is a constant tendency to change, the old combinations being broken up and new combinations being formed. One of the most active of the elements in this respect is oxygen.

4. Chemical Combination.—When two or more elements are brought into contact under favorable circumstances, they will combine and form a substance unlike either of the elements. For example, hydrogen, an exceedingly light gas, will burn in the air with a light blue flame. This burning is a form of chemical combination, in which the hydrogen unites with another gas, the oxygen of the air, and forms water, a substance with which we are all familiar.

Chemical combination produces heat; chemical separation, on the other hand, absorbs heat. Thus, if carbon and oxygen are brought together at high temperature, they will combine and form carbon dioxide; hydrogen and oxygen combine to form water; hydrogen, nitrogen, and oxygen, when combined in certain proportions, form nitric acid; a given volume of nitrogen and three times that volume of hydrogen combine and form ammonia, a gas that differs greatly from both nitrogen and hydrogen. In each of these combinations a certain amount of heat is produced; to separate the elements again an amount of heat will be absorbed exactly equal to that produced by their combination.

5. It is supposed that the *molecules* of most elements, such as hydrogen or oxygen, are composed of two *atoms*. A **molecule** is the physical unit, or the smallest portion which, so long as the substance is chemically unchanged, keeps together without complete separation of its parts, while an **atom** is the chemical unit, or the smallest mass of an element that can exist. It is further supposed, by chemists, that equal volumes of all gases, whether simple or compound, contain the same number of molecules; thus, a cubic foot of hydrogen, a cubic foot of air, a cubic foot of steam, contain the same number of molecules.

Suppose, now, that a cubic foot of hydrogen gas is allowed to come into contact with a cubic foot of chlorine gas; if the mixture is exposed to heat or light, the gases combine. The process of combination is explained as follows: There is a certain attraction or affinity between the hydrogen and the chlorine atoms. Under the influence of heat or light, this attraction becomes so strong that the two atoms composing the molecule of hydrogen are torn apart. Likewise, the atoms composing a molecule of chlorine separate. Each atom of chlorine seizes upon an atom of hydrogen and forms a molecule of an entirely new gas, viz., hydrochloric acid gas. Since each atom of chlorine takes one atom of hydrogen, it is plain that the number of molecules of each gas must be the same. In other words, a cubic foot of chlorine requires a cubic foot of hydrogen to combine with it; these gases cannot be made to combine in any other proportion. For example, if 3 cubic feet of chlorine were placed in contact with 2 cubic feet of hydrogen, 4 cubic feet of hydrochloric acid gas would be formed, and the extra cubic foot would still remain chlorine. The symbol for hydrochloric gas is HCl.

Suppose, now, that hydrogen and oxygen are placed in contact and heated; they will combine and form steam (or water). But it will be found that each atom of oxygen seizes two atoms of hydrogen to form a molecule of water, and, therefore, the volume of hydrogen must be double the volume of the oxygen with which it combines. This is shown by the symbol for water, H_2O; that is, two parts or atoms of hydrogen to one of oxygen. Similarly, the symbol for ammonia is NH_3; that is, three parts of hydrogen to one of nitrogen. Again, hydrogen and carbon form a compound; each atom of carbon seizes four atoms of hydrogen and forms a molecule of marsh gas; the symbol for marsh gas is therefore CH_4.

The symbol of any compound indicates how the atoms of the elements combine to form the compound. Thus, the symbol H_2O shows that two atoms of hydrogen and one of oxygen unite to form a molecule of water. The symbol H_2SO_4 (sulphuric acid) shows that a molecule of the sulphuric acid contains two atoms of hydrogen, one of sulphur, and four of oxygen.

6. Combination by Weight.—A cubic foot of hydrogen combines with just 1 cubic foot of chlorine. On weighing each gas, it is found that the cubic foot of chlorine weighs 35.5 times as much as the cubic foot of hydrogen. A cubic foot of oxygen weighs 16 times as much as a cubic foot of hydrogen.

As like volumes of gases contain equal numbers of molecules, and oxygen and hydrogen molecules contain two atoms each, like volumes of the two gases must contain equal numbers of atoms. Now, since the former weighs 16 times as much as the latter, it follows that an atom of oxygen weighs 16 times as much as an atom of hydrogen. Similarly, an atom of chlorine weighs 35.5 times as much as an atom of hydrogen. This ratio between the weight of an atom of any element and the weight of an atom of hydrogen is called the **atomic weight,** or the **chemical equivalent,** of the element. The atomic weight of any element (or compound) may be found by dividing the weight of a given volume, say a cubic foot, of the element, when in a gaseous state, by the weight of a cubic foot of hydrogen. The atomic weight is, therefore, about the same thing as specific gravity, hydrogen being the base of comparison instead of water or air. Elements combine with other elements or parts of compounds in proportion to their atomic weight, or some simple multiple of it.

7. The atomic weights of some of the elements considered in this Section are as follows:

Hydrogen, *H*	1.0
Oxygen, *O*	16.0
Nitrogen, *N*	14.0
Carbon, *C*	12.0
Sulphur, *S*	32.0
Chlorine, *Cl*	35.5

By the aid of these atomic weights, the composition of any substance, by weight, can be found when its symbol is known. For example, take water, H_2O; multiply the number of atoms of each element by the atomic weight of the atom. Thus,

$2 \times 1 = 2$ parts, by weight, of hydrogen
$1 \times 16 = 16$ parts, by weight, of oxygen
18 parts, by weight, of water

Water, then, is composed of $\frac{2}{18} = 11.11$ per cent. of hydrogen, and $\frac{16}{18} = 88.89$ per cent. of oxygen.

As another example, take carbon dioxide, CO_2. We have

1 atom of $C \times$ atomic weight, $12 = 12$ parts, by weight, of C
2 atoms of $O \times$ atomic weight, $16 = 32$ parts, by weight, of O
44 parts, by weight, of CO_2

Hence, CO_2 contains $\frac{12}{44} = 27.27$ per cent. of carbon and $\frac{32}{44} = 72.73$ per cent. of oxygen. From these examples it is plain that the atomic weight of water is 18, and of carbon dioxide.44.

8. Mixtures.—Two or more elementary substances may be mixed together and yet not combine to form a new substance; they are then said to form a **mixture.** The mixture has the properties of the elements composing it. The most familiar example of a mixture is ordinary air, which is composed of 23 parts, by weight, of oxygen and 77 parts, by weight, of nitrogen. The two gases are not combined chemically; they are simply mixed.

9. Specific Gravity.—This value is the ratio of the weight of a given substance to that of a standard substance. For solids and liquids, the standard is pure water at a temperature of 62° F.; while for gas, the standard is usually hydrogen gas, though sometimes air. Frequently, however, the specific gravity of gases is measured with water as a standard.

The weight of a cubic foot of pure water at 62° F. is 62.355 pounds; so that in order to find the weight per cubic foot of any substance, the specific gravity of the substance is multiplied by 62.355; while to find the specific gravity of a substance, its weight per cubic foot is multiplied by the reciprocal of 62.355 or .016037. To ascertain the weight per cubic inch of a substance, its specific gravity is multiplied by .036085. With each element, briefly treated in the following description, are given the chemical symbols and the specific gravity. The values given for the gaseous elements are based on hydrogen as the standard, while those for solids have pure water at 62° F. for the basis of the value.

GASEOUS ELEMENTS

10. Oxygen, *O*, specific gravity 16, is a transparent, colorless, odorless gas. The word *oxygen* is derived from two Greek words that literally mean sour generator, or acid former, as it was formerly believed that oxygen was a constituent of all acids. The chemist Davy, however, showed that hydrogen and not oxygen was necessary for an acid.

Oxygen is found in both a combined and a free state. In the air it is mixed, but not combined, with nitrogen in the proportions of about 23 per cent. of oxygen and 77 per cent. of nitrogen. It is an important constituent of all animal and vegetable substances, and is contained in all mineral substances. In granite, slate, clay, limestone, and other rocks, except coal, the quantity of oxygen is nearly 50 per cent. It combines with all other elements except fluorine, which is a gaseous element never found uncombined. Water dissolves about 4 per cent. of oxygen; about 47 per cent. of the earth's crust is composed of it, while it constitutes about 86 per cent. of the water. Decay and the ordinary forms of decomposition are simply the result of oxidation.

11. Hydrogen, *H*, specific gravity 1, is the lightest of all known substances. As one of the two elements constituting water, it is found everywhere present in the crust of the earth. In combination with water, many minerals form the group of hydrous minerals. It exists in combination everywhere, as a constituent of water, of all plants and animals, and in numerous minerals, abundantly in coal, petroleum, bitumen, and in a lesser degree in rocks. The amount of water chemically combined in the rocks or mechanically held in the crevices is estimated to many times exceed the amount of water in the oceans.

Hydrogen, in combination with carbon, forms the *hydrocarbons* that are found in the rocks that contain the remains of animal and plant life, such as the rocks from which petroleum and natural gas are obtained.

Hydrogen, combined with the minerals, forms but .21 per cent. of the crust of the earth. The amount of it present in the air and water, however, increases the importance of this element.

12. Chlorine, *C*, specific gravity 35.5, is a nearly transparent gas, of a greenish-yellow color. Combined with magnesium, potassium, or sodium, it forms the chloride of which common salt is the most familiar example. It is never found in a free state. Chlorine is one of the strongest oxidizing agents. It has a great affinity for hydrogen and most metals, of which it forms chlorides. Its most abundant compounds are chlorides, of which sodium chloride, $NaCl$, and magnesium chloride, $MgCl_2$, are the most common and widely distributed. Sodium and magnesium chlorides are found in sea-water, and all soils of the earth contain some sodium chloride.

Chlorine forms about .01 per cent. of the crust of the earth, but about 2 per cent. of the oceans.

SOLID ELEMENTS

13. Silicon, *Si*, specific gravity 2.49, is never found in a free state but is found chiefly in the form of the oxide, SiO_2, commonly called *silica* or *silicon dioxide*, and also in combination with oxygen and several of the common metals, particularly with sodium, potassium, aluminum, and calcium, in the form of *silicates*, which are combinations of silicon with oxygen and other elements, as, for instance, feldspar, which is a complex silicate of aluminum and potassium. Silicon in one or the other of these forms is found in nearly all the rocks of the earth. In sandstone, it constitutes the greater part of the rock, while in coal it is only present in small quantities.

Next to oxygen, silicon is the most abundant element. Some extensive mountain ranges consist almost entirely of silicon dioxide, SiO_2, in the forms known as *quartz* and *quartzite*. Other ranges are made up of silicates that are compounds formed by the combination of silicon dioxide and bases; that is, the greater number of rocks of the earth's surface are combinations of complex silicates, limestone, $CaCO_3$, being the only important exception. The clay of valleys, river beds, etc. also contains silicon in large quantities, while the sand found so abundantly on the seashore is mostly silicon dioxide, or pulverized quartz.

While it is difficult to decompose the oxide in such a way as to obtain the element, under proper conditions the silicon can be obtained in the form of crystals that are of a gray color and are harder than glass, although they are only a chemical curiosity. Crystalline silica is found as quartz, rock crystal, and sand in variously colored varieties, and the amorphous (not crystalline) forms of flint, chalcedony, opal, and others. The specific gravity of quartz is about 2.6. It is very hard but not so hard as ruby, corundum, Al_2O_3, and diamond. All forms of silica are more or less soluble in alkali by prolonged heating, but crystalline quartz is the least attacked. Silicon will not burn or dissolve in any acids except a mixture of nitric and hydrofluoric acids.

About 27 per cent. of the crust of the earth is composed of silicon, and therefore about 74 per cent. of the rocks are composed of oxygen and silicon.

14. Aluminum, *Al*, specific gravity 2.6, is a light, silver-white metal that is never found uncombined; its compounds are mostly silicates. It is the most abundant and widely distributed of all metals. Combined with oxygen it forms the oxide of aluminum, or *alumina*, Al_2O_3. Aluminum is commonly found in combination with both oxygen and silicon and forms the group of minerals known as the silicates of alumina. Alumina is found in all rock formations, except limestone, and in nearly 200 minerals. Corundum, Al_2O_3, is the richest of its ores. Emery is a mixture of the oxides of aluminum and iron.

The compounds of aluminum are difficult of reduction. The ruby and sapphire are Al_2O_3, colored by impurities. About 8 per cent. of the crust of the earth is composed of aluminum and about 80 per cent. of the crust of the earth is composed of oxygen, silicon, and aluminum. The greater part of the clay soil and rocks is composed of these three elements.

15. Iron (*ferrum*), *Fe*, specific gravity 7.8, is rarely found in an uncombined state, but is usually found as an oxide. It is widely disseminated throughout the earth's crust, of which it constitutes about 5 per cent., and its presence is shown in the red and yellow colors in the soil and rocks.

Iron has a great affinity for oxygen and, combined with it, forms what is commonly known as *rust*, which is the hydrated oxide of iron. It commonly exists in the soil in the form of the highest oxide and gives it the red and yellow color, but in the presence of an excess of organic matter, the ferric oxide is reduced to ferrous carbonate, and its red or yellow color is destroyed. The clay of swamps is always bluish, but if this clay is burned to brick the organic matter is destroyed and the iron peroxidized colors the brick red.

16. Calcium, *Ca*, specific gravity 1.578, is a brass-yellow lustrous metal that is never found in an uncombined state. In combination it is very widely distributed and is found in enormous quantities. It is found principally as *carbonate*, $CaCO_3$; in the form of limestone, marble, and chalk, as *sulphate*, $CaSO_4$; in the form of gypsum, as *phosphate*, Ca_3PO_4. Calcium combined with carbon dioxide, CO_2, and with another molecule of oxygen forms *calcite*, $CaCO_3$, which is present in some marbles and most limestones. Calcium is slightly soluble in water, and in solution forms the source from which fish, shell fish, and reef-building corals derive the materials from which to build their shells, cells, etc. Calcium in combination with sulphur and oxygen forms *gypsum*, $CaSO_4+2H_2O$.

Calcium enters into the composition of many of the complex silicates, and therefore forms a constituent of many of the rocks. In the form of carbonate, such as chalk and the various limestones, it forms an enormous rock mass in different parts of the earth. In some districts the sulphate or gypsum is found in considerable quantities. The chief varieties of native calcium carbonate, or carbonate of lime, are the amorphous forms, chalk, and the various kinds of limestone, the crystalline forms of marble, and the well-crystallized forms of calcite, or Iceland spar, and aragonite. Nearly 4 per cent. of the crust of the earth is composed of calcium.

17. Magnesium, *Mg*, specific gravity 1.7, is a less abundant element than calcium, but is almost invariably associated with it. It is a silver-white metal with a high luster, and is never found in a free state. Magnesium carbonate is found as *magnesite*, $MgCO_3$, but usually occurs with calcium carbonate in dolomite and magnesian limestones. The sulphate and carbonate may be found in many natural waters. As silicate, magnesium enters into the composition of numerous minerals, as soapstone, serpentine, and meerschaum, while asbestos, hornblende, and many other silicates, contain this element. Magnesium forms about 2½ per cent. of the crust of the earth.

18. Potassium (*kalium*), *K*, specific gravity .875, is a white metal with a bright, metallic luster. It is so light as to float on water. It is a constituent of many minerals, particularly feldspar, which is a complex silicate of aluminum and potassium. It is never found in a free state, but is present as a silicate in the older Plutonic rocks, which are rocks formed from molten material at a great depth and pressure and including the granites, and in a small quantity in most soils. It is also present in sea-water. Potassium forms about 2½ per cent. of the crust of the earth.

19. Sodium (*natrium*), *Na*, specific gravity .973, is abundant, very widely distributed, and found in large quantities, principally as *sodium chloride*. It is found in a number of silicates and enters into the composition of an enormous number of rocks, being present in traces almost everywhere. It is a silver-white metal with a bright metallic luster, with a very light pink tinge, and is so light as to float on water. Sodium forms about 2½ per cent. of the crust of the earth.

20. Carbon, *C*, specific gravity 3.52, is commonly found in combination with oxygen. In an uncombined state it is found in the form of diamond and graphite. It also forms part of the tissue of all animals and plants, and when these tissues decay or burn the carbon unites with oxygen to form *carbon dioxide*, CO_2, which in combination with other elements forms the carbonates of which calcite (calcium carbonate) is an example.

Carbon is the central element of organic nature; every living thing contains it as an essential constituent. The number of compounds that it forms is almost infinite, and they present such peculiarities that they are commonly treated under the head of "organic chemistry."

Carbon is present in a number of rocks, having been placed there by the agency of organic life, as in coal, where it has been taken from the air, and in limestone, where it has been taken from the water. .22 per cent. of carbon is estimated to be in the crust of the earth. The *diàmond* is pure crystallized carbon.

21. Phosphorus, *P*, specific gravity 1.84 to 2.01, is a yellowish, translucent, crystalline, waxy solid that is never found in a free state because of its affinity for oxygen and other elements. Combined with oxygen it forms the phosphates, of which phosphate of lime is the most common. Its compounds are quite abundant and widely diffused, the principal one being calcium phosphate, which occurs in certain minerals as *phosphorite* and *apatite*.

Phosphorus, in minute quantities, enters into the bones and tissues of many animals and the tissues of many plants. It constitutes but about .1 per cent. of the crust of the earth.

22. Sulphur, *S*, specific gravity 7.086, is found particularly in the neighborhood of volcanoes. It occurs also as sulphide, in combination with many metals, as in *iron pyrites*, FeS_2, *copper pyrites*, $FeCuS_2$, etc. As sulphate, it occurs in combination with metals and oxygen, as *calcium sulphate*, or *gypsum*, $CaSO_4 + 2H_2O$.

Sulphur is found in the tissues of many animals and plants, and is found in the rocks in small quantities. It forms but about .03 per cent. of the crust of the earth.

23. Manganese, *Mn*, specific gravity 7.14 to 7.20, is said to be the hardest of metals. It is brittle, almost infusible, and looks like cast iron. It is easily oxidized in the air and dissolves in diluted acids. It forms alloys and imparts valuable properties to steel, perhaps by combining with such impurities as sulphur and phosphorus, but this is uncertain. Manganese is widely disseminated and its presence is often shown by black and purple stains on the surface of the rocks. It forms less than 1 per cent. of the crust of the earth.

COMMON MINERALS

24. The elements, when they unite, follow definite laws and produce definite results, uniting themselves in regular proportions, forming molecules, which, if free to act under favorable conditions, often assume regular geometrical forms or crystals. These combinations produce *minerals*, which (with the exception of mercury) may be defined as homogeneous solids of definite chemical composition, found in nature, but not of apparent organic origin.

A mineral may be composed of a single element, such as gold, silver, copper, lead, etc., or it may be made up of two or more elements chemically combined, as is usually the case. These elements may be only slightly bound together, as in the case of sulphide of mercury, which may be disunited by the action of heat, or they may be so firmly united and their chemical affinity so strong that, like silica, they may resist almost any effort to separate them. Therefore, some minerals are enduring and others are unstable, and while the chemical composition is definite, it is subject to change whenever the conditions are favorable. These changes are the cause of the decay and crumbling of rocks.

Minerals are frequently formed under conditions that do not permit sufficient room for the perfect development of the crystals; therefore, the perfect crystal is rare. When the crystal is not perfect, though the internal structure shows a crystal formation, the mineral is termed *crystalline.*

Some minerals, like diamonds and quartz, are exceedingly hard, while others, like graphite or black lead, are soft. Some are brittle and others are plastic. Most of the minerals are crystalline, though sometimes built up in irregular shapes, to which is applied the term *amorphous*, which means not crystalline.

The minerals that make up the greater part of the earth's crust, named approximately in the order of their importance, are:

1. Quartz.
2. The group of feldspars.
3. The group including calcite, dolomite, and siderite.
4. The group of micas.
5. The amphibole group.
6. The pyroxene group.
7. The group of iron minerals.
8. Gypsum.
9. Salt.
10. Ice and its liquid form, water.

25. Quartz.—Quartz (*silica*, SiO_2) is the most abundant mineral in the crust of the earth. In color it varies from a transparent rock crystal to a jet-black glassy mass. It is often colorless, though sometimes topaz-yellow, amethystine, rose, smoky, and other tints occur, also quartz of various shades of red, yellow, green, blue, or brown are found. In some varieties, the colors are in bands, stripes, or clouds, of all degrees of transparency to opaque, of a vitreous luster, with brilliant crystals, sometimes dull and often waxy.

The common mineral impurities of quartz are chlorite, rutile, asbestos, actinolite, tourmaline, hematite, and limonite. Hematite, or the red oxide of iron, is usually the red coloring matter; limonite, another oxide of iron, is the yellow coloring matter. Chlorite and actinolite give the green color, and an oxide of silicate of nickel an apple-green tint. Manganese gives an amethystine color, and carbonaceous matter, such as color marsh waters, the smoky-brown shades. Quartz crystals often contain liquid in cavities, either water, petroleum, or liquid carbon dioxide. Clear quartz is sometimes speckled with scales of mica or rendered opaline by means of asbestos. Flint or chert are often colored by mixture with the materials of the enclosing rock.

Quartz is sometimes amorphous, but usually crystalline, and is often found in perfect crystals; the varieties of the latter include agate, amethyst, and jasper. Quartz is light in weight, brittle, and so hard that it cannot be scratched with a knife. Its chemical composition is strongly fixed and silica, or quartz, once formed, remains as silica throughout all the changes to which rocks are ordinarily subjected. By reason of its hardness and chemical strength, quartz may give a great durability to rock. There is no common mineral that resists destruction so well.

Quartz is found in various forms and colors, but may be distinguished by the lack of true cleavage, by hardness, infusibility before the blowpipe, and insolubility with the common acids.

It is found nearly everywhere on the earth's surface, but is particularly abundant in granite rocks, sandstone, and most of the soils. Wherever found, it is fresh, pure silica, and in this it differs from many of the common minerals, which are so liable to decay.

The varieties of quartz may be classified as *vitreous*, distinguished by their glassy fracture; *chalcedonic*, having a subvitreous or waxy luster and generally translucent; *jaspery cryptocrystalline*, having little or no luster, and opaque. The vitreous varieties comprise rock crystal, amethyst, rose quartz, false topaz, smoky quartz, prase, aventurine quartz, and ferruginous quartz. The chalcedonic varieties comprise chalcedony, chrysoprase, carnelian, sard, agate, onyx, cat's-eye, flint, rhomb spar, chert, and plasma. The jaspery varieties comprise jasper, bloodstone, Lydian stone, touchstone, basanite, besides other varieties arising from the structure.

26. Feldspar Group.—The name *feldspar* is given to all silicates of alumina with some other elements, such as magnesium, potassium, and calcium. Each species differs slightly from the other in chemical composition and mineral form. The two most common divisions are **orthoclase** and **plagioclase**. The former name signifies that the mineral has the lines of cleavage at right angles to each other, while the latter designation means that there are two prominent lines of cleavage oblique in direction.

Feldspar is variable in color, light in weight, and almost as hard as quartz. While quartz may be distinguished by its lack of cleavage, feldspar has distinct cleavage planes. Feldspar is always crystalline, though good crystals are not common. It is not as soluble as the nearly insoluble quartz, but when exposed to the weather it begins to change and crumble and in time changes from a clear, hard, glassy mineral to a dull, opaque substance that can be scratched with a knife, and finally becomes a powdery white clay known as *kaolin*, from which fine chinaware is made. In changing to kaolin, some of the original sodium, calcium, or potassium enters into combination with other elements, producing a soluble salt that can be removed by solution in water; therefore, when exposed to the weather, rocks that contain feldspar decay and crumble.

27. Calcite Group.—The mineral in the calcite group is carbonate of lime, $CaCO_3$, and though commonly white, may be any color. It is usually crystalline but frequently amorphous. It is easily scratched with a knife, light in weight, and cleaves readily in two or three directions. Its principal varieties are Iceland spar, dog-tooth spar, satin spar, and limestone (a general name for a massive calcite as well as for massive dolomite), as well as granular or crystalline limestone, which includes all the fine marbles, compact limestone, chalk, hydraulic limestone, oolite, pisolite, argentine, etc,

Calcite is the main constituent of limestone and is one of the most abundant of the common minerals. It is present in nearly all waters on or in the earth, and is being constantly formed by the destruction of minerals that contain calcium. Its presence in water makes it possible for many animals and some plants to take it from solution and build it into their skeletons or substance, from the remains of which great beds of limestone are deposited in the sea.

Calcite, not being chemically strong, is not a durable substance, and hence rocks made of it do not resist the weather well.

28. Dolomite, or Magnesian Limestone.—Dolomite is calcium magnesium carbonate, $CaMgCO_3$, and its usual colors are white, or white tinged with yellow, red, green, brown, and sometimes black. Iron and manganese are often present, replacing part of the magnesium or calcium.

Iron-bearing varieties become brown on exposure, and magnesia-bearing varieties, black.

The principal kinds of dolomite are the white, crystalline, granular, and the white, massive varieties, of which the former resembles granular limestone and the latter is extensively used as a marble. The latter species include pearl spar, rhomb spar, and brown spar. Dolomite resembles calcite in that it burns to quicklime but is somewhat more resistant. It forms great beds of magnesian or dolomitic limestone, which is closely associated with calcite.

29. Siderite, or Spathic Iron.—Spathic iron, iron carbonate, or chalybite, FeO,C or $FeO + CO_2$, often contains some manganese oxide or magnesia and lime replacing some of the *iron protoxide*, FeO. The iron, on exposure to air, becomes *hydrous sesquioxide*, that is, an iron oxide having three parts oxygen to two parts of iron, and gives the siderite a brown or brownish-yellow color. The crystallized or foliated variety is called *spathic* or *sparry iron* because the mineral has the aspect of a spar or crystallized mineral. The cleavage of siderite resembles calcite, calcinite, and dolomite.

The argillaceous variety occurs in nodular or lumpy forms and is called clay ironstone, and is abundant in coal measures.

Siderite occurs in rocks of various ages and often accompanies other ores. Large deposits exist in gneiss and mica schist, clay, slate, also in some limestone and in the coal formations, principally in the form of clay ironstone.

30. Mica Group.—Mica is a general name applied to a large number of minerals that are complex silicates of aluminum and some other metals, as potassium, lithium, and magnesium. There are numerous species of mica, depending on differences in chemical composition. The one characteristic feature of them all, however, is the cleavage, which is such that the mineral splits readily into thin elastic plates.

Micas vary in color from light brown to deep black, and are very soft. They decay readily, forming soluble and insoluble products, the soluble part passing off in the water, the insoluble parts usually remaining as clayey remnants. Some micas are not so easily decayed, however, and are frequently found in the soil, on the beach, and in beds that have been made from the decay of other rocks.

Mica is common in lava, granites, and many other rocks. The common mica is called *muscovite*, which varies in color from white to green, yellow, and brownish shades, rarely rose red or reddish violet. Mica possesses a pearly luster and is transparent or translucent. It is a constituent of granite, gneiss, and mica schist.

31. Amphibole and Pyroxene Groups.—The minerals of these groups possess distinct chemical and crystalline characteristics, which can be distinguished in the crystal, but which resemble each other so closely that they cannot readily be distinguished by the naked eye. *Hornblende* is the common representative of the amphibole group and *augite*, of the pyroxene group.

These minerals are found in many of the lavas and granitic rocks and are commonly dark-colored with jet-black grains. They are complex silicates, and iron is often present. They easily decay, forming reddish or yellowish stains of iron rust as the iron becomes an oxide. Much of the iron coloring matter of the soil is formed by their disintegration.

32. Ores of Iron.—In addition to siderite, which is a carbonate of iron, several oxides of iron and the sulphide of iron are quite common. Of the oxides, the most common is *magnetite*, Fe_3O_4, a black mineral, frequently crystalline, and found in many of the volcanic rocks, in ore beds, and elsewhere. When magnetite rusts, it forms *hematite*, Fe_2O_3, which is the commonest of the iron ores, and forms the red coloring matter of the rocks. When hematite is further oxidized, it becomes *limonite*, $2Fe_2O_33H_2O$. The sulphide of iron or pyrites, FeS_2, is found in many of the rocks in the form of cubical or other crystals, and is of a brownish-yellow color and commonly known as *fool's gold*.

33. Gypsum (Hydrous Calcium Sulphate).—Gypsum, $CaSO_4 + 2H_2O$, is sulphate of lime, and is formed by the decomposition of lime-bearing minerals or by alteration of the carbonate of lime by the sulphate. It is soluble in water. Gypsum, in appearance, resembles calcite somewhat and has a cleavage nearly as perfect as that of some of the micas, but the cleavage plates are not elastic. While present all through the crust of the earth gypsum is not abundant. It is rarely found in large quantities, and when it is thus found it is mined for plaster of Paris. Its principal varieties are selenite, radiated and fibrous gypsum, and alabaster.

34. Salt, or Halite.—Salt is *chloride of sodium*, $NaCl$. Chlorine, which is one of the strongest of the oxidizing agents, has such strong affinities that it never occurs in a free state. Its most abundant compound is sodium chloride, which is very soluble. The salt in the ocean has been originally derived from the rocks reaching it in solution.

35. Ice.—Ice is a substance that plays a very important part in the changes that occur in the earth's crust. Water in freezing and producing ice expands with great force; this expansion is one of the most destructive agents in nature, and one with which the engineer has frequently to combat.

FORMATION AND CHARACTERISTICS OF ROCKS

36. Definition.—The term **rock** is commonly defined as a hard mass of mineral matter, composed of one or more kinds of minerals, having, as a rule, no definite external form and liable to vary considerably in chemical composition. According to geology, however, rock includes all the consolidated materials forming the crust of the earth, such as sand, gravel, and clay, as well as the fragmental or detrital beds that have been derived from it.

37. Classification.—Rocks are divided into two principal kinds, according to their structure and origin, namely, *stratified* and *unstratified*. **Stratified rocks** are more or less consolidated sediments and are therefore aqueous in origin and earthy in structure. **Unstratified rocks** have been more or less fused and are therefore igneous in origin and either crystalline or glassy in structure.

The rocks of the earth's crust owe their formation to five different causes:

1. The solidification of molten rock, as, for instance, the lavas.

2. The chemical precipitation from water, as illustrated by the beds of salt.

3. The action of animals or plants, as exemplified in the case of coral or coal strata.

4. The mechanical destruction of other rocks, as in the sand and clay beds.

5. The alteration or metamorphism of one of these classes of rocks, as exemplified in the case of marble.

Rocks that owe their formation to the first cause are termed *igneous*. Rocks that owe their formation to the second, third, and fourth causes are termed *sedimentary* or *stratified*; the fifth class is termed *metamorphic*.

UNSTRATIFIED, OR IGNEOUS, ROCKS

38. Unstratified, or **igneous, rocks** can be distinguished by the absence of true stratification or the lamination by sorting of material, by the absence of fossils, by crystalline or glassy texture in place of an earthy texture, and by their mode of occurrence, all of which characteristics are due to their mode of origin. The igneous rocks have consolidated from a state of fusion or semifusion instead of being deposited as sediments. Their original fused condition is shown by the crystalline or glassy texture, and by their occurrence in fissures, which shows that they were injected in a molten state; also by their effects, due to their heated condition, on the stratified rocks with which they come in contact.

Igneous rocks occur in three principal positions: underlying stratified rock and appearing on the surface in large masses, particularly in the mountain regions; in vertical sheets intersecting the stratified rocks or other igneous rocks;

in streams or sheets overlying the stratified rocks or sometimes between the strata, sometimes as veins connected with the underlying masses of igneous rocks. They occupy but a small portion of the surface of the earth, estimated to be about one-tenth of the land surface, but underneath the stratified rocks they are supposed to form the great mass of the earth.

Igneous rocks are classified into two groups: the *Plutonics*, or the *granitics*, and the *volcanics*, or the true *eruptives*. The **Plutonics** are coarse-grained and occur only in great masses either underlying the stratified rocks or appearing on the surface over wide areas, especially in the axes of mountain ranges. Usually, the **granitics** have not been erupted at all, although they often form the reservoirs from which the eruptive rocks are derived. The **volcanic, or eruptive,** rocks are fine-grained, sometimes glassy, and are found in sheets injected among the strata or as streams and sheets outpoured on the surface. The term *trappean* has been applied to the eruptive rocks that are injected among the stratified rocks.

These rocks may be briefly summarized as the *granitic*, which occur beneath, the *trappean*, which are injected among, and the *volcanic*, which are outpoured on, the stratified rocks.

39. Granitic, or Plutonic, Rocks.—The granitic groups are found in great masses and never in sheets or streams. They are very coarse-grained in texture and have a mottled or speckled appearance, due to the crystals of which they are formed being of considerable sizes and of various colors aggregated. These crystals mainly consist of quartz, feldspar, mica, and hornblende. The quartz crystals are bluish, glassy, transparent spots. The feldspar crystals are opaque, whitish, greenish, or rose-colored crystals with fluted surfaces. The hornblende crystals are usually black spots. The mica may be distinguished by its thin, scaly structure, the color of which is pearly or black. The term *granitic* has been applied to this group because granite is its best type; in fact, all these rocks are commonly called *granite* by those unfamiliar with mineralogy. Thus, true granite consists of quartz, feldspar, and mica, with hornblende and talc sometimes appearing as impurities. Syenite is feldspar and hornblende frequently associated with quartz, mica, etc.

Igneous rocks, whether Plutonic or volcanic, are divisible into two subgroups: the *acidic rocks* and the *basic rocks*. In the **acidic rocks,** quartz and potash feldspar (orthoclase) predominate; in the **basic rocks,** hornblende or augite and soda lime feldspar (plagioclase) predominate. The acidic rocks are light-colored and not so dense as the basic group; the basic rocks are darker and heavier than the acidic group. These two groups, while sometimes sharply defined, frequently grade into each other. Granite is the best type of the acidics, and diorite, particularly gabbro and diabase, of the basics.

40. Trappean, or Intrusive, Rocks.—This group of rocks is intermediate between the Plutonics and the volcanics. They occur in sheets intruded among the strata, particularly of the older rocks. They are finer grained than the Plutonics and more crystalline than the volcanics, presumably because they cooled more rapidly than the Plutonics and less rapidly than the volcanics.

The trappean rocks are also divisible into acidics and basics. *Felsite* and *porphyry* are among the acidics, and *diorite* and *diabase* are among the basics. **Diorite and diabase,** when occurring among the intrusive rocks, are finer grained than the massive varieties. **Felsite** is a fine-grained, light-grayish rock, consisting essentially of orthoclase and quartz. **Porphyry** consists of a fine-grained, feldspar, ground mass, with large crystals of feldspar disseminated through it. Any rock is said to be porphyritic, however, if it consists of a fine-grained ground mass with large crystals of any kind disseminated through it, as porphyritic diorite and porphyritic granite.

41. Volcanic, or Eruptive, Rocks.—The volcanic, or eruptive, rocks are distinguished from the Plutonic and trappean by their texture and mode of occurrence. In texture they are characterized by a crystalline structure, though some of the fine-grained varieties are microcrystalline and are in a more or less uncrystalline or glassy base or cement, showing that the fused mass has cooled too quickly to permit of complete crystallization. Frequently these rocks are in a wholly glassy condition.

These rocks are also divisible into the acidic and the basic. **Trachyte** may be taken as a type of the acidics. It is a light-colored rock and rough to the touch, consisting essentially of orthoclase with more or less quartz. When the quartz grains are conspicuous it becomes *rhyolite*. **Phonolite** is a dense variety, of light grayish color, that splits into slabs on weathering and has a metallic ring under the hammer. **Obsidian** and **pumice** are glassy scoriaceous (that is, of cinder or slaglike foundation) varieties of trachyte. **Basalt** is a type of the basics. It is a very dark, almost black, heavy rock, with an almost invisible fine-grained texture, and consists of plagioclase with augite, olivine, and magnetite. **Dolorite** has a similar composition, but has a more distinctly crystalline texture and is therefore dark grayish in color.

42. The volcanic or eruptive rocks have been formed in two ways: through the craters of volcanoes, in which the fused mass comes up through the opening and flows off in streams or is thrown out as cinders and ashes; or, it comes up through great fissures in the crust of the earth, frequently hundreds of miles long, and spreads as extensive sheets. These two methods of formation may be called *crater eruption* and *fissure eruption*. The greater part of the rocks on the surface of the earth is due to the latter method.

The fissure eruptions may be subdivided into three classes: *dikes*, *overflow sheets*, and *intercalary beds*.

Dikes are vertical sheets, filling great fissures in other igneous or stratified rocks and are the most common of all methods of occurrence of eruptives and intrusive rocks.

Overflow sheets are masses of lava that have come up, in liquid form, through great fissures and spread out on the surface as extensive sheets. These sheets are often outpoured one on the other until they reach a total thickness of 2,000 to 3,000 feet. Some of these lava floods are of great thickness and extent. The whole of Southern California, Northwestern Nevada, and a great part of Oregon, Washington, and Idaho, and part of Montana and British Columbia, are covered with these lava floods, which are supposed to have come up in fissures in the Cascade and Blue Mountains, and spread as sheets covering the whole intervening space. The most extensive overflow sheets are usually basalt. The basic lavas, like basalt, were very liquid and spread out in thin sheets, while the acidic lavas were stiffly viscous and were squeezed out in dome shape.

Intercalary beds are sheets of rock found between strata. As such they may have been poured out on the bed of the sea or lake, and afterwards covered with sediment, or they may have broken through the strata for a certain distance and then have spread between the separated strata.

SEDIMENTARY AND METAMORPHIC ROCKS

43. Sedimentary.—Some rocks disintegrate through mechanical or chemical agencies more readily than others. For instance, in granite, the quartz will waste very slowly and then only as it is dissolved, for it will not be altered chemically; the feldspar and hornblende, being more complex, are less durable and soon commence to change, finally becoming the clay from which some of the elements go off in solution; the quartz grains, left without cementing material, fall out and the granite crumbles. From the original minerals of this rock, three different products result: soluble salts, fine clayey fragments, and larger grains of pure quartz. What is true of granite is also true, in a certain degree, of all rocks.

44. According to the way in which these mineral products are gathered into layers or strata, they may be classified into three groups of sedimentary rocks: the *fragmental*, or *clastic;* the *chemical precipitates;* the *organic*. The chemical precipitates and some of the fragmental and organic rocks are of aqueous formation. Nearly all are stratified and some of each group are truly sedimentary.

45. Fragmental, or clastic, rocks are composed of distinct fragments of other rocks, gathered into layers by means of wind, ice, water, or volcanic eruption. Sandstones are formed of particles of sand that have been gathered into beds by one of these agencies and consolidated. Strata in which the material is transported by wind must necessarily be composed of fine grains or dust; ice, water, and volcanoes can transport particles of any size. The size of the particles transported by the wind or water will be determined by the velocity with which they are moved; therefore, by the varying action of these transporting agencies, the fragments are assorted into layers, according to their size, the coarsest being moved only by the stronger currents, while the finer particles settle in quiet air or water, giving rise to banding or stratification. This is also true of rocks transported by means of volcanic action, the larger pieces falling first and the finer particles or ash being carried great distances.

46. The pebbly rocks formed at the base of a cliff may be transported by water and rounded, forming pebbly or granite beds which, when consolidated into hard rock, become known as a **conglomerate.** The matrix is usually sand that contains a cement which binds the fragments together and consolidates the rock. Among the various kinds of conglomerate rock are limestone conglomerates, shale conglomerates, quartz-pebbly conglomerates, granite-pebbly conglomerates, etc. There are also volcanic conglomerates, composed of larger fragments of volcanic pumice and ash.

47. The **sandstones** may be coarse in texture, according to the size of the conglomerate, or very fine in grain, almost like clay, as exemplified in bluestone. Sandstones are usually composed of grains of quartz. There are also shale sands, magnetite sands, and garnetiferous sands.

Sandstones have been given different names, depending on the cement that consolidates the sand grains. *Argillaceous sandstone* is applied to rock that has a clayey cement. If the grain is fine it is sometimes called an *arenaceous clay*. In *calcareous sandstone*, the cementing material is lime. If the cementing material is of iron, the rock is *ferruginous*, as in the brown and red sandstone. In *siliceous sandstone*, the cementing material is silica. In some sandstones, there are many small angular fragments giving the rock the character of a grit, which is used for grindstones. Sandstone rocks that split easily in every direction are called *freestones*. Sometimes, owing to the presence of many mica flakes, sandstone cleaves readily in only one direction and it is then said to be *shaly* or *micaceous sandstone*.

48. The **clay rocks** are so constructed that they frequently split into layers. Such rock is called a *shale*, the cleavage depending on many minute flat particles of mineral, often mica. Near the coral islands, the grinding action of the waves on the beach often wears the coral into fine clay, which, in settling to the bottom, forms a limy mud, afterwards becoming transformed into limestone. There are many other kinds of clay rocks, such as the *kaolin clay*, formed by the decay of feldspar; and *fireclay*, which has lost its alkalies by having them extracted by the plants that grew upon it, leaving the clay so free of alkali that it resists the action of fire. There are also sandy or *arenaceous clays*, containing considerable fine sand, and *carbonaceous clays* that contain fragments of plants.

49. **Chemically precipitated rocks** are interesting but not important. Chemical precipitation, however, has much to do with the destruction of rocks.

Water passing through rocks often takes a mineral in solution from one place and places it in another; this is one of the ways in which rocks are cemented. The water often produces chemical reactions and sometimes completely changes entire beds of rock, as, for instance, certain limestones have been changed to magnesian limestone or dolomite, while others have been changed to iron beds by the precipitation of siderite or some other salt of iron from some solution of iron in water. Other limestones have been changed to gypsum.

Pure rain water exerts very little power as a solvent in the dissolving of most minerals, but from decaying vegetation and from the air it derives impurities, including carbonic-acid gas, or it may encounter alkaline substances that are easily dissolved. These impurities transform the water, in some cases, into a weak acid, and in other cases, into a weak alkali, in which condition it may attack the minerals directly.

50. Organic rocks comprise calcareous, silicious, and phosphate rocks, and plant deposits.

1. *Calcareous Rocks.*—The greater part of the limestone beds of the earth have been formed through the agency of animals. In the oceans, there are reefs built of coral fragments, which are made of carbonate of lime that the coral animals have abstracted from the ocean water. On a great part of the ocean floor, an ooze or limestone mud is now forming deposits like the chalk, which originated in a similar manner. In past geological ages, a species of animal, now very rare in the ocean, built limestone beds that are known as *crinoidal limestone.* Limestones are composed of carbonate of lime, sometimes accompanied by numerous impurities that color them. When clay is present in large quantities, the rock is called an argillaceous limestone, which grades into a calcareous clay rock. They usually contain many fossil fragments of shells or corals.

2. *Silicious Rocks.*—These rocks, of animal origin, are not common. **Infusorial earth** is composed of the skeletons of microscopic animals and is found in shallow lakes and beneath some of the swamps. **Diatomaceous earth** is a silicious rock, geologically, containing large numbers of silicious cells of a plant belonging to the group of diatoms.

3. *Phosphate Rocks.*—In some places in the United States, particularly near Charleston, S. C., and Florida, and in certain sections of Tennessee, the bones of marine and land animals of great size have accumulated into bone beds.

These rocks are phosphate of lime. They are known as **phosphate rocks** and are used to form the phosphoric acid of fertilizers. These phosphate rocks have no value for structural purposes.

4. *Plant Deposits.*—Vegetation takes carbon from the air and mineral substances from the water. With these substances they form their structure, which, on decaying, are returned in large part to the air of the earth. In swamps, where decay is retarded, the plant remains may accumulate in beds of peat which may later on become transformed into coal or mineral fuel, which is mostly composed of carbon. Peat, lignite or brown coal, bituminous coal, anthracite, and graphite are successive transformations of organic matter.

51. There are other forms of rock that are directly or indirectly of organic origin, while nearly all the sedimentary strata have some sedimentary animal or plant remains. Flint and chert are dense, hard layers of nodules of silica, some of which have been formed by silicious animals, although most of them appear to be of chemical origin.

52. **Oolite** is usually a limestone, although occasionally it is an iron or even a silicious rock. It is made up of minute grains resembling bunches of fish eggs, whence its name. Each of the rounded layers is made up of concentric layers like an onion. Oolite may be formed in three ways: by the action of lower forms of plant life, which build up the grains; by chemical deposit in water at the surface, as in the geyser origin of the Yellowstone; and by a chemical change, which causes a rock of a different origin to assume the conditions of an oolite. Oolite grains are now accumulating on many shores, as those of Florida and the Great Salt Lake.

Sedimentary rocks, particularly those that owe their origin to mechanical and organic agencies, are of the greatest importance to man, as they furnish him with most of his building stones. At present sedimentary rocks are forming, all over the globe, on land and in the sea. More than one-half of the earth's crust that is visible is made of sedimentary strata.

53. Metamorphic Rocks.—These rocks are an intermediate series between the stratified and unstratified rocks. They are stratified, banded, or foliated rocks, but crystalline in texture, like the igneous rocks, and are usually destitute of fossils. This banding, however, is quite different from stratification, for it is an arrangement of crystalline minerals, while in sedimentary rocks it is usually a banding of fragments arranged according to size or color, etc. The metamorphic strata result from complex changes of other rocks, in which the elements are often made to combine in a new manner. Given the same assemblage of elements, the effects of metamorphism will produce the same result; thus, a schist may be formed from either a shale or a lava.

Metamorphic rocks are supposed to have been formed from sediments, like stratified rocks, but to have been subsequently changed by (1) heat, (2) water, (3) alkali, (4) pressure, (5) crushing. To produce metamorphism by heat alone, that is, dry heat, requires a temperature of 2,500° to 3,000° F., but in the presence of water an incipient change begins at 400° F., while complete hydrothermal fusion takes place at 800°. If any alkaline carbonate be present in the water, these effects occur at a still lower temperature. Pressure is necessary because without it it is impossible to have even such moderate heat in the presence of water.

Metamorphic rocks may grade into stratified rocks on the one hand and into igneous rocks on the other. They cover a large area; the older rocks especially, are found along the axes of great mountain chains. The whole of Labrador, the larger portion of Canada, the whole eastern slope of the Appalachian system, and also the axes of the Colorado and Sierra ranges consist of them. In Canada, they are supposed to be 40,000 to 50,000 feet thick. They are very much crumbled.

54. Metamorphism is nearly always associated with great thickness and crumbling. Some rocks, such as the igneous, were originally solid; many, such as the fragmental and organic, were unconsolidated. Metamorphism solidifies these rocks and in some places changes them entirely. As a result of these changes, the strata are so altered that their original condition cannot be told at a glance. Sandstone becomes a dense quartz rock, called *quartzite*, in which the sand grains may be no longer visible to the eye. A peat bog may be changed to anthracite or even to graphite. A dense, apparently structureless, limestone, may become transformed to a white or variegated marble composed of many crystals of calcite, or a clay stratum may be metamorphosed into a slate, in which the dense rock becomes harder, although at the same time an ability to split easily in one direction is introduced. This is called slaty cleavage and is one of the features of metamorphism. The cleavage is developed because of many plates of micaceous material formed in the rock as a result of heat or the mashing together of the whole rock mass in a direction at right angles to the cleavage plane. In this change one of the conditions is great pressure, and as the minerals develop they grow in a plane at right angles to the direction of the pressure because this is the plane of the least resistance, and because of the many cleavage planes of the newly formed micaceous material, slate splits easily along these lines.

55. Metamorphism of slate rock continued further would develop other minerals and would soon be altered in character and become a **schist,** in which the various minerals are arranged in bands having a cleavage no less marked than that of slate. A schist will split into layers much less easily and uniformly than the slate. Schist is known by various names, depending on the minerals present, such as mica schist, hornblende schist, chlorite schist, etc. All these rocks are characterized by the banding and the ease of splitting, which is due to the schistose structure arising from the banding of the minerals along these planes. As in the case of slate, these bands are at right angles to the direction of the pressure that was present during the metamorphism.

A final metamorphic change is one in which the original condition is hidden. The last stage before actual melting takes place produces **gneiss,** which resembles granite, except that it has its minerals more or less perfectly banded, while granite is massive and without layers. In granite and gneiss, the minerals are frequently the same in kind and in general formation.

56. Where metamorphism has taken place in sedimentary rocks, the changes are sometimes so pronounced that the stratification is destroyed and the original nature of the rocks is so changed that it is impossible to tell whether they were originally sedimentary or igneous. Metamorphic rocks are found in the earth's crust in three positions: first, near some mass of intruded igneous rock; second, in the core of mountains deep below the original surface; third, among the most ancient rocks, particularly the Archæan, which are the lowest and the oldest of all the rocks. In the latter place the metamorphic rocks occur in great masses.

It was formerly believed that metamorphic rocks were part of the original crust of the earth and represented the most ancient strata, but this belief has been proved erroneous and it can now be shown that the rocks that have been very much altered belong to later ages and are transformations of sediments belonging to the same general age as deposits of shales, limestone, and sandstone in adjoining regions.

JOINTS AND FISSURES IN ROCKS

57. All rocks—igneous, sedimentary, and metamorphic—are divided into separable blocks of different sizes and shapes by cracks in various directions; these cracks are called **joints.** In stratified rocks one of the division planes is between the strata and the other two nearly at right angles to this. The different kinds of rocks have blocks of characteristic shapes and sizes. In sandstone, for instance, the blocks are usually very large and roughly prismatic; in limestone, they are usually very regularly cubic; in shale, oblong rhomboidal; in slate, small and sharply rhombic; in granite, sometimes large and roughly cubic, sometimes scaling in concentric shells producing domes; in eruptives, of many shapes, such as roughly cubic, roughly spherical columnar, etc. These joints are supposed to have been formed by the shrinkage of the rocks; in stratified rocks, in consolidating from sediments; in igneous or metamorphic rocks, in cooling from a state of fusion or semifusion. In stratified rocks, the joints are usually confined to the stratum, though some of the larger joints (*master joints*) run through several strata.

Fissures are undoubtedly formed by the movements of the earth's crust. Joints are cracks in the individual strata, while fissures are fractures of the earth's crust extending through many formations and continuing for great distances. Owing to the great horizontal pressure, due to the shrinkage of the crust of the earth, the crust is sometimes thrown into ridges and hollows, thereby producing enormous fractures parallel to the axis of bending; thus mountain ranges are produced parallel to mountain ranges, while sometimes a system is at right angles to the main system.

Great fissures occur in systems, usually parallel to the axis of elevation or length, and frequently extend for hundreds of miles, and are miles deep. When filled, at the moment of formation, with fused matter from below, they form *dikes*, and all great dikes and igneous overflows have been through such fissures. If these fissures are not filled at once with fused matter, but are afterwards slowly filled with mineral matter, they form *fissure veins*.

PHYSICAL PROPERTIES OF BUILDING STONE

58. The structural manner in which the constituent parts of the rocks are grouped together bears a greater relation to the value or quality of the rock than the character of the minerals composing it; or in other words, the **physical characteristics** may be, and frequently are, more important than the chemical qualities.

59. Density.—The weight, strength, and absorptive properties of stone are dependent on the **density**. Thus, among rocks having the same mineral composition but differing as to structure, generally the strongest will be the densest, and the heaviest will be the least absorptive.

60. Hardness.—The manner in which the mineral constituents of a rock are cemented to each other and the individual hardness of such mineral constituents determine the **hardness** of the rock as a structure. The minerals composing a rock may be hard but the rock itself as a structure will be soft if the particles do not strongly adhere to one another. Thus, some of the softest sandstones are composed of quartz, which is a hard mineral, but the grains are so weakly cemented together that the stone as a whole is soft.

61. Structure.—The structure of a rock depends on the form, size, and arrangement of its component minerals. All rocks may be approximately classified as *crystalline*, *vitreous* or *glassy*, and *fragmental*. Granite and crystalline limestone may be taken as types of the crystalline group; obsidian and pitchstone may be taken as types of the vitreous group; while the sandstones are types of the fragmental group.

62. Though all rocks have some common structural characteristics, certain peculiarities are found only in single types of rock. If the structure can be recognized by the unaided eye, the rock is said to have a *macroscopic structure*, and such rocks may be said to be granular, massive, polished, stratified, porphyritic, and concretionary.

The term *granular*, as its name implies, is applied to rocks built up of distinct grains of crystalline, or fragmental and water-worn character.

The term *massive*, or *unstratified*, is applied to rocks that are not arranged in any definite form in layers, or strata, but have the constituent parts mingled together, as in diabase and granite.

The term *polished*, or *schistose*, is applied to rocks that have their constituents arranged in definite planes nearly parallel to each other.

The term *stratified* is applied to rocks composed of parallel layers or beds, as is frequently seen in limestone and sandstone. When the strata are thin and fine, the rock structure is said to be scaly or laminated.

The term *porphyritic* is applied to rocks that consist of a ground mass of fine or compact and evenly crystallized material, with larger crystals of feldspar scattered through it. A granite fragment has a porphyritic structure, but it is difficult to distinguish owing to the similarity of color existing between the crystals and the ground mass. In such rocks as the felsites, it is quite noticeable. In the porphyries of Eastern Massachusetts, the ground mass is of a black or no color and very compact and dense, while the large white crystal feldspars are in marked contrast.

The porphyritic structure is so noticeable that any rocks possessing this characteristic in a marked degree are commonly termed *porphyries* without regard to the mineral composition. The word *porphyry* is now commonly applied as an adjective, because any rock may possess this structure, whatever may be its origin or composition.

The term *concretionary* is applied to rocks composed of concretions or rounded particles built up by the collections of mineral matter around a center forming a rounded mass of concentric layers like the coating of an onion. When the concretions are small, like the roe of a fish, the structure is called *oolitic*, or if large like a pea, the structure is called *pisolitic*. The Bedford, Ind., limestones are examples of the oolitic type. The concretionary structure is rarely found in crystalline rocks.

63. Aggregation of Particles.—The hardness of the rock depends largely on the aggregation of the particles; therefore, the working qualities of the rock are fixed by the character of this aggregation. If the grains are loosely coherent in a rock composed of hard minerals it may work readily, while a rock consisting of softer materials may be worked with difficulty because the particles tenaciously adhere to each other.

The durability of a stone is, to a great extent, a matter of texture If the grains adhere closely, the stone will be less absorbent and less durable than one in which the adhesion is not so great, as in the friable and loose-textured rocks.

The kind of fracture shown by a rock is determined by the fineness or the coarseness of the grain and the relation of the particles to themselves or their state of aggregation. Such rocks as flint, obsidian, and some varieties of limestone have a compact fine grain and show, on fracture, a concave or convex shell-like face of conchoidal form, and are difficult to dress. Other stones show, on fracture, a jagged surface or split along certain planes, all dependent on the aggregation of the particles.

64. Rift and Grain.—Rocks that do not possess rift and grain cannot be worked into rectangular form without great difficulty, unless they are of a very soft nature, but with these qualities the hardest rocks can be readily worked; for instance, the South Dakota quartzite, which is one of the hardest rocks known, can be as easily broken into pieces for paving as a soft sandstone or a granite.

The **rift** of a rock is a line of cleavage parallel to the bed and is visible in such rocks as mica schist, gneiss, and other sedimentary rocks. It is along these lines that the rock can be readily split. Rift, however, is commonly found in massive rocks, although it is not so easily discerned as in the examples cited. The **grain** of a rock is always at right angles to the plane of the rift or bed.

65. Color.—The chemical properties of a rock, as a rule, determine its **color**. The color of granites, however, is affected by the action of the light on the feldspars, which when clear and glassy, absorb the light, making the rock apparently darker than when the feldspars are white and opaque and reflect the light.

Iron, the principal coloring matter in rocks, may be found in chemical composition with other minerals or in such simpler compounds as the sulphides and carbonates, or as an oxide distributed throughout the mass of rock. The brownish or reddish hues are due to the free oxides of iron while the bluish or grayish hues are caused by the carbonates or the sulphides. The absence of iron in any of its forms is usually indicated by the white, or nearly white, color of the rock. The permanency of the color of the rock depends on the form in which the iron is found. Oxidation is likely to result if it is in the form of a sulphide, carbonate, or other protoxide compound. Therefore, stone containing these forms of iron is apt to fade and turn yellowish and stain on exposure. The sesquioxide, being in the last stages of oxidation, can undergo no further change from oxidation and is therefore a permanent color; hence, the decidedly red color may be considered permanent. The blue and the black colors of marbles and limestones are largely caused by the presence of carbonaceous matter, usually of vegetable origin.

SILICIOUS STONES

66. The granitic group of igneous rocks is richest in silica and therefore its members are known as **silicious stones.** Of this class, granite, syenite, gneiss, greenstone, and trap and the harder varieties of sandstone are most commonly used for structural purposes.

67. These stones compose the **primary rocks,** which are those rocks supposed to have been formed from the slow cooling of the incandescent earth. The granites are also unstratified, eruptive rocks, and underlie the stratified rocks. They are composed of an aggregation or assemblage of crystals of feldspar, quartz, and mica, the principal impurities being hornblende and talc: Quartz (pure silica, SiO_2) has a hardness of 7; feldspar (silica and aluminum, together with potash), a hardness of 6; hornblende, a hardness of from 5 to 6; the small scales of mica, a hardness of 3.

The colors of granite are white, grayish white, yellowish, reddish, rose, flesh color, or deep red, but rarely green. It is distinguished by its even and brilliant fracture, its pearly luster, and its outline, which is seldom regular, but in which may be recognized rectangles and parallelograms.

Granite varies in quality according to the proportions of its components and their method of aggregation. Stone of the greatest durability and hardness contains a greater proportion of quartz and a less proportion of feldspar and mica. Hornblende renders the stone tough and heavy. Feldspar renders it lighter in color, easier to cut, and more susceptible to decomposition by the solution of potash contained in it. Mica renders it friable.

The granites are among the most valuable of the building stones and are extensively used in important works. They can be readily quarried and by reason of the lack of grain in the stone, blocks can be obtained of any size. On account of its great hardness, granite is difficult to work and therefore very costly to use if the stone is required to be cut. It weighs about 166 pounds per cubic foot.

Granite is found in the eastern part of the United States, in Canada, in many parts of the Alleghany and Rocky Mountains, and, as a rule, wherever the later rock formations and the underlying beds have been left exposed. It is generally classified into gray and red. Gray granite is found throughout New England, the border states, and in Virginia. Red granite is composed of red orthoclase (aluminum potassium silicates), bluish quartz, and a little hornblende, with very little mica. It is hard and takes a fine polish. It is found on the Bay of Fundy, in the islands on the St. Lawrence River, Virginia, Lake Superior, Maine, and many points in the Rocky Mountains.

68. Syenite.—This stone derives its name from Syene in Egypt. It consists of feldspar and hornblende, frequently associated with mica and quartz; is of a granular texture closely resembling granite; and is hard and tough, somewhat coarse-grained, and will not take a polish. It is one of the most durable of the granitic rocks when its feldspar constituent is not too readily decomposed by the removal of its potash when open to the weather. For this reason it should be carefully tested before it is used.

69. Gneiss and Mica Slate.—These are similar to granite in composition but differ from it in being stratified. Granite, syenite, and gneiss resemble one another so closely that they are all frequently called granite by those not familiar with their characteristics.

Gneiss is not as valuable a stone as granite on account of its stratification, which will not permit it to be split in any direction. It is, however, a good building material and often answers as well as granite.

70. Greenstone, Trap, and Basalt.—These stones are igneous, unstratified rocks, consisting of hornblende and feldspar. The term *trap* has been suggested as a generic name for these rocks. The greenstone is not as coarse-grained as granite, and in the trap and basalt the granular structure is not apparent. The greenstone and trap break into blocks and the basalt into columns of prismatic form. They are found in veins and dikes and injected among the stratified rock of all ages.

These rocks vary in color, from nearly white in some varieties of greenstone, to nearly black, as in basalt, the difference of color being determined by variation in the proportions of hornblende, which gives a dark color, and feldspar, which gives a light color. The green is due to chromium. These stones, while making very durable building material, cannot be obtained in large blocks and are difficult to cut. Trap rock forms one of the best aggregates for use in making concrete.

SEDIMENTARY STONES

71. Sandstones.—Such material as sandstone consists of fragmentary rocks, composed mostly of grains of silica (or quartz), cemented together by a deposition of silica, carbonate of lime, oxide of lime, and aluminous matter. Sandstone is a stratified rock and belongs to the later geological periods.

If the cementing material is silica, the rock is very durable, but difficult to work. Iron oxide in the cementing material, consisting of carbonate of lime, and clayey matter, gives the stone a reddish or brownish color. Lime renders the stone particularly liable to disintegration when exposed to an atmosphere containing gases, or when used for foundations where the soil is impregnated with acid water. The presence of clay or the oxide of iron is also deleterious.

Sandstones are variable in character, some being nearly as valuable as granite, while others are practically useless for permanent construction. The best stone is characterized by small grains with a small proportion of cementing material, and when broken has a bright, clear, sharp fracture. It is usually found in thick beds and shows slight evidences of stratification. Water can readily penetrate between the layers of this stone; therefore, in foundations it should be laid on its natural bed so that the penetration of moisture and possible disintegration by freezing may be prevented as far as it may be possible.

Sandstone of good quality possesses strength and durability and can be readily cut and dressed, which qualities make it one of the most frequently used of our common building stones. When the grains are extremely small, it is termed a "freestone" because of the ease with which it can be quarried, cut, and dressed.

Sandstones vary much in color: The Ohio and Nova Scotia varieties are yellowish and cream color and sometimes nearly white; the Missouri sandstone is of a yellowish drab color and possesses durability; the Portland, Conn., Newark, N. J., Marquette, Mich., and Bass Island, in Lake Superior, sandstones are of a dark brownish-red color, which is due to the presence of iron, and are termed *brownstones*. The Potsdam, N. Y., red sandstone is durable, hard, highly silicious, and of a reddish color. The Hummelstown, Pa., sandstone has a brownish color.

72. Soapstone.—This stone is the silicate of magnesia and is found in many places in the United States. It possesses valuable qualities where a stone capable of resisting high temperature is required.

73. Argillaceous Stone.—Stones of this nature are generally weak and soft and are not durable when exposed to the weather. They are therefore of little value as building stones. Clay slate is a sedimentary argillaceous rock, fine-grained, compact, and of a laminated structure. Its colors are usually dark purple, blue, and light green. The best varieties of clay slate are used for roofing and flagging.

CALCAREOUS STONES

74. Calcareous stones are composed largely of lime; therefore, limestones and marbles are the most familiar examples. Limestone is a carbonate of lime and effervesces when attacked by acids that are stronger than the carbonic acid in its composition, the weaker acid being rejected and new lime salts being formed. The carbonic acid can also be expelled by heat, in which case the product is caustic lime, commonly termed *quicklime*, which is uncombined lime. All varieties of calcareous stone are found in the United States. Extensive deposits are found in a line parallel with the Atlantic Coast; another deposit underlies the Middle States. The marbles are mostly confined to the mountainous districts, while the common limestones are frequently found in immense strata that have been deposited on the bed of an ancient ocean.

75. Limestones.—These common building stones are of various qualities, some being hard and strong while others are soft and friable. There are two sorts, the granular and the compact, from either of which excellent stones may be obtained. As they are usually easily worked, they are among the lowest-priced dressed stones in the building.

76. Marble.—Metamorphosed limestone gives masonry material known as **marble,** which is easily dressed to a smooth surface and polished. The granular varieties are generally superior to the compact for building purposes. The impure carbonates of lime are sometimes of great value as marble. The magnesian limestones, or the dolomites, are usually of excellent quality.

White marble is found in the Laurentian rocks, Canada, but much of that used in the Northern Atlantic States is obtained from the Green Mountains, which extend through

Vermont, Western Massachusetts, Western Connecticut, and Southeastern New York. Quarries exist at Granden, Rutland, Danby, Dorset, and Manchester, in Vermont; at Lanesborough, Lee, Stockbridge, Great Barrington, and Sheffield, in Massachusetts; at Canaan, in Connecticut; and at Pleasantville and Tuckahoe, in New York. The snowflake marble is obtained from the Pleasantville quarries, and a fine grade of statuary marble from Rutland, Vt. From this place toward the south, the marbles become coarser and harder and more suitable for building purposes. Dolomitic marbles are found in the southeastern part of New York and in Delaware. White dolomite marble is found in Maryland.

The colored marbles used in building construction are of several varieties and are found in Vermont, Connecticut, New York, Pennsylvania, and Tennessee. **Brecciated marbles,** that is, those in which the conglomerate fragments are angular instead of water-worn, are found in Vermont on the shores of Lake Champlain, and a dove-colored marble with greenish veins is found at Rutland. Black marbles are found at Shoreham, Conn., and Williamsport, Pa. Black Trenton limestone is found at Glen Falls, N. Y. The Warwick marble, found in Orange County, N. Y., is beautifully colored with carmine, with white veins. The Knoxville marble is of a reddish-brown color with lines of blue. Tennessee marble is brown and white mottled. The foreign marbles are largely imported from Italy, Spain, and Belgium. The Bardiglio of Italy is of a gray color shaded with black; the Siena of Spain is a pale yellow color; the Lisbon of Portugal a pale reddish color; and the Belgian, of Belgium, is black. Verde antique is composed of bands of serpentine and white marble.

77. Chalk.—Soft limestone in which the minute shells composing it have not been entirely destroyed by the pressure to which it has been subjected in early geological times is called **chalk.** It is not suitable for constructive purposes but is very useful in making lime and cement.

78. Quicklime.—This material is obtained by *calcination* from various limestones and is the basis of common mortar; the act, or operation, of calcination is the expelling, by heat, of some white substance by which the stone or the cementing material is broken down and reduced to a friable state. Limestones are necessary in the manufacture of iron, as they afford an alkaline base that unites chemically with the silica, alumina, and other impurities of the ores and allows the metal to separate into a state of approximate purity.

79. Gypsum, alabaster, or **plaster of Paris** is a sulphate of lime containing the water of crystallization. The term "plaster of Paris" is due to the fact that large deposits of this stone underlie the city of Paris. This natural sulphate of lime, when raised to a high temperature, loses its water of crystallization and is then ground into a fine powder. This becomes the plaster of Paris of commerce, which is used for molds, ornaments, and casts, as well as in wall plaster and staff. Gypsum is found in many parts of the United States, great quantities being found in the state of New York.

MANUFACTURED STONES

80. The artificial stones include *brick*, *firebrick*, *concrete*, and *terra cotta*. While there are several ornamental stones manufactured, they have little structural significance, and only brick, firebrick, and concrete, which are entirely materials of construction, will be discussed in this Section. Terra cotta is much used in modern construction to obtain decorative and architectural effect, but its chief structural use is in the construction of fireproof floors, partitions, and coverings.

81. Bricks.—As is generally known, **bricks** are made of burnt clay and are used extensively in all classes of building operations of permanency. They can be cheaply made, readily handled, and formed into structures of almost any desired form. Brick, if properly made, burned hard, and laid up with Portland cement mortar, is one of the most durable building materials in use.

The clay of which the common brick is made consists principally of silicate of alumina, but usually also contains lime, magnesia, and oxide of iron; the oxide of iron gives the brick strength and hardness; silicate of lime renders the clay easily fusible and causes the brick to become distorted in the burning. Uncombined silica is beneficial, if there is not too much of it, as it preserves the shape of the brick at high temperature. If it is in excess, however, it renders the brick weak and brittle, because it destroys cohesion; 20 to 25 per cent. of silica makes a good proportion.

Bricks are made by hand and machine, the machine-made ones usually being the denser. They are burnt for about 2 weeks, first at a moderate heat, until all the moisture has been expelled, and then at a slowly increasing temperature until, at the end of 24 hours, the "arch bricks" attain a white heat, when the temperature is slightly lowered, but a constant high temperature is kept up until the end of the allotted time, after which the openings in the kiln are closed, the fire is drawn, and the kiln is allowed to cool very slowly. There are various kinds of kilns, some temporary and some permanent; the permanent kilns are the best and most economical.

Three kinds of bricks are usually taken from the kilns: those forming the top and sides of the arches in which the fire is placed are overburned and partly vitrified; the lower bricks in the arch are usually overburned on one end and underburned on the other and are called **arch bricks;** they are hard, brittle, and have little strength. Bricks from the interior of the pile are usually of the best quality in the kiln and are termed **body bricks,** sometimes **hard** or **cherry bricks.** Bricks from the exterior of the kiln are usually underburned and are called **soft** or **salmon bricks.** Salmon bricks are too soft for use in important places, as they are deficient in strength and will not resist the weather.

Bricks of good quality should be of regular shape with parallel surfaces, plane faces, and sharp square edges, of uniform texture, and should be burned hard. They should be thoroughly sound, free from cracks, and should ring clearly when struck a sharp blow. They will fail under a compressive stress of about 10,000 pounds per square inch; soft bricks will not resist more than about one-tenth of this stress. Pressed bricks will bear about twice as much as good hard bricks. Hard, well-burned brick should not absorb more than 6 per cent. of its weight in water. Brickwork in masses will crush under a very much smaller load than a single brick, presumably because of the combined stresses due to bending, etc. The Watertown Arsenal tests of brick piers are worth careful study.

82. Firebrick.—Firebrick is usually made of a very pure clay with clean sand, or sometimes of pure silica cemented with a small proportion of clay. The clay should be silicate of alumina. Oxide of iron in the clay is very injurious, and if it reaches 6 per cent. the brick is not suitable for the purpose. Specifications for firebrick should require that the oxide of iron shall be less than this amount, and that the aggregate of lime, soda, potash, and magnesia shall be less than 3 per cent. The sulphide of iron or pyrites has a harmful effect on the fireclay, and brick containing it should not be accepted. An excess of silica in the brick makes it refractory in extremely high temperatures. Where the brick has to resist the action of metallic oxides, which would have a tendency to unite with silica, alumina should be in excess.

83. Concrete.—This name is given to any mixture of gravel, slag, cinders, or broken stone with cement and sand. The best concrete for heavy structural purposes has for its base broken stone or slag, while gravel and cinders are used for such work as filling in fireproof construction and for the foundation of cement pavements. An excellent concrete is made of cement and sand with a mixture of broken stone and gravel.

The broken stone, slag, cinders, or gravel forming the base of the concrete is known as the *aggregate*, while the cementing material, composed of the cement and sand, is known as the *matrix*. The combination of the aggregate and the matrix forms, when the cement has attained its final set, an artificial conglomerate rock that is superior in strength and durability to many of the natural rocks or stones used for building purposes. The purpose of the aggregate is to provide strength and solidity to the mass and to make up the bulk as cheaply as possible, so that the amount of the cement mortar, which is the expensive material, may be reduced to a minimum. The matrix holds the aggregate together by its adhesion, and in a complete concrete must fill all *voids*, as the spaces that would exist between the parts of the aggregate are called.

In mixing concrete, water should be judiciously used, for the mixture should be pulverulent, or nearly so, rather than wet. An excess of water tends to weaken the concrete and prevents the proper distribution of the matrix throughout the mass of the aggregate.

Concrete should be put in place in layers of from 6 to 10 inches in depth and should be tamped until the moisture begins to appear on the top of the layer. In placing concrete, layer upon layer, the upper surface of the bed should be free from dirt and dust and moderately wet.

84. Stone and Gravel Concrete.—When concrete is composed of broken stone and gravel as a base, the stone, or gravel, and sand must be sharp and thoroughly clean in order that the concrete may attain its maximum strength. Sand containing loam, dirt, or dust, which will soil the hands when rubbed, should not be used in the manufacture of concrete where dependence must be placed on it for structural stability.

The proportion of parts for the ingredients composing stone and gravel concrete depend entirely on the purpose or use for which the material is required. In many instances, a *poor concrete*, that is, one in which a small quantity of cement is used in proportion to the stone, gravel, and sand, will fulfil the requirements and promote economy, while it is frequently imperative that a *rich concrete* be used, or one in which considerable cement is employed in the mixture.

For structural work requiring strength and durability, engineers favor a mixture consisting of one part of Portland cement, two parts of sand, and five parts of broken stone that will pass through a 2- or 2½-inch ring. An excellent concrete is also made by using 5½ cubic feet of cement, 7 cubic feet of sand, and 27 cubic feet of broken stone. These proportions are just sufficient to make 1 cubic yard of concrete.

When gravel is used in making the concrete it assists in filling the voids and consequently decreases the amount of cement mortar required. Good proportions for concrete of this character consist of one part of Portland cement, two parts of sand, three parts of gravel, and four parts of broken stone. This formula, which is easily remembered, has been used by the engineer corps of the United States army in the construction of foundations. Rosendale or natural-rock cements may be substituted for Portland, if desired. Concrete composed of one part of natural-rock or Rosendale cement to two parts of sand and four parts of chips or broken stone is specified by many architects and was used in the construction of the East River Bridge. The mass of the foundation for the Statue of Liberty in New York harbor was made of two parts of Rosendale cement, two parts of sand, and seven parts of broken trap stone.

The modulus of elasticity of good stone concrete is about 700,000 pounds, while of neat cement it is approximately 3,000,000 pounds. The unit adhesive strength of stone concrete to rough iron or steel is about 600 pounds, while the adhesion of cement to brick or stone, in pounds per square inch, is approximately 15.

85. Cinder Concrete.—This material has fire-resisting qualities superior to stone concrete, owing to the fact that its mass is more porous and that the air spaces confined within are non-conductors and also because the cinders are more refractory than the broken stone commonly used in stone concrete. Besides, it adapts itself more quickly to sudden changes in temperature and therefore is not as liable to crack, split, or disintegrate when rapidly cooled by streams of water used in the extinguishing of a fire.

Cinder concrete is light in weight, compared with stone concrete, and is used extensively for fireproof-floor construction both as a filling material and as a structural element of the construction, though the laying of any material in which the matrix is Portland cement should not be conducted when the thermometer is below 28° or 30° F. Cinder concrete, in case of necessity, can be laid at a much lower temperature, especially if it is being used in poor construction where the centers are so open as to allow the surplus water to drain away before freezing can take place in the body of the concrete.

86. Experience teaches that the rapid and complete destruction of iron or steel when in contact with damp ashes or cinders is certain. Even cast-iron pipes, which are less liable to corrosion than steel, running through cinder banks, oxidize rapidly and in a short time are unfit for use. It is only natural, therefore, that on this account cinders as aggregate for concrete should be looked on with some disfavor by those who contemplate using that material in conjunction with steel or iron.

The active agents producing corrosion are water, carbonic acid, and sulphuric acid. The water and carbonic acid occur in the atmosphere, and if allowed to come in contact with the bare metal will cause corrosion. Where the metal is embedded in the concrete, however, there is little chance for this to occur, and no more opportunity is offered by the use of cinder concrete than in that in which the principal ingredient is stone, the cement in any case being the protective material.

That cement will protect iron and steel from corrosion indefinitely is well known from observation and experiments; and if it protects the metal against the action of moisture and carbonic acid, it is reasonable to suppose that, if the cinders are embedded in a sufficient matrix of cement mortar, there will be no action on the metal due to the sulphuric acid, or to sulphur as sulphide, in the cinders. Sufficient matrix in order to completely protect bare iron or steel, can only be had in complete cinder concrete, which is obtained by the proportions of 1 part of cement, 2.75 parts of sand, and 4.75 parts of cinders. Such a concrete is not so good a fire- and water-resisting material as that made of 1 part of cement, 2 parts of sand, and 5 parts of cinders, and is not as practicable, for it is heavier, and is not in accordance with the building ordinances, as usually framed in the principal cities.

Though the usual concrete in the proportion of 1, 2, and 5, will not entirely protect bare iron or steel, and some oxidation will take place more or less rapidly, depending on the absence or presence of moisture, it is hardly likely that the corrosion will be serious with the iron or steel used in the ordinarily dry atmosphere between the floors of a building. The concrete, however, contains considerable moisture when it is first laid, and an initial corrosion is likely to take place before the concrete has dried out. In order to avoid this initial corrosion, which is serious, for when it once starts it is liable to continue owing to chemical action and reaction, the iron or steel should be painted with some good paint, preferably one free from oils that are likely to be affected by acids and alkalies. Some paint of the asphaltum variety would probably be best.

Since meager concrete in the proportion of 1, 2, and 5 fails to entirely protect bare iron or steel, it is good practice, where possible, to cover the metal with a thin coat or layer of cement mortar of 1 part of Portland cement and 2.75 parts of sand, and then follow with the concrete. If this is done, the concrete will not come in contact with the metal, and the steel or iron will be fully protected from corrosion, which, though slight and probably locally confined, is likely to be instituted from the several causes previously cited.

FIRE-STONES

87. Fire-stones are stones capable of resisting the action of great heat without fusing, exfoliating, or cracking. Lime and magnesia, except in the form of silicates, are prejudicial to the quality of fire-stones; potash, also, is very injurious because it increases the fusibility of the stone, which, on melting, causes the formation of a fusible glass. Quartz and mica alone or in combination make the most refractory stones. Mica, slate, and gneiss make an excellent combination. Gneiss is particularly refractory when it contains a considerable portion of arenaceous quartz; that is, quartz in which the particles partake of the nature of sand.

Limestones do not stand well in the presence of high temperatures, as they sometimes explode, owing to the rapid expulsion of the carbonic-acid gas.

Granitic and other primary rocks usually contain some water, which, in the presence of fire, causes them to crack and sometimes explode.

Sandstones, if somewhat porous, uncrystallized, and free from feldspar, are the most refractory of the common building stones.

Firebrick is perhaps the most fire-resisting material now known, while common hard-burned brick is more refractory than any of the building stones.

Concrete made of Portland cement and stone is a fire-resisting medium of only fair value.

DURABILITY OF BUILDING STONE

88. In the structural use of building stones, it is seldom that the full safe strength of the stone is required to resist the stresses imposed, and consequently the range of choice is not limited by this consideration so much as by the factor of *durability*. While in architectural work color is of great importance, for on it the architect depends to a large extent for the success of his design, it is exceptional in purely structural work for the color of the stone to be a deciding factor.

The **durability** of a building stone depends not only on the physical and chemical formation of the stone, but to a considerable extent on the climate in which the structure is to be built and also the method employed in quarrying the material. Where a selection is to be made of two stones equally durable and structurally fit for their purpose, the economic consideration influences the choice. The cost of structural building stones is regulated by the difficulties of quarrying, the refractory nature of the stone in finishing, and the distance it must be transported.

89. Physical Structure.—The most durable building stones are generally of a compact and uniform texture and show a clean fracture free from earthy or soluble mineral matter. Stones showing lamination or layers are not likely to prove as durable as those of a more homogeneous structure especially when laid with the laminations perpendicular to the bed of the wall, or on *edge*.

Non-porosity is not always a quality synonymous with durability, for many stones that absorb moisture also permit of its rapid evaporation; such stones are likely to prove more durable than those that absorb less moisture and part with it more reluctantly.

Stones showing a streaked appearance and lack of uniformity in color are usually composed of several minerals of various degrees of hardness, and in some instances one of them may be slightly soluble. Such stones are not likely to weather well, for the softer or more soluble mineral will be corroded and washed away, leaving the harder substance to protrude. When the less durable substance is in small pockets or spots, the stone will, on long exposure, be pitted; while if the softer mineral is in streaks or veins the material will be grooved, fissured, or channeled.

Small fossils or shells embedded in the substance of a building stone have usually a deleterious influence on its durability and weathering qualities. Such fossils and shells are calcareous in nature and generally soft and partially soluble under atmospheric influences.

90. Climate and Environment.—Building stones of the most durable character are required in climates where the diurnal changes in temperature are great and where there is much moisture in the atmosphere. The structure of a stone consists of minute particles that are surrounded by a matrix that forms the cementing material of the mass or are closely attached to each other by cohesion. In either case, changes in temperature cause these particles to expand and contract with considerable force, thus loosening particles from the matrix or from each other, causing deterioration and the ultimate destruction of the rock. The freezing of water in the pores of the stone or in the crevices and the spaces between the laminations in stratified stones, is the primary cause of the rapid destruction of some building stones. Water in freezing expands about one-tenth of its bulk and is said to exert a pressure of about 150 tons per square foot, which is sufficient, under favorable conditions, to split and rend the strongest rocks. The freezing of moisture within the pores of the stone is very deleterious to the stability of its structure, especially if there is not contained in the rock reserve pore space sufficient to accommodate the increased bulk of the water when frozen.

When water freezes in the crevices or spaces between the laminations, its action is that of an adjustable wedge tending to split the rock and widening the crevice more and more with each repetition of the freezing process. By this means, stones of laminated or stratified structure are particularly liable to disfigurement, when laid on edge by *exfoliation* or the scaling of the surface. In the large cities, it is not uncommon to observe balusters and carved details partially destroyed from this cause. The damages from the freezing of water in the spaces and crevices between the laminations is not as great when the stone is laid on its bed as when laid on edge, because there is not the same opportunity for the space lying in a horizontal plane to collect the moisture and also because the pressure on the stone from the superimposed masonry nullifies, to some extent, the wedging or bursting action of the freezing water.

The severest atmosphere on building stone is one that frequently and for long periods contains great quantities of suspended moisture in the shape of fogs and is also subjected, by environment, to much smoke and gas from the bituminous coals of manufactories. Such atmospheres are likely to contain carbonic-acid gas and sulphurous fumes, which have an appreciable effect on limestones and marbles. The actions from these sources affecting the durability of building stones are especially marked where the atmosphere is extremely moist.

91. Effect of Quarrying and Finishing.—Before stones are used in an important structure they must be thoroughly seasoned. When detached from the rock, stone is generally saturated with quarry water. It should, therefore, be exposed for some months, preferably under cover, to allow this water to evaporate. If the stone is not seasoned before it is placed in the wall of the structure, it is likely to remain peculiarly damp, and the excess of moisture, in freezing, will influence the durability of the material.

The use of heavy explosives for detaching dimensioned stone is detrimental to the quality of durability, from the fact that the severe concussion is likely to jar the particles and partially destroy their cementation and cohesion, producing incipient cracks and flaws that make the face of the stone more permeable to moisture and thus facilitating the deterioration of the stone by freezing and chemical action. For the same reasons stones sawed to size are more durable than those hammered and broken; and stones taken from the quarry by channeling or cutting are preferable to those procured by wedging.

92. Effect of Fire.—The fierce conflagrations that occur in the large cities frequently subject the stone walls of structures to intense heat. While stone is an excellent non-conductor, it is not as a rule as durable, when subjected to intense heat, as brick.

The severest test to which a stone can be subjected in a fire is for it to be heated intensely and then cooled by the sudden application of water from the fire-hose. This rapid change of temperature causes the exterior heated layer of the stone to contract more rapidly than the mass, and from many stones under this condition, large pieces will crack and break off; the process being several times repeated results in the entire destruction of the stone.

The silicious sandstones are the least destructible by fire, while the granites and conglomerates are probably the most affected by intense heat and the sudden cooling incident to the application of water. Limestones are very refractory in temperatures less than 1,000°, and at this temperature are not liable to deterioration by sudden cooling, though above this temperature, they may be reduced to quicklime, which crumbles and falls away after a few weeks' exposure to the air.

STRENGTH OF STONES AND MASONRY

93. The resistance of stones to stress varies greatly, and the strength of masonry depends not only on the material of which it is composed but on the manner in which these materials are handled; that is, on the workmanship. Stones that are the densest usually possess the greatest resistance, and masonry composed of squared stones with close joints is the strongest.

Many tables, based on the results of tests, give the strength values of building stones, but they differ widely, the discrepancy being due to the following causes:

1. Samples are taken from different quarries or from different parts of the same quarry.
2. The pieces of stone used for testing are not uniformly seasoned.
3. Test pieces are of different sizes.
4. They are not uniformly dressed or finished.
5. Variations exist in the method of placing the test specimens in the machine.

Frequently stones quarried from different parts of the same bed will show 20 or 30 per cent. difference in their crushing resistance, and stones that have been quarried some time and exposed will show a different resistance from those lately detached. The larger the test piece, the greater will be the unit stress developed, for small cubes do not develop as great a unit resistance as large ones, and within certain limits the unit stress that test cubes of the same material will sustain varies directly as the cube of the sides.

The method of finishing the test pieces and the accuracy and fineness with which the sides are dressed have much to do with the results of the test. Specimens that have been sawed to shape test higher than those that have been finished with a tool or chisel. Microscopic examination of the surface finished with a chisel reveals numerous minute cracks, caused by the excessive jars, that tend to reduce the crushing strength by starting fractures. The fineness of the surface finish also affects the result, from the fact that when the bearing surfaces are rough, transverse stresses that tend to disrupt the specimen are created.

94. It is well determined that from lack of homogeneousness or uniformity of texture, building stones and masonry of the same material have variable strength values. This uncertainty regarding the exact strength of masonry materials, together with their usually rapid deterioration, necessitates the use of a high factor of safety, so that in all work of this class minimum safety factors ranging from 10 to 20 are employed. When, therefore, the average strength values of commercial masonry materials are known and a high factor of safety is used, the basis on which the design is made is assuredly safe.

95. The average strength values for masonry materials, which are sufficiently conservative for good engineering practice, are given in the following tables:

TABLE I

STRENGTH OF BUILDING STONES AND MASONRY

Materials		Weight per Cubic Foot Pounds	Compressive Strength Pounds per Square Inch	Tensile Strength Pounds per Square Inch	Transverse Strength Pounds per Square Inch
Granite,	Colorado	166	15,000		
	Connecticut	166	14,000		1,500
	Massachusetts	165	16,000		
	Maine	165	15,000		
	Minnesota	166	25,000		
	New York	166	16,000	600	1,800
	New Hampshire	166	12,000		
Sandstone,	bluestone	160	15,000	1,400	2,700
	Connecticut, Middletown	148	7,000	590	1,000
	Massachusetts { Longmeadow, brown	142	10,000	450	
	Massachusetts { Longmeadow, red	149	12,000	450	
	New York { Hudson River		12,000		
	New York { Little Falls, brown		10,000		
	Ohio	139	8,000	100	479
	Pennsylvania, Hummelstown, brown		12,000		
Limestone,	New York { Kingston	168	12,000		
	New York { Garrison Station	164	18,000	Average	Average
	Indiana, Bedford, oolitic	146	8,000	1,000	1,500
	Michigan, Marquette	146	8,000		
	Pennsylvania, Conshohocken		15,000		
Marble,	Pennsylvania, Montgomery County		11,000		
	Massachusetts, Lee, dolomite		22,800	Average	Average
	New York, Pleasantville, dolomite		22,000	700	1,200
	Italian	168	12,000		
	Vermont	167	10,000		
Slate		160–180	10,000	10,000	5,000
Rubble, in lime mortar		150	500		

TABLE II

STRENGTH OF BRICKS, BRICKWORK, AND RUBBLE

Materials	Weight per Cubic Foot Pounds	Compressive Strength Pounds per Square Inch	Tensile Strength Pounds per Square Inch	Transverse Strength Pounds per Square Inch
Bricks, soft, inferior	100	1,000	40	
good, common	120	10,000	200	600
best, hard	125	12,000	400	800
paving	130	5,000		
Philadelphia, pressed	150	6,000	200	600
Brickwork, common, in lime mortar	120	1,000	50	
stretchers, in cement and lime	125	1,500	100	
best, hard in cement mortar	130	2,000	300	
Terra cotta	110	5,000		
Terra-cotta work	112	2,000		
Rubble	150	500		

TABLE III

STRENGTH OF CEMENTS, MORTARS, AND CONCRETE

Materials		Weight per Cubic Foot Pounds	Compressive Strength Pounds per Square Inch	Tensile Strength Pounds per Square Inch	Transverse Strength Pounds per Square Inch
Neat cement,	Portland, 1 month old		2,000	300	400
	Portland, 1 year old		3,000	550	800
	natural rock, 1 month old		1,200	150	200
	natural rock, 1 year old		2,000	400	400
	lime	59	600		200
	plaster of Paris	79	600	70	
Concrete,	Portland cement, 1 month old		1,000	200	100
	Portland cement, 1 year old	120–140	2,000	400	150
	natural cement, 1 month old	120–140	500	100	75
	natural cement, 1 year old				
Mortar,	lime	98	400	50	125
	natural cement and lime	100	600	75	200
	natural cement	102	1,000	145	102
	Portland cement	109	2,000	300	700
	plastering	86	400		

TABLE IV

ALLOWABLE UNIT STRESSES FOR MASONRY MATERIALS

Description of Material		Compressive Strength Pounds per Square Inch	Transverse Strength Pounds per Square Inch
Capstones,	bluestone	700	300
Templets,	granite	700	180
Monoliths,	limestone	500	150
	marble	400	120
	sandstone, other than bluestone	350	100
	slate	700	400
Squared-Stone	bluestone	350	
Masonry,	granite	350	
	limestone	250	
	sandstone, other than bluestone	175	
Rubble,	laid in Portland cement mortar	150	20
	laid in natural cement mortar	120	
	laid in lime-and-cement mortar	100	
	laid in lime mortar	80	
Brickwork,	laid in Portland cement; cement 1, sand 3	250	50
	laid in natural cement; cement 1, sand 3	150	40
	laid in lime and cement; cement 1, lime 1, sand 1	125	30
	laid in lime; lime 1, sand 4	100	15
Concrete,	Portland cement; cement 1, sand 2, stone 4	200	30
	Portland cement; cement 1, sand 2, stone 5	150	20
	natural cement; cement 1, sand 2, stone 4	125	16
	natural cement; cement 1, sand 2, stone 5	100	10

96. The values given in Tables I, II, and III are the average ultimate and breaking loads for the different materials. They are the results of tests made at different times on specimens prepared for the purpose. It will be noticed that the strength values of squared masonry are not given, and though conservative practice recommends that masonry of squared stone may be considered as having an ultimate strength equal to four-tenths of the strength of the stone, this is only an assumption that has not been substantiated by tests. The scarcity of reliable tests on masonry piers and walls is due to the fact that in order to obtain accurate results of the test, specimens must be of full-size dimensions and when thus built their strength is so great as to resist the ultimate power of the testing machine.

97. In using the values given in Tables I, II, and III, factors of safety of not less than 10 for compression and 15 and 20 for tension and transverse stress, respectively, should be employed. The usual practice in structural and architectural engineering is to use the allowable unit values for masonry and masonry materials given in Table IV. These values are considered good practice, and, in most materials, correspond with values recommended by the building laws of several cities.

SELECTION OF BUILDING STONES

98. In the building of important masonry structures, it is of primary importance that the stone employed shall be of sufficient strength and durability. Probably nothing in engineering construction is so neglected as the inspection of the building stone that is to be used.

Where it is necessary to employ great quantities of building stone in important situations, with reference to the stability of the structure, an inspection of the quarry from which the stone is to be obtained should be made. Besides, it should be the effort of the person who is to decide on the merits of the stone to inspect some building or structure that has been erected of the same material for a considerable length of time.

It is well, however, not to depend wholly on either inspection at the quarry or at the building but to subject the stone to laboratory tests, when it should be tested both chemically and physically as well as subjected to microscopic inspection.

99. The inspection at the quarry, when conservatively made, will frequently reveal the durability of the stone as well as its uniformity. Exposed quarry faces will sometimes show the weathering properties of the stone, besides its liability to disintegration through moisture and running water containing deleterious acids and alkalies. Such inspection will also determine whether there is sufficient stone of a uniform texture and color in sight to supply the amount of material required for the work. By quarry inspection likewise, the several grades of stone are known, and in first-class work it is imperative that the best grade or run of the quarry be insisted on. Frequently, the third-grade stone is employed in the structure, and on showing deterioration and poor weathering qualities causes otherwise excellent building stone, when of first-class cuttings, to be condemned.

100. By the inspection of stone that is in place in a building or structure for a considerable length of time, an excellent idea may be had of its durability as to structure, color, and weathering properties. If, after years of exposure in the severe atmosphere of an industrial city situated in a temperate zone, the building stone shows no disintegration or exfoliation and has retained its original luster and color, but for the soil of dust and smoke stains, due to its environments, it certainly can be considered of the best structural value for building purposes.

101. While the quarry and building inspections of stone are of the utmost practical importance, they should, wherever possible, be augmented by laboratory tests; in fact, these tests are absolutely necessary. When the stone to be used is from a new quarry, the characteristics of the product are little known. The laboratory tests usually consist of chemical analysis, microscopic examinations, and physical tests.

The *chemical analysis* determines both qualitatively and quantitatively the chemical constituents of the stone. Examined qualitatively, the mineral elements and chemical combinations of compressing the stone, together with the impurities and original matter, are determined; while the quantitative analysis shows the proportion of the different elements and chemical combinations. When the chemical composition of a stone is in this way determined, conclusions can usually, though not always, be drawn as to the quality of durability and the weathering properties of the stone.

The *microscopic examination* of building stone is of even more importance and is less expensive to conduct than the chemical analysis, for by it is revealed the structure of the stone. By the microscope may be observed whether the stone is igneous, metamorphic, or unaltered sedimentary rock. It also shows the size and shape of the particles or crystals composing the stone, their relative closeness, and the character and compactness of the cementing material holding them together. Usually the mineral constituents of the stone may be determined likewise by microscopic examination and frequently their proportions may be estimated, together with the percentage of impurities contained in the stone. Likewise by the microscope may be detected any flaws in the structure, such as cracks, cavities, incipient fractures, and gas bubbles.

The *physical tests* of stone are of great practical value in determining its strength and durability. When the strength values of a building stone are not known, it is frequently necessary to make tests in order to determine the crushing and transverse unit stress of the material, for though the unit crushing strength of building stone is usually greatly in excess of that required, the transverse strength is frequently barely sufficient to sustain the load when a fair factor of safety is allowed. An estimation of the durability of a building stone may be deduced from data collected from physical tests made to ascertain the specific gravity, porosity, and weight of the stone and from such tests as will demonstrate the effects of extreme heat and cold and the actions of carbonic and sulphurous acids. In making these last tests, the object is to impose on the stone, as nearly as possible, conditions that in a few hours' time will approach the effect produced by climatic changes and vitiated atmosphere through a lapse of years.

MORTARS

CHARACTERISTICS OF MATERIALS

102. Mortars for structural purposes are composed of lime or cement and sand mixed to the proper consistency with water. When lime and sand are used, the mixture is known as *lime mortar;* when the mixture is of cement and sand, it is designated as *cement mortar.* Frequently in building work, lime mortar is mixed with cement mortar, when the term *lime-and-cement mortar* is used. Small percentages of other materials, such as salt, sugar, brick, and volcanic dust, are sometimes mixed with mortar in order to accomplish specific purposes, such as the prevention of freezing, hastening of setting, increasing of strength, and the creating of hydraulic properties, though with doubtful effectiveness. In mixing mortar, it is usual to designate the amount of such ingredients by a notation, such as 1 to 1, 1 to 2, or 1 to 3, which signifies that the mortar is composed of one part of cement or lime to one, two, or three parts of sand, respectively. The first number of the notation always indicates the amount of cement, which for convenience is taken at unity. In proportioning the mixture, the parts of the ingredients are generally measured by volume and just sufficient water supplied to work the mortar to the proper consistency.

103. Sand.—This material is an important part of all mortar mixtures, for in lime mortar it prevents excessive shrinkage and adds crushing resistance and tenacity, while in cement mortar it is principally used for economic reasons and to provide additional resistance to crushing, though it decreases tenacity of cement.

It is important that sand for mortars shall be clean and sharp. *Clean sand* when rubbed between the fingers will not soil them, but sand containing impurities will leave an earthy stain. When clean sand is dampened and squeezed in the palm of the hand, on the release of the pressure, it will fall apart, while if it contains soil, clay, or earthy substance, the particles will be held together by being cemented by the impurities. Sand for structural purposes should be *sharp;* that is, the grains should be cubical and angular rather than globular or rounded. The sharpness is best tested by rubbing or rolling the sand between the hands and if there is a distinct grating sound, it indicates that the sand grains are angular. Sand that is sharp and angular lies closer together and offers more surface of adhesion for the cement composing the matrix of the mortar.

Much importance has unwarrantedly been placed on the size of the grains of sand for structural purposes. Very fine sand and sand so coarse as to contain grains larger than the joints of masonry should not be used. A sand containing a mixed size of grain is the most economical and the best for structural purposes. When the usual specification is complied with, namely, that the sand shall be sifted through a screen having a mesh $\frac{1}{8}$ inch square, grains of not more than $\frac{1}{16}$ inch on a side will be passed, and sand of such size gives excellent results.

The usual sources from which building sand is obtained are the seashore, the river bank, and sand deposits or pockets. The first, which is known as *sea sand*, is usually objected to on account of the salt that it contains and should not be used where the appearance of the work is a factor unless it is well washed. The second, which is termed *river sand*, is generally composed of rounded particles rather than angular and may be either clean or dirty. The third, or *pit sand*, is usually sharp, though it is likely to contain clayey and earthy substances, which destroy its cleanliness.

104. Lime.—By calcining or burning calcium carbonate or limestone at a high temperature, **quicklime** or

commercial lime is produced. In the process of burning, the water that the stone contains is driven off, together with carbonic-acid gas or carbon dioxide. Lime has a great affinity for water and should be used soon after being made, as it is likely to become *air-slaked*, when it is unfit for use in making lime mortar. When lime is brought in contact with water, it takes it up rapidly, swelling and generating considerable heat, the lime falling apart and producing a fine white powder; this process is known as *slaking*. When more water is added to this powder, a *lime paste* is formed, which by the addition of sand produces lime mortar. The best limes slake vigorously and completely, the *lime flour*, or precipitate, being fine and free from refractory lumps and cores. The presence of such unslaked fragments indicates that the limestone was not pure or that the process of calcination was imperfect.

In slaking lime, a volume of water equal to two or three times the volume of the lime is used. The entire amount of water necessary for slaking should be used at one time, as it is deleterious to turn in cold water after the lime has commenced to slake.

105. Cement.—Commercial cements used for making mortar for structural purposes may be classified as *Portland* and *natural-rock cements*. The difference in their manufacture consists more essentially in the method of procuring the raw material rather than in the process of preparation. The name **Portland cement** is given to any cement that is manufactured from the several necessary materials carefully gathered and proportioned, while the term **natural,** or **natural-rock, cement** is given to those cements that are manufactured direct from a natural rock containing the necessary ingredients.

Portland cement is obtained by mixing lime or marl with clay in the proper proportion. The mixture is made into balls, or nodules, which are calcined in a furnace and ground to a fine powder. The natural-rock cements are manufactured from a natural rock that contains lime, alumina, and silica.

Rocks containing the proper materials for making these cements are of a mixed limestone and clayey nature. These rocks are calcined at a high temperature approaching vitrification and are broken, pulverized, and finally ground to form the commercial cement. Both Portland and natural-rock cements are *hydraulic*, that is, will set under water, and are consequently invariably used for work in damp situations and below the water-line.

When cement is brought in contact with water, it hardens rapidly even when air is not present. This hardening or *setting*, as it is called, is caused by a chemical reaction that takes place and forms with the water of crystallization double silicates of lime and alumina. Some cements set more rapidly than others and cements that take an initial set in from 5 to 30 minutes, or even 1 hour after mixing, are known as *quick-setting cements*, while those that require from several hours to a day or longer, in order for the initial set to take place, are known as *slow-setting cements*.

Usually natural-rock cements set in less time than the Portland cements; besides they are lighter in weight and do not obtain the same degree of strength in as short a time after mixing. A barrel of natural-rock cement weighs about 300 pounds, while Portland cements are usually one-third heavier.

Portland cement is considered, in many respects, superior to natural-rock cements for structural purposes, yet in this country great quantities of natural cement have been used for the most important engineering work and have been found, after years of service, to have been exceedingly durable, attaining great strength.

The natural-rock cements are found and manufactured in New York, Pennsylvania, Maryland, Virginia, and Kentucky, as well as in several of the western states. The principal brands are many. Those having an extended reputation are known as Rosendale, Louisville, Cumberland, James River, and Round Top. The Portland cements that are used in this country are both domestic and imported. Some of the principal domestic Portland cements are known as the

Vulcanite and Alpha, manufactured in New Jersey; the Dexter, Lehigh, Atlas, and Whitehall, manufactured in Pennsylvania. The imported cements come from Germany, France, and England. Lagerdorfer German cement is used to some extent, it being a strong, finely ground, and uniform material. Vicat cement, which is imported from France, is a high-priced cement and owes its use in this country to the fact that it is not as likely to stain fine ashlar work as some of the domestic cements. English Portland cement is also used to some extent, though it is doubtful whether it possesses any qualifications of excellency over the best American Portland cements. The distinction that the New York building laws make with regard to brands of cements sold under the general nomenclature of Portland and natural is interesting, though it does not determine actually whether the cement is Portland or natural.

The requirements of the building laws are as follows:

"Cement classified as Portland cement shall be considered to mean such cement as will, when tested neat, after 1 day set in air, be capable of sustaining, without rupture, a tensile stress of 120 pounds per square inch, and after 1 day in air and 6 days in water be capable of sustaining, without rupture, a tensile stress of at least 300 pounds per square inch.

"Cements other than Portland cement shall be considered to mean such cement that when tested neat, after 1 day set in air, be capable of sustaining, without rupture, a tensile stress of at least 120 pounds per square inch."

The building laws continue to state that the test shall be made under the supervision of the Commissioner of Buildings, and that those tests shall be made at such a time as he shall deem advisable. They also require that records of the tests shall be kept for public information.

106. Lime Mortar.—Common mortar of quicklime and sand, though formerly used for all classes of work, is now supplanted by cement mortar for the better class of structural work. In mixing **lime mortar,** the lime is thoroughly slaked to a powder and wetted to a paste when the sand is mixed and thoroughly incorporated with the paste by working and kneading. While lime on exposure to the atmosphere will become air-slaked and unfit for use, the slaked lime, when mixed with sand, is not so impaired and it is usually considered beneficial to mix up large batches of the mortar some time before using it, when it only requires tempering with water and some working to be ready for immediate use.

Sand is mixed with lime to form lime mortar in the proportions of one part of lime to from three to six parts of sand. With rich limes, that is, those that slake freely and completely, the proportion of three parts of sand to one of lime is not sufficient, and four parts of sand makes a better mixture. The proportion of six parts of sand to one of lime is never used except in the poorest work, and in fact the building ordinances of the several cities require that mortar shall be made of one part of lime and not more than four parts of sand, the lime to be in all cases thoroughly burnt, of good quality, and properly slaked before it is mixed with the sand.

Lime mortar sets by the drying out of the mortar and by a chemical action produced through the absorption of carbon dioxide, or carbonic-acid gas, from the air. By taking up this gas, the slaked lime is again converted to calcium carbonate, which surrounds each particle of sand, thus converting the mortar into artificial stone. Since the absorption of carbon dioxide is necessary, in order that lime mortar may harden, this mortar will not set in water and should not be used in foundation work below the water-line or in extremely damp situations. Lime mortar used in the interior of very thick walls has been known to remain soft for years; consequently, it cannot be used in concrete.

Though lime mortar will not harden in wet situations, the process of crystallization caused by the absorption of carbon dioxide in the air is facilitated by keeping the work damp for a short time after laying. For this reason, brickwork should not be laid in very hot or dry weather unless the bricks are thoroughly soaked in water.

When lime mortar does not retain sufficient moisture to properly promote the chemical action of hardening, it forms a chalky mass that is easily pulverized between the fingers, showing that it has lost most of its properties of adhesion and cohesion.

107. Cement Mortar.—Mortar composed of either Portland or natural-rock cement is being used extensively for all structural purposes, owing to the reduction in the price of these cements, and their excellent quality.

In this mortar, only cement and sand are used with just sufficient water to mix it to a mealy consistency; an excess of water is detrimental to its qualities of strength. The usual proportions of sand and cement, when natural-rock cements are used, are one part of cement and one or two parts of sand, the usual practice being to use the latter proportion. In mixing cement mortar with Portland cement, it is customary to mix the sand and cement in the proportion of 1 to 3, for this mixture is considered equal in strength and durability to a mixture of one part of sand and two parts of natural-rock cement.

The New York building laws specify that cement mortar shall be made of cement and sand in the proportion of one part of cement and three parts of sand and shall be used immediately after being mixed. The cement and sand are to be measured and thoroughly mixed before adding water. They also stipulate that the cements must be finely ground and free from lumps. The usual practice in making cement mortar is to mix the sand and cement together dry, measuring the parts by volume, adding the water as required in order to reduce the mixture to work consistently, care being exercised in supplying the water that the amount added is not excessive.

Owing to the fact that cement mortar sets rapidly, it should never be mixed, except immediately before being used; and after a batch of cement mortar has once taken its initial set no attempt should be made to retemper it, but the whole mass should be discarded. The natural-rock cement mortars set more quickly than Portland cement mortars, but with care can be conveniently used in any class of work. Both natural-rock and Portland cement mortars have hydraulic properties and set rapidly and completely under water. This property of hydraulicity is not destroyed by the addition of lime in the proportion of not more than one part of lime to two of cement. Because of the hydraulic properties of cement mortar and on account of the great strength that it attains when it sets in water, this mortar should always be used for masonry in damp situations or for foundations that are likely to be immersed in water. The usual cement mortars are durable even when submerged in sea-water.

One barrel of cement to two barrels of sand will make about 8 cubic feet of cement mortar and will lay 1 cubic yard of brickwork or rubble, while one barrel of cement and three barrels of sand will make about 12 cubic yards of mortar, which will lay 1½ cubic yards of ordinary masonry, or provide sufficient matrix for 1½ cubic yards of concrete. The quantity of water that is required in mixing cement mortar varies considerably with the temperature and with the dryness of the materials. Ordinarily, one-third of a barrel of water mixed with one barrel of cement, will make two-thirds of a barrel of stiff paste.

In making cement mortar, sometimes the cement is mixed to a paste with the water and the sand added, but the usual practice is to mix the cement and sand dry and to introduce sufficient water in order to work it to a mealy and pasty condition. In the manufacture of all cement mortar, the sand and cement must be thoroughly mixed, as otherwise the substance will not have a uniform texture and its mass will incorporate spots of dry cement, and frequently there will be pockets of sand without the matrix of cement.

There is much difference of opinion as to the advisability of using cement mortar in freezing weather. Undoubtedly, the best practice is to suspend all masonry operations when the thermometer is below freezing, but when necessary, however, cement mortar may be used in temperatures several degrees below freezing. It is well, however, in laying work with cement mortar in freezing weather to heat the sand, brick, and stone, if possible, and to use hot water in mixing. With ordinary precautions and by covering the work when operations are suspended at night, cement mortar may be used in low temperatures with safety. The frost may have some action on the exterior edges of the joints, but the action will not penetrate to the interior of the wall.

Some authorities recommend the introduction of salt in cement mortar when it is used in freezing weather. This practice is not to be recommended, however, for the salt retards the setting and is liable to stain the masonry or brickwork.

108. Cement-and-Lime Mortar.—Frequently mortar for building work is composed of a mixture of both cement and lime with sand. The lime introduced with cement in mortar makes the material more tractable by increasing the plasticity of the mortar, and thus facilitates the work of laying the masonry or brickwork. It also retards the setting, which is frequently requisite, and lowers the freezing point of the mortar. Besides, it does not shrink, like lime mortar, in setting and is therefore desirable in laying the backing of ashlar walls. Its use also promotes economy, for lime is much less costly than cement.

The proportion of the ingredients used in mixing this mortar varies considerably with the class of work and the judgment of the designer. For brick backing, piers, and walls, one part of Portland cement, two parts of good, fresh, wood-burned lime, and three parts of clean, sharp sand will give satisfaction. When this mixture is to be used for ashlar facing one part of Vicat Portland cement, two parts of good, fresh, wood-burned lime, and three parts of clean, sharp sand, mixed together as soon as prepared, have been found to give excellent results. Frequently only a small percentage of lime is added to the cement mortar to give it plasticity, and in this instance the specifications usually require that the mortar shall be composed of one part of Portland cement, three parts of clean, sharp sand, and just sufficient lime to make the mortar pasty. Lime-and-cement mortar is admirably adapted to the construction of brick arches used in fireproof-floor construction.

The New York building laws stipulate that cement-and-lime mortar mixed shall be made of one part of lime, one part of cement, and not more than three parts of sand to each.

In mixing lime-and-cement mortar, the slaked lime is mixed in bulk with the sand and the cement added, after tempering it with water, immediately before being required. The lime mortar and the cement mortar made in this way must be thoroughly incorporated by working, though this must be done rapidly before the cement has had time to set.

109. Comparison of Lime and Cement Mortars.—It is interesting to compare lime and cement mortars so as to see wherein one is more advantageous than the other. The points in favor of lime are: its cheapness; the large quantity of mortar that can be made from a given quantity of lime; the fact that if, before use, it is kept damp, it remains good indefinitely; and the ease and simplicity of mixing. Against its use may be mentioned these points: in damp situations, it will not harden; in the interior of walls of any considerable thickness, it hardens (if at all) but very slowly; and it is much weaker than cement.

Among the advantages of cement mortar are: its hydraulic quality, in virtue of which it hardens even quicker in water than in air; its greater adhesive power; its rapidity of setting; and the great ultimate strength that it attains. The points against it are: the extra care required in mixing and handling; and its greater cost, which is, of course, the controlling factor. Summing up, it is evident that cement is greatly preferable to lime, and the quality of masonry laid with it is so much superior as to more than counterbalance the difference in cost over lime mortar.

Cement mortar possesses an advantage over lime mortar, in that in setting it does not shrink. Because of this property, and because of the fact that cement mortar sets rapidly, walls can be laid up in cement mortar much more rapidly than they can in lime mortar and with greater safety.

A mixture of cement and lime mortars is frequently used, and is a considerable improvement on ordinary lime mortar, in that, if the quality of lime is not greater than one-fourth or one-fifth that of the cement, the strength will be practically that of cement mortar, and the mixture will set quickly in damp places, while the cost will be materially less than that of cement mortar.

It may be interesting to compare the relative cost of masonry laid in lime and in cement mortar. For this purpose the following figures are given, which, being obtained from a very reliable source, will undoubtedly be found quite accurate. To work from a common basis, the estimates are made per 1,000 brick, and per perch of rubble masonry, all laid in 1-to-3 mortar. The quantities given, of course, remain constant, while the prices vary according to local rates. These figures represent actual cost.

BRICKWORK IN LIME MORTAR

1,000 brick	$6.00
3 bushels lump lime, at $.25	.75
½ load (½ cubic yard) sand, at $1.50	.75
Bricklayer, 7 hours, at $.35	2.45
Helper, 7 hours, at $.15	1.05
Cost, per 1,000 laid	$11.00

BRICKWORK IN CEMENT MORTAR

1,000 brick	$6.00
1½ barrels Rosendale cement, at $1.50	2.25
½ load sand, at $1.50	.75
Labor, same as before	3.50
Cost	$12.50

NOTE.—If Portland cement is used, the cost will be about $13.63 per thousand brick.

RUBBLE MASONRY IN LIME MORTAR

1 perch stone	$1.25
1 bushel lime	.25
⅙ load sand, at $1.50	.25
Mason, ⅓ day, at $3.00	1.00
Helper, ¼ day, at $1.50	.38
Cost, per perch	$3.13

RUBBLE MASONRY IN CEMENT MORTAR

1 perch stone	$1.25
½ barrel Rosendale cement, at $1.50	.75
⅙ load sand, at $1.50	.25
Labor, same as before	1.38
Cost	$3.63

NOTE.—Using Portland cement, at $2.25 per barrel, the cost per perch will be $4.01.

From these figures, it is evident that a considerable difference exists between the cost of lime and cement mortars, in favor of the former. A large part of this difference may be overcome by replacing some of the cement by lime, which will cheapen the mortar, and at the same time give it the valuable qualities possessed by both materials.

MATERIALS OF STRUCTURAL ENGINEERING

(PART 3)

TIMBER AND METALS

CHARACTERISTICS OF TIMBER

1. Wood, as a building material, is divided into three general groups; namely, the *evergreen*, the *tropical*, and the *hardwood*. In the first of these are classed pine, spruce, hemlock, cedar, cypress, etc.; in the second, palm, rattan, bamboo, etc.; in the third, oak, chestnut, walnut, locust, maple, hickory, ash, boxwood, whitewood, and a number of others. Each of these woods has peculiarities and characteristics which render it fit and useful for some building purposes, and utterly unfit and useless for others.

2. White pine, commonly known as *pine*, or sometimes referred to as *northern pine*, to distinguish it from the species described below, is a tree common in the northern part of the United States and in Canada. It furnishes a light, soft, and straight-grained wood of a yellowish color, but is not so strong as other woods of the same class, and in building is used principally as a finishing material, where a good, durable, but inexpensive job is required. As a material for patternmaking it has no equal, and its power of holding glue renders it invaluable to the cabinetmaker and joiner.

3. Georgia pine, also known as *hard pine*, *pitch pine*, and occasionally as *long-leafed pine*, which is really the best name for it, is a large forest tree growing along the southern coast of the United States, from Virginia to Texas, and extending only about 150 miles inland. Its annual rings are smaller than those of the white pine, and have a dense, dark-colored, resinous summer growth, which gives the wood a well-marked grain.

The wood is heavy, hard, strong, and, under proper conditions, very durable. For heavy framing timbers and floors it is most desirable, and on account of its grain is sometimes used for the trim of unimportant rooms. It rapidly decays in a damp location, and therefore cannot be used for house sills, or as sleepers or posts that are in contact with the ground, but if situated in a dry, well-ventilated place, it will remain practically unchanged for over a century.

Great care should be exercised in obtaining Georgia pine, as in many localities this wood is confused with another material variously known as *Carolina pine* and *Northern yellow pine*, which is greatly inferior to it in every respect.

The *Carolina pine* is not a long-leafed pine, and is neither so strong nor so durable as the Georgia or Southern pine. In appearance it is somewhat lighter than the long-leafed pine, and the fiber is softer and contains less resin than the regular hard, or pitch, pine.

4. Spruce is a name given to all the wood furnished by the various species of the spruce fir tree. There are four varieties of the wood, known as *black spruce*, *white spruce*, *Norway spruce*, and *single spruce*.

Black spruce grows in the northern half of the United States and throughout British America. Its wood is light in weight, reddish in color, and, though easy to work, is very tough in fiber and highly desirable for joists, studs, and general framing timber. It is also greatly used for piles and submerged cribs and cofferdams, as it not only preserves well under water but also resists the destructive action of parasitic crustacea, such as barnacles and mussels, longer than any other similar wood.

White spruce is not so common as the black variety, though, when sawed into lumber, it can scarcely be distinguished from it. Its growth is confined to the extreme northern part of the United States and to British America. Another variety of white spruce is a large-sized tree growing in the central and southern parts of the Rocky Mountains, from Mexico to Montana.

Norway spruce is a variety growing in Central and Northern Europe and in Northern Asia, and its tough, straight grain makes it an excellent material for ships, masts, spars, etc., as well as the more ordinary purposes of house building. Under the name of *white deal*, it fills the same place in the European woodworking shops as white pine does in America.

Single spruce grows in the central and the western part of the United States. It is lighter in color, but otherwise its properties are similar to the black and the white spruce.

5. Hemlock is similar to spruce in appearance, though much inferior as a building material. The wood is very brittle, splits easily, and is liable to be *shaky*. Its grain is coarse and uneven, and though it holds nails much more firmly than pine, the wood is generally soft and not durable.

Some varieties of it are better than others, but in commerce they are so mixed that it is difficult to obtain a large quantity of even quality. Hemlock is used almost exclusively as a cheap, rough framing timber.

6. White cedar is a soft, light, fine-grained, and very durable wood, but lacks both strength and toughness. Its durability makes it a desirable material for shingles, and also for tanks in which water is stored; these are about the only purposes for which it is used in building construction, though it is used largely in boat building, cigar-box manufacture, and cooperage.

7. Red cedar is a smaller tree than white cedar, and of much slower growth. The wood is very similar in texture to white cedar, but even more compact and durable. It is of a reddish-brown color, and possesses a strong, pungent odor, which repels insects. Its extreme durability makes it valuable for posts, sills, sleepers, etc. in contact with the ground, and its strong odor renders it extremely serviceable as shelving for closets and linings for chests and trunks, where the exclusion of moths and other insects is desired.

8. Cypress is a wood very similar to cedar, growing in Southern Europe and in the southern and western portions of the United States. It is one of the most durable woods, and is well adapted for outside use.

In the northern part of the United States its use is confined almost exclusively to shingles, but in the South it is used as extensively as pine is in the North.

9. Redwood is the name given to one of the species of giant trees of California, and is the most valuable timber grown in that state. It grows to a height of from 200 to 300 feet, and its trunk is bare and branchless for one-third of its height. The color is a dull red, and while the wood resembles pine and is used generally in the West for the same purposes as pine is in the East, it is inferior to pine on account of its peculiarity of *shrinking lengthwise* as well as crosswise. It is used largely for railroad ties, fence posts, telegraph poles, and other purposes where durability under exposure is required. As an interior finishing material it is highly prized, as it takes a high polish, and its color improves with age.

10. The **hardwood** group is headed by the **oak** as typical of its class, nearly all others being compared with it in regard to hardness, durability, and strength.

White oak is the hardest of the several American species of the oak tree, and it grows in abundance throughout the eastern half of the United States. It furnishes a wood that is heavy, hard, cross-grained, strong, and of a light yellowish-brown color. It is used where great strength and durability are required, as in framed structures, ship building, cooperage, and carriage making.

Red oak is similar in nearly every respect to white oak, except in its grain and color, the grain usually being coarser and the color darker and redder. It is also about 12 per cent. softer.

English oak is similar to the American oaks in color, texture, and appearance, but is superior to them for such structural purposes as ship building and house framing.

The structure of the fiber, and the large, thick, and numerous medullary rays, make oak especially prized as a material for cabinetwork and furniture when the log is quarter-sawed. The silver grain and the high and durable polish that the wood is capable of receiving, make it one of the most beautiful used in joinery and cabinetwork.

11. Ash, the wood of a large tree growing in the colder portions of the United States, is heavy, hard, and very elastic. Its grain is coarse, and its color is very similar to that of red oak, which it also resembles in strength and hardness.

Ash is sometimes used for furniture and cabinetwork, making a good imitation of oak, but it is never so strongly marked in the silver grain as oak, and its tendency, after a few years, to become decayed and brittle renders it unfit for structural work.

12. Hickory is the heaviest, hardest, toughest, and strongest of all the American woods. The medullary rays are very numerous and distinct, and produce a fine effect in the quarter-sawed plank. The flexibility of the wood, together with its toughness and strength, render it valuable in the manufacture of carriages, sleighs, and implements requiring bent-wood details.

As a building material, it is unfit for use: first, on account of its extreme hardness and difficulty of working; and second, on account of its liability to the attacks of boring insects, even after the fibers have been filled and varnished.

13. Locust is one of the largest forest trees in the United States, and furnishes a wood that is as hard as white oak. It is composed of very wide annual layers, in which the vessels are few, but very large, and are arranged in rows, giving the wood a peculiar striped grain. Its principal use is in exposed places where great durability is required, while for posts for buildings and fences in damp locations it has no superior. Its hardness increases with age, and on this account it is used for turned ornaments and occasionally in cabinetwork.

14. Maple is a large-sized timber tree that furnishes a light-colored, fine-grained, hard, strong, and heavy wood. The annual growth is narrow and close, but on careful examination small vessels may be seen scattered through it. The medullary rays are small and distinct, giving to the quarter-cut lumber a clearly defined silver grain. Two other characteristics of the grain are observed, especially in old trees, and are known as *curly maple* and *bird's-eye maple.* The former is a waviness of the grain similar to the burl obtained from the root timber of the walnut tree, while the bird's-eye is an effect produced in old trees by the circular inflexion of the fibers. The plank appears to be covered with numerous small spots, similar to minute knots, and strongly resembling birds' eyes, whence it derives its name. Though the appearance of both the curly maple and the bird's-eye maple is practically due to distorted fibers, which materially reduce the strength of the wood, they are highly prized in the cabinetmaker's art, as they lend to the polished surface a variegation and impart a beauty equaled by few other materials.

15. Chestnut, a large forest tree common to the eastern part of the United States, produces a comparatively soft, coarse-grained wood that, though very brittle, is exceedingly durable when exposed to the weather. It will not stand variations of slowly evaporating moisture as well as locust, and is therefore not so well suited for fence posts and sills laid in contact with the earth; but for exposed structures and sleepers laid in concrete or sandy soil, it affords a material much more easily worked than locust and nearly as durable as cedar.

At the age of 50 years the tree is in fine condition for cutting, previous to which the wood is likely to be composed of large cells filled with moisture, that do not dry out without impairing the quality of the timber. On the other hand, if the tree is not cut at 50 years, it is almost sure to become decayed in the heart wood, and thereby rendered unfit for use.

16. Beech is the wood of a large forest tree growing in the eastern part of the United States, and in Europe. It is used but slightly in building, owing to its tendency to rot in damp situations, but it is often used, especially in European countries, for piles, in places where it will be constantly submerged. It is very hard and tough, and of a close, uniform texture, which renders it a desirable material for tool handles and plane stocks, a use to which it is often put. It is occasionally used for furniture on account of its susceptibility to a high polish, but is too brittle for very fine work requiring strength.

17. Whitewood, so called from the purity of its color, is the lumber of the tulip tree, a large, straight forest tree abundant in the United States. It is light, soft, very brittle, and shrinks excessively in drying. When thoroughly dry it will not split with the grain, and in even slight atmospheric changes will warp and twist exceedingly. Its cheapness, ease of working, and the large size of its boards cause it to be used in carpentry and joinery, in many places where it is utterly unsuited.

18. Buttonwood, also called **sycamore,** is the name given in the United States to the wood of a species of tree generally known as **plane tree.** The wood is heavy and hard, of a light brown color, and very brittle. Its grain is fine and close, but, though susceptible to a high polish, it is not much used in general carpentry or joinery, as it is very hard to work and has a strong tendency to warp and twist under variations of temperature. In damp places it will soon show signs of decay and is therefore unfitted for any but the most protected positions.

19. Lignum vitæ is an exceedingly heavy, hard, and dark-colored wood, with an almost solid annual growth. It is very resinous, difficult to split, and has a soapy feeling when handled. Its color is dark brown, with lighter brown markings, and it is used mostly for small turned articles, tool handles, and the sheaves of block pulleys.

STRENGTH VALUES

20. The **ultimate strength** of any material is that unit stress that is just sufficient to break it.

The **ultimate elongation** is the total elongation produced in a unit of length of the material having a unit of area, by a stress equal to the ultimate strength of the material.

21. Modulus of Rupture.—When a simple beam breaks, the fibers at the top are in *compression*, and those at the bottom in *tension*, as shown in Fig. 1. By actual tests, it has been found that though some of the different fibers of materials under transverse stresses are in compression and some in tension, the ultimate resistance of the material does not agree with the ultimate resistance of the fibers to either tension or compression. Though many attempts have been made to account for it, this fact remains; hence, it becomes necessary to obtain some constant, or value, more closely agreeing with the strength of materials under transverse stress. It is usual, therefore, where the cross-section of the beam is uniform, to obtain, by actual tests, the constant or value for each material. This value is called the **modulus of rupture** and is generally expressed in pounds per square inch.

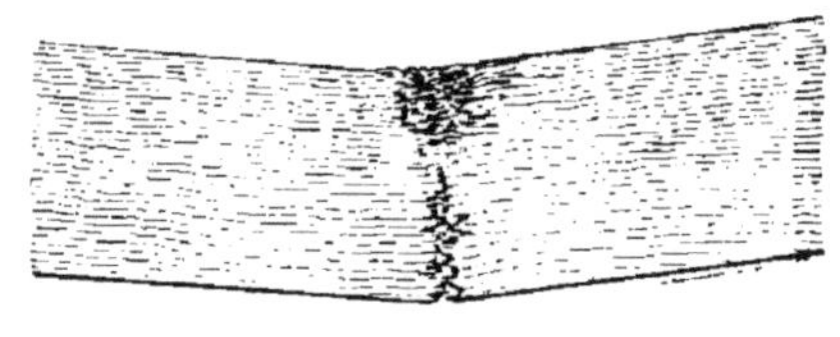

Fig. 1

22. Table I gives values of the strength of timber when subjected to different stresses. The column in the table, headed Compression, With Grain, will be found useful in computing the strength of columns. The values in the column headed Compression, Across Grain, are used in cases similar

TABLE I

AVERAGE ULTIMATE BREAKING UNIT STRESSES, IN POUNDS, PER SQUARE INCH

Kind of Timber	Tension		Compression			Transverse		Shearing	
			With Grain						
	With Grain	Across Grain	End Bearing	Columns Under 15 Diams.	Across Grain	Modulus of Rupture	Modulus of Elasticity	With Grain	Across Grain
White oak	10,000	2,000	7,000	4,500	2,000	6,000	1,100,000	800	4 000
White pine	7,000	500	5,500	3,500	800	4,000	1,000,000	400	2,000
Southern long-leaf or Georgia yellow pine	12,000	600	8,000	5,000	1,400	7,000	1,700,000	600	5,000
Douglas, Oregon, and yellow fir	12,000		8,000	6,000	1,200	6,500	1,400,000	600	
Washington fir or pine (red fir)	10,000					5,000			
Northern or short-leaf yellow pine	9,000	500	6,000	4,000	1,000	6,000	1,200,000	400	4,000
Red pine	9,000	500	6,000	4,000	800	5,000	1,200,000		
Norway pine	8,000		6,000	4,000	800	4,000	1,200,000		
Canadian (Ottawa) white pine	10,000			5,000				350	
Canadian (Ontario) red pine	10,000			5,000		5,000	1,400,000	400	
Spruce and eastern fir	8,000	500	6,000	4,000	700	4,000	1,200,000	400	3,000
Hemlock	6,000			4,000	600	3,500	900,000	350	2,500
Cypress	6,000		6,000	4,000	700	5,000	900,000		
Cedar	8,000		6,000	4,000	700	5,000	700,000		1,500
Chestnut	9,000			5,000	900	5,000	1,000,000	600	1,500
California redwood	7,000			4,000	800	4,500	700,000	400	
California spruce				4,000		5,000	1,200,000		
Factor of safety	10	10	5	5	4	6	2	4	4

to Fig. 2, and are such as will not produce an indenture of more than $\frac{1}{100}$ inch in the surface of the timber, a value well within the safe limit. The values under the heading, Shearing, With Grain, are used in computing the strength of the end of the tie-beam at the heel of the main rafter in a

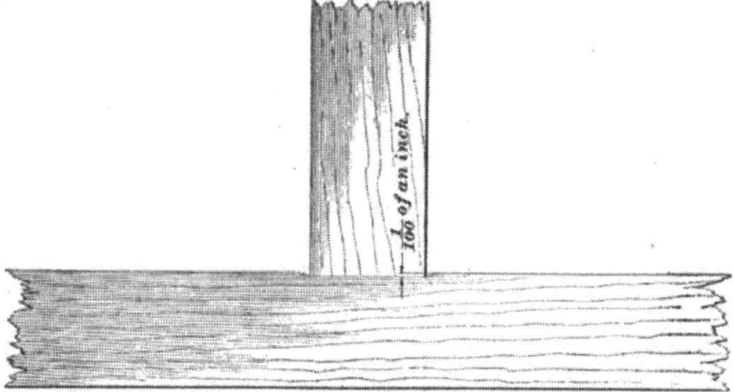

Fig. 2

roof truss, as shown in Fig. 3. The tendency is to shear off the piece *h* parallel to the grain along the line *a b*. The figures in the column headed Transverse, Modulus of Rupture, are the constants, or values, used when computing the strength of beams.

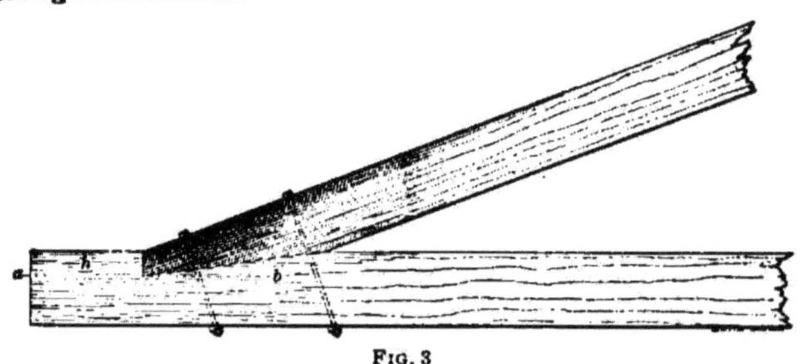

Fig. 3

The values in Table I were recommended by the committee on Strength of Bridge and Trestle Timbers of the Association of Railway Superintendents of Bridges and Buildings in its fifth annual convention, which was held in New Orleans, October, 1895.

CHARACTERISTICS OF IRON AND STEEL

23. Iron is very widely distributed in nature and its compounds are abundant. Probably no portion of the earth's crust is free from it, yet it occurs native only in very small quantities, and the iron thus found is probably of meteoric origin and is always alloyed to some extent with other metals, as nickel, cobalt, copper, etc. Its strong affinity for non-metals explains its infrequent occurrence in the native condition; and the dissimilarity between the metal and its ores may explain why it was among the later useful metals to be discovered, if, as is generally believed, such is the case. It may be mentioned, however, that some writers think that iron was known and used at a much earlier period in the world's history than is generally believed, but that its tendency to corrode has destroyed all traces of its use in ancient times, while instruments of brass and bronze remain.

Chemically pure iron is valuable only for experimental purposes and as a curiosity, as it has no use in the arts except, perhaps, in medicine. It may be obtained on a small scale in several ways, among which may be mentioned the reduction of pure ferric oxide by heating it in a current of hydrogen, and by the electrical decomposition of a solution of pure ferrous sulphate or chloride. While pure iron is devoid of value, when it contains small quantities of other elements it is the most useful and widely used of all metals; in fact, it is almost impossible to overestimate its importance in the arts.

24. The manufacture of iron from its ores depends on chemical principles with which we are familiar. As iron does not occur native, it is necessary to reduce its compounds, and this is done in such a manner that the resulting metal will contain the elements necessary to give it the properties that have made it so valuable. The method almost universally employed is to charge in the ore, together with the fuel—which at present is nearly always either coke, coal, or charcoal—at the top of a tall furnace; as the ore always contains extraneous matter, a flux is added in the proper amount to form a fusible slag with these impurities. Hot air blown into the furnace near the bottom, on coming in contact with the highly heated fuel in excess, forms carbon monoxide, which passes up through the descending charge of ore, fuel, and flux. At the temperature of the furnace, both the carbon of the fuel and the carbon monoxide thus formed act as reducing agents on the ore, removing the oxygen and leaving metallic iron, which, at the intense heat near the bottom of the furnace, melts and drops to the bottom, taking up some carbon from the fuel, and silicon, sulphur, phosphorus, and manganese from the ore, fuel, and flux. At the same time, the silica, alumina, lime, and magnesia of the ore, fuel, and flux unite, forming a fluid slag that, being lighter than iron, floats on the molten metal in the bottom. The iron and slag thus formed are drawn out at proper intervals through openings provided for them in the bottom of the furnace.

When the ascending gas reaches the top of the furnace it contains considerable carbon monoxide, which is very combustible. It passes through an opening near the top of the furnace and is led through the "down-comer" to a feedpipe. Part of it is then used to heat the stoves that are employed to heat the blast of air blown in near the bottom of the furnace by means of blowing engines. The part of the gas not used in the stoves is burned under the boilers that produce steam to run the blowing engines.

25. Definition of Steel.—While at first thought it seems to be a simple matter to define *steel* properly, the more familiar one becomes with the subject, the more perplexing is it to write a concise definition that will apply to the wide range of steels produced, or even to the greater part of them. Before the introduction of the modern methods of manufacture, the distinction between steel and wrought iron was sharp and well marked, and steel could be defined as any alloy of iron with carbon that would take a temper on quenching. Wrought iron does not sensibly harden on sudden cooling in water from a red heat. Modern methods of manufacture, however, have produced a metal that largely partakes of the nature of wrought iron, though it is made by the same processes that give a metal that hardens on quenching. For this reason such a classification as the above would now throw out the greater amount, or at least a very large tonnage, of the material classed and accepted by the metallurgical and commercial world as steel. The Bessemer converter and open-hearth furnace early showed an adaptability to produce a soft metal having great strength, elasticity, and ductility, capable of displacing wrought iron, and, for most purposes, far superior to it. It is impossible, therefore, to offer a thoroughly comprehensive definition of steel that is not easily assailable and whose inapplicability is shown from some standpoint.

Steel may be defined as a metal produced by the complete fusion of materials in a bath, the necessary properties being given, after conversion, by additions of carbon or carbon alloys. **Wrought iron** may be defined as a metal produced by the partial fusion, or bringing to a pasty condition, of materials on a hearth.

Blister, or *cementation*, *steel*, which is made by soaking bars of iron, at or above a red heat, in charcoal or carbon, would seem to be a notable exception; but as this is mainly an intermediate product for remelting in crucibles, and as its production is of little importance, it will be disregarded in this treatment of the subject.

The question of the proper classification of steels is one to which much attention has been given in the past, an international committee at one time having been selected from the metallurgical and technical societies of the principal steel-producing countries to adopt a universal classification. While much good came of their work, and strenuous efforts were made to adopt their classification, it was never generally used metallurgically nor commercially.

Many theories have been advanced as to what steel is. One that is held by many practical metallurgists is that the ideal steel is an alloy of pure iron and carbon only, all other elements being regarded as impurities. From this point of view, all grades of steel can be produced by simply varying the amount of carbon; but as impurities are necessarily present, all steel contains varying, and usually very small, amounts of sulphur, phosphorus, silicon, metallic oxides, and gases, which require other additions for their neutralization or elimination. Again, special alloys are required for giving steels characteristic qualities for particular purposes; such are the nickel, tungsten, chrome, manganese, and molybdenum steel.

26. Processes of Manufacture.—There are only three processes for the manufacture of steel: The *crucible*, the oldest of present methods; the *Bessemer;* and the *open-hearth.* The last two were developed almost simultaneously. The **Bessemer** was first perfected, and for the first 35 years, or up to about 1890, led the open-hearth, both as to tonnage produced and in the perfection of methods and appliances—both metallurgical and mechanical. While the Bessemer process still produces the greater tonnage, this is the only direction in which it can claim superiority over the open-hearth. In the order of their metallurgical and commercial importance today the processes rank: first, the open-hearth; second, the Bessemer; and third, the crucible.

While the **crucible process** is of the least consequence, it holds the most distinctive field metallurgically, and one from which the others seem unlikely to crowd it out. Given the same composition, it is well established that crucible steel is superior to either of the others, but owing to the much higher cost of production, its use is now restricted mainly to the making of high-grade tools, certain mining drills, parts of intricate machines, and, in general, where the first cost of the steel can be ignored.

The **open-hearth process** can claim as its own a larger field than the Bessemer. Open-hearth steel is now used for the better grades of plate steel, forgings, car axles, and structural steel. The basic open-hearth process is used where an extra-soft, pure steel is required, as in plates, sheets, rods, wires, etc. Bessemer steel is used for rails, nails, tin plate, light axles, in fact, for those articles where cheapness is desired. It is, however, being rapidly replaced by steel produced by the basic open-hearth process. The basic process, by cheaper production than was possible in the acid open-hearth, makes this a formidable rival of the Bessemer and seems practically sure to largely supplant it in the next few years. Owing to lower cost of production, the Bessemer process held undisputed sway for years in all lines using a large tonnage of steel. The open-hearth gradually demonstrated its superior fitness for special lines. While both the crucible and open-hearth processes have distinctive fields, held from the cheaper metal by the superior quality of their product, the Bessemer has no field the open-hearth cannot fill, and only by lower cost does it still produce the greater tonnage. Practically all rails are as yet made of Bessemer metal; also, most of the "billets" and "slabs" for merchant bar, tin plate, sheets, nails, light axles, and some ship and tank plates, etc.

27. Table II gives the average ultimate strengths of the various metals employed in building construction.

EXAMPLE 1.—What pull will be required to break a 2-inch diameter rod of wrought iron?

SOLUTION.—The area of the rod is equal to the area of a 2-in. circle, which is $2^2 \times .7854 = 3.14$ sq. in.; the ultimate tensile, or breaking, strength of wrought iron, according to Table II, is 48,000 lb. per sq. in. Therefore, the ultimate strength of the rod in question is $3.14 \times 48,000 = 150,720$ lb. Ans.

EXAMPLE 2.—What length of wrought-iron bar, if hung by one end, will break of its own weight, assuming the weight of 1 cubic inch of wrought iron to be .227 pound.

SOLUTION.—Assume any size of bar; say, 1⅛ in. in diameter. The area of this bar is .99 sq. in., which may, for convenience, be called 1 sq. in. Now, as there is just 1 cu. in. in each lineal inch in the rod, a length of 1 ft. weighs $.277 \times 12 = 3.32$ lb. The tensile strength of wrought iron being 48,000 lb. per sq. in., and 1 ft. of its length weighing 3.32 lb., the length of rod required is $\frac{48,000}{3.32} = 14,458$ ft. Ans.

TABLE II

AVERAGE ULTIMATE STRENGTHS OF MATERIALS IN POUNDS PER SQUARE INCH

	Compression	Tension	Elastic Limit	Shearing	Modulus of Rupture	Modulus of Elasticity
Brass, Bronze, and Copper						
Brass, cast	(30,000)	24,000	6,000	36,000	20,000	9,000,000
Brass wire, annealed		50,000				
Brass wire, unannealed		80,000	16,000			14,000,000
Bronze, aluminum	120,000	75,000				
Bronze, gun metal	(20,000)	32,000	10,000		53,000	10,000,000
Bronze, manganese	120,000	60,000	30,000			
Bronze, phosphor		50,000	24,000			14,000,000
Bronze, Tobin		66,000	40,000			4,500,000
Copper, bolts	30,000	30,000				
Copper, cast	(40,000)	24,000	6,000	30,000	22,000	10,000,000
Copper wire, annealed		36,000				15,000,000
Copper wire, unannealed		60,000	10,000			18,000,000
Cast and Wrought Iron						
Iron, cast	80,000	15,000	6,000	18,000	30,000	12,000,000
Iron wire, annealed		60,000				15,000,000
Iron wire, unannealed		80,000	27,000			25,000,000
Iron, wrought, shapes	46,000	48,000	26,000	40,000	44,000	27,000,000
Iron, wrought, rerolled bars	48,000	50,000	27,000	40,000	48,000	26,000,000
Cast and Structural Steel						
Steel, castings	70,000	70,000	40,000	60,000	70,000	30,000,000
Steel, structural, soft	56,000	56,000	30,000	48,000	54,000	29,000,000
Steel, structural, medium	64,000	64,000	33,000	50,000	60,000	29,000,000
Steel wire, annealed		80,000	40,000			29,000,000
Steel wire, unannealed		120,000	60,000			30,000,000
Steel wire, crucible		180,000	80,000			30,000,000
Steel wire, for suspension bridges		200,000	90,000			30,000,000
Steel wire, special tempered		300,000				

Note.—Compression values enclosed in parentheses indicate loads producing 10 per cent. reduction in original lengths.

FACTOR OF SAFETY

28. The **factor of safety,** or, as it is sometimes called, **the safety factor,** is the ratio of the breaking strength of the structure to the load that, under usual conditions, it is called on to carry. Suppose that the load required to break, dismember, or crush a structure is 5,000 pounds, and that the load it is called on to carry is 1,000 pounds, then the factor of safety may be obtained by dividing 5,000 pounds by 1,000 pounds, or $\frac{5000}{1000} = 5$, the factor of safety in this structure.

The factor of safety depends on the conditions, circumstances, or materials used; in other words, it is the *factor of ignorance.* When a piece of steel, wood, or cast iron is used in a building, the engineer does not know the exact strength of that particular piece of steel, wood, or cast iron. From his own experience, and that of others, he knows the approximate tensile strength of structural steel to be 60,000 pounds per square inch, and that it varies more or less from this value. In regard to timber, the uncertainty is much greater, because of knots, shakes, and interior rot, not always evident on the surface. Cast iron is even more unreliable, on account of almost indeterminable blowholes, flaws, and imperfections in the castings.

29. Deterioration.—Another factor to be considered is **deterioration** in the material, due to various causes. In metals there is corrosion on account of moisture and gases in the atmosphere, especially noticeable in the steel trusses over railroad sheds, where the sulphur fumes from the stacks of the locomotives unite with the moisture in the air, forming free sulphuric acid, which attacks the steel vigorously, and it demands constant painting to prevent its entire destruction. Wood is subject to decay from either dry or wet rot, caused by local conditions; it may, like iron and steel, be subjected to *fatigue*, produced by constant stress due to the load it is required to sustain. Cast iron does not deteriorate to any great extent, its corrosion not being as rapid, possibly, as that of steel or wrought iron. However, there are internal strains produced in cast iron by the irregular cooling of the metal in the mold. Under the slightest blow, castings will sometimes, owing to these internal stresses, snap and break in a number of places.

These reasons are, in truth, sufficiently cogent to require the factor of safety now adopted in all engineering work. Table III gives the factor of safety commonly employed by conservative constructors for various materials used in building work.

TABLE III

FACTOR OF SAFETY FOR DIFFERENT MATERIALS USED IN CONSTRUCTION

Materials	Factor of Safety
Structural steel and wrought iron	3 to 4
Wood	4 to 5
Cast iron	6 to 10
Stone	10 at least

In this table, the factor of safety generally used for structural steel is 3 to 4, which simply means that the steel structure should not break until it bears a load three or four times greater than it is designed to carry.

EXAMPLE.—If the breaking strength of a cast-iron column is 200,000 pounds, what safe load will the column sustain if a factor of safety of 6 is used?

SOLUTION.— $200,000 \div 6 = 33,333$ lb. Ans.

EXAMPLES FOR PRACTICE

1. Provided a factor of safety of 3 is adopted, what will be the safe working strength of a 2-inch diameter tension rod of medium structural steel? Ans. 67,020 lb.

2. The pull on a 2-inch eyebolt, passing through a piece of Southern yellow pine, is 40,000 pounds. What should be the diameter of the washer, if the bolt hole through it is 2¼ inches in diameter? Ans. 12¼ in.

3. The bottom of the notch in a spruce timber 10 inches wide and 12 inches deep, forming the tie-member in a roof truss, is 18 inches from the end. What resistance will the end of the tie offer to the thrust of the rafter? Ans. 72,000 lb.

4. A short block of Northern yellow pine, 10 inches by 10 inches in section, standing on end, supports 50,000 pounds. What is its factor of safety? Ans. 12

BEAMS AND GIRDERS

(PART 1)

ELEMENTS OF BEAMS

DEFINITIONS

1. Any bar resting on supports and liable to be subjected to transverse stresses is called a **beam.**

A large beam that carries smaller, or secondary beams, is a **girder.**

A beam resting on two supports very near its end is a **simple beam.**

A beam resting on one support at its middle, or having one end fixed (as in a wall) and the other end free, is a **cantilever.**

A beam that has both ends firmly secured is a **fixed beam.**

A beam that rests on more than two supports is a **continuous beam.**

The distance between the supports of a simple beam is its **span.**

MOMENTS OF FORCES

2. In order to calculate the stresses produced in beams under different conditions, and the forces at the points of support, it is necessary to understand the theory of the *moments of forces.*

In Fig. 1, W is a weight that acts downwards with a force of 10,000 pounds. If some fixed point a, not in the line along which the weight W acts, is connected with the line of action of W by a rigid arm, so that W pulls on one end of this arm, while the other end is firmly held at a, experience teaches that the pull of W tends to turn, or rotate, the arm around the point a. The tendency of a force to produce rotation around a given point is called the **moment of the force** with respect to that point.

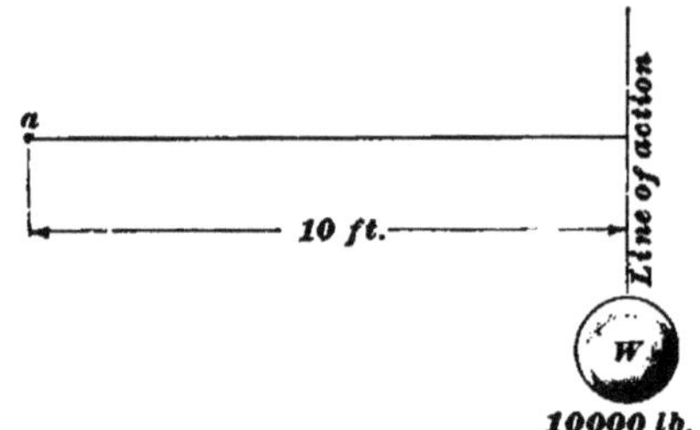

Fig. 1

The point a that is taken as the center around which there is a tendency to rotate, is called the **center of moments.**

The perpendicular distance from the center of moments to the line along which the force acts, is the **lever arm** of the force, also called the **leverage** of the force.

3. The measure of the moment of a force, that is, of the tendency of the force to produce rotation around a given center, is the product of the magnitude of the force multiplied

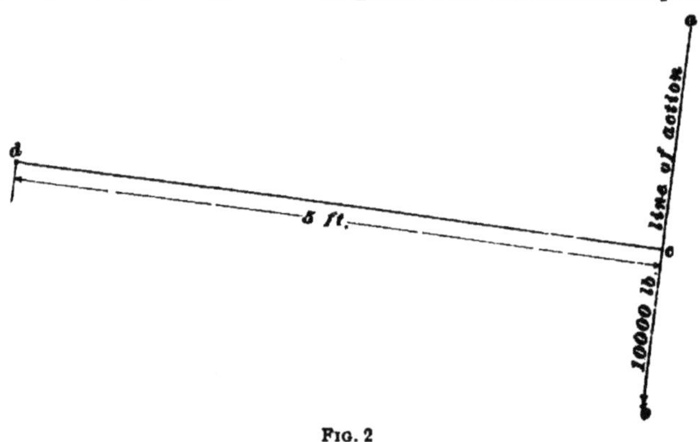

Fig. 2

by the length of its lever arm. In Fig. 1, for example, the magnitude of the force is equal to the weight of 10,000 pounds, and the length of the lever arm is 10 feet; therefore, the moment of the force W, with respect to the center a, is

10,000 × 10 = 100,000, which, since the factors by which it is produced represent feet and pounds, is called **foot-pounds.** Thus, the moment of a force of 10,000 pounds, whose lever arm is 10 feet long, is 10,000 × 10 = 100,000 foot-pounds.

EXAMPLE.—What is the moment of the force of 10,000 pounds whose line of action is *a b*, Fig. 2, the center of moments being at *d*?

SOLUTION.—The perpendicular distance *c d* from the line of action of the force to the center of moments being 5 ft., and the magnitude of the force 10,000 lb., the moment is 10,000 × 5 = 50,000 ft.-lb. Ans.

4. In Fig. 3, the line of action of the force of 10,000 pounds passes directly through the point *d*; consequently, the

d 10000 lb.

FIG. 3

perpendicular distance from the line of action to the point *d* is *zero* and there is no tendency to rotate around that point; therefore, there is no motion.

5. The moment of a force may be expressed in *inch-pounds*, *foot-pounds*, or *foot-tons*, depending on the unit of measurement used to designate the magnitude of the force and the length of its lever arm. For instance, if the magnitude of a force is measured in pounds, and the lever arm through which it acts, in inches, the moment will be in *inch-pounds;* if a force of 10 tons acts through a lever arm of 20 feet, the moment of the force is 10 × 20 = 200 *foot-tons.*

EXAMPLE.—What is the moment, in inch-pounds, of a force of 8,000 pounds, if the length of the lever arm is 13 feet?

SOLUTION.—Since the moment is to be in inch-pounds, the length of the lever arm must be in inches. 13 ft. = 13 × 12 = 156 in., and the moment is 8,000 × 156 = 1,248,000 in.-lb. Ans.

6. Equilibrium of Moments.—When a body is at rest, the forces that act on it must balance one another; the forces are then said to be in **equilibrium.** That there may be perfect balance among the forces, it is necessary that there be not only no unbalanced force tending to move the body along some given line, but that there be, also, no unbalanced moment, the effect of which would turn the body about some point.

Fig. 4 shows a beam or lever, resting on the support *c*, on the right-hand end of which a force *b* of 5 pounds acts downwards tending to turn it around the point of support *c* in the direction traveled by the hands of a clock, that is, to produce **right-hand** rotation. The measure of this tendency is $5 \times 10 = 50$ foot-pounds. Another force *a* acts downwards on the left-hand end of the lever, tending to produce **left-hand** rotation, or to turn the lever in the direction opposite to that traveled by the hands of a clock. Since the force *a* is 10 pounds, and it acts with a lever arm of 5 feet, its moment is $10 \times 5 = 50$ foot-pounds, the same as the moment of the force *b*. There are thus two equal moments, one tending to turn the lever to the right and the other to the left;

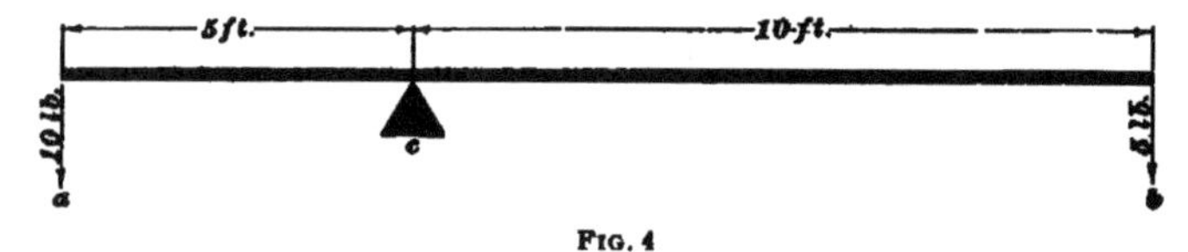

Fig. 4

as a result, the effect of one is neutralized by the effect of the other, and the second condition of equilibrium is fulfilled; that is, there is *equilibrium of moments*.

7. Positive and Negative Moments.—It is customary to distinguish between the directions in which there is a tendency to produce rotation by the use of the signs + and −. Thus, if a force tends to produce right-hand rotation, its moment may be called **positive** and be given the plus sign, while a force that tends to produce rotation in the opposite direction is called **negative**, and its moment is given the minus sign. That there may be equilibrium of moments, the above considerations show that the difference between the sum of the positive moments and the sum of the negative moments must be zero; this difference is called the **algebraic sum** of the moments. The following principle is then evolved: *In order that there may be equilibrium, the algebraic sum of the moments of all the forces acting on a body must be zero.*

8. Resultant and Reactionary Moments.—In Fig. 5 is shown a lever composed of two arms at right angles to each other and free to turn about the center *c*. A force *a*, whose moment is $10 \times 5 = 50$ foot-pounds, acts on the horizontal arm in such a manner that it tends to produce left-hand rotation. Another force *b*, whose moment with respect to the center *c* is $12 \times 3 = 36$ foot-pounds, tends to produce right-hand rotation. These two forces, therefore, will not produce equilibrium, but their combined efforts will be equal to the algebraic sum of the moments. This sum is the **resultant moment** of the two forces *a* and *b*. Thus, $-50 + 36 = -14$ foot-pounds, which is the resultant moment of the forces considered, because it produces the same effect as the two moments combined.

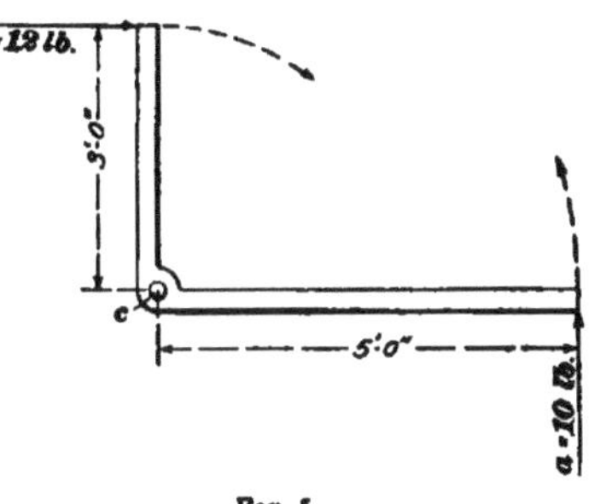

Fig. 5

In order to secure equilibrium, there must be another force acting in such a manner as to overcome the algebraic sum of these two moments, and consequently its moment must be equal and opposite in effect to the resultant moment. The name given to the moment of this force is **reactionary moment.**

If the length of the lever arm of the force that acts to produce the reactionary moment is known, the magnitude of the force may be readily found. Thus, in the present case, the resultant moment is -14 foot-pounds; let it be required to find the force to produce equilibrium, when acting with a lever arm 7 feet long. Since the moment is the product of the force multiplied by its lever arm, it follows that the required force may be found by dividing the given moment by the length of the lever arm; consequently, the required force is $14 \div 7 = 2$ pounds.

If, instead of the two forces just considered, we have a body acted on by a number of forces whose moments about a given center are known, the reactionary moment of these forces, that is, the moment of the force required to produce equilibrium, is the algebraic sum of the moments of the given forces with the sign changed; and, further, if the length of the lever arm of the reactionary moment is known, the magnitude of the required force can be found by dividing the moment by the length of the lever arm.

9. The above principles may be expressed as follows:

Rule.—*To find the force required to produce equilibrium of moments, when the moments of any number of given forces and the lever arm of the required force are given, divide the algebraic sum of the given moments by the length of the given lever arm. If the algebraic sum is positive, the required force must tend to produce left-hand rotation; if negative, the force must tend to produce right-hand rotation.*

EXAMPLE.—In Fig. 6, a system of forces, shown by the arrows, acts in various directions and at various distances from the center O. The force F' is 25 pounds and its lever arm Op' is 8 feet, F'' is 16 pounds with a lever arm Op'' of 12 feet, F''' is 40 pounds with a lever arm Op''' of 6 feet, and the force F^{v} is 100 pounds, acting directly through the center O. If the distance Op^{iv} is 12 feet, what must be the magnitude of the force F^{iv} in order to produce equilibrium of moments?

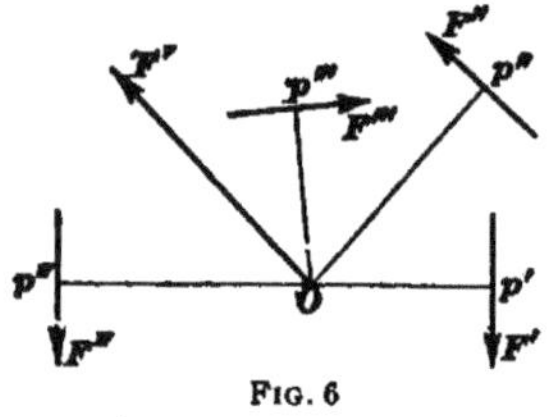

FIG. 6

SOLUTION.—As shown by the arrows, the forces tending to produce right-hand rotation are F' and F''', and their moments, called positive, are, respectively, $25 \times 8 = +200$ ft.-lb., and $40 \times 6 = +240$ ft.-lb. The lever arm of the force F^{v} is zero; consequently, it has no moment with respect to the center O. The force F'' tends to produce left-hand rotation and its moment is $16 \times 12 = -192$ ft.-lb. The algebraic sum of the moments of the given forces is $+ 200 + 240 - 192 = +248$ ft.-lb.; therefore, according to the rule, the force F^{iv} must be $248 \div 12 = 20\frac{2}{3}$ lb., which, since the algebraic sum of the given moments is positive, must tend to produce left-hand rotation, as shown by the arrow. Ans.

10. The principles involved in the theory of moments are among the most simple in mechanics, and, at the same time, of the greatest practical importance in the solution of problems relating to the strength of beams, girders, and trusses.

EXAMPLE.—In Fig. 7, the lower tie-member in the roof truss has been raised to secure a vaulted ceiling effect in the upper story of the building that the truss covers. The weight transmitted through this member to the pier wall is 30,000 pounds; there is, consequently, an equal upward force due to the reaction of the wall. This force of 30,000 pounds tends to break the truss by producing rotation about the point *b*. What is its moment around the point *b*?

SOLUTION.—Since the perpendicular distance from the line of action of the force is 3 ft., the moment of the force *a* around the point *b* is $30,000 \times 3 = 90,000$ ft.-lb. Ans.

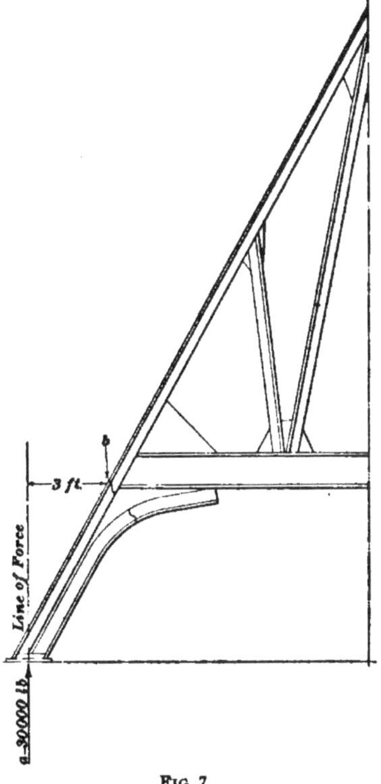

FIG. 7

THE LEVER

11. A **lever** is a bar capable of being turned about a pin, pivot, or point, as in Figs. 8, 9, and 10.

The object W to be lifted is called the **weight**; the force P used is called the **power**; and the point or pivot F is called the **fulcrum**.

That part of the lever between the weight and the fulcrum, or Fb, is called the **weight arm**, and the part between the power and the fulcrum, or Fc, is called the **power arm**.

Take the fulcrum, or point F, as the center of moments; then, in order that the lever shall be in equilibrium, the moment of P about F, or $P \times Fc$, must equal the moment of W about F, or $W \times Fb$. That is, $P \times Fc = W \times Fb$, or, in other words, the power multiplied by the power arm equals the weight multiplied by the weight arm.

If F be taken as the center of a circle, and arcs be described through b and c, it will be seen that, if the weight arm is moved through a certain angle, the power arm will

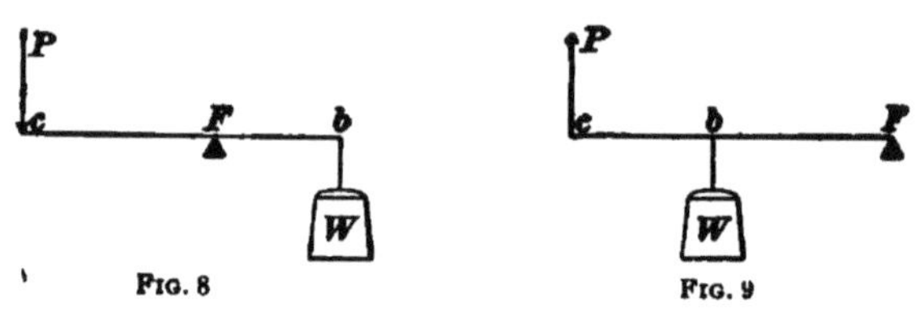

FIG. 8 FIG. 9

move through the same angle; also, that the vertical distance that W moves will be proportional to the vertical distance that P moves. From this it is seen that the power arm is proportional to the distance through which the power moves, and the weight arm is proportional to the distance through which the weight moves.

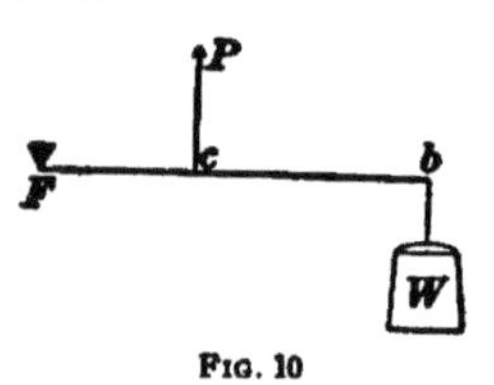

FIG. 10

Hence, instead of writing $P \times Fc = W \times Fb$, it might have been written $P \times$ distance through which P moves $= W \times$ distance through which W moves. This is the general law of all machines, and can be applied to any mechanism, from the simple lever to the most complicated arrangement. Stated in the form of a rule, it is as follows:

Rule.—*The power multiplied by the distance through which it moves equals the weight multiplied by the distance through which it moves.*

EXAMPLE 1.—If the weight arm of a lever is 6 inches long and the power arm is 4 feet long, how great a weight can be raised by a force of 20 pounds at the end of the power arm?

SOLUTION.— 4 ft. = 48 in. Hence, $20 \times 48 = W \times 6$, or $W = 160$ lb. Ans.

EXAMPLE 2.—(*a*) What is the ratio between the power and the weight in the last example? (*b*) In the last example, if P moves 24 inches, how far does W move? (*c*) What is the ratio between the two distances?

SOLUTION.—(*a*) 20 : 160 = 1 : 8; that is, the weight moved is eight times the power.

(*b*) $20 \times 24 = 160 \times x$. $x = \frac{480}{160} = 3$ in., the distance that *W* moves. Ans.

(*c*) 3 : 24 = 1 : 8, or the ratio is 1 : 8. Ans.

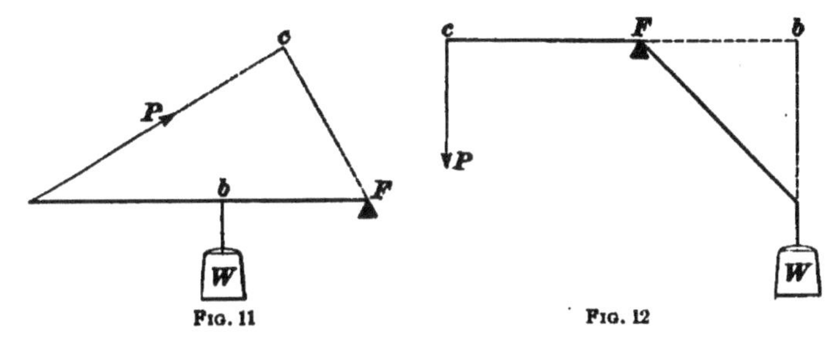

FIG. 11 FIG. 12

The law that governs the straight lever also governs the bent lever; but care must be taken to determine the true length of the lever arm, which is in every case the

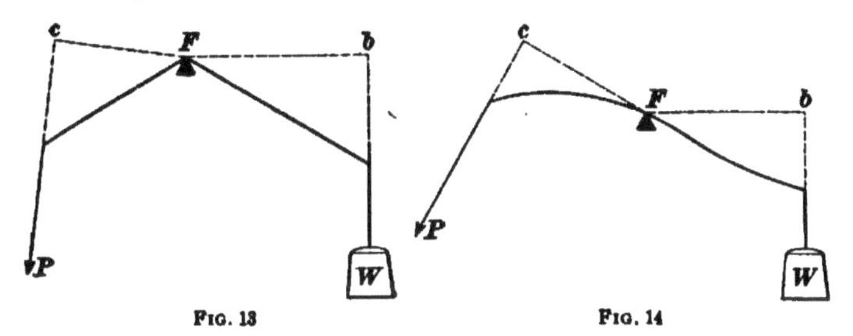

FIG. 13 FIG. 14

perpendicular distance from the fulcrum to the line of direction of the weight or power.

Thus, in Figs. 11, 12, 13, and 14, *Fc* in each case represents the power arm, and *Fb* the weight arm.

EXAMPLES FOR PRACTICE

1. A lever arm has a length of 10 feet; the load acting on the end of the lever is 6,000 pounds. What is the moment of this load, in inch-pounds? Ans. 720,000

2. A piece of timber 20 feet long is balanced at a point 8 feet from one end, the load at this end being 9,000 pounds. What is the load at the other end? Ans. 6,000 lb.

3. The one support of a beam 20 feet long is 8 feet from the left-hand end; at this end is a load of 25 pounds; at the right of the support, 3 feet distant, is a load of 5 pounds; and at 7 feet to the right of the support is a load of 10 pounds. What load is required, and at which end should it be placed, to produce balance, or equilibrium, in the beam? Ans. 9.58 lb. at right-hand end

4. A steel I beam that extends 6 feet outside of the center of a building wall, and 3 feet inside, is required to support a load on the outside end of 4,000 pounds. What load on the inner end will keep the beam from tilting? Ans. 8,000 lb.

REACTIONS

12. Since one condition of equilibrium requires that the sum of all the forces acting on a body in one direction must be balanced by an equal set of forces acting in the opposite direction, it follows that, in order that any body may be kept from falling, there must be an upward pressure, or thrust, against it, just equal to the downward pressure due to its weight; this upward thrust is called a **reaction.**

In accordance with this principle, it is evident that the simple beam shown in Fig. 15 is supported by the sum

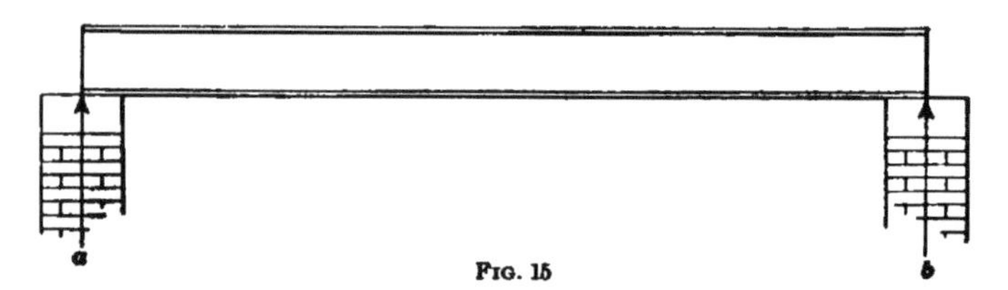

Fig. 15

of the upward pressures exerted on it by the two brick piers on which it rests; also, that this sum is equal to the weight of the beam plus the weight of any load it may carry. This is expressed by the statement: The sum of the reactions at the supports of any beam is equal to the sum of the loads.

13. Relation Between the Reactions.—If the load on a simple beam is either uniformly distributed over the entire length of the beam, applied at the center of the span, or symmetrically placed on each side of the center of the span, the reaction at each support is equal to one-half of the total load. When, however, the loads are not symmetrically

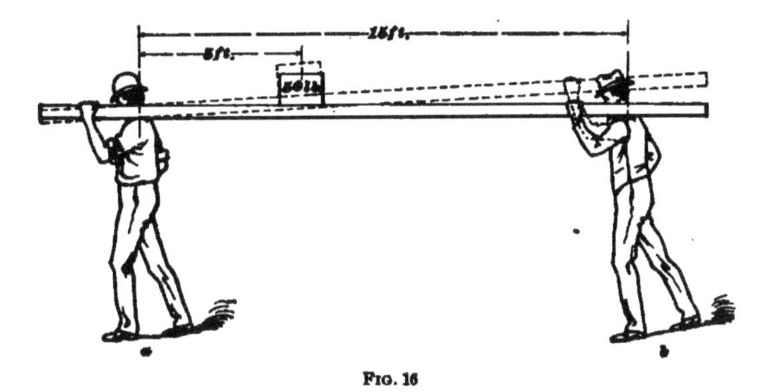

FIG. 16

placed, the reactions are unequal and must be determined before the first step toward obtaining the strength of the beam can be taken. The reactions at the points of support

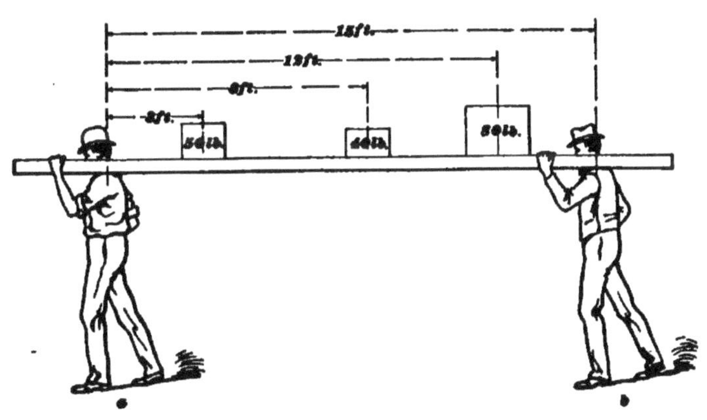

FIG. 17

of a beam carrying a number of loads irregularly placed, are determined by applying the principle of moments, as shown in the following illustrative examples:

14. Two men, *a* and *b*, 15 feet apart, carry a 50-pound weight between them on a plank, as shown in Fig. 16. What part of the load does each man carry?

If the load had been placed midway between them, it is quite evident that each man would have half the weight of the plank and load to support. But, since the load is moved until within 5 feet of *a*, he must support a greater proportion of the load than *b*. If *b* raises his end of the plank, as shown in dotted lines, it is evident that *a* simply acts as a hinge while *b* raises the weight with a lever arm 15 feet long. The weight of 50 pounds acts down with a leverage of 5 feet; its moment about *a* as a center is, therefore, $50 \times 5 = 250$

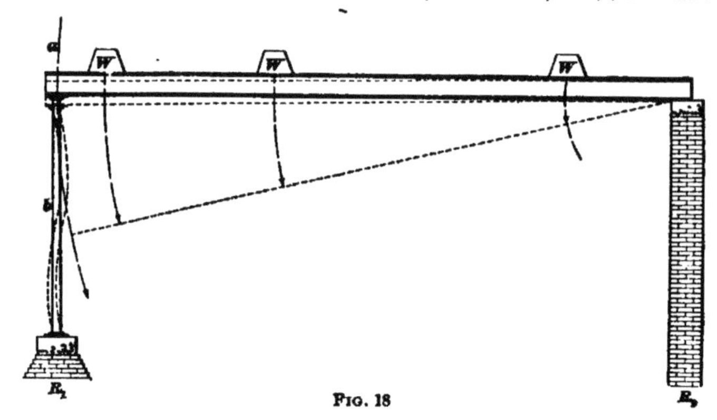

FIG. 18

foot-pounds. That there may be equilibrium of moments, it is evident that the man *b* must exert an upward force whose moment with a lever arm of 15 feet equals 250 foot-pounds; that is, he must exert a force of $250 \div 15 = 16\frac{2}{3}$ pounds to support his share of the weight. Since the sum of the reactions must equal the sum of the loads, it follows that if *b* supports $16\frac{2}{3}$ pounds, *a* must support the difference between the load of 50 pounds and $16\frac{2}{3}$ pounds, or $33\frac{1}{3}$ pounds.

15. Fig. 17 shows the men *a* and *b* supporting three loads of 50, 40, and 80 pounds, respectively. It is desired to estimate the force that each must exert to sustain the weights, leaving the weight of the plank out of the question. Assuming the center of moments to be at *a*, find the resultant moment of all the weights about this point, as follows:

$$
\begin{aligned}
50 \times 3 &= 150 \text{ foot-pounds} \\
40 \times 8 &= 320 \text{ foot-pounds} \\
80 \times 12 &= 960 \text{ foot-pounds} \\
\text{Total,} \quad &\ 1430 \text{ foot-pounds}
\end{aligned}
$$

This is the moment of all the loads on the beam about the point *a* as a center. Hence, the force that *b* must exert, in order to produce equilibrium, is $1,430 \div 15 = 95\frac{1}{3}$ pounds. The part of the load that *a* supports is the difference between the total load, $50 + 40 + 80 = 170$ pounds, and the part of the load supported by *b*; that is, $170 - 95\frac{1}{3} = 74\frac{2}{3}$ pounds.

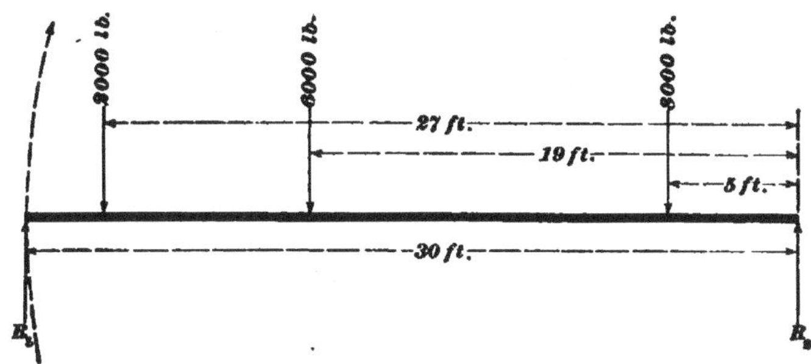

FIG. 19

16. Take a more practical example. In Fig. 18, let it be required to find the reactions R_1 and R_2. (In all the subjoined problems, R_1 and R_2 represent the reactions.) The center of moments may be taken at either R_1 or R_2. Taking R_2 as the center in this case, construct a diagram as in Fig. 19. The three loads are forces acting in a downward direction; the sum of their moments, with respect to the assumed center, may be computed as follows:

$$
\begin{aligned}
8,000 \times 5 &= 40000 \text{ foot-pounds} \\
6,000 \times 19 &= 114000 \text{ foot-pounds} \\
2,000 \times 27 &= 54000 \text{ foot-pounds} \\
\text{Total,} \quad &\ 208000 \text{ foot-pounds}
\end{aligned}
$$

The magnitude of the reaction R_1 acting in an upward direction with a lever arm of 30 feet is therefore 208,000 ÷ 30 = 6,933⅓ pounds. The sum of all the loads is 2,000 + 6,000 + 8,000 = 16,000 pounds. Then, 16,000 − 6,933⅓ = 9,066⅔ pounds, the reaction at R_2.

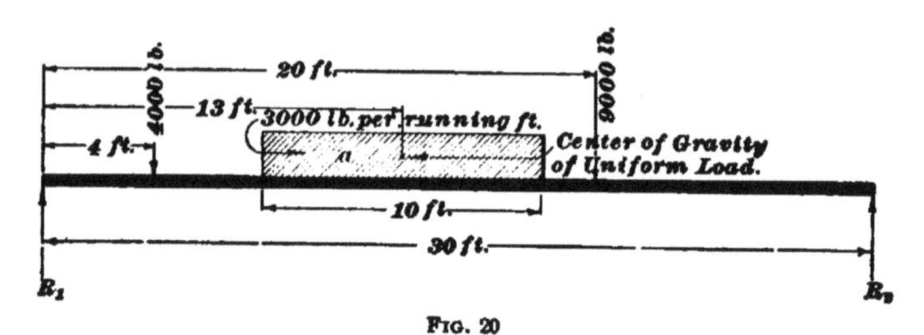

FIG. 20

EXAMPLE 1.—What is the reaction at R_2 in Fig. 20?

SOLUTION.—In computing the moment due to a uniform or evenly distributed load, as at *a*, the lever arm is always considered as the distance from the center of moments to the center of gravity of the load. The amount of the uniform load *a* is 3,000 × 10 = 30,000 lb., and the

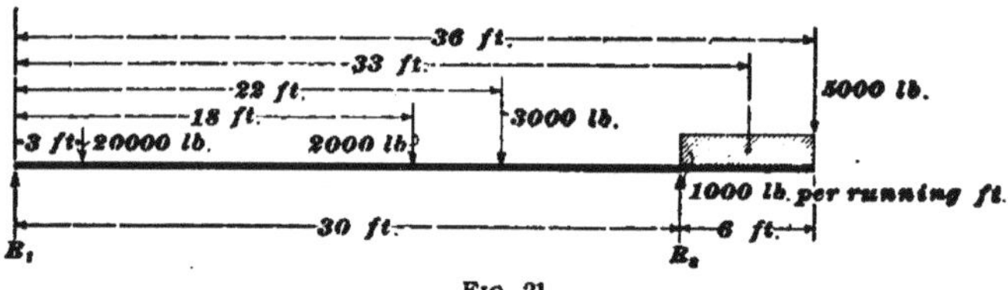

FIG. 21

distance of its center of gravity from R_1 is 13 ft. The moments of the loads on this beam may then be seen from the following:

$$\begin{aligned} 30{,}000 \times 13 &= 390000 \text{ ft.-lb.} \\ 4{,}000 \times 4 &= 16000 \text{ ft.-lb.} \\ 9{,}000 \times 20 &= 180000 \text{ ft.-lb.} \\ \text{Total,} &\quad 586000 \text{ ft.-lb.} \end{aligned}$$

This is the sum of the moments of all the loads about R_1 as a center. The leverage of the reaction R_2 is 30 ft. Hence, 586,000 ÷ 30 = 19,533⅓ lb., the reaction at R_2. Ans.

EXAMPLE 2.—A beam is loaded as shown in Fig. 21. What is the amount of each of the reactions R_1 and R_2?

SOLUTION.—Considering R_1 as the center of moments, the moments of the loads about it are:

$$\begin{aligned} 20{,}000 \times 3 &= 60000 \text{ ft.-lb.} \\ 2{,}000 \times 18 &= 36000 \text{ ft.-lb.} \\ 3{,}000 \times 22 &= 66000 \text{ ft.-lb.} \\ 5{,}000 \times 36 &= 180000 \text{ ft.-lb.} \\ 1{,}000 \times 6 \times 33 &= 198000 \text{ ft.-lb.} \\ \text{Total,} \quad & 540000 \text{ ft.-lb.} \end{aligned}$$

This, divided by 30, the length of the lever arm of the reaction R_2, equals 18,000 lb., the reaction at R_2. The sum of the loads is 20,000 + 2,000 + 3,000 + 5,000 + 6,000 = 36,000 lb.; and 36,000 − 18,000 = 18,000 lb., the amount of the other reaction, or R_1. Ans.

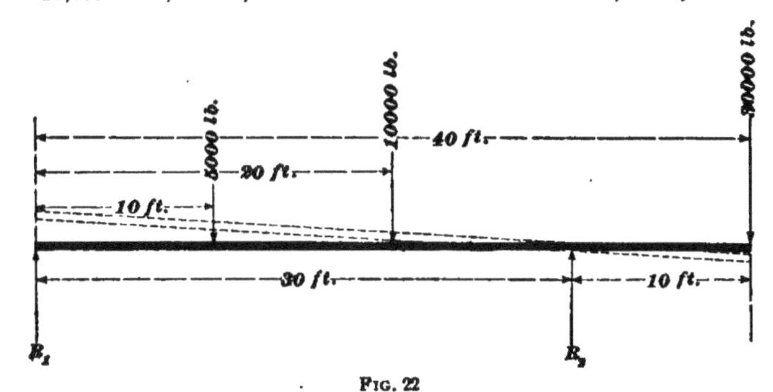

FIG. 22

EXAMPLE 3.—Compute the reactions at the supports R_1 and R_2 in a beam loaded as shown in Fig. 22.

SOLUTION.—Letting R_1 represent the center of moments, the moments of the loads are:

$$\begin{aligned} 5{,}000 \times 10 &= 50000 \text{ ft.-lb.} \\ 10{,}000 \times 20 &= 200000 \text{ ft.-lb.} \\ 30{,}000 \times 40 &= 1200000 \text{ ft.-lb.} \\ \text{Total,} \quad & 1450000 \text{ ft.-lb.} \end{aligned}$$

Now, 1,450,000 ÷ 30, the distance between the supports, equals 48,333⅓ lb., the required reaction at R_2. The sum of the loads is 5,000 + 10,000 + 30,000 = 45,000 lb.; therefore, the reaction R_2 is greater than the sum of the loads. This shows that the force at R_1 must act in a downward direction in order that the sum of the downward forces may equal the upward force at R_2. Since this is opposite to the usual direction, the reaction at R_1 is called negative or minus. In other words, instead of an upward reaction at R_1, there must be a downward force at this point, or the beam will, as shown by the dotted lines, rotate around the support R_2. The magnitude of this downward force is the difference between the upward reaction at R_2 and the sum of the downward pressures due to the loads; that is, $48,333\frac{1}{3} - 45,000 = 3,333\frac{1}{3}$ lb. Compute the reaction at R_1 by taking the center of moments at R_2 and applying the rule in Art. **9** to find the magnitude and direction of action of the force at R_1 whose moment is the resultant of the moments of the loads on the beam. The load of 30,000 lb. tends to produce right-hand rotation around the center R_2; hence, its moment, $30,000 \times 10 = 300,000$ ft.-lb., is positive. The 10,000-lb. load is 10 ft. to the left of R_2 and its tendency is to produce left-hand rotation about R_2; consequently, its moment is negative and equal to $10,000 \times 10 = 100,000$ ft.-lb. In a similar manner, the moment of the 5,000-lb. load is found to be negative and equal to $5,000 \times 20 = 100,000$ ft.-lb. These results may be collected thus:

Positive moment:

$30,000 \times 10 = \qquad\qquad + 300000$ ft.-lb.

Negative moments:

$10,000 \times 10 = -100000$ ft.-lb.

$5,000 \times 20 = -100000$ ft.-lb.

$\qquad\qquad - 200000$ ft.-lb.

Difference, $+100000$ ft.-lb.

the resultant of the moments of the three loads. Since the positive moment is greater than the sum of the negative moments, the force at R_1 must tend to produce left-hand rotation; that is, it must act downwards; its lever arm being 30 ft. long, its magnitude must be $100,000 \div 30 = 3,333\frac{1}{3}$ lb., the same result as was obtained before.

EXAMPLES FOR PRACTICE

1. The span of a simple beam is 25 feet; at distances of 9 feet, 16 feet, and 18 feet from the left-hand end are placed concentrated loads of 8,000, 4,000, and 16,000 pounds, respectively. What is the amount of the left reaction? Ans. 11,040 lb.

2. The two reactions supporting a beam are 2,500 and 3,000 pounds; what is the amount of a single concentrated load necessary to produce these reactions? Ans. 5,500 lb.

3. A 30-foot beam overhangs the right-hand support 6 feet; on this end is a weight of 6,000 pounds; 10 feet, 12 feet, and 18 feet from the left-hand support are loads of 8,000, 6,200, and 7,800 pounds, respectively. What is the amount of the right-hand reaction? Ans. $19,783\frac{1}{3}$ lb.

4. If for a distance of 10 feet from the left-hand end of a beam there is distributed a load of 1,000 pounds per running foot, and at the center of the beam is located the concentrated load of 16,500 pounds, what is the amount of the left-hand reaction, provided that the beam is supported at both ends and is 30 feet long? Ans. $16,583\frac{1}{3}$ lb.

STRESSES IN BEAMS

17. It has been seen that a beam is a body acted on by various external forces so related as to be in a condition of equilibrium; so far, however, the effect of these forces on the beam itself has not been considered.

In a body subjected to a direct pull or thrust, as a rope or a column, the external forces are directly opposed to each other, and the resultant stresses in all sections are of the same kind, tension, or compression. In a beam, however, the external forces, while they generally act in parallel lines, are not directly opposed to each other, and it is the function of the beam to transfer these forces from one line of action to another. Take, for example, the case of a weight suspended from a pin driven in a wall, as shown in Fig. 23. The downward force of 20 pounds due to the action of the weight is balanced by the upward pressure, or reaction, of the pin on the rope; the rope is thus subjected to the action of two directly opposing forces, the result being a tensile stress that is the same for each section of the rope between the weight and the pin. The pin acts as a cantilever beam that transfers the downward pressure, due to the pull of the rope, horizontally to the wall, where it is balanced by an equal upward pressure, or reaction. The pin is thus subjected to two opposing forces that, however, act in different lines; these forces produce a set of opposing forces, or stresses, in the pin itself, which are different in kind for different parts of the pin, and vary in magnitude for each section between the rope and the wall.

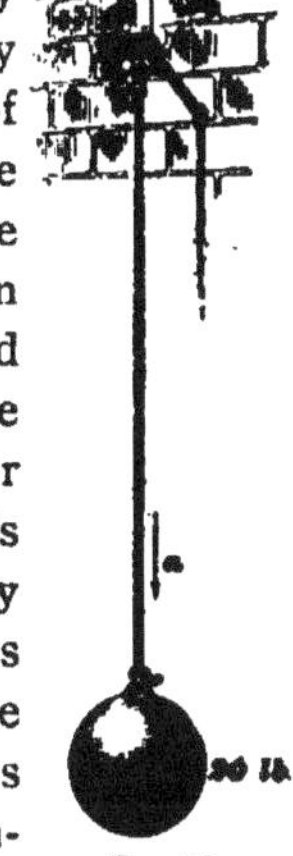

FIG. 23

SHEAR

18. An inspection of Fig. 23 shows that if a vertical plane is passed through any point between the rope and wall, the part of the pin between this plane and the rope will be acted on by a downward force, due to the pull of the rope, while the other part is subjected to an equal upward force due to the reaction of the wall; the action of these two forces tends to slide the two parts of the pin past each other, along the section formed by the cutting plane. The pin is thus subjected to a stress which, from its similarity to a shearing action, is called **shear.**

19. Shear in a Simple Beam.—By observation of Fig. 24 (*a*), which shows a simple beam having a concentrated load at the center, the only apparent stress in the beam is that of bending. However, if the fibers of the beam were not continuous, and instead of being one piece it were composed of short blocks, the tendency would be for the blocks to slip past each other, as shown at (*b*). This tendency would exist in the entire beam, producing vertical shear, which must be considered in calculating its strength. If the beam were laminated, the effect would be as shown at (*c*); that is, one layer would slide on the next one as the beam deflects. In a solid beam there is the same tendency to shear the beam longitudinally and this horizontal shear is equal to the vertical shear.

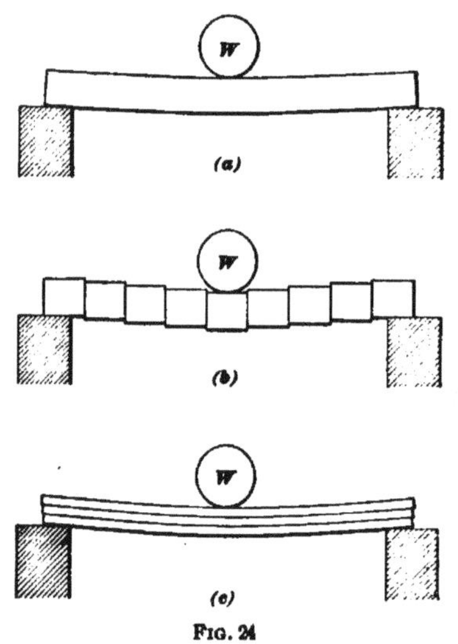

Fig. 24

Consider now the simple beam shown in Fig. 25. Since the loads are symmetrically applied, each reaction is equal to 40 pounds, one-half of the total load on the beam. Beginning at the left reaction R_1, there is an upward force of 40 pounds acting on the beam; since the forces are in equilibrium, this upward force is balanced by an equal downward force, which is the vertical resultant of the loads and the reaction R_2. Considering, therefore, any section of the beam between R_1 and the point of application of the load *n*, it is seen that the part of the beam at the left of this section is subjected to an upward thrust of 40 pounds, while the part at the right is subjected to an equal downward thrust; the result is a shearing stress on this section, whose magnitude is equal to the reaction R_1.

When the point of application of *n* is reached, the effect of the upward force R_1 is partly balanced by the downward force of 10 pounds due to the load *n*; considering, therefore, any section of the beam *a b* between the points of application of the loads *n* and *m*, it is seen that the part of the beam at the left is acted on by the vertical resultant of the reaction R_1 and

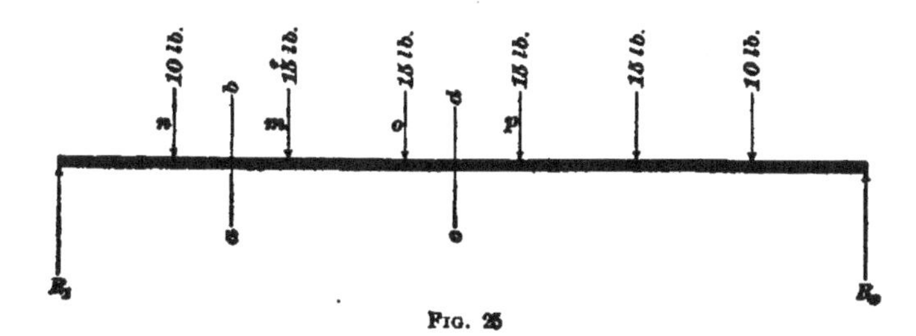

FIG. 25

the load *n*; that is, by an upward force of 40 − 10 = 30 pounds, while the part at the right is acted on by an equal downward force, which is the vertical resultant of the remaining loads and the reaction R_2. Any section between the points of application of *n* and *m* is therefore subject to a shearing stress equal to the difference between the reaction R_1 and the load *n*; that is, to 40 − 10 = 30 pounds. In the same way, it follows that the shearing stress for any section between *m* and *o* is 40 − (10 + 15) = 15 pounds. For any section *c d* between the points of application of *o* and *p*, the shearing stress is 40 − (10 + 15 + 15) = 0; in other words, on each side of this section the downward forces and the reactions are equal, and their resultant is zero; it is, therefore, a section in which there is no shear.

For convenience, it is customary to call the reactions, or forces, acting in an upward direction, positive, and the loads, or downward forces, negative; since the difference between the sums of the positive and negative numbers representing a given set of values is called their algebraic sum, it follows that *the shear for any section of a beam is equal to the algebraic sum of either reaction and the loads between this reaction and the given section.*

In nearly all cases the external forces—loads and reactions—act on a beam along vertical lines; the shearing stress just considered is called the vertical shear, because it is the resultant of these forces along a section formed by an imaginary vertical cutting plane.

20. Maximum Shear.—From what has been said, it is evident that the shear in any simple beam is always greatest between the reactions and the nearest loads, and that in any case the **maximum shear** is equal to the greater reaction.

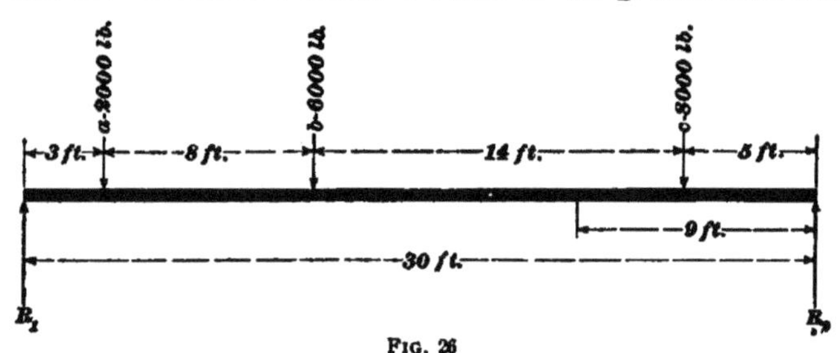

Fig. 26

21. Positive and Negative Shear.—If a section of the beam near the left reaction is taken and the forces acting on the part of the beam at the left are considered, it is seen that their resultant is positive; the shear at this section is therefore called **positive shear.** If, however, a section near the right reaction is taken, the resultant of the forces at the left is found to be negative, and in consequence the shear is called **negative.** It is also evident that there is a section between the two, where the resultant of the forces changes from positive to negative; at such a section the shear is said to **change sign.**

Example 1.—(*a*) What is the maximum shear on the beam shown in Fig. 26? (*b*) What is the shear at a point 9 feet from the right support? (*c*) What is the shear at a point 18 feet from the right support?

Solution.—(*a*) First estimate the reactions as follows: Taking the center of moments at the left support, the moments of the loads are:

$$
\begin{aligned}
2{,}000 \times 3 &= 6000 \text{ ft.-lb.}\\
6{,}000 \times 11 &= 66000 \text{ ft.-lb.}\\
8{,}000 \times 25 &= \underline{200000} \text{ ft.-lb.}\\
\text{Total,}\quad & \ 272000 \text{ ft.-lb.}
\end{aligned}
$$

$272{,}000 \div 30 = 9{,}066\frac{2}{3}$ lb., the reaction at R_2. The sum of the loads equals $2{,}000 + 6{,}000 + 8{,}000 = 16{,}000$ lb.; $16{,}000 - 9{,}066\frac{2}{3} = 6{,}933\frac{1}{3}$ lb., the reaction at R_1. The maximum shear is therefore $9{,}066\frac{2}{3}$ lb. Ans.

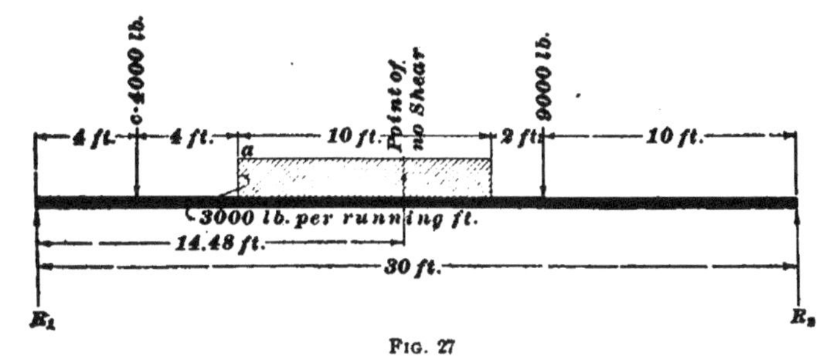

Fig. 27

(*b*) As the reaction R_2 at the right support is equal to $9{,}066\frac{2}{3}$ lb., and as there is only the one load *c* of 8,000 lb., between R_2 and a point 9 ft. away, the shear at this point must equal $9{,}066\frac{2}{3} - 8{,}000 = 1{,}066\frac{2}{3}$ lb. Ans.

(*c*) The shear at 18 ft. from the reaction R_2 is also $1{,}066\frac{2}{3}$ lb., because there is no other weight occurring between this point and R_2.

Example 2.—At what point in the beam loaded as shown in Fig. 27 does the shear change sign?

Solution.—Compute the reaction R_1 as follows: With the center of moments at R_2 the moments of the loads are:

$$
\begin{aligned}
9{,}000 \times 10 &= 90000 \text{ ft.-lb.}\\
4{,}000 \times 26 &= 104000 \text{ ft.-lb.}\\
3{,}000 \times 10 \times 17 &= \underline{510000} \text{ ft.-lb.}\\
\text{Total,}\quad & \ 704000 \text{ ft.-lb.}
\end{aligned}
$$

$704{,}000 \div 30 = 23{,}466\frac{2}{3}$ lb., the reaction at R_1. The first load that occurs, working out on the beam from R_1, is *c* of 4,000 lb. Then, $23{,}466\frac{2}{3} - 4{,}000 = 19{,}466\frac{2}{3}$ lb. The next load that occurs on the beam is the uniform load of 3,000 lb., per running ft. There being altogether

30,000 lb. in this load, it is evident that it will more than absorb the remaining amount of the reaction R_1; the point where the change of sign occurs must consequently be somewhere in that part of the beam covered by the uniform load. The load being 3,000 lb. per running ft., if the remaining part of the reaction, 19,466⅔ lb., be divided by the 3,000 lb., the result will be the number of feet of the uniform load required to absorb the remaining part of the reaction, and this will give the distance of the section, beyond which the resultant of the forces at the left becomes negative, from the edge of the uniform load at a; thus, 19,466⅔ ÷ 3,000 = 6.48 ft. The distance from R_1 to the edge of the uniform load is 8 ft. The entire distance to the section of change of sign of the shear is therefore 8 + 6.48 = 14.48 ft. from R_1. Ans.

EXAMPLES FOR PRACTICE

1. The uniformly distributed load on a beam supported at both ends is 40,000 pounds; what is the maximum shear on the beam?

Ans. 20,000 lb.

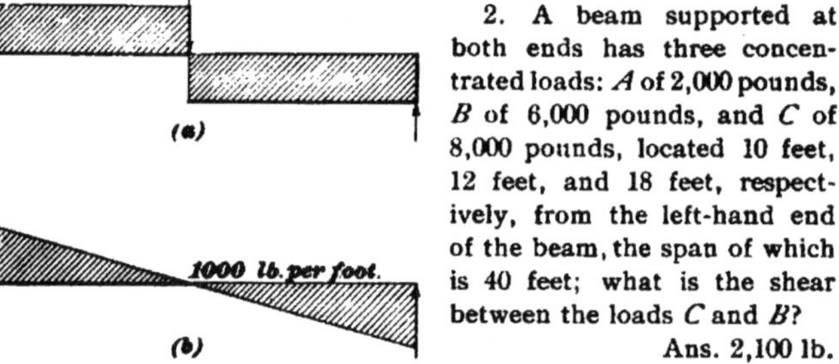

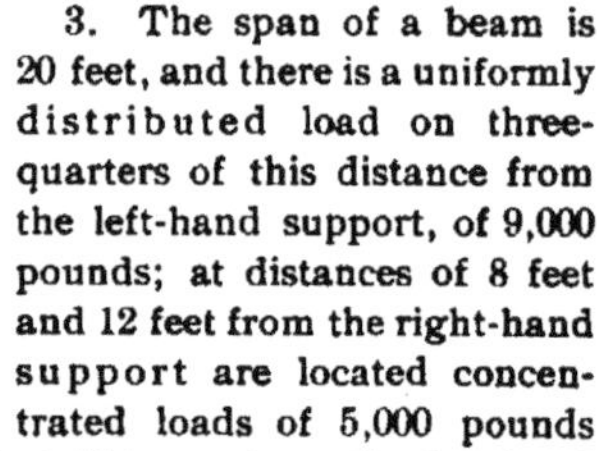

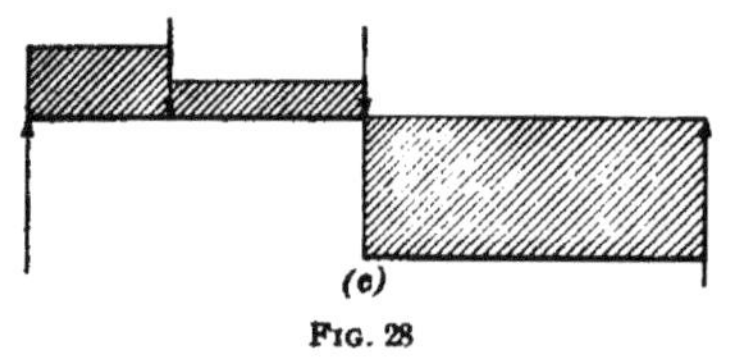

FIG. 28

2. A beam supported at both ends has three concentrated loads: A of 2,000 pounds, B of 6,000 pounds, and C of 8,000 pounds, located 10 feet, 12 feet, and 18 feet, respectively, from the left-hand end of the beam, the span of which is 40 feet; what is the shear between the loads C and B?

Ans. 2,100 lb.

3. The span of a beam is 20 feet, and there is a uniformly distributed load on three-quarters of this distance from the left-hand support, of 9,000 pounds; at distances of 8 feet and 12 feet from the right-hand support are located concentrated loads of 5,000 pounds and 6,000 pounds, respectively. At what distance from the left-hand end of the beam does the shear change sign? Ans. 8 ft. 8½ in.

---

22. The shear on a beam may be represented graphically as shown in Fig. 28, in which the shaded portion designates the amount of the shear along the beam. The portion above the horizontal line represents the positive shear, while the negative shear is indicated by the portion below the line; (*a*) shows the shear on a simple beam loaded at the center; the shear is uniform at all points except directly under the load, where it is zero. The shear on a beam uniformly loaded is illustrated at (*b*); in this case the shear is maximum at the supports and decreases gradually until it becomes zero at the center. The shear on a beam having concentrated loads may be represented as shown at (*c*). These diagrams are simply representations of the results obtained by analysis, but a method of obtaining the amount of shear graphically will be given later.

BENDING STRESSES

23. Bending Moment.—If, in a cantilever loaded as in Fig. 29, any point x on the center line $a\,b$ is taken as a center of moments, and a section made by a vertical plane $c\,d$ through this center is considered, it is evident that the moment of the force due to the downward thrust of the load tends to turn the end of the beam to the right of $c\,d$, around the center x; the measure of this tendency is the product of the weight W and its distance from $c\,d$; and, since it is the moment of a force that tends to bend the beam, it is called the **bending moment.**

24. Resisting Moment.—An inspection of Fig. 29 shows that if the end of the beam turns around the center x until it takes the position shown by the dotted lines, the parts of the two surfaces formed by the cutting plane $c\,d$ that are above the center x must be pulled from each other, while those below are pushed closer together. Thus, it is seen that if a vertical section is considered through any point on the center line $a\,b$ between the load and the point of support, the tendency of the load is to separate the particles in this section above the center line, and to push those below the center line closer together; in other words, through the bending action of the load, the upper part of the beam is subjected to a tensile stress, while the lower part is subjected to a compressive stress.

Fig. 29 also shows that the greater the distance of the particles in the assumed section above or below the center x, the greater will be their displacement; since the stress in a loaded body is directly proportional to the strain, or relative displacement, of the particles, it follows that the stress in a particle of any section is proportional to its distance from the center line, and that the greatest stress is in the particles composing the upper and lower surfaces of the beam.

In accordance with the conditions of equilibrium, the algebraic sum of the moments of all the forces tending to produce rotation around a given center must be zero; it is evident that the weight of the load is a force that tends to produce right-hand rotation around the center x; therefore, if the beam does not break under the action of the load, there must be forces acting whose moments, with respect to the center x, balance the moment of the load. These forces are the resistances with which the particles of the beam oppose any effort to change their relative positions. The tensile stresses in the particles above the center x and the compressive stresses in those below it, are a set of forces that resist the tendency of the load to turn the end of the beam, and, when the effect of the load is just balanced by the effect of these forces, it is evident that the sum of the moments of these resisting stresses is equal to the moment of the load. The sum of the moments of the stresses of all the particles composing any section of a beam is called the **resisting moment, or moment of resistance**, of that section.

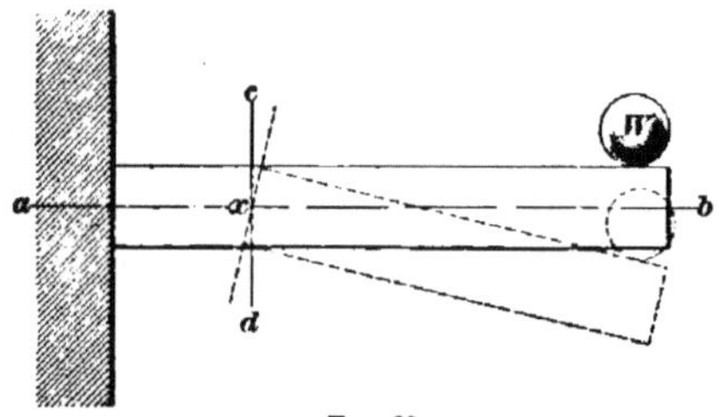

Fig. 29

25. The relations between the effect of a load and the resulting stresses in a beam have been thoroughly proved, both by mathematical investigations and numerous experiments. The results of these experiments on beams may be briefly expressed by the following:

Experimental Law.—*When a beam is bent, the horizontal elongation or compression of any fiber is directly proportional to its distance from the neutral surface, and, since the strains are directly proportional to the horizontal stresses in each fiber, they are also directly proportional to their distances from the neutral surface, provided the elastic limit is not exceeded.*

26. Bending Moments in Simple Beams.—Referring to the simple beam shown in Fig. 26, take the center of moments on the neutral axis directly under the load a, and consider the effect produced on a vertical section of the beam through this center, by the reaction R_1. It was shown in Art. **21**, that the reaction R_1 is an upward force of $6,933\frac{1}{3}$ pounds; it therefore has a tendency to turn the end of the beam upwards around the assumed center with a moment of $6,933\frac{1}{3} \times 3 = 20,800$ foot-pounds. It is evident that, to prevent this turning from actually taking place, the positive moment of the reaction must be balanced by a negative moment that can be produced only by a set of internal stresses. The condition that the moment of the stresses must be negative makes it plain that the upper fibers must be in compression and the lower in tension, a result exactly opposite to the effect produced by the bending moment on the fibers in the cantilever.

27. Effect of the Moments Due to Loads.—The only force acting on the beam at the left of the section considered in the last article was the reaction R_1. The load a acted downwards directly through this section, but its lever arm, and consequently its moment, with respect to the assumed center, was zero. Take now a point on the center line of the beam directly under the positive load b. The reaction has a moment, with respect to this center, of $6,933\frac{1}{3} \times 11 = 76,266\frac{2}{3}$ foot-pounds, while the load a, which acts downwards with a lever arm of 8 feet, has a negative moment of $2,000 \times 8 = 16,000$ foot-pounds. The bending moment at the assumed section is the algebraic sum of these moments, that is, $76,266\frac{2}{3} - 16,000 = 60,266\frac{2}{3}$ foot-pounds. Again, taking the center of moments on a section 9 feet from the right reaction R_2, the moments are as follows:

Positive moment:

Reaction R_1, 6,933⅓ × 21 = 145600 ft.-lb.

Negative moments:

Load *a*, 2,000 × 18 = 36,000 ft.-lb.

Load *b*, 6,000 × 10 = 60,000 ft.-lb. 96000 ft.-lb.

Difference, 49600 ft.-lb.

This resultant moment is the bending moment at the given section.

28. The illustrations show that the bending moment varies from point to point in a beam, and depends on the length of the beam, and on the size as well as position of the loads. Since the stresses in the beam, and consequently its ability to carry its loads, depend directly on the bending moment, it follows that it is important to find not only the bending moment for any assumed section but also the section where the bending moment is greatest. It is, in this connection, useful to note the relation between the bending moment and the shear.

29. The shear in a simple beam is always greatest at the greater reaction, being equal to that reaction. In passing along the beam from either reaction, there is no change in the shear until a load is reached; at each point where a load is added, the shear is diminished by an amount equal to the load. At the point where the sum of the added loads equals or exceeds the reaction, the shear is said to change sign. The section where the change in sign in the shear takes place, depends on the method of loading. With a uniformly distributed load, the shear diminishes uniformly from each reaction, and the section where the sign changes is the section of the beam midway between the supports. With a single concentrated load, the shear is equal to each reaction at all sections between that reaction and the point where the load is applied, and the section where the shear changes sign is directly under the load. With any system of loading, the section where the shear changes sign can be found by adding the successive loads from either reaction toward the center of the beam, until a sum is obtained that equals or exceeds the reaction; the section where the shear changes sign is under the point of application of the last load added.

30. The bending moment in a simple beam increases as the shear decreases; it is zero at either reaction and increases toward the center, becoming greatest at the section where the shear changes sign. With a uniformly distributed load, the greatest bending moment is at the section of the beam midway between the supports; with a single concentrated

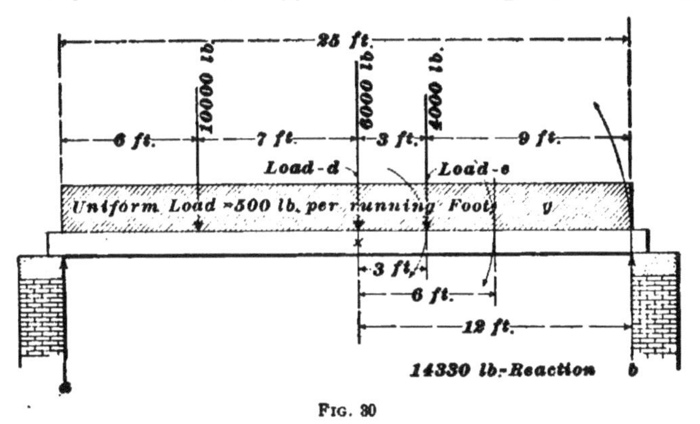

Fig. 30

load, the greatest bending moment is directly under the load; and with any system of loading, the greatest bending moment occurs at the section where the shear changes sign. Having located the section of greatest bending moment, the magnitude of this moment can be readily computed, by taking the center of moments, on the section of greatest bending moment and computing the resultant moment of either reaction and all the loads between it and the center in question.

Example.—A wooden beam, supported on two brick piers, is loaded as shown in Fig. 30: (*a*) What is the greatest shear? (*b*) Where does the shear change sign? (*c*) What is the greatest bending moment, in inch-pounds?

SOLUTION.—(*a*) Since the greatest shear is equal to the greater reaction, the reactions will first be computed. Take the center of moments at the edge of pier *a*, and as the moment of the uniform load is the same as the moment of an equal concentrated load acting at the center of gravity of the uniform load, the moments of the loads are:

10,000-lb. load	10,000 × 6 =	60000 ft.-lb.
Load *d*	6,000 × 13 =	78000 ft.-lb.
Load *e*	4,000 × 16 =	64000 ft.-lb.
Uniformly distributed load . .	12,500 × 12½ =	156250 ft.-lb.
	Total,	358250 ft.-lb.

This also equals the moment of the reaction of the pier *b*. The reaction at *b* is therefore 358,250 ÷ 25 = 14,330 lb. The total load is 10,000 + 6,000 + 4,000 + 12,500 = 32,500 lb.; the reaction at *a* is therefore 32,500 − 14,330 = 18,170 lb. This, being the greater reaction, is the greatest shear. Ans.

(*b*) Beginning at the left reaction and adding the loads in succession toward the right, the load of 10,000 lb., plus the uniformly distributed load between the reaction and the point of application of the load *d*, is 10,000 + 500 × 13 = 16,500 lb. This is less than the left reaction, but, when the load *d* is added, the sum of the loads is greater than the reaction; consequently, the shear changes sign under the load *d*. Ans.

(*c*) Taking the center of moments under the load *d*, and considering the forces at the left, the moments are:

Positive moment:		
18,170 × 13 =		236210 ft.-lb.
Negative moments:		
6,500 × 6½ =	42250 ft.-lb.	
10,000 × 7 =	70000 ft.-lb.	
		112250 ft. lb.
Difference (bending moment in ft.-lb.),		123960 ft.-lb.,

The bending moment, in inch-pounds, is therefore 123,960 × 12 = 1,487,520 in.-lb. Ans.

31. Formulas for Maximum Bending Moments and Safe Loads.—The following table gives formulas for obtaining the maximum safe loads that may be supported by beams under different conditions and the maximum bending moments produced by these loads. In these formulas, W is the weight, in pounds; S, the section modulus; s_a, the safe unit fiber stress; and l, the span in inches.

In Case XVI, the greatest load can be supported and consequently the least bending moment occurs when the distance x is .207 l, and the formulas given apply only to this condition. The bending moment for any point is found by the formula $M = \frac{Wc}{2}\left(\frac{c}{l} - 1 + \frac{x}{c}\right)$, in which the value c is the distance of the point at which the amount of the bending moment desired is located from the left-hand end of the beam.

When the end of a beam extends into the wall it is considered as being fixed, if the masonry is built around it tightly; while if the end rests on a wall or other support it is considered as being simply supported. When the load is represented by cross-sectioning, it denotes that it is a distributed load; the round weight represents a concentrated load. A load such as shown in Cases III and X is sometimes encountered, for instance, in a case where a beam is required to support the side walls or curbing of a stairway. A triangular load, as shown in Case IX, is considered when the beam is to support a solid brick wall.

32. The rule given in Case VII is that most used, as it applies to a beam uniformly loaded, such as floor joists, girders, and, in some cases, the rafters of a roof. The rule in Case IX is convenient in calculating the bending moment on lintels supporting brickwork or masonry over openings. It will be observed that if the beam supporting a concentrated load at the center is firmly fixed or fastened at both ends, as in Case XIII, instead of being simply supported as designated in Case IV, the bending moment under the same load will be only half as much. Also, in Case XIV, where the ends of the beam are firmly fixed, the uniformly distributed load that may be supported is one and one-half times as great as where the ends merely rest on supports, as in Case VII. It is seldom advisable, in ordinary building practice, to consider the ends of a beam fixed, it being good practice to assume the ends of the beam as simply bearing on the wall, using the rules and formulas in Table I. However, it should be understood that all of these rules and formulas apply to static loads. The

TABLE I

FORMULAS FOR MAXIMUM SAFE LOADS AND MAXIMUM BENDING MOMENTS ON BEAMS

Case	Method of Loading	Maximum Bending Moment	Maximum Safe Load
I	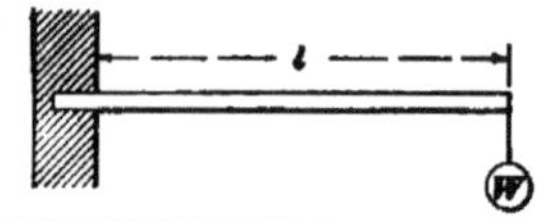	Wl (1)	$\frac{Ss_a}{l}$ (17)
II	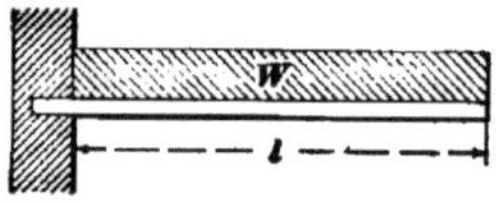	$\frac{Wl}{2}$ (2)	$\frac{2Ss_a}{l}$ (18)
III		$\frac{Wl}{3}$ (3)	$\frac{3Ss_a}{l}$ (19)
IV	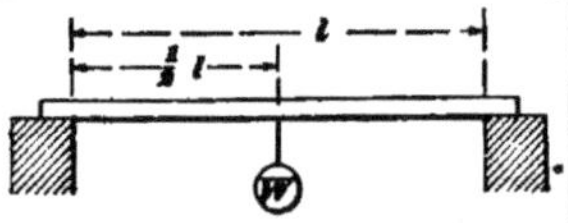	$\frac{Wl}{4}$ (4)	$\frac{4Ss_a}{l}$ (20)
V	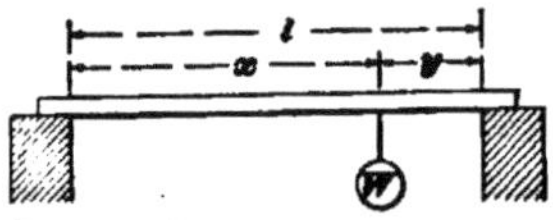	$\frac{Wxy}{l}$ (5)	$\frac{lSs_a}{xy}$ (21)
VI	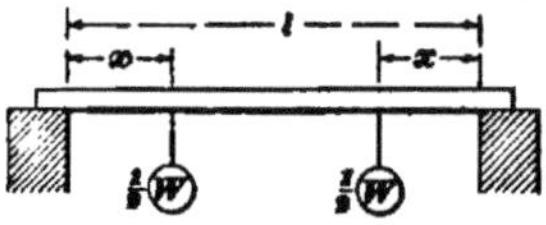	$\frac{Wx}{2}$ (6)	$\frac{2Ss_a}{x}$ (22)
VII	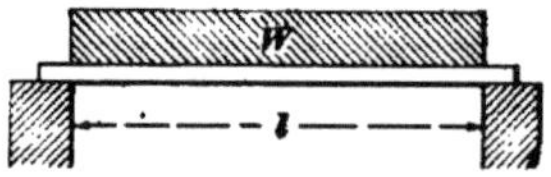	$\frac{Wl}{8}$ (7)	$\frac{8Ss_a}{l}$ (23)
VIII	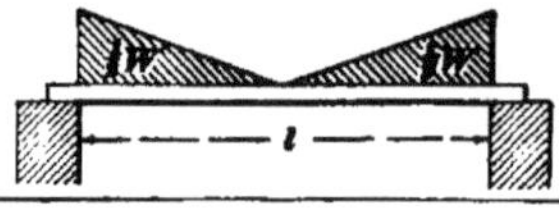	$\frac{Wl}{12}$ (8)	$\frac{12Ss_a}{l}$ (24)

TABLE I—(Continued)

Case	Method of Loading	Maximum Bending Moment	Maximum Safe Load
IX		$\frac{Wl}{6}$ (9)	$\frac{6 S s_a}{l}$ (25)
X	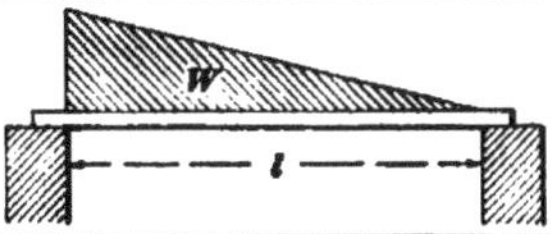	$\frac{104\, Wl}{108}$ (10)	$\frac{810\, S s_a}{104\, l}$ (26)
XI	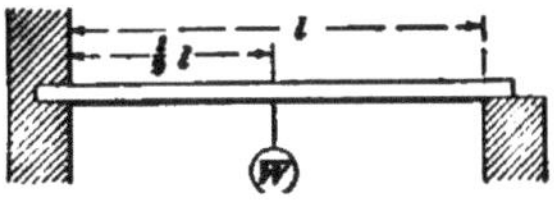	$\frac{6\, Wl}{32}$ (11)	$\frac{16\, S s_a}{3\, l}$ (27)
XII	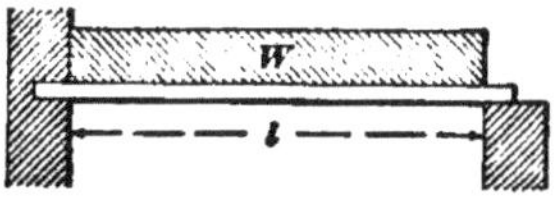	$\frac{Wl}{8}$ (12)	$\frac{8\, S s_a}{l}$ (28)
XIII	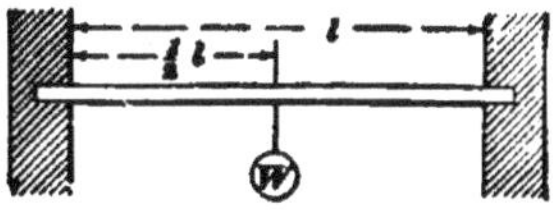	$\frac{Wl}{8}$ (13)	$\frac{8\, S s_a}{l}$ (29)
XIV	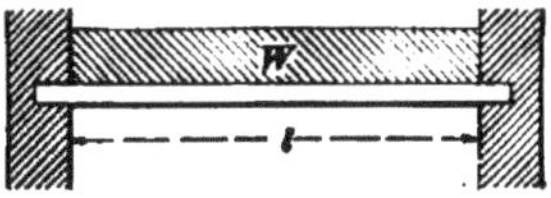	$\frac{Wl}{12}$ (14)	$\frac{12\, S s_a}{l}$ (30)
XV	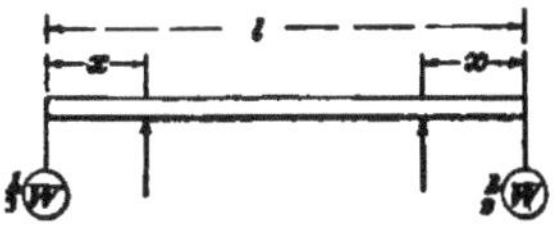	$\frac{Wx}{2}$ (15)	$\frac{2\, S s_a}{x}$ (31)
XVI	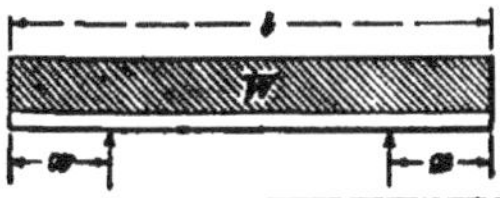	$\frac{3\, Wl}{140}$ (16)	$\frac{140\, S s_a}{3\, l}$ (32)

same load suddenly applied produces a stress in the beam twice as great as that of a static load. The safe suddenly applied load is therefore only half as much.

A graphical method for determining the bending moment on a beam is explained in the following article:

EXAMPLE.—What will be the bending moment, in inch-pounds, on a wooden girder supporting a floor area of 150 square feet, the dead and live load being 100 pounds per square foot, and the span of the girder 20 feet?

SOLUTION.—The total uniformly distributed load is $150 \times 100 = 15{,}000$ lb.; therefore, by applying the formula in Case VII, Table I, the bending moment

$$M = \frac{W l}{8} = \frac{15{,}000 \times 20 \times 12}{8} = 450{,}000 \text{ in.-lb. Ans.}$$

EXAMPLES FOR PRACTICE

1. A beam has a span of 20 feet, and is loaded with a uniformly distributed load of 2,500 pounds per lineal foot; what is the greatest bending moment, in inch-pounds, on the beam? Ans. 1,500,000 in.-lb.

2. What is the bending moment, in inch-pounds, on a cantilever beam, securely fastened into a wall, extending from the point of support 10 feet and loaded with a uniformly distributed load of 1,000 pounds per lineal foot? Ans. 600,000 in.-lb.

3. What is the bending moment, in inch-pounds, on a girder having a span of 30 feet, if there is a uniformly distributed load of 1,500 pounds per lineal foot and a load of 20,000 pounds concentrated at the center? Ans. 3,825,000 in.-lb.

4. A plate girder in a building is required to support a uniformly distributed load of 2,000 pounds per lineal foot, extending 20 feet each side of the center of the girder; in addition, it is required to support a load of 30,600 pounds, concentrated 10 feet from one end of the girder, and another load of 43,000 pounds, located 22 feet from the same end. What will be the greatest bending moment on the girder, in foot-pounds, if the span is 60 feet? Ans. 1,528,933 ft.-lb.

GRAPHICAL METHOD OF OBTAINING BENDING MOMENT AND SHEAR

33. The graphical method of obtaining bending moment and shear on a beam can best be explained by assuming an example and solving it.

Fig. 31 (*a*) shows a simple beam having three concentrated loads, W_1, W_2, and W_3. First lay out, as in (*b*), the *force polygon*, or *stress diagram*, as it is sometimes called; the

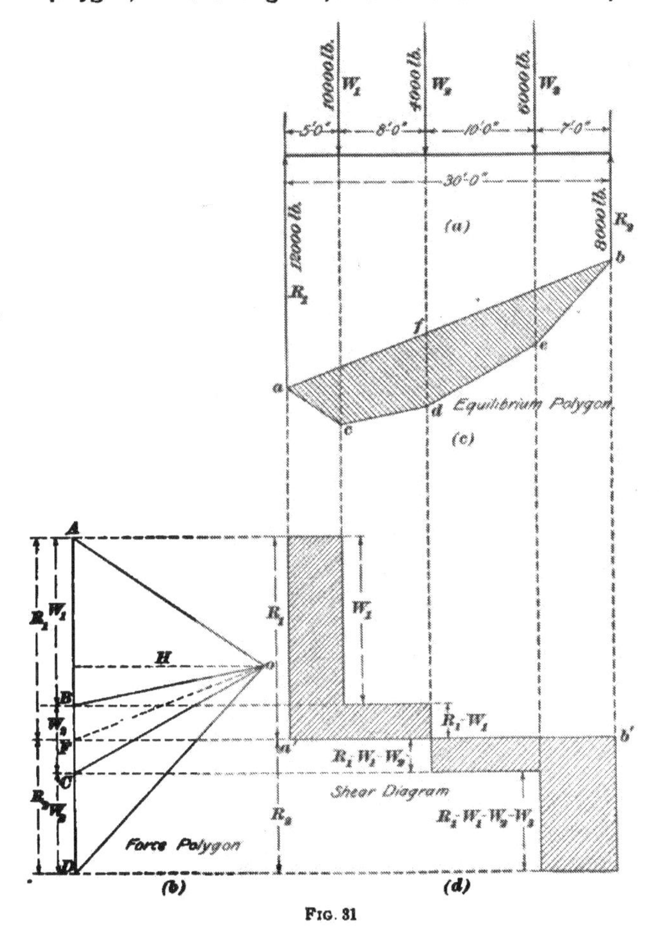

Fig. 31

loads are laid off to scale at *A B*, *B C*, and *C D*. From any point *o*, outside of the load line, draw lines to *A*, *B*, *C*, and *D*. If a cord is supported at two points, *a* and *b*, in Fig. 31 (*c*), and the loads W_1, W_2, and W_3 are hung at the points *c*, *d*, and *e*, respectively, the cord will be in equilibrium, provided that the portions *a c*, *c d*, *d e*, and *e b* are parallel to *A o*, *B o*, *C o*, and *D o*, respectively.

The figure in (*c*) is called the *equilibrium*, or *funicular, polygon*, and is a representation of the variation of the bending moment along the beam. Therefore, the bending moment at any point is proportional to the ordinate in the funicular polygon below that point, or the vertical line intercepted between the sides of the polygon. For instance, the bending moment under the load W_2 is proportional to the ordinate *f d*. The reactions R_1 and R_2 may be obtained from the force polygon by drawing the line *o F* parallel to *a b* in the equilibrium polygon; *A F* and *F D* represent the reactions R_1 and R_2, respectively.

In the force polygon, a horizontal line drawn from the point *o* to the load line represents the horizontal component of the tensions *a c*, *c d*, *d e*, and *e b* in the funicular polygon, and is designated as *H*. The bending moment at any point is equal to the product of the vertical ordinate in the funicular polygon at that point and the force *H*. The ordinate should be measured by the scale to which the beam was drawn and the line *H* by the scale to which the load line was laid off. The explanation of this is given in Art. **35**.

34. To construct the shear diagram shown in (*d*), project downwards the points at which the loads and reactions occur. The point *F* in the force polygon is projected horizontally, thus giving *a' b'*, the base line of the polygon. Project the point *A* until it intersects the line drawn from the load W_1 in (*a*), and the point *B* until it intersects the line drawn from the load W_2. The shear at this point changes sign and consequently the remainder of the diagram will be below the base line *a' b'*. Project the points *C* and *D* until they intersect the lines projected from the load W_3 and the reaction R_2, respectively. This completes the shear diagram, as shown by the shaded section in (*d*). The shear at any point is found by measuring the ordinate at that point by the scale used in laying off the load line *A D*.

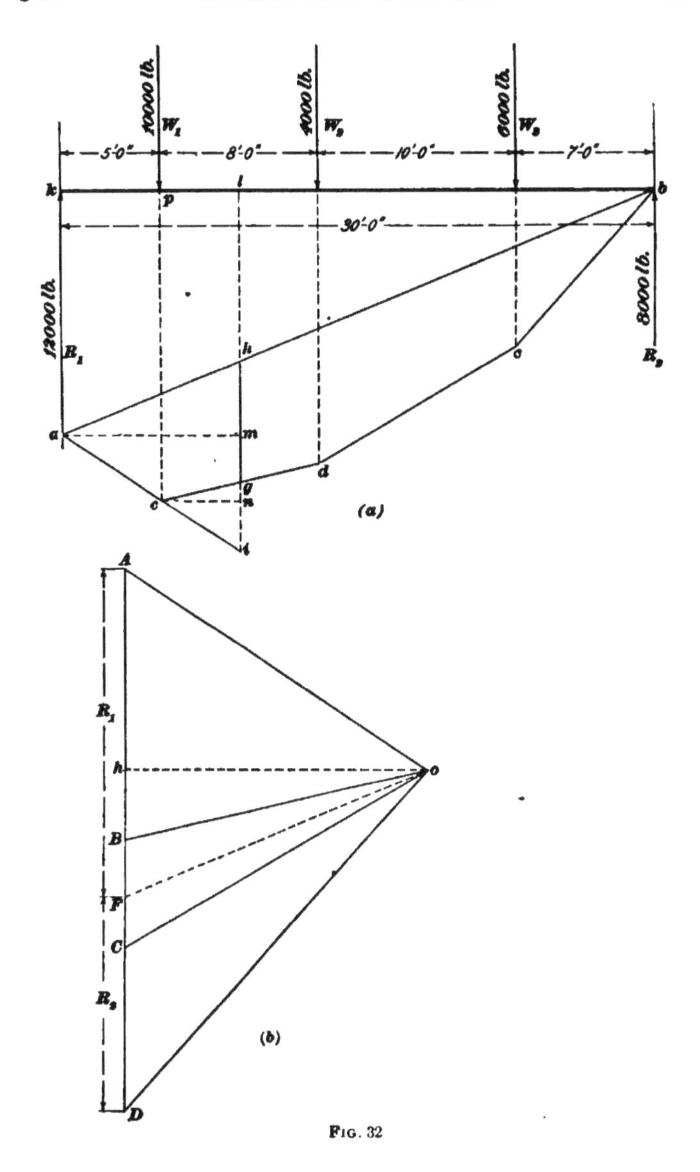

10000 lb.
4000 lb.
6000 lb.
W_1
W_2
W_3
5'-0"
8'-0"
10'-0"
7'-0"
30'-0"
12000 lb.
8000 lb.
R_1
R_2
k
p
l
b
h
e
a
m
d
g
c
n
i
(a)
A
h
o
B
F
C
D
(b)

FIG. 32

35. The graphical determination of the bending moment may be explained by means of Fig. 32, in which the beam shown in Fig. 31 is considered.

On the beam $k b$ it is desired to find the bending moment at the point l, 9 feet from k. Let the bending moment be considered first with reference to the reaction R_1 alone, omitting the effect of the load W_1. Construct the triangle $a h i$ by producing the lines $g h$ and $a c$ until they intersect at i. As the triangle $a h i$ in (*a*) is similar to $A o F$ in (*b*), it follows that $a h : h i = F o : A F$, and as the altitudes of the two triangles are proportional to their bases, the following proportion may be written: $a m : h i = h o : A F$; therefore, $A F \times a m = h o \times h i$. But $a m = k l$ and $A F = R_1$, hence, $R_1 \times k l = h o \times h i$.

As the load W_1 acts in opposition to the reaction R_1, its moment must be subtracted from that of R_1. In the similar triangles $c g i$ and $A o B$, $c g : g i = o B : A B$ and consequently $c n : g i = o h : A B$, but $A B = W_1$ and $c n =$ its moment $p l$; therefore, $W_1 \times p l = g i \times o h$. The resulting moment is then found by subtracting that of W_1 from R_1, as follows: $h i \times h o - g i \times h o$, or $h o (h i - g i) = h o \times h g$. Thus, it is proved that the bending moment at any point on a beam is equal to the product of the line representing the horizontal force $o h$ and a vertical line intercepted between the sides of the funicular polygon and passing through the point considered.

In Fig. 32 (*a*) the line $g h$ measures 6.14 feet and in (*b*) the line $h o$ measures 11,000 pounds, according to the scales to which the two figures have been drawn; the product $6.14 \times 11{,}000 = 68{,}154$ foot-pounds is the bending moment at l. By calculation, the bending moment is found to be $9 \times 12{,}000 - 4 \times 10{,}000 = 68{,}000$ foot-pounds. The difference is caused by inaccuracies in the drawing by reason of the small scale to which it is drawn.

CONTINUOUS BEAMS

36. When beams or girders extend over three or more supports in one piece, that is, are not jointed over the supports, they are said to be *continuous*, and the strains produced are very different from those in ordinary beams.

Fig. 33 shows the action of a beam fixed at each end and having one intermediate support. The points *b*, *c*, *e*, and *f*, where the direction of the curve changes and, consequently, where the stress in the upper and lower portions of the beam changes, are called the points of **contraflexure.** The portions *a b*, *c d*, *d e*, and *f g* act as cantilevers and the upper edges of the beam are in tension while the lower are in compression.

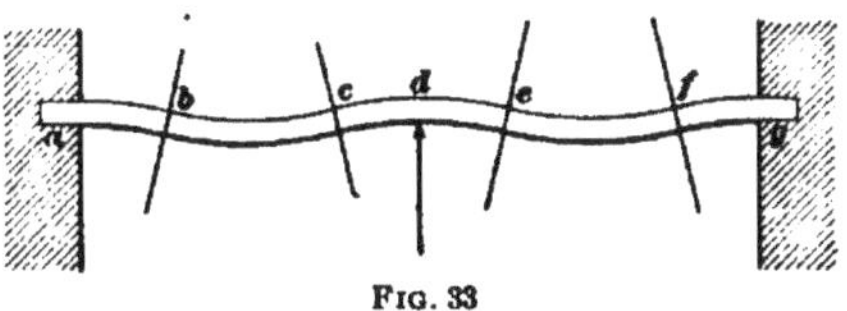

FIG. 33

The portions *b c* and *e f* resemble simple beams and their upper edges are in compression while the lower are in tension. If the ends of the girders are not fixed, but are simply supported, the curves will be as shown in Fig. 34. The curves of each span produced by a uniform load resemble those of a beam fixed at one end and supported at the other. The location of the points of contraflexure and the value of the bending moments are affected by the distribution of the load and the section of the girder.

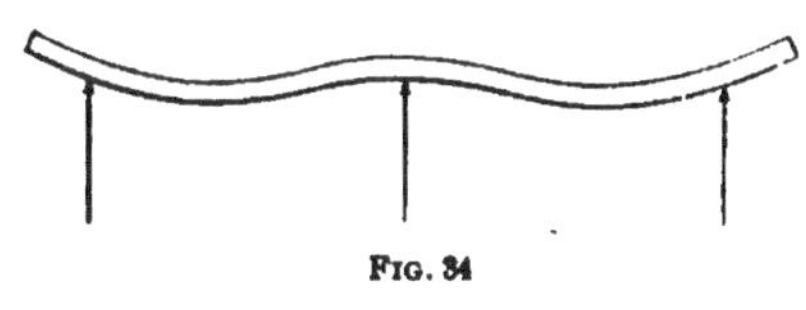
FIG. 34

The calculations in regard to continuous girders are very complicated and consequently will not be discussed here, but the following formulas will give all the information required in proportioning such beams to the load they are required to carry.

Tables II and III give the reactions and bending moments at the supports for continuous beams uniformly loaded and extending over a number of equal spans. In these tables, l equals the length of each span and w the weight per unit of

TABLE II

REACTIONS FOR CONTINUOUS BEAMS OVER EQUAL SPANS

Coefficients of w l

Number of Span	Number of Each Support									
	1st	2d	3d	4th	5th	6th	7th	8th	9th	10th
1	$\frac{1}{2}$	$\frac{1}{2}$								
2	$\frac{3}{8}$	$\frac{10}{8}$	$\frac{3}{8}$							
3	$\frac{4}{10}$	$\frac{11}{10}$	$\frac{11}{10}$	$\frac{4}{10}$						
4	$\frac{11}{28}$	$\frac{32}{28}$	$\frac{26}{28}$	$\frac{32}{28}$	$\frac{11}{28}$					
5	$\frac{15}{38}$	$\frac{43}{38}$	$\frac{37}{38}$	$\frac{37}{38}$	$\frac{43}{38}$	$\frac{15}{38}$				
6	$\frac{41}{104}$	$\frac{118}{104}$	$\frac{100}{104}$	$\frac{106}{104}$	$\frac{100}{104}$	$\frac{118}{104}$	$\frac{41}{104}$			
7	$\frac{56}{142}$	$\frac{161}{142}$	$\frac{137}{142}$	$\frac{143}{142}$	$\frac{143}{142}$	$\frac{137}{142}$	$\frac{161}{142}$	$\frac{56}{142}$		
8	$\frac{153}{388}$	$\frac{440}{388}$	$\frac{374}{388}$	$\frac{392}{388}$	$\frac{386}{388}$	$\frac{392}{388}$	$\frac{374}{388}$	$\frac{440}{388}$	$\frac{153}{388}$	
9	$\frac{209}{530}$	$\frac{601}{530}$	$\frac{511}{530}$	$\frac{535}{530}$	$\frac{529}{530}$	$\frac{529}{530}$	$\frac{535}{530}$	$\frac{511}{530}$	$\frac{601}{530}$	$\frac{209}{530}$

TABLE III

BENDING MOMENTS FOR CONTINUOUS BEAMS OVER EQUAL SPANS

Coefficients of $w l^2$

Number of Span	Number of Each Support									
	1st	2d	3d	4th	5th	6th	7th	8th	9th	10th
1	0	0								
2	0	$\frac{1}{8}$	0							
3	0	$\frac{1}{10}$	$\frac{1}{10}$	0						
4	0	$\frac{3}{28}$	$\frac{2}{28}$	$\frac{3}{28}$	0					
5	0	$\frac{4}{38}$	$\frac{3}{38}$	$\frac{3}{38}$	$\frac{4}{38}$	0				
6	0	$\frac{11}{104}$	$\frac{8}{104}$	$\frac{9}{104}$	$\frac{8}{104}$	$\frac{11}{104}$	0			
7	0	$\frac{15}{142}$	$\frac{11}{142}$	$\frac{12}{142}$	$\frac{12}{142}$	$\frac{11}{142}$	$\frac{15}{142}$	0		
8	0	$\frac{41}{388}$	$\frac{30}{388}$	$\frac{33}{388}$	$\frac{32}{388}$	$\frac{33}{388}$	$\frac{30}{388}$	$\frac{41}{388}$	0	
9	0	$\frac{56}{530}$	$\frac{41}{530}$	$\frac{45}{530}$	$\frac{44}{530}$	$\frac{44}{530}$	$\frac{45}{530}$	$\frac{41}{530}$	$\frac{56}{530}$	0

length; hence, the load on each span is equal to $w\,l$. The reactions are expressed in terms of $w\,l$ and the bending moments in terms of $w\,l^2$. Only the fractional coefficients are given.

To illustrate the method of using the tables, the following example will be assumed:

EXAMPLE.—The lower tie-member of an A-shaped truss, *a a* Fig. 35, supported by the four reactions R_1, R_2, R_3, and R_4, sustains a uniformly distributed load of 2,000 pounds per lineal foot. (*a*) What will be the theoretical amount of the reactions? (*b*) What is the amount of the greatest bending moment from this uniformly distributed load?

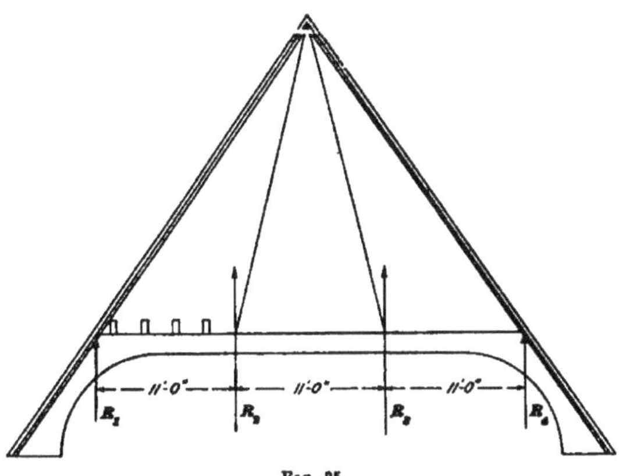

FIG. 35

SOLUTION.—(*a*) From Table II, the reactions at the ends are equal to $\frac{4}{10}$ of $w\,l$, and for the intermediate supports, $\frac{11}{10}$ of $w\,l$. Substituting these values, the reactions at the ends are equal to $\frac{4}{10} \times 2{,}000 \times 11 = 8{,}800$ lb. Ans.

The reactions at the intermediate supports are equal to $\frac{11}{10} \times 2{,}000 \times 11 = 24{,}200$ lb. Ans.

(*b*) The greatest bending moment occurs at the intermediate supports; from Table III, this moment is equal to $\frac{1}{10}\,w\,l^2$, or $\frac{1}{10} \times 2{,}000 \times 11 \times 11 = 24{,}200$ ft.-lb. Ans.

EXAMPLES FOR PRACTICE

1. The first floor of a building used as a store was supported on 12-inch I beams having a clear span of 30 feet. A subsequent tenant desired to use the first floor of the building as a storage warehouse and it was decided to place a row of piers down the center of the

basement to further support these girders. By what percentage is the carrying capacity of the floor increased by the introduction of the central pier? Ans. 400%

2. What will be the load from the girder on the central pier mentioned in the above problem, if the load is 2,000 pounds per lineal foot? Ans. 37,500 lb.

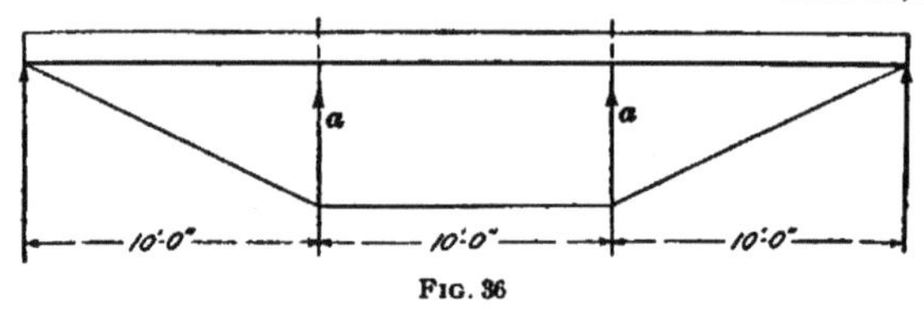

FIG. 36

3. A trussed girder, Fig. 36, is loaded with a uniformly distributed load of 1,500 pounds per foot; what will be the amount of the compressive stress in the straining posts *a*, *a*. Ans. 16,500 lb.

DEFLECTION OF BEAMS

37. Elasticity is that property which a body possesses of returning to its original form, after being strained or distorted by the application of a stress. This property is possessed by all bodies in a greater or less degree. If, after being distorted, a body does not perfectly resume its original form, it is said to have a **permanent set.** It is believed that the elasticity of all solids is more or less imperfect, and that the slightest strain produces a corresponding permanent set. It is customary, however, to consider the elasticity of all building materials as practically perfect within certain limits. Under this assumption, stresses, up to a certain limit, may be applied and removed, and the resulting strain or alteration of form will be only temporary, with no appreciable permanent set. However, stresses above this limit will cause permanent sets.

38. The *elastic limit* of any material is the maximum unit stress that may be applied to it without causing any apparent permanent set; or, it is that point at which the strain ceases to be proportional to the stress.

To illustrate: Consider a piece of steel wire supported at one end and loaded by a weight suspended from the other.

The wire is found to stretch under the action of the load, and by varying the weight or stress, the strain in each case is found to vary in the same proportion, so long as the weight is not greater than one-fourth of the breaking strength of the wire; within this limit it is found, on removing the weight, that the wire resumes its original length.

If the load is made considerably greater than one-fourth of the breaking strength of the wire, it is found that when the load is removed the wire has taken a permanent set; in other words, it will not return to its original length. If the wire remains permanently longer than it was before the load was applied, it has been strained beyond the limit of elasticity.

Suppose that a weight of 2,000 pounds is hung from the end of a wrought-iron rod having a sectional area of 1 square inch, and that the rod stretches about $\frac{1}{13000}$ of its original length. When the weight is removed, the bar resumes its original length, as far as can be measured by ordinary instruments. Now, instead of 2,000 pounds, attach a weight of 24,000 pounds to the rod and it stretches about $\frac{1}{1000}$ of its length; when this weight is removed, we find that the bar does not return to its original length, but that it is slightly longer than it was before; that is, the bar has a permanent set.

The unit stress where the weight on the rod is just sufficient to produce the least permanent set is called the **elastic limit.**

39. The **modulus of elasticity** is the ratio of the unit stress to the unit strain for loads within the elastic limit.

For example, if the weight of 2,000 pounds on an iron bar, whose section is 1 square inch, produces an elongation of $\frac{1}{13000}$ of the original length of the bar, the unit stress is 2,000 pounds per square inch; the unit strain is $\frac{1}{13000}$; and the modulus of elasticity is $2{,}000 \div \frac{1}{13000} = 26{,}000{,}000$ pounds per square inch.

In most building materials, the modulus of elasticity for tension and the modulus for compression may be considered as practically equal. The moduli of elasticity of some of the principal building materials are: yellow pine, 1,200,000 pounds per square inch; cast iron, 17,000,000 to 20,000,000; steel, 28,000,000 to 30,000,000; neat cement, 3,000,000; concrete, 700,000.

40. The stresses of tension and compression created in a loaded beam cause elongation and shortening of the longitudinal elements above and below the neutral plane, the result of which is a curvature of the beam. The amount of this curvature depends on the amount and distribution of the load, the material of which the beam is composed, its span and manner of support, and on the dimensions and form of the cross-section.

Deflection is the name applied to the distortion or bending produced in a beam when subjected to transverse stresses. The measure of the deflection at any point on a beam is the perpendicular displacement of the point from its original position. If, on the removal of the transverse stresses or loads on the beam, it returns to the straight or original form, the material in the beam has not been strained beyond the elastic limit. On the other hand, if the internal stresses exceed the elastic limit of the material, a permanent set will be given the beam.

41. Stiffness is a measure of the ability of a body to resist bending; this property is very different from the strength of the material or its power to resist rupture.

The stiffness of a structure does not depend so much on the elasticity of the material of which it is composed as on its arrangement and form; for example, a floor may be built of shallow and wide joists that will be sufficiently strong to carry a given load, but it will not be nearly so stiff as a floor of equal strength built of narrow and deep ones. This property of stiffness is as important in building construction as mere strength, and the two should be considered together; thus, the floor joists of a building may be strong enough to resist breaking, but so shallow as to lack stiffness, in which case the floor will be springy and vibrate from people walking on it. If there is a plastered ceiling on the under side of the joists of such a floor, the deflection of the joists may cause the plaster to crack and fall into the room below. Where stiffness is lacking in the rafters of a roof, they will be liable to sag, thereby causing unsightly hollows in the surface of the roof, in which moisture and snow may lodge, which would be very detrimental to the roof covering.

42. From the foregoing, it is evident that not only must the strength of the beams composing a structure be calculated to withstand rupture, but the beams must be stiff or rigid enough to resist bending. It is, therefore, important to be able to calculate the deflection of any beam under its load, and if found excessive, the size of the beam may be increased and the deflection reduced to working limits.

The amount of deflection that exists in beams loaded and supported in different ways may be calculated by the formulas given in Table IV. In using these formulas, all the loads should be expressed in pounds and the lengths in inches. The modulus of elasticity is denoted by E, and the moment of inertia of the section by I.

EXAMPLE 1.—A 10-inch steel I beam, supported at the ends, must sustain a uniformly distributed load of 10,000 pounds. The span of the beam is 20 feet, and its moment of inertia is 146.4; there is to be a plastered ceiling on its under side, the allowable deflection of which is $\frac{1}{30}$ inch for each foot of span. Will the deflection of the beam be excessive?

SOLUTION.—The formula for the deflection of a beam of this character, from the table, is $\frac{5\,W\,l^3}{384\,E\,I}$. From Art. **39**, the modulus of elasticity of structural steel is 28,000,000 to 30,000,000; taking the average value, and substituting the other values of the example in the formula, the deflection equals

$$\frac{5 \times 10{,}000 \times 240^3}{384 \times 29{,}000{,}000 \times 146.4} = .42, \text{ or about } \tfrac{7}{16} \text{ in.}$$

Since the allowable deflection for each foot of span is $\frac{1}{30}$ in., the total allowable deflection is $\frac{1}{30}$ of $20 = \frac{2}{3}$ in. This is greater than the calculated deflection, and the beam therefore satisfies the required conditions. Ans.

EXAMPLE 2.—A $12'' \times 16''$ yellow-pine girder must support a symmetrically placed triangular piece of brickwork, which weighs about 12,000 pounds. What will be the deflection of the timber if the span is 20 feet?

TABLE IV

FORMULAS FOR DEFLECTION OF BEAMS

Case	Method of Loading	Deflection Inches	
I	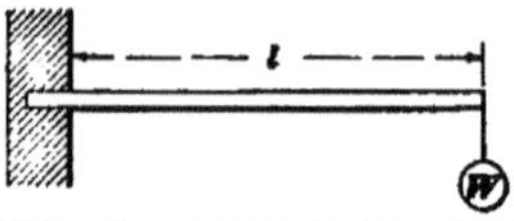	$\dfrac{Wl^3}{3EI}$	(33)
II	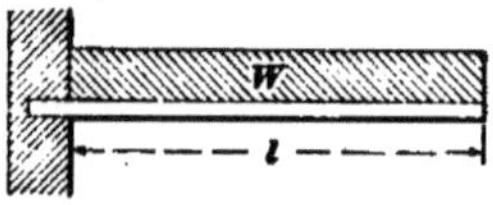	$\dfrac{Wl^3}{8EI}$	(34)
III		$\dfrac{Wl^3}{15EI}$	(35)
IV	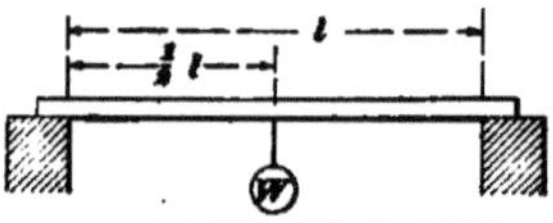	$\dfrac{Wl^3}{48EI}$	(36)
V	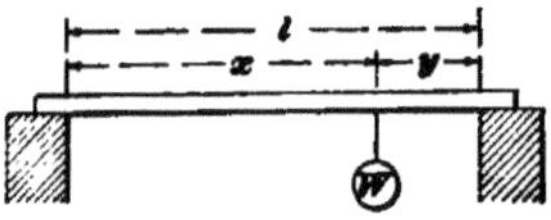	$\dfrac{Wxy(2l-x)\sqrt{3x(2l-x)}}{27lEI}$	(37)
VI	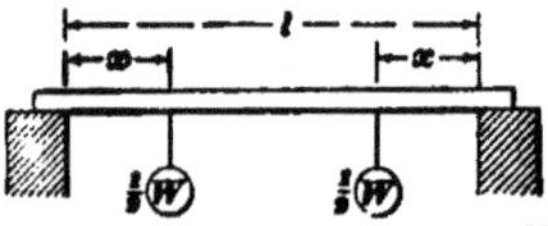	$\dfrac{Wx}{48EI}(3l^2-4x^2)$	(38)
VII		$\dfrac{5Wl^3}{384EI}$	(39)
VIII		$\dfrac{3Wl^3}{320EI}$	(40)

TABLE IV—(Continued)

Case	Method of Loading	Deflection Inches	
IX		$\frac{Wl^3}{60EI}$	(41)
X		$\frac{47\,Wl^3}{3{,}600EI}$	(42)
XI	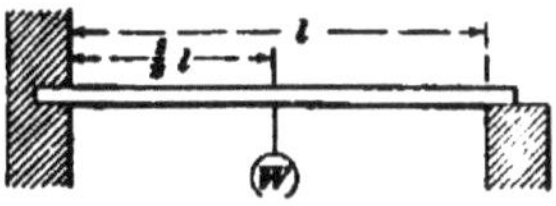	$\frac{3\,Wl^3}{322\,EI}$	(43)
XII	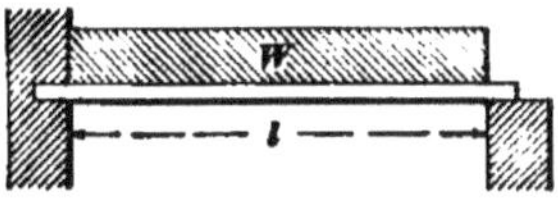	$\frac{5\,Wl^3}{926\,EI}$	(44)
XIII	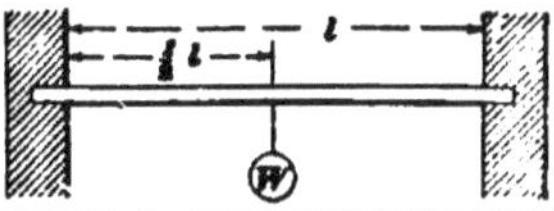	$\frac{Wl^3}{192\,EI}$	(45)
XIV	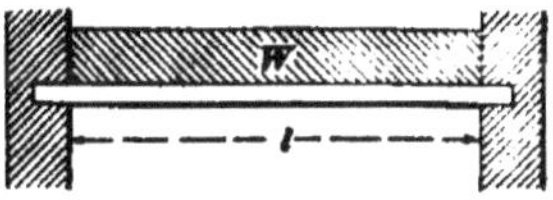	$\frac{Wl^3}{384\,EI}$	(46)
XV		For overhang: $\frac{Wx}{12\,EI}(3\,xl-4\,x^2)$	(47)
		For part between supports: $\frac{Wx}{16EI}(l-2\,x)^2$	(48)
XVI	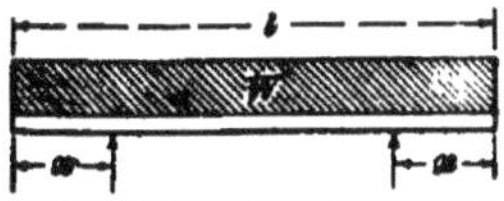	Variable	

SOLUTION.—The formula for the deflection, in this case, from the table, is $\frac{W l^3}{60 E I}$. From Art. **39**, the value of the modulus of elasticity is found to be 1,200,000. The moment of inertia of the section, from the formula $I = \frac{b d^3}{12}$, is $I = \frac{12 \times 16^3}{12} = 4,096$. Then, by substituting the given values, the deflection is

$$\frac{12,000 \times 240^3}{60 \times 1,200,000 \times 4,096} = .56, \text{ about } \tfrac{9}{16} \text{ in. Ans.}$$

EXAMPLES FOR PRACTICE

1. The moment of inertia of a 12-inch steel I beam is 228.3, and its span is 25 feet. If the ends of the beam are simply supported, what will be its deflection under a concentrated load of 10,000 pounds suspended from its center? Ans. .85 in.

2. A cantilever beam of 12″ × 16″ yellow pine extends from a building wall 10 feet, and is loaded on the end with a concentrated load of 12,500 pounds; what will be the greatest deflection of the beam? Ans. 1.46 in.

3. The span of a 15-inch steel I beam is 30 feet, and the moment of inertia of its section is 455.8; the load on the beam is uniformly distributed and amounts to 3,000 pounds per lineal foot. If the ends of the beam are firmly fixed, what will be its deflection? Ans. .83 in.

BEAMS AND GIRDERS

(PART 2)

WOODEN, STEEL, AND REENFORCED CONCRETE BEAMS

FORMULA OF GENERAL APPLICATION

1. In every section of a beam under stress, each fiber offers against rupture a resistance whose moment is equal to the resisting force of the fiber multiplied by its perpendicular distance from the neutral axis of the section. The sum of these fiber moments is called the **resisting moment** of the section, and is equal to the bending moment producing the stress. By higher mathematics, it is proved that the resisting moment is equal to the product of the greatest unit stress in any part of a section multiplied by a factor, called the **section modulus,** or the **resisting inches,** which depends on the shape of the section; the latter term is now rarely used.

If the greatest unit stress is assumed to be the **modulus of rupture** or ultimate fiber stress of the material composing the beam, the following rule may be stated:

Rule I.—*To find the ultimate resisting moment of a beam, multiply the section modulus by the modulus of rupture of the material of which the beam is composed.*

The moduli of rupture for the materials used in building construction may be obtained from the table in *Materials of Structural Engineering*, Part 3.

From rule I; the ultimate resisting moment, which is represented by M_1 is equal to the product of the section modulus and the modulus of rupture, or Ss. If M represents the bending moment, the beam is loaded to its maximum capacity when M and M_1 are equal; that is, the load applied to produce this condition would be the breaking load. For a uniformly distributed load the formula for the bending moment on a beam is $M = \frac{WL}{8}$, in which W equals the total load in pounds, and L the span of the beam in feet.

The bending moment is expressed in foot-pounds while the resisting moment is in inch-pounds; therefore, in equating these values it is necessary to reduce both values to the same denomination, which is accomplished by dividing Ss by 12. This gives the equation

$$\frac{WL}{8} = \frac{Ss}{12}, \text{ then, } W = \frac{8\,Ss}{12\,L}, \text{ or}$$

$$W = \frac{2\,Ss}{3\,L} \qquad (1)$$

in which S = section modulus;
s = modulus of rupture;
L = span of beam, in feet;
W = total load, in pounds.

The uniformly distributed load that will break a beam whose size is known is determined by formula **1**, which may be stated by the following rule:

Rule II.—*To determine the uniformly distributed load, in pounds, that will break a beam, multiply twice the section modulus by the modulus of rupture and divide this product by three times the span, in feet.*

EXAMPLE.—Taking the modulus of rupture of hemlock at 3,500 pounds and using a factor of safety of 4, what uniformly distributed load can be safely carried by a hemlock beam 8 inches by 12 inches, the span being 25 feet?

SOLUTION.—Section modulus $= \frac{b\,d^2}{6} = \frac{8 \times 12^2}{6} = 192$. The formula for the breaking load is $W = \frac{2\,Ss}{3\,L}$; hence, using a factor of safety of 4, the safe uniformly distributed load will be $\frac{2 S s}{4 \times 3 L}$. Substituting the values in this formula, the breaking load is

$$\frac{2 \times 192 \times 3{,}500}{4 \times 3 \times 25} = 4{,}480 \text{ lb. Ans.}$$

WOODEN BEAMS

2. In practice, it is often necessary to determine what size beam will be required to carry a given load; the formula to be used in such a case is obtained by transposing the values in formula **1.** Thus,

$$W = \frac{2 S s}{3 L}, \; 3 W L = 2 S s, \; \frac{3 W L}{2 s} = S, \text{ or}$$

$$S = \frac{3 W L}{2 s} \qquad (2)$$

This formula will give the value of the section modulus that the beam must have in order to furnish the necessary resistance. Then, for a rectangular beam, the formula $S = \frac{b d^2}{6}$ is used, in which the value of S found from the previous formula is substituted and a width or depth is assumed. The other dimension may be readily determined. The following example will illustrate this method:

EXAMPLE.—What size yellow-pine beam is required to support a uniformly distributed load of 500 pounds per foot over a span of 10 feet, the modulus of rupture of yellow pine being 6,000, and a factor of safety of 4 being used?

SOLUTION.—The safe strength of the wood, using 4 as a factor of safety, is $6{,}000 \div 4 = 1{,}500$ lb. per sq. in. Substituting the values of W, L, and s in the formula $S = \frac{3 W L}{2 s}$ gives

$$S = \frac{3 \times (500 \times 10) \times 10}{2 \times 1{,}500} = \frac{150{,}000}{3{,}000} = 50$$

the required section modulus. Assuming 6 in. as the width of the beam, the values of S and b may be substituted in the formula $S = \frac{b d^2}{6}$; thus, $50 = \frac{6 d^2}{6}$; $d^2 = 50$; $d = \sqrt{50} = 7.07$ in. It will therefore be necessary to use a 6″ × 8″ beam. Ans.

3. When a beam is designed to resist the bending moment produced by the loads on it, it will usually be strong enough to resist the vertical and horizontal shear, so that it is not necessary to take into account these stresses in proportioning the size of the beam. However, they must be considered to some extent in designing the beam at its bearing, and care should be taken that no material is cut away at the bottom. If the section is reduced at the bearing by cutting out the under side, the resistance to the horizontal shear at this point will be partially destroyed.

GENERAL NOTES RELATIVE TO WOODEN BEAMS

4. In the framing of floors it is frequently necessary to frame around chimney breasts, elevator shafts, and similar projections. This construction is generally accomplished by framing or hanging the joists from a beam or girder that extends in front and parallel to the face of the projection. This girder is known as a *header* and is, in timber framing, composed of two or more joists placed side by side and spiked together; it is supported at the ends by similar girders extending parallel with the joists, which are termed **trimmers**; the joists that are framed into the header are called **tail-beams.**

Wooden beams or any other timbers entering a party wall of a building constructed of stone, brick, or iron should be separated from the beam or timber entering in the opposite side of the wall by at least 4 inches of solid masonry.

A header or trimmer more than 4 feet long should be hung in stirrup irons of suitable thickness for the size of the timbers. Patent hangers are also frequently used. All beams, except headers and tail-beams, should have one end resting 4 inches in the wall, or on a girder.

The ends of all wooden floorbeams and roof beams resting on brick walls should be cut to a bevel sloping away from the vertical toward the top, this slope being not less than 3 inches for their depth. Except in framed buildings, a floorbeam or roof beam should not be supported on stud partitions. Also, all wooden beams should be bridged with cross-bridging placed not more than 8 feet apart.

Each tier of beams should be anchored to the side, front, rear, or party walls at intervals of not more than 6 feet, with good, strong, wrought-iron anchors not less than 1½ inches by ⅜ inch; these anchors should be fastened to the side of the beams by two or more wrought-iron nails at least ¼ inch in diameter. Where beams are supported by girders, the latter should be anchored to the walls and fastened to each other by iron straps.

Where the ends of wooden beams rest on girders, they should be butted together and strapped by wrought-iron straps, those beams being strapped that were anchored into the wall. The straps are secured to the beams in the same manner as the anchors. If the beams are not butted on the girder, they should lap each other at least 12 inches and should be well spiked or bolted together.

It is necessary to anchor front and rear walls and all piers to the beams of each story. The same size anchors as are required for the side walls should be used and should extend over four beams.

TABLE I

Wood	Factor of Safety
Hemlock	70
Spruce	90
Oak	120
White pine	90
Yellow pine	140

5. The safe uniformly distributed load that a floorbeam will support, as determined by the New York building laws, is found by multiplying the area, in square inches, by its depth, in inches, and dividing this product by the span of the beam, in feet. This result is then multiplied by the value given in Table I for the wood of which the beam is composed.

EXAMPLE.—What will be the safe uniformly distributed load, in pounds, for a yellow-pine beam 3 inches by 10 inches and having a span of 20 feet?

SOLUTION.—Substituting the values stated in the example and solving according to the method explained above gives

$$\frac{3 \times 10 \times 10}{20} \times 140 = 2{,}100 \text{ lb.}$$

the safe uniformly distributed load. Ans.

FLITCH-PLATE GIRDERS

6. A **flitch-plate girder** is composed of a steel plate having a wooden beam bolted on each side. In designing a girder of this kind, it is necessary to proportion it so that the different parts will deflect equally. The load distribution may be disregarded, as it is the same for the wood and steel; also, the span is the same for each portion of the beam, and hence this value and all constants may be eliminated when considering the tendency of the materials to deflect under the same conditions.

By referring to the deflection formula and disregarding the values just mentioned, it will be observed that in order to have equal deflection in the two materials, the ratio of the load to the product of the modulus of elasticity and the moment of inertia must be the same for each material. When the depth of the wooden beams and the plate are the same, the moment of inertia will vary with the breadth; the other values used in determining this property may be disregarded, as they are constant. Then the ratio will be the load divided by the product of the modulus of elasticity and the width of the section for each material.

As the bending moment M is equal to the section modulus multiplied by the modulus of rupture, the value of the latter may be found by dividing the bending moment by the section modulus. The cross-section of a flitch-plate beam is rectangular; therefore, the section modulus is obtained from the formula $\frac{b\,d^2}{6}$; as the width b is the only value that varies, the others may be disregarded, and it may be said that the modulus of rupture, or unit fiber stress, for each material varies as the ratio of the load to the width of the section for each. When s represents the unit fiber stress; W, the load; and b, the width of the section for one material; and s', W', and b' represent the same values for the other material, the relation stated above may be expressed by the proportion:

$$s : s' = \frac{W}{b} : \frac{W'}{b'}$$

It has been stated that the ratio of the load to the product of the modulus of elasticity and the width of the section for each material must be equal; or, when E and E' represent the moduli of elasticity for the materials, $\frac{W}{E\,b} = \frac{W'}{E'\,b'}$; but, as the unit fiber stress for each material varies in the ratio of the load to the width, the former may be substituted for this ratio in the proportion. Thus, $\frac{s}{E} = \frac{s'}{E'}$, or $s : s' = E : E'$; hence, the unit fiber stress varies as the modulus of elasticity for each material. This proportion must always be true in order to provide for a uniform deflection in the two materials.

EXAMPLE.—If a flitch-plate girder 12 inches in depth is required to resist a bending moment of 300,000 inch-pounds, what thickness of spruce planks will be required, the steel plate being $\frac{1}{2}$ inch thick? The modulus of elasticity for structural steel is 29,000,000 and for spruce 1,200,000, while the safe unit stress for steel may be taken at 15,000 pounds.

SOLUTION.—The ratio of the moduli of elasticity for the two materials is $\frac{1,200,000}{29,000,000} = \frac{12}{290}$. Then the safe unit stress that may be employed for spruce is $\frac{12}{290}$ of 15,000 = 620 lb. per sq. in. The bending moment is equal to the modulus of rupture multiplied by the section modulus, or $M = Ss$. As in this instance there are two separate beams, there are also two section moduli, viz.: $\frac{b\,d^2}{6}$ and $\frac{b'\,d^2}{6}$, that have to be added together; hence, in this case, $M = \frac{b\,s\,d^2}{6} + \frac{b'\,s'\,d^2}{6}$. By transposing, it is found that $b = \frac{6\,M - b'\,s'\,d^2}{s\,d^2}$. Considering s and b as the values for wood, and substituting in the formula gives

$$b = \frac{6 \times 300,000 - .5 \times 15,000 \times 144}{620 \times 144} = 8+ \text{ in.}$$

Therefore, each plank should be 8 ÷ 2 = 4 in. in width. Ans.

7. Flitch-plate girders should have two lines of bolts, as shown in Fig. 1, the distance between centers on each line being twice the depth of the girder, while at each end there should be two bolts in the same vertical line. When a flitch-plate girder is connected to a column and it is desired to have the column flanges concealed, the girder may be notched out to fit over the flange. The floor may then be laid up to the column and the bolts in the girder arranged as shown in Fig. 2, which is a sectional plan and elevation of a column

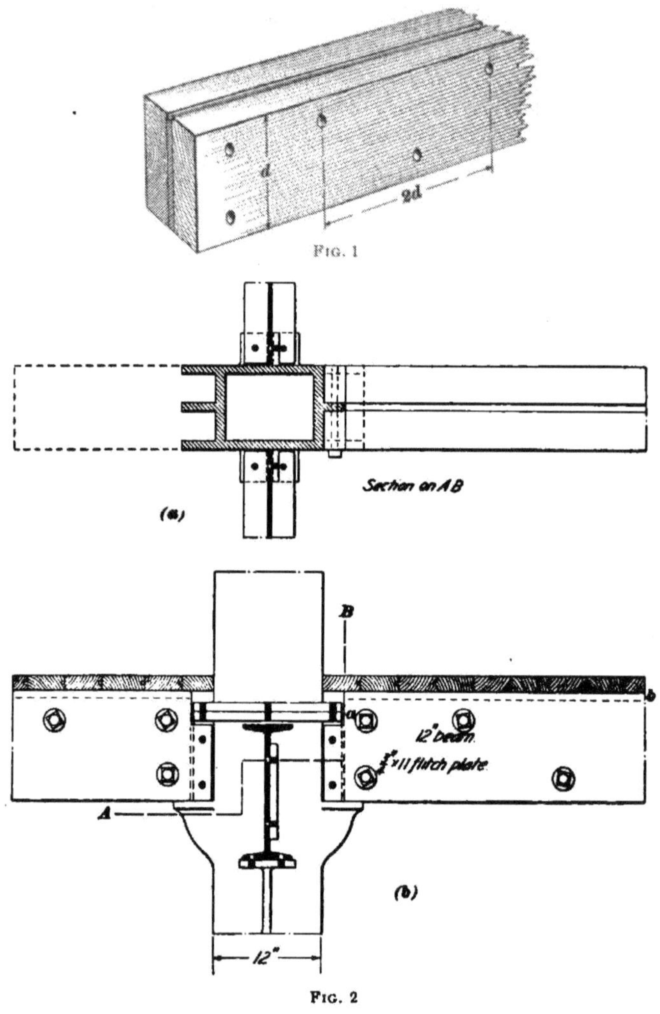

FIG. 1

FIG. 2

with two steel and two flitch-plate girders. The flooring shown in the elevation is omitted in the plan.

If a lug projects from the column, the iron plate may be cut away sufficiently to permit the lug to slip between the wooden beams, which should be bolted to it. Where a flitch-plate beam is connected to an I beam or plate girder, an angle may be riveted to the web of the girder to support the flitch plate, and if the floor is to be raised above the top flange of the I beam, it will be necessary to cut the flitch plate in the same manner, as illustrated in Fig. 2 (*b*).

BUILT-UP WOODEN BEAMS

8. Difficulty is frequently experienced in obtaining timbers of the size required for very heavy work; therefore, the method of building up a beam of smaller pieces is sometimes resorted to. The most efficient built-up beam, and the form commonly used, is constructed by securing narrow pieces of material together so that the joints are vertical, as shown in Fig. 3. They may be fastened with through

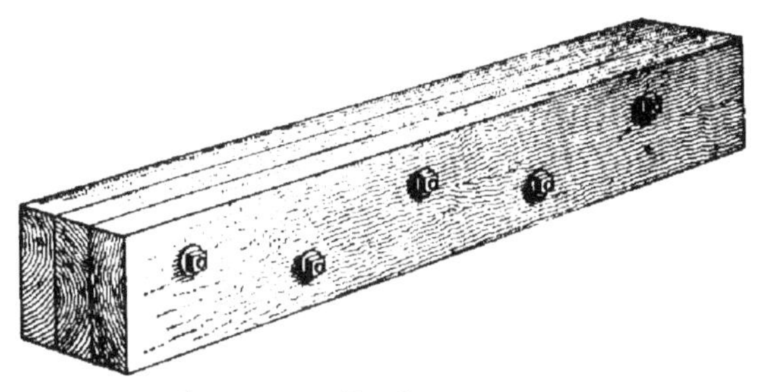

FIG. 3

bolts, which give the more efficient beam, or with lagscrews; in some cases they are simply spiked together. These built-up members are of advantage in trusses for church roofs, especially when it is desired to have the lower member of the truss curved. Fig. 4 shows a portion of a church roof truss, in which the rafter member, as well as the lower chord member, is built up of three pieces $1\frac{7}{8}$ inches thick, secured by lagscrews. Theoretically, a beam built up in this manner is as strong as a solid beam of the same dimensions, and it is found in many cases to be even stronger. This is due to the fact that a solid beam of large dimensions is liable to have defects in the center of the timber or to be composed partly of the heart wood, which is not so strong as the outer layers. Better material

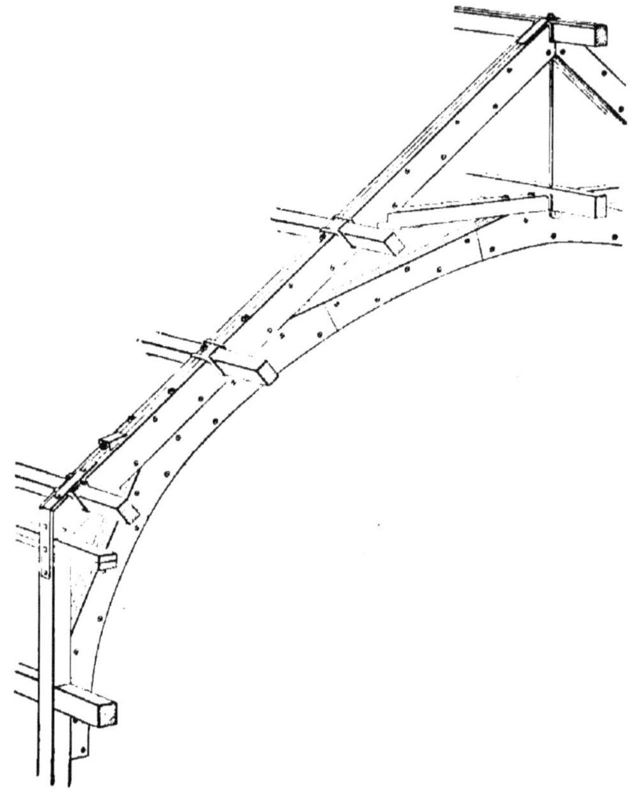

Fig. 4

can be obtained in small pieces and they are more likely to be well seasoned than the larger timbers. By having the lower chord member made up of sections it is also possible to cover up the tension bar entirely, which is sometimes desirable.

9. Another form of built-up beam is one in which the pieces are placed so that the joints are horizontal, as shown in Fig. 5 (*a*). The longitudinal shear produced in this beam

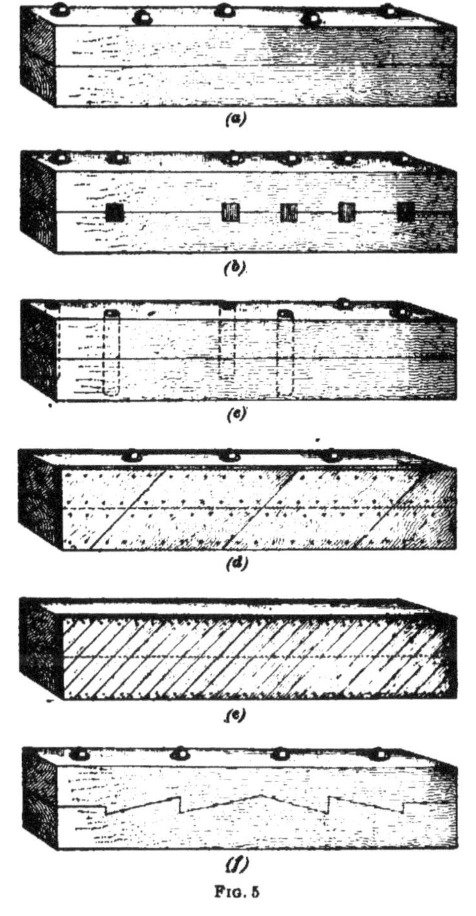

FIG. 5

when it is subjected to transverse stress is resisted only by the bolts. Probably a better construction is one in which hardwood or iron keys are used, in addition to the bolts. This is known as a *keyed beam*, and is shown in (*b*). Another method of resisting the longitudinal shear is by inserting pieces of pipe in the beam, as shown in (*c*). This is a fairly cheap form, but is inferior to the keyed beam.

10. Clark's design for a built-up beam consists of two pieces, one above the other, fastened by narrow strips on each side, running diagonally, those on opposite sides of the beam running in opposite directions. Sometimes the diagonal strips are wide pieces and the heavy timbers are bolted together. These two forms of the Clark beam are shown in Fig. 5 (*d*) and (*e*), respectively. What is known as the *indented beam*, shown at (*f*), is sometimes used. The notches are intended to resist the horizontal shear, or the tendency of the two pieces to slide on each other, but when shrinkage occurs, the indented surfaces are not held tightly together and the efficiency of the beam is greatly diminished.

11. Efficiency of Built-Up Beams.—Table II gives the average efficiency of several kinds of built-up beams. These results were obtained from tests and the efficiency is the ratio of the strength of a built beam to a solid

TABLE II

EFFICIENCY OF BUILT-UP WOODEN BEAMS

Kind of Beam	Deflection	Efficiency Per Cent.
Indented	2.00	69.5
Clark's	2.00	76.0
Piped	1.70	84.6
Oak keys	1.25	90.6
Flat iron keys	1.50	78.6
Square iron keys	1.50	89.5

one of the same size and quality of material. The deflection is expressed in the same ratio, the deflection of a solid beam being the unit of comparison, so that from the table it may be observed that Clark's and the indented beams have a deflection equal to twice that of a solid beam under the same condition of loading, while the deflection of the other beams tested and listed varies from 1¼ to 1¾ times the deflection of a solid beam.

DETAILS OF DESIGN

12. Beam Hangers.—Various devices for supporting the ends of beams where they are carried by headers and trimmers or where they run into the wall, have been manufactured; some of these give very good results, while others are not entirely satisfactory. When the **Goetz hanger,** shown in Fig. 6, is employed the beam is held in place by a nail driven up through the hole *a*; therefore, when the beams are placed opposite on each side of the girder, this arrangement ties the beams together. The pins, or lugs, that enter

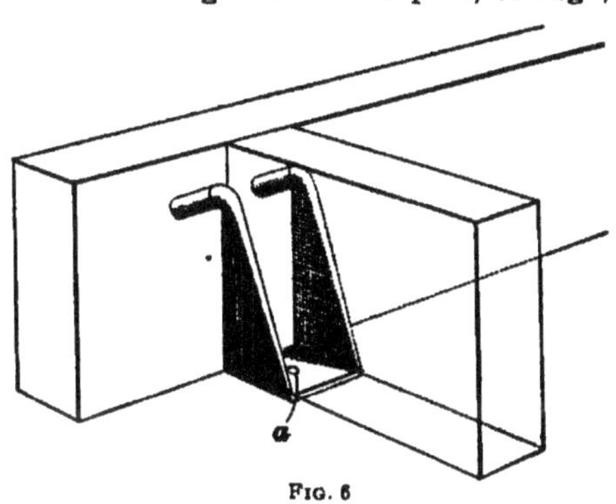

FIG. 6

the main beam are bent at an angle so that the hanger will lock against the beam when the weight is applied to the joist.

The **Van Dorn hanger** for wooden beams is shown in Fig. 7 (*a*), while the double and single hangers used on I beams are shown in (*b*) and (*c*), respectively. The double hanger is formed by bolting two hangers to an iron plate that rests on the top of the I beam; a block of wood is placed on each side of the web of the beam to keep the hanger in position. Fig. 8 shows a special Van Dorn hanger arranged to support joists on the opposite sides of a beam or girder.

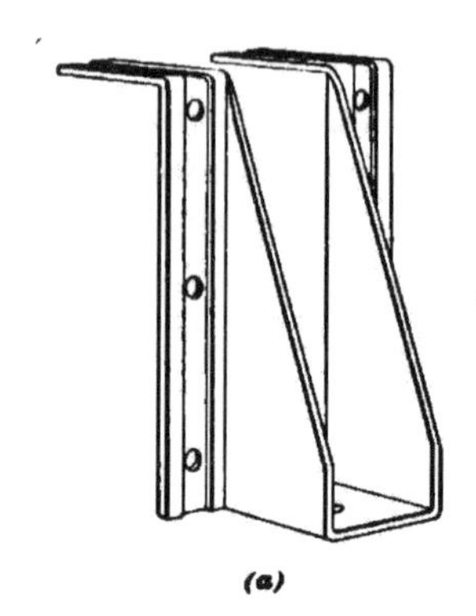

(a)

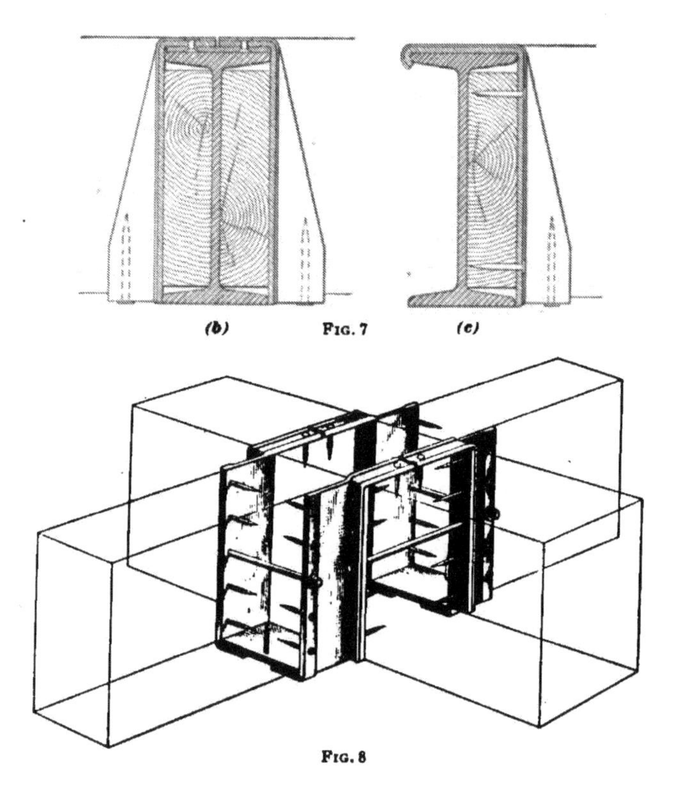

(b) Fig. 7 (c)

Fig. 8

A **stirrup iron hanger** is made by bending an iron strap to fit over the top of the beam and form a loop to hold the joist. Some special stirrup iron hangers are shown

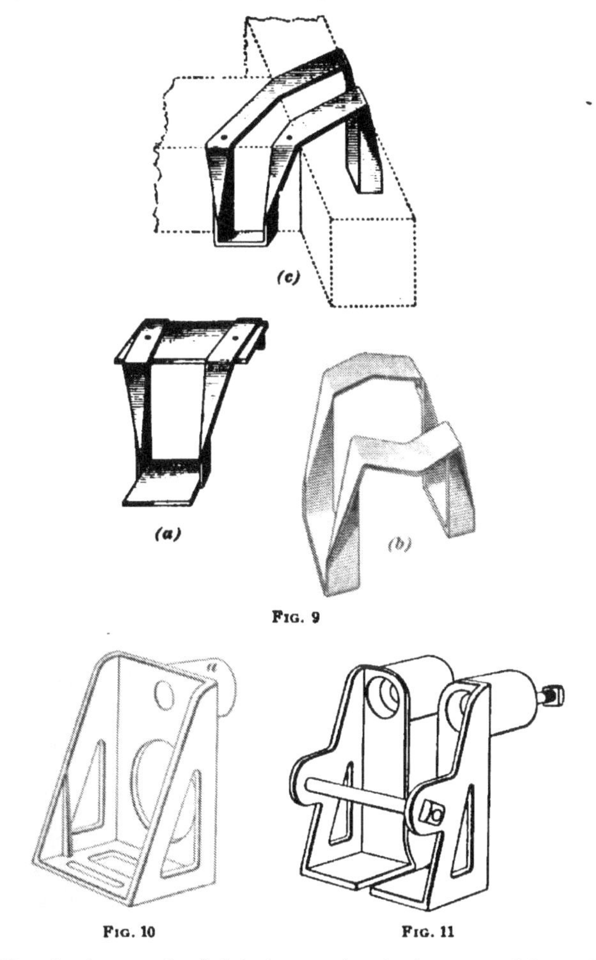

FIG. 9

FIG. 10

FIG. 11

in Fig. 9, the one in (*a*) being a simple hanger with a plate riveted in the loop of the hanger to carry the joist and one on the top to rest on the beam or to be built into the wall.

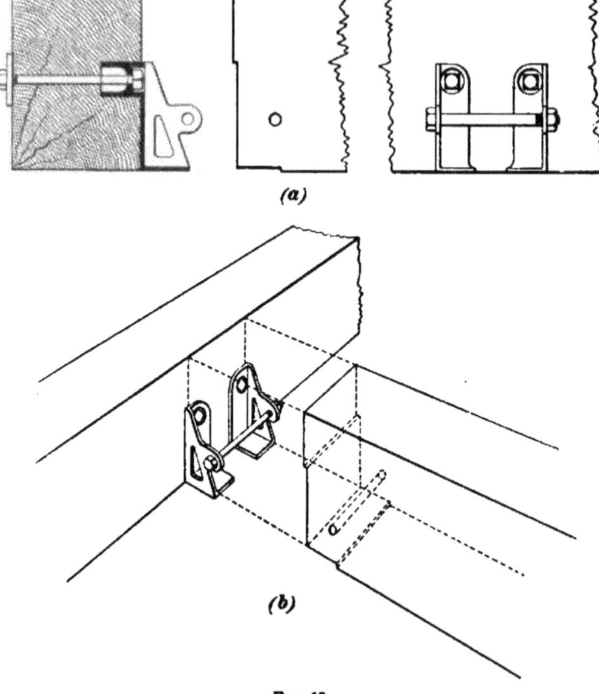

FIG. 12

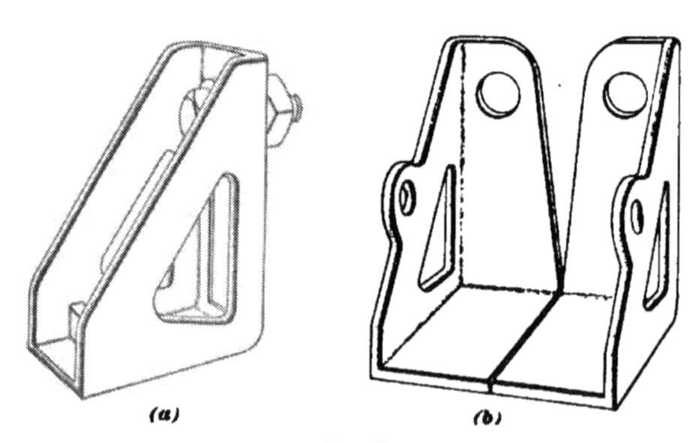

FIG. 13

In (*b*) is shown a double hanger having one side shorter than the other, while (*c*) illustrates a double hanger that is used where the beams are not opposite each other and do not run in the same direction. These hangers are generally held in place by being nailed to the beams that support them.

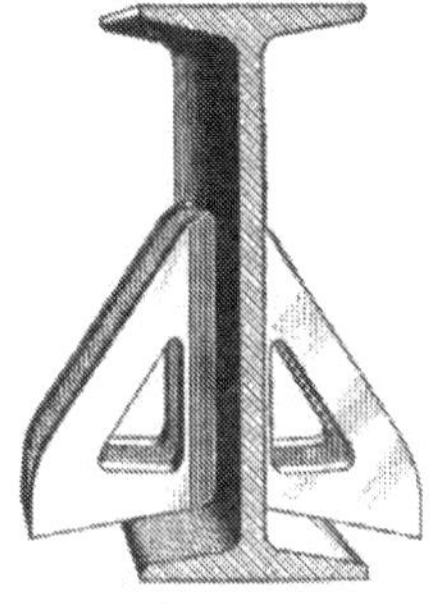

Fig. 14

The **Duplex hanger** for a small joist or beam is shown in Fig. 10; the girder on which it is hung has a hole drilled in it to receive the lug *a*. For wide beams, a double hanger, shown in Fig. 11, is used. The beam is bolted to the hanger, the latter being secured to the girder by bolts run through the lugs that enter the beam. Fig. 12 (*a*) shows a side and front elevation of a double hanger in position, while (*b*) shows a perspective of a

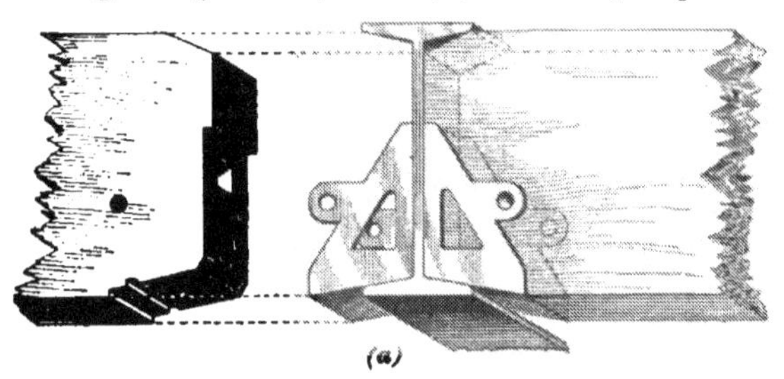

(*a*)

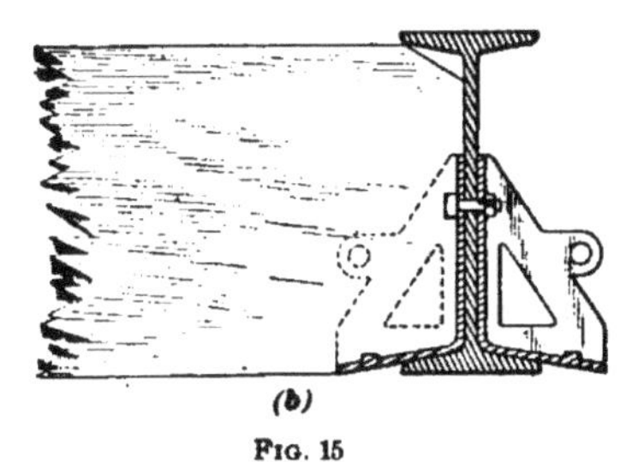

(*b*)

Fig. 15

connection made with such a hanger, the beam being moved out of position and shown in full lines, while the dotted lines indicate its position when held by the hanger. In Fig. 13 (*a*) and (*b*) are shown the hangers used for **I** beams; these are made to fit exactly the flange of the **I** beam and are bolted through the web of the beam. Fig. 14 represents the single hanger, used for narrow beams, in position on the **I** beam. The double hanger used for heavier construction is shown in position in Fig. 15 (*a*) and (*b*), (*a*) showing the manner in which the end of the wooden beam should be cut to fit in the hanger, while (*b*) illustrates the method of bolting the hangers to the **I** beams, and also shows the rib over which the joists must be cut and which serves as a tie when the latter are in place. The Duplex wall hangers, shown in Fig. 16 (*a*) and (*b*), are used extensively and give satisfactory results.

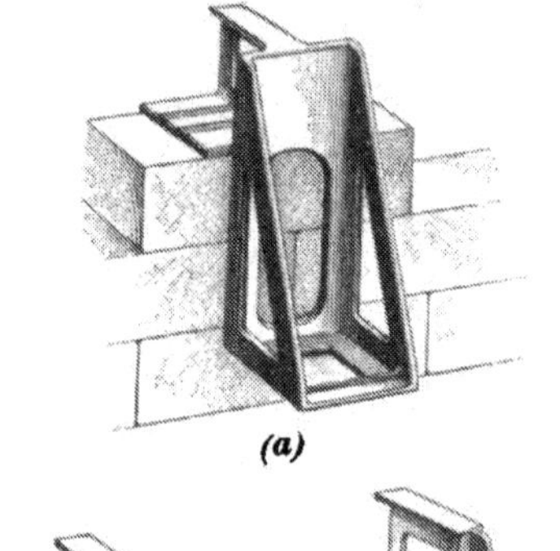

(*a*)

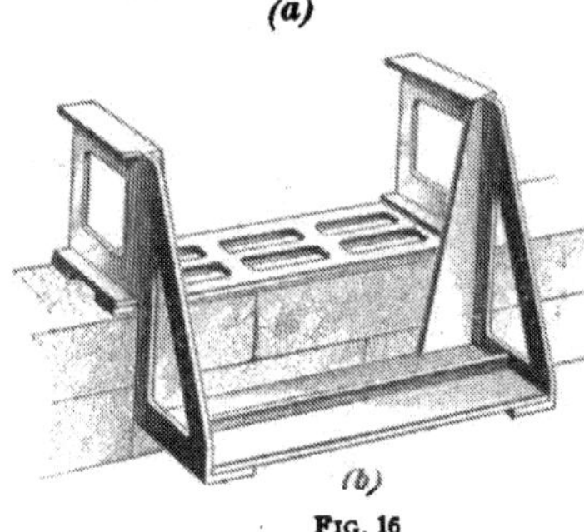

(*b*)

Fig. 16

Fig. 17

13. Tests Made on Beam Hangers.—The results of a test made on a Van Dorn hanger, a stirrup iron hanger, and a Duplex hanger are shown in Figs. 17, 18, and 19. In each case, one end of the **I** beam to which the load was applied rested on a pine block 7 inches thick, that was securely fastened in the hanger; the other end was supported on an iron bar. When 13,300 pounds had been applied to the beam carried by the Van Dorn hanger, shown in Fig. 17, the hanger began to straighten out, and failed at a load of 18,750 pounds. The stirrup iron hanger failed at a load of 13,750 pounds, by pulling off from the header, as shown in Fig. 18; the crushed parts of the header are the points where the stirrup was hung for the test. The beam supported by the Duplex

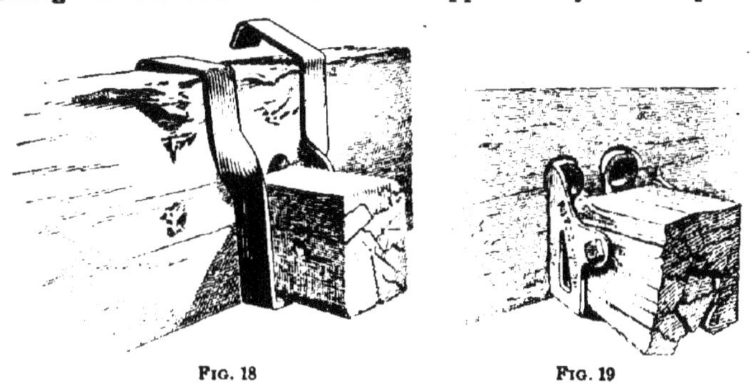

Fig. 18 Fig. 19

hanger, Fig. 19, was loaded to 20,000 pounds, when the wood began to show signs of failing, but the hanger held until a load of 39,550 pounds was applied. At this point, one side of the hanger broke off short under the lug projecting into the header. The Duplex hanger was bolted to the header with two $\frac{3}{4}$-inch bolts through the lugs.

14. Heavy Wooden Construction.—In heavy wooden construction, the wooden beams are carried by iron girders and are sometimes supported on beam ledges or blocks bolted to the girder. Fig. 20 shows a 15-inch **I** beam that carries 8″ × 12″ wooden beams placed opposite each other. These beams rest on beam ledges *a*, of which only one is shown in the illustration. They extend the full length of the girder and are bolted to it with $\frac{3}{4}$-inch bolts. The ledges are notched out to receive the wooden beams and are also cut out to fit around the lower flange of the **I** beam, extending $\frac{3}{8}$ inch below it. The wooden beams, which are cut out to fit the upper flange of the girder, are notched at the top and a $1\frac{7}{8}$-inch anchor plate is set in the full width of the beams; this arrangement ties the beams together. For additional stiffness, a $\frac{7}{8}$-inch cleat b is nailed on each side of the beam, extending from the beam ledge to within $\frac{3}{8}$ inch of the top of the wooden beam; the cleat is fastened to the anchor plate and the beam, thus forming a rigid connection. A $\frac{7}{8}$-inch casing is rabbeted into the beams and extends from one beam to the next, thus enclosing the girder and making a neat finish. The only part of the girder that is exposed is the lower flange, but even this is protected to some extent by the beam ledge. Fig. 21 shows an elevation of the girder with the finished casing at (*a*) and without it at (*b*).

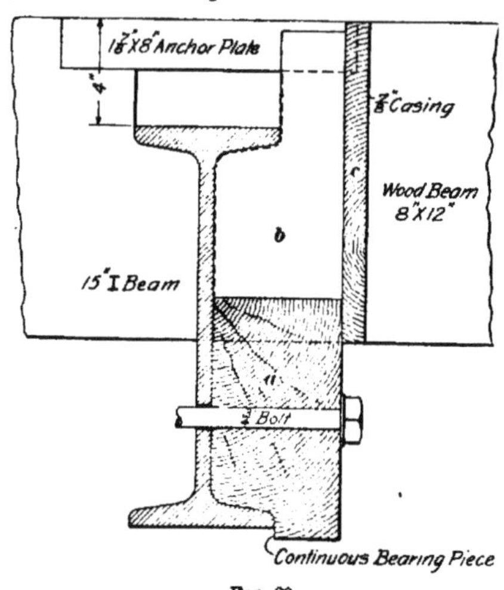

FIG. 20

The arrangement for the casing of a 20-inch girder carrying 8″ × 12″ wooden beams is shown in Fig. 22, in which *a*, *a* are the blocks that are placed under each beam and bolted through the web of the girder in two places, the blocks being counterbored to allow the bolt heads to sink below

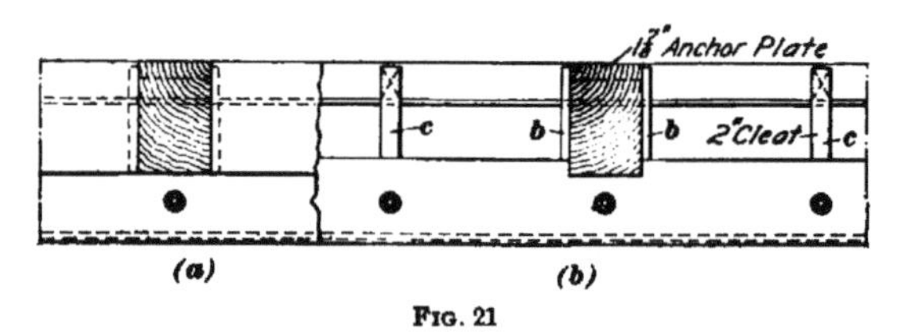

FIG. 21

the surface. The beams are notched and an anchor plate is set in as shown; a cleat, extending from the bottom flange of the **I** beam almost to the top of the wooden beam, is spiked to each side, thus keeping the block, beam, and anchor plate in line and securing them rigidly together. The casings *b*, *c*, and *d* and the molding *e* are then put in place, the piece marked *d* being rabbeted into the beam. Though the arrangements shown in Figs. 20 and 22 can hardly be considered as strictly slow-burning construction, they are especially strong and rigid and may be employed with good results in buildings where it is necessary to provide against the effects of vibration.

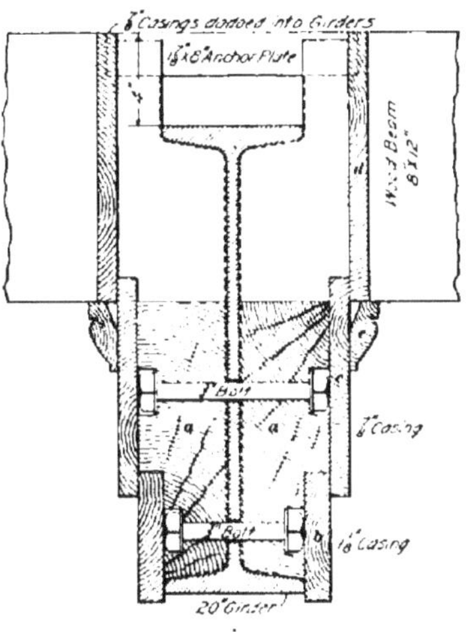

Fig. 22

STEEL BEAMS

15. The **steel beams** used in building construction are generally either channels or I beams. When the word *beam* is used, it is understood that an I beam is meant. For instance, a 12-inch 40-pound beam means an I beam 12 inches in depth, weighing 40 pounds to the lineal foot; a channel of the same size is expressed as a 12-inch 40-pound channel. In designating rolled shapes on working drawings, various systems of abbreviations are used. A 12-inch 40-pound beam may be expressed as 12″ **I** 40 #, or a channel as 12″ **[** 40 #. This is entirely a matter of judgment with the draftsman, or is governed by the practice used in the particular drafting room. As long as the size, character, and weight of the beam are given, it matters little how they are expressed, if intelligibly written.

TRANSVERSE STRENGTH

16. In calculating the strength of steel beams, it is first necessary to find the bending moment, using the methods and rules already given. Then the section modulus required in the beam may be obtained by dividing the bending moment, in inch-pounds, by the quotient obtained by dividing the modulus of rupture by the factor of safety. Assume, for example, the bending moment on a beam to be 50,000 foot-pounds. Reduce it to inch-pounds by multiplying it by 12, which gives 600,000 inch-pounds. The modulus of rupture for structural steel is 60,000 pounds. If a factor of safety of 4 is used, the safe working value of this material will be 60,000 ÷ 4 = 15,000 pounds per square inch. Then 600,000 ÷ 15,000 = 40, the section modulus required.

The approximate section modulus of an I beam, or a channel, may be found by the following rules:

Rule I.—*To obtain the approximate section modulus of an I beam, multiply the sectional area of the beam, in square inches, by the depth, in inches, and divide by the constant 3.2.*

Rule II.—*To obtain the approximate section modulus of a channel, multiply the sectional area of the channel, in square inches, by the depth, in inches, and divide by the constant 3.67.*

Letting A equal the sectional area, in square inches, of an I beam or a channel, and d its depth, in inches, the approximate section moduli S_i and S_c of an I beam and channel, respectively, may be found from the formulas

$$S_i = \frac{A\,d}{3.2} \qquad (3)$$

and

$$S_c = \frac{A\,d}{3.67} \qquad (4)$$

EXAMPLE.—What is the section modulus of a 12-inch I beam, the sectional area of which is 9.01 square inches?

SOLUTION.—Applying the formula,

$$S_i = \frac{A\,d}{3.2} = \frac{9.01 \times 12}{3.2} = 33.8. \quad \text{Ans.}$$

17. Finding the Dimensions of a Steel Beam.—To illustrate the method of calculating the dimensions of a steel beam, let it be required to find what size steel I beams are necessary to support the floor of an office building; this floor rests on brick arches sprung between the beams, and weighs complete 110 pounds per square foot. The building is designed to carry a live load of 40 pounds per square foot. The span of the beams is 20 feet and they are spaced 5 feet on centers. The owner requires that the building have a large factor of safety, and suggests that for the floorbeams a safety factor of 5 be used. The total dead and live load on the floor is 110 pounds + 40 pounds = 150 pounds per square foot. The floor area supported by one beam is 20 × 5 = 100 square feet. Then the total load on one beam is 100 × 150 = 15,000 pounds. The load being uniformly distributed, the formula for the bending moment is $M = \frac{WL}{8}$; substituting the values for W and L, $M = \frac{15{,}000 \times 20}{8} = 37{,}500$, the bending moment in foot-pounds, which, being multiplied by 12, equals 450,000 inch-pounds.

The modulus of rupture for structural steel is 60,000 pounds per square inch, and, since a factor of safety of 5 is required, the safe working value will be 60,000 ÷ 5 = 12,000 pounds per square inch. The bending moment in inch-pounds is 450,000, which, divided by 12,000, gives a section modulus of 37.5.

From the table Properties of Standard I Beams in *Properties of Sections*, it is found that the section modulus of a 12-inch 35-pound beam is 38; hence this size beam should be used.

In selecting beams from the table, care should be taken to obtain the deepest beam of the least weight with the required section modulus. Thus, from the table, it is found that the section modulus of a 10-inch beam, weighing 40 pounds, is 31.7, while a 12-inch beam of 40 pounds, or the same weight as the 10-inch beam, has a section modulus of 41, and, in consequence, possesses nearly one-third more strength, making it, therefore, the more economical beam to use.

18. By way of general review of the subject of beams, the following practical example will be considered: Fig. 23

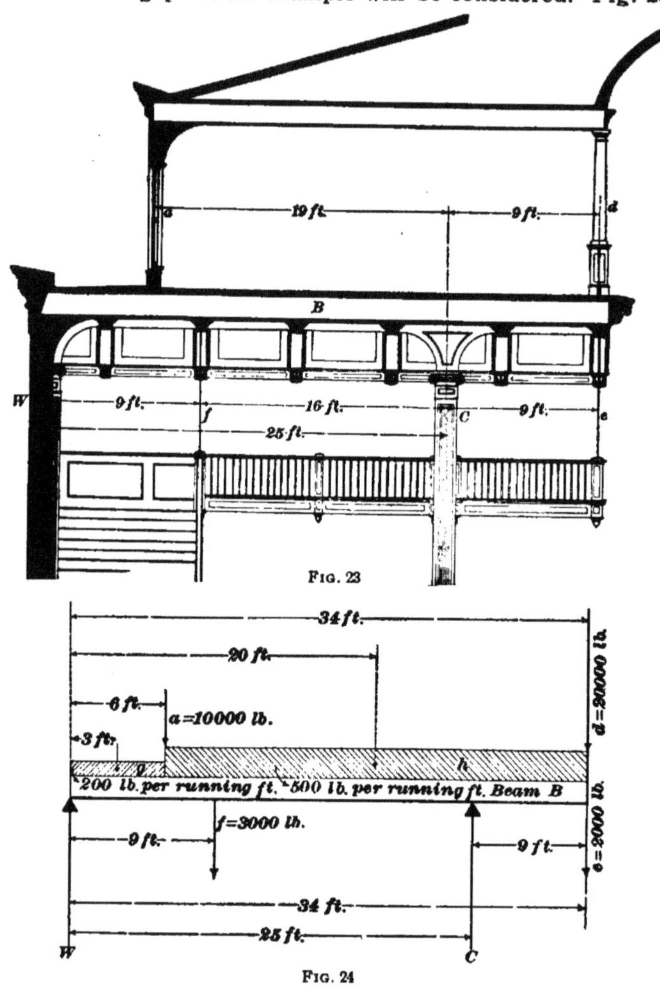

FIG. 23

FIG. 24

shows the transverse sectional elevation of a large department store, in which the girder B is made up of two I beams. What is the size and weight of these steel beams?

Before commencing the calculations, draw the outline diagram, as shown in Fig. 24; this is called a frame diagram. The two supports for the girder are the wall *W* and the column *C*. The loads *g*, *h* on the girder are uniform; the load *h* is due to the weight of the floor, girder, and the ceiling, together with the live load on the floor due to the people, furniture, etc. This load has been assumed to amount to 500 pounds per running foot of the girder. The load *g*, being due only to the ceiling and a portion of the roof, and there being no floor load on it, has been considered as amounting to 200 pounds per running foot.

The girder is also loaded with four concentrated loads: *a* of 10,000 pounds, due to the weight of the light wall and a portion of the roof; *d* of 20,000 pounds, due to the load coming down the small column from a portion of the roof; and two hanging loads *f* and *e*, of 3,000 and 2,000 pounds, respectively, from the weight of the stair landing or hall.

The reactions may now be calculated. The moments about *W* are as follows:

Due to		Foot-Pounds
Load *g* (200 × 6 = 1,200 lb.) .	1,200 × 3 =	3600
Load *h* (500 × 28 = 14,000 lb.) .	14,000 × 20 =	280000
Load *a*	10,000 × 6 =	60000
Load *f*	3,000 × 9 =	27000
Load *d*	20,000 × 34 =	680000
Load *e*	2,000 × 34 =	68000
Total moments	=	1118600

This, divided by the distance between the supports, or the span, 25 feet, will give 44,744, the load, in pounds, coming on the column *C*; or, in other words, the reaction at *C*. The loads are as follows:

Load *g* =	1200	pounds
Load *h* =	14000	pounds
Load *a* =	10000	pounds
Load *f* =	3000	pounds
Load *d* =	20000	pounds
Load *e* =	2000	pounds
Total load =	50200	pounds

Then, the reaction at *W* is 50,200 − 44,744 = 5,456 pounds.

Find the point between the two supports *W* and *C*, where the shear changes sign.

Working out on the beam from *W*, the first load encountered and to be deducted from the reaction *W* is the uniform load *g*, equal to 200 × 6 = 1,200 pounds. Then, 5,456 (reaction at *W*) − 1,200 (load *g*) = 4,256 pounds. The next load on the beam is the concentrated load *a* of 10,000 pounds, which is much more than the remaining portion of the reaction *W*. The greatest bending moment occurring between the column and the wall is, therefore, at the point *a*, and is equal to 5,456 (reaction at *W*) × 6 = 32,736 foot-pounds, from which is to be taken the moment of the load *g* of 1,200 × 3 = 3,600 foot-pounds. Then, 32,736 − 3,600 = 29,136 foot-pounds, the greatest bending moment between the supports *C* and *W*.

Again referring to the diagram, Fig. 24, it is seen that there is quite a bending moment directly over the column *C*, due to the two concentrated loads *d*, *e* on the end of the beam and the portion of the uniform load *h* overhanging the support *C*. This portion of the beam may be considered as a cantilever; the bending moment at *C* is equal to the sum of the moments of all the loads on the overhanging portion of the beam, which are:

Load		Foot-Pounds
d	20,000 × 9.0 =	180000
e	2,000 × 9.0 =	18000
h (500 × 9 = 4,500 lb.) .	4,500 × 4.5 =	20250
Total moments	=	218250

218,250 × 12 = 2,619,000 inch-pounds. This, divided by 20,000, the safe working value of structural steel (using the modulus of rupture of 60,000 pounds ÷ 3, the safety factor used in this case) = 131, the required section modulus in the two beams. Then, 131 ÷ 2 = 65.5, section modulus required in one of the beams.

From the table in *Properties of Sections*, it is found that the section modulus of a 15-inch 55-pound beam is 68.1. The 18-inch beam of the same weight has a section modulus of 88.4, and while this is in excess of the required amount, it is the preferable beam to use, two of the kind being required.

19. Relation Between Span and Depth of Beam.—In order to select beams that will not deflect too much under the load that they are required to sustain, the depth of the beam, in inches, should never be less than half the span of the beam, in feet. Thus, if the span of the beam be 20 feet, a beam not less than 10 inches in depth should be used to avoid excessive deflection.

20. Separators for I Beams.—In building construction, it frequently happens that a single I beam is insufficient to carry the imposed load. Where heavy loads, such as brick walls, vaults, etc., are to be supported, a single I beam is inadequate; therefore, two or more beams are placed side by side and bolted together, with steel **separators** between, as shown in Fig. 25, or with cast-iron separators, as shown in Fig. 26 (*a*) and (*b*). In (*a*) is shown a type of cast-iron separator so formed that the bolts pass through a hole in the

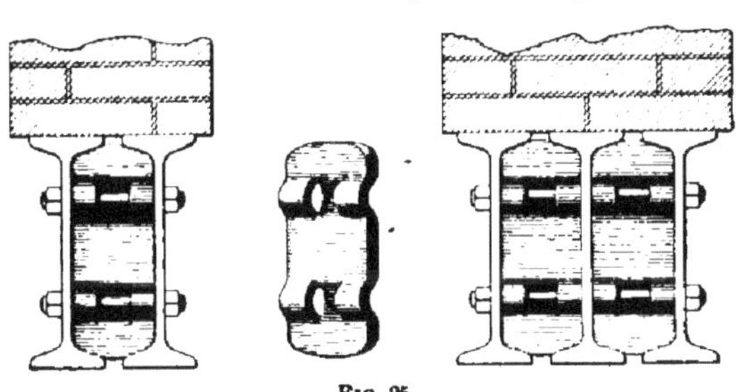

Fig. 25

swollen portion of the casting. Another pattern of cast-iron separator is shown in (*b*), in which the connecting bolt passes through a lug cast on the separator. A common use for a double beam secured together by bolts and separators is shown at *a*, *a*, Fig. 26 (*c*). These separators hold the compression flanges of the beams in position, preventing deflection sideways, and also, in a measure, cause the beams to act together, distributing the load uniformly on both. Separators should be spaced from 6 to 7 feet apart throughout the length of the beam; they should also be provided at the supports and at the points where heavy loads are concentrated.

Standard separators may be obtained in such widths that the inner edges of the flanges of the two beams connected are about $\frac{1}{4}$ inch apart. For beams 10 inches in depth and under, one $\frac{3}{4}$-inch bolt is used through the separator, while

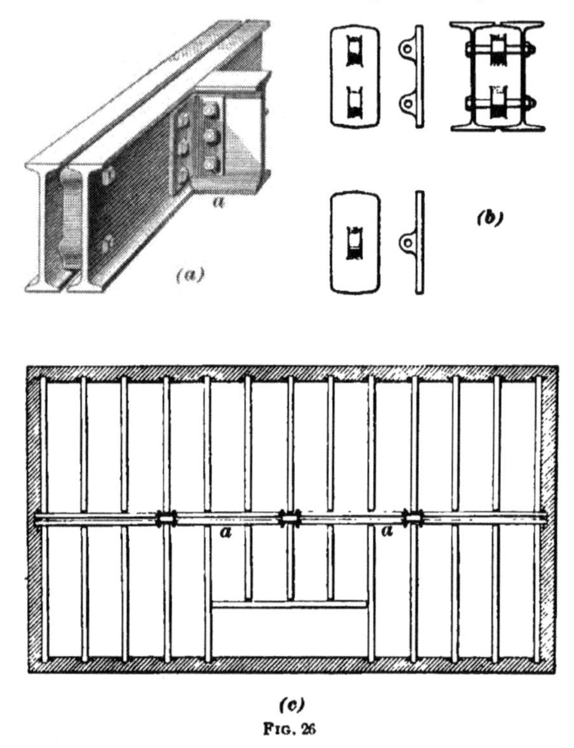

Fig. 26

two $\frac{3}{4}$-inch bolts are used for connecting beams over 10 inches in depth. Where two bolts are used, the distance between centers is usually made 10 inches for beams 18 and 20 inches in depth, 7 inches for 15-inch beams, and 6 inches for 12-inch beams. Separators for 18-inch and 20-inch beams weigh about 20 pounds each; for 15-inch beams, about 12 pounds;

for 12-inch beams the weight is from 8 to 10 pounds; for 8-, 9-, and 10-inch beams, from 5 to 7 pounds; and for 4-, 5-, and 6-inch beams, from 1 to 4 pounds.

21. I Beam Girders.—In designing floors of buildings it is desirable to have a minimum number of interior supporting columns, consistent with economy. A beam girder consisting of a pair of I beams is frequently advantageous for supporting the steel floorbeams, as shown at *a* in Fig. 26 (*a*).

Girders composed of two or more I beams are commonly used to span openings in brick walls. If the wall to be supported is thoroughly seasoned and without openings, the

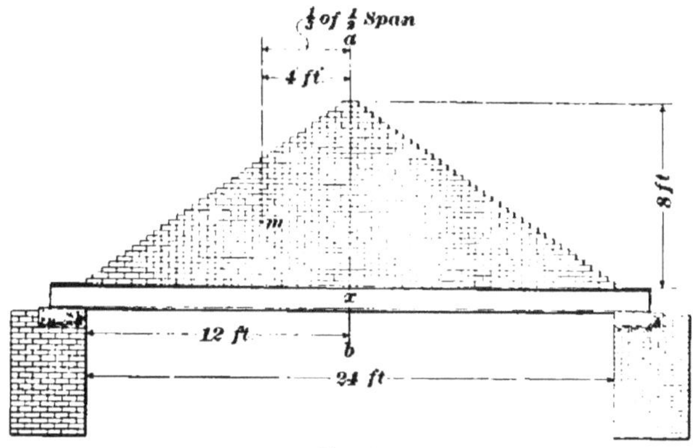

FIG. 27

weight carried by the girder can safely be assumed as the weight of a triangular piece of brickwork, whose altitude is one-third of the span of the girder. If the wall is newly built, or has openings for windows or other purposes, the girder must be designed to carry the entire wall above it between the supports.

EXAMPLE.—Required, the size of a steel I-beam girder to carry a wall 12 inches thick, made of hard brick laid in lime mortar; there are no openings in the wall above the girder, nor does the wall support floor joists or roof beams, while the span of the opening is 24 feet.

SOLUTION.—Draw the diagram as shown in Fig. 27. The area of the triangular piece of brickwork is $24 \times 4 = 96$ sq. ft., since the area of a triangle is equal to the base, multiplied by one-half the altitude.

As the wall is 1 ft. thick, there are 96 cu. ft. in this triangular piece. The weight of brickwork in lime mortar per cubic foot is 120 lb. Then, 96 × 120 = 11,520 lb., the load on the girder.

The bending moment may be determined by the formula or rule given in *Beams and Girders*, Part 1, for a beam carrying a triangular load, or it may be determined by calculating the moments, as follows: The reactions at the two supports are each equal to half the load, or 11,520 ÷ 2 = 5,760 lb. The greatest bending moment is at the center of the beam. Then, the moment of the reaction about the point x is 5,760 × 12 = 69,120 ft.-lb. But, counteracting this, and to be deducted from it, is the moment of the load at the left of x, equal to half of the triangular piece of brickwork. The moment of this load about the point x is equal to the product of its weight, multiplied by the horizontal distance from a vertical line through its center of gravity to the point x. Take the line $a\,b$ as the base of a triangle, remembering that a line drawn parallel to the base line of a triangle, at a distance of one-third of the altitude from it, always passes through its center of gravity. Now, the distance from the point x to the vertical line through the center of gravity m of the triangle is 4 ft., and the moment due to the triangular piece of brickwork to the left of the center is 5,760 × 4 = 23,040 ft.-lb. Deducting this from the moment of the reaction already found, the calculation is: 69,120 − 23,040 = 46,080 ft.-lb., the bending moment on this beam or girder; or, 46,080 × 12 = 552,960 in.-lb. This calculation may be checked by applying the formula $\frac{WL}{6}$. The bending moment, in inch-pounds, being 552,960, using a safe working value, or fiber stress, of 15,000 lb., the section modulus required would be 552,960 ÷ 15,000 = 36.86. From the table in *Properties of Sections*, it is seen that the section modulus of a 12-in. 35 lb. beam is 38, which gives the required strength in this case. It may be found to be better practice to use two channels instead of one I beam, for the top flange of the I beam may be too narrow to properly support the brick wall, while the two channels placed side by side, with separators between, could be made of the same thickness as the wall.

22. Connections.—The standard connections for the principal sizes and weights of steel I beams are illustrated in Fig. 28. These connections are based on an allowable shearing stress of 10,000 pounds per square inch, a bearing stress of 20,000 pounds per square inch on the bolts, and an extreme fiber stress of 16,000 pounds. They may be used for beams whose spans are not less than those given in Table III.

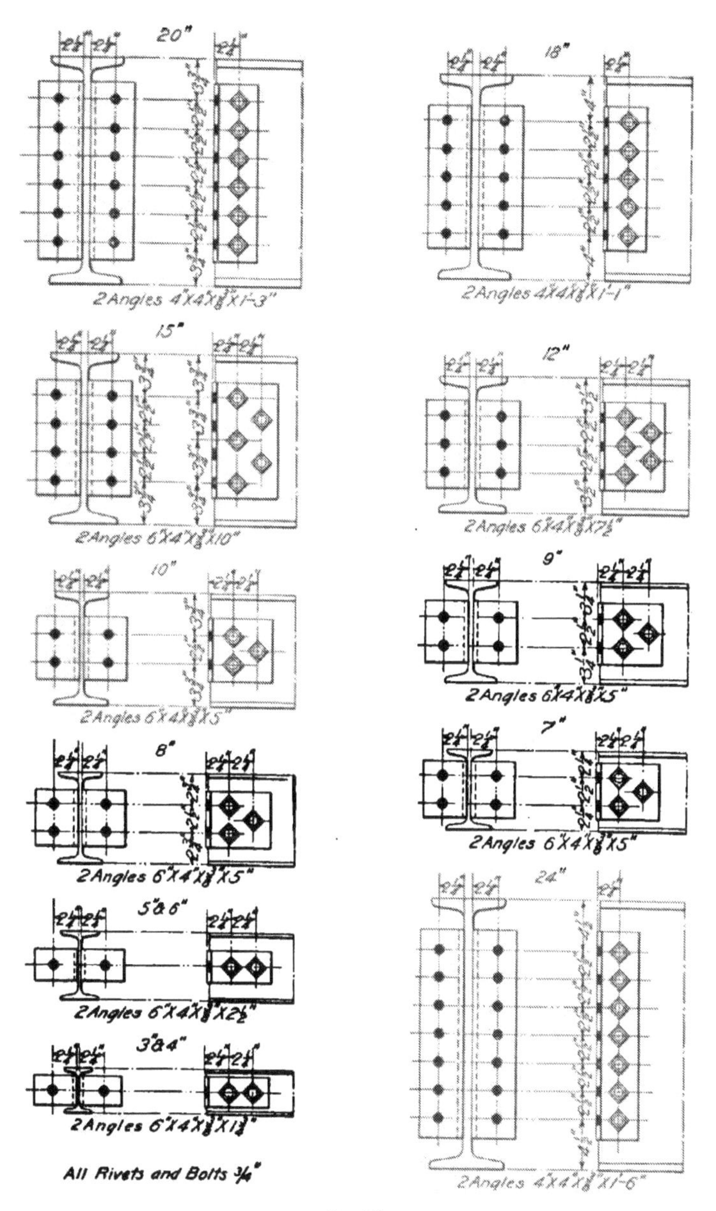
20"
2 Angles 4"X4"X3/8"X1'-3"
18"
2 Angles 4"X4"X3/8"X1'-1"
15"
2 Angles 6"X4"X3/8"X10"
12"
2 Angles 6"X4"X3/8"X7½"
10"
2 Angles 6"X4"X3/8"X5"
9"
2 Angles 6"X4"X3/8"X5"
8"
2 Angles 6"X4"X3/8"X5"
7"
2 Angles 6"X4"X3/8"X5"
5"&6"
2 Angles 6"X4"X3/8"X2½"
24"
2 Angles 4"X4"X3/8"X1'-6"
3"&4"
2 Angles 6"X4"X3/8"X1½"
All Rivets and Bolts ¾"

FIG. 28

TABLE III

MINIMUM SPANS FOR I BEAMS, WITH STANDARD CONNECTIONS, THAT ARE UNIFORMLY LOADED TO REALIZE A UNIT STRESS OF 16,000 POUNDS

Depth of Beam	Weight per Foot	Minimum Safe Span	Depth of Beam	Weight per Foot	Minimum Safe Span	Depth of Beam	Weight per Foot	Minimum Safe Span
Inches	Pounds	Feet	Inches	Pounds	Feet	Inches	Pounds	Feet
3	5.50	1.7	9	30.0	6.8	15	80.0	15.9
3	6.50	1.2	9	35.0	7.5	15	85.0	16.4
3	7.50	1.2	10	25.0	9.3	15	90.0	17.0
4	7.50	2.8	10	30.0	8.1	15	95.0	17.5
4	8.50	2.2	10	35.0	8.8	15	100.0	18.1
4	9.50	2.0	10	40.0	9.6	18	55.0	13.7
4	10.50	2.2	12	31.5	7.3	18	60.0	11.9
5	9.75	4.1	12	35.0	7.7	18	65.0	11.8
5	12.25	3.3	12	40.0	8.2	18	70.0	12.4
5	14.75	3.7	12	45.0	9.6	20	65.0	13.9
6	12.25	5.6	12	50.0	10.2	20	70.0	12.5
6	14.75	4.8	12	55.0	10.8	20	75.0	12.8
6	17.25	5.3	15	42.0	10.2	20	80.0	14.8
7	15.00	4.9	15	45.0	9.4	20	85.0	15.2
7	17.50	3.8	15	50.0	9.7	20	90.0	15.7
7	20.00	3.6	15	55.0	10.3	20	95.0	16.2
8	18.00	6.2	15	60.0	10.8	20	100.0	16.7
8	20.25	5.1	15	65.0	12.8	24	80.0	17.7
8	22.75	4.8	15	70.0	13.4	24	85.0	16.1
8	25.25	5.1	15	75.0	13.9	24	90.0	16.1
9	21.00	7.7	15	80.0	14.5	24	95.0	16.6
9	25.00	6.2				24	100.0	17.1

Beams that are shorter than this are, as a consequence, able to support greater loads without exceeding the unit stress of 16,000 pounds on which the standard connections are based. The joints illustrated would, in such cases, be exposed to a stress greater than that for which they are intended. To avoid such excessive stresses the minimum safe spans have been specified in Table III for the various beams. The larger beams to which these connections are made are supposed to have webs not less than $\frac{9}{16}$ inch in thickness. If these beams or girders are framed opposite one another, into another beam or girder whose web is less in thickness than $\frac{9}{16}$ inch, the length of the minimum spans should be increased. The reason for this is that a thinner web cannot carry so large a load. To prevent the overloading of the connections it is necessary to increase the length of the spans of the adjoining beams in the ratio of the $\frac{9}{16}$-inch web to the thickness of the thinner web.

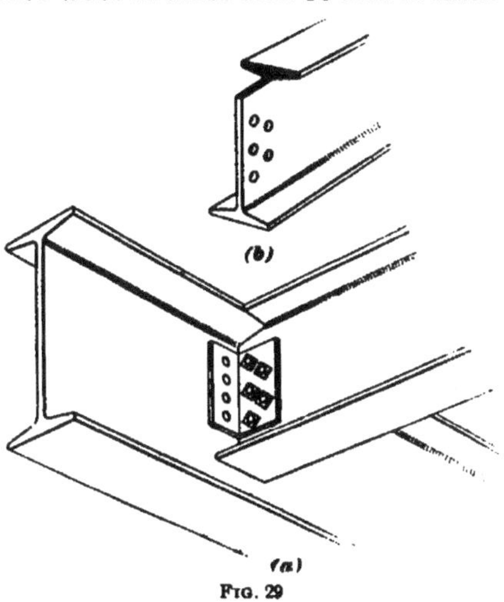

FIG. 29

For instance, a beam 7 inches deep, weighing 20 pounds per foot should, according to Table III, have a minimum span of 3.6 feet when connected to a beam with a $\frac{9}{16}$-inch web. In case the latter web is only $\frac{3}{8}$ inch thick, the following proportion will give the length x of the longer beam required:

$$x : 3.6 = \frac{9}{16} : \frac{6}{16}, \text{ or } x = \frac{3.6 \times \frac{9}{16}}{\frac{6}{16}} = 5.4 \text{ feet}$$

The standard connections given are designed for $\frac{13}{16}$-inch holes and $\frac{3}{4}$-inch diameter rivets or bolts. Connection angles may, if so specified, be riveted instead of bolted to the beams; but, unless otherwise ordered, bolted connections are generally used.

When beams having very short spans are loaded to their full capacity, the shear at the end, which must be transmitted through the connections, becomes so great that connections stronger than the standard must be used. Table III gives the limits of lengths below which the standard connections cannot be used and below which special designs should be made.

Where floorbeams of different depths are to be joined to a larger girder and the tops or bottoms of the beams are to be level, a special construction is required. An example of this is given in Fig. 29 (*a*), in which two beams are joined with their tops flush. It was necessary to cut the flange of one enough to receive half the flange of the other, as shown in (*b*). This method of cutting the beam is called *coping*.

LATERAL STRENGTH OF BEAMS

23. Obtaining the Fiber Stress of Beams.—Where beams are subjected to a horizontal force in addition to the vertical load, some means must be provided to resist the lateral flexure. The beams in a floor composed of arches receive a horizontal thrust in addition to the vertical load. It is assumed that, for the interior beams, the thrusts on opposite sides, due to the dead load, balance each other; therefore, the horizontal thrust is calculated for the live load only. However, in beams which receive a thrust on one side only, as end beams, and beams around shafts and wells, the total load is used in figuring the horizontal thrust. The total stress, due to both the horizontal and vertical loads, should not exceed the allowable fiber stress. The stress due to the lateral force is obtained by the formula

$$s_h = \frac{T c x^2}{I'} \qquad (5)$$

in which s_h = stress, in pounds per square inch, due to lateral forces;

T = thrust of arch, in pounds per lineal foot;

c = distance of extreme fiber from neutral axis, in inches;

x = distance between tie-rods or lateral supports, in feet;

I' = moment of inertia about vertical axis of section, or axis at right angles to line of application of lateral forces.

When the web of the I beam is placed vertically, as usual, c is equal to $\frac{b}{2}$, b representing the width of the flange in inches. Then the preceding formula becomes

$$s_h = \frac{T b x^2}{2 I'} \qquad (6)$$

The formula is derived from the equation, bending moment M = resisting moment M_1.

The portion of the beam between two adjacent tie-rods may be regarded as a beam with fixed ends; the bending moment for a uniform load will then be $\frac{W L}{12}$. The distance between supports is represented by x in formula 5; therefore, x will be used in this formula instead of L, while instead of W, the thrust per lineal foot, or T, will be used. The load is equal to $T x$ and the bending moment created by it is $\frac{T x \times 12 x}{12}$, or $T x^2$, considering the span in inches, as usual.

The resisting moment is equal to the section modulus multiplied by the safe unit fiber stress, or $S s$. As explained in *Properties of Sections*, the section modulus is equal to the moment of inertia divided by the distance from the neutral axis to the farthest edge of the section, or $S = \frac{I}{c}$. In this case, the moment of inertia is taken with respect to the neutral axis parallel to the web; consequently, c equals half the width of the flange, or $\frac{b}{2}$, and $S = \frac{2 I}{b}$. Hence, substituting the values thus obtained for M and M_1 gives $T x^2 = \frac{2 I s}{b}$. Then, $s = \frac{T b x^2}{2 I}$.

24. Obtaining the Horizontal Thrust.—The horizontal thrust is found by the formula

$$T = \frac{3 w L^2}{2 r} \qquad (7)$$

in which T = pressure or thrust, in pounds per lineal foot of arch;

w = load on arch, in pounds per square foot, uniformly distributed;

L = span of arch, in feet;

r = rise of arch, in inches.

The thrust of an arch having a concentrated load at the center equal to W, is found from the formula

$$T = \frac{3WL}{r} \qquad (8)$$

If the floor arch is flat, the value of r may be taken as the effective depth of the arch, as will be explained later.

25. Obtaining the Resultant Stress.—In a simple beam a vertical load produces a maximum compressive stress of s_v at the extreme top fibers and a maximum tensile stress of s_v in the extreme bottom fibers. A horizontal load, as the thrust of an arch, will produce a compressive stress of s_h in the extreme fibers on the side toward the thrust and a tensile stress of s_h on the fibers farthest from the thrust. Therefore, the compressive stress due to both vertical and horizontal loads, on the upper surface of the beam at the corner toward the horizontal thrust, is $s_v + s_h$. On the corner of the beam diagonally opposite, there is a tensile stress of $s_v + s_h$. On the other corners the resultant stresses are $s_v - s_h$ and $s_h - s_v$.

26. Therefore, $s_v + s_h$ should equal the safe working stress per square inch of the material in the beam. It is customary when combining stresses due to vertical loading and horizontal arch thrust, to use a little higher safe fiber stress than is usual under other circumstances, because the mortar in the joints of the arch takes up a considerable part of the horizontal thrust. For interior beams, the safe stress is taken at 20,000 pounds per square inch and for exterior beams, 18,000 pounds is allowed; hence, the formula for interior beams may be written

$$s_v + s_h = 20{,}000 \text{ pounds} \qquad (9)$$

and for exterior beams,

$$s_v + s_h = 18{,}000 \text{ pounds} \qquad (10)$$

In each of the above formulas, however, s_v must not exceed 16,000 pounds, which is the same quantity that is usually taken for the safe working unit tensile stress.

EXAMPLE.—Determine whether 15-inch 42-pound I beams have sufficient strength to support a brick-arch floor over a span of 15 feet. The rise of the arches is assumed to be 6 inches, the span is taken at 4 feet, and it is decided to place the tie-rods 4 feet apart. The dead load is 150 pounds per square foot and the live load 200 pounds per square foot.

SOLUTION.—The thrust of the arch, due to the live load, is obtained from formula **7**, $T = \frac{3wL^2}{2r}$ where w = 200 pounds per square foot, L = 4 feet, and r = 6 inches; therefore,

$$T = \frac{3 \times 200 \times 4 \times 4}{2 \times 6} = 800 \text{ lb. per ft.}$$

From the table giving the properties of standard I beams in the paper on *Properties of Sections*, the moment of inertia with respect to an axis parallel to the web of a 15-in. 42-lb. I beam is 14.62 and the width of the flange b is 5.50. Then substituting in formula **6** for the live load only, $s_h = \frac{800 \times 5.50 \times 5 \times 5}{2 \times 14.62} = 3{,}762$ lb., and for the total load, $s_h = \frac{1{,}400 \times 5.50 \times 5 \times 5}{2 \times 14.62} = 6{,}583$ lb.

The section modulus of this beam, obtained from the same table in *Properties of Sections*, is 58.9; therefore,

$$s_v = \frac{Wl}{8S} = \frac{350 \times 4 \times 15 \times 15 \times 12}{8 \times 58.9} = 8{,}022 \text{ lb.}$$

Now for the inside beams s_v = 8,022, which is less than the allowable stress of 16,000, and $s_v + s_h$ = 8,022 + 3,762 = 11,784, which is less than the allowable stress of 20,000; therefore, the inside beams are sufficiently strong. For the outside beams $s_v + s_h$ = 8,022 + 6,583 = 14,605, which is less than the allowable stress of 18,000; hence, these beams are also amply strong.

SPACING OF TIE-RODS

27. When an arch is placed between two beams, tie-rods should be supplied to resist its thrust. The conditions are shown in Fig. 30, in which $abcd$ represents a flat tile arch sprung between the I beams e and f. In this case the nominal depth of the arch is considered as 10 inches. The dotted line *dgc* represents the theoretical line of pressure in the flat arch while *gh* may be considered as the theoretical rise, or as it is called, the ***effective depth*** of the arch. The point *g* may be

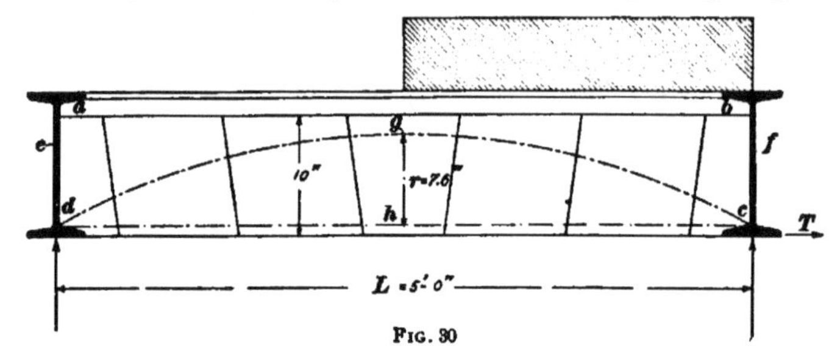

FIG. 30

taken as a center of moments, and the thrust of the arch to be resisted by the tie-rods may be represented by the force T. For a section 1 foot in width, the load on one-half of the arch is equal to $\frac{wL}{2}$ when w equals the load per square foot of surface and L the span of the arch, in feet. The moment of the force T about the point g must equal the moment of the load about the same point; and as the lever arm r of the force T is in inches, the lever arm of the load must also be in inches in order to equate the expressions; hence, the lever arm of the load will be $\frac{12L}{4}$ and its moment $\frac{wL}{2} \times \frac{12L}{4} = \frac{3wL^2}{2}$. The moment of the force T about the point g is Tr. Then, as these moments must be equal, $Tr = \frac{3wL^2}{2}$. The strength of one rod is equal to its area A multiplied by its safe unit fiber stress, and this value divided by the resistance T required in 1 foot of length of beam will give the distance between rods. Hence, assuming the safe unit fiber stress to be 15,000 pounds, the spacing of the rods is obtained from the formula

$$x = \frac{A \times 15{,}000}{T} \qquad (11)$$

From the formula $Tr = \frac{3wL^2}{2}$, the value of T is found to be $\frac{3\,w\,L^2}{2\,r}$. Substituting this value for T in formula **11** gives $x = \frac{A \times 15{,}000}{\frac{3\,w\,L^2}{2\,r}}$, or $x = \frac{A \times 15{,}000 \times 2\,r}{3\,w\,L^2}$. By cancelation the formula becomes

$$x = \frac{10{,}000\,A\,r}{w\,L^2} \qquad \textbf{(12)}$$

In the formulas just given, the load w includes the weight of the arch as well as the load on it. Ordinarily, $\frac{3}{4}$-inch tie-rods are used in floor construction; these should not be placed farther apart than 6 feet. In calculating the spacing of the rods the area should be taken at the root of the thread.

EXAMPLE.—In the example in Art. **26**, what size rods will be required if they are placed 5 feet apart?

SOLUTION.—Transposing formula **12**, $A = \frac{x\,w\,L^2}{10{,}000\,r}$; then substituting the values obtained from the example in Art. **26** in this formula gives

$$A = \frac{5 \times 350 \times 16}{10{,}000 \times 6} = .466 \text{ sq. in.}$$

A 1-in. rod, which has an area of .550 sq. in. at the root of the thread is the nearest commercial bar that can be used, for a $\frac{7}{8}$-in. rod has an area at root of thread of only .420 sq. in. Ans.

28. The effective depth and nominal depth of flat tile arches, in inches, are given below:

Nominal depth . .	6.0	7.0	8.0	9.0	10.0	12.0
Effective depth . .	3.6	4.6	5.6	6.6	7.6	9.6

EXAMPLE.—What should be the greatest distance between tie-rods that resist the thrust of an 8-inch tile arch having a span of 4 feet? The diameter of the rods is $\frac{3}{4}$ inch, and the arch carries a load of 120 pounds per square foot.

SOLUTION.—The area of the rod at the root of the thread is found to be .302 sq. in., and the weight of the arch, including the filling, flooring, and ceiling is found to be 63 lb. per sq. ft.; therefore, the total load per square foot is 120 + 63 = 183 lb. From the above table, the effective depth of the arch is found to be 5.6 in. Substituting these values in formula **10** gives

$$x = \frac{.302 \times 5.6 \times 10{,}000}{183 \times 4 \times 4} = 5.77, \text{ or 5 ft. 9 in. Ans.}$$

29. In designing a floor system for the support of fireproof arches of brick, it is advisable to provide tie-rods throughout the system. Though the beams supporting many portions of the floor will not be subjected to a material thrust, from the fact that adjacent arches react against each other, nevertheless, in actual construction it is often convenient to lay the arches in alternate panels, and the wooden centers are frequently removed from one panel to the adjacent one. Where a single arch is supported on two **I** beams, as shown in Fig. 31 (*a*), each beam is subjected to a lateral force equal to the thrust of the arch, and in proportioning such beams for the floor load, this lateral thrust should be considered in conjunction with the vertical load on the beam, in order to determine the ultimate unit stress.

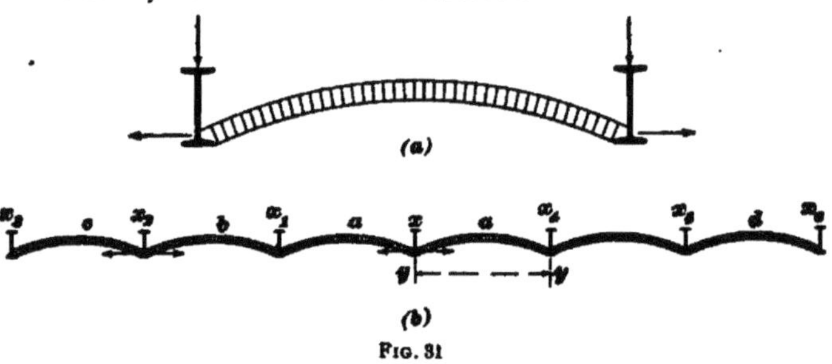

(*a*)

(*b*)

Fig. 31

Considering the system of arch construction shown in Fig. 31 (*b*) from the center beam x outwards each way, it will be noticed that the arches a, a react on the beam x and their thrusts are balanced. Likewise, the thrust of the arches a and b are balanced on the beam x_1, and in a similar manner b and c react against each other. The horizontal thrust of the arch c must be provided for in the beam x_3 and sufficient transverse lateral resistance must be obtained either by the use of a larger beam, or tie-rods must be placed at frequent intervals. If tie-rods connect all the beams in the series, the effect is the same as though a continuous tie-rod extended from the beam x_3 to the beam x_6, and the outward thrusts of the system due to the arches c and d are resisted by the series of tie-rods. If the tie-rods in the bay yy

toward the center of the system are omitted for reasons of economy or design, the thrust of the arch *a* is resisted by the transverse strength of the three beams x_4, x_5, and x_6. In large floor systems, however, if it is possible to contruct all the arches at once, the tie-rods may be omitted in central portions of the system and provided only for the exterior beams and three or four adjacent panels.

30. It is interesting to observe the effect of the horizontal force on the bending moment in a beam supporting

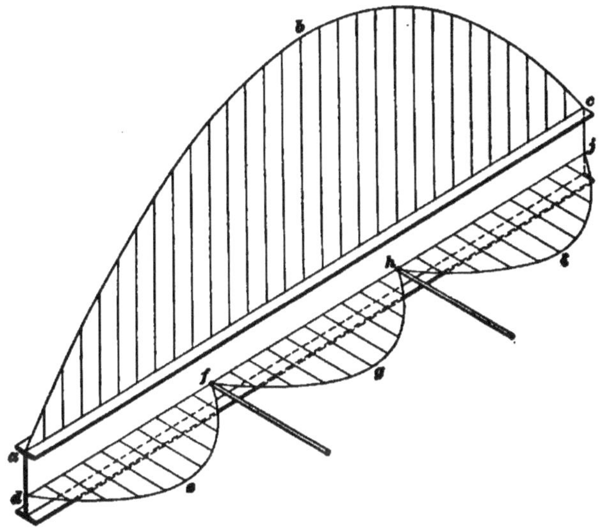

FIG. 32

arches and tied to the adjacent beam by tie-rods. Such a beam is subjected to two bending moments—one vertical and the other horizontal. In each case the load is uniform and the bending moment may be represented by a parabola. Fig. 32 shows the conditions that exist, the parabola *a b c* representing the bending moment due to the vertical load, and *d e f*, *f g h*, and *h i j* the bending moments between tie-rods or support and tie-rod. The resultant bending moment is found as shown in Fig. 33, in which *a b* represents the length of the beam and *a c* the bending moment for one-half of the beam. The tie-rods are located at the points *d* and *e*, and *a f* shows the bending moment for one-half of the distance between the end support and the first tie-rod.

31. The bending moment at any point on the beam is equal to the resultant of the vertical and horizontal bending moments; hence, the bending moment at *g* is represented by the resultant of *g h* and *g i* or the line *h j*. This is obtained by laying off *g j* at right angles to *h g* and equal to *g i*, and joining *h* and *j*. The divisions from *x* to *b* are made equal to those from *a* to *x* and the line *h j* is laid off at *j' h'*. Likewise, the bending moment at *k* is equal to the

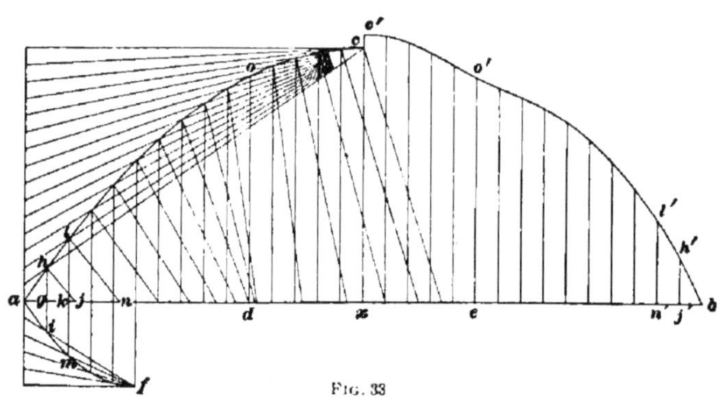

FIG. 33

resultant of *k l* and *k m*, or *l n*, and this distance is laid off at *n' l'*. The bending moment at the succeeding points is found in a similar manner until the first tie-rod is reached. The horizontal thrust is counteracted at this point and hence no horizontal bending moment is produced. Therefore, the distance *d o* is laid off at *e o'* and the resultant bending moments at the remaining points are laid off as explained. A curve drawn through the points thus determined, as *b h' l' o' c'*, represents the resultant bending moment on one-half of the beam.

EXAMPLE 1.—What size I beam is required to support a tile arch on one side for a distance of 20 feet, the conditions being as follows: The span of the arch is 5 feet; its nominal depth, 10 inches; and the weight, including filling, flooring, etc., is 69 pounds per square foot; the live load is 150 pounds per square foot. The tie-rods are $\frac{3}{4}$ inch in diameter and the allowable unit fiber stress of the beam is 18,000 pounds.

SOLUTION.—The spacing of the rods is found from formula **12**; thus, $x = \frac{.302 \times 7.6 \times 10{,}000}{219 \times 5 \times 5} = 4.19$ ft.

The lateral thrust of the arch is determined from formula **7**,

$$T = \frac{3 \times 219 \times 5 \times 5}{2 \times 7.6} = 1{,}080.59, \text{ or, say, } 1{,}081 \text{ lb.}$$

Assuming a 12-in. 31.5-lb. beam, the values of b and l' are 5 and 9.5, respectively. Substituting these values in formula **5**, gives

$$s_h = \frac{1{,}081 \times 5 \times 4.19 \times 4.19}{2 \times 9.5} = 4{,}994 \text{ lb. per sq. in.}$$

Hence, 18,000 − 4,994 = 13,006 lb. per sq. in. Now the section modulus for this beam under a vertical load is found from *Properties of Sections* to be 36. Therefore,

$$s_v = \frac{Wl}{8S} = \frac{219 \times 5 \times 20 \times 20 \times 12}{2 \times 8 \times 36} = 9{,}125 < 13{,}006 \text{ lb. per sq. in.}$$

Hence, a 12-in. 31.5-lb. beam is sufficiently strong. A 10-in. 30-lb. beam by a similar calculation would be found too light; consequently, a 12-in. 31.5-lb. beam is the lightest one that is safe under the given conditions.

EXAMPLE 2.—What size I beam will be required to support a tile arch on each side, for a distance of 18 feet? The span of the arch is 5 feet; the nominal depth, 9 inches; and the spacing of the tie-rods is 4 feet. The live load is 160 pounds per square foot, and the dead load is 40 pounds per square foot.

SOLUTION.—From formula **7**, $T = \frac{3 \times 160 \times 5 \times 5}{2 \times 6.6} = 909$ lb. per ft.

Assuming a 12-in. 31.5-lb. I beam, the values of b and l' are 5 and 9.5, respectively. Substituting these values in formula **6**,

$$s_h = \frac{909 \times 5 \times 4 \times 4}{2 \times 9.5} = 3{,}827 \text{ lb. per sq. in.}$$

The section modulus of a 12-in., 31.5-lb. beam is 36. Then the vertical stress s_v is $\frac{Wl}{8S}$, or

$$s_v = \frac{(160 + 40) \times 5 \times 18 \times 18 \times 12}{8 \times 36} = 13{,}500 \text{ lb. per sq. in.}$$

Then, as $s_v = 13{,}500 < 16{,}000$, and $s_v + s_h = 13{,}500 + 3{,}827 = 17{,}327 < 20{,}000$, the beam assumed is sufficiently strong.

EXAMPLES FOR PRACTICE

1. What size I beam will be required to support a tile arch on one side over a distance of 18 feet? The nominal depth of the arch is 9 inches

and it has a span of 4 feet. The weight of the arch, including filling, flooring, and ceiling, is 65 pounds per square foot and the load to be supported by it is 135 pounds per square foot. The only lateral support is the usual tie-rods, in this case ¾-inch rods, spaced according to formula. Ans. 10-in. 25-lb. beam

2. The outside beam of a floor system supporting brick segmental arches is connected with the other beams by the usual ¾-inch tie-rods, and it is to have a safe unit fiber stress of 18,000 pounds. The span of the arch is 4 feet, the rise or effective depth 6 inches, and its weight 130 pounds per square foot. The live load is 200 pounds per square foot and the span of the beam is 22 feet. What size I beam will be required? Ans. 12-in. 31½-lb. I beam

---

BEARING PLATES

32. Beams resting on masonry walls or piers will usually require bearing plates to distribute the load over an area such that the safe bearing value of the masonry will not be exceeded. These bearing plates are usually of steel or cast iron, though stone templets are sometimes used. The method of computing the size required is explained in *Statics of Masonry*, Part 1.

Table IV gives the bearing value, in pounds, for plates of various sizes bearing on the different kinds of masonry. As the thickness of the plate is not stated, it must be computed for each case. It depends on the allowable load and unit stress and the width of the beam or channel resting on it, and may be determined by the following formula:

$$t = .866\,(l - b)\sqrt{\frac{R}{s_a\,b'\,l}} \qquad (13)$$

in which t = thickness of plate, in inches;
l = length of plate perpendicular to axis of beam, in inches;
b = width of flange of beam or channel, in inches;
R = reaction at point of support, in pounds;
b' = width of plate, in inches, in direction of axis of beam or channel;
s_a = allowable stress, in pounds per square inch, on extreme fiber of plate.

TABLE IV

BEARING PLATES FOR I BEAMS AND CHANNELS

Bearing on Wall Inches	Size of Plate Inches	Safe Bearing Value of Plates, in Pounds				
		Ordinary Stone Masonry	Good Stone Masonry	Brick in Lime Mortar	Brick in Rosendale Cement Mortar	Brick in Portland Cement Mortar
4	4 × 4	2,880	4,800	1,600	2,400	3,200
4	4 × 6	4,320	7,200	2,400	3,600	4,800
4	4 × 8	5,760	9,600	3,200	4,800	6,400
6	6 × 6	6,480	10,800	3,600	5,400	7,200
6	6 × 8	8,640	14,400	4,800	7,200	9,600
6	6 × 10	10,800	18,000	6,000	9,000	12,000
8	8 × 8	11,520	19,200	6,400	9,600	12,800
8	8 × 10	14,400	24,000	8,000	12,000	16,000
8	8 × 12	17,280	28,800	9,600	14,400	19,200
10	10 × 10	18,000	30,000	10,000	15,000	20,000
10	10 × 12	21,600	36,000	12,000	18,000	24,000
10	10 × 14	25,200	42,000	14,000	21,000	28,000
12	12 × 12	25,920	43,200	14,400	21,600	28,800
12	12 × 14	30,240	50,400	16,800	25,200	33,600
12	12 × 16	34,560	57,600	19,200	28,800	38,400
12	12 × 18	38,880	64,800	21,600	32,400	43,200
14	14 × 14	35,280	58,800	19,600	29,400	39,200
14	14 × 16	40,320	67,200	22,400	33,600	44,800
14	14 × 18	45,360	75,600	25,200	37,800	50,400
14	14 × 20	50,400	84,000	28,000	42,000	56,000
16	16 × 16	46,080	76,800	25,600	38,400	51,200
16	16 × 18	51,840	86,400	28,800	43,200	57,600
16	16 × 20	57,600	96,000	32,000	48,000	64,000
16	16 × 22	63,360	105,600	35,200	52,800	70,400
18	18 × 18	58,320	97,200	32,400	48,600	64,800
18	18 × 20	64,800	108,000	36,000	54,000	72,000
18	18 × 22	71,280	118,800	39,600	59,400	79,200
18	18 × 24	77,760	129,600	43,200	64,800	86,400
20	20 × 20	72,000	120,000	40,000	60,000	80,000
20	20 × 22	79,200	132,000	44,000	66,000	80,000
20	20 × 24	86,000	144,000	48,000	72,000	96,000
20	20 × 26	93,600	156,000	52,000	78,000	104,000

If the allowable unit stress in the extreme fiber for steel is considered as 16,000 pounds, the formula may be written:

$$t = .00685(l - b)\sqrt{\frac{R}{b' l}} \qquad (14)$$

EXAMPLE.—What size steel bearing plate is required in a brick wall laid in lime mortar, to support the end of a 12-inch 31.5-pound standard I beam having a flange width of 5 inches, which carries a uniformly distributed load of 600 pounds per lineal foot over a span of 25 feet? The allowable unit stress in the extreme fiber of the plate may be considered as 16,000 pounds.

SOLUTION.—The bearing value required is equal to the reaction, or $\frac{15000}{2} = 7{,}500$ lb. From Table IV, it is seen that an 8″ × 10″ plate will give the required bearing value for a plate on brickwork laid in lime mortar. Substituting the values in the formula $t = .00685(l - b)\sqrt{\frac{R}{b' l}}$ gives

$$t = .00685(10 - 5)\sqrt{\frac{7{,}500}{8 \times 10}} = .332 \text{ in.}$$

The nearest size above this is $\frac{3}{8}$ in.; hence, an 8″ × 10″ × $\frac{3}{8}$″ bearing plate will be required. Ans.

EXAMPLES FOR PRACTICE

1. An 18-inch 65-pound I beam having a flange width of 6.18 inches supports a uniformly distributed load of 1,000 pounds per lineal foot. The span of the beam is 32 feet and its ends rest on piers built of ordinary stone masonry. Provided that the allowable unit stress in the extreme fiber of the plate is 16,000 pounds, what size bearing plate will be required?

Ans. { .515 in. thick
8″ × 12″ × $\frac{9}{16}$″ plate

2. A steel lintel composed of two 15-inch 35-pound channels placed flange edge to flange edge supports the wall above and also a floor load, the combined weight amounting to 4,000 pounds per lineal foot. The span is 15 feet, and the over-all width of flange is 8 inches. Assuming an allowable unit stress of 16,000 pounds in the extreme fiber of the plate, what size bearing plate will be required to support the end of the lintel on a brick wall laid in Portland cement mortar?

Ans. 12″ × 14″ × $\frac{9}{16}$″ plate

REENFORCED CONCRETE BEAMS

33. Adaptability.—Reenforced concrete in building construction is beginning to occupy an important place in fireproof and slow-burning mill construction. Concrete is one of the best fire retardants known and is very strong under compression, but when subjected to a tensile stress it is weak and soon cracks and fails. For this reason it is necessary to reenforce it with some material that is strong under tension, such as steel or iron.

The reliability of reenforced concrete depends, to a great extent, on the workmanship; that is, reenforced concrete constructed under thorough and conscientious supervision will be as safe as any other form of construction, while if the work is carelessly executed there is a liability of its failing when least expected.

The increase in weight may sometimes be advanced as an argument against this construction. However, when the saving in cost over other methods of fireproof construction is considered, the additional weight is a question of little importance. That this work can be erected with considerable speed has been demonstrated in several building operations of magnitude. In the instance of a building 50 feet wide by 160 feet long, having a basement and four stories, the concrete construction of the entire building was completed in 12 weeks, the plastering was finished in 1 month, and the entire structure was completed in 4½ months, which was 34 days before the time of completion stated in the contract. As the concrete work included the laying of about 35,000 square feet of floor and roof, in addition to the columns, girders, etc., the progress was remarkable.

34. Durability.—That reenforced concrete will develop fire-resisting qualities has been proved by numerous tests and by its behavior in conflagrations of considerable magnitude. A test was made on reenforced concrete composed of one part of Portland cement, three parts of gravel, and five parts of trap rock, in which the average temperature was about 1,800°; when it was further subjected to the severe condition of rapid cooling by water from a hose under a pressure of about 65 pounds it revealed: First, no separation between the reenforced concrete girders and the concrete

FIG. 34

slab; second, no spalling of the exposed surfaces of the floor system; third, only slight, superficial cracks in the under surface of the concrete composing the system; fourth, a fair condition of the reenforced concrete columns, though shearing breaks and cracks were observed and one of the columns was buckled about $\frac{5}{8}$ inch. This test was made under a load of 150 pounds per square foot, which was afterwards increased.

It was further proved by the Baltimore fire of 1904 that reenforced concrete can at least partially withstand the severe destructive effects of a great fire. A survey of the United States Fidelity and Guaranty building of that city, which was composed of floor systems carried by reenforced concrete columns and girders, showed that though the cast-iron front and the brick party wall of the building had been destroyed, the concrete floor systems and their supporting columns remained in place. Fig. 34 shows the condition of the building several weeks after the fire, some of the party wall having been removed. When in this condition, the floors of the building were tested without failure, the load being 400 pounds per square foot. Even the overhanging portion *a* sustained a load of 200 pounds per square foot without damage. The reenforced concrete of which the structural members of this building were constructed, was composed of crushed granite, sand, and American Portland cement. Most of the plaster in the building had fallen from the concrete, thus making it evident that the former afforded no adequate protection.

35. Rigidity of Reenforced Concrete Construction. As there are no joints, such as the column and girder connections, used in structural steel work, which sometimes prove inefficient, the structure is practically a monolith and is, consequently, extremely rigid. This construction is particularly desirable for buildings that contain looms, presses, rotary mills, etc., or any machinery liable to run in unison and produce considerable vibration, in buildings as ordinarily constructed.

36. Comparison of the Cost.—An investigation of the cost per square foot of area of reenforced concrete when compared with other recognized types of good construction, such as slow-burning and steel-frame construction, shows that it occupies a middle position between the two, as may be observed from the following tabulation:

Slow-burning construction	$1.07
Reenforced concrete	1.45
Steel floor and roof, fireproofed	1.85

As a fireproof construction, the reenforced concrete is cheaper than the steel frame fireproofed with terra cotta, and is cheap enough to be used in mills and factories.

37. Details of Reenforced Concrete Construction. In Fig. 35 (*a*) is shown a system of reenforced concrete construction that corresponds to the usual framed construction executed in wood. The longitudinal girder is made a few inches deeper than the cross-girders to permit the bottom rods in the latter to rest on top of those in the longitudinal girder. By this means a direct bearing is secured and the straining of the concrete is avoided. Wooden planks are bolted to the concrete beams, to which the sleepers supporting the shafting may be secured. The columns are reenforced by rods placed in each corner and braced by horizontal bars. In view (*b*) is shown the method of anchoring concrete beams to the wall. The mason builds a dovetailed pocket in the brick wall into which the concrete is poured at the same time that the beam is formed. All girders should rest on cast- or wrought-iron bearing plates placed in the wall. An isometric view of the steel reenforcement is shown in Fig. 35 (*c*).

A type of reenforced concrete construction very similar to the one just described is illustrated in Fig. 36. View (*a*) shows an elevation, partly in section, of the reenforced concrete girder, while (*b*) represents an enlarged cross-section taken on the line *c d*. In this case, the reenforcement is composed of the plain round steel rods *a*, part of which are straight and part bent, as shown in (*a*), these rods being held in place by a number of **U** bars or stirrups *b* of hoop steel. A comparison between a steel girder and the reenforced concrete girder may be made by referring to Fig. 37. The lower flange of the steel girder is replaced by the rods;

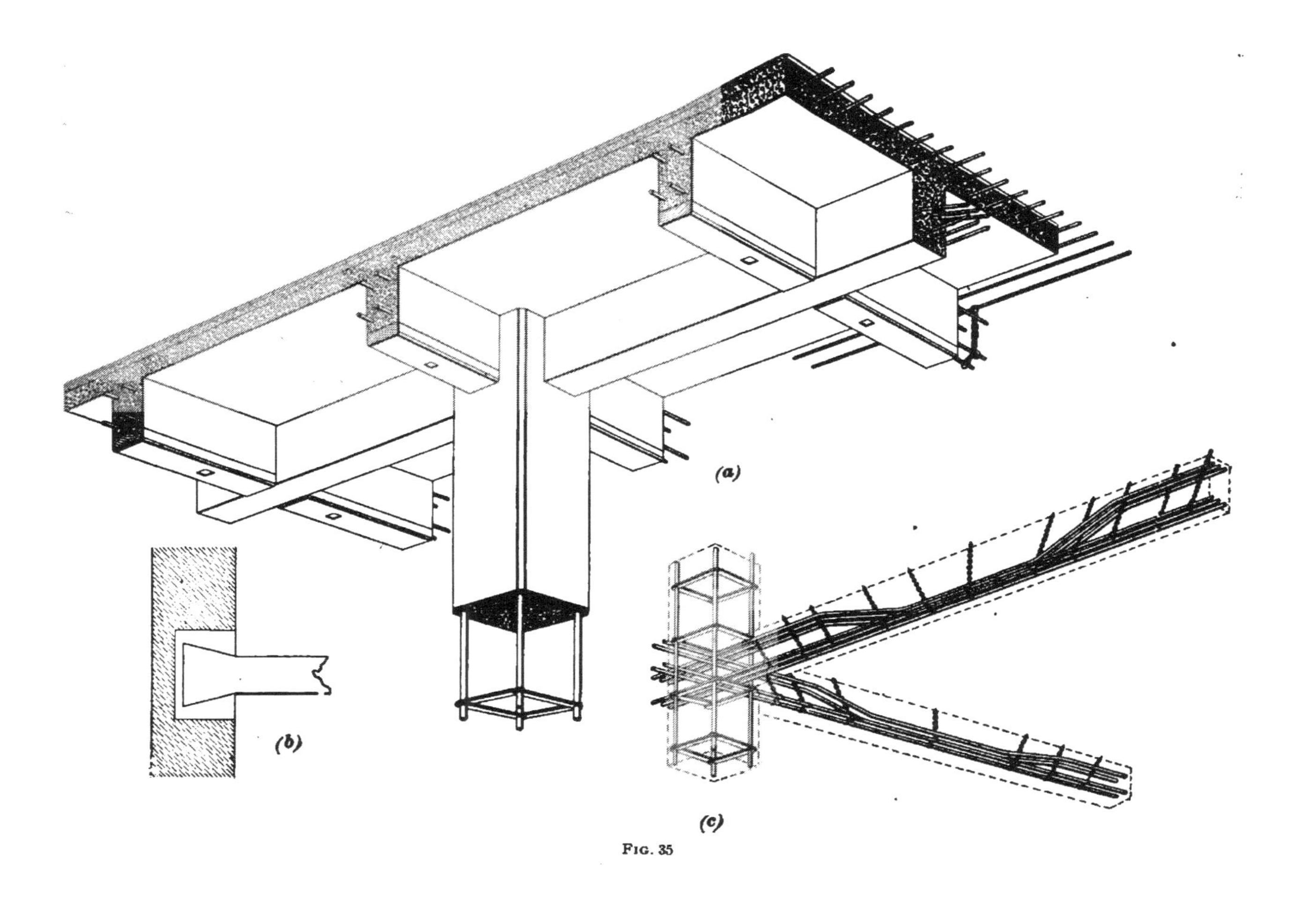

FIG. 35

the upper flange, by the concrete floor slab; and the web, by the **U** bars and the concrete rib.

Experience has proved that concrete ribs that are reenforced with a large percentage of metal, as is generally the case, show the first signs of failure near the supports. Diagonal cracks appear, caused by the combined tensile and

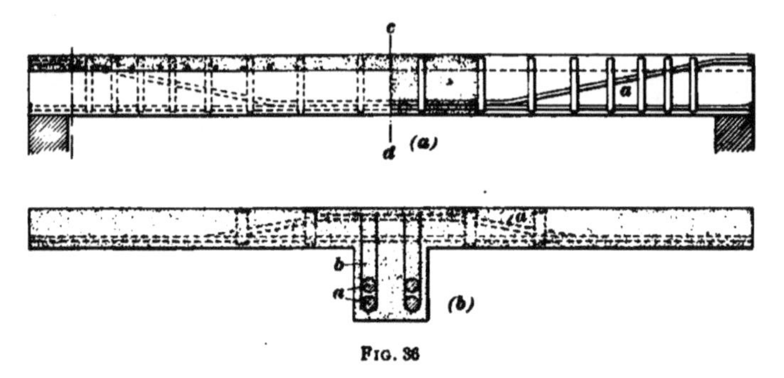

FIG. 36

shearing stresses. In order to resist this tendency to fail, some of the rods are bent up similar to the construction of a cambered beam, as shown in Fig. 36, the inclined portion of the bent rods resisting the stress that produces the diagonal cracks. The **U** bars hold the rods in position and prevent

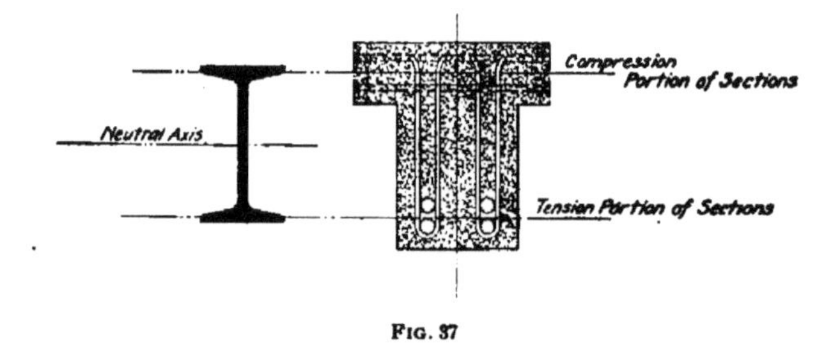

FIG. 37

their breaking out of the concrete; they are usually placed closer together near the supports, as the shearing stresses are maximum in that portion of the beam.

Concrete girders are seldom constructed with the ends free, but are generally connected in a monolithic manner with other girders, making them continuous and, consequently, much stronger than girders freely supported at the ends. The deflection is also considerably less than in girders with free ends.

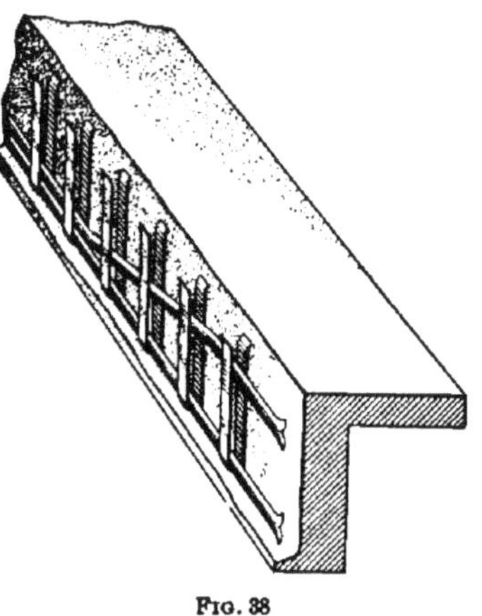
FIG. 38

Fig. 38 shows a perspective view of the longitudinal section of a reenforced, or armored concrete, girder. A connection between the reenforcing material in a continuous girder and a column is shown in Fig. 39 (*a*). The column is composed of four rods braced at intervals by pieces of sheet metal, as shown in (*b*). If a stronger column is required, more rods are used, as shown in Fig. 40, where they are tied together by wires at intervals of not more than the diameter or least side of the column. The rods are placed near the surface of the column in order to give the largest radius

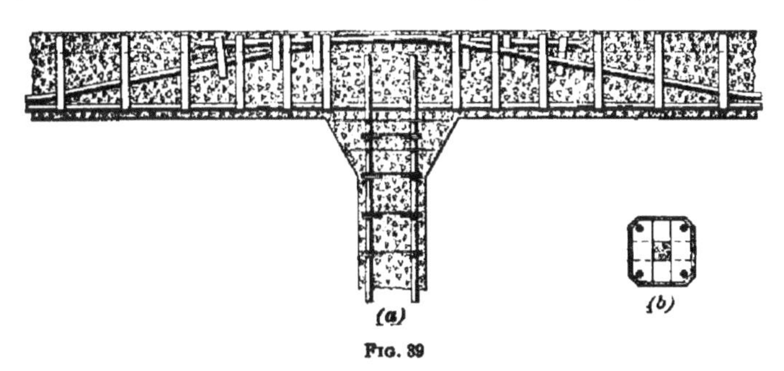

FIG. 39

of gyration and to resist the tensile stresses produced by eccentric loading, wind pressure, pull of beltings, etc.

38. Size of Girders.—The economical depth of reenforced concrete girders is practically the same as the depth of steel girders having the same carrying capacity. As a rule, the depth should not exceed one-twentieth of the span.

The following tabulation shows the widths of concrete beams that correspond to steel beams of the same strength:

6 inches for girders corresponding to 12-inch I beams.
8 inches for girders corresponding to from 12-inch to 18-inch beams.
10 inches for girders corresponding to from 20-inch to 24-inch beams.
12 inches for girders corresponding to riveted girders from 30 to 40 inches in height.
From 12 to 24 inches for girders corresponding to very heavy box girders.

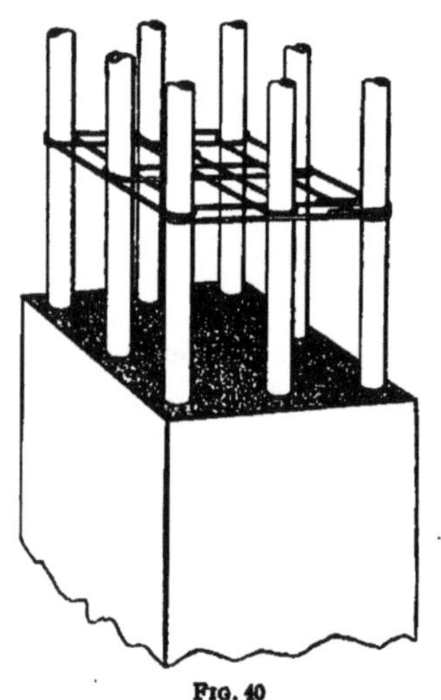

FIG. 40

JOHNSON'S METHOD OF CALCULATION

39. The strength that the reenforcing bars are required to furnish may be determined by the following method:

When the steel bar is rigidly attached to the concrete its distortion is the same as that of the concrete in the same horizontal plane. But the stresses, or the resistances of the materials to distortion, vary as their rigidities, or as their moduli of elasticity. The modulus of elasticity of steel is usually taken at about 28,000,000, but that of concrete varies with the quality of the material, and in this discussion will be considered as 1,000,000. Hence, if the steel is twenty-eight times as rigid as concrete, it resists twenty-eight times as much as the same area of concrete for equal distortions; and in making the calculations for the strength of a concrete beam, the area of the steel bars may be replaced by an area of concrete twenty-eight times as great, provided that it is placed in the same horizontal plane as the bars. Thus, in Fig. 41 (*b*) the area of the bars shown in (*a*) has been replaced by concrete of an area twenty-eight

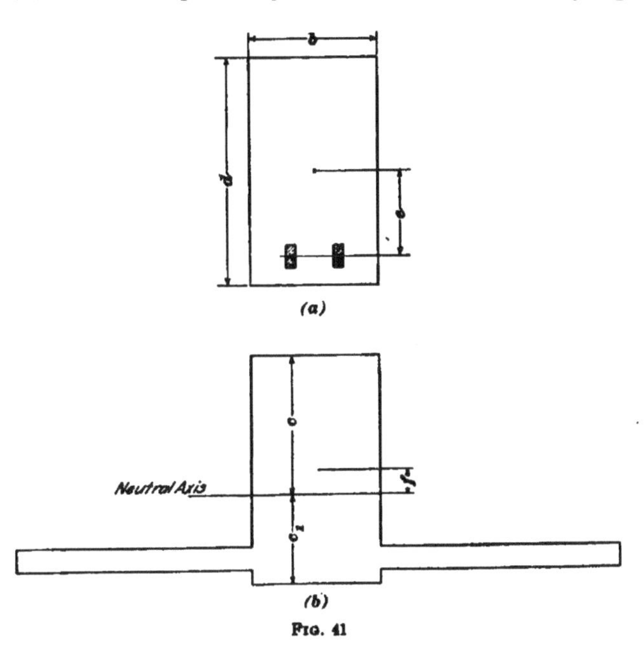

FIG. 41

times the combined area of the bars. Considering this new section, the strength of the beam to resist the tension in the concrete is determined from the equation

$$M_1 = \frac{s_t I}{c_1} \qquad (15)$$

in which s_t = ultimate unit tensile strength of concrete;
I = moment of inertia of transformed cross-section;
c_1 = distance from neutral axis to tension side;
M_1 = moment of resistance of actual beam.

If the reenforcement were not used, the equation would be $M = \frac{s_t b d^2}{6}$ $\left(\text{since } \frac{I}{c_1} = \text{section modulus, or } \frac{b d^2}{6}\right)$. In this formula s_o represents the stress on the outer portion of the beam if no steel bars are used; then, $s_o = \frac{6M}{bd^2}$.

The tensile stress in the concrete at the bottom of the beam may be determined from the formula $s_t = \frac{Mc_1}{I}$; but the moment of inertia for the transformed section is

$$I = bd\left(\frac{d^2}{12} + \frac{e^2}{1+m}\right) \qquad (16)$$

in which e = distance from center of beam to center of steel bars;

m = a value used for convenience $= \frac{bd}{A} \times \frac{E_c}{E_s}$.

(A = area of steel bars; E_c = modulus of elasticity of concrete; E_s = modulus of elasticity of steel.)

Also, $c_1 = \frac{d}{2} - f$; but $f = \frac{e}{m+1}$, and, therefore, $c_1 = \frac{d}{2} - \frac{e}{m+1}$. Substituting these values in the original formula gives

$$s_t = s_o\left(\frac{1 + m - 2\frac{e}{d}}{1 + m + 12\frac{e^2}{d^2}}\right) \qquad (17)$$

The formula for the compressive stress in the concrete at the top is

$$s_c = s_o\left(\frac{1 + m + 2\frac{e}{d}}{1 + m + 12\frac{e^2}{d^2}}\right) \qquad (18)$$

The steel bars are assumed to stretch the same as the concrete adjacent to them, and for any distance e from the center of the beam, the total stress on the steel bars is

$$\frac{2s_o e b}{1 + m + 12\frac{e^2}{d^2}} \qquad (19)$$

40. The following example will make clear the application of the above formulas:

EXAMPLE.—A concrete beam in a metal and concrete factory building has a span of 14 feet and is to be 15 inches in width and 24 inches in depth. Ten ½" × 1" bars are embedded in the bottom, the center of the bars being 2 inches from the edge of the beam, and a load of 1,000 pounds per lineal foot is to be supported. Taking the safe unit compressive stress of the concrete at 250 pounds, the safe tensile stress at 100 pounds per square inch, and the unit stress of the steel that may be developed before the concrete fails at 3,000 pounds, determine whether the beam has sufficient strength to carry its load.

SOLUTION.—The greatest bending moment $M = \frac{WL}{8} = \frac{1{,}000 \times 14 \times 14}{8}$ $= 24{,}500$ ft.-lb. $= 294{,}000$ in.-lb.; $s_c = \frac{6M}{bd^2} = \frac{6 \times 294{,}000}{15 \times 24 \times 24} = 204\frac{1}{6}$ lb. per sq. in.; $m = \frac{bd}{A} \times \frac{E_c}{E_s}$, and when $\frac{E_c}{E_s}$ is taken as $\frac{1}{28}$, $m = \frac{15 \times 24}{10 \times 1 \times \frac{1}{2}}$ $\times \frac{1}{28} = 2.57$.

Substituting these values in the formula $s_t = s_c\left(\frac{1 + m - 2\frac{e}{d}}{1 + m + 12\frac{e^2}{d^2}}\right)$

gives $s_t = 204\frac{1}{6}\left(\frac{1 + 2.57 - \left(2 \times \frac{10}{24}\right)}{1 + 2.57 + \left(12 \times \frac{10^2}{24^2}\right)}\right) = 98$ lb. per sq. in. Also,

$$s_c = 204\tfrac{1}{6}\left(\frac{1 + 2.57 + \left(2 \times \frac{10}{24}\right)}{1 + 2.57 + \left(12 \times \frac{10^2}{24^2}\right)}\right) = 159 \text{ lb. per sq. in.}$$

The total stress on the steel bars is equal to $\frac{2 \times 204\frac{1}{6} \times 10 \times 15}{1 + 2.57 + \left(12 \times \frac{10^2}{24^2}\right)}$ $= 10{,}841$ lb. The unit stress on the steel will then equal the result just obtained divided by the entire sectional area of the reenforcing metal. There being ten bars ½ inch by 1 inch the area of reenforcing metal equals 5 and the unit stress is $10{,}841 \div 5 = 2{,}168$ lb. As these values are all within the safe limit, the beam is considered as having sufficient strength to support its load. Ans.

RANSOME SYSTEM

41. In the **Ransome system** of reenforced concrete construction, the square steel bars have been twisted cold, by being gripped at the ends, until the angle made by the edge with the axis of the bar is about 20°. This makes the bar like a long screw and the mortar of the concrete embeds itself in the concave whorls; in order to remove the bar an area of concrete must be sheared off equal to the superficial area of the cylinder enclosing the bar. The bar is held rigidly throughout its entire length, so that it cannot draw out or become loosened from the concrete when the load is applied to the beam. The bond may be considered as practically perfect, and the composite beam can be figured accurately by assuming that the steel resists the tensile stress while the concrete takes care of the compression. A formula in common use, and one that is sufficiently accurate for practical work, is obtained by equating the external force, or the bending moment on the beam, with the resistance that the bars offer. The point about which the bars tend to revolve is considered as the center of gravity of the section of concrete supplying the resistance to compression; this section is assumed to be the upper third of the concrete beam, or $abcd$, in Fig. 42. Disregarding the distance from the bars to the lower edge of the beam, which is only 2 or 3 inches, and considering the center line of the bars as the lower line of the beam section, their lever arm about the point x is $\frac{5}{6}d$. Then, if $s_a A$ represents the strength of all the steel bars, the equation is

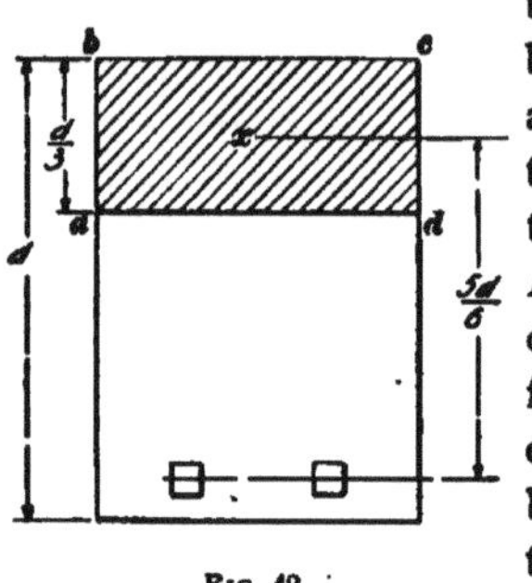

Fig. 42

$$\frac{Wl}{8} = s_a A \times \frac{5d}{6}. \quad s_a A = \frac{Wl}{8} \times \frac{6}{5d} = \frac{Wl}{6\frac{2}{3}d}$$

or

$$A = \frac{Wl}{6\frac{2}{3}d\,s_a} \qquad (20)$$

in which s_a = safe unit fiber stress of bars;
A = area of steel bars, in square inches;
W = uniformly distributed load on beam;
l = span of beam, in inches;
d = depth of beam, in inches.

It is claimed that cold twisting of mild steel increases the tensile strength and raises the elastic limit from 10 to 15 per cent.; therefore, it will do no harm to increase the denominator 5 per cent., changing $6\frac{2}{3}$ to 7. This gives the formula

$$A = \frac{W l}{7 d s_a} \qquad (21)$$

HENNEBIQUE SYSTEM

42. A formula used by M. Hennebique for determining the area of the reenforcing material is based on the assumption that the resistance to compression in the concrete takes care of one-half the bending moment on the beam while the other half is resisted by the tension in the steel. The concrete is not considered as offering any resistance to the tension in the beam. This engineer assumes that the neutral axis is at a distance of two-thirds of the depth of the beam from the top, or at $a\,d$, Fig. 43, and that the portion $a\,b\,c\,d$, acting about the axis $a\,d$, provides the resistance to compression. This resistance is assumed to be concentrated at the point x located at a distance of one-fourth $a\,b$ or one-sixth d from this axis. The resistance to tension, supplied by the steel bars, is assumed to act about the same axis and therefore its lever arm is $d\,e$. Considering d as the depth and b the width of the beam, and taking 350 pounds per square inch as the compressive stress in the concrete, the equation that represents the resistance of the beam to compression is

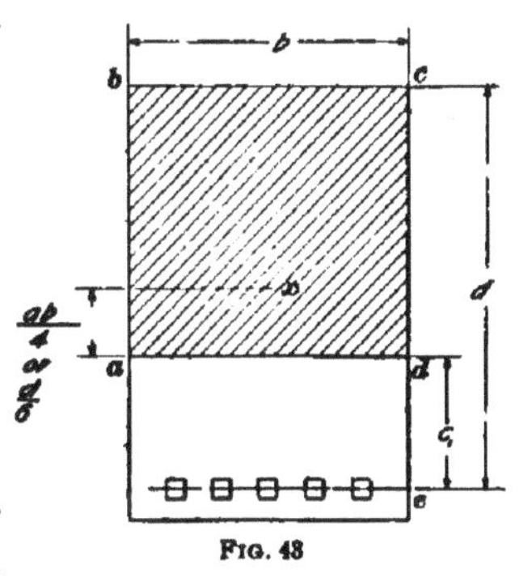

FIG. 43

$$\frac{2\,b\,d}{3} \times 350 \times \frac{d}{6}$$

or

$$\frac{350\,b\,d^2}{9} = \frac{M}{2} \qquad (22)$$

When A equals the area of steel required and the unit tensile stress is taken at 14,000 pounds, the following equation expresses the resistance to tension: $14{,}000\ A \times \frac{d}{3} = \frac{M}{2}$, or

$$A = \frac{3\ M}{28{,}000\ d} \qquad (23)$$

In this formula, the distance from the center of the bars to the lower line of the beam is disregarded and the lever arm is considered as $\frac{d}{3}$.

It can be seen that the assumptions made by Hennebique are not entirely correct, but the results obtained from this reasoning would probably be sufficiently accurate.

METHOD USED IN COMMON PRACTICE

43. The following formulas are derived by Mr. Edwin Thatcher and published by the Concrete Steel Engineering Company. They are probably the ones in most common use in reenforced concrete design.

Let s_c = stress per square inch on concrete (compression), in pounds;

s_s = stress per square inch on steel (tension), in pounds;

s_s' = stress per square inch on steel (compression), in pounds;

E_c = modulus of elasticity of concrete;

E_s = modulus of elasticity of steel;

A = area of steel in tension for 1 inch width of beam;

$n A$ = area of steel in compression for 1 inch width of beam;

M_1 = resisting moment, in foot-pounds, for 1 inch width of beam.

Fig. 44 represents, in section, a concrete beam reenforced both top and bottom. The dimensions c, h, a, b, d, x, and y are inches.

In order to design a reenforced concrete beam, let values of s_s, s_c, E_s, E_c, n, d, and a be assumed.

$$y = \frac{d}{\left(\frac{s_c}{s_s}\frac{E_s}{E_c} + 1\right)}$$

$$x = d - y$$

$$A = \frac{s_c x y}{2 s_s [y - n(x - a)]}$$

$$s_s' = \frac{(x - a)}{y} s_s$$

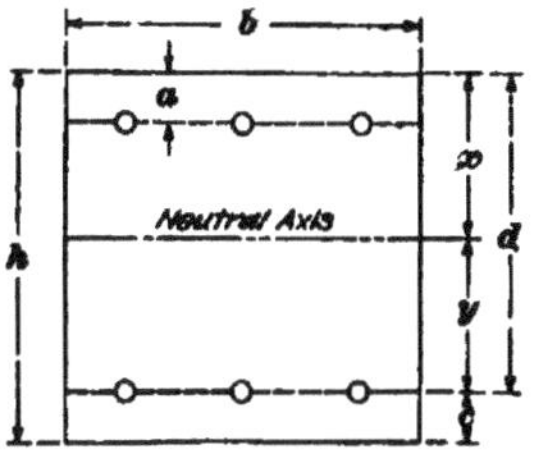

Fig. 44

$$M_1 = \frac{s_s}{36}\left[\frac{E_c}{E_s}\frac{x^2}{y} + \frac{3nA(x-a)^2}{y} + 3Ay\right]$$

To review a beam already built, assume values of s_s, E_s, and E_c.

$$x = \sqrt{2A\frac{E_s}{E_c}(d + na) + \left[A\frac{E_s}{E_c}(n+1)\right]^2} - \left[A\frac{E_s}{E_c}(n+1)\right]$$

$$y = d - x$$

$$A = \frac{\text{area of bars in tension}}{b}$$

$$s_s' = \frac{(x - a)s_s}{y}$$

$$s_c = s_s \frac{E_c}{E_s}\frac{x}{y}$$

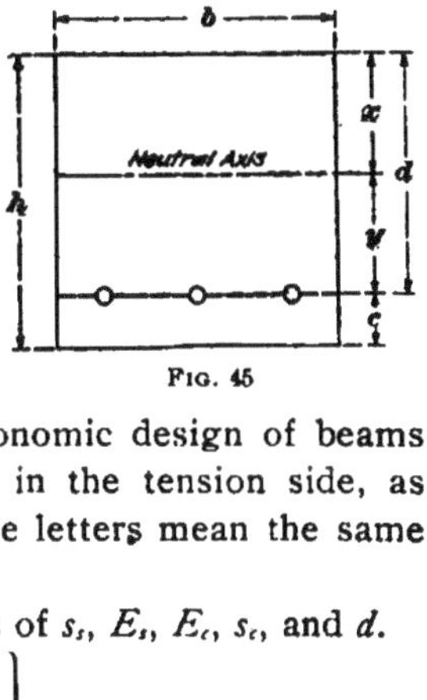

M_1 = same as above

Fig. 45

However, the most usual and economic design of beams is that which has steel rods only in the tension side, as illustrated in section in Fig. 45; the letters mean the same as before.

To design a beam, assume values of s_s, E_s, E_c, s_c, and d.

$$\left.\begin{aligned} y &= \frac{d}{\left(\frac{s_c}{s_s}\frac{E_s}{E_c} + 1\right)} \\ x &= d - y \\ A &= \frac{d}{2\left[\frac{s_s}{s_c} + \left(\frac{s_s}{s_c}\right)^2 \frac{E_c}{E_s}\right]} \\ M_1 &= \frac{s_s}{36}\left(\frac{E_c}{E_s}\frac{x^2}{y} + 3A\right) \end{aligned}\right\} \quad (24)$$

To review a beam, assume values of s_s, E_s, and E_c.

$$\left.\begin{array}{l} x = \sqrt{2A\frac{E_s}{E_c}d + \left(A\frac{E_s}{E_c}\right)^2} - A\frac{E_s}{E_c} \\ y = d - x \\ s_c = s_s\frac{E_c}{E_s}\frac{x}{y} \\ M_1 = \text{same as above} \end{array}\right\} \quad (25)$$

If the beams are continuous over two or more supports, make A = 0.8 of above value, and reenforce the beams at the top over each support.

There are different opinions in regard to how c should be proportioned. Some engineers state that it should be at least 1½ inches for girders, while others say that it should be made equal to one-fifth of d, which is the same thing as one-sixth of the total depth of the beam. So long as c is great enough to keep the steel free from moisture and to fireproof it, it is sufficient because the formulas given above do not consider that the concrete in tension, which is an uncertain quantity at best, adds to the strength of the beam in resisting bending and its only function below the neutral axis is to protect the embedded rods. Mr. Thatcher uses the figures given in the following table for finding c which will be found safe and yet economical:

h	d	c	h	d	c	h	d	c
1.5	1.0	0.5*	8.0	7.0	1.0	20.0	18.25	1.75
2.0	1.5	0.5	8.5	7.5	1.0	22.0	20.0	2.0
2.5	1.75	0.75	9.0	7.75	1.25	24.0	22.0	2.0
3.0	2.25	0.75	10.0	8.75	1.25	26.0	24.0	2.0
3.5	2.75	0.75	11.0	9.75	1.25	28.0	26.0	2.0
4.0	3.25	0.75	12.0	10.75	1.25	30.0	28.0	2.0
4.5	3.5	1.0	13.0	11.5	1.5	32.0	30.0	2.0
5.0	4.0	1.0	14.0	12.5	1.5	34.0	32.0	2.0
5.5	4.5	1.0	15.0	13.5	1.5	36.0	33.75	2.25
6.0	5.0	1.0	16.0	14.5	1.5	42.0	39.5	2.5
6.5	5.5	1.0	17.0	15.5	1.5	48.0	45.5	2.5
7.0	6.0	1.0	18.0	16.5	1.5	54.0	51.0	3.0
7.5	6.5	1.0	19.0	17.25	1.75	60.0	57.0	3.0

To reduce formula **25**, Mr. Thatcher substitutes values for the stress in the steel and the concrete and the moduli of elasticity of these two materials as follows:

$$E_s = 30{,}000{,}000$$
$$s_s = 64{,}000 \text{ (ultimate)}$$
$$E_c \text{ (1–2–4, 1 month old)} = 1{,}460{,}000$$
$$E_c \text{ (1–2–4, 6 months old)} = 2{,}580{,}000$$
$$E_c \text{ (1–3–6, 1 month old)} = 1{,}220{,}000$$
$$E_c \text{ (1–3–6, 6 months old)} = 1{,}860{,}000$$
$$s_c \text{ (1–2–4, 1 month old)} = 2{,}400 \text{ (ultimate)}$$
$$s_c \text{ (1–2–4, 6 months old)} = 3{,}700 \text{ (ultimate)}$$
$$s_c \text{ (1–3–6, 1 month old)} = 2{,}050 \text{ (ultimate)}$$
$$s_c \text{ (1–3–6, 6 months old)} = 3{,}100 \text{ (ultimate)}$$

By substituting these values just given in formula **25**, the following table may be derived with slight approximations which can be used in the design of all ordinary beams and slabs:

Mixture of Concrete	Age of Concrete	Area of Steel in Square Inches Required for 1-Inch Width of Beam-A	Ultimate Resisting Moment, Foot-Pounds for 1-Inch Width of Beam-M_1
1–2–4	1 month	$\frac{d}{142}$	$35.62d^2$
1–2–4	6 months	$\frac{d}{100}$	$51.25d^2$
1–3–6	1 month	$\frac{d}{165}$	$30.62d^2$
1–3–6	6 months	$\frac{d}{109}$	$46.25d^2$

In obtaining these formulas, it will be noticed that the ultimate and not the safe working strength of the concrete has been used. It is, therefore, necessary that the bending moment due to the imposed loads should be multiplied by a factor of safety. That is to say, the beam is designed to break under a load several times, depending on the factor used, greater than the load it is intended to carry.

If a factor of safety is used in the design of the beam based on the strength of concrete 1 month old, there will not be sufficient steel to develop the full strength of the concrete when the latter is 6 months old, as the strength of almost all hydraulic cement concrete increases with its age, as has been noted in a previous paper. It is better practice, therefore, to use such a factor of safety for concrete 6 months old as will give a minimum factor of safety for concrete 1 month old. That is to say, a factor of safety of 5 for 6 months reenforced concrete will give a factor of safety of only about 3.5 when the concrete is one month old, which will, however, usually be ample for ordinary engineering work. A factor of safety of 4 in 6 months reenforced concrete beams will give a factor of safety of about 2.75 in 1 month, which will be sufficient for such uniform loads without impact, such as earth fills, water pressure, etc.

EXAMPLE 1.—A certain floor in an office building is to be supported by beams of reenforced concrete 1-2-4 mixture. The span of these beams is 30 feet and they are on 10-foot centers. The dead load is 50 pounds per square foot and the live load is 150 pounds per square foot. Design the beam.

SOLUTION.—The maximum moment on the beam, in foot-pounds, is, $\frac{200 \times 10 \times 30 \times 30}{8} = 225{,}000$ ft.-lb. Using a factor of safety of 5, the beam must be designed to break under a moment of $225{,}000 \times 5 = 1{,}125{,}000$ ft.-lb. Assume a value for h of 36 in.; this will make d equal to 33.75 in., according to the table given; therefore, the bending moment for 1 in. width of beam is $51.25d^2 = 51.25 \times 33.75^2 = 58{,}380$ ft.-lb. Therefore,

$$b = \frac{1{,}125{,}000}{58{,}380} = 19.27 \text{ in., say 20 in.}$$

$$A = \frac{d}{100} = \frac{33.75}{100} = .3375 \text{ sq. in. per in. of width}$$

Therefore, the total area of steel required is $.3375 \times 20 = 6.75$ sq. in.; this is about the area of sixteen ¾-in. round rods. The beam required is, therefore, 36 in. deep and 20 in. wide, and is reenforced by sixteen ¾-in. steel bars placed 2¼ in. from the bottom of the beam.

EXAMPLE 2.—Design a reenforced concrete floor which is on an 8-foot span to carry a total load of 250 pounds per square foot, if the concrete to be used is a 1-3-6 mixture.

SOLUTION.—The maximum moment on a strip of floor 1 ft. wide is $\frac{250 \times 8 \times 8}{8} = 2{,}000$ ft.-lb. Using a factor of safety of 5, the beam must be designed to break under a moment of $2{,}000 \times 5 = 10{,}000$ ft.-lb. Now, since the section of the floor under consideration is 12 in. wide instead of 1 in., we must take twelve times the moment given in the table. That is $46.25 \times d^2 \times 12 = 10{,}000$. $d = 4.245$, say 4.25 in. Therefore, $h = 4.25 + 1.00 = 5\frac{1}{4}$ in. $A = \frac{4.25}{109} = .03899$, say .04 sq. in. per in. of width. Now a ⅝-in. round bar has an area of .3068 sq. in. Therefore, if ⅝-in. bars are used, they will be $\frac{.3068}{.03899} = 7.869$, say 7⅞ in. apart. Ans.

THE KAHN SYSTEM OF REENFORCED CONCRETE

44. The **Kahn system** of reenforced concrete construction is illustrated in Fig. 46, which shows the shape of the reenforcing metal for the beams and columns. This system

FIG. 46

is based on the assumption that all the tensile stresses should be resisted by the steel reenforcement, which is constructed on the principle of a Pratt trussed beam. For instance, in

Fig. 47, which represents a Pratt truss, the diagonal members and the lower chord are in tension while the vertical members and the upper chord are in compression. The Kahn standard bar is shown in section in Fig. 48. It consists of a central rib *c* with a web *d* on either side and may be so formed as to correspond to the lower chord and oblique members of a Pratt truss. This is done by shearing the webs *d* along the rib *c* for a certain distance and bending them into the positions indicated at *a*, *a* in Fig. 46. It is evident that a bar bent up as described cannot possibly slip through the concrete and, therefore, it must resist all of the tension produced in the beam, provided that the concrete is so proportioned that it will not fail. When a load is applied on top of the beam, the concrete throughout the whole length of the beam tends to arch itself, the steel bar in the bottom acting as a

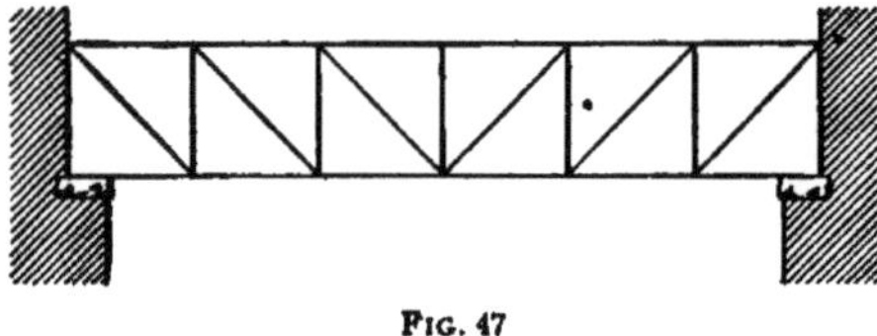

Fig. 47

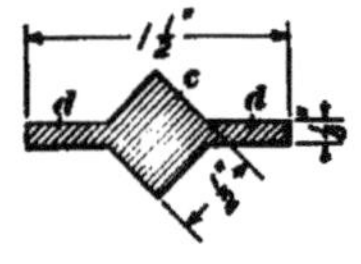

Fig. 48

tie. In addition to this, a series of concentric arches, gradually decreasing in length, may be supposed to be produced. For instance, one arch will extend from the prongs *e* to the corresponding prongs at the other end of the beam, which are not shown in the illustration; another arch from the prongs *f* to prongs similarly placed at the other end, and so forth, the arches gradually decreasing in size until the central arch at *c* is reached. In each of these arches the prongs receive the weight and carry it upwards, distributing it on the other arches of larger span, which are tied together by the horizontal tension bar. It is evident then that, theoretically, the tension bar is subjected to a stress equal to the horizontal component of the stresses in the oblique members, and it would be strong enough to resist this stress, if it were placed on the outside of the concrete. However, it is more advisable to embed the steel in the concrete to prevent rusting and secure a fireproof construction. When a beam composed of good materials and properly proportioned is tested to destruction, either the concrete crushes at the top or the steel bar pulls in two; as the resistance to compression in the top of the concrete beam is largely supplied by the floor itself, it is usually impossible for the beam to fail in compression. It is evident, therefore, that a large amount of steel can be placed in the bottom of the beam to balance the compression.

45. Formulas.—The formulas and values used in the calculations made by the company manufacturing the Kahn system of reenforced concrete are given below:

E_s = modulus of elasticity of steel, or 30,000,000;
E_c = modulus of elasticity of concrete, or 2,000,000;
s_s = ultimate unit tensile strength of steel, or 64,000;
s_t = ultimate unit tensile strength of concrete, or 200;
s_c = ultimate unit compressive strength of concrete, or 3,000;
A = area of metal.

The values for b, d, x, and y are shown in Fig. 49, which shows the bar in section with part of the web bent up to form a pair of prongs. The location of the neutral axis is determined by the formula

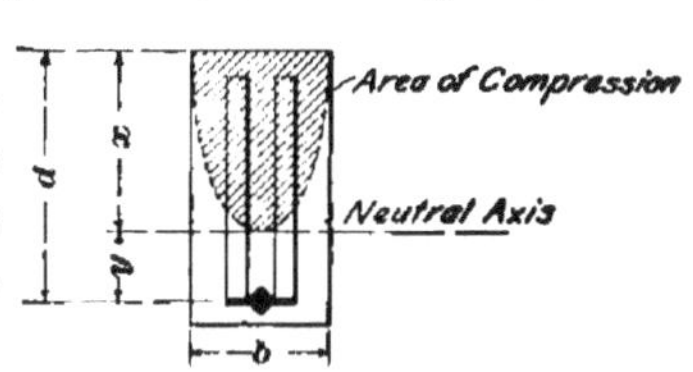

Fig. 49

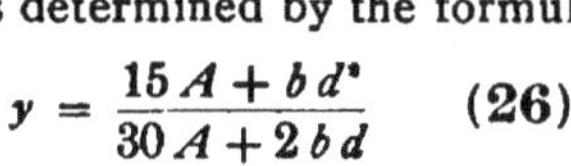

$$y = \frac{15A + bd^2}{30A + 2bd} \qquad (26)$$

When the tensile strength of the concrete is taken into consideration, the moment of resistance is

$$M_1 = (\tfrac{7}{8}x + y)As_s + \frac{s_t b y^2}{3} \qquad (27)$$

If the concrete is disregarded in calculating the resistance to tension, the formula for the resisting moment is

$$M_1 = (\tfrac{7}{8}x + y)As_s \qquad (28)$$

The value of b should never be less than $\frac{As_s}{1,800x}$. A

factor of safety of 4 or 5 should be used in determining the strength of reenforced concrete. The values used in these calulations are somewhat higher than those given in *Beams and Girders*, Part 1.

46. Safe Loads.—Table V gives the safe uniformly distributed loads for concrete beams reenforced with Kahn trussed bars.

Fig. 50 shows a section of the ordinary beam, and in Case I of Table V the bars are ½ inch by 1½ inches, the area being .76 square inch and the weight per lineal foot 2.8 pounds; the length of the diagonals, shown at *d*, Fig. 46, is 6 inches, and *b* equals 8 inches. The batter of the sides of the beam should not be less than 1 inch in 6 inches for beams up to 12 inches in depth, but above that depth the sides may be made vertical.

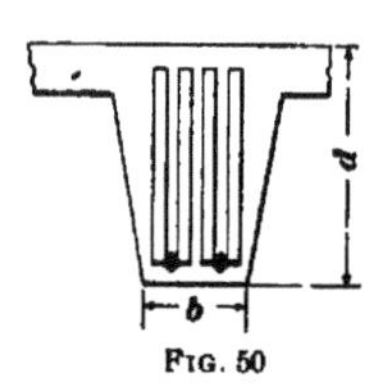

Fig. 50

In Case II, $b = 10$ inches; the size of the bars is ¾ inch by 2 3/16 inches; their area and weight per lineal foot are 1.56 square inches and 5.4 pounds, respectively, and the length of the diagonals is 12 inches. 1″ × 3″ bars are used in Case III, the area being 2.84 square inches, and the weight per lineal foot 9.6 pounds. The length of the diagonals is 12 inches for beams under 24 inches in depth and 18 inches for those over this depth, and *b* equals 12 inches.

In Case IV, $b = 14$ inches, the bars are 1¼ inches by 3¾ inches, weighing 13.8 pounds per square foot and having an area of 4 square inches. The length of the diagonals is 18 feet. In Cases III and IV, the beams should have a batter of 1 to 6 up to a depth of 16 inches, but for a greater depth they may have vertical sides. The loads given in the table were figured for a fiber stress in the steel of 16,000 pounds per square inch. If the beams are continuous across two or more supports, having inverted reenforcing bars built in as required, the safe loads may be increased by ¼.

47. Bars.—Fig. 51 shows the bars kept in stock and having standard cuts; these may be obtained in any length.

Fig. 52 illustrates the method of placing the reenforcing bars for the different conditions of loading and support. At (*a*) is shown a cantilever having a concentrated load at the end, and as the tension is at the upper side of the beam, the bars are placed at the top with the diagonals pointing downwards. Under these conditions the safe load will be one-eighth of that given in the table. (*b*) shows a cantilever supporting a uniform load; in this case the safe load is

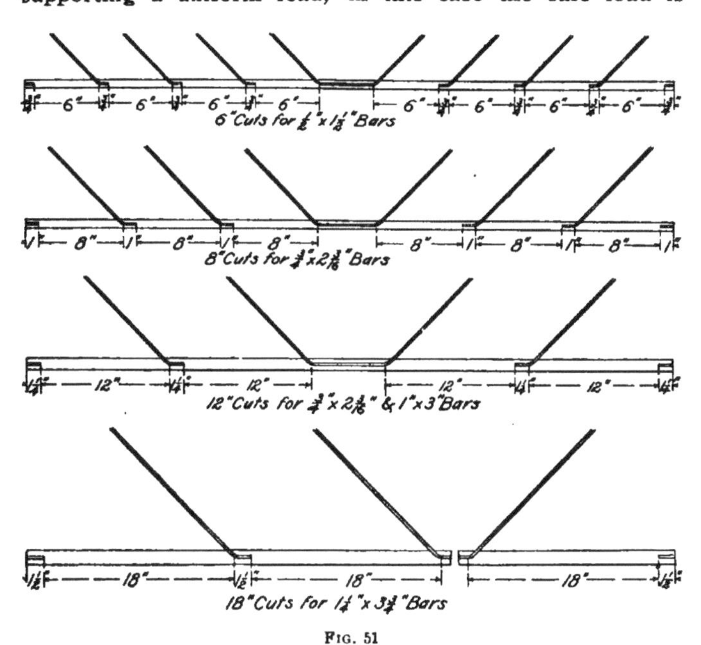

FIG. 51

one-fourth of that given in the table. As the load in (*c*) is concentrated at the center, the safe load will be one-half of that given in the table, while under the conditions shown in (*d*), it will be the same as stated. At (*e*) is shown a continuous beam uniformly loaded and supported at three points, and as there will be tension in the upper part of the beam over the central support, an inverted bar is placed at this point; the safe load under these conditions is the

TABLE V

SAFE UNIFORMLY DISTRIBUTED LOAD, IN HUNDREDS OF POUNDS

	Depth Inches	Distance Between Supports, in Feet																						
		8	9	10	11	12	13	14	15	16	17	18	19	20	21	22	23	24	25	26	27	28	29	30
Case I	6	40	35	32	29	27	25	23	21	20	19	18	17	16										
	8	56	50	45	41	38	35	32	30	28	27	25	24	23	21	20								
	10	74	66	59	53	49	45	42	40	37	35	33	31	30	28	27	26	25	24	23				
	12	93	82	74	67	62	57	53	50	46	44	41	39	37	35	34	32	31	30	28	27			
	14	110	98	88	80	73	68	63	59	55	52	49	46	44	42	40	38	37	35	34	33	31		
	16	123	110	98	89	82	75	70	65	61	58	55	52	49	47	45	43	41	39	38	36	35	34	
	18	138	122	110	100	92	85	79	73	69	65	61	58	55	52	50	48	46	44	42	41	39	38	37
	20	154	137	123	113	102	95	88	82	77	72	68	65	61	58	56	54	51	49	47	46	44	42	41
Case II	8	110	98	88	80	73	68	63	59	55	52	49	46	44	42	40	38	37	35	34	33	31		
	10	147	130	117	106	98	90	84	78	73	69	65	62	59	56	53	51	49	47	45	43	42	40	
	12	190	169	152	138	127	117	108	101	95	90	85	80	76	72	69	66	63	61	58	56	54	52	51
	14	210	188	169	154	141	130	121	113	106	99	94	89	85	81	77	74	71	68	65	63	60	58	56
	16	248	220	198	180	165	152	142	132	124	117	110	104	99	94	90	86	82	79	76	73	71	68	66
	18	280	249	224	203	186	172	160	149	140	132	124	118	112	107	102	97	93	90	86	83	80	77	75
	20	314	279	251	228	209	193	179	168	157	148	139	132	125	120	114	109	105	100	97	93	90	87	84
	22	347	309	278	252	232	214	198	185	174	163	154	146	139	132	126	121	116	111	107	103	99	96	93

Case III	10	244	217	195	177	163	150	140	130	122	115	108	103	98	93	89	85	81	78	75	72	70	68	65
	12	328	290	261	237	218	200	186	174	163	153	145	137	130	124	118	113	107	104	100	97	93	90	87
	14	394	350	315	286	263	242	225	210	197	185	175	166	157	150	143	137	131	126	121	117	112	108	105
	16	448	400	358	326	298	276	256	239	224	211	199	188	179	170	163	155	149	143	138	132	128	123	119
	18	508	452	406	369	338	312	290	270	254	239	226	214	203	193	184	176	169	162	156	150	145	140	135
	20	568	504	453	412	378	348	324	302	283	266	252	238	226	216	206	197	189	181	174	168	162	156	151
	22	634	564	507	461	423	390	362	338	317	298	281	266	254	242	230	220	211	203	195	186	182	175	169
	24	680	605	544	395	455	418	388	362	340	320	302	286	272	259	247	236	226	217	209	201	194	188	181
Case IV	12	452	402	361	328	300	277	258	240	226	212	200	190	182	172	164	157	150	144	139	134	129	125	121
	14	540	480	432	392	360	332	308	288	270	254	240	227	216	206	196	188	180	173	166	160	154	149	144
	16	625	555	499	455	416	382	357	333	312	294	277	263	250	238	227	217	208	200	192	185	178	172	167
	18	707	630	566	515	472	435	405	377	352	333	314	298	283	270	257	246	236	226	218	210	202	195	188
	20	788	700	630	572	525	485	450	420	394	370	350	331	315	300	286	273	262	252	242	233	225	217	210
	22	867	770	694	630	577	535	495	462	434	407	386	365	346	330	315	302	289	278	267	257	248	239	231
	24	960	854	768	700	640	590	550	512	480	452	427	404	382	366	349	334	320	307	295	284	276	265	256
	26	1,040	925	832	757	695	640	595	555	520	490	462	438	415	396	378	361	346	333	320	308	297	287	277
	28	1,125	1,000	900	820	750	692	642	600	563	530	500	474	450	428	410	392	375	360	346	334	321	310	300
	30	1,210	1,075	968	878	805	744	680	645	605	568	536	508	483	460	440	420	404	387	372	358	345	333	321

same as that given in the table. View (*f*) shows a continuous beam supported at four points, that is, carried across two supports, and in this case the safe load is equal to one and one-fourth times that given in the table.

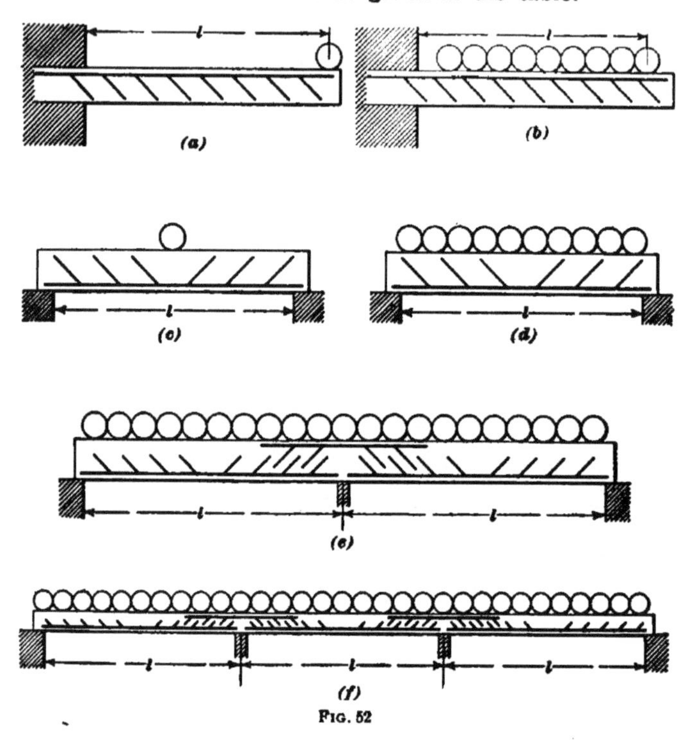

FIG. 52

THE UNIT CONCRETE STEEL REENFORCEMENT

48. The most modern type of reenforced concrete construction is illustrated in Fig. 53. This construction is known as the **unit system,** and takes its name from the fact that all the metal necessary for the actual reenforcement of the concrete girder is in one piece, differing in this respect from other reenforced concrete systems. The steel reenforcement of the unit-girder frame is made from a rolled section consisting of four round bars connected with a thin web, as shown in Fig. 53 (*b*). For a considerable distance from each end, the web is sheared through the section at *a a* and the round bars with the attached portion of web are bent up in a similar form to a trussed girder, as shown at *b*, *b*, Fig. 53 (*a*). The web remaining on the two central bars of the four-bulbed rolled section is punched and bent downwards, as shown at *c*, *c*, while the web between the two inside bars of the rolled section is punched out at intervals with slotted holes, as at *d*, *d*. The purpose of these slotted holes is best understood by referring to Fig. 54.

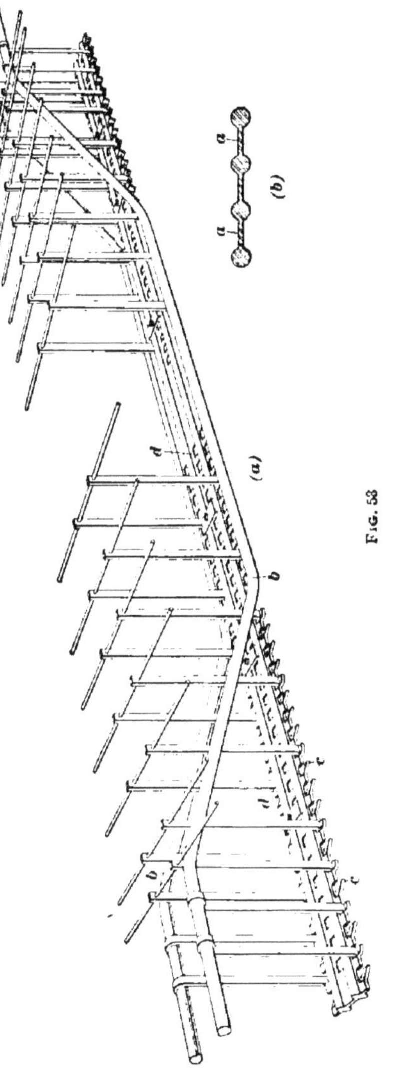

Fig. 53

In Fig. 54 (*a*) a sectional elevation of a unit-girder frame is shown, while cross-sections through the girder are shown at (*d*) and (*e*). From these three views it will be seen that especially designed bolts are placed through the slotted holes punched in the central web of the rolled section.

These bolts consist of an annealed cast-steel socket *b*,

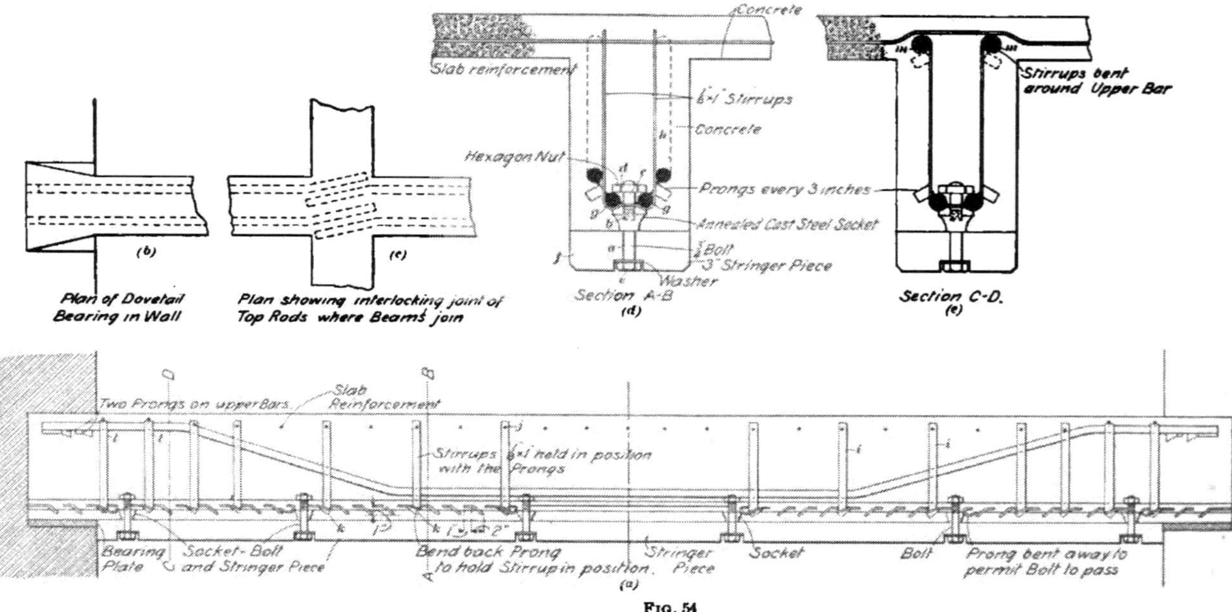
Plan of Dovetail Bearing in Wall
(b)
Plan showing interlocking joint of Top Rods where Beams join
(c)
Concrete
Slab reinforcement
Stirrups
Concrete
Hexagon Nut
Prongs every 3 inches
Annealed Cast Steel Socket
Bolt
3" Stringer Piece
Washer
Section A-B
(d)
Stirrups bent around Upper Bar
Section C-D.
(e)
Two Prongs on upper Bars
Slab Reinforcement
Stirrups held in position with the Prongs
Bearing Plate
Socket-Bolt and Stringer Piece
Bend back Prong to hold Stirrup in position.
Stringer Piece
Socket
Bolt
Prong bent away to permit Bolt to pass
(a)

Fig. 54

Fig. 54 (*d*), arranged to fit into the space between the two center bars. Both ends of this casting are capped and supplied with stud bolts, as at *a* and *c*, the stud bolts being furnished with the nuts *d* and *e*. The purpose of these especially designed bolts is threefold: First, when put in place and securely fastened, by the nut *d*, to the steel girder, they form a means by which the bottom of the mold into which the concrete for the beam is tamped may be held in place. Second, when the bottom of the mold is secured in this manner the trussed unit of steel is held in the correct position within the mold and there is no danger of disturbing the reenforcing metal, as there is in some systems by the tamping of the concrete. Third, the nut *e*, when the concrete has set and the bottom of the mold is removed, may be used to secure wooden stringers, as at *f*. These stringers are used in mill construction for the purpose of holding up the shafting by lagscrewing the hangers to them.

Again referring to Fig. 54 (*a*), it is observed that the trussed frame is supplied with stirrups *i*, *i* consisting of $\frac{1}{8}'' \times 1''$ bars punched at the upper end, as at *j*, to allow the steel reenforcing members of the concrete floor between the girders to pass through them. These bars are **U**-shaped, as shown at *h* in view (*d*), and are bent at the bottom, as at *g*, *g* to fit the shape of the rolled section. They are securely held in place at the bottom by bending back the prongs, as illustrated at *k*, *k* in view (*a*). In this manner, the prongs, or clips, answer the double purpose of securing the stirrups in position and acting as a key for the concrete at the bottom of the girder, thus preventing it from spalling or breaking off with exposure to fire or water. The stirrups toward the end of the bar are arranged somewhat differently at the top, for these are bent around the upper bar as shown at *l*, *l* in view (*a*), and at *m*, *m* in view (*e*). The stirrups are necessary throughout the girder to take care of the shearing stresses, and as the shear increases toward the abutments they are spaced closer together at the supports, in a similar manner to the stiffeners in a plate girder. The girders are so arranged in plan at the ends as to be interchangeable and interlocking; that is, where the girders meet on the supporting wall, pier, or column the bars are bent to one side, as shown in view (*c*), and where they enter a wall of masonry the dovetailed end construction is employed, simulating the Goetz-Mitchel wall-box construction, as shown in view (*b*).

The advantages claimed for the unit-girder system of reenforced concrete are that it is designed, built, delivered, and erected as a unit, or in one piece; that the principal members are made of one rolled section and all other parts are securely held together so that none will be inadvertently omitted when the concrete is put in place; further, that the entire frame is placed in position at once, and that it is the only system in which the concrete can be thoroughly tamped without disturbing the reenforcing members; also, that the slab reenforcement is braced through the stirrups. The cost of this system of reenforced concrete is a little more than slow-burning construction, but it will not exceed the combined cost of slow-burning construction and the usual fire-equipment, such as sprinkler system, stand pipes, etc., that is necessary with the latter.

SPECIFICATION FOR REENFORCED CONCRETE

49. In connection with reenforced concrete construction, the following specification is interesting and covers most of the necessary stipulations:

The work is to be executed entirely in reenforced concrete and the system employed must be one that provides for the shearing stresses at the ends of girders and beams. The floors are to be made sufficiently strong to carry their own weight and the superimposed load of . . . pounds; the ceilings, to carry . . . pounds; and the roof . . . pounds. The live load on the floor will include the finish used on top of the concrete construction. The floor shall be able to carry, without failure, a superimposed load of four times the amount stated in the specification, and the right is reserved to test any unit area of the floor construction to failure. If it fails before the required load has been placed on it, the damage must be made good and all work must be strengthened so as to meet the requirements of these specifications, without expense to the owner.

The reenforced concrete beams must be in accordance with an approved system of construction. They must provide resistance to bending moment and also to shearing, by being reenforced in the vertical plane as well as in the horizontal. The members resisting the shear shall be rigidly connected to the main horizontal tension member, and shall make an angle of about 45° with this member. In calculating the resistance to shear, the adhesion of the concrete to steel will not be assumed at more than 50 pounds per square inch and the vertical reenforcement shall therefore be proportioned to take care of the remaining stresses not provided for by the concrete.

The metal in the floor system must be protected from fire by making the distance from the lower surface of the concrete slab to the edge of the metal at least 1 inch, while in girders and beams this thickness shall be 1½ inches.

The concrete shall be composed of sand, broken stone or gravel, and an approved quality of Portland cement. If broken stone is used, it must be free from fine dust, and small enough to pass through a 1-inch screen, while if gravel is employed, it must be clean and vary in size from ¼ to 1 inch. For beams, the proportion of concrete shall be one part of cement, two parts of sand, and four parts of broken stone. For floor slabs, the proportion shall be one part of cement, two and one-half parts of sand, and five parts of broken stone. In general, one layer of concrete shall be entirely completed before another is commenced and the layers shall not exceed about 6 inches in thickness. The concrete shall be deposited as closely as possible to the place where it is to be used, in order to avoid rehandling while in the mold, and each layer shall be thoroughly rammed until moisture appears on the surface. The mixture shall be as nearly uniform as possible in character, and shall be mixed by one of the approved standard batch machine mixers; it shall be of such consistency that when thoroughly rammed it will quake slightly.

General sketches must be furnished by each bidder, showing his system of reenforced concrete and the method of applying it to the work in hand, the thickness of floor slabs, general location of beams, etc., and he must state the tensile strength, elastic limit, elongation, etc. of the steel bars he intends to use. The steel must come up to the standard structural material adopted by the Association of Steel Manufacturers, which provides an ultimate strength of from 55,000 to 60,000 pounds per square inch. Before the work is commenced, detail working drawings and specifications must be furnished by the successful bidder and approved by the architect. The centering used must be strong enough to hold the plastic concrete to the true line and shape.

The architect shall have power to condemn any concrete improperly mixed or proportioned and prevent its incorporation in the work, and whenever, in his judgment, the concrete is liable to be injured from freezing, work shall be suspended. The reenforcing bars used in the concrete will not be painted, but any bar on which rust scales have begun to form will be rejected. No bids will be considered except from persons or firms with recent and extensive experience in the form of work called for.

GENERAL NOTES ON REENFORCED CONCRETE

50. Examples of Construction.—In Fig. 55 is shown a section through an eight-story warehouse constructed entirely of reenforced concrete. The size of the building is 92 feet by 125 feet, and it was designed to carry floor loads

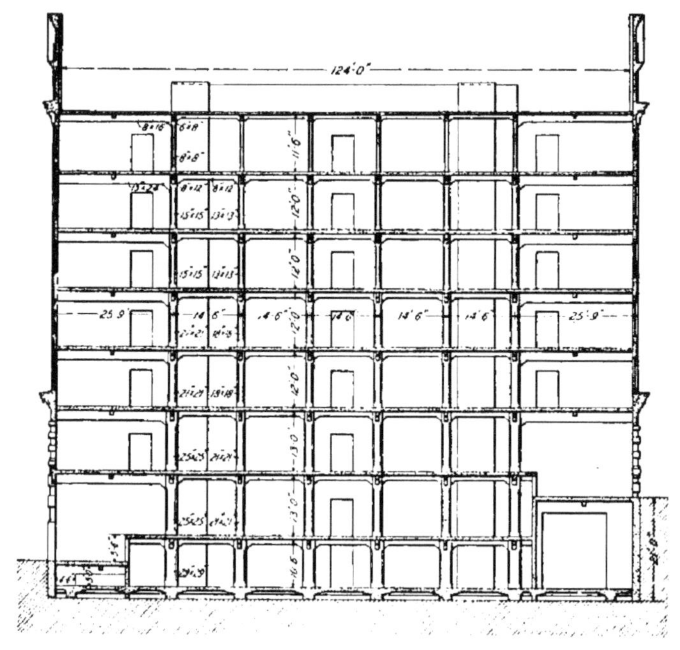

FIG. 55

up to 800 pounds per square foot. The ground on which the building was located was of the worst description imaginable for foundations, and it was necessary to support the entire structure on a reenforced concrete raft. This raft measures 2 feet 6 inches in its thickest part and only 7 inches in

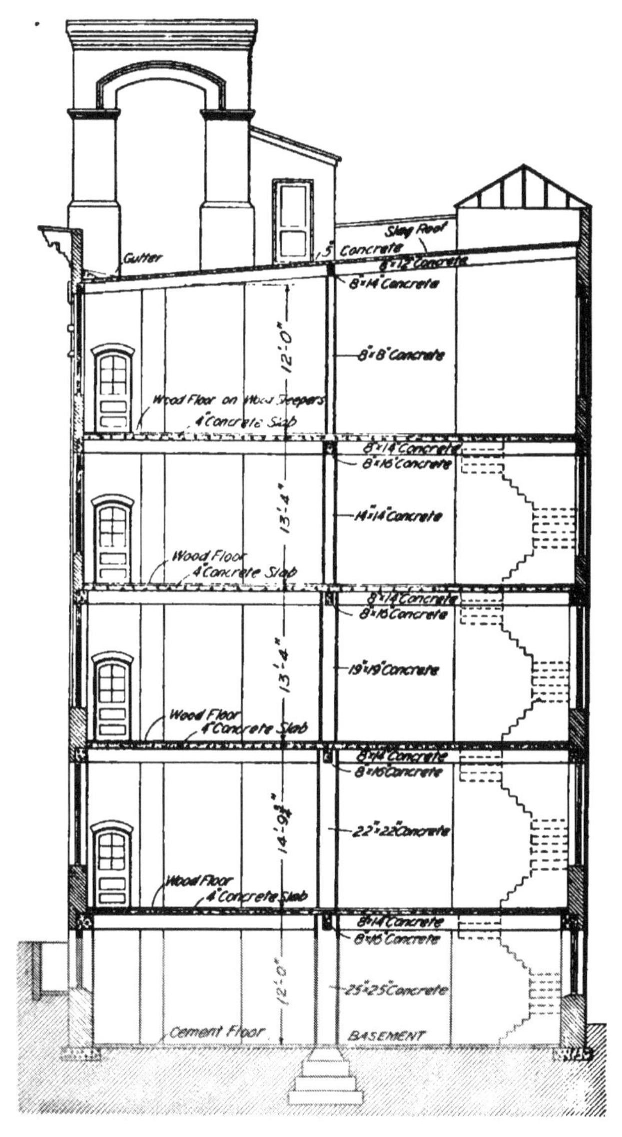
Slag Roof
5" Concrete
8"×12" Concrete
Gutter
8"×14" Concrete
12'-0"
8"×8" Concrete
4" Concrete Slab
8"×14" Concrete
8"×16" Concrete
13'-4"
14"×14" Concrete
Wood Floor
4" Concrete Slab
8"×14" Concrete
8"×16" Concrete
13'-4"
19"×19" Concrete
Wood Floor
4" Concrete Slab
8"×14" Concrete
8"×16" Concrete
22"×22" Concrete
Wood Floor
4" Concrete Slab
8"×14" Concrete
8"×16" Concrete
12'-0"
25"×25" Concrete
Cement Floor
BASEMENT

Fig. 56

its thinnest part. It is laid out in panels and is arranged so that each column will be supported at the intersection of two beams. Between the date of construction of the footings and the date of construction of the first floor, this building settled 3½ inches in the front and 3 inches in the rear, but no further settlement occurred. It is not advisable to use this character of foundation for supporting a building where it is possible to employ isolated piers.

Fig. 56 shows a section through a laundry; this structure, including the roof, is composed entirely of concrete. It is interesting to note the sizes of the beams, girders, and columns used in this building, which may be considered as a fair sample of reenforced concrete construction applied to factory buildings.

51. The Practice of Construction.—It is customary to leave the centering for the floors in place at least 2 weeks. Where two sets of forms are used, rapid progress can be made. After the concrete is laid on the first set of forms, the centering for the second floor may be built up, for by the time the concrete for the second floor is in place the first floor will be sufficiently hardened to permit the removal of the forms supporting it, which may be used for the third floor, and so on. In fact, reenforced concrete construction can be executed as fast as the material can be put in place. Concrete may be laid in freezing weather, provided that the temperature does not fall below 26° F., but if freezing has occurred, the centering should not be removed until every indication of frost has disappeared and the concrete has thoroughly set. Many contractors use only 1-inch material for the centerings, but it is more advisable to use 2-inch. The forms should be put together so that they can be readily taken down. This may be accomplished by not driving the nails to their full length.

Limestone should never be used for the aggregate in concrete, as it renders it worthless as a fireproof material. The limestone will disintegrate under heat and water and the pieces will swell and break out of the slab of concrete. It is very important for the cement to be uniform and of the best quality. In many cases, the mixtures for foundation work may have a high percentage of stone, sometimes as much as eight parts of stone to one part of cement and four parts of sand, but on an average, the best results for foundation work are obtained from one part cement, three parts of sand, and five parts of broken stone. It is not advisable to attempt to color exterior concrete walls by mixing the coloring matter in the concrete, as the cement tends to combine chemically with the coloring matter. If it is desired to color the wall, the best method is to paint the concrete afterwards. However, the most satisfactory method of obtaining a color scheme is to use sand that has considerable color, as red or yellow.

GOVERNMENTAL REGULATIONS FOR REENFORCED CONCRETE CONSTRUCTION

52. Municipalities and some national governments have adopted laws and ordinances regulating the construction of reenforced concrete. The following is an extract from the latest of these, with some alterations, and adherence to its stipulation cannot but insure excellent results.

53. Materials.—Only Portland cement, delivered in original packages, shall be used in the making of reenforced concrete construction. The sand used in mixing the concrete shall be clean and sharp while the other ingredients shall be of such a suitable size as experience has found to be satisfactory.

54. Mixing Concrete.—The concrete shall be mixed only in quantities required for immediate use. It shall be put in place immediately after mixing and be tamped uniformly; if laid in the consistency of moist earth, it shall be tamped until water appears on the surface. Suitably shaped tampers of proper weight shall be used in tamping.

55. Testing.—In the case of untried methods of construction, the building authorities may first require preliminary trial constructions and load tests.

The building authorities may also determine the qualities of the building materials that are employed, by the aid of an official testing laboratory or in any other way they may deem suitable, and they also may test the strength of the concrete made.

At the final inspection for acceptance, the structure shall be uncovered and accessible at the places indicated by the representative of the building authorities, to show the quality and character of construction. The building authorities have also the right to test the quality of the construction, the degree of hardening obtained, and the sustaining strength, by special tests.

If loading tests are deemed necessary, they shall be made according to the directions of the representative of the building authorities. The owner and the contractor shall be invited to be present at these tests.

If a strip of a floor panel be cut out and tested by a trial load, this load shall be distributed uniformly over the whole strip and not exceed the weight of the floor and double the live load it is computed to sustain. If such a strip is tested without being cut out of the panel, the test load shall be increased by one-half. Thus, if W_d denotes the dead load and W_l the live load, the test load will be, for the former case, $W_d + 2\ W_l$ and for the latter, $1.5\ W_d + 3\ W_l$.

56. Regarding Work in Freezing Weather.—No work shall be done in freezing weather, except when the influence of frost is excluded. After prolonged freezing weather, the work shall not be taken up again in warmer weather until the approval of the building authorities has been obtained.

Until sufficient hardening of the concrete has occurred, the structural parts shall be protected against the effect of freezing and premature drying as well as against vibrations and loads. If freezing weather occurs during the hardening period, the time ordinarily allowed for setting shall be extended by the time of freezing because of the retardation in the hardening caused by the latter.

A daily account of the work shall be kept on the ground and be always open for inspection. Freezing weather shall be especially noted, and the hour and temperature recorded.

57. Molding the Concrete.—Special care shall be exercised in placing the reenforcing metal in its correct position and in surrounding it tightly with cement mortar.

The concrete shall be laid in layers that shall have a thickness not exceeding 6 inches and each layer shall be thoroughly tamped in place.

Continuous walls shall be laid uniformly for their entire length, and care shall be taken to obtain good connections with the adjacent transverse walls. Layers that form the top surfaces of a story must be leveled off.

The molding boards must have sufficient resistance to bending as well as to shocks and vibrations due to tamping, and they shall be arranged to be safely removable after taking away their supports.

The side molds of concrete beams and the molds of floor slabs up to spans of 5 feet shall not be removed before 3 days, and the remaining molds and the supports not before 14 days from the completion of the tamping. In removing the molds and supports all jar and vibration shall be avoided.

If the tamping has been completed only a short time before the occurrence of the freezing weather, special care shall be taken in removing the molds and supports.

In placing a new layer of concrete on a fresh one it will suffice to wet the old surface thoroughly, though in building on hardened concrete the old surface shall be roughened, cleaned, and wet.

In constructing walls and piers for buildings of several stories, the upper stories shall not be started before the molds of the lower story shall have been removed.

58. Rules for Static Computation.—The weight of concrete, inclusive of the reenforcing metal, shall be taken at 150 pounds per cubic foot, unless another weight be definitely ascertained, and in the computation for floors the weight of the flooring material shall be added to the weight of the sustaining parts.

59. Determination of External Forces.—For the members subject to bending, the end moments and reactions shall be computed, according to the character of loading and support, by the formulas for freely supported or continuous beams.

For freely supported slabs, the clear opening plus the depth of the slab shall be taken as the span; for continuous slabs, the distance between centers of supports.

For slabs continuous over several spans, the bending moment at the centers of slabs may be taken as four-fifths the value of the moment for a freely supported beam, unless the actually occurring moments and reactions be ascertained by computation or tests.

The same rule holds true for beams, tee-formed beams and girders, with the exception, however, that no end moment shall be taken into account unless special structural details to insure the fixed end be provided. The span shall be taken as the free opening plus the length of one bearing.

For T-formed beams, the flange shall not be considered for a width of more than one-third of the length of the beam.

For columns, consideration shall be given to possible eccentric loading.

60. Allowable Working Stresses.—In the members subject to bending, the compressive stress in the concrete shall not exceed one-fifth of its ultimate resistance; the tensile and compressive stresses in the steel shall not exceed 17,000 pounds per square inch. The following loads shall be provided for: (*a*) For structural parts subject to moderate impact, such as floors of dwellings, offices, and warehouses; the actual dead and live loads. (*b*) For parts subject to higher impact or widely varying loads, such as floors of assembly rooms, dancing halls, factories, and storehouses; the actual dead load and one and one-half times the live load. (*c*) For parts subject to heavy shocks, such as roofs of vaults, under passageways and yards; the actual dead load and twice the live load.

In columns, the concrete shall not sustain a stress above one-tenth its breaking strength. In computing the steel reenforcing for column flexure, a factor of safety of 5 shall be provided.

The shearing stress in the concrete shall not exceed 64 pounds per square inch. In any instance, the shearing stress shall not exceed one-fifth of the ultimate resistance. The adhesive stresses shall not exceed the allowable shearing stress.

61. Determination of Internal Forces.—The modulus of elasticity of steel shall be taken as fifteen times that of concrete, unless another ratio be shown.

The stresses in a section of a body subject to bending shall be computed on the assumption that the elongations are proportional to their distances from the neutral axis and that all the tensile stresses are taken up by the steel reenforcement.

Shearing stresses shall be computed, unless the form and design of the members show at once their insignificance. If no allowance is made for them in the design of the member, they must be taken up by suitably shaped steel reenforcement.

The reenforcing steel shall, as far as possible, be so formed that its displacement in the concrete shall be prevented by its form. As far as this is lacking, the adhesive stress shall be computed.

Computations for the flexure of columns shall be made if the height exceeds eighteen times the least diameter. Transverse connections that shall hold the embedded steel rods in their positions relative to each other shall be placed at distances apart not more than thirty times the diameter of the rods.

For the computation of the columns for flexure, Euler's formula shall be used.

62. Method of Computation.—If the beam is a simple one reenforced on the tension side only, the computation for the stresses in the concrete and steel may be made by formulas **27** and **28**, respectively.

The following formulas are those stipulated in the government regulations introduced in Art. **52**. In these formulas,

f_s = sectional area of reenforcing steel for width b of beam;
n = ratio of coefficient of elasticity of steel to that of concrete;
M = bending moment of beam;
h = total depth of beam;
x = distance of neutral axis from upper surface;
a = distance from center of steel to bottom of beam;
b = width of beam;

$$x = \frac{n f_s}{b}\left[\sqrt{1 + \frac{2b(h-a)}{n f_s}} - 1\right] \qquad \textbf{(29)}$$

From the equation of moments, the greatest compressive stress in the concrete s_c and the greatest tensile stress in the steel s_s will be:

$$s_c = \frac{2M}{b\left(h - a - \frac{x}{3}\right)} \qquad \textbf{(30)}$$

$$s_s = \frac{M}{f_s\left(h - a - \frac{x}{3}\right)} \qquad \textbf{(31)}$$

For **T**-formed beams when the width of the top part or flange is called b and its thickness or depth is called d, the same formulas may be used if the neutral axis lies in the flange or at the junction of flange and web. If the neutral axis passes through the web, the slight compressive stresses in the web may be neglected. The resulting formulas are:

$$x = \frac{(h-a) n f_s + \frac{b d^2}{2}}{b d + n f_s} \qquad \textbf{(32)}$$

$$y = x - \frac{d}{2} + \frac{d^2}{6(2x - d)} \qquad \textbf{(33)}$$

$$s_s = \frac{M}{f_s(h - a - x + y)} \qquad \textbf{(34)}$$

$$s_c = s_s\left(\frac{x}{n(h - a - x)}\right) \qquad \textbf{(35)}$$

63. Centric Pressure on Columns.—If F be the cross-sectional area of the concrete in compression, and f_s that of the total reenforcing steel, the allowable load will be

$$P = s_c(F + n f_s) \qquad (36)$$

The greatest stress in the concrete will be

$$s_c = \frac{P}{F + n f_s} \qquad (37)$$

The greatest stress in the steel will be

$$s_s = n s_c = \frac{n P}{F + n f_s} \qquad (38)$$

64. Eccentric Pressure on Columns.—The computation is the same as for a homogeneous material, except that in the expressions for the cross-sectional area and the moment of inertia of the reenforcing steel, n times the value of an equivalent section of concrete is substituted. Any tensile stresses that may be produced must be taken up by reenforcing steel provided for this purpose.

BEAMS AND GIRDERS

(PART 3)

BUILT-UP GIRDERS

GENERAL CONSTRUCTION

1. Definitions.—A **plate girder** is one built up of a number of plates and angles securely riveted together. The names given to its different parts may be understood by referring to Fig. 1, in which *a* is the **flange plate,** of which there may be one or more on each flange, depending on the strength required. These plates are the principal elements that resist the bending stresses in the girder. The **flange angles** *b*, *b* are the means of connecting the flange plates to the **web-plate** *c*. When the load on the girder is small, the flange plates may be omitted, in which case the flange angles are the members that chiefly offer resistance to the bending stresses.

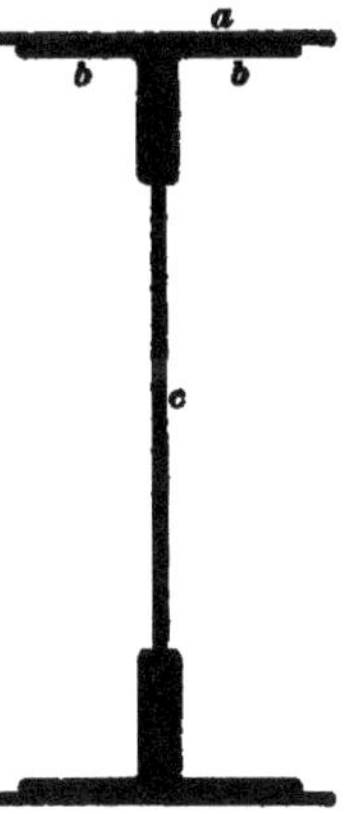

FIG. 1

On account of the construction of a plate girder, there is very little stiffness in the web-plate; consequently, there is always a strong tendency for it to fail by buckling and twisting under the load imposed on the girder. This tendency to buckle is greatest at the supports or abutments of the girder and at the points where concentrated loads are applied. Because of this buckling tendency it becomes necessary to reenforce the girder by riveting to it at stated intervals **stiffeners** generally made of angles.

The most common and cheapest form of stiffener is shown in Fig. 2 (*a*). This is simply a straight piece of angle riveted to the web-plate and flange angles. The space between the stiffeners and web-plate, due to the thickness of the flange angles, is filled with a piece of bar iron or plate, as shown at *d*; this is called a **filler** or **packing piece.**

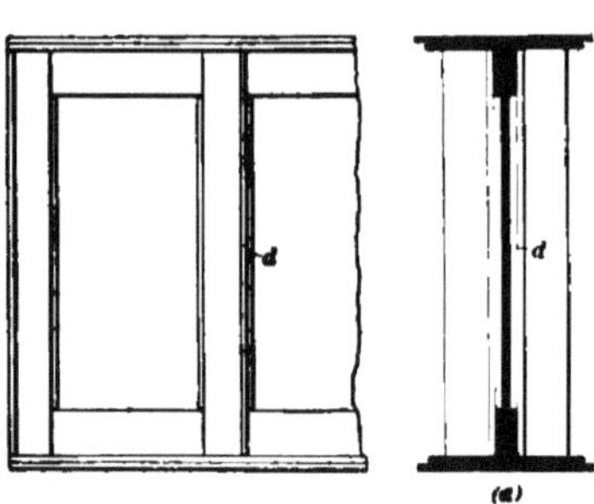

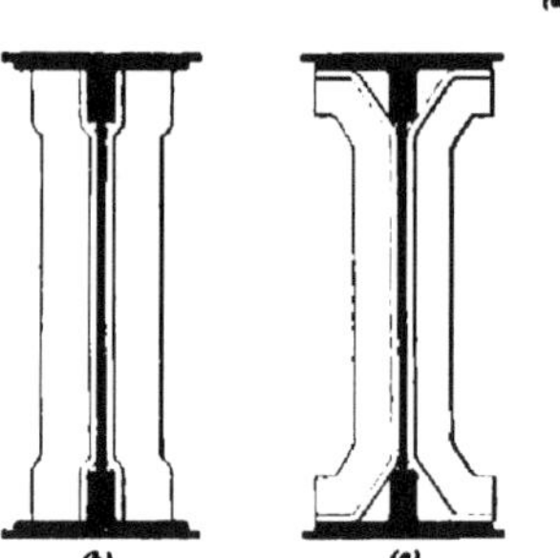

Fig. 2

Another form of stiffener for plate girders is shown in Fig. 2 (*b*). The angle is swaged out, to allow it to fit over the flange angles, and is riveted directly to the web-plate, thus doing away with the filler or packing piece. This construction does not require as much material as that shown in (*a*), but, unless there are a large number of girders of the same dimensions to be built, in which case dies, in connection with a power or hydraulic press, may be used for swaging the ends, the labor required is so much greater that it makes the girder more expensive.

The stiffeners shown at (*c*) are sometimes used, but are subject to the same general criticism in regard to cost of manufacture as those shown at (*b*). Stiffeners of this shape possess a possible advantage in the fact that they stiffen the flanges considerably more than either of the other styles.

2. Usual Forms of Sections.—The four principal sections used in plate-girder construction are shown in Fig. 3.

A simple plate girder with a web-plate and two flange angles, but with no flange plate, is shown in (*a*); this section is used for short spans or light loads. In (*b*) is shown a similar girder with one flange plate; this girder is used to support heavier loads and to clear longer spans. The girder in (*c*), which may have two or more flange plates at each flange, may, if the conditions require, be made as heavy as is necessary in order to carry great loads over long spans. In fact, the strength of a girder of this character may be increased almost indefinitely by the addition of flange plates. The section, Fig. 3 (*d*), is a plate girder of box section. It is stiffer laterally than the other forms shown, but the difficulty of reaching the interior for painting and inspecting and

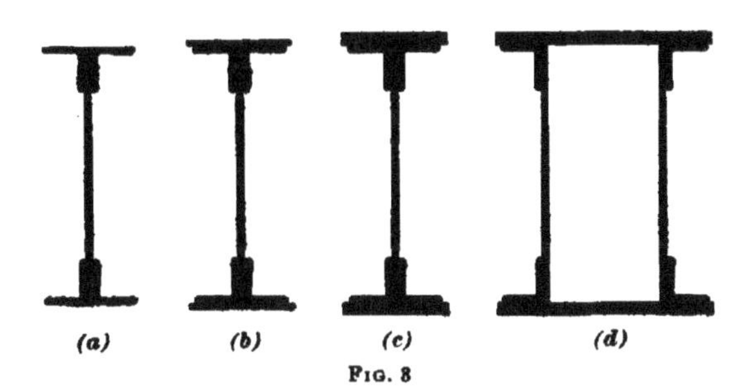

Fig. 3

the excessive amount of labor required in its construction, are such serious objections that it is much less used than the sections shown at (*a*), (*b*), and (*c*). On account of their open construction, the latter are especially good forms to use in buildings where the objection in regard to lateral stiffness does not hold good, as when the girder is used in the position in which it is usually found, it is generally prevented from deflecting laterally by the floorbeams; any lack of stiffness in comparison with the box girder is more than compensated for by the simplicity of construction and easy access on all sides for painting and inspecting.

PRINCIPLES OF DESIGN

STRESSES

3. The external forces, loads, and reactions produce the same kind of **stresses** in a plate girder as in an ordinary beam, but, on account of its special construction, the distribution of these stresses in the girder is assumed to be somewhat different from that in a beam made of a single piece. In the girder, the shear is generally assumed to be borne wholly by the web-plate, while the bending moment is assumed to be resisted by the stresses in the flange members. The method of calculating the magnitude of the shear and bending moment is the same as that for beams; owing, however, to the different assumption in regard to the distribution of these forces, a different method of calculation is used in determining the relations between them and the stresses in the girder.

4. Shearing Stresses in Web-Plate.—In discussing the methods of calculating the dimensions of a plate girder for a given purpose, we will first consider the shear, which is the principal factor that determines the thickness of the web-plate and the number and size of stiffeners required. The greatest shear in a beam occurs at the point of support at which the reaction is greatest, and the magnitude of the shear is equal to the reaction at that point; consequently, in a simple plate girder, the greatest shear occurs at a point of support, and is equal in amount to the reaction at that point.

WEB-PLATES AND STIFFENERS

5. Depth of Girder.—Having calculated the shear, the depth of the girder is assumed in accordance with practical rules that fix the relation between the depth and span. In accordance with the best practice, the depth should not be less than one-fifteenth of the span, though some authorities consider one-twentieth ample. The latter proportion, however, gives an exceedingly shallow girder, and cannot be recommended except where the loads are very light and the span short, or where it is absolutely necessary that an extremely shallow girder be used, on account of decorative features, or lack of space in regard to headroom, in which case the girder should be so proportioned that when fully loaded its deflection will not be excessive.

6. Thickness of Web-Plate.—Knowing the depth of the girder, and the shear at the points of support, the thickness of the web-plate is proportioned so as to give it sufficient area to resist the maximum shear. It is always necessary to stiffen the plates over the supports, as shown in Fig. 5; these stiffeners are riveted to the plate and transfer the shearing stress from it to the supports.

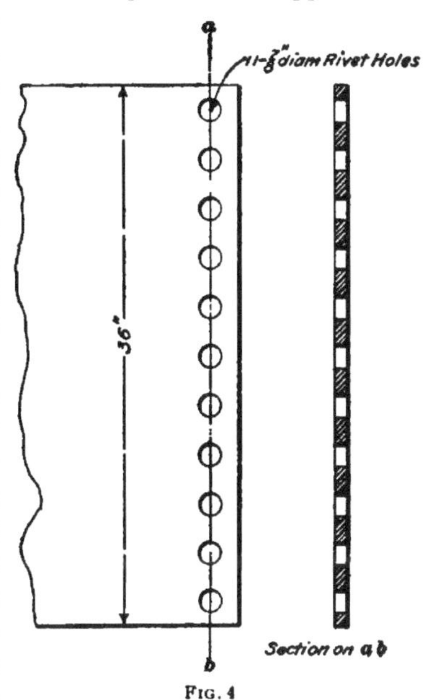

FIG. 4

A considerable portion of the plate is cut away by the holes for the rivets by which it is fastened to the stiffeners; hence, the least strength of the plate is along the line of the rivet holes. It can be seen, by referring to Fig. 4, which shows the end of a plate with the holes punched for riveting to the stiffener, that the net or efficient depth of the plate is equal to the actual depth minus the sum of the diameters of the rivet holes.

The following rule may be used for calculating the thickness of a web-plate so that it will have sufficient strength to resist the shearing stress:

Rule.—*From the total depth of the web-plate, deduct the sum of the diameters of the rivet holes, which will give the net* *or efficient depth of the web-plate; multiply the net depth by the safe resistance of the material to shear, and divide the maximum shear, in pounds, by the product; the quotient will be the required thickness of the metal in the web of the girder.*

This rule may be expressed by the formula

$$t = \frac{R}{d \times s} \qquad (1)$$

in which t = thickness of web-plate;
R = greatest reaction or maximum shear;
s = safe shearing resistance of material per square inch;
d = net depth of web-plate after all rivet holes have been deducted.

7. The safe resistance of the material to shear is of course governed by the factor of safety required in the girder. For example, the ultimate shearing strength of structural steel being 52,000 pounds per square inch, if a factor of safety of 4 is required, the safe resistance of the metal will be 52,000 ÷ 4 = 13,000 pounds, while if a factor of safety of 5 is desired, the safe strength will be 52,000 ÷ 5 = 10,400 pounds.

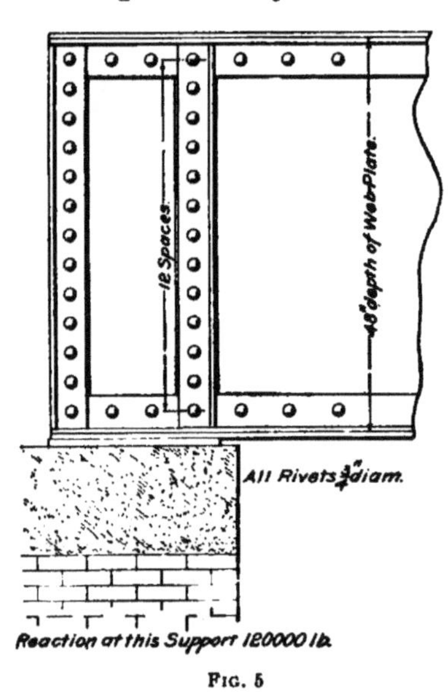

Fig. 5

In deducting the metal for the rivet holes in order to ascertain the net depth of the web-plate, the holes should always be considered as being $\frac{1}{8}$ inch larger than the nominal diameter of the rivet; this allowance is made because the holes are always made $\frac{1}{16}$ inch larger in diameter than the rivet so that the rivet may be inserted easily, and another $\frac{1}{16}$ inch should also be allowed in the diameter of the hole to compensate for any injury that the metal immediately around it may suffer from the punch.

It will often be found that the calculated thickness of the web-plate is less than is allowable for practical reasons. The thinnest plate that should be used for any case is $\frac{5}{16}$ inch.

EXAMPLE.—Fig. 5 shows the end of a plate girder in which the greatest reaction is 120,000 pounds. The girder is made of structural steel, the safe fiber stress of which is assumed to be 11,000 pounds per square inch for shear. What should be the thickness of the web-plate?

SOLUTION.—The width of the plate is 48 in., and there are thirteen holes punched for $\frac{3}{4}$-in. rivets. The deduction to be made for each rivet hole is $\frac{1}{8}$ in. + $\frac{3}{4}$ in. = $\frac{7}{8}$ in.; therefore, the net depth of the plate is $48 - 13 \times \frac{7}{8} = 48 - 11\frac{3}{8} = 36\frac{5}{8}$ in. Applying formula **1**, the thickness of the plate is $t = \frac{120{,}000}{36\frac{5}{8} \times 11{,}000} = .297$ in. In no case, however, should the web-plate of a girder be less than $\frac{5}{16}$ in. in thickness. Hence, as .297 is less than $\frac{5}{16}$, the thickness of the web-plate in this girder should be $\frac{5}{16}$ in. Ans.

8. Buckling of Web-Plate and Distribution of Stiffeners.—The shearing stresses in a web-plate, in addition to their tendency to shear the plate, are liable to cause it to fail by buckling; therefore, in order to properly resist the vertical shearing stresses and prevent them from buckling the web-plate before its full shearing strength is realized, it is necessary to provide the stiffeners.

The shearing stresses in a simple beam are always greatest at the points of support, and diminish toward the center of the beam until a point is reached such that the sum of the loads between it and the support is equal to the reaction; at such a point the shear is said to change sign. Therefore, the stiffeners should be most numerous at the points of greatest vertical shear, and should decrease in number as the shear decreases. Theoretically, this would be a correct method of locating the stiffeners, but practically they are spaced at equal distances along the length of the girder, except at the points of support, where several are placed near together in order to give the end of the girder more nearly the character of a column and enable it to successfully resist the great vertical shear, due to the reaction at this point. In no case should the stiffeners at the end of a plate girder be omitted, even if the conditions make the intermediate ones unnecessary.

It is good practice to place stiffeners directly under any concentrated load that may be placed on the girder. These not only stiffen the web-plate at the point of application of the load, and thus prevent buckling, but also, through the medium of the rivets, assist in distributing the load on the web-plate and other members of the girder.

The end stiffeners of a plate girder may be considered as columns subjected to a compressive stress equal to the reaction, and calculated by the rules and formulas for columns. For safety, the stress on the end stiffeners should never exceed 15,000 pounds per square inch of section.

9. Practice in regard to the placing of stiffeners on plate girders varies considerably, being more a matter of judgment and experience than of calculation. Some engineers determine the resistance of the web-plate to buckling by the formula

$$b = \frac{11{,}000}{1 + \frac{d^2}{3{,}000\, t^2}} \qquad (2)$$

in which b = safe resistance of web to buckling, in pounds per square inch;

d = depth of web-plate, in inches;

t = thickness of web-plate, in inches.

If the value of b given by this formula is less than the unit shearing stress, the girder should be stiffened.

EXAMPLE.—The allowable shearing stress on the web of a plate girder is 11,000 pounds per square inch. The stiffeners at the end supports are riveted by nine $\frac{7}{8}$-inch rivets to the web-plate, which is 36 inches wide. The end reaction on the girder is 100,000 pounds. (*a*) What should be the thickness of the web-plate? (*b*) Will it be sufficiently strong without the addition of stiffeners?

SOLUTION.—(*a*) Since $\frac{7}{8}$-in. rivets are used, the allowance to be made for one rivet hole is $\frac{7}{8}+\frac{1}{8}=1$ in., and the effective depth of the plate along the line of rivets is $36-9\times 1=27$ in. Applying formula **1**, the thickness of the plate is found to be $t=\frac{100{,}000}{27\times 11{,}000}=.33$ in. The thickness of the nearest standard-size plate above this is $\frac{3}{8}$ in., which will be the thickness used. Ans.

(*b*) By formula **2**, the safe unit resistance of the plate to buckling is $b=\dfrac{11{,}000}{1+\dfrac{36^2}{3{,}000\times(\frac{3}{8})^2}}=2{,}700$ lb. per sq. in.; which, since it is much less than the unit shearing stress on the web-plate, shows that stiffeners are required. Ans.

10. Practical Rule for Spacing Stiffeners.—It is not the general practice to make the above calculations to determine whether stiffeners are required; according to the best engineering practice, stiffeners should be provided, unless the thickness of the web-plate is at least one-fiftieth of the clear distance between the vertical legs of the flange angles.

EXAMPLE 1.—Assume the girder in the previous problem to be provided with 6″ × 6″ flange angles; the depth and thickness of the plate, as shown, are 36 inches and $\frac{3}{8}$ inch, respectively. According to the above rule, does this girder require stiffeners?

SOLUTION.—The unsupported depth of the plate between the flange angles is $36-2\times 6=24$ in.; $\frac{1}{50}$ of 24 in. $=.48$, say, $\frac{1}{2}$ in. As the thickness of the web-plate is only $\frac{3}{8}$ in., the girder must be provided with stiffeners. Ans.

Another rule, which gives nearly the same result, is as follows:

Rule.—*Provide stiffeners whenever the thickness of the web-plate is less than one-sixtieth of its total depth.*

This rule is modified by some authorities so as to allow a thickness of one-eightieth of the total depth of the plate as being amply safe without stiffeners. The more conservative rule, however, which requires the thickness to be at least one-fiftieth of the unsupported depth of the web or the distance between the flange angles, is the one to be recommended, and will be used in this Section.

The spacing and size of stiffeners to be used on a plate girder is almost entirely a matter of experience and judgment. As a general rule, it may be said that stiffeners should be provided at the ends of all plate girders over the supports or abutments, and they should be so proportioned that they will take care of the entire reaction at these points. The stiffeners between the abutments or supports should be of such size that they will best suit the general requirements of the design of the girder. The practice in spacing intermediate stiffeners is to make the distance between their center lines equal to the depth of the girder, thus dividing the girder into equal square panels. Under no conditions, however, should stiffeners be placed more than 5 feet apart from center to center of line of rivets.

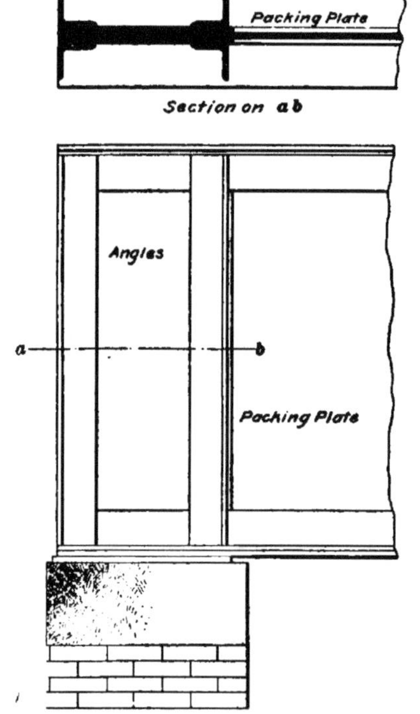

FIG. 6

Having proportioned the stiffeners at the abutments to take the entire reaction, it is good practice, when possible, to make the intermediate stiffeners of the same size as the end ones. In general, the angles used for stiffeners should not be less than 3 inches by 3 inches by $\frac{5}{16}$ inch, though on shallow girders, with extremely light loads, it might be economical to use angles as light as $2\frac{1}{2}$ inches by $2\frac{1}{2}$ inches by $\frac{5}{16}$ inch. Sizes smaller than this should certainly never be used for this purpose.

Stiffeners should always extend over the vertical legs of the flange angles; they should always be either swaged out to fit over the flange angles, or be provided with a filling piece, as illustrated in Fig. 2.

EXAMPLE 2.—The end reaction on a plate girder is 300,000 pounds. If a compressive fiber stress of 13,000 pounds per square inch is allowed on the stiffeners, and four stiffeners are used, as shown in Fig. 6, what should be the size of the angles?

SOLUTION.—The reaction or greatest shear being 300,000 lb., and the allowable stress 13,000 lb. per sq. in., the area of the stiffeners required must be 300,000 ÷ 13,000 = 23 sq. in.; this sectional area, divided among four angles, gives 23 ÷ 4 = 5.75 sq. in. as the area required for each angle.

By referring to the list of Angles With Equal Legs, in the table in *Properties of Sections*, it is seen that a $5'' \times 5'' \times \frac{5}{8}''$ angle has a sectional area of 5.86, while in the list of areas of the angles with unequal legs, a $5'' \times 4'' \times \frac{11}{16}''$ is shown to have an area of 5.72 sq. in.; therefore, either of these angles may be used. Ans.

FLANGES

11. The **flanges** of a riveted girder include all the metal at the top and bottom of the girder, and are sometimes called the top and bottom chords, though this term is more frequently applied to lattice or open girders, such as are more often used for railroad and highway bridges.

In building construction, it is customary to include in the flange the two flange angles, the flange plates, and one-sixth or one-eighth of the web-plate included between the flange angles.

The resisting moment, or the resistance to bending, of the section may be calculated as follows:

Taking the center of gravity of one flange section as a center, the moment of resistance of the other flange is equal to the area of the flange section multiplied by the distance between the centers of gravity of the flange areas and the safe stress per square inch of section. The resistance of the web is equal to its section modulus multiplied by the safe stress per square inch. Assuming d as the depth of the web which is considered here as being the distance between the centers of gravity of the two flanges, the resisting moment of the entire section may be expressed by the formula

$$M_1 = A\,d\,s + \frac{s\,b\,d^2}{6} \qquad (3)$$

in which A = area of flange section;
d = height of girder between centers of gravity of flanges;
s = allowable stress per square inch;
b = thickness of web.

Considering $b\,d$ as equal to the area of the web, or A', we have $M_1 = A\,d\,s + \frac{A'\,s\,d}{6}$, or $M_1 = d\,s\left(A + \frac{A'}{6}\right)$. Thus, it may be seen that the entire resistance of the section includes about one-sixth of the area of the web, and consequently, one-sixth of the web may be considered as assisting to resist the bending moment on the girder.

The reduction of the moment of resistance of the web by rivet holes must be taken into consideration, however, and consequently it would be more nearly correct to assume one-eighth of the depth of the web as acting with the flange to resist the bending moment.

The building ordinances of some of the large cities in the United States, however, will not allow any portion of the web-plate to be included as part of the flange. In this Section, the web-plate will not be considered in calculating the flange area.

If, because of economic considerations, one-sixth or one-eighth of the web-plate must be included as part of the flange, it should be remembered that the plate should never be spliced near the center, when the girder is uniformly or symmetrically loaded, or directly under the point of greatest bending moment, when the load on the girder is unsymmetrically placed. Especial care must also be taken to insure that any splice made on the length of such a web-plate is so designed as to furnish the greatest possible percentage of strength of the solid plate included within one-sixth or one-eighth of the depth of the web.

The best practice dictates that where flange plates are used, the sectional area of the flange angles should equal the sectional area of the flange plates. This, however, is not possible in heavy work, where the best that can be done is to use the heaviest sections obtainable for the flange angles.

12. Flange Stresses.—In a simple girder, the top flange is subjected to compression and the bottom flange to tension. Nevertheless, it is customary in practice to make the two flanges equal and composed of the same size of rolled plates and angles.

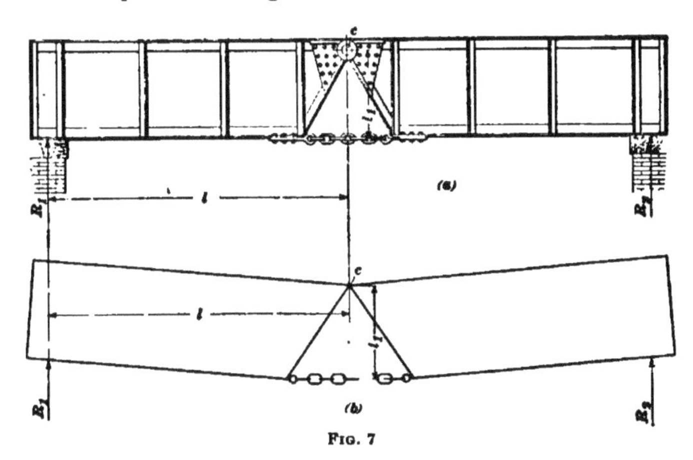

Fig. 7

In proportioning the flanges of a plate girder, the lower flange is calculated for tension; the areas of the rivet holes cut out of the flanges are deducted from the total area, so as to give the net or actual area of the flange at the point of least strength.

The stresses in the flanges are assumed to be produced wholly by the bending moment on the girder, and the moments of these stresses are assumed to be equal to the moments of the external forces.

The principles on which the flange stresses of a plate girder are calculated will be made clear by reference to Fig. 7, which shows a girder in two sections joined by a hinge pin c at the upper flange, and a chain at the lower flange. The resultant moment of the loads and reactions tends to produce rotation about the center c, which, however, is taken at a point in the upper flange instead of on the neutral axis, as was done in the case of the beam composed of a single section; in reality, owing to the fact that the web is entirely neglected in calculating the resistance of the girder to the bending stresses, there is no neutral axis, in the sense in which that term was used in connection with ordinary beams. Strictly speaking, there may be said to exist a neutral axis in the web, similar to that in an ordinary beam, but, on account of the thinness of the web, the additional tensile and compressive resistance offered by it is so small that it may be neglected, as far as the longitudinal stresses in the flanges are concerned. The centers of moments of the tensile and compressive stresses are therefore supposed to be shifted to the upper and lower flanges, respectively. The vertical stresses have still to be transmitted by the web to the points of support, the compressive stresses being taken up by the stiffeners and the tensile ones by the web.

The stress in the chain, which represents the lower chord or flange of the girder, resists the tendency to rotation about the center of moments c, with a lever arm l_1, which is the perpendicular distance from the chain to the point c. It is evident, then, that the strength of the girder depends on two factors: the tensile strength of the lower chord, and its distance from the center of the hinge c, the latter of which represents the depth of the girder.

If the center of moments is taken on the center line of the chain, directly under the point c, it is evident that the resultant moment of the external forces, with respect to this center, is the same as when the center was taken at c; it is also evident that the force in the beam whose moment, with respect to this center, balances the resultant moment of the external forces, is the compression on the pin c. Since the moment and the lever arm of the compressive stress on the pin are respectively equal to the moment and the lever arm of the tensile stress in the chain, it follows that these two stresses are equal; in other words, the compressive stress in the top flange of the girder is equal to the tensile stress in the bottom flange.

13. Proportioning the Flanges.—Having determined the principles on which the bending strength of a plate girder is calculated, it remains to show a method for proportioning the metal in the flanges. The usual process is as follows:

First calculate the maximum bending moment on the girder; in calculating the bending moment on a plate girder it is customary to express the moment in foot-pounds, the depth of a girder being generally given in feet and not in inches, as in solid beams of shallow depth. If, however, the depth of the girder is expressed in inches, the bending moment must be calculated in inch-pounds.

Having found the maximum bending moment on the girder, it is necessary to assume an allowable fiber stress for the material of which the flanges are composed. The following rule may be used to calculate the sectional area of either flange:

Rule.—*Divide the bending moment on the girder, in foot-pounds, by the product obtained by multiplying the depth of the girder, in feet, by the safe fiber stress.*

The safe fiber stress for a given case is obtained by dividing the ultimate fiber stress per square inch of the material by the factor of safety required in the girder.

The rule may be expressed by the formula

$$A = \frac{M}{D \times s_a} \qquad (4)$$

in which A = net area of one flange, in square inches;
D = depth of girder, in feet;
s_a = safe fiber stress per square inch of material;
M = bending moment on girder, in foot-pounds.

EXAMPLE.—The depth of a plate girder is 6 feet, the span is 80 feet, and the load on the girder is 3,000 pounds per lineal foot: (*a*) What will be the required net flange area for structural steel, if a factor of safety of 4 is used? (*b*) Of what size rolled sections should the flange be composed?

SOLUTION.—(*a*) The span being 80 ft. and the load 3,000 lb. per lineal ft., the entire load on the girder will be $80 \times 3{,}000 = 240{,}000$ lb. Substituting in the formula $M = \frac{WL}{8}$ for the bending moment on a simple beam, gives $M = \frac{240{,}000 \times 80}{8} = 2{,}400{,}000$ ft.-lb. The net area of the flange, from formula **4**, is

$$A = \frac{2{,}400{,}000}{6 \times 15{,}000} = 26.7 \text{ sq. in.} \quad \text{Ans.}$$

(*b*) In order to determine the size of the flange plates and angles, it is useful to assume some particular size of angles and plates, and make a detail sketch of the flange, as shown in Fig. 8, marking on it the size of the respective plates and angles that have been assumed. The rivets should also be shown, so that the metal cut out of the rivet holes may be deducted from the sectional area of the flange in order to determine that area.

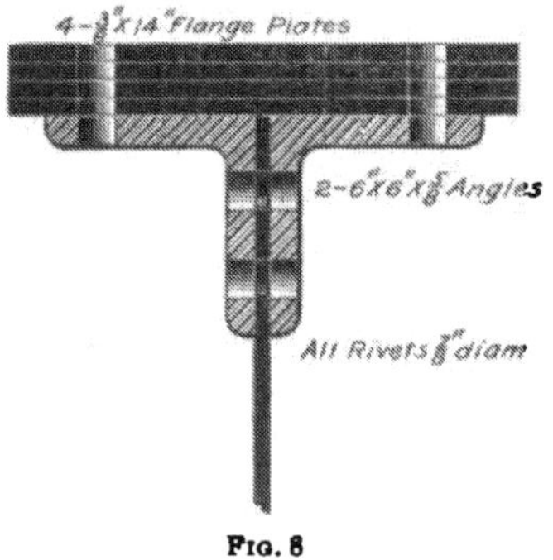

FIG. 8

It is assumed in the section under consideration that there are two rows of rivets through the vertical legs of the angles, each pair of these rivets being placed in the same vertical plane, in consequence of which the amount to be deducted from the net section is double the area cut out for one rivet. The rivets in the two rows through the horizontal legs of each angle are staggered, and consequently only one rivet hole in each horizontal leg affects the area of the flange section.

According to the table Properties of Standard Angles, in *Properties of Sections*, the area of a 6" × 6" × ⅝" angle is 7.11 sq. in.; therefore, the total area of the metal in the flange is

Two 6" × 6" × ⅝" angles,	7.11 × 2 =	14.22 sq. in.
Four 14" × ⅜" plates,	4 × 14 × ⅜ =	21.00 sq. in.
	Total,	35.22 sq. in.

From the total area of the flange it is necessary to deduct the metal cut out for the rivet holes. As ⅞-in. rivets are used, the rivet holes are considered to be ⅛ in. larger, or 1 in. in diameter.

There are four 1-in. holes through $\frac{5}{8}$ in. of metal to be deducted from the vertical legs of the angles, and in the plates and the horizontal legs of the angles there are two 1-in. holes through $2\frac{1}{8}$ in. of metal. The areas to be deducted for the rivet holes are, therefore,

Four 1-in. holes through $\frac{5}{8}$ in. of metal	$= 2\frac{1}{2}$ sq. in.
Two 1-in. holes through $2\frac{1}{8}$ in. of metal	$= 4\frac{1}{4}$ sq. in.
Total,	$6\frac{3}{4}$ sq. in.

and the net area of the flange is $35.22 - 6.75 = 28.47$ sq. in. Since the calculations showed that the net area required in this flange is 26.6 sq. in., it is evident that the assumed flange is amply strong. Ans.

While plate girders are more economical than box girders, the latter are stiffer in a lateral direction and hence should be used where a wide top flange is required in order to furnish the necessary lateral stiffness for a long span. If the girder is not held in place laterally the top flange should have a width equal to at least $\frac{1}{20}$ of the span. Otherwise, the gross area of the top flange may be found by the following formula:

$$A' = A\left(1 + \frac{e^2}{5,000}\right) \qquad (5)$$

in which A = gross area required in top flange, with girder supported laterally;

A' = gross area in top flange, girder unsupported laterally;

e = span ÷ width of flange, both in inches.

14. Lengths of Flange Plates.—Since the bending moment in a simple beam varies along the entire length of the beam, the location of the maximum bending moment depending on the distribution of the load, it would seem that in order to design an economical girder, the area of the flange should vary with the bending moment. Where flange plates are used, this condition may be partially fulfilled by the use of plates of different lengths, each extending only as far as may be required in order to provide the flange section demanded by the bending moment.

Reference to Fig. 9 will make this construction more clear. In this figure, it is seen that the top plate of the top flange is the shortest, and extends over a small portion of the girder only, each successive plate under this one being longer than the one above it. The third plate from the top is the longest and extends nearly the full length of the girder, while the angles extend from end to end.

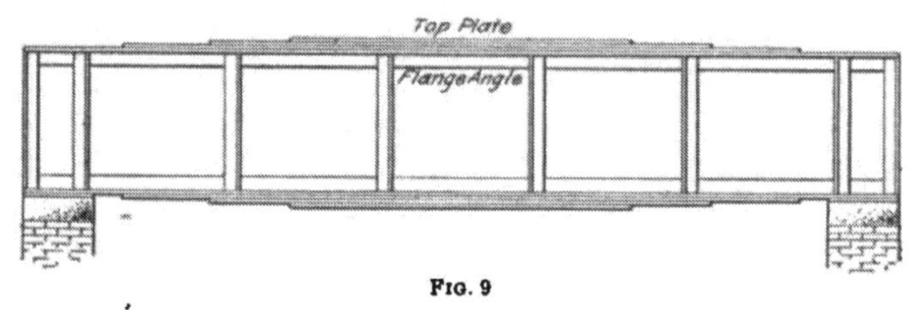

Fig. 9

Where the beam is uniformly loaded, the following method may be used to obtain the theoretical length of each of the flange plates:

Commencing with the outside plate of the flange, find the sum of all the net areas, in square inches, of the plates to and including the plate in question. Thus, in Fig. 10, if it be required to obtain the length of the third plate from the top, find the sum of the areas of the first, second, and third plates. If the length of the second plate is required, then the sum of the areas of the first and second plates is to is taken. Divide the area so obtained by the net area of the whole flange, in square inches, and multiply the square root of this quotient by the length of the girder, in feet; the product will be the theoretical length of the plate in feet.

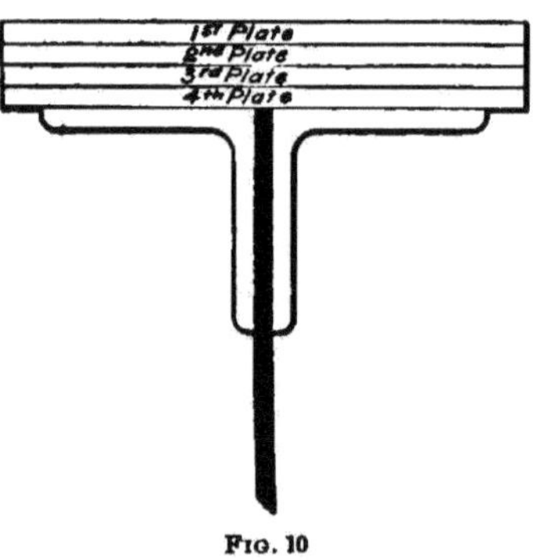

Fig. 10

Having obtained the theoretical length of the plate, it is necessary to add from 12 to 16 inches to each end, in order that the plate in question may be carried sufficiently past the point of bending moment that governs the area of the flange at its ends to be securely riveted to the plates and angles making up the flange from there on to the abutment.

The method for determining the length of flange plates where the beam is uniformly loaded may be expressed by the formula

$$L_p = L\sqrt{\frac{a}{A}} \qquad (6)$$

in which L_p = theoretical length, in feet, of plate in question;

L = length of girder, in feet;

a = net area of all plates to and including plate in question, beginning with outside plate;

A = total net area of entire flange.

EXAMPLE.—In Fig. 11 is shown a section through the flange of a plate girder the span of which is 60 feet. What is the theoretical length of each of the three flange plates?

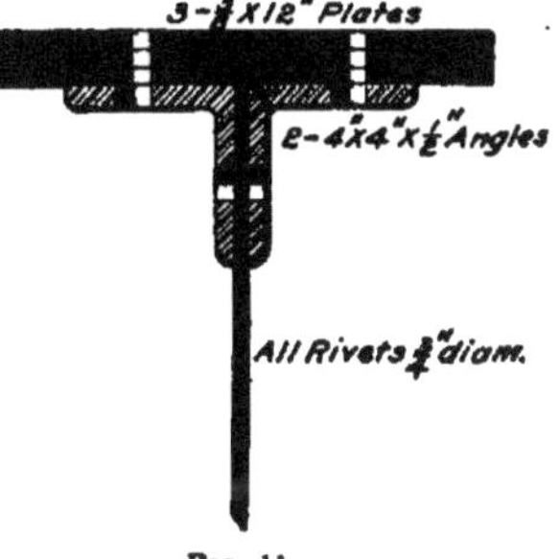

FIG. 11

SOLUTION.—The area of a 4″ × 4″ × ½″ angle, according to the table Properties of Standard Angles, in *Properties of Sections*, is 3.75 sq. in. The area of each plate is ⅜ × 12 = 4.5 sq. in. The diameter to be deducted for the rivet holes is ¾ + ⅛ = ⅞ in.

The area cut out by a ⅞-in. hole through a ⅜-in. plate is .875 × .375 = .328 sq. in. Then, as there are two rivet holes in each plate, its net area is 4.5 sq. in. − (.328 sq. in. × 2) = 3.844 sq. in.

The net area of the angles is (3.75 sq. in. × 2) − (.4375 sq. in. × 4) = 5.75 sq. in.

The net area of the flange section is, therefore,

Three plates 3.844 × 3	=	11.532 sq. in.
Two angles	=	5.750 sq. in.
Total,		17.282 sq. in.

Now calculate the length of the outside plate. Substituting in the formula gives

$$L_p = 60\sqrt{\frac{3.844}{17.282}} = 28.29 \text{ ft.}$$

By substituting the proper values, the theoretical length of the second or middle plate is

$$L_p = 60\sqrt{\frac{7.688}{17.282}} = 40.0 \text{ ft.}$$

The length of the third or last plate in the flange, that is, the one next to the flange angles, is next to be calculated, though some engineers prefer to run this the entire length of the girder, as it stiffens the girder laterally and assists in preventing any tendency toward side deflection. The theoretical length of this plate is

$$L_p = 60\sqrt{\frac{11.532}{17.282}} = 49.12 \text{ ft.}$$

15. Graphic Method of Determining Length of Flange Plates.—The graphic method for determining the theoretical length of flange plates in built-up girders is more convenient than the analytic method previously given. In order to explain this method, a section through the flange of a plate girder will be assumed and the lengths of the several flange plates determined.

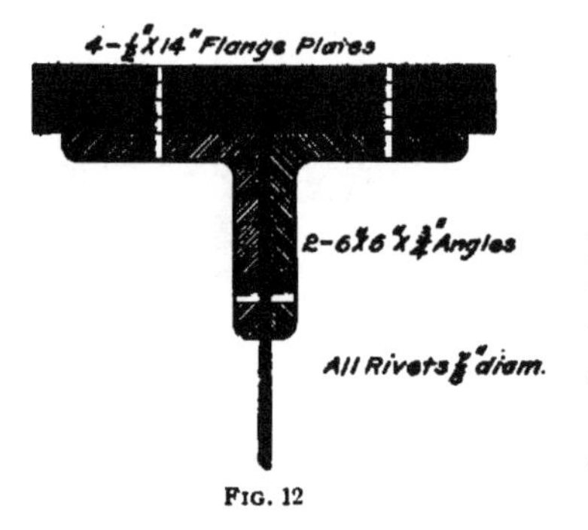

FIG. 12

Fig. 12 shows a section through the flange of a girder, built up of four ½″ × 14″ flange plates, the span of the girder being 90 feet. It will be noticed that there are two rows of rivets in the flange, and two rows in the vertical leg of the angles, but as the latter are staggered, there will be but one rivet hole to be deducted from the vertical leg of each angle.

The sectional area of a 6″ × 6″ × ¾″ angle is found from the table Properties of Angles, in *Properties of Sections*, to be 8.44 square inches; from this deduct 1½ square inches, the area cut out by the two rivet holes, making the net area of each flange angle 6.94 square inches.

The sectional area of a ½″ × 14″ flange plate is 7 square inches, from which there is to be deducted 1 square inch for the sectional area cut out by the two rivet holes. Hence, the net area of one flange plate is 7 square inches − 1 square inch = 6 square inches.

The net area of the entire flange will, therefore, be

Two 6″ × 6″ × ¾″ angles	=	13.88 square inches
Four ½″ × 14″ flange plates	=	24.00 square inches
Total,		37.88 square inches

Since the load is uniformly distributed, the flange plates extend equally on each side of the center; consequently, the diagram for only one-half of the girder will be drawn. Draw, to any scale, a horizontal line *a b*, Fig. 13, equal to one-half of the span; divide this line into any number of equal parts (in the figure, twelve parts have been used). Upwards from the points of division, draw indefinite perpendicular lines. On the perpendicular from *b*, lay off to some scale a distance that represents the entire net section of the flange, thus locating the point *r*. For example, the net area of the flange in this case is 37.88 square inches; letting $\frac{1}{16}$ inch represent 1 square inch, the distance *b r* must be $37.88 \times \frac{1}{16} = 2.37$ inches, nearly.

Lay off to the same scale on the line *b r* a distance *b n*, which represents 13.88 square inches, the net area of the two flange angles, also the distances *n o*, *o p*, *p q*, and *q r*, each representing 6 square inches, the net area of each of the flange plates. From the point *r*, draw a horizontal line cutting the vertical line erected at *a*, thus locating the point *b′*. Divide the vertical line *a b′* into the same number of equal parts as the line *a b*, thus locating the points *c′*, *d′*, *e′*, *f′*, etc.; and from these points, draw the lines *c′ r*, *d′ r*, etc. Draw the curve *a s v r* through the points where the vertical lines from *c*, *d*, *e*, etc. intersect the corresponding lines *c′ r*, *d′ r*, *e′ r*, etc. Now from the points *q*, *p*, *o*, and *n* draw horizontal lines as shown, cutting the curve in the points *s*, *t*, *u*, and *v*, and from each of these points of intersection, draw a perpendicular, extending it until it intersects the horizontal line next above. The rectangles *v′ r q v*, *u′ q p u*, *t′ p o t*, etc. thus formed, represent the flange plates and angles.

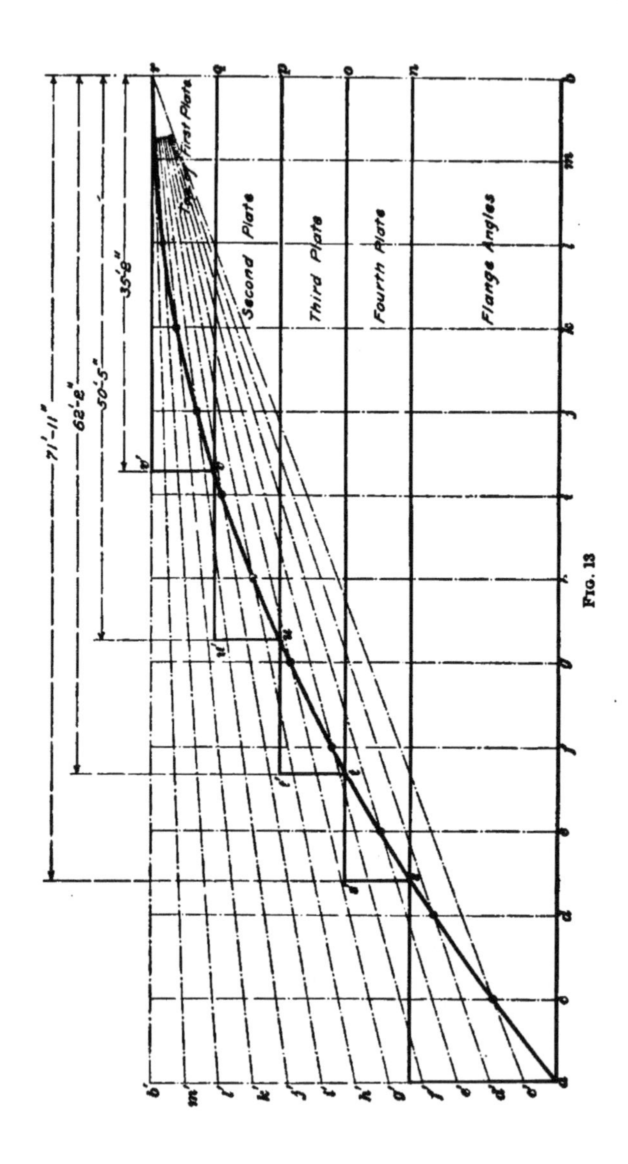
71'-11"
62'-2"
50'-5"
35'-8"
First Plate
Second Plate
Third Plate
Fourth Plate
Flange Angles

Fig. 13

In order to obtain the theoretical length of any flange plate, measure the length of the corresponding rectangle by the scale to which the half span was laid out on the line *a b*; this length multiplied by 2 gives the length of the plate in question. For example, if it is desired to obtain the length of the first or top plate with the scale to which the half span was laid out, measure the length of the line *v′ r*; as only one-half of the diagram is drawn, this gives one-half of the length of the top or first plate, and by doubling this the entire theoretical length of the plate in question is obtained.

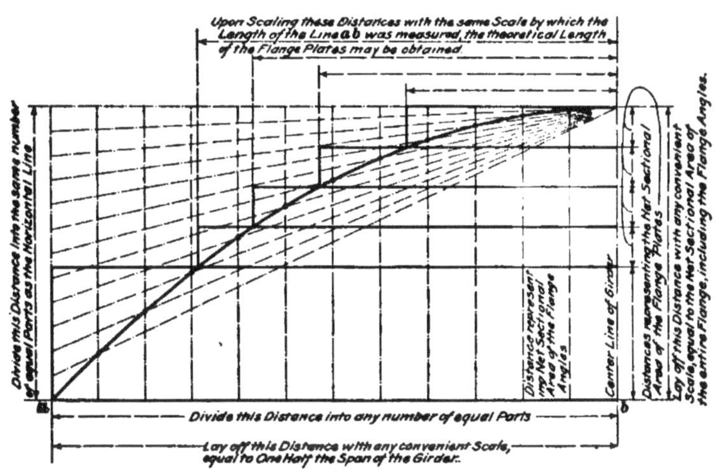

FIG. 14

The length of the other plates may be determined in like manner. It is to be borne in mind that to the theoretical length given by the diagram it is necessary to add a length of about 1 foot at each end of the plate. From the diagram, the theoretical lengths of the flange plates of the girder shown in Fig. 13 are found to be 35 feet 2 inches, 50 feet 5 inches, 62 feet 2 inches, and 71 feet 11 inches, respectively.

The student will find on checking these lengths by formula **6** that they are approximately correct. Fig. 14, in which all the different steps are indicated, is presented in order that the student may always have a guide for laying out this diagram.

16. Application of Graphic Method to Girders With Concentrated Loads.—The graphic method for determining the theoretical length of flange plates when the girder is loaded with concentrated loads, which is similar to that given for a uniformly loaded girder, will be illustrated by constructing a diagram for the lengths of the four flange plates required for a girder with a span of 80 feet and a depth of 6 feet, carrying a concentrated load of 185,000 pounds at 30 feet from one end.

The bending moment on the girder may be calculated by the formula

$$M = W \times \frac{xy}{L} \qquad (7)$$

in which M = bending moment;
W = load on girder;
L = span, in feet;
x = distance that load is located from one abutment;
y = distance that load is located from other abutment.

In Fig. 15, the load is located at the point f, 30 feet from R_1 and 50 feet from R_2. Substituting these values in the formula, the bending moment is found to be $M = 185,000 \times \frac{30 \times 50}{80} = 3,468,750$ foot-pounds.

From this bending moment, the required net flange area, assuming a safe unit stress of 15,000 pounds per square inch, is found to be, approximately, 38 square inches, which is provided by the use of four $\frac{1}{2}'' \times 14''$ plates and two $6'' \times 6'' \times \frac{3}{4}''$ angles. The flange therefore has the section shown in Fig. 12.

To construct the diagram, which is shown in Fig. 15, draw the base line $c\,d$, to any scale, equal to the span of the girder, in feet. (In this case, owing to the fact that the load is not symmetrically placed, the center line of the girder will not divide the lengths of the plates in halves, and it will be necessary to draw the entire diagram.) Locate the point of application of the concentrated load at 30 feet from R_1,

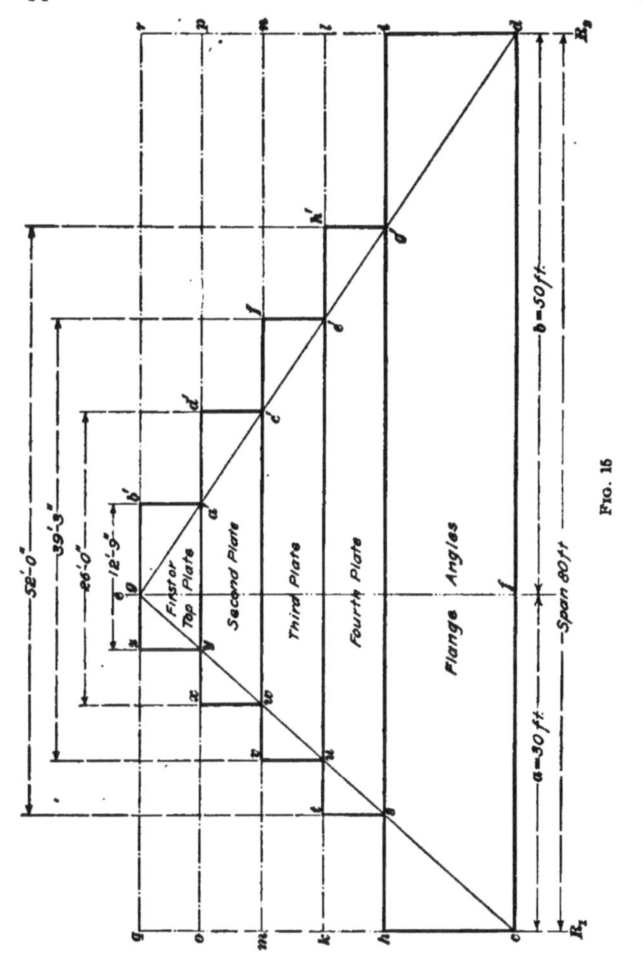

FIG. 15

and draw the perpendicular line ef. On this line, lay off, to any scale, a distance that represents the bending moment. For example, in this case, a scale has been used on which

50,000 foot-pounds is represented by $\frac{1}{32}$ inch; the distance to be laid off is, therefore, $\frac{3468750}{50000} \times \frac{1}{32} = 2.168$ inches. This locates the point g, which is then connected by straight lines to the points c and d. Draw vertical lines from the points c and d until they meet a horizontal line through the point g, at q and r.

In the flange section, the angles have a combined net area of 13.88 square inches, and each flange plate a net area of 6 square inches. The combined net sectional area of the flange is 37.88 square inches.

Place the zero mark of any convenient scale at the point f, and slant the scale until the marking on the scale that represents 37.88 square inches, the net area of the flange, falls on the horizontal line $q\,r$. Thus, in this case, a $\frac{1}{16}$-inch scale has been used, and the zero mark is placed at f, while the mark on the scale representing the division 37.88 is placed on the horizontal line $q\,r$. With the scale still in this position, begin at the zero mark and lay off a distance 13.88 to represent the net sectional area of the two angles; then lay off four distances, each equal to 6, to represent the net sectional area of each of the flange plates.

Through the points thus found, draw the horizontal lines $h\,i$, $k\,l$, $m\,n$, and $o\,p$. At the points s, u, w, y, etc., where these horizontal lines cut the oblique lines $g\,c$ and $g\,d$, draw the perpendicular lines $s\,t$, $u\,v$, $w\,x$, $y\,z$, $a'\,b'$, etc., extending each until it meets the next horizontal line above. Then the rectangles enclosed by the horizontal and vertical lines, shown heavy in the diagram, represent the cover-plates and flange angles. By measuring the lengths of these rectangles with the scale to which the span was laid off on the line $c\,d$, the theoretical length of the plates may be determined.

In this case the length of the angles is equal to the span, 80 feet, as marked on the diagram. The length of the first, or top, plate measures 12 feet 9 inches; of the second plate, 26 feet 0 inches; of the third plate, 39 feet 3 inches; and of the fourth, or last, plate, 52 feet 0 inches. It may be that when the student lays out the diagram for himself, he will obtain results that will vary slightly from those given.

However, a variation of a few inches in the length of a flange plate on a girder need not be considered.

17. Graphic Diagram for Several Concentrated Loads.—In order to illustrate this method still further, a more complicated problem, in which there are three concentrated loads, will now be presented.

Assume that a girder having a span of 80 feet with a depth of 6 feet is loaded as shown in Fig. 16. What flange area is required, provided that a safe unit fiber stress of 16,000 pounds per square inch is used, and of what should the flange be constructed; also, what will be the length of the several flange plates?

The reactions at R_1 and R_2 are found to be 142,500 and 117,500 pounds, respectively. It will first be necessary to

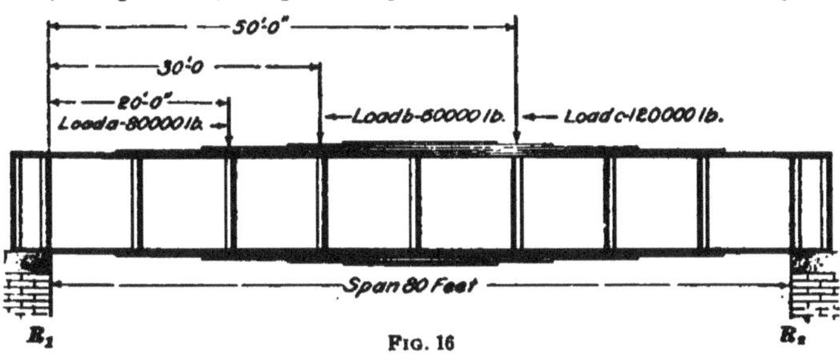

Fig. 16

calculate the bending moment, in foot-pounds, at the points where the loads *a*, *b*, and *c*, are concentrated. The bending moment under load *a* is $142,500 \times 20 = 2,850,000$ foot-pounds; under load *b*, the bending moment is $(142,500 \times 30) - (80,000 \times 10) = 3,475,000$ foot-pounds; and the bending moment under load *c* is $(142,500 \times 50) - [(80,000 \times 30) + (60,000 \times 20)] = 3,525,000$ foot-pounds.

From this it is seen that the greatest bending moment is under the load *c*, and its magnitude is 3,525,000 foot-pounds.

From this the flange area required may be calculated by applying formula **4**, as follows:

$$A = \frac{3,525,000}{6 \times 16,000} = 36.71 \text{ square inches}$$

We will now select a flange section that will have the required net sectional area; referring to the section shown in Fig. 12, the area is found to be 37.88 square inches; this flange will therefore satisfy the requirements.

18. Having determined the bending moments and the greatest net flange area, begin the diagram shown in Fig. 17 by drawing to any scale the horizontal line *de* equal in length to the span of the girder; with the same scale locate the points of application *a*, *b*, and *c* of the concentrated loads. Upwards from the points *d*, *a*, *b*, *c*, and *e* draw indefinite perpendicular lines, and on the perpendiculars from *a*, *b*, and *c* lay off to some convenient scale distances *af*, *bg*, and *ch*, which represent the respective bending moments at these points.

For example, the bending moment at *a* is 2,850,000 foot-pounds; at *b*, it is 3,475,000 foot-pounds; and at *c*, it is 3,525,000 foot-pounds; therefore, assuming a scale on which each $\frac{1}{32}$ inch represents 50,000 foot-pounds of bending moment, the respective bending moments at the points *a*, *b*, and *c* are represented by lengths *af* of 57 thirty-seconds, *bg* of 69½ thirty-seconds, and *ch* of 70½ thirty-seconds.

Draw straight lines connecting the points *d*, *f*, *g*, *h*, and *e*. Through the highest point *h*, representing the greatest bending moment, draw the horizontal line *jk*.

The net area of the flange being 37.88 square inches, place the zero mark of any convenient scale on the line *de*, and slant the scale until the mark that represents 37.88 falls on the line *jk*.

Starting from the zero mark on the scale, lay off a distance that represents the net area of the two flange angles, in this case 13.88 square inches, then divide the remaining distance into equal parts, each of which represents 6 square inches, the net sectional area of the several flange plates.

Through the points just found, draw the horizontal lines *lm*, *no*, *pq*, and *rs*. Where these horizontal lines intersect the oblique lines at the points *t*, *u*, *v*, *w*, *x*, etc., draw vertical lines until they intersect the next horizontal line above.

Then draw in, with heavy lines, the rectangles representing the flange plates and flange angles.

The theoretical length of the flange plates may now be

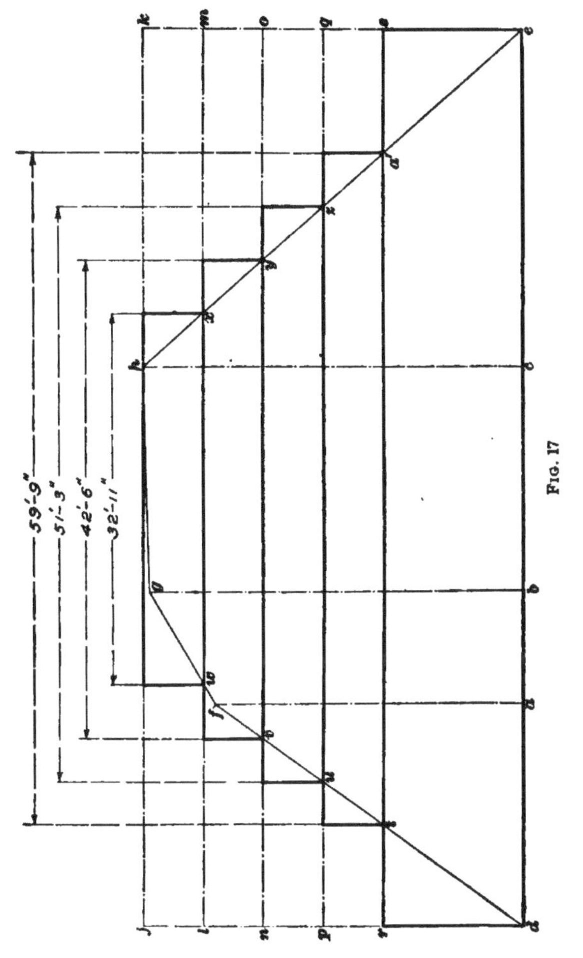

Fig. 17

determined by measuring with the scale to which the span was laid off on the line *de*. The length of the top, or first, plate in this case is found to be 32 feet 11 inches; of the second plate, 42 feet 6 inches; of the third plate, 51 feet 3 inches; and of the fourth, or last, plate, 59 feet 9 inches.

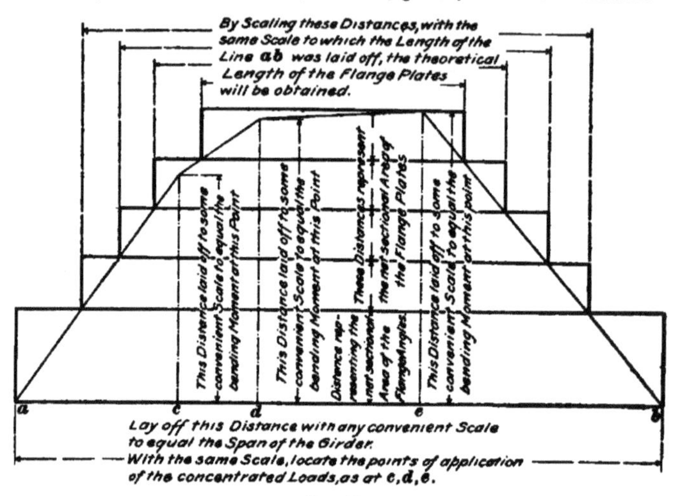

FIG. 18

In Fig. 18 is shown a diagram that will serve as a general rule for determining the length of the several flange plates of a girder loaded with several concentrated loads.

19. Diagram for a Combination of Concentrated Loads With a Uniformly Distributed Load.—There is another condition of girder loading that is frequently

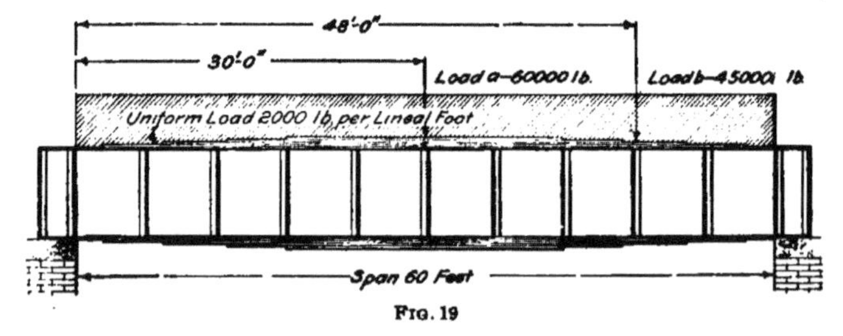

FIG. 19

encountered in practical work in which it is necessary to determine the length of the several flange plates by the graphic method; this condition is produced by a combination of a uniformly distributed load with several concentrated loads located at different points along the girder.

In order to explain the method for obtaining the length of the flange plates in a girder loaded in this manner, the following problem will be assumed and the diagram will be constructed as was done in previous cases.

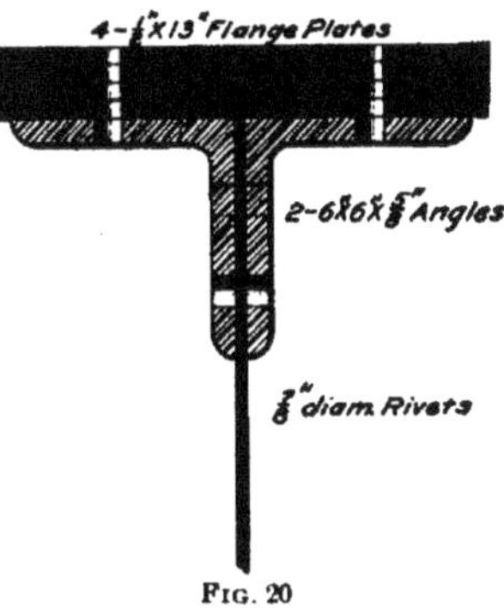

Fig. 20

Assume the girder to be loaded, as shown in Fig. 19, with a uniformly distributed load and the two concentrated loads. The flange section shown in Fig. 20 is sufficient to resist the bending moments due to these loads. It is required to determine by the graphic method the theoretical lengths of the several cover-plates making up the flange section.

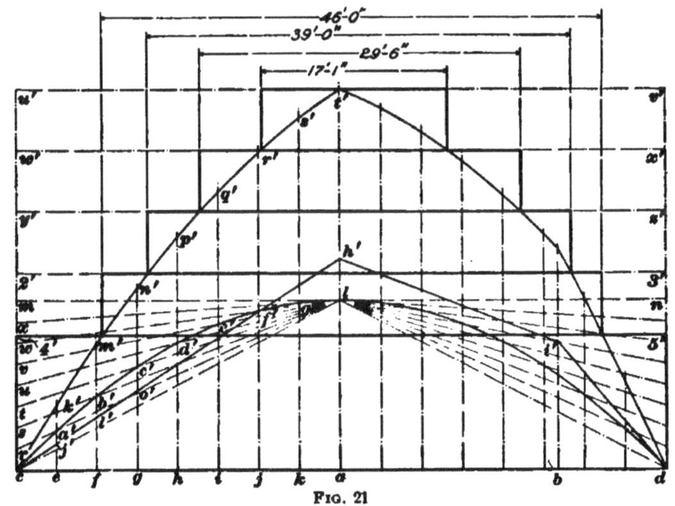

Fig. 21

Before starting to draw the diagram shown in Fig. 21, it is necessary to make the calculations for the following: The greatest bending moment; the maximum bending moment due to the uniformly distributed load; and the bending moment under each of the concentrated loads, neglecting the uniformly distributed load. These bending moments should be expressed in foot-pounds. The flange area required to resist successfully each of these bending moments should also be calculated.

The calculations in this case have been made in the usual manner, and the results are as follows:

	Foot-Pounds
Greatest bending moment . . . . =	2,070,000
Bending moment due to a uniform load . . . =	900,000
Bending moment under concentrated load *a*, considering the concentrated loads only . . =	1,170,000
Bending moment under concentrated load *b*, considering the concentrated loads only . . =	792,000

Since the depth of the girder is 4 feet, if a unit fiber stress of 15,000 pounds is used, the flange area required to resist the greatest bending moment is $A = \frac{2,070,000}{4 \times 15,000} = 34.5$ square inches.

The flange area required to resist the bending moment due to the uniform load is $A = \frac{900,000}{4 \times 15,000} = 15$ square inches.

The flange area required to resist the bending moment at the point on the girder where the concentrated load *a* is situated, considering the concentrated loads only, is $A = \frac{1,170,000}{4 \times 15,000} = 19\frac{1}{2}$ square inches.

The flange area required to resist the bending moment at the point on the girder under the concentrated load *b* is $A = \frac{792,000}{4 \times 15,000} = 13.2$ square inches, considering, as before, only the concentrated loads.

20. As the loads on the girder are not symmetrically placed with regard to the center, it will be necessary to draw the complete diagram. Begin the diagram by drawing, to any convenient scale, the horizontal line *c d*, Fig. 21, equal in length to the span of the girder, and locate the points of application of the concentrated loads at *a* and *b*; upwards from the points *c*, *a*, *b*, and *d* draw indefinite vertical lines.

Now, in accordance with the method explained in Art. **15**, make the construction to determine the curved line representing the bending moment due to the uniform load as follows:

At the center of the girder draw a vertical line (in this case the center of the girder is found to be at the point where the load *a* is concentrated); divide half the span into any number of equal parts, as at *e*, *f*, *g*, *h*, *i*, etc., and from the points so obtained draw perpendiculars. Lay off on the vertical line passing through the center a distance *a l*, which may represent either the greatest bending moment at this point due to the uniform load, or the flange area required to resist this bending moment, as they are proportional. In this case the net area required in the flange will be used; hence, as the area of the flange required for the uniform load is 15 square inches, if $\frac{1}{16}$ inch is assumed to represent 1 square inch of flange area, the distance *a l* will be $\frac{15}{16}$ inch.

Through the point *l*, draw the horizontal line *m n*. Divide the distance *m c* into the number of equal parts into which the half of the span was divided, and from the points *r*, *s*, *t*, *u*, *v*, etc. thus obtained, draw converging lines to the point *l*; where these oblique lines intersect the vertical lines, mark the points *a'*, *b'*, *c'*, *d'*, etc., and through these points draw the curve *c l*. Draw the other half of the curve *l d* in the same manner, thus completing the diagram for the uniform load.

21. Now draw the diagram for the concentrated loads. On the vertical line *a l*, extended, lay off the distance *a h'*, equal to the net flange area required to support the concentrated load *a*, and on the perpendicular line erected at *b* lay off the distance *b i'* equal to the flange area required at the point *b* to support the concentrated loads.

It must be remembered that the same scale is to be used as that with which the flange area required for the uniform load diagram was laid off; also, that if the vertical distances are laid off to represent the bending moment in the one case, the bending moment should be used in the other, while if the flange area required is used in the one case, it is evident that it should also be used in the other. The student will understand the importance of this fact when he proceeds further with the diagram. Having located the points *h'* and *i'*, complete the diagram by connecting, with straight lines, the points *c*, *h'*, *i'*, and *d*, as was done in the previous diagrams of concentrated loads.

The next step in the process is to measure the distance *e j'* with a pair of dividers, and from the point *a'* on the curve representing the uniform load, lay off on the vertical line the distance *a' k'* equal to *e j'*; also lay off from the point *b'* the distance *b' m'* equal to *f l'*, and from the point *c'* lay off on the vertical line the distance *c' n'* equal to *g o'*; continue in this manner through the entire diagram. Having determined the points *k'*, *m'*, *n'*, *p'*, etc. through the entire diagram, draw in the curve *c k' m' n'*, etc. The point *t'* is the highest point in the diagram, and its distance from the horizontal line *c d* represents the entire flange area required in the girder to resist the greatest bending due to both the uniform and the concentrated loads.

Through the point *t'*, draw the horizontal line *u' v'*, and lay off between the horizontal lines *u' v'* and *c d* the several distances representing the net area of the flange plates and flange angles. Through the points of these divisions, draw the horizontal lines *w'*–*x'*, *y'*–*z'*, *2'*–*3'*, and *4'*–*5'*. Where these horizontal lines intersect the curved line representing the net area required for the combined uniform and concentrated loads, draw short vertical lines to the next horizontal line above; draw, with heavy lines, the rectangles representing the flange plates and flange angles; scale the length of the flange plates with the scale to which the span *c d* was laid off, and the theoretical length of the flange plates will be found.

In the girder under consideration, the theoretical length of the top, or first, plate is found to be 17 feet 1 inch; the length of the second plate, 29 feet 6 inches; the length of the third plate, 39 feet 0 inches; and the length of the last, or fourth, plate is 46 feet 0 inches.

RIVET SPACING

22. Rivets in End Angles or Stiffeners Over Abutments.—First, the allowable safe load on the rivet should be determined. Whether the double shear of the rivet or the bearing value of the plate around the rivet hole is the greater, should be found as previously explained. Having obtained the safe allowable load for each rivet, a sufficient number should be placed in the end angles or stiffeners to take care of the entire shear at that point. Assume the reaction at the end of a girder to be 100,000 pounds; $\frac{7}{8}$-inch rivets are used and the web is $\frac{3}{8}$ inch thick. Using an allowable double shearing stress of 12,000 pounds per square inch, the value of the rivet in double shear is 7,216 pounds, while the web-bearing value is 5,251 pounds; as the latter is the smaller, it is the allowable load on the rivet. The number of rivets required in the two pair of end angles is, therefore, $100{,}000 \div 5{,}251 = 19.04$, say 20, or 10 rivets in each pair.

23. Rivets in Stiffeners Between Abutments.—If possible, the rivets in the intermediate stiffeners are usually spaced the same as in the end stiffeners. It is hardly possible to make any calculation of practical value in regard to the number and spacing of these rivets, and in fact no calculation is required; a practical rule is that the pitch of these rivets should never exceed 6 inches, nor should it exceed sixteen times the thickness of the leg of the angle.

24. Rivets Connecting Flange Angles With Web. When a plate girder is loaded, the tendency of the flanges and angles is to slide horizontally past the web; this tendency to slide induces a horizontal flange stress. The rivets connecting the angles to the web resist this tendency, and there must be a sufficient number of rivets to do it safely.

The stress that is transmitted horizontally from the web to the flange at any point is equal to the increment of the flange stress at that point. When the web is not considered as resisting any portion of the bending moment, this increment is found by the formula

$$f_i = \frac{S}{h} \qquad (8)$$

in which S = maximum shear at point considered;
h = height, in inches, between center lines of rivets;
f_i = increment of stress per inch of run.

The increment of stress divided into the resistance of one rivet, gives the distance between centers of rivets, or their pitch. Hence,

$$p = \frac{r}{\frac{S}{h}} = \frac{r\,h}{S} \qquad (9)$$

in which p = pitch of rivets;
r = resistance of one rivet;
h and S = same as in formula **8.**

Where the maximum bending moment is considered as being resisted by the flange area and one-eighth of the web area, the pitch of the rivets is increased as the ratio of the flange area and the combined area, including the flange area and one-eighth of the web. Therefore,

$$p = \frac{A + \frac{A'}{8}}{A} \times \frac{r\,h}{S} \qquad (10)$$

when a portion of the web is included in the area resisting the shear. In this formula, A has the same value as in formula **4;** A' is equal to the area of the web, while p, r, h, and S are the same as in formula **9.** The application of formula **10** is shown in the following example:

EXAMPLE.—A plate girder having a span of 40 feet supports a uniform load of 5,000 pounds per lineal foot. The distance from the outside reenforcing angle of the girder over the abutment to the first stiffener is 4 feet, while the depth of the girder from center to center of the flange rivets is 3 feet. The thickness of the web-plate is $\frac{5}{16}$ inch and its depth is 40 inches; it is perforated on a single section with ten

$\frac{13}{16}$-inch rivet holes. Provided that the safe unit flange stress is 15,000 pounds, what will be the theoretical pitch of the $\frac{3}{4}$-inch rivets in the vertical legs of the flange angles of the first panel?

SOLUTION.—The bending moment at the first panel point is equal to the moment of the reaction about this point minus the moment of the load on the first panel; hence, $M = (100,000 \times 4) - (5,000 \times 4 \times 2) = 360,000$ ft.-lb. In this example, the depth of the girder will be considered as the distance between centers of rivets; then substituting in formula **4**, the net flange area equals $\frac{360,000}{3 \times 15,000} = 8$ sq. in. The net area of the web equals $(40 \times \frac{5}{16}) - (10 \times \frac{13}{16} \times \frac{5}{16}) = 9.961$ sq. in. The area of the flange, not including any portion of the web, is therefore equal to $8 - \frac{9.961}{8} = 6.755$ sq. in.

From the table Shearing Value of Rivets, in *Details of Construction*, the shearing value of a $\frac{3}{4}$-in. rivet in double shear is 6,627 lb. The maximum shear at the point under consideration is $100,000 - (5,000 \times 4) = 80,000$ lb. Then substituting these values in formula **10**,

$$p = \frac{A + \frac{A'}{8}}{A} \times \frac{r\,h}{S} = \frac{6.755 + \frac{9.961}{8}}{6.755} \times \frac{6,627 \times 36}{80,000} = 3.53 \text{ in., or } 3\frac{1}{2} \text{ in.}$$

Ans.

25. The example given below illustrates the method of finding the pitch of the rivets when no portion of the web-plate is considered as resisting the bending moment, or when formula **9** is applied.

Assume a girder of 40 feet span, as shown in Fig. 22, with a depth of 4 feet, and a uniformly distributed load of

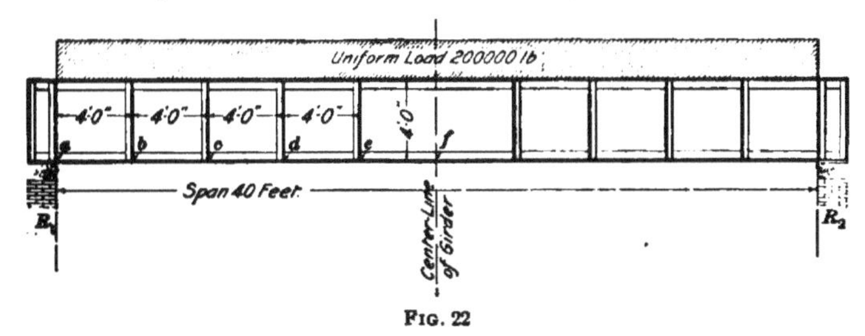

FIG. 22

200,000 pounds. The shearing stress in the girder at the left reaction, or point a, is equal to R_1, in this case 100,000 pounds. At b, 4 feet from R_1, the vertical shear in the girder is $100,000 - (5,000 \times 4) = 80,000$ pounds; at c, 8 feet from R_1, the vertical shear is 100,000 − (5,000 × 8) = 60,000 pounds; at *d*, it is 100,000 − (5,000 × 12) = 40,000 pounds; at *e*, the shear is 20,000 pounds, and at *f* it is zero.

By substituting the above results in formula **8**, the rate of increase, per inch of length, or the increment of the horizontal stress in the flange at the several points *a*, *b*, *c*, *d*, *e*, and *f* may be obtained. Thus, at the end *a* of the girder the increase in the horizontal flange stress is $\frac{100,000}{4 \times 12} = 2,083$ pounds per inch of run; at *b*, $\frac{80,000}{4 \times 12} = 1,667$ pounds per inch of run; at *c*, $\frac{60,000}{4 \times 12} = 1,250$ pounds per inch of run; at *d*, $\frac{40,000}{4 \times 12} = 833$ pounds per inch of run; and at *e*, $\frac{20,000}{4 \times 12} = 417$ pounds per inch of run.

If $\frac{7}{8}$-inch rivets are used, the safe load for one rivet, at 12,000 pounds per square inch, in a $\frac{3}{8}$-inch plate, is 3,938 pounds, from the table, Bearing Value of Riveted Plates, in *Details of Construction*, and in web bearing, the strength of one rivet is 3,938 × $1\frac{1}{3}$ = 5,251 pounds per square inch. At the end, where the increase in stress is 2,083 pounds per inch of run the pitch of the rivets should be 5,251 ÷ 2,083 = 2.52 inches, from center to center. At *b*, the maximum allowable pitch of the rivets is 5,251 ÷ 1,667 = 3.75 inches; at *c*, the pitch may be 5,251 ÷ 1,250 = 4.20 inches; and at *d*, 5,251 ÷ 833 = 6.30 inches. Since, for practical reasons, the rivets in the vertical leg of the flange are spaced the same in both the upper and lower chords, and since the greatest allowable pitch of rivets in a compression member is 6 inches, it is needless to carry the calculation further.

Hence, the pitch of the rivets between *a* and *b* should be $2\frac{1}{2}$ inches; between *b* and *c*, $3\frac{3}{4}$ inches; between *c* and *d*, $4\frac{1}{4}$ inches; since the theoretical pitch between *d* and *e* is more than 6 inches, which for practical reasons is not allowable, all the rivets between *d* and the center of the girder should be spaced 6 inches from center to center.

26. Effect of Vertical Stress.—Sometimes the vertical as well as the horizontal stress in the flange is taken into account in spacing the rivets, in which case the resultant of the two stresses is the stress that must be provided for. The vertical stress is due directly to the load resting on the flange of the girder, which, through the rivets, is transmitted to the web-plate.

In the plate girder shown in Fig. 22, the increase in the horizontal flange stress at the end is, as previously calculated, 2,083 pounds per inch of run; the load on the girder being uniformly distributed, the vertical stress on the flange, per lineal inch, is equal to the entire load on the girder divided by the span of the girder in inches; it is, therefore, $200,000 \div (40 \times 12) = 416$ pounds per inch of run.

The total stress to be resisted by the rivets is, therefore, equal to the resultant of 2,083 pounds—due to the increase in the horizontal stress on the flange—and the vertical stress of 416 pounds; this resultant is $\sqrt{2,083^2 + 416^2} = 2,124$ pounds per inch of run. The pitch of the rivets at the end of the girder would then be $5,251 \div 2,124 = 2.47$, approximately, $2\frac{1}{2}$ inches.

At *b*, 4 feet from the end of the girder, the horizontal increment of stress on the flange, as previously calculated, is 1,667 pounds per inch of run, while the vertical stress remains the same; the combined action of these two forces produces a resultant stress on the rivets of $\sqrt{1,667^2 + 416^2} = 1,717$ pounds per inch of run, and this divided into the value of one rivet gives $5,251 \div 1,717 = 3.06$, or about $3\frac{1}{4}$ inches. Similar calculations may be made for each panel point to the center of the girder, or until the pitch exceeds the allowable limit of 6 inches.

The above results show that the values of the pitch in which the vertical stress due to the load is taken into account, are nearly the same as those first obtained; the effect of the vertical stress has, therefore, little influence on the pitch of the rivets, and it is hardly necessary to go into such refinement in the design of an ordinary plate girder.

27. Rivets Spaced According to Stress Produced by Bending Moment.—The rivets that connect the flange angles with the web-plate may also be spaced according to the stresses produced on the flanges by the bending moment.

The horizontal stress on the flanges diminishes either way from the point of greatest bending moment toward the end reactions, where it becomes zero, and for any point this stress may be calculated by the application of the principle of moments.

If the bending moment is obtained at any panel point and is divided by the depth of the girder, the stress on the flange at that point will be obtained; and, if this stress is divided by the allowable load on one rivet, the number of rivets required between that point and the end reaction will be obtained.

For example, in the girder used in the previous illustration, Fig. 22, the span being 40 feet and the load 200,000 pounds, the bending moment at the center is equal to $\frac{WL}{8} = \frac{200,000 \times 40}{8} = 1,000,000$ foot-pounds; then the depth of the girder being 4 feet, the flange stress at this point is $1,000,000 \div 4 = 250,000$ pounds. The allowable load on each rivet being 5,251 pounds, the number of rivets between the center and the end reaction is $250,000 \div 5,251 = 48$ rivets, approximately.

Now, although the number of rivets between the end reaction and the center of the girder has been obtained, the pitch of these rivets is still unknown. Since the horizontal stress in the flange varies, being greatest at the ends and least under the position of maximum bending moment, it follows that the rivets should be spaced nearer together at the ends, with an increase in the spacing toward the point of greatest bending moment.

In practical work, the rivet spacing is seldom varied in any one panel; if, however, the flange stress is obtained at each of the stiffeners b, c, d, e, and f, Fig. 22, the number of rivets required between each of these points and the end reaction may be obtained; by finding the difference between these numbers for any two consecutive stiffeners, the number of rivets required in the panel between those stiffeners is arrived at. For example, the stresses on the flange at each of the stiffeners of the girder shown in Fig. 22 are as follows:

	Bending Moment Foot-Pounds		Depth of Girder Feet		Flange Stress Pounds
At *b*,	360,000	÷	4	=	90,000
At *c*,	640,000	÷	4	=	160,000
At *d*,	840,000	÷	4	=	210,000
At *e*,	960,000	÷	4	=	240,000
At *f*,	1,000,000	÷	4	=	250,000

The approximate number of rivets between each stiffener and the reaction R_1 is as follows:

Between *b* and R_1, 90,000 ÷ 5,251 = 18 rivets
Between *c* and R_1, 160,000 ÷ 5,251 = 31 rivets
Between *d* and R_1, 210,000 ÷ 5,251 = 40 rivets
Between *e* and R_1, 240,000 ÷ 5,251 = 46 rivets
Between *f* and R_1, 250,000 ÷ 5,251 = 48 rivets

Then the number of rivets required is:

Between *b* and *a*, 18 − 0 = 18 rivets
Between *c* and *b*, 31 − 18 = 13 rivets
Between *d* and *c*, 40 − 31 = 9 rivets
Between *e* and *d*, 46 − 40 = 6 rivets
Between *f* and *e*, 48 − 46 = 2 rivets

Consequently, the pitch between the stiffeners will be as follows:

Between *b* and *a*, 48 ÷ 18 = 2.67 inches
Between *c* and *b*, 48 ÷ 13 = 3.69 inches
Between *d* and *c*, 48 ÷ 9 = 5.33 inches
Between *d* and *e*, 48 ÷ 6 = 8.00 inches

Between *d* and *e* the theoretical pitch exceeds 6 inches, the limit allowable for a compression member.

By the first method, the pitch at each stiffener or panel point is determined, while by the second the average pitch between two consecutive panel points is obtained; in order to compare the two, we will reduce the results obtained by the first to the basis of the second. In the first method the pitch at the several stiffeners or panel points was found to be:

At a = 2.52 inches At c = 4.20 inches
At b = 3.75 inches At d = 6.30 inches

From these the average pitch between the several points would be:

Between a and b, $(2.52 + 3.75) \div 2 = 3.14$ inches
Between b and c, $(3.75 + 4.20) \div 2 = 3.98$ inches
Between c and d, $(4.20 + 6.30) \div 2 = 5.25$ inches

These, on comparison, are found to correspond approximately with the values 2.67, 3.69, and 5.33 inches obtained by the second method.

28. Rivets Spaced According to Direct Vertical Shear.—This is the method much used in practical work, and will be found to give safe results, corresponding favorably with those obtained by the previous methods. The method is based on the assumption that at any point the horizontal shear between the flange angles and the web-plate is equal to the vertical shear on the girder; for example, the vertical shear at the end stiffener or point a, Fig. 22, is 100,000 pounds; then, according to this method, the shearing stress between the flange angles and the web-plate is 100,000 pounds, distributed over the space between the panel points a and b, and sufficient rivets should be placed between these points to safely sustain this shear.

The allowable web-bearing load on a $\frac{7}{8}$-inch rivet in a $\frac{3}{8}$-inch plate being 5,251 pounds, the number of rivets required between a and b is $100{,}000 \div 5{,}251 = 20$, approximately; the vertical shear at b is 80,000 pounds, and $80{,}000 \div 5{,}251 = 16$, approximately, the number of rivets to be used between b and c; the shear at c is 60,000 pounds, and $60{,}000 \div 5{,}251 = 12$, approximately, the number of rivets to be used between c and d; similarly, the number of rivets required between d and e is found to be 8. According to these results, the pitch of the rivets between a and b should be $48 \div 20$

= 2.4 inches; between *b* and *c*, 48 ÷ 16 = 3 inches; between *c* and *d*, 4 inches; and from there on, 6 inches.

Rivet spacing in plate girders is governed so largely by practical considerations, that this method is to be recommended on account of its convenience. It gives safe results that agree closely with those obtained by the more cumbersome methods.

29. Graphic Method of Determining Number of Rivets in Vertical Leg of Flange Angles.—Besides the several analytical methods of determining the number of

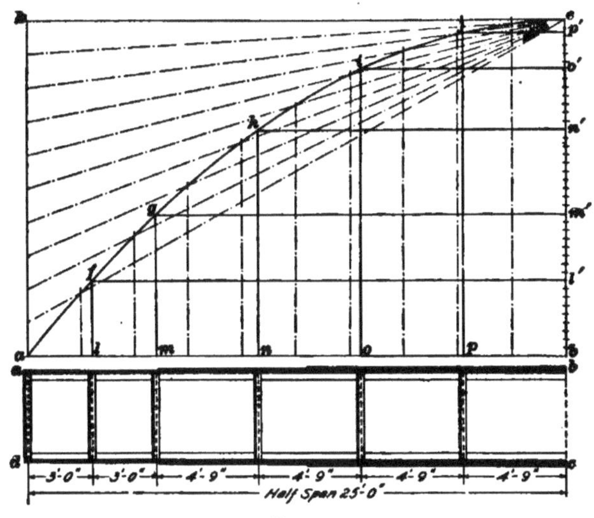

FIG. 23

rivets through the vertical legs of the flange angles in the several panels of the plate girder, a convenient graphic method that gives approximate results sufficiently accurate for all practical purposes is illustrated in Fig. 23. In this method, a diagrammatic drawing of one-half of the plate girder with the stiffeners properly placed, is made to scale, as shown at *a b c d*. The number of rivets required through the vertical leg of the top flange of the plate girder, from the center line of the girder to the abutment, is next calculated by the method described in Art. **27**, which consists in first calculating the bending moment, in foot-pounds, due to the uniformly distributed load and dividing by the depth of the girder, in feet, the result being the horizontal flange stress. This stress divided by the allowable resistance of one rivet will give the number of rivets between the center of the plate girder and the abutment, or between *a* and *b* in the sketch. Having determined the number of rivets required between these points, lay off any vertical distance, as *b e*, and through the points *a* and *e* describe a parabola by the method explained in Art. **15**. Having drawn in the parabola as shown in the figure, extend upwards the center lines of the angle stiffeners on the plate girder until they intersect the parabola, as at *f*, *g*, *h*, *i*, and *j*. Through the points of intersection, extend horizontal lines until they intersect the line *b e*, the greatest ordinate of the parabola, which is coincident with the center line of the plate girder. Having proceeded thus far, divide the length of the greatest ordinate of the parabola, or the distance *b e*, into the same number of equal parts as there are rivets required between *a* and *b*. This may conveniently be done by applying the scale obliquely between the horizontal lines *k e* and *a b*, as explained in connection with the method for determining the length of the flange plates in compound riveted girders.

Assuming that in the diagram forty rivets are required between *a* and *b*, and that in consequence there are forty divisions on the scale, the number of rivets between *a* and *l* will equal the number of spaces between *b* and *l′*, or nine. The number of rivets between *l* and *m* will equal the spaces between *l′* and *m′*, or eight, while between *m* and *n*, from the portion included between *m′* and *n′*, there would be required ten rivets. In the panel *n o*, the theoretical number of rivets will equal eight, while between *o* and *p*, the theoretical requirements will be fulfilled by placing four rivets through the vertical leg of the flange angles. In all cases where there is a fraction of a rivet in a panel, one additional rivet should be used. Since between *a* and *l* nine rivets are required, and the distance from the center of the end stiffener to the center of the stiffener *l* is 36 inches, the theoretical pitch between *a* and *l* will equal 36 ÷ 9 = 4 inches.

In the panel *l m*, since there are eight rivets required, the pitch will be 36 ÷ 8 = 4.5 inches. Between *m* and *n*, there are ten rivets required and the theoretical pitch will be 57 ÷ 10 = 5.70 inches. The number of rivets required between *n* and *o* is eight, and consequently their pitch will be 57 ÷ 8 = 7.13 inches. This last distance is greater than the maximum allowable pitch for rivet spacing in plate girders; therefore, it is needless to calculate the theoretical pitches in the remaining panels. From these calculations it is probable that in designing the plate girder a pitch of 4 inches will be adopted for the panels *a l* and *l m*, provided that a single row of rivets is used, while throughout the remaining panels of the girder the rivets will be spaced at the maximum pitch of 6 inches.

30. The student will observe in this method the application of the principles involved in the graphic method for determining the length of the flange plates of the girder supporting the uniformly distributed load. The parabola always represents graphically the bending moment created in a simple beam by a uniformly distributed load. Since the horizontal flange stress varies directly with the amount of the bending moment throughout the girder, it is evident that the parabola likewise truly represents the flange stress between the abutments and the center line of the girder. Likewise, if the middle or greatest ordinate of the parabola represents the horizontal flange stress, to scale, at the center of the girder, the ordinates of the parabola at any point will equal the horizontal flange stress at that point. If the heights of the ordinates projected from the center line of each stiffener are laid off on the center ordinate and the center ordinate is reduced from horizontal flange stress to the number of rivets whose resistance will equal the amount of flange stress, then the ratio between the several ordinates and the greatest ordinate will be the same as the ratio of the number of rivets required between the ordinate and the abutment of the girder and the number of rivets required between the center of the girder and the abutments; that is, the length mg is to the length eb as the number of rivets between a and m is to the number of rivets between a and b; thus, if the length of the ordinate mg were equal to $1\frac{19}{32}$ inches and the length of the ordinate eb were $3\frac{3}{4}$ inches, while the number of rivets required between a and b is 40, the number of rivets required between m and a would equal $\frac{mg \times 40}{be}$; or, substituting the values, $\frac{1.59375 \times 40}{3.75} = 16.92$, or 17. Figuring in the same way, if lf were $\frac{27}{32}$ inch, the number of rivets between l and a would equal $\frac{.84375 \times 40}{3.75} = 9$. The difference between the number of rivets required between the points a and m and the points a and l equals $17 - 9 = 8$, which accurately expresses the number of rivets called for by the space $l'm'$.

EXAMPLE 1.—A plate girder 60 feet long and 4 feet in depth is loaded with a uniformly distributed load of 2,000 pounds per lineal foot. Required, by the graphic method, the theoretical pitch of rivets through the vertical leg of the flange angle in the several panels, assuming that the allowable resistance of each rivet is 4,800 pounds. The first two panels are 3 feet 9 inches, and the remainder 5 feet in length.

SOLUTION.—The maximum bending moment, which occurs at the center of the girder, is expressed by the formula $M = \frac{WL}{8}$. The value of W, according to the problem, is $2,000 \times 60 = 120,000$, while L or the length of the girder equals 60 ft. By substitution, the bending moment, $M = \frac{120,000 \times 60}{8} = 900,000$ ft.-lb. Since the depth of the girder is 4 ft., the horizontal flange stress at the center of the girder is $900,000 \div 4 = 225,000$ lb. The strength of each rivet equals 4,800 lb.; hence, the number of rivets required between the center and the end is $225,000 \div 4,800 = 46.875$, or approximately 47. Having obtained this result, lay out a diagrammatic elevation of one-half of the plate girder to scale, as shown in Fig. 24, locating the stiffeners where required. On this plate girder draw one-half of the parabola, the center ordinate of the parabola coinciding with the center line of the girder. By placing a convenient scale obliquely between the horizontal lines ab and cd, divide the greatest ordinate into 47 equal parts. Extend upwards from the points *k*, *l*, *m*, *n*, *o*, and *p* the several ordinates, as shown, and project the points *e*, *f*, *g*, *h*, *i*, and *j* horizontally to the greatest ordinate of the parabola.

From the diagram, Fig. 24, the number of rivets required between the several panel points is as follows:

Between *a* and *k*, 11 rivets
Between *k* and *l*, 10 rivets
Between *l* and *m*, 11 rivets
Between *m* and *n*, 8 rivets
Between *n* and *o*, 5 rivets
Between *o* and *p*, 3 rivets
Between *p* and *b*, 1 rivet

The distance between the panel points or the center line of the stiffeners for the first two panels is equal to 45 in., while for the

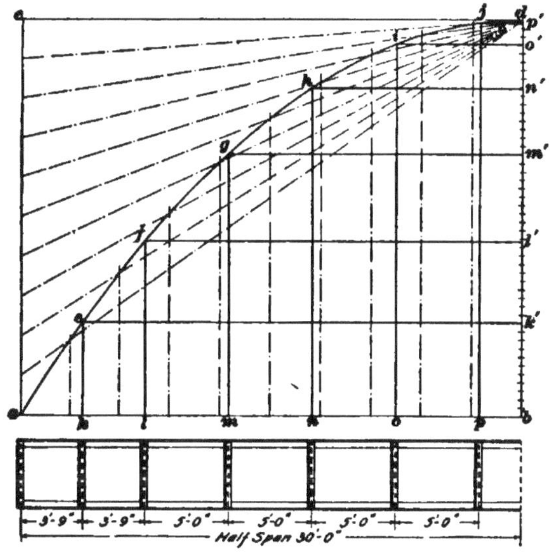

Fig. 24

remaining panels the stiffeners are spaced 60 in. on center lines; hence, the pitch between the several theoretical panel points from the abutment toward the center is as follows:

Between *a* and *k* = 45 ÷ 11 = 4.09 in.
Between *k* and *l* = 45 ÷ 10 = 4.5 in.
Between *l* and *m* = 60 ÷ 11 = 5.45 in.
Between *m* and *n* = 60 ÷ 8 = 7.5 in.

The calculations need not be carried further, for the theoretical pitch of the rivets in the next panel beyond *lm* exceeds the maximum allowable pitch for rivets in plate girders. Ans.

EXAMPLE 2.—Find the pitch of the rivets, by the graphic method, in the several panels of a plate girder having a span of 75 feet and a depth of 5 feet. The length of the panels throughout the girder is 5 feet and the load supported is 2,500 pounds per lineal foot. The strength of one rivet may be taken at 5,000 pounds, and the rivets should be placed in two rows.

SOLUTION.—As in the previous examples, the diagram should be laid out and the parabola representing the bending moment drawn, as shown in Fig. 25. The number of rivets required between the

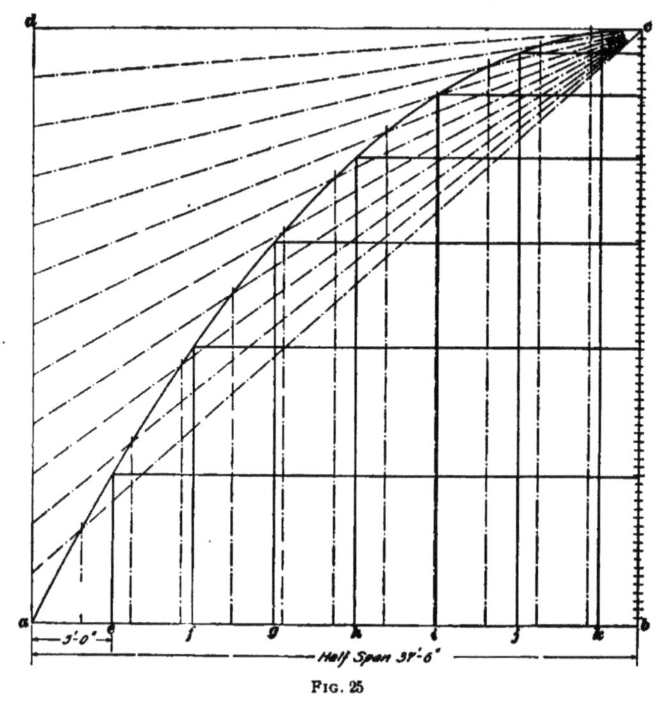

FIG. 25

center and the end is found by dividing the strength of one rivet into the quotient obtained by dividing the bending moment by the depth of the girder. Thus, $\frac{2{,}500 \times 75 \times 75}{8 \times 5 \times 5{,}000} = 70.3$, or 71 rivets.

Divide the greater ordinate bc into seventy-one equal spaces and project horizontal lines from the points where the lines drawn from the panel points e, f, g, h, etc. intersect the parabola, thus determining the number of rivets required in each panel. The first requires eighteen; the second, fifteen; the third, thirteen; etc. Then the spacing in the first panel is $60 \div 18 = 3.3$ in.; for the second panel the pitch is $60 \div 15 = 4$ in.; for the third and fourth panels the pitch is found to be 4.6 in. and 6 in., respectively. As 6-in. angles would undoubtedly be used in this girder, it is advisable to place the rivets in two rows; in the first two panels they may be spaced 6 in. in each row, while in the remaining panels the pitch will be 8 in., which is the maximum pitch for rivets placed in a double row. Ans.

31. Pitch of Rivets in Girders Supporting Concentrated Loads.—In order to illustrate the method of finding the pitch of the rivets in a girder that supports concentrated loads, the conditions shown in Fig. 26 (*a*) will be

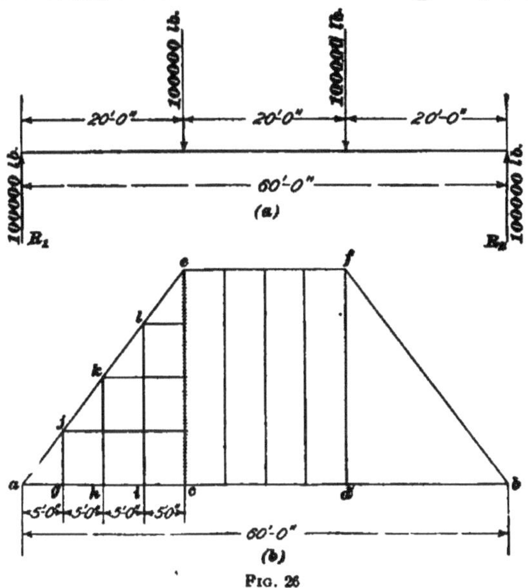

Fig. 26

assumed. The greatest bending moment occurs under the loads and along the length of the girder between them. Lay off *a b*, in (*b*), equal to the span of the girder, and mark the points *c* and *d* where the concentrated loads occur. At these points erect perpendiculars and lay off any convenient distance on each, as *c e* and *d f*, to represent the bending moment. Draw *a e*, *e f*, and *f b*, and from the panel points erect perpendiculars intersecting these lines at *j*, *k*, and *l*. The maximum bending moment, occurring at *c*, is equal to the moment of the reaction about that point, or 100,000 × 20 = 2,000,000 foot-pounds. The flange stress equals 2,000,000 ÷ 5 = 400,000 pounds. Assuming the strength of one rivet to be 6,750 pounds, the number required is 400,000 ÷ 6,750 = 59.26, or 60. Divide *ce* into sixty equal parts and from

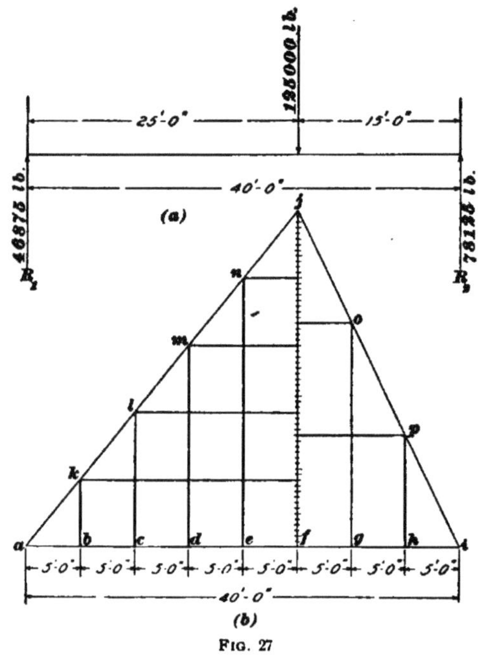

FIG. 27

the points *j*, *k*, and *l* draw horizontal lines cutting *ce*. These lines show that fifteen rivets are required for each panel up to the point *c* and from *c* to *d* no rivets are necessary, theoretically. The pitch of the rivets in the first four panels is 60 ÷ 15 = 4 inches, while for the panels from *c* to *d*, they will be spaced at the maximum pitch, or 6 inches.

EXAMPLE.—In Fig. 27 (*a*) is shown a girder having a span of 40 feet and carrying a concentrated load of 125,000 pounds at a distance of 15 feet from one end. If the girder is 4 feet deep and the panels are 5 feet in length, what will be the pitch of the rivets, assuming the strength of 1 rivet to be 5,000 pounds?

SOLUTION.—Lay off the span of the girder at $a\,i$ in view (b), marking the panel points, as at b, c, d, etc., and at the location of the concentrated load erect a perpendicular, as $f\,j$, of any convenient height. Draw $a\,j$ and $j\,i$, and from the panel points draw perpendiculars to $a\,i$, intersecting these lines at k, l, m, n, o, and p. The maximum bending moment is under the load and is equal to $46{,}875 \times 25 = 1{,}171{,}875$ ft.-lb.; $1{,}171{,}875 \div 4 = 292{,}968.75$ lb. flange stress; $292{,}968.75 \div 5{,}000 = 58.59+$, or, say, sixty rivets are required between the concentrated load and each end of the girder. Divide $f\,j$ into 60 equal parts and draw horizontal lines from the points k, l, m, etc. until they intersect $f\,j$. This shows that twelve rivets are required in each panel to the left of the load, while twenty are needed in each panel to the right. The pitch will then be $60 \div 12 = 5$ in. in the first five panels from the left, and $60 \div 20 = 3$ in. in the remaining panels.

32. Rivet Spacing in Flange Plates.—In spacing the rivets that bind the several flange plates together, a sufficient number of rivets, spaced from $2\frac{3}{4}$ to 3 inches on centers, should be used at the ends of each plate to transmit the allowable stress in it to the members below. For the remainder of the plate, the rivets should have the greatest allowable pitch for a compression member; that is, sixteen times the thickness of the thinnest outside plate, provided that such a distance does not exceed 6 inches. To illustrate:

An intermediate flange plate in a certain girder is $\frac{3}{8}$ inch by 12 inches, the sectional area thus being $4\frac{1}{2}$ square inches. From this area is to be deducted the section cut out by two $\frac{7}{8}$-inch rivet holes, $(1 \times \frac{3}{8}) \times 2 = \frac{3}{4}$ square inch; then the net area of the cover-plate is $4\frac{1}{2} - \frac{3}{4} = 3\frac{3}{4}$ square inches. Assuming that a safe fiber stress of 15,000 pounds was used in calculating the strength of the girder, the safe strength of the cover-plate is $3\frac{3}{4} \times 15{,}000 = 56{,}250$ pounds. Now the safe load on a $\frac{7}{8}$-inch rivet depends, in this position, on the ordinary bearing value of a $\frac{3}{8}$-inch plate, which, calculated on the basis of a fiber stress of 12,000 pounds, is 5,119 pounds. Hence, the number of rivets required in the end of this cover-plate is $56{,}250 \div 5{,}119 = 10.9$, say, 11; but in order to have them symmetrical, there should be six on each side of the web, and they should be spaced about 3 inches from center to center. The remaining rivets in this plate may have the greatest allowable pitch until the next cover-plate is reached.

EXAMPLES FOR PRACTICE

1. Determine, by the graphic method, the theoretical pitch for the rivets in the vertical legs of the flange angles in a plate girder 45 feet long from center to center of bearing plate. The depth of the girder is 36 inches and the load per lineal foot is 1,400 pounds. The safe resistance of the rivets is determined to be 4,200 pounds; the distance between the stiffeners or panel points is 5 feet throughout the girder.

Ans. { First panel, 5 in.
Second panel, 6.67 in.
Third panel, 10 in., etc

2. Theoretically, how many rivets will be required in the vertical legs of the flange angles between the several panel points or stiffeners of a plate girder having a span of 80 feet and a depth of 5 feet, assuming that the load on the girder is uniformly distributed and is equal to 3,000 pounds per lineal foot; also, that the resistance one rivet offers is 5,200 pounds.

Ans. { First panel, 22 rivets
Second panel, 19 rivets
Third panel, 16 rivets
Fourth panel, 13 rivets
Fifth panel, 10 rivets

3. Find the pitch of the rivets in the vertical legs of the flange angles in a girder 4 feet in depth and having a span of 50 feet, which supports a concentrated load of 90,000 pounds located at a distance of 20 feet from one end of the girder. The value of one rivet is 4,500 pounds and the length of each panel is 5 feet.

Ans. { 4-in. pitch in four panels at end nearest load
6-in. pitch in remainder

PRACTICAL DESIGN

33. In order to illustrate the application of the rules and formulas previously given, the following practical problem will be assumed and worked out:

The floor of a building used for light manufacturing purposes is to be supported by three plate girders, as shown at *a*, *a*, *a*, Fig. 28. The floor is composed of 1-inch yellow-pine flooring laid on 3″ × 12″ hemlock joists, spaced on 16-inch centers; these joists are to carry a plastered ceiling on the under side. The live load on the floor will be

80 pounds per square foot. The girder itself is to extend below the surface of the ceiling and is to be painted. A detail of the construction is shown in Fig. 29.

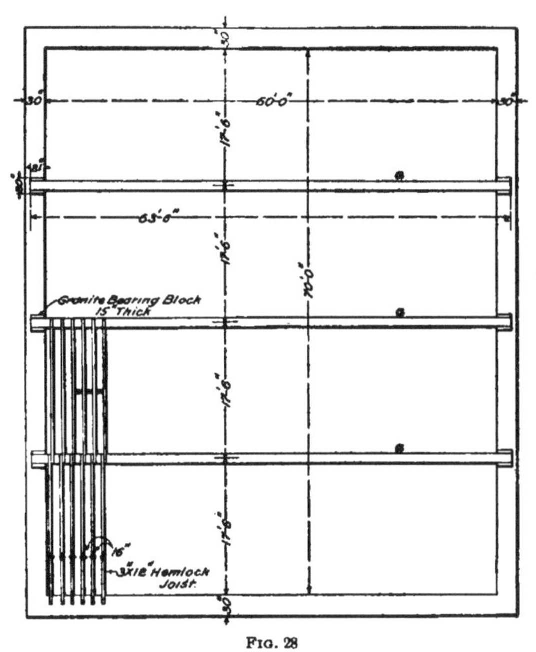

FIG. 28

The total load on each square foot of floor surface is as follows:

Live load, per square foot of floor surface .	80 pounds
Lath and plaster, per square foot of floor surface	8 pounds
1-inch yellow-pine flooring, per square foot of floor surface	4 pounds
Hemlock joist flooring, per square foot of floor surface	6 pounds
Girder (assumed), per square foot of floor surface	8 pounds
Total	106 pounds

The floor area to be supported by one girder is 60 × 17.5 = 1,050 square feet; and the total uniformly distributed load on the girder is 1,050 × 106 = 111,300 pounds.

The greatest bending moment on the girder is

$$M = \frac{WL}{8} = \frac{111{,}300 \times 60}{8} = 834{,}750 \text{ foot-pounds}$$

The depth of the girder is 4 feet, and the allowable unit fiber stress to be used is 15,000 pounds; therefore, the

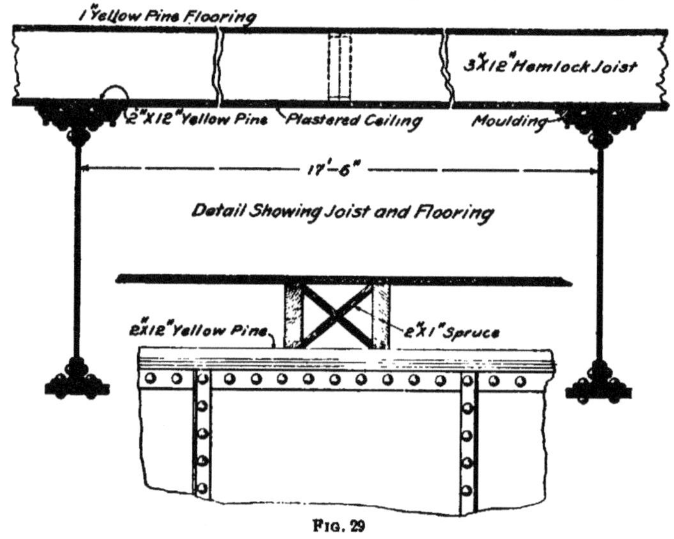

Fig. 29

required flange area may be determined by formula **4.** Substituting the proper values in the formula gives

$$A = \frac{834{,}750}{4 \times 15{,}000} = 13.91 \text{ square inches}$$

Assume a flange composed of two $5'' \times 5'' \times \frac{7}{16}''$ angles and two $\frac{3}{8}'' \times 12''$ flange plates; a sketch of the section with the location of the rivets is shown in Fig. 30. The entire area of the flange is:

Two $\frac{3}{8}'' \times 12''$ plates	=	9	square inches
Two $5'' \times 5'' \times \frac{7}{16}''$ angles	=	8.36	square inches
Total,		17.36	square inches

The sectional areas cut out for rivet holes are:

Four $\frac{7}{8}$-inch holes through $\frac{3}{8}$-inch plate = 1.3 1 2 sq. in.
Four $\frac{7}{8}$-inch holes through $\frac{7}{16}$-inch angles = 1.5 3 1 sq. in.
Total, 2.8 4 3 sq. in.

The net area of the flange is, therefore, 17.36 − 2.84 = 14.52 square inches, which, since the required area is 13.91 square inches, is ample, and this section will be adopted.

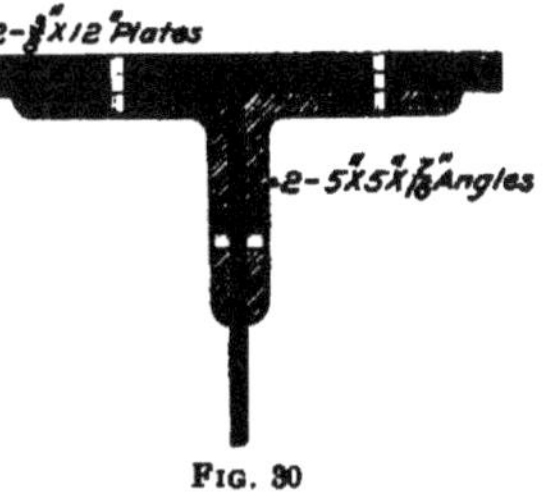

FIG. 30

We will now determine the thickness of the web-plate. The reaction at either end is equal to one-half of the load, or 55,650 pounds. Assuming that there are eleven $\frac{7}{8}$-inch holes cut in line through the web-plate, the net depth of the plate will be $48 - 11 \times .875 = 38.375$ inches. Using an allowable unit shearing stress of 11,000 pounds, the theoretical thickness of the web-plate, from formula **1**, is $t = \frac{55,650}{38.375 \times 11,000} = .132$ inch.

However, it is not practicable to use this thickness of metal for a web-plate, since it would not provide sufficient bearing value for the rivets. As it is never good practice to use a web-plate less than $\frac{5}{16}$ inch in thickness, this size will be adopted.

34. The lengths of the flange plates are now required; they may be determined either by the graphic method or by formula **6**; using the latter method, the theoretical length of the outside plate is found to be $l = 60\sqrt{\frac{3.844}{14.52}} = 30.87$ feet, or about 30 feet 10 inches, to which is to be added 1 foot at each end to allow for riveting. The total length of the plate is therefore 32 feet 10 inches, say, 33 feet.

Applying the formula again, the length of the second flange plate is $l = 60\sqrt{\frac{7.688}{14.52}} = 43.66$ feet, or about 43 feet 8 inches; adding a foot at each end gives us 45 feet 8 inches.

Consider now the size of the four stiffeners at the end of the girder. The reaction at the end of the girder is 55,650 pounds, and the allowable compressive strength of the material in the girder will be taken at 13,000 pounds. Then the sectional area required in the four angles composing the stiffeners on the plate girder over the abutments is 55,650 ÷ 13,000 = 4.28 square inches. Since it would not be advisable to use smaller than a 4″ × 4″ × $\frac{5}{16}$″ angle in this position, the sectional area of which is 2.4 square inches (see table Properties of Angles, in *Properties of Sections*), it is evident that there will be ample strength in the four stiffeners. The other stiffeners may be made of 3″ × 3″ × $\frac{5}{16}$″ angles, which is the smallest size that should be used for any girder requiring intermediate stiffeners.

35. The rivet spacing, etc. needs no explanation; it would be well, however, for the student to calculate the number of rivets for the several parts and compare the results with the number actually used, as shown by the detail drawing, Fig. 31. He will undoubtedly find that more rivets are used than are actually required, but he must bear in mind that there are always practical considerations that influence more or less the design of structural work.

In Fig. 31 it will be noticed that the web-plate is spliced at the point *a*. The shear at this point is equal to 55,650 pounds, the reaction at R_1 minus the load on the girder between R_1 and the point *a* under consideration, that is, 55,650 − 37,150 = 18,500 pounds. A sufficient number of rivets must be placed on the two sides of the joint to take care of this shear safely.

Fig. 32 shows the design of a heavily loaded girder with a long span; this girder was designed to carry a uniformly distributed load of 2,400 pounds per lineal foot and two concentrated loads of 60,000 pounds, placed one on each side of the center of the girder and 12 feet 6 inches therefrom. The unit fiber stress allowable in calculating the flange section was taken at 15,000 pounds.

The student should note particularly the splices on this

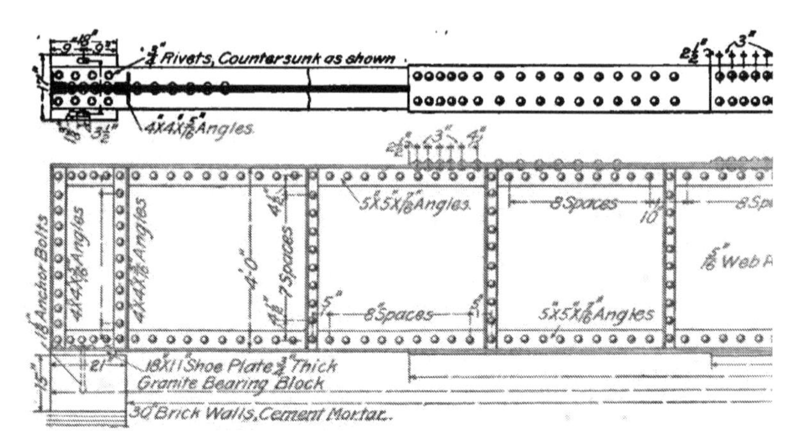

Rivets, Countersunk as shown
4X4X5/16 Angles.
5X5X7/16 Angles.
8 Spaces
Web P
Anchor Bolts
4X4X5/16 Angles
7 Spaces
4'-0"
8 Spaces
5X5X7/16 Angles
18X11 Shoe Plate 3/4 Thick
Granite Bearing Block
30 Brick Walls, Cement Mortar.

FIG.

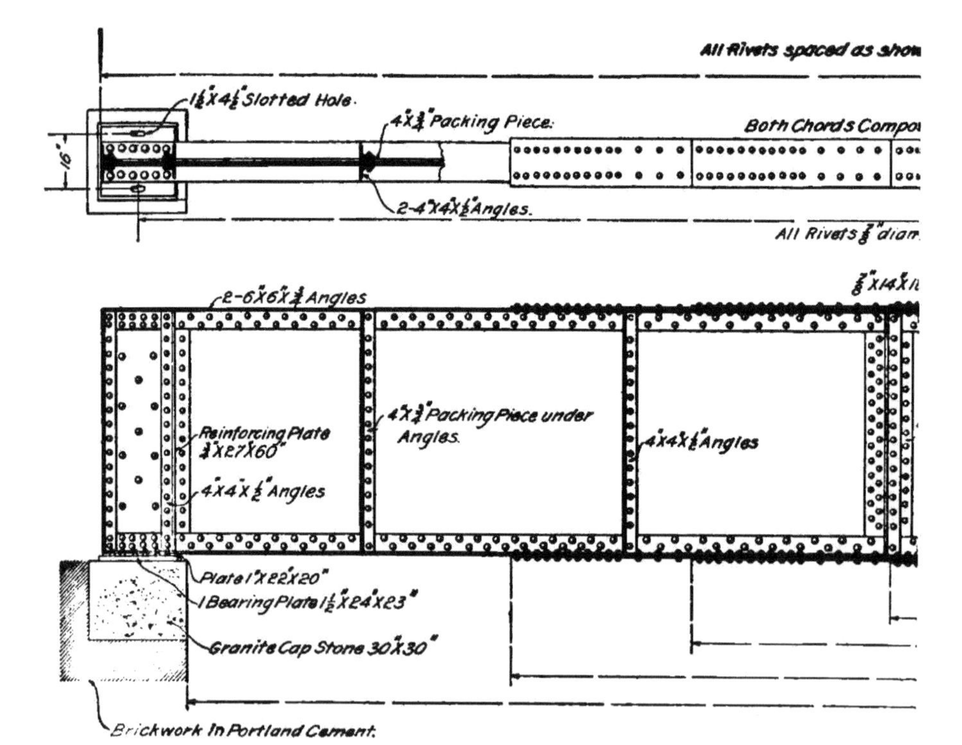

All Rivets spaced as show
1½X4½ Slotted Hole.
4X¾ Packing Piece.
Both Chords Compo
16"
2-4X4X½ Angles.
All Rivets ⅞ diam
2-6X6X¾ Angles
Reinforcing Plate ¾X27X60"
4X¾ Packing Piece under Angles.
4X4X½ Angles
4X4X½ Angles
Plate 1"X22X20"
1 Bearing Plate 1½X24X23"
Granite Cap Stone 30X30"
Brickwork In Portland Cement.

FIG.

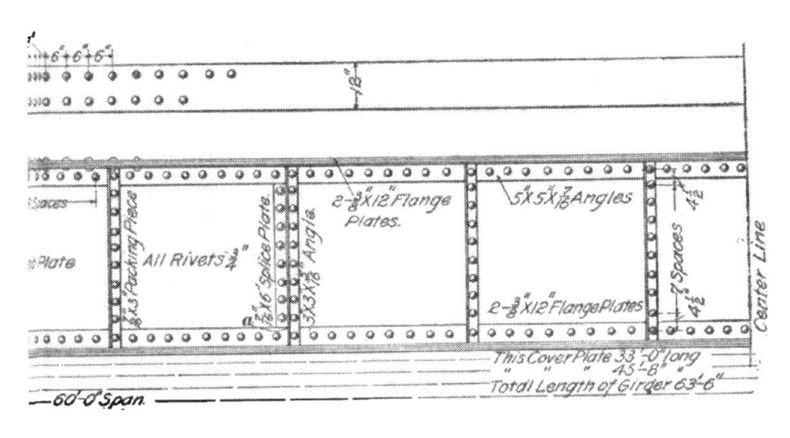
2-3/8"X12" Flange Plates.
5"X5"X7/16" Angles
All Rivets 3/4"
2-3/8"X12" Flange Plates
7 Spaces
Center Line
This Cover Plate 33'-0" long
45'-8"
Total Length of Girder 63'-6"
60'-0" Span.

FIG. 81

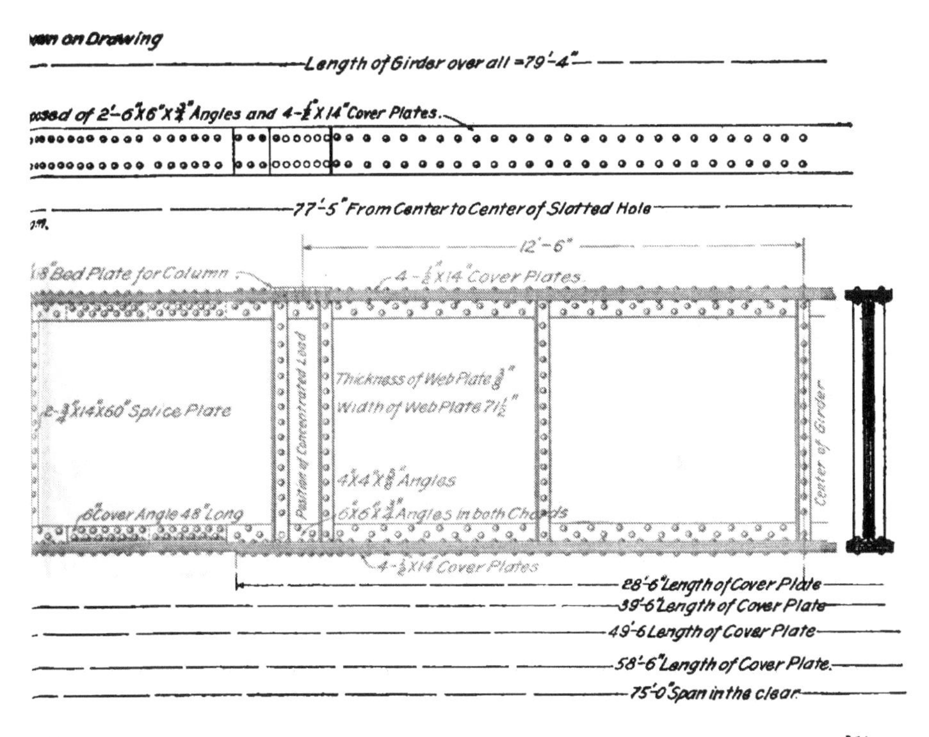
Length of Girder over all = 79'-4"
Composed of 2'-6"X6"X3/4" Angles and 4-1/2"X14" Cover Plates.
77'-5" From Center to Center of Slotted Hole
12'-6"
Bed Plate for Column
4-1/2"X14" Cover Plates.
2-1/2"X14"X60" Splice Plate
Position of Concentrated Load
Thickness of Web Plate 3/8"
Width of Web Plate 71 1/2"
4"X4"X3/8" Angles
6"X6"X3/4" Angles in both Chords
6" Cover Angle 48" Long
Center of Girder
4-1/2"X14" Cover Plates
28'-6" Length of Cover Plate
39'-6" Length of Cover Plate
49'-6 Length of Cover Plate
58'-6" Length of Cover Plate.
75'-0" Span in the clear.

FIG. 82

girder. In this case it was necessary to splice the flange angles; when this is done, care must be taken to weaken the sectional area of the angle as little as possible by the punching of the rivet holes, and a sufficient number of rivets must also be placed each side of the joint, so that the resistance of the rivet section may equal that of the net section of the flange angles. A careful study should be made of the various other details on this drawing, which represents excellent modern practice.

CAMBERED GIRDERS

ANALYTIC METHOD OF COMPUTING STRESSES

GIRDERS WITH ONE STRUT

36. When wooden girders of great span are heavily loaded, it becomes necessary to strengthen them with iron or steel camber rods, as in Figs. 33 and 34, which show a girder with one and with two supports, respectively. The span of the beam or girder may be considered, in each case, as the distance between the supports, the strength of the girder being thereby materially increased.

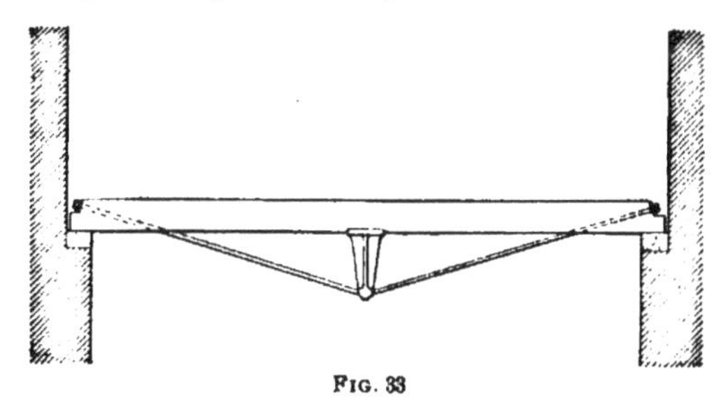

FIG. 33

37. In Fig. 35, let W represent the load concentrated at D. Then the stress in the member DC is equal to W. The stress in the other members may be found by applying the following rules:

Rule.—I. *To find the stress in AC or BC, divide the length of the line AC by the length of the line DC, and then multiply this result by one-half of the load W.*

II. *To find the stress in the beam AB, divide the length of the line AD by the length of the line DC, and multiply this result by one-half of the load W.*

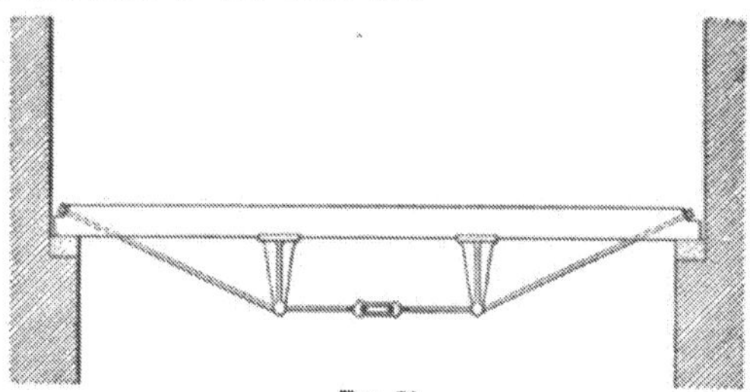

FIG. 34

In the diagram, Fig. 35, the members represented by the solid lines are in compression, and those shown dotted are in tension.

The length of the members in the above rules may be taken in feet or inches, but all lengths should be taken in the same

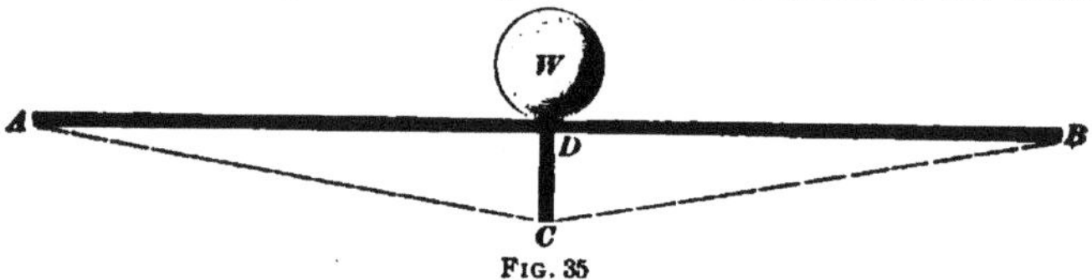

FIG. 35

unit of measurement. The rules may be expressed by the formulas:

$$\text{Stress } DC = +W \qquad (11)$$

$$\text{Stress } AC \text{ or } BC = -\frac{AC}{DC} \times \frac{W}{2} \qquad (12)$$

$$\text{Stress } AB = +\frac{AD}{DC} \times \frac{W}{2} \qquad (13)$$

The + and − signs in the formulas indicate compression and tension, respectively. The + sign denotes that the result obtained is a *compressive stress.* The − sign means that the result is a *tensile stress.*

EXAMPLE.—What is: (*a*) the tension in the camber rod; (*b*) the compression on a trussed beam of the dimensions and loads shown in Fig. 36?

SOLUTION.—The load W coming on the strut DC must first be computed. The load is, in this case, usually considered equal to one-half

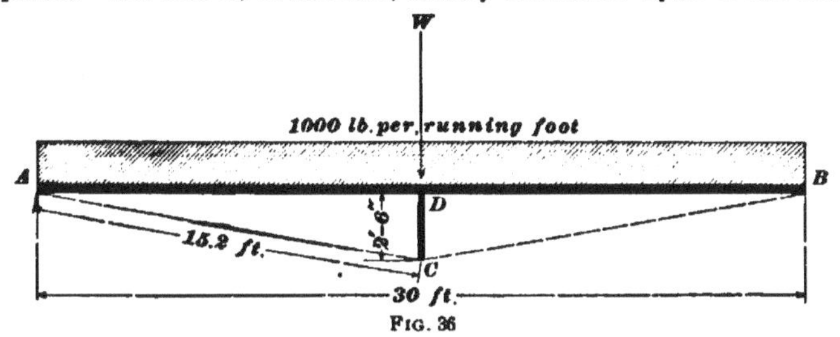

FIG. 36

the entire load on the beam. But as the beam is composed of one length of timber, and is not hinged at D, being, in effect, a continuous beam, it is more correct to consider the load on the center strut as being five-eighths of the entire load on the beam. The entire load on the beam is equal to $30 \times 1{,}000 = 30{,}000$ lb.; five-eighths of $30{,}000 = 18{,}750$ lb., the load W acting on the beam directly over the strut DC.

(*a*) The tension in the camber rod AC is equal to the length $AC \div DC$ multiplied by one-half of W, or, substituting the given dimensions, $15.2 \div 2.5 = 6.08$; and $6.08 \times (\frac{1}{2}$ of $18{,}750) = 57{,}000$ lb., the tensile stress in the rod AC. Ans.

(*b*) To determine the stress in beam AB, divide the length AD by DC, and multiply by one-half of W. Thus, $15 \div 2.5 = 6$; $6 \times (\frac{1}{2}$ of $18{,}750) = 56{,}250$ lb., the compressive stress in the beam AB. Ans.

GIRDERS WITH TWO STRUTS

38. In Fig. 37, the calculations for the stresses in the various members are similar to those given for the trussed

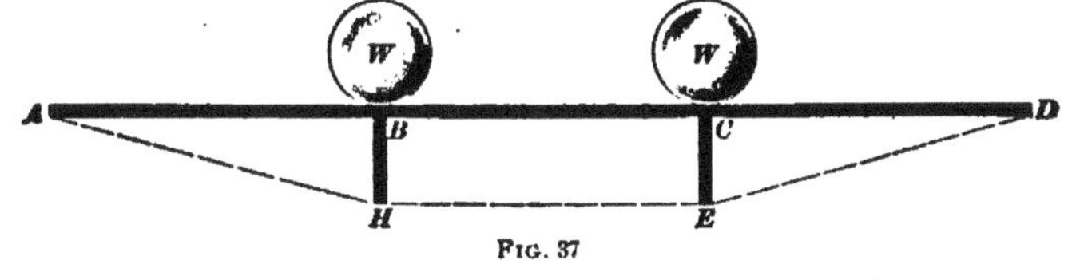

FIG. 37

beam with one support. In the two trussed beams, the stress in BH or $CE = W$. The stresses in the other members may be expressed by rule, as follows:

Rule.—I. *To obtain the stress in AH or DE, divide the length of AH by the length of BH, and multiply this result by the amount of the load W.*

II. *To find the stress in AD or HE, divide the length of the line AB by the length of the line BH, and multiply this result by the amount of the load W.*

The above may be expressed in formulas:

$$\text{Stress } BH \text{ or } CE = +W \qquad (14)$$

$$\text{Stress } AH \text{ or } DE = -\frac{AH}{BH} \times W \qquad (15)$$

$$\text{Stress } HE = -\frac{AB}{BH} \times W \qquad (16)$$

$$\text{Stress } AD = +\frac{AB}{BH} \times W \qquad (17)$$

As previously noted, compression is indicated by the + sign, and tension by the − sign.

EXAMPLE.—A beam is trussed, as shown in Fig. 38. What is: (*a*) the stress in camber rod HE; (*b*) the compression in the beam AD?

SOLUTION.—The entire load on the beam AD is $30 \times 1{,}000 = 30{,}000$ lb. The loads W, W could be considered equal to one-third of the entire load on the beam. But the beam, being a con-

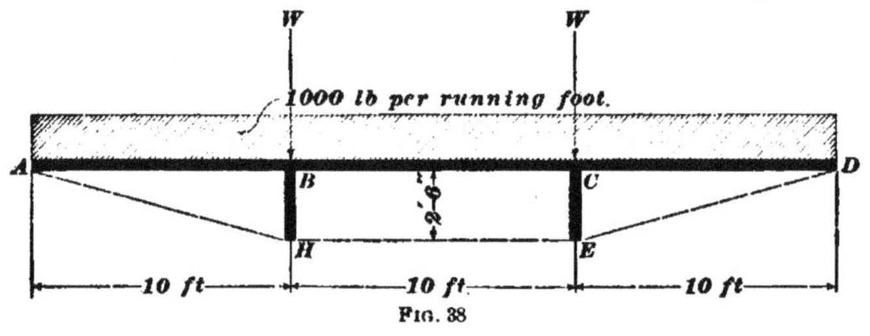

FIG. 38

tinuous girder, as in the previous example, it is better practice to consider it equal to eleven-thirtieths of the entire load. Hence, W is equal to eleven-thirtieths of 30,000, or 11,000 lb.

(*a*) Applying formula **16**, we have

$$\text{Stress } HE = \frac{10}{2.5} \times 11{,}000 = 44{,}000 \text{ lb. Ans.}$$

(*b*) Applying formula **17**, we have

$$\text{Stress } AD = \frac{10}{2.5} \times 11{,}000 = 44{,}000 \text{ lb. Ans.}$$

GRAPHIC METHOD OF DETERMINING STRESSES

39. The stress in the various members of a trussed beam may be obtained by means of a graphic method that is simply an application of the principles of the resolution of forces. Although not as exact in its results as the mathematical method, it is probably more satisfactory, there being, under it, less chance of errors creeping into the calculation. This method is fully explained in the subjoined example:

A floor is to be supported by yellow-pine girders, each composed of two 4″ × 12″ beams, trussed with a wrought-iron rod, as shown in Fig. 39. The span of the girders is 24 feet, and they are spaced 8 feet from center to center. The load is light, amounting to only 40 pounds per square

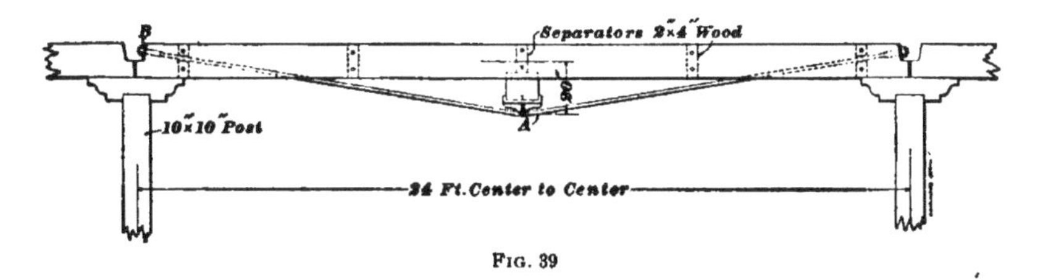

FIG. 39

foot of floor surface. Required, to determine whether the two yellow-pine beams are sufficiently strong, and what should be the size of the wrought-iron camber rod; also, to design the detail construction for the parts A and B.

The floor area supported by each girder is 24 × 8 = 192 square feet; therefore, the total load on a girder is 192 × 40 = 7,680 pounds. To find the stress produced in the different members of the truss by this load, first draw to some convenient scale, as in Fig. 40, the lines ab, ac, bc, and dc corresponding, respectively, to the center lines of the girder, the wrought-iron camber rod, and the strut; thus, the line ab represents the center line of the pine beams, its length being equal to 24 feet on the assumed scale; while dc, drawn perpendicular to ab at its middle point, represents, on the same scale, 20 inches, the length of the strut.

In accordance with the principles stated in Art. **37**, the load carried by the strut may be taken as five-eighths of the total load on the girder; therefore, the force *f*, Fig. 40, acting downwards on the frame, and borne directly by the strut *dc*, is $7{,}680 \times \frac{5}{8} = 4{,}800$ pounds. This force is held in equilibrium by the stresses in the members of the truss, represented by the center lines *ad*, *db*, *ac*, and *cb*, one-half of it, or 2,400 pounds, being held by each of the pairs *ad*

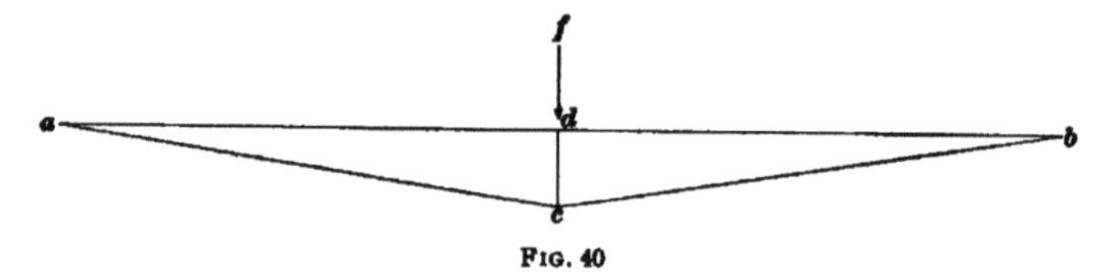

FIG. 40

and *ac*, *db* and *bc*. Considering the half of the load carried by the pair *ad* and *ac*, we have a downward force of 2,400 pounds, which it is required to resolve into two components, one acting along the line *ac* and the other along *ad*. Assuming a scale of forces, one, for example, in which a line 1 inch long represents a force of 800 pounds, draw the line *dc*, Fig. 41, parallel to the center line *dc*, Fig. 40, of the strut, and make its length correspond to a force of 2,400 pounds, the part of the total load on the strut that is

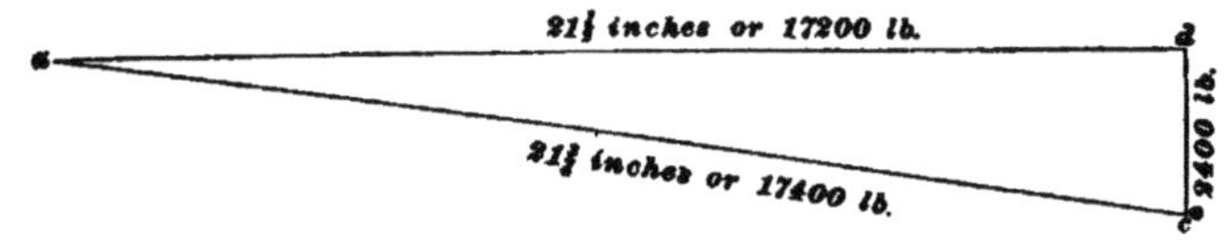

FIG. 41

borne by the members *ad* and *ac*. From the upper extremity of *dc*, Fig. 41, draw the line *da* parallel to the line *da* of Fig. 40, and from the lower extremity draw the line *ca* parallel to *ca* of Fig. 40, prolonging these two lines until they meet at the point *a*. The lines *da* and *ca* of Fig. 41 represent, on the scale of forces to which the line *dc* was drawn, the stresses in the corresponding members of the girder. With the assumed scale of 1 inch = 800 pounds, the line dc must be 2,400 ÷ 800 = 3 inches long; by measurement, the lines da and ca are found to be $21\frac{1}{2}$ and $21\frac{3}{4}$ inches long, respectively; therefore, the stress represented by the line da is $21\frac{1}{2} \times 800 = 17{,}200$ pounds, and that represented by ca is $21\frac{3}{4} \times 800 = 17{,}400$ pounds.

DETAIL DESIGN

40. The stress of 17,200 pounds is the total compressive stress produced in the two yellow-pine beams through the action of the downward thrust on the strut. The ultimate resistance to compression of yellow pine per square inch is usually taken at about 4,400 pounds; and as wood is not so reliable as iron, it is considered advisable to use a factor of safety of 6, as against a factor of safety of 4 for the camber rods. Since the trussed girder is secured against lateral deflection by the floor joist, and as it is secured from deflection in an upward direction at the center by the load on the floor, and by the camber rod and strut, the length of the wooden girder, which may be considered as a column under compressive stress, is only one-half the span, or 12 feet. The sectional dimension of the girder is so great in comparison with its length, that it is not necessary to apply the column formula, and its strength may be considered as its resistance to direct compression. Hence, 4,400 ÷ 6 = 733 pounds, which is the allowable compressive strength of the girder per square inch of section. Then, 17,200 pounds (the compression) ÷ 733 pounds (the allowable unit stress) = 23 square inches required to take care of the compressive stress. As the girder is known to be 12 inches in depth, it is readily seen that this compressive stress will require a section of the timber girder equal to 2 inches by 12 inches.

There is, in addition to this, a transverse stress on one-half of the girder produced by the uniformly distributed load. To find the amount of this bending stress, consider the left-hand half of the girder as a simple beam sustaining a uniformly distributed load equal to one-half of the total load on the girder, that is, a load of 7,680 ÷ 2 = 3,890 pounds.

Applying formula **1**, the bending moment due to this load is $M = \frac{3,890 \times 12 \times 12}{8} = 70,020$ inch-pounds. The section modulus is obtained by the formula $S = \frac{M}{s_a}$, where S equals the section modulus, M the bending moment in inch-pounds, and s_a the allowable unit fiber stress of the material, which is equal, in this case, to 7,300 (the modulus of rupture of yellow pine) ÷ 6 (the factor of safety) = 1,216 pounds. Substituting the values in the above formula, $S = \frac{70,020}{1,216}$ = 57.6, the section modulus required to resist the transverse stress. The bending moment might also be obtained by considering the beam as continuous, having three supports.

41. Since the section modulus of a rectangular beam may be obtained by the formula $S = \frac{b d^2}{6}$, b being the width of the beam in inches, and d the depth, and as S is already known to be 57.6 and the depth of the beam to be 12 inches, the width of the beam required to resist the transverse stress may be obtained by transposing the formula to $b = \frac{S \times 6}{d^2}$; the values substituted give $b = \frac{57.6 \times 6}{12 \times 12}$ = 2.4 inches, which is the width of the required beam. Then, adding the size of the timber required to resist compression and the size of timber required to resist the transverse stress, we have a timber 2 inches wide by 12 inches deep, added to a timber 2.4 or, say, 2½ inches by 12 inches, which equals a piece 4½ inches by 12 inches. In the girder, there are two 4″ × 12″ timbers, and, as only a single 4½″ × 12″ timber is required, it is evident that the girder is nearly twice as strong as is necessary. However, it must be borne in mind that the theoretical dimensions of members do not always agree with those required in practical rules; for instance, in the above case it would not be good practice to make the combined sectional area of the girder equivalent to that of a 4½″ × 12″ timber, as obtained by the calculation, because this would make each timber a little larger than 2″ × 12″, and no timber or girder, especially where rafter or flooring is spiked to it, should be less than 3 inches wide.

42. The ultimate tensile strength of wrought iron is usually taken at 50,000 pounds per square inch; hence, if we use a factor of safety of 4, the safe working fiber stress in the rod must be 50,000 ÷ 4 = 12,500 pounds per square inch.

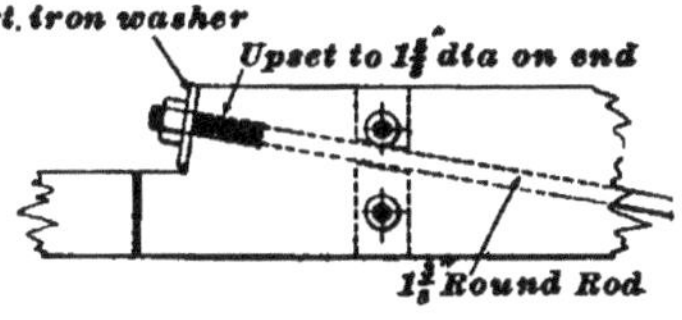

Fig. 42

According to the results given by the diagram, the total stress in the rod is 17,400 pounds; therefore, the rod must have a net section of 17,400 ÷ 12,500 = 1.39 square inches. The area of a $1\frac{3}{8}$-inch round rod is 1.48 square inches, and as this is the nearest standard size having the required sectional area, it will be used. As the area at the bottom of the thread of a $1\frac{3}{8}$-inch bolt is, however, only 1.06 square inches, it will be necessary to upset or enlarge the ends of the rod to a diameter of $1\frac{5}{8}$ inches, in order to get the requisite strength in the threaded portion. The washer at B, Fig. 39, must be large enough to distribute the pressure due to the pull of the rod over a sufficient area of the end of the beams to prevent danger of crushing the wood. The allowable compressive strength of yellow pine, parallel to the grain, may be taken as 800 pounds per square inch; this requires a washer whose area is 17,400 ÷ 800 = 22 square inches, nearly. Using a washer 6 inches wide, extending across the ends of the two beams, we get a bearing area of 2 × 4 × 6 = 48 square inches. In order to

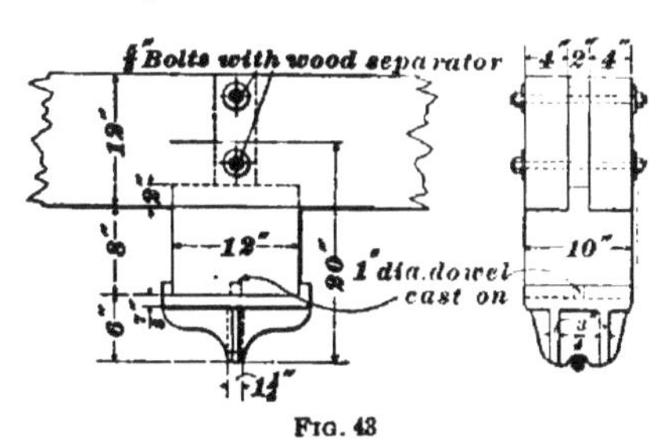

Fig. 43

resist the bending stress due to the pull of the rod, the washer should be from $\frac{3}{4}$ inch to 1 inch in thickness.

Figs. 39, 42, and 43, which are so clearly drawn as to require no further explanation, show excellent details for the different parts of the trussed stringer under consideration.

EXAMPLES FOR PRACTICE

1. It is found necessary to truss the yellow-pine purlins supporting a roof, with a wrought-iron camber rod on each side of the purlin; the length of the purlin is 20 feet, the depth of the truss from the center of the rods to the center of the purlin is 14 inches, and the load on the central strut is 3,200 pounds. What should be the diameter of the camber rods if the ends of the rods are upset, and a safety factor of 4 is desired? Ans. $\frac{7}{8}$ in. diam.

2. A girder of 24-foot span is trussed at the center by a camber rod and strut; the depth of the truss from the center of the girder to the center of the rod is 2 feet; if the beam is loaded with a uniformly distributed load of 2,000 pounds per lineal foot: (*a*) what is the stress on the rod? (*b*) what is the compressive stress on the beam? (*c*) what is the stress on the central strut?

Ans. (*a*) 91,200 lb.; (*b*) 90,000 lb.; (*c*) 30,000 lb.

TRUSSED GIRDERS

43. In some buildings it is necessary to use **trussed girders** to support the weight over a large space, such as a stock exchange, ballroom, etc. They are also sometimes required when it is desired to place a runway or bridge between two adjacent buildings.

In the first case, the girder may support only the weight of the floor above and its live load or it may carry the weight from several stories above, in which case the girder will probably have concentrated loads due to columns, in addition to the uniform load produced by the dead and live loads.

The trussed girders supporting a bridge between two buildings should be proportioned to carry the entire dead load of the bridge and also a moving or rolling load of considerable weight

A trussed girder may be constructed of wood with iron or steel tension members, or it may be entirely of steel.

THE HOWE TRUSS

44. A form of truss known as the **Howe** is shown in Fig. 44. The stresses in the different members may be determined analytically or graphically. As the latter method is explained in *Graphical Analysis of Stresses*, Parts 1 and 2, only the analytic method will be considered here.

Suppose the truss shown in Fig. 44 to be divided along the dotted line de, thus cutting the members a, b, and c; it is evident that if the section at the left were considered as a separate piece, it would be held in equilibrium by applying to each bar that is cut, a force of the same direction and magnitude as the stress existing in the bar before the

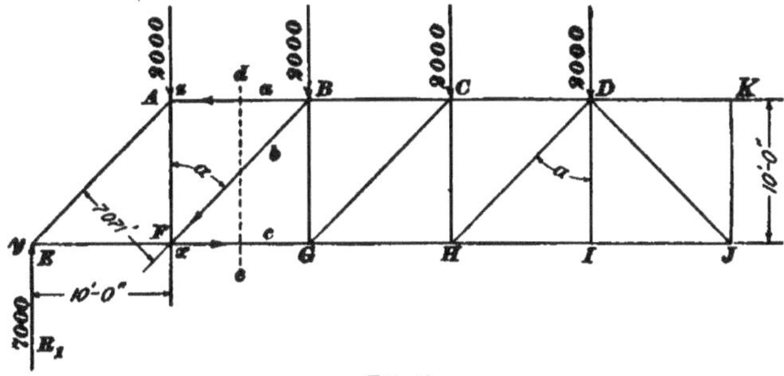

Fig. 44

section was made. Therefore, the forces R_1, A, a, b, and c acting on this portion of the truss are in equilibrium, and the sum of their moments is zero. To find the stress in a, moments will be taken about the point x. As the other unknown forces b and c and the vertical force A pass through this point, they have no effect on the equilibrium of the forces taken about this point, and consequently only the forces R_1 and a need be considered. The moment of R_1 is $7,000 \times 10 = 70,000$ foot-pounds, and the opposing force a is equal to this amount divided by its lever arm, or $70,000 \div 10 = 7,000$ pounds. Regarding the moment of the reaction as positive, the moment of the force a will be considered as negative, because it tends to revolve around the point x

in the opposite direction from the reaction R_1, or in the direction of the arrow. Since the arrow points toward the joint z, the stress in a is compressive.

The stress in b is found by taking the center of moments at the point y. As R_1 and c pass through this point, they are not taken into consideration, but only A, a, and b. The moment of a is $7{,}000 \times 10 = 70{,}000$ foot-pounds, while that of A is $2{,}000 \times 10 = 20{,}000$ foot-pounds. $70{,}000 - 20{,}000 = 50{,}000$ foot-pounds; then the value of b is $50{,}000 \div 7.071 = 7{,}071$ pounds.

Taking z as the center of moments, the forces to be considered are R_1, b, and c. The positive moments of R_1 and b are

	$7{,}000 \times 10 =$	70000 foot-pounds
and	$7{,}071 \times 7.071 =$	50000 foot-pounds
	Total,	120000 foot-pounds

Then the force c is equal to $120{,}000 \div 10 = 12{,}000$ pounds. The stress in b is compressive, as designated by the arrow, and, in this case, produces a positive moment, while that in c is tensile and produces a negative moment. In the same manner, a section may be taken at any point and thus the stress in any member may be obtained. This method is called the *method of sections.*

45. As stated above, the stress in the three members a, b, and c must be in equilibrium with the external forces on the left of the section; and therefore the algebraic sum of the vertical components of these forces equals zero the same as the horizontal components a and c. Then, as $b \cos \alpha$ is the vertical component of the oblique force, $R_1 - A - b \cos \alpha = 0$; but $R_1 - A$ is equal to the vertical shear in the panel AB. Hence, shear $- b \cos \alpha = 0$, and transposing, $- b \cos \alpha = -$ shear, or $b \cos \alpha =$ shear. Then $b = \frac{\text{shear}}{\cos \alpha} = \text{shear} \times \sec \alpha$, and the following rule may be deduced:

Rule.—*The stress in any oblique web member is equal to the vertical shear in the panel multiplied by the secant of the angle that the member makes with the vertical.*

It is assumed that the angle a is equal to 45°. After obtaining the secant of a, as explained in Art. **46,** and applying this rule to determine the stresses in the remaining web members, it is found that

Stress in CG = (7,000 – 2,000 – 2,000) × sec a = 3,000 × 1.4142 = 4,242.6 pounds.

Stress in DH = (7,000 – 2,000 – 2,000 – 2,000) × sec a = 1,000 × 1.4142 = 1,414.2 pounds.

Stress in AE = 7,000 × sec a = 7,000 × 1.4142 = 9,899.4 pounds.

46. The same results may be obtained by the following method: The load at the center of the truss or the point D is equal to 2,000 pounds and half of this load goes to each of the compression members DH and DJ, since DI does not take any of the load, but is simply a rod to prevent the chord HJ from sagging, or to assist in supporting any load placed on the lower chord. Therefore, the stress in each of these members is equal to one-half of the load at D multiplied by the secant of the angle HDI. The secant of the angle may be found from a table of secants, if the angle is known. If not, it may be calculated by considering HDI as a right triangle and ascertaining the ratio between the hypotenuse DH and the side DI adjacent to the angle, which ratio is equal to the secant of the angle a. Therefore, $\sec a = \frac{DH}{DI}$; but $DH = \sqrt{\overline{DI}^2 + \overline{HI}^2}$; therefore, $\sec a = \frac{\sqrt{\overline{DI}^2 + \overline{HI}^2}}{DI}$. This formula, so as to apply to any of the other panels, may be stated as follows:

$$\sec a = \frac{\sqrt{(\text{height of panel})^2 + (\text{width of panel})^2}}{\text{height of panel}}.$$

As the width and height in this instance are equal, the angle a is 45°, the secant of which is 1.4142. The horizontal thrusts of the members HD and DJ are resisted by the tension member HJ. The vertical thrusts must also be resisted at the points H and J by the tension members CH

and KJ, which transmit the downward forces to the points C and K. At the point C, there is applied a load of 2,000 pounds in addition to the force of 1,000 pounds transmitted by the tension bar CH from the member HD. Consequently, the stress in CG is $1,000 + 2,000 = 3,000$ pounds multiplied by the secant of the angle HDI, or a, since the compression web members are all slanted at the same angle. The stresses in the other oblique web members may be found in a similar manner and be expressed in a tabulation as follows:

$$DH = 2,000 \times \tfrac{1}{2} \times \sec a = 1,000 \times 1.4142 = 1,414.2 \text{ pounds}$$
$$CG = 2,000 \times 1\tfrac{1}{2} \times \sec a = 3,000 \times 1.4142 = 4,242.6 \text{ pounds}$$
$$BF = 2,000 \times 2\tfrac{1}{2} \times \sec a = 5,000 \times 1.4142 = 7,071.0 \text{ pounds}$$
$$AE = 2,000 \times 3\tfrac{1}{2} \times \sec a = 7,000 \times 1.4142 = 9,899.4 \text{ pounds}$$

Theoretically, there is no stress in the vertical web DI, but practically there would be a slight tensile stress. In the web CH, there is a tensile stress equal to the vertical component of the stress in HD, or 1,000 pounds, which stress may be found by means of the formula $CH = \frac{DH}{\sec a} = \frac{1,414.2}{1.4142} = 1,000$. The stress in BG is equal to the vertical component of the stress in CG and is found, by means of the preceding formula, to be 3,000 pounds. The other stresses may be found in a similar manner and are as given in the following tabulation:

$$DI = 0$$
$$CH = 2,000 \times \tfrac{1}{2} = 1,000 \text{ pounds}$$
$$BG = 2,000 \times 1\tfrac{1}{2} = 3,000 \text{ pounds}$$
$$AF = 2,000 \times 2\tfrac{1}{2} = 5,000 \text{ pounds}$$

47. The stresses in the horizontal members, or chords, may be obtained when the stresses in the web members are known. Consider the compression member AE separately, as shown in Fig. 45. Then the forces acting on it are the reaction R_1 of 7,000 pounds and a vertical load at A of 2,000 pounds. The stress in AF was found to be 5,000 pounds, so that the total downward force acting on the end of the strut

AE is 2,000 + 5,000 = 7,000 pounds, which just counteracts the reaction, or upward force. The stress in AF equals the vertical component of that in AE less the load at A, while the stress in EF is the horizontal component of AE. It may also be observed that in order to keep the end of the strut from being pushed upwards by the reaction, the force EF must act away from the joint E, or in the direction of the arrow; this indicates a tensile stress in the chord. In order that the strut may be in equilibrium, it is necessary to have a horizontal force at the point A equal and opposite in direction to the force EF. Therefore, the stress in AB is equal to that in EF, but it is a compressive stress, as indicated by the arrow. This stress is equal to the horizontal component of the stress in AE, which may be expressed as (AF + load at A) tan α; similarly, the stress in BC or FG, Fig. 44, is equal to the stress in EF plus the horizontal component of the stress in BF, or EF + (BG + load at B) tan α. The tangent of α is obtained from the formula $\tan \alpha = \frac{\text{width of panel}}{\text{height of panel}}$.

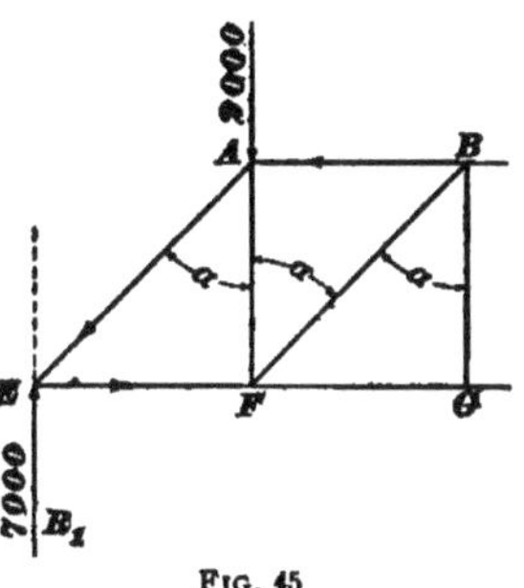

FIG. 45

Substituting these values for the problem under consideration, the value of the tangent is $\frac{10}{10} = 1$. The following notation of the chord stresses may now be made:

$AB = EF = 7{,}000 \times \tan \alpha = 7{,}000 \times 1 = 7{,}000$ pounds.

$BC = FG = (7{,}000 + 5{,}000) \times \tan \alpha = 12{,}000 \times 1 =$ 12,000 pounds.

$CD = GH = (7{,}000 + 5{,}000 + 3{,}000) \times \tan \alpha = 15{,}000 \times 1 = 15{,}000$ pounds.

$HI = IJ = (7{,}000 + 5{,}000 + 3{,}000 + 1{,}000) \times \tan \alpha = 16{,}000 \times 1 = 16{,}000$ pounds.

It may be noticed that the stress in the outside panel of the lower chord, or EF, is equal to the shear in that panel multiplied by the tangent of the angle EAF, that is, equal to the reaction $\times$ tan α, while in the next panel, or FG, the stress is equal to the sum of the shears in EF and FG multiplied by tan α. In GH, the stress is equal to shear in EF + shear in FG + shear in GH multiplied by tan α. Hence, the following rule may be given for the stress in a chord member of a Howe truss:

Rule.—*To find the stress in any chord member, multiply the sum of the shears in all panels from the reaction up to and including the one in question, by the tangent of the angle that the inclined web member makes with the vertical.*

The portions of the upper chord are all in compression, while those of the lower chord are all in tension. If it is desired to check the results obtained by either of the preceding methods, the stress diagram may be drawn for the truss and the stresses measured on it.

MAXIMUM STRESSES PRODUCED BY LIVE LOAD

48. The live and dead loads are generally figured separately in structures where it is necessary to provide for a moving load, such as a bridge. The greatest chord stress is produced when the live load covers the entire truss, while the greatest stress is created in the web member when the live load covers the portion of the truss from the web member in question to the remote abutment and the portion to the other abutment is unloaded. This fact may be demonstrated by considering the following example: The live load per panel in Fig. 44 is 5,000 pounds. When the live load covers the entire truss, the reactions are 17,500 pounds and the chord and web stresses are as follows:

WEB STRESSES

$AF = 17{,}500 - 5{,}000 = 12{,}500$ pounds

$BG = 17{,}500 - 5{,}000 - 5{,}000 = 7{,}500$ pounds

$CH = 17{,}500 - 5{,}000 - 5{,}000 - 5{,}000 = 2{,}500$ pounds

$DI = 0$

$AE = 17{,}500 \times \sec\alpha = 17{,}500 \times 1.4142 = 24{,}748.5$ pounds

$BF = 12{,}500 \times \sec\alpha = 12{,}500 \times 1.4142 = 17{,}677.5$ pounds

$CG = 7{,}500 \times \sec\alpha = 7{,}500 \times 1.4142 = 10{,}606.5$ pounds

$DH = 2{,}500 \times \sec\alpha = 2{,}500 \times 1.4142 = 3{,}535.5$ pounds

CHORD STRESSES

$AB = EF = 17{,}500 \times \tan \alpha = 17{,}500 \times 1 = 17{,}500$ pounds
$BC = FG = (17{,}500 + 12{,}500) \times \tan \alpha = 30{,}000 \times 1 = 30{,}000$ pounds
$CD = GH = (17{,}500 + 12{,}500 + 7{,}500) \times \tan \alpha = 37{,}500 \times 1 = 37{,}500$ pounds
$HI = (17{,}500 + 12{,}500 + 7{,}500 + 2{,}500) \times \tan \alpha = 40{,}000 \times 1 = 40{,}000$ pounds

49. Assuming the truss to be loaded at every panel point except A, as shown in Fig. 46, the left-hand reaction R_1 is found by taking moments about R_2. Considering the width of the panel as the unit of measurement, the moments of the loads are $(5{,}000 \times 1) + (5{,}000 \times 2)$, etc., and as the load is the same at each panel point and the lever arms increase

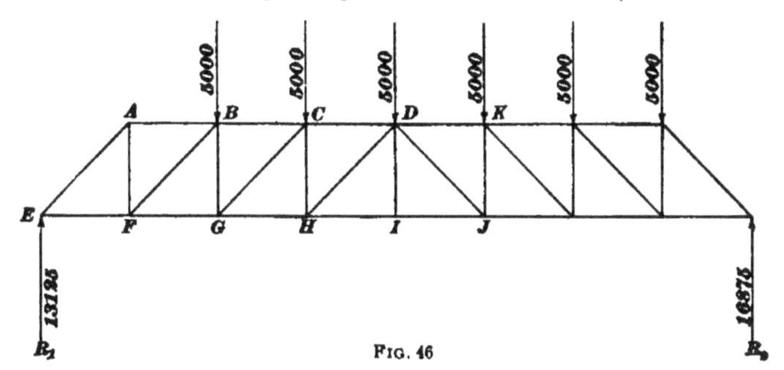

FIG. 46

in arithmetical progression, the sum of the lever arms, or the terms of the progression, is equal to the sum of the first and last terms multiplied by half the number of terms; therefore, the sum of the moments about R_2 is $5{,}000 \times \frac{6(1 + 6)}{2} = 105{,}000$, and the reaction at R_1 is equal to this amount divided by the number of panels, or $105{,}000 \div 8 = 13{,}125$ pounds. The stresses now existing are as follows:

WEB STRESSES

$AF = 13{,}125$ pounds
$BG = 13{,}125 - 5{,}000 = 8{,}125$ pounds
$CH = 13{,}125 - 5{,}000 - 5{,}000 = 3{,}125$ pounds
$DI = 0$

$AE = 13{,}125 \times \sec\alpha = 13{,}125 \times 1.4142 = 18{,}561$ pounds
$BF = 13{,}125 \times \sec\alpha = 13{,}125 \times 1.4142 = 18{,}561$ pounds
$CG = 8{,}125 \times \sec\alpha = 8{,}125 \times 1.4142 = 11{,}490$ pounds
$DH = 3{,}125 \times \sec\alpha = 3{,}125 \times 1.4142 = 4{,}419$ pounds

CHORD STRESSES

$AB = EF = 13{,}125 \times \tan\alpha = 13{,}125 \times 1 = 13{,}125$ pounds
$BC = FG = (13{,}125 + 13{,}125) \times \tan\alpha = 26{,}250 \times 1 = 26{,}250$ pounds
$CD = GH = (13{,}125 + 13{,}125 + 8{,}125) \times \tan\alpha = 34{,}375 \times 1 = 34{,}375$ pounds
$HI = (13{,}125 + 13{,}125 + 8{,}125 + 3{,}125) \times \tan\alpha = 37{,}500 \times 1 = 37{,}500$ pounds

50. Consider the load as covering all points except A and B. The reaction at R_1 is equal to $\left[\frac{5(1+5)}{2} \times 5{,}000\right] \div 8 = 9{,}375$ pounds, and the stresses are as shown in the following tabulation:

WEB STRESSES

$AF = 9{,}375$ pounds
$BG = 9{,}375$ pounds
$CH = 9{,}375 - 5{,}000 = 4{,}375$ pounds
$DI = 0$
$AE = 9{,}375 \times \sec\alpha = 9{,}375 \times 1.4142 = 13{,}258$ pounds
$BF = 9{,}375 \times \sec\alpha = 9{,}375 \times 1.4142 = 13{,}258$ pounds
$CG = 9{,}375 \times \sec\alpha = 9{,}375 \times 1.4142 = 13{,}258$ pounds
$DH = 4{,}375 \times \sec\alpha = 4{,}375 \times 1.4142 = 6{,}187$ pounds

CHORD STRESSES

$AB = EF = 9{,}375 \times \tan\alpha = 9{,}375 \times 1 = 9{,}375$ pounds
$BC = FG = (9{,}375 + 9{,}375) \times \tan\alpha = 18{,}750 \times 1 = 18{,}750$ pounds
$CD = GH = (9{,}375 + 9{,}375 + 9{,}375) \times \tan\alpha = 28{,}125 \times 1 = 28{,}125$ pounds
$HI = (9{,}375 + 9{,}375 + 9{,}375 + 4{,}375) \times \tan\alpha = 32{,}500 \times 1 = 32{,}500$ pounds

The results of the preceding calculations, arranged in a convenient form, are given below:

	Fully Loaded	All Joints Loaded Except A	All Joints Loaded Except A and B
AE =	24,748.5	18,561	13,258
BF =	17,677.5	18,561	13,258
CG =	10,606.5	11,490	13,258
DH =	3,535.5	4,419	6,187
$AB = EF$ =	17,500.0	13,125	9,375
$BC = FG$ =	30,000.0	26,250	18,750
$CD = GH$ =	37,500.0	34,375	28,125
HI =	40,000.0	37,500	32,500

It may be observed from these results that the greatest stress is produced in the chord members when the entire truss is loaded; also, that the greatest stress in AE is when the truss is fully loaded. In BF the greatest stress is created when the truss is loaded to the right of the panel, including this member, that is, all joints are loaded except A, while CG has the greatest stress when all joints except A and B are loaded. If loads are considered at every point except A, B, and C, it will be found that the greatest stress occurs in DH under that condition.

51. Counterbracing is required in the panels where the shear created by the dead load is exceeded by the shear of opposite kind produced by the live load. This may be readily ascertained by drawing the shear diagrams; for instance, Fig. 47 (*a*) represents the frame diagram of the truss having only the dead load on it. The live load is 5,000 pounds at each panel point and will be considered first as being applied at only the first panel point to the left, or AB. Then the reaction for the live load is $\frac{5,000 \times 70}{80}$ = 4,375 pounds at R_1 and 5,000 − 4,375 = 625 pounds at R_2. In the shear diagram (*b*), the shaded portion represents the shear, to scale, created by the dead load along the truss; the portion above the line will be called positive and that below negative. The heavy line designates the shear produced by the live load, which for the first panel is equal to the left reaction, while for the remainder of the truss it is equal to the right reaction, or 625 pounds. It is observed from this diagram that the negative shear from the live load does not exceed the positive shear produced by the dead load and, consequently, no counterbraces are required.

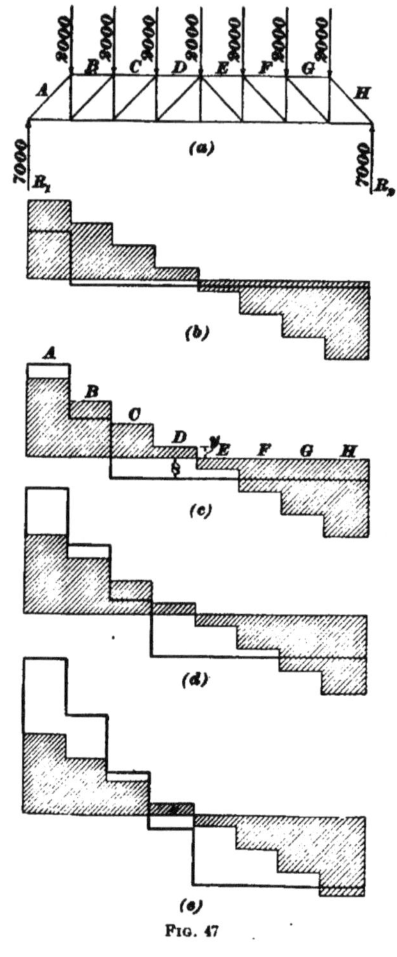

Fig. 47

The shear shown by the heavy line in the diagram (*c*) is produced by placing live loads on two panel points of the truss, at *AB* and *BC* in (*a*). As will be observed, the negative shear in the panel *D*, created by the live load and represented by the heavy line, exceeds the positive shear produced by the dead load, that is, x is greater than y; therefore, it will be necessary to counterbrace this panel. Also, the diagrams (*d*) and (*e*) show that counterbracing is required in the panel *D*, but not in any other, though in practice the panel *C* would be braced as an additional precaution. Then, theoretically, the only panels to be braced are those at *D* and *E*.

THE PRATT TRUSS

52. The **Pratt truss** has its vertical members in compression and the inclined members in tension, while the top chord is in compression and the lower one in tension. This fact may be readily determined by considering a portion of the truss as being separated from the remainder by a line, as at de, Fig. 48, cutting three members of the truss. The kind of stress in each of these members and its amount may then be found by taking moments about different points, as x, y, and z. To ascertain the stress in the member c, moments will be taken about the point x as a center. The positive moment is $7,000 \times 24 = 168,000$ foot-pounds, while the negative moment is $2,000 \times 12 = 24,000$ foot-pounds.

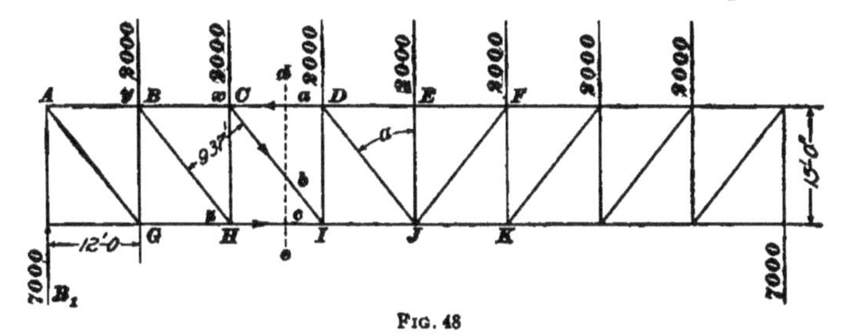

FIG. 48

The algebraic sum is $168,000 - 24,000 = 144,000$ foot-pounds, which is the moment of the stress in c. The lever arm of this force is 15 feet; hence, the stress is equal to $144,000 \div 15 = 9,600$ pounds. The surplus of the moment $R_1 \times 24$ is therefore equal to 144,000 foot-pounds and is positive. To create equilibrium, the moment of c should equal this amount, but be of opposite sign, therefore negative, and must act in the direction shown by the arrow, or away from the joint, which indicates that it is a tensile stress.

The stress in b is determined by considering the moments about the point y as a center. The positive moments are:

$$7,000 \times 12 = 84000 \text{ foot-pounds}$$
$$2,000 \times 12 = 24000 \text{ foot-pounds}$$
$$\text{Total, } 108000 \text{ foot-pounds}$$

The negative moment of *c* is 9,600 × 15 = − 144,000 foot-pounds; hence, the excess of this moment over the positive one is 108,000 − 144,000 = − 36,000 foot-pounds. To produce equilibrium, the moment of *b* should be of opposite sign, that is, positive, and as its lever arm is 9.37 feet, its stress is equal to 36,000 ÷ 9.37 = 3,842 pounds. The arrow points away from the joint, and the stress is therefore tensile.

The stress in *a* may be obtained by considering *z* as the center of moments. The positive moments are:

7,000 × 24	=	168000 foot-pounds
3,842 × 9.37	=	36000 foot-pounds
Total,		204000 foot-pounds

The negative moment is 2,000 × 12 = 24,000 foot-pounds. As the positive moments exceed the negative by 204,000 − 24,000 = 180,000 foot-pounds, the moment of *a* must equal this amount in order to establish equilibrium and it must be of opposite sign, or negative. As its lever arm is 15 feet the stress is equal to 180,000 ÷ 15 = 12,000 pounds. In order to produce a negative moment, this force must act in the direction of the arrow shown at *a*, or toward the joint, and consequently the stress is compressive. The stress in any member may be obtained by this method.

53. The method of determining the stresses by a notation, as explained in connection with the Howe truss, may also be applied to this form of truss. The stress in *EJ* is compressive and equal to the amount of the load at *E*, or 2,000 pounds. One-half of this stress must be taken care of by each of the rods, *DJ* and *FJ*; thus, 1,000 pounds is transmitted to the point *D* by *DJ*, so that the compression in *DI* is 2,000 + 1,000 = 3,000 pounds. This amount is carried up the oblique member *IC*, and consequently the stress in *CH* is 2,000 + 3,000 = 5,000 pounds. Hence, the stress in the vertical web members may be tabulated as follows:

Stress in EJ = 2,000 pounds
Stress in DI = 3,000 pounds
Stress in CH = 5,000 pounds
Stress in BG = 7,000 pounds

The stress in DJ is equal to 1,000 multiplied by the secant of the angle DJE, or a; in CI it is equal to 3,000 multiplied by sec a, etc. Hence, the following notation may be made:

Stress in $DJ = 1{,}000 \times \sec a = 1{,}000 \times 1.28075 = 1{,}281$ pounds.

Stress in $CI = 3{,}000 \times \sec a = 3{,}000 \times 1.28075 = 3{,}842$ pounds.

Stress in $BH = 5{,}000 \times \sec a = 5{,}000 \times 1.28075 = 6{,}404$ pounds.

Stress in $AG = 7{,}000 \times \sec a = 7{,}000 \times 1.28075 = 8{,}965$ pounds.

The stress in the chord members may be obtained as explained in connection with the Howe truss. For instance, the stress in AB is equal to the horizontal component of that in AG and is a compressive stress, while the stress in GH is equal to the same amount, but is tensile. In this problem, the horizontal component of AG is equal to $7{,}000 \times \tan a = 7{,}000 \times .800196 = 5{,}601$ pounds. The stress in the members BC and HI is equal to $(7{,}000 + 5{,}000) \times \tan a = 12{,}000 \times .800196 = 9{,}602$ pounds. Then the chord stresses may be written as follows:

$AB = GH = 7{,}000 \times \tan a = 7{,}000 \times .800196 = 5{,}601$ pounds.

$BC = HI = (7{,}000 + 5{,}000) \times \tan a = 12{,}000 \times .800196 = 9{,}602$ pounds.

$CD = IJ = (7{,}000 + 5{,}000 + 3{,}000) \times \tan a = 15{,}000 \times .800196 = 12{,}003$ pounds.

$DE = (7{,}000 + 5{,}000 + 3{,}000 + 1{,}000) \times \tan a = 16{,}000 \times .800196 = 12{,}803$ pounds.

THE WARREN TRUSS

54. The **Warren truss** is usually built of iron or steel, and while it is not much used in this country, in England it is frequently employed for comparatively short spans. The stresses in the members may be obtained analytically, as explained in connection with the Howe and Pratt trusses. The chord stresses may also be determined by the *method of chord increments*, which is explained as follows: Referring

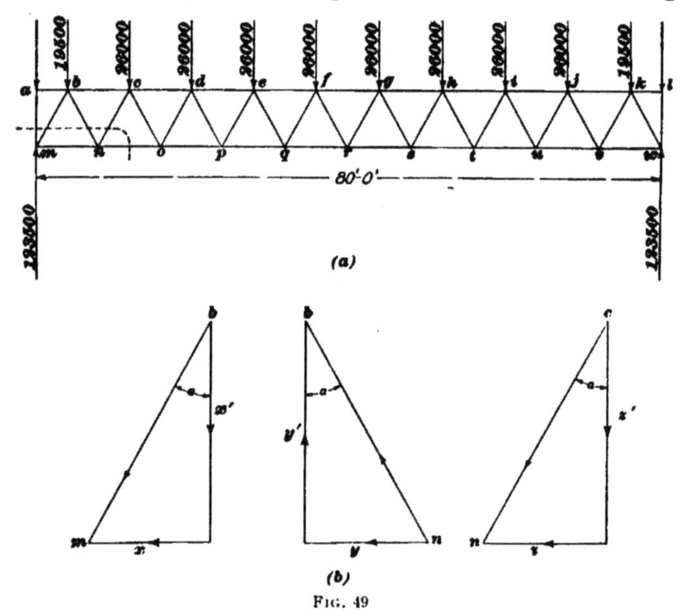

FIG. 49

to Fig. 49 (a), assume that it is desired to find the stress in the chord member $n\,o$. Draw a curved section, as shown by the dotted line, cutting this member and all the web members to the left. The vertical shears in the web members may be designated by s_1, s_2, s_3, etc. Then in order to have the truss in equilibrium, the sum of the horizontal components must equal zero. The action of the stresses in the web members $b\,m$, $b\,n$, and $c\,n$ may be observed from Fig. 49 (b), which shows the horizontal components of these stresses. The members bm and cn are in compression; hence, their vertical components x' and z' act downwards and their horizontal components x and z toward the left, as shown by the arrows. The member bn is in tension and its vertical component y' is an upward force, while its horizontal component y acts toward the left. Therefore, the sum of the horizontal components, since all act in the same direction, is $x + y + z$, and the stress in no must be equal to this sum and act in the opposite direction. Then, as the horizontal component of any web member is equal to the vertical shear in that member multiplied by the tangent of the angle which that member makes with the vertical, the following rule may be stated:

Rule.—*The stress in any chord member is equal to the sum of the products of the vertical shear in each web member between the chord in question and the nearer reaction, multiplied by the tangent of the angle that the web member makes with the vertical.*

Hence, the stress in no is equal to $s_1 \tan a_1 + s_2 \tan a_2 + s_3 \tan a_3$, but as a_1, a_2, and a_3 are all equal, the equation becomes, stress in $no = \tan a(s_1 + s_2 + s_3)$.

55. The stress in the web member is equal to the vertical component of its stress multiplied by the secant of the angle that the web makes with the vertical, or, referring to Fig. 49 (*b*) $x' \times \sec a$; but as the vertical component of the oblique force represents the shear in that member, its stress is equal to the shear multiplied by the secant of its angle, or $s \times \sec a$. Hence, the following rule may be given:

Rule.—*The stress in any web member is equal to the shear in that member multiplied by the secant of its angle with the vertical.*

Assume a Warren truss to be of the dimensions given in Fig. 49; the sum of the live and dead loads is taken at 26,000 pounds per panel. It will be noticed that at the points b and k there is a load of only 19,500 pounds. This may be explained by supposing that the extent of a panel is that part of the truss included between two adjoining apexes and that between each of these there is a distributed load of 26,000 pounds. Between the points a and b there is only one-half panel with a load of 13,000 pounds, one-half of which is supported directly at the point m through the member $a\,m$. The other half, or 6,500 pounds, is supported at the point b, which point also supports one-half of the total panel load $b\,c$, or 13,000 pounds. Adding these loads, 13,000 + 6,500, gives a total load of 19,500 pounds, as indicated in the diagram. The small load at a is disregarded and, in fact, the members $a\,m$ and $a\,b$ are not considered in determining the stresses in the members of the truss. Applying the rule given in this article, the stress in the member $b\,m$ is equal to the shear in that member, which is 123,500 pounds, since no load is applied until the point b is reached, multiplied by the secant of α; as this angle is 30°, its secant is 1.1547. Hence,

Stress in $b\,m$ is 123,500 × 1.1547 = 142,605 pounds.

Shear in $b\,n$ is 123,500 − 19,500 = 104,000 pounds.

Stress in $b\,n$ is 104,000 × 1.1547 = 120,089 pounds.

The shear in $c\,n$ is the same as in $b\,n$ and consequently the stress is the same amount, but a different kind, the stress in $c\,n$ being compressive, while that in $b\,n$ is tensile. The stresses in the remaining web members may be obtained similarly. The chord stress, as previously stated, is equal to the sum of the shears in all the web members to the left, multiplied by the tangent of α. Then,

Stress in $m\,n = s_1 \times \tan\alpha$ = 123,500 × .57735 = 71,303 pounds.

Stress in $n\,o = (s_1 + s_2 + s_3) \times \tan\alpha$ = (123,500 + 104,000 + 104,000) × .57735 = 191,391 pounds.

Stress in $o\,p = (s_1 + s_2 + s_3 + s_4 + s_5) \times \tan\alpha$ = (123,500 + 104,000 + 104,000 + 78,000 + 78,000) × .57735 = 281,458 pounds.

Considering the upper chord, the stress in $b\,c = (s_1 + s_2) \times \tan\alpha$ = (123,500 + 104,000) × .57735 = 131,347 pounds.

Stress in $c\,d = (s_1 + s_2 + s_3 + s_4) \times \tan\alpha$ = 123,500 + 104,000 + 104,000 + 78,000) × .57735 = 236,425 pounds.

The stresses in all the chord members may be obtained similarly, and those in the lower chord are always in tension while those in the upper are in compression.

THE LATTICE TRUSS

56. The **lattice truss,** or **double Warren,** as it is sometimes called, is shown in Fig. 50. It consists of two systems of triangular bracing, one of which is represented in the figure by full lines and the other by dotted lines. The truss is equivalent to the two trusses welded into one and the loads at *c*, *e*, and *g*, in view (*a*), are considered as being carried to the abutments by the diagonals shown dotted, while the full-line diagonals transfer the loads at *b*, *d*, *f*, and *h* to the abutments. The chord stresses are found by the method of chord increments, as in the Warren truss. For instance, the stress in *a b* is equal to the vertical shear in the web *a k* multiplied by the tangent of the angle *j a k*, or α, which in this case is 45°, and consequently its tangent is 1. Taking the dead load at 350 pounds per foot, the panel loads are $350 \times 8 = 2{,}800$ pounds. The system of diagonals to which *a k* belongs receives three panel loads; then half of the center load will be carried to the left reaction down the diagonal *e m* and up *m c*, at which point another load is added, so that one and one-half panel loads are carried down to *k* and up to *a*. Hence, the vertical shear in *a k* is $1\frac{1}{2} \times 2{,}800 = 4{,}200$ pounds, and the stress in the chord *a b* is shear in $a b \times \tan \alpha = (1\frac{1}{2} \times 2{,}800) \times 1 = 4{,}200$ pounds.

The stress in the chord *b c* is equal to the sum of the shears in all the web members cut by the curved section *x x*, multiplied by tan α. The whole load at *d* supported by the full-line diagonals is transferred to the left abutment by passing down *d l* and up *l b*, where it encounters another panel load; hence, the vertical shear in *b j* is equivalent to two panel loads, while that in *b l* is equal to one panel load. As *a k* is equal to one and one-half panel loads, the stress in the chord *b c* equals $(1\frac{1}{2} + 2 + 1) \times 2{,}800 \times 1 = 12{,}600$ pounds. In the same manner, the stresses in all the chord members may be determined, those in the upper chord always being compressive and in the lower tensile.

The stress in any web member is equal to the vertical shear in that member multiplied by the secant of its angle with

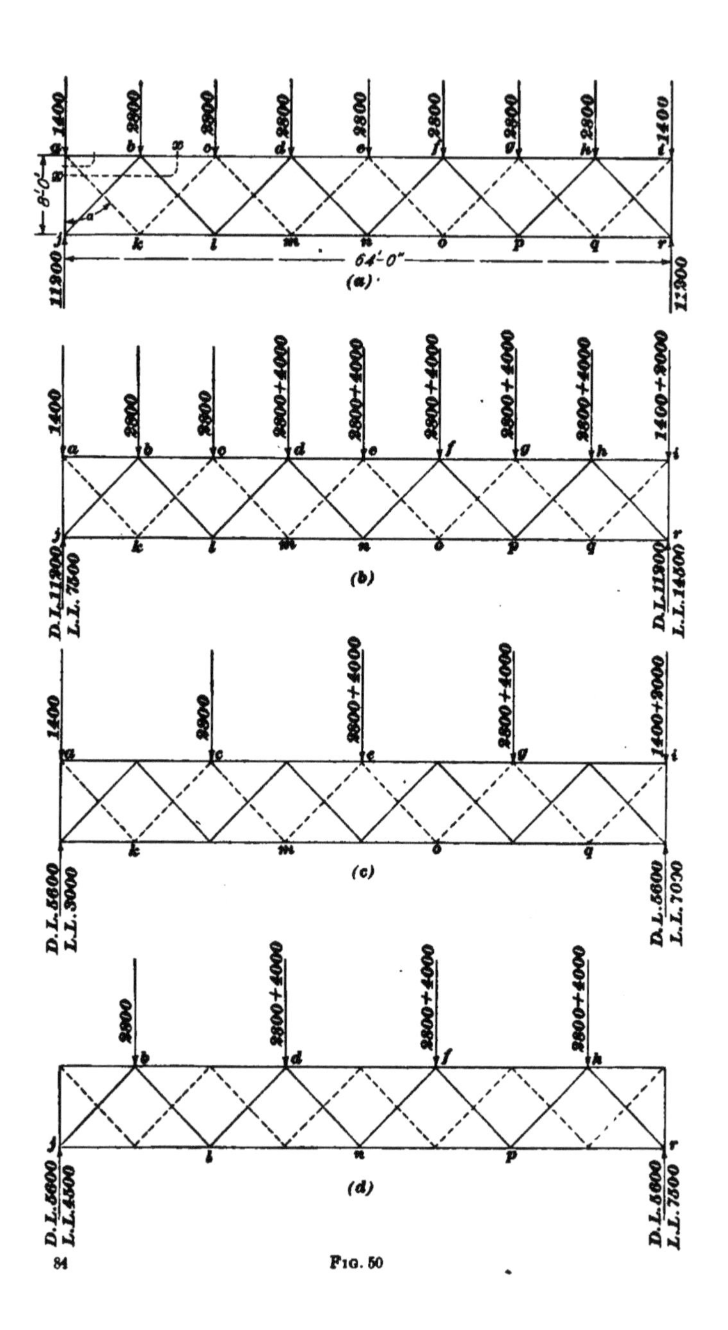
1400
2800
2800
2800
2800
2800
2800
2800
1400
a
b
c
d
e
f
g
h
i
x
8'-0"
a
j
k
l
m
n
o
p
q
r
64'-0"
11900
11900
(a)
1400
2800
2800
2800+4000
2800+4000
2800+4000
2800+4000
2800+4000
1400+9000
D.L.11900
L.L.7500
D.L.11900
L.L.14500
(b)
1400
2800
2800+4000
2800+4000
1400+9000
D.L.5600
L.L.3000
D.L.5600
L.L.7000
(c)
2800
2800+4000
2800+4000
2800+4000
D.L.5600
L.L.4500
D.L.5600
L.L.7500
(d)

Fig. 50

the vertical, which, in this truss, is 1.4142; hence, the stress in $a\,k$ produced by the dead load is $1\frac{1}{2} \times 2{,}800 \times 1.4142 = 5{,}940$ pounds. To find the maximum stresses in the webs, it is necessary to consider the truss as only partly covered by the live load; therefore, it will be assumed that the live load extends over the entire portion of the truss to the right of the panel c, as shown in Fig. 50 (b). As stated previously, the two systems of diagonals must be considered separately in determining the web stresses, the loads supported by the dotted diagonals and the resulting reactions being shown in Fig. 50 (c). The dead-load reaction is marked *D. L.*, while that produced by the live load is designated by *L. L.* The stress in $a\,k$, under the condition of loading shown in (c), is $(3{,}000 + 5{,}600 - 1{,}400) \times \sec a = 7{,}200 \times 1.4142 = 10{,}182$ pounds. The same stress exists in $k\,c$, while in $c\,m$ and $m\,e$ it is $(3{,}000 + 5{,}600 - 1{,}400 - 2{,}800) \times \sec a = 4{,}400 \times 1.4142 = 6{,}222$ pounds. In $e\,o$ and $o\,g$, the stress is $(3{,}000 + 5{,}600 - 1{,}400 - 2{,}800 - 2{,}800 - 4{,}000) \times \sec a = -2{,}400 \times 1.4142 = -3{,}394$ pounds. The stress in $g\,q$ and $q\,i$ is $(3{,}000 + 5{,}600 - 1{,}400 - 2{,}800 - 2{,}800 - 4{,}000 - 2{,}800 - 4{,}000) \times \sec a = -9{,}200 \times 1.4142 = -13{,}011$ pounds.

The $-$ sign in the last two results does not denote a tensile stress, but simply indicates that the shear in these members is negative, when the shear at the left-hand reaction is considered as positive; in other words, the shear changes sign at the point e, for the system of dotted diagonals.

Referring to Fig. 50 (d), the stresses in the full-line diagonals are determined as follows:

Stress in $j\,b = (5{,}600 + 4{,}500) \times \sec a = 10{,}100 \times 1.4142 = 14{,}283$ pounds.

Stress in $b\,l$ and $l\,d = (5{,}600 + 4{,}500 - 2{,}800) \times \sec a = 7{,}300 \times 1.4142 = 10{,}324$ pounds.

Stress in $d\,n$ and $n\,f = (5{,}600 + 4{,}500 - 2{,}800 - 2{,}800 - 4{,}000) \times \sec a = 500 \times 1.4142 = 707$ pounds.

Stress in $f\,p$ and $p\,h = (5{,}600 + 4{,}500 - 2{,}800 - 2{,}800 - 4{,}000 - 2{,}800 - 4{,}000) \times \sec a = -6{,}300 \times 1.4142 = -8{,}909$ pounds.

Stress in *hr* = (5,600 + 4,500 − 2,800 − 2,800 − 4,000 − 2,800 − 4,000 − 2,800 − 4,000) × sec *a* = − 13,100 × 1.4142 = − 18,526 pounds.

The same method may be pursued to determine the web stresses under the different conditions of loading, and the maximum stress in each member may thus be ascertained. The members *ak*, *cm*, *go*, *iq*, and *bl*, *dn*, *fn*, *hp* are in tension, while *ck*, *em*, *eo*, *gq*, and *bj*, *dl*, *fp*, *hr* are in compression.

57. In Fig. 51, the stress diagrams for the different conditions of loading are shown. The stresses may be scaled from these diagrams, computed by the method of sections, or determined analytically, as explained below.

When the dead load only is considered, the reactions produced by the loads on the full-line diagonals are each 5,600 pounds; then the stress in *hr*, Fig. 51 (*a*), is compressive and is equal to 5,600 × sec *a*. At the point *h*, a load of 2,800 pounds counteracts half of the force, thus leaving 5,600 − 2,800 = 2,800 pounds of vertical force that creates tension in the member *hp*. The same force is carried up the web *fp*, producing a compressive stress in that member, and at *f* it meets the load of 2,800 pounds, which is just equal to it and consequently entirely neutralizes it. Hence, under a uniform load, there is no stress in *fn* and *dn*; that is, the truss would be stable without these members. This condition is changed, however, when the live load of 4,000 pounds is applied at *HI*. The right-hand reaction is then equal to 9,100 pounds and the left-hand reaction to 6,100 pounds. The stress in *hr* is 9,100 × sec *a*; at *h*, the load of 4,000 + 2,800 = 6,800 pounds is encountered and a force of 9,100 − 6,800 = 2,300 pounds is left to pass down *hp* and up *pf*. The load of 2,800 pounds at *f* must be taken care of, but as *fp* can take only 2,300 pounds, the remaining 500 pounds must be resisted by *fn*; hence, the stress in *fn*, which is compressive, is equal to 500 × sec *a*. The same amount of stress is created in *dn*, but it is of opposite character, or tensile. At *d*, another load of 2,800 pounds is applied, so that *dl* must carry 2,800 + 500 = 3,300 pounds.

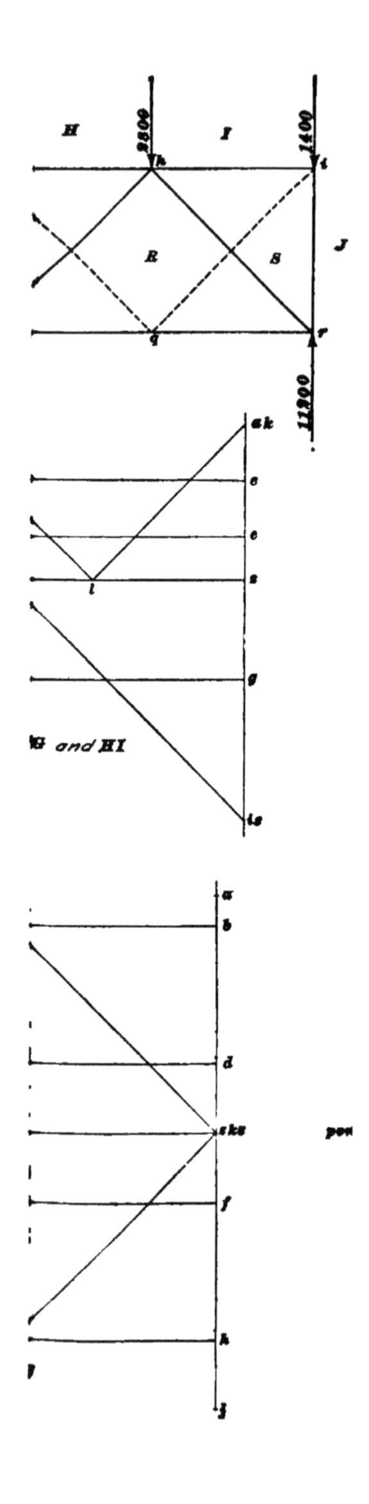

9500
1400
11200
G and HI

The stress in this member is, of course, compressive and the same amount of tension exists in bl. At b, the load of 2,800 pounds is added, thus producing a force of (3,300 + 2,800) = 6,100 pounds to be resisted by bj; therefore, the compression in this member is 6,100 × sec a. If the live load is applied at the points h and f, the right-hand reaction is 11,600 pounds and the left-hand, 7,600 pounds. Then the stress in hr is 11,600 × sec a, while in hp and fp the stress is (11,600 − 6,800) × sec a. At f, another load of 6,800 pounds is to be resisted and as fp can take care of 4,800 pounds, the remainder, or 2,000 pounds, must be resisted by fn. Hence, the compression in fn and the tension in dn are equal to 2,000 × sec a. In dl and bl, the stress is (2,000 + 2,800) × sec a, while the compression in bj is (2,000 + 2,800 + 2,800) × sec a. In this manner the web stresses may be determined for all conditions of loading. It will be observed that when the live load comes from the right, as considered in the calculations given above, the stress in fn is compressive, while in dn it is tensile, but when the live load is applied at the left-hand end of the truss, the stresses in these members will be reversed. Therefore, it will be necessary to make dn and fn capable of resisting both tension and compression. In the dotted diagonals, the kind of stress does not change under the different conditions of loading.

LATERAL BRACING

58. When a trussed girder is used to support a tramway between two buildings or in any similar position, it is necessary to take into consideration the wind pressure on the girder, and to provide a system of **lateral bracing** to resist this pressure. When the girder is not enclosed, as in a bridge, the wind acts on both sides of it, that is, on both trusses forming the girder; but the area to be regarded as exposed to the wind is a question that the designer must decide. If the girder is enclosed, of course the surface exposed to the wind is equal to the entire area of one covered side.

In the following example the stresses in the members of the trusses may be determined as previously explained; therefore, only the bracing required to resist the wind pressure will be considered.

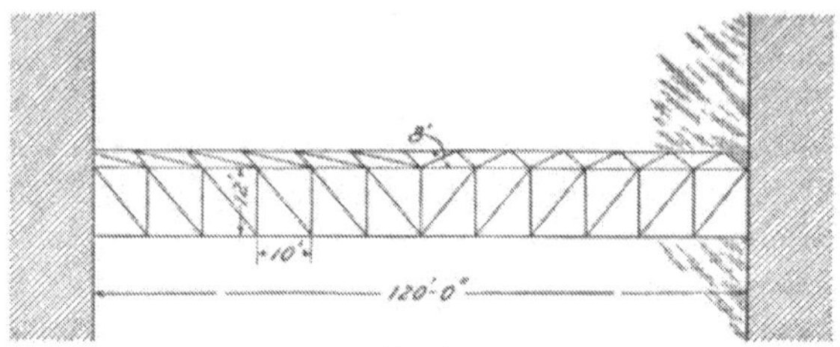

Fig. 52

Example.—A covered runway, or bridge, connecting two wings of a large factory at the fifth story has a span of 120 feet and a depth of 12 feet. The trusses are composed of twelve panels and the distance between trusses is 8 feet. The floor of the bridge, which is supported on the lower chords of the trusses, is composed of two layers of 2½-inch tongued-and-grooved spruce plank laid diagonally, one layer crossing the other, and a 1-inch finished floor; this construction furnishes sufficient bracing against the wind at the lower chords, but it is necessary to introduce a system of diagonal bracing between the upper chords, as shown in Fig. 52. Considering the wind pressure as 30 pounds per square foot, what will be the stresses created in the system of lateral bracing?

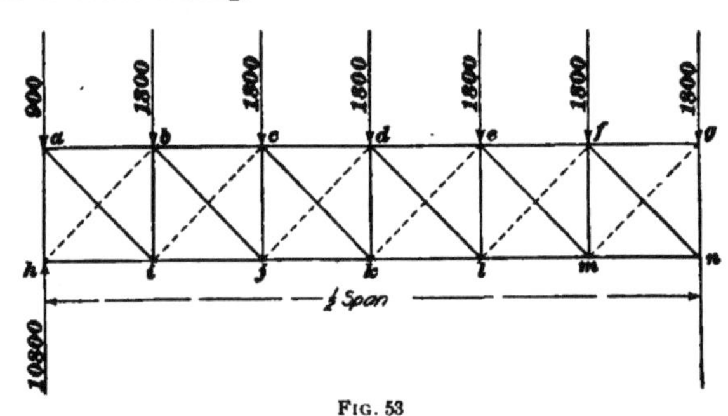

Fig. 53

Solution.—As the ends of the trusses are to be built into the walls, the horizontal reactions due to the wind load are transmitted to the masonry instead of being resisted by the transverse strength of the end struts. (The transverse strength of columns or struts against lateral forces is treated in *Wind Bracing*, and consequently it is not necessary to consider it here.) The area of one panel is 12 × 10 = 120 sq. ft. and therefore the wind load per panel is 120 × 30 = 3,600 lb., but as one-half of this amount is to be resisted by the floor, the panel loads acting on the system of bracing between the upper chords are 1,800 lb. each. A plan of the top of the girder showing the wind loads applied is given in Fig. 53. The members *a h*, *b i*, *c j*, etc. are heavy wooden beams and hence should always be in compression. The tension rods designated by the heavy lines are in use when the load is applied as shown, but when the wind comes from the other side the dotted diagonals provide the required tension and the heavy diagonals are not in use. The stress in *g n* is 1,800 lb. and half of this force reacts on *f n*, while the other half goes to the tension member on the right of *g n*. Then the tensile stress in *f n* is 900 × sec *f n g* = 900 × 1.601 = 1,441 lb. The stress in *f m* is 900 + 1,800 = 2,700 lb., compression, and the tension in *e m* is 2,700 × 1.601 = 4,323 lb. The remaining stresses may be analyzed and the following tabulation made:

g n = 1,800 +	*c k* = 6,300 × 1.601 = 10,086−
f n = 900 × 1.601 = 1,441−	*c j* = 6,300 + 1,800 = 8,100+
f m = 1,800 + 900 = 2,700+	*b j* = 8,100 × 1.601 = 12,968−
e m = 2,700 × 1.601 = 4,323−	*b i* = 8,100 + 1,800 = 9,900+
e l = 2,700 + 1,800 = 4,500+	*a i* = 9,900 × 1.601 = 15,850−
d l = 4,500 × 1.601 = 7,205−	*a h* = 9,900 + 900 = 10,800+
d k = 4,500 + 1,800 = 6,300+	Ans.

DETAILS OF TRUSS CONSTRUCTION

WOODEN TRUSSED GIRDERS

59. It is frequently desirable to support roofs on trussed girders; that is, trusses with horizontal top and bottom chords. A truss of this kind constructed of timber is illustrated in Fig. 54 by a plan (*a*), an elevation (*b*), and with details (*c*), (*d*), (*e*), and (*f*). The elevation shows that the timber trusses *A* and *B* have a theoretical span of 47 feet 11 inches and the plan shows that they are placed 16 feet 11½ inches between centers. The elevation, Fig. 54 (*b*), shows the general outline of the truss. Its structural elements consist of a rigid central panel counterbraced with 1-inch diagonal rods that insure this portion of the frame against any distortion produced by unsymmetrical loads, and the two triangles

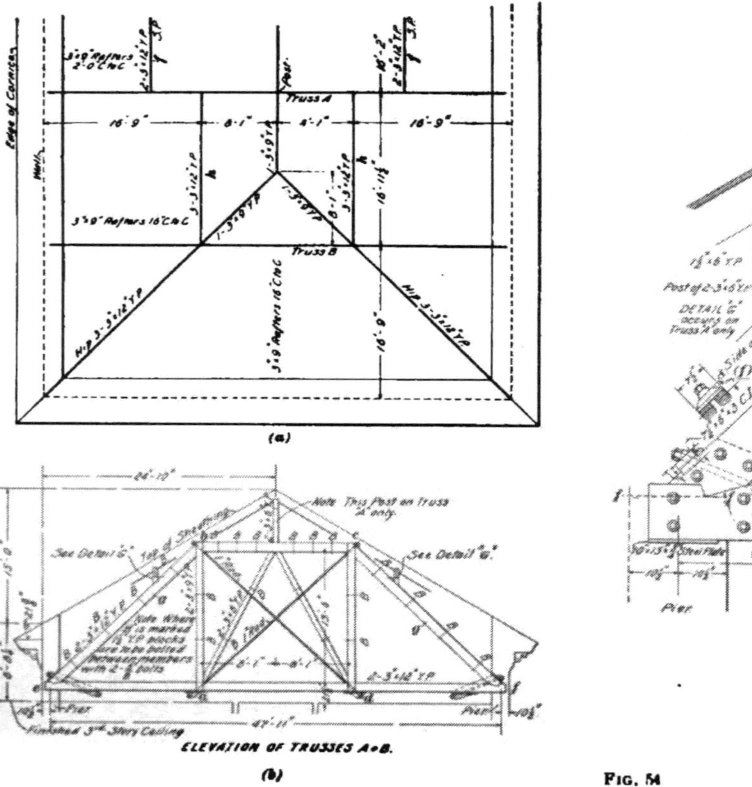

Edge of Cornice
Wall
Truss A
Truss B
Post
16'-9"
8'-1"
4'-1"
(a)
24'-10"
Note This Post on Truss "A" only.
See Detail "G"
See Detail "G"
2-3"x12" Y.P.
47'-11"
Pier
Finished 3rd Story Ceiling
ELEVATION OF TRUSSES A+B.
(b)
Note Other Half of Bracing the same Reversed
Purlin of 2-3"x12"
Post of 2-3"x6" Y.P.
DETAIL "G" occurs on Truss "A" only
(d)
Dotted Lines show 1½" Y.P. Cleats between 3"x8" and 3"x6"
6" Spikes
Note: All Bolts must have Washers
Center Line of Truss
C.I. Washers
Pier
(c)
PLAN
(e)
PLAN
DETAILS OF TRUSSES A+B.

Fig. 54

e b a and *d c f* form a rigid frame incapable of distortion. The chord members and end posts of the truss *g*, *g*, Fig. 54 (*b*), are each composed of two 3″ × 12″ pieces of yellow pine, placed side by side and separated by yellow-pine blocks 1½ inches thick. The position of these blocks is shown in the elevation (*b*) at *B*, *B*, *B*.

Referring to the details shown in Fig. 54 (*c*), it will be seen that the diagonal counterbracing, consisting of 1-inch rods, extends through the space left between the timbers making up the upper and lower chords and that they are secured at the ends by especially designed check-washers *a*, *a*. The strut members *b*, *b* in the middle panel of the truss are each made up of two 3″ × 6″ yellow-pine timbers. These are cut to the shape shown in the detail at the ends and are secured in place by being spiked to cleats *c*, *c* 1½ inches thick that fit between the middle space of the connecting members and form what would be termed in steel construction, a gusset plate. The vertical members *d*, as designated in the drawing of the details, are composed of two 3″ × 9″ yellow-pine pieces fitted tightly between the upper and lower chord members and further held in place and reenforced by two 1″ × 9″ yellow-pine pieces nailed on each side of the heavier pieces of the member. These side pieces extend over the top and bottom chords and are bolted to them with ¾-inch bolts. The required strength in the heel connection of the truss is obtained by the introduction of the 1½-inch through bolt *e* and by 3″ × 9″ and 3″ × 6″ cleats e_1 and e_2, one being placed above the bolt and the other below it, both being inserted between the pieces composing the end post and the lower chord of the truss. The presence of the lower cleat e_2 is not quite evident in the drawing, because its outline coincides with that of the lower chord and end post. The connection is secured to the cleats by through bolts, and the 1½-inch bolt forming the main resistance of the joint is provided with special cast-iron washers, details of which are shown at (*e*) and (*f*). Additional resistance is obtained at this important joint by notching the end post into the lower chord member and thus obtaining the shearing resistance of the timber along the line *ff*. From the plan, Fig. 54 (*a*), it is noticed that purlins *h*, *h* composed of 3″ × 12″ pieces of yellow pine extend between the trusses *A* and *B*. These purlins are supported from the truss by wrought-iron stirrup irons *g*, as shown in (*c*), and carry the bulk of the weight of the roof between the trusses *A* and *B*. The other purlins, marked *f* in the plan, are supported on the end posts of the truss in the manner shown in (*d*), which is so clear as to need no explanation.

60. Another type of timber truss, which is frequently used in mill construction or temporary work, is that shown

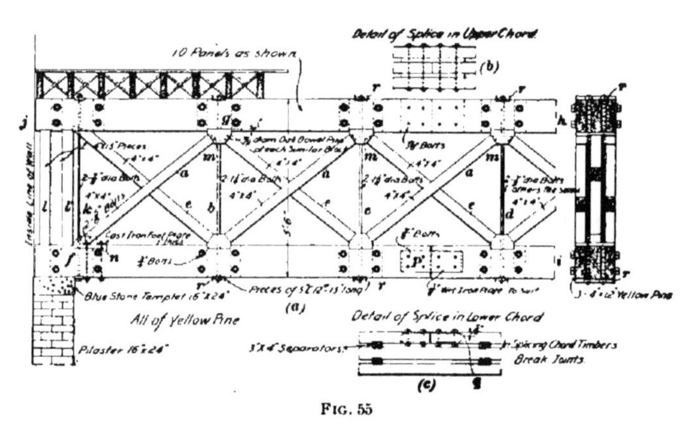

FIG. 55

in Fig. 55. This trussed girder was designed for a span of 45 feet and a total live and dead load of 50,000 pounds, uniformly distributed. It is constructed of Georgia yellow pine and is 5 feet 6 inches in depth. The members and details of construction are designed for a factor of safety of at least 4. Both chords of the trusses are composed of three pieces of timber placed side by side and separated with 3″ × 4″ separators composed of the same material as the truss. These separators are gained into the chord timbers about ¼ inch. While this is apt to weaken the chord in tension somewhat, the rigidity gained by the insertion of the separators in this manner, laterally and vertically, more than compensates for the loss in strength. From the figure, it will be noticed that the web compression members are all composed of 4″ × 4″ pieces. Those marked *a*, *a*, *a* are doubled, for with the rods *b*, *c*, and *d*, they form the principal members of the web, while the members *e*, *e*, *e* are composed of single pieces of timber. These pieces would not be necessary if it were certain that the load would always be uniformly distributed. The possibilities are, however, that with a girder of this span and supporting the great floor area that it necessarily must, the floor load will frequently be concentrated on portions of the girder, and consequently the counterbraces are necessary to insure the truss against distortion. The structural outline of this truss extends along the lines *f g h i* and the upper chord is extended from *g* to *j* only in order to support the portion of the floor adjacent to the wall. It is evident from this that the frame diagram of the truss is as shown in Fig. 56, or in the form of a Howe truss. Consequently, the rods *k* are of little use except to rigidly tie the extended portion of the upper chord to the lower one at the abutments, while the struts *l*, *l* are introduced to take up the reaction at the end *j* of the extended portion of the upper chord. The timber strut members of the web are held in place on bearing blocks by means of oak dowels, as shown at *m*, special care being taken to secure a rigid and adequate connection at *n* by the introduction of a cast-iron foot-plate.

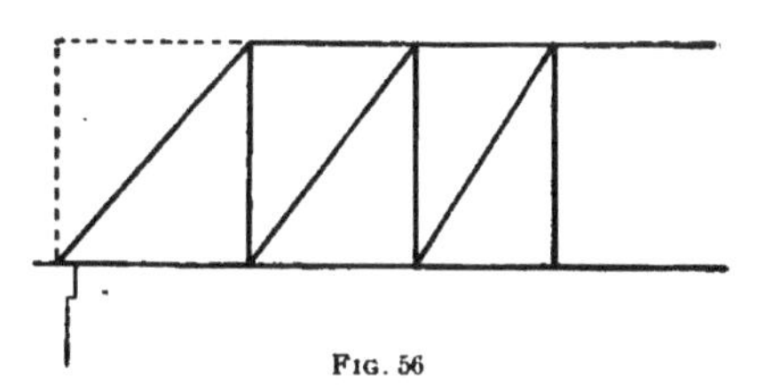

Fig. 56

61. Should it be necessary to splice the timbers making up both the top and bottom chord, the splices should be kept as near as possible to the abutments, for, in the chords, the stresses decrease toward the supports. For the splice in the upper, or compression, chord it is only necessary to observe that the spliced timbers are cut true and square and abut, their alinement being secured by pieces of timber bolted on each side. Such a detail as described is shown in view (*b*). The splice for the chord in tension is shown at *p* in view (*a*). It is necessary to use wrought-iron splice plates having a lug *q*, Fig. 55 (*c*), on each end. All such splice connections should be analyzed for shear parallel with the grain and for bearing on the end wood, and, if possible, a strength greater by 25 per cent. than the strength of the net section of the member spliced should be realized. The washers *r*, *r* are subjected to considerable bending stress and are therefore made of pieces of 5-inch channel iron equal in length to the width of the chord members. When channels are thus used for washers it is generally necessary to provide a socket wrench to tighten up the bolts.

62. In Fig. 57 is shown a trussed girder having a span of over 50 feet. At the ends, it is built in a manner similar to a plate girder with a web *a* extending between the flange angles. Toward the center, however, for economy and also because it was desired to provide a construction that would allow the diffusion of light from a skylight, the girder was made open, as shown. The girders are stiffened laterally at the lower flange by I beams extending to the walls and at the top flange by a brace, secured at *c*, extending to a horizontal steel beam member resting on the walls. At the center of the span, the girder is further reenforced laterally by the beams for the support of the diffusing sash shown at *d*. There are no peculiarities of the design requiring detailed explanation, as the drawing is sufficiently clear with reference to the constructive details.

63. In Fig. 58 is shown a detail design of a trussed through bridge, such as would be used to connect two wings or two buildings of a manufacturing plant. It must be designed to carry, besides its own weight, the truck loads of the heaviest merchandise that would be likely to be sent across it. The bridge has a span of 120 feet, is 10 feet from center to center of trusses across, and is 15 feet 6 inches

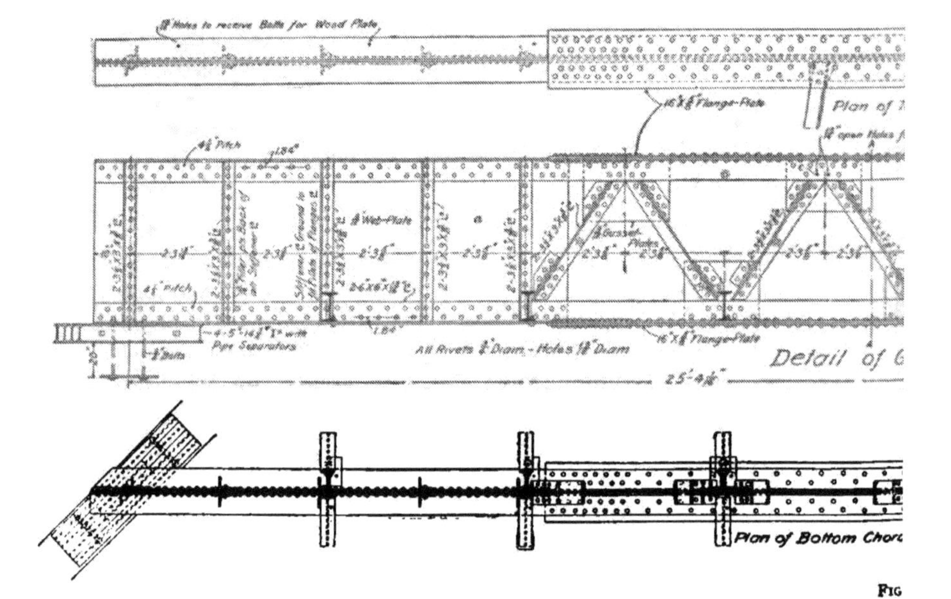
Plan of Bottom Chord
Detail of G
All Rivets ⅞" Diam. - Holes 1⅛" Diam
16"x⅝" Flange-Plate
25'-4⅛"

FIG

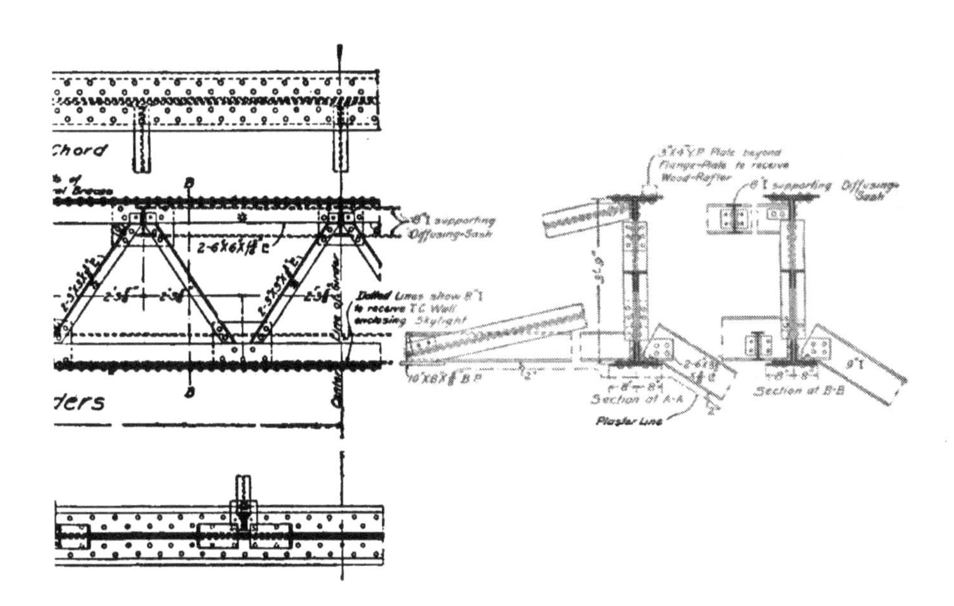
Chord
ders
8" I supporting Diffusing-Sash
Dotted Lines show 8" I to receive T.C. Wall enclosing Skylight
Wood-Rafter
Section at A-A
Plaster Line
Section at B-B
9" I

§ 14

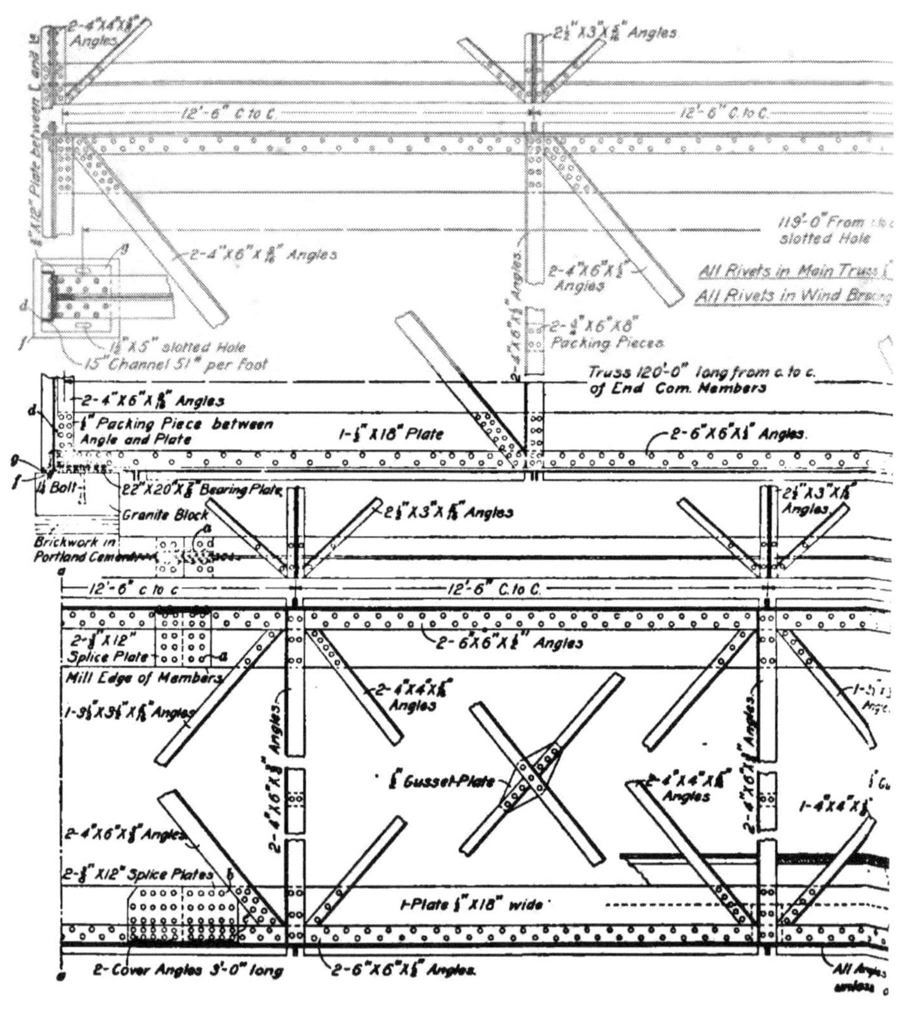

2-4"X4"X3/8" Angles
2½"X3"X5/16" Angles
12'-6" C to C
12'-6" C. to C.
2-4"X6"X9/16" Angles
2-4"X6"X½" Angles
2-4"X6"X½" Angles
2-½"X6"X8" Packing Pieces
119'-0" From
slotted Hole
All Rivets in Main Truss
All Rivets in Wind Bracing
1½"X5" slotted Hole
15" Channel 51# per foot
Truss 120'-0" long from c. to c. of End Com. Members
2-4"X6"X9/16" Angles
½" Packing Piece between Angle and Plate
1-½"X18" Plate
2-6"X6"X½" Angles.
1¼ Bolt
22"X20"X⅞" Bearing Plate
Granite Block
Brickwork in Portland Cement
2½"X3"X5/16" Angles
2½"X3"X5/16" Angles
12'-6" c to c
12'-6" C. to C.
2-½"X12" Splice Plate
2-6"X6"X½" Angles
Mill Edge of Members
1-3½"X3½"X5/16" Angles
2-4"X4"X5/16" Angles
2-4"X6"X⅜" Angles
½" Gusset-Plate
2-4"X4"X5/16" Angles
2-4"X6"X⅜" Angles
1-4"X4"X5/16"
2-4"X6"X½" Angles
2-½"X12" Splice Plates
1-Plate ½"X18" wide
2-Cover Angles 3'-0" long
2-6"X6"X½" Angles.

Fig. 5

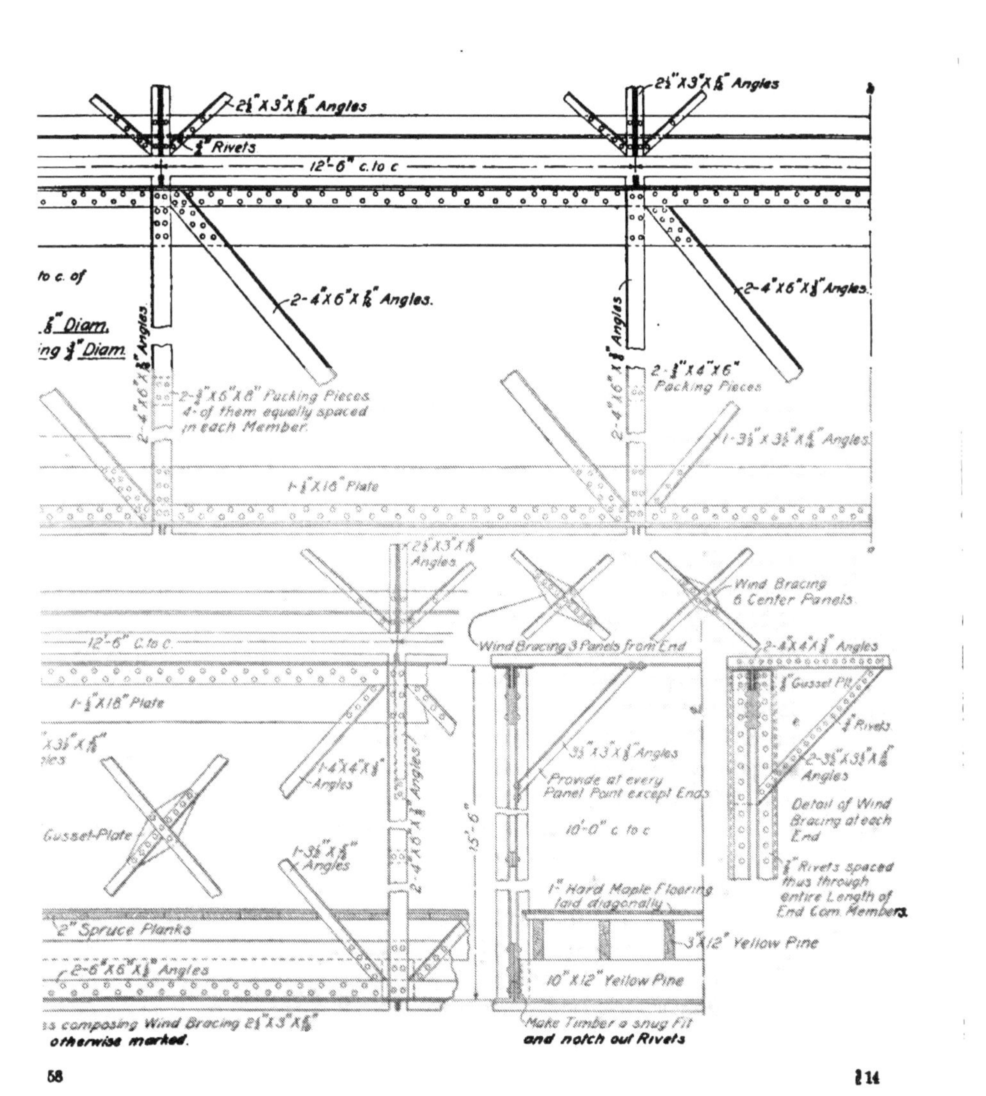

2½" X 3" X 5/16" Angles
¾" Rivets
12'-6" c. to c
2-4" X 6" X 7/16" Angles.
2-4" X 6" X 3/8" Angles
2-⅜" X 4" X 6" Packing Pieces
1-3½" X 3½" X 5/16" Angles
1-½" X 18" Plate
12'-6" C. to C.
Wind Bracing 3 Panels from End
Wind Bracing 6 Center Panels
2-4" X 4" X ⅜" Angles
⅜" Gusset Plt.
3½" X 3" X ⅜" Angles
Provide at every Panel Point except Ends
10'-0" c. to c
Gusset-Plate
15'-6"
1" Hard Maple Flooring laid diagonally
2" Spruce Planks
3" X 12" Yellow Pine
10" X 12" Yellow Pine
2-6" X 6" X ½" Angles
Detail of Wind Bracing at each End
Make Timber a snug Fit and notch out Rivets
otherwise marked.

from out to out of chord members in height or depth. The two main views are elevations supposed to join along the lines *a a*. The chord members of the truss are composed of two 6″ × 6″ × ½″ angles riveted to a ½″ × 18″ web-plate. By the adoption of a built-up **T** section of this kind the necessity of gusset plates is obviated, as the web members of the truss can be riveted through the vertical parts of the chord members. Owing to the fact that it is impossible to obtain such angles and plates as are shown in the drawing longer than 50 or 60 feet, the upper and lower chord members must be spliced as shown at *a*, *b*.

In the upper splice, it is only necessary to provide sufficient splice plates and rivets to hold the ends of the upper chord member in alinement, but in the design of the splice *b* for the lower chord, long splice plates must be used and sufficient rivets to realize, by their shearing resistance, the strength of the net section of the chord member. To stiffen the lower chord laterally at the splice, it is advisable to use cover angles that fit into the chord angles, as at *c*. Where possible, it would be well to splice the angles of the lower chord at a different place from the splice of the web plate. The central panels are counterbraced so that there will be no distortion of the frame from a moving load. The end posts of the truss, as at *d*, should be strongly constructed so as to withstand the bending stress produced by the gusset-plate brace *e*, shown in the view placed in the lower right-hand corner. This gusset-plate brace transmits considerable stress from the upper lateral wind bracing, and the only way to resist the stress is to provide a sufficient section modulus in the end members so as to supply the necessary resisting moment. While the floor system will offer sufficient lateral resistance at the bottom of the truss, it is good practice to provide lateral bracing at the botton as well as at the top. As the truss is of steel, and of considerable span, it is necessary to provide for the expansion and contraction of the frame. For this purpose the steel bearing plate *f* is supplied and the anchor bolts are passed through slotted holes in the heel plate *g*.

COLUMNS AND STRUTS

(PART 1)

INTRODUCTION

1. With the exception of the foundations, the **columns** are the most important element in the constructive design of a building. To them should be devoted the most careful study of the architect and engineer, for the failure of one of them means the falling of a portion of the building, or its entire demolition. They may be classified according to the material of which they are constructed, and according to the shape or form of their cross-section.

Under the first classification are included wooden, cast-iron, and structural steel columns and monoliths of stone subjected to compression. In making the second classification, the first must be considered. Wooden columns are either square, rectangular, or round in section. Cast-iron columns are usually hollow, and are square, rectangular, or round in section, though frequently necessity requires the use of a variety of sections. Steel columns are made in all forms that can be built up of the usual rolled shapes, and in some instances special shapes are made in order to secure a column of desirable section.

It is always a matter of some nicety for the engineer to make a selection of the column that will be best adapted to the work under consideration, particularly where the material to be used is structural steel. In selecting wooden and cast-iron columns, the choice is limited and is generally decided by the matters of appearance, availability of material, and the practicability with which the girders and columns can be made.

2. In selecting steel columns that will best fulfil the conditions imposed by the structure, the following considerations should be carefully studied:

1. Economy in both material and cost of construction. The first is attained when the metal in the column section is distributed at a maximum distance from the central axis of the column, which will also fulfil the condition that the radii of gyration shall be the same, or nearly so, on all the axes of the column section. Economy in cost of construction is principally accomplished by so designing the column section that it can be readily built up of stock-rolled shapes, arranged in such a manner that the column can be readily fitted up and the riveting done with facility, by power. The availability of the material should be embodied in this consideration; that is, patented sections controlled by a distant manufacturer should be avoided unless prompt shipment is assured, as the delay occasioned by failure to ship, and the impossibility of getting the material elsewhere, would cause pecuniary loss.

2. That column section should be considered the best constructively in which it is arranged that the loads may be transmitted directly to the center of the column. Thus eccentric loading and its attending dangers are avoided.

3. The column section should be such that all beam and girder connections thereto can be made conveniently without the use of swaged or bent plates of peculiar shape, and that all fitting in the field is avoided, in order that the work can be rapidly and securely assembled.

4. Where, as in buildings of many stories, the columns extend from the basement to the roof in one continuous length, and for reasons of economy it is necessary to reduce the column section at every several stories, that section should be used that will allow the reduction of the sectional area of the rolled-steel shapes employed, without materially reducing the general dimensions of the column section. Should the general dimensions of the column be materially reduced at the several floors, it is evident that economy will require a column of such small dimensions in the upper tiers

that the difficulty of making good constructive details at the beam and girder connections will be manifest.

5. The column section should be of such form as to be easily accessible for inspection and painting; though when completely fireproofed this consideration is of less importance, as the covering will tend to prevent the accumulation of moisture on the columns and lessen the liability of deterioration.

COLUMNS IN GENERAL

STRESS IN COLUMNS

3. The stresses in a tension member always tend to straighten it, but in a column a very slight inequality at the ends, or a weakness on one side, will produce a bending moment and cause the column to bend. This is shown exaggerated in Fig. 1. The stress in the plane *A B* may be considered as the algebraic sum of the diagrams shown in Fig. 2, in which the line *c d* may be considered as the zero line, its length representing the total thickness of the column, based on any

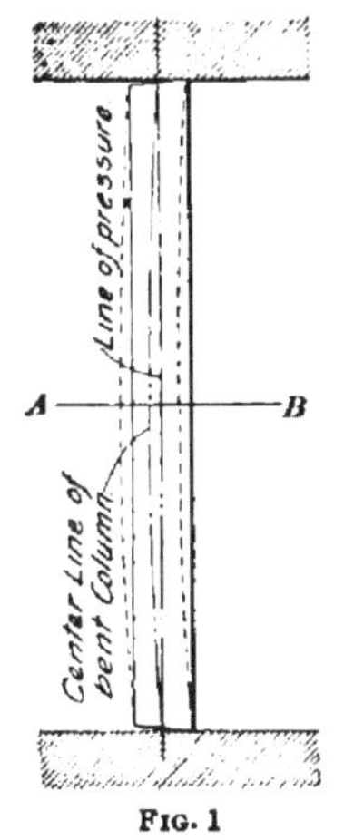

FIG. 1

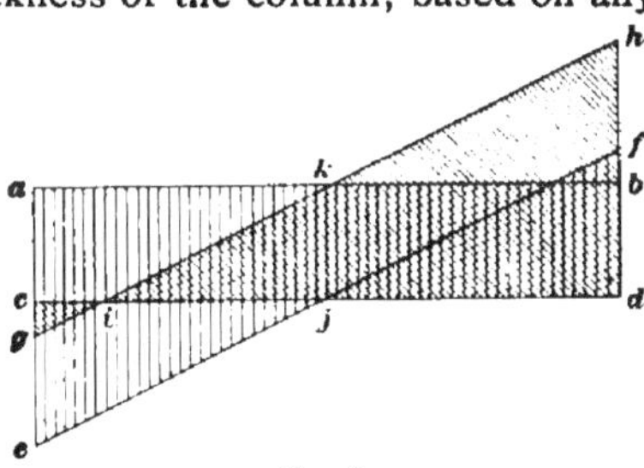

FIG. 2

assumed unit length, while the ordinates *a c*, *k j*, *b d*, or any of the intermediate verticals, represent the value of the longitudinal pressure in the column, in pounds per square inch, drawn to any convenient scale. Ordinates above *c d* represent positive stress, that is, compression; those below, negative or tension. When the column remains straight, the pressure is uniform throughout the section, as seen by the uniform length of the ordinates in the rectangle *abcd*. As soon as the pressure on the column becomes one-sided, the length and signs of these ordinates are changed, some changing from positive to negative, as in the triangle *cej*, while those in the triangle *jfd* remain positive, increasing from zero, at *j*, to maximum compressive, at *d*. The triangles *cgi* and *ihd* represent the algebraic sums of the two stresses, showing a very large amount of compression from *i* to *d* and tension from *i* to *c*. At the extreme left of the figure, the algebraic sum of the stresses is $ac - ce$ or *cg*, a negative or tensile stress. At *i*, the compression and tension are equal and counterbalance each other, the result being zero. From *i* to *j*, the algebraic sum of the tension and compression is equal to a compressive stress *ikj*. At *d*, the stress is all compression and is equal to the sum of *fd* and *bd*, or *hd*. In this way the total algebraic sum of *abdc* and *cejfd* is found to equal *cgihd*. It will be seen that the maximum compression *hd* should not exceed the safe compressive strength of the material.

If we consider the area *abdc* to be represented by the letter *A* and *cejfd* by the letter *B*, the strength of the column depends on the sum of *A* and *B*. The compressive strength *A* is very easily found, being equal to the sectional area of the column multiplied by the ultimate unit compressive strength of the material, but the value of *B* varies considerably and increases as the center line of the bent column is farther from the line of pressure, Fig. 1. On this elementary formula, most formulas for the strength of columns are founded.

As a general rule, the shorter the column the lower is the value of *B*; and, vice versa, the longer the column the greater is the value of *B*. It is for this reason that two formulas are generally adopted, one for short columns and one for long. *Short columns* are those in which the length is small compared with the width of the shortest side; *long columns* are those in which the length is great compared with the shortest side. The limit or point of change between short and long columns will be given under the proper heading.

4. Method of Failure of Short Columns.—Materials under compression fail in two distinct ways, depending on the nature of the material. Those that are malleable, such as steel, flatten and are deformed. Those that are comparatively brittle shear along a plane. In some of the first tests made, the test pieces being cubical, it was found that the pieces sheared along planes extending diagonally across the sides from one top edge to the opposite lower edge, making an angle of 45° with the horizontal. This was supposed to be the angle of rupture and was the same for all materials. Later tests, however, proved that this angle varies, and with most materials is about 60° with the horizontal. It is for this reason that in making compression tests with cast iron, or any other brittle material, the height of the test piece is made at least 1½ times the diameter, or shortest side. These test pieces are actually diminutive columns and show the method of failure of short columns. Where the height is not greater than the least dimension, the angle of rupture will be along planes of 45° with the horizontal, but as the length is increased in proportion to the least side, this angle increases until it reaches about 60°, where it remains the same. A test piece of cast iron showing the plane of rupture of this material is shown in Fig. 3.

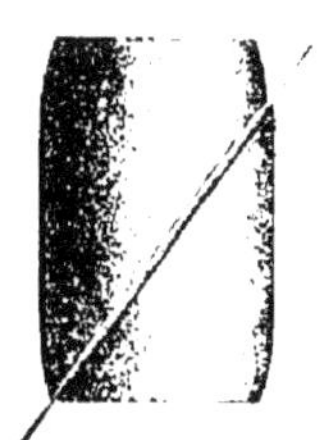

Fig. 3

5. Method of Failure of Long Columns.—Materials in compression develop more strength if the surfaces are true and level. Under this condition the tendency is to resist compression equally, and not to crush in one place before the balance of the material can be brought under compression, as would be the case if the bearing surfaces were uneven and rough.

The theoretical column is assumed to be hinged on a knife edge so as to be free to bend sidewise, as illustrated in Fig. 4 (*a*). This is never met with in practice, the nearest approach to it being the compression members of a pin-connected truss where the pin is comparatively small. If, however, the pin is large, the strength of the column is increased by the friction existing in the connection. In (*b*) is shown a column with rounded ends. Here the results are different from those of a knife-edge column, because the bending of the column changes the point of application of the load and consequently the line of pressure. (*c*) shows a column with flat ends. Here the column must be considerably bent out of line in order to overcome the effect of the flat

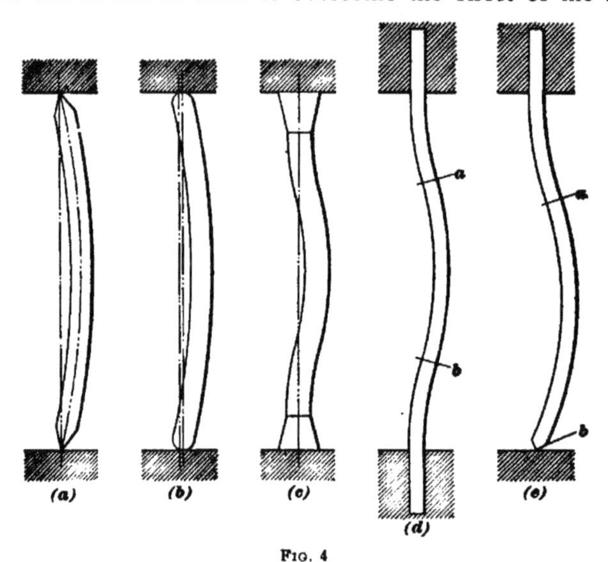

Fig. 4

ends. This case approaches the conditions shown in (*d*) where the ends are held tight, and the curve assumed by the column is similar to that of a beam uniformly loaded and fixed at the ends. The portion considered as the actual column lies between the points *a* and *b*, for at these two points the bending moment changes sign and compression alone exists. In (*e*) is shown a column fixed at one end and movable at the other, the portion between *a* and *b* acting as the column. The entire column assumes an **S** shape under the load.

COLUMN FORMULAS

6. The **column formulas** in general use do not give a direct method of calculating the dimensions of a column that will safely support a given load. The usual method of obtaining these is to assume values for the dimensions of the column, substituting these values in the following formulas and solving for the ultimate average compression u_l of this column. If the assumed size gives a value of u_l that is satisfactory for the given conditions, it is correct. If, however, the resulting value of u_l is smaller than the load per square inch, the assumed size is too small and it will be necessary to choose a larger size and solve again. If, on the contrary, the value of u_l is greater than the load per square inch of column section, a smaller size of column is assumed and a new value of u_l obtained. After a few trials, a size that gives a satisfactory stress for the given conditions will be found.

The four formulas in general use for calculating the strength of columns are: *Gordon's* or *Rankin's*, *Euler's*, the *straight-line*, and the *parabolic*.

7. Gordon's or Rankin's Formula.—This formula is based on the assumption that the column is subjected to three distinct stresses: (1) a direct compressive stress uniform throughout the section of the column; (2) a stress due to any slight eccentricity of the applied load; (3) a transverse stress created by the initial deflection of the column under the action of the applied load. Assuming these conditions, we have the following rule:

Rule.—*The load, per square inch, on any column should not exceed the compressive strength of the material at the elastic limit divided by 1 plus a value representing the eccentricity of the load plus a value representing the bending of the column.*

This rule is expressed by the formula

$$u_l = \frac{f_c}{1 + \frac{v c}{r^2} + \frac{f_c - u_l}{10 E}\left(\frac{l}{r}\right)^2} \qquad (1)$$

in which u_l = average compression, per square inch, of column section when maximum stress, per square inch, equals elastic limit of material;

f_c = elastic limit of material under compression;

E = modulus of elasticity of material;

l = length of column, in inches;

r = least radius of gyration, in inches;

v = eccentric displacement of load;

c = distance of center of gravity of section from extreme fiber.

This formula is the most rational of the theoretical formulas, its results giving a safe unit value for the compressive strength of the material. Its form, however, is indeterminate, since the value u_l appears on both sides of the equation; besides, the second term of the denominator is very difficult to obtain. In the formula in general use, this term has been neglected and a constant a substituted for the value of $\frac{f_c - u_l}{10E}$. Also, the ultimate strength s per square inch of the material in the column is used instead of f_c, the elastic limit, the result obtained being the ultimate strength, per square inch, of the column section. Thus,

$$u = \frac{s}{1 + a\left(\frac{l}{r}\right)^2} \qquad (2)$$

in which u = ultimate strength, per square inch, of column section;

s = ultimate strength, per square inch, of material in column;

a = constant varying with material;

l = length, in inches;

r = radius of gyration.

8. Euler's Formula.—This formula is based on the assumption of an ideal column; that is, a column that is centrally loaded and pivoted at the ends so as to be free to bend sidewise, and will fail by compression and bending alone, no eccentric load being taken into consideration. It also assumes that the strength of the column is realized when the elastic limit of the material is reached. The values given by this formula are correct only for long columns, or columns whose lengths are great in comparison with their radii of gyration. Its general form is

$$u_l = \frac{\pi^2 E}{\left(\frac{l}{r}\right)^2} \qquad (3)$$

in which u_l = average compression, per square inch, of column section;
E = modulus of elasticity of material;
l = length of column, in inches;
r = least radius of gyration.

For fixed ends, the constant 4 is introduced into the numerator, and for one end pivoted and one end fixed, a constant of $\frac{9}{4}$ is used.

9. Straight-Line Formula.—This formula was suggested by Thos. H. Johnson, C. E., in 1886, and was derived by plotting, on squared paper, similar to that shown in Fig. 7, the values of the breaking loads of columns of different materials. From these results the values in the following formula were obtained. It was found that an approximately straight line would pass through the plottings of the tests and hence the name of the formula is derived. The general form of this formula is

$$u = a - b\frac{l}{r} \qquad (4)$$

in which u = ultimate strength, per square inch, of column section;
a = constant, representing strength of material under compression, whose value is determined by test;
b = constant;
l = length of column, in inches;
r = radius of gyration.

This formula, while fairly accurate for the general sizes, is not good in cases where $\frac{l}{r}$ falls below a certain value, which varies for different materials and will be given later.

10. Parabolic Formula.—This formula is similar to the straight-line formula in that it is very convenient to apply and agrees closely with the actual tests on full-size columns. Its general form is

$$u_f = f_c - b\left(\frac{l}{r}\right)^2 \qquad (5)$$

in which u_f = average compression, per square inch, of column section;

f_c = elastic limit for compression of material composing column;

b = constant determined by experiment;

l = length of column, in inches;

r = radius of gyration, in inches.

This formula is also limited in its application to columns in which the ratio of $\frac{l}{r}$ does not exceed a certain value. This value is so high, however, that it is beyond the general practice, and if a case occurs in which the ratio of $\frac{l}{r}$ is greater than the limit of the parabolic formula, the strength may be calculated by Euler's formula. These limitations of the formulas are fully explained later.

TIMBER COLUMNS AND POSTS

11. Timber, although not generally used in modern buildings, is a valuable constructive material for columns. Timber columns or posts are adapted to mill construction, owing to the fact that they are economical and when of considerable size are not easily destroyed by fire. The kinds of timber usually employed for this purpose are the long-leaf and short-leaf yellow pines, white pine, red pine, white oak, spruce, hemlock, cypress, cedar, and redwood. The timber generally preferred is the yellow pine, the long-leaf variety being stronger and more durable than the short-leaf. The disadvantage of this wood is that the resinous sap makes it rather more inflammable than the other woods.

SHORT COLUMNS

12. The column may, in its first stage of development, be considered a cubical or rectangular block, as shown in Fig. 5. If the length of the column does not exceed from six to about ten times the width of the least side, the load it can safely carry may be estimated by multiplying its sectional area, in square inches, by the safe resistance to compression of the material parallel to the grain. To obtain the safe resistance of a short column, or block, divide the

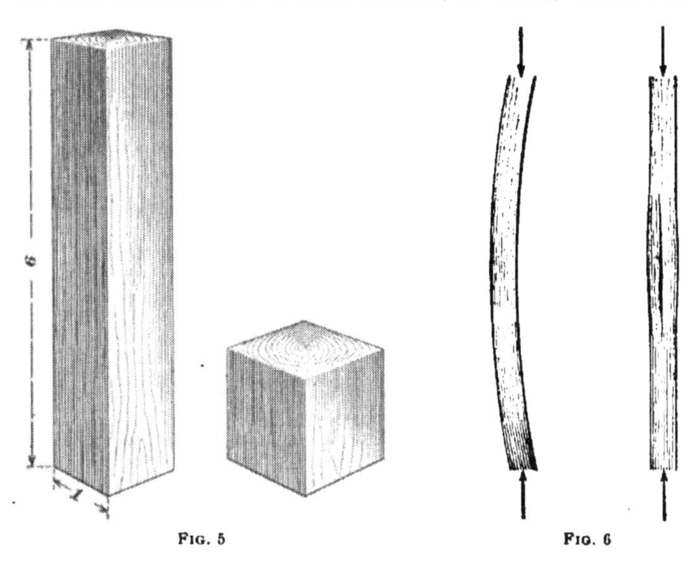

Fig. 5 Fig. 6

ultimate resistance to compression of the material parallel to the grain, by the factor of safety, which, for wooden columns, is from 4 to 5, though it may be, in some instances, good practice to use 6; it is all a matter of judgment, governed by the conditions to be met with. After having obtained the safe resistance of the material to compression, multiply by the sectional area of the column, in square inches, and the result will be the safe resistance of the short column to compression.

EXAMPLE.—What safe load will be supported by a short yellow-pine block, 12 inches square and 6 feet long, standing on end, a safety factor of 5 being used?

SOLUTION.—Assuming that the ultimate compressive strength of yellow pine parallel to the grain, per square inch, is 5,000 lb., with a factor of safety of 5, the safe load, per square inch, on this short column will be 5,000 ÷ 5 = 1,000 lb. per sq. in. The area of the column is 12 in. × 12 in. = 144 sq. in., and therefore the safe resistance to compression of the column = 144 × 1,000 = 144,000 lb. Ans.

LONG COLUMNS

13. When **long columns**, or those over ten times the width of the smallest side, are under compression, and not secured against yielding sidewise, it is evident that they are liable to bend before breaking. To ascertain the exact stress in such pieces is sometimes quite difficult. Hence, we must have a formula making due allowance for this tendency in the column to bend, or to split, and spread from the center, as shown in Fig. 6.

GORDON'S FORMULA FOR RECTANGULAR POSTS

14. Square-End Columns.—The **Gordon formula** for the different timbers usually employed in building construction is expressed in the following equations:

Southern yellow pine, $$u_s = \frac{1,125}{1 + \frac{l^2}{1,100 d^2}} \quad (6)$$

White oak, $$u_s = \frac{925}{1 + \frac{l^2}{1,100 d^2}} \quad (7)$$

White pine and spruce, $$u_s = \frac{800}{1 + \frac{l^2}{1,100 d^2}} \quad (8)$$

in which u_s = safe unit load, or safe bearing strength of column, in pounds per square inch of cross-section;

l = length of column, in inches;

d = dimension of least side, in inches.

The factor of safety varying from 4 to 5 gives results of from $\frac{1}{4}$ the ultimate strength in the case of small columns, to $\frac{1}{5}$ the ultimate strength in long columns.

By comparing these formulas with formula **2**, it will be observed that d^2 has been substituted for r^2. This is done for convenience in calculation, since the dimension of the least side is known, while r, or the least radius of gyration, would have to be calculated. In making this substitution, the constant a in formula **2** does not represent the same value as the constant of 1,100 given in formulas **6**, **7**, and **8**, for, to introduce the value of r^2 or $\frac{d^2}{12}$ as given in *Properties of Sections*, a different value of a than that in formula **2** is substituted in the denominator.

EXAMPLE.—What will be the safe compressive strength of a white-oak post 8 inches by 10 inches by 15 feet long?

SOLUTION.—In this case, l equals 15×12, or 180 in., and d equals 8 in. By substituting these values in formula **7**, which gives the safe unit stress for a white-oak column, we have $u_s = \dfrac{925}{1 + \dfrac{180^2}{1,100 \times 8^2}} = 634$ lb. per sq. in. This result multiplied by the area of cross-section of the column, which is 8 in. $\times$ 10 in., equals $634 \times 80 = 50,720$ lb. Ans.

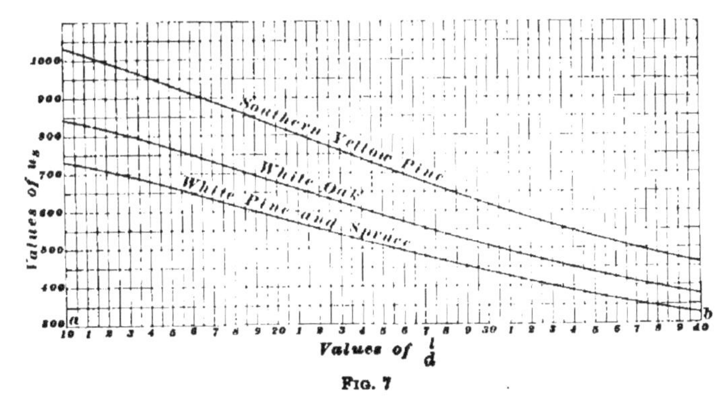

FIG. 7

15. Diagram Chart for Safe Load.—Fig. 7 gives the curves for these formulas for the different materials, by which the values of u_s, in pounds per square inch, may be obtained. These curves are described by plotting the values of u_s for corresponding values of $\frac{l}{d}$ that have been laid off on a horizontal line. From the points thus established, vertical lines are drawn, on which are laid off, to some convenient scale, vertical heights equal to the resulting values of u_s. The use of this chart is shown in the following example:

To find the value of u_s for any value of $\frac{l}{d}$, divide the length, in inches, by the smallest dimension of the column or post and find the corresponding value on the line *a b*. By following up the vertical line through this point and reading on the vertical scale, the value of u_s is found. In the example in Art. **14**, $\frac{l}{d}$ equals $\frac{180}{8}$ or $22\frac{1}{2}$. The value of u_s corresponding to the value of $22\frac{1}{2}$ for white oak is 635. This is very nearly the same value as that found in the previous example, the slight difference being due to the impossibility of accurate reading of the chart.

EXAMPLE.—What size square column of white oak, 18 feet long, will be required to sustain 30 tons?

SOLUTION.—It will be necessary to assume some size column and substitute its value in formula **7**, solving for the safe strength of the column; if this equals or exceeds slightly the load to be sustained, the assumed size is satisfactory. Assuming in this case a 10-in. square column, and substituting the values in formula **7**, the safe unit load of a white-oak column of the dimensions named is

$$u_s = \frac{925}{1 + \frac{216^2}{1,100 \times 10^2}} = 649 \text{ lb.}$$

and the entire safe load is $649 \times 100 = 64,900$ lb., which exceeds the load of 30 T. to be sustained, and the assumed size is therefore satisfactory. Ans.

EXAMPLES FOR PRACTICE

1. What is the safe strength of a $10'' \times 12''$ Southern yellow-pine post 20 feet long? Ans. 88,560 lb.

2. Find the size of a square post of white pine 14 feet 6 inches long required to support a load of 20 tons. Ans. $8\frac{1}{4}$ in.

STRAIGHT-LINE FORMULA

16. The formula mostly used for long, square or rectangular wooden columns, with square ends, is the following, deduced from elaborate tests made on full-length columns at Watertown arsenal:

$$u = s_c - \left(\frac{s_c\, l}{100\, d}\right) \qquad (9)$$

in which u = ultimate strength per square inch of sectional area of column;

s_c = ultimate compressive strength of material per square inch parallel to grain;

l = length of column in inches;

d = length of least side of column, in inches.

EXAMPLE.—What safe load will a white-pine column 10 inches square and 20 feet long, support, using a factor of safety of 6?

SOLUTION.—The ultimate compressive strength of white pine parallel to the grain is 3,500 lb. per sq. in. Therefore, by substituting in the formula, we have $u = 3{,}500 - \left(\frac{3{,}500 \times 240}{100 \times 10}\right) = 2{,}660$ lb., the ultimate bearing value of the column per square inch. As the factor of safety required is 6, the safe bearing value per square inch of sectional area is $2{,}660 \div 6 = 443\frac{1}{3}$ lb. The area of the column being 100 sq. in., the safe load is $100 \times 433\frac{1}{3} = 44{,}333$ lb. Ans.

PARABOLIC FORMULAS

17. The **parabolic formulas** for various materials are given as follows, in which u_f is the same value as defined in Art. **8**.

For white pine,

$$u_f = 2{,}500 - .6\left(\frac{l}{d}\right)^2 \qquad (10)$$

For short-leaf yellow pine,

$$u_f = 3{,}300 - .7\left(\frac{l}{d}\right)^2 \qquad (11)$$

For white oak,

$$u_f = 3{,}500 - .8\left(\frac{l}{d}\right)^2 \qquad (12)$$

For long-leaf yellow pine,

$$u_r = 4{,}000 - .8\left(\frac{l}{d}\right)^2 \qquad \textbf{(13)}$$

These formulas apply to posts with flat ends when the value of $\frac{l}{d}$ is less than 60.

18. The following rules and tables are based on a system of figuring recommended by the Department of Agriculture, Bureau of Forestry, and on the values of the ultimate strength of timber recommended by the committee on strength of bridges and trestle timbers, of the fifth annual convention of the Association of Railway Superintendents of Bridges and Buildings in 1895.

The strength of timber varies with the amount of moisture it contains and hence on its exposure to the elements. On account of this fact, the following classification is made of timber in general use:

Class *a*, moisture contents, 18 per cent., is used for structures freely exposed to the weather, such as railway trestles, uncovered bridges, etc.

Class *b*, moisture contents, 15 per cent., is used for structures under roof but without side shelter, freely exposed to outside air but protected from rain, such as roof trusses of open shops and shèds, covered bridges over streams, etc.

Class *c*, moisture contents, 12 per cent., is used for structures in buildings unheated but more or less protected from outside air, such as roof trusses or barns, enclosed shops and sheds.

Class *d*, moisture contents, 10 per cent., is used for structures in buildings at all times protected from the outside air, heated in winter, such as roof trusses in houses, halls, churches, etc.

The difference in strength due to exposure in the above classes of buildings is accounted for in the factor of safety used.

19. Factor of Safety.—Table I gives the factors of safety for these different classes of structures.

TABLE I

FACTOR OF SAFETY TO BE USED WITH THE DIFFERENT CLASSES

Classes	Yellow Pine	All Others
Class *a*	.20	.20
Class *b*	.23	.22
Class *c*	.28	.24
Class *d*	.31	.25

20. A. L. Johnson, Civil Engineer in the United States Department of Agriculture, Division of Forestry, suggests the following formula:

$$u = s_c \frac{700 + 15g}{700 + 15g + g^2} \qquad (14)$$

in which u = ultimate strength of post, in pounds per square inch;

s_c = ultimate crushing strength of material, in pounds per square inch;

$g = \frac{l}{d}$.

EXAMPLE.—Apply this formula to the example given in Art. **16.**

SOLUTION.—In this case, $g = \frac{20 \times 12}{10} = 24$. Substituting in formula **14,** $u = 3{,}500 \times \frac{700 + 15 \times 24}{700 + 15 \times 24 + 24^2} = 2{,}267.7$ lb. This, divided by 6, the factor of safety, gives a safe bearing value per square inch of sectional area of $2{,}268 \div 6 = 378$ lb. The area of the column being 100 sq. in., the safe load is $100 \times 378 = 37{,}800$ lb. This is less than the result given by the formula in Art. **16,** but it is reliable and goes to show the variation in different formulas.

21. Table II gives values of u for various values of g. Its use greatly shortens the work, as it is only necessary to find g for any desired column. The table is divided into five classes:

Class *A* includes all woods whose ultimate crushing

TABLE II

ULTIMATE STRENGTH PER SQUARE INCH FOR TIMBER POSTS

Value $x = \frac{l}{d}$	Class A $s_c = 3{,}500$	Class B $s_c = 4{,}000$	Class C $s_c = 4{,}500$	Class D $s_c = 5{,}000$	Class E $s_c = 6{,}000$
2	3,481	3,978	4,475	4,973	5,967
3	3,458	3,952	4,446	4,940	5,928
4	3,428	3,918	4,407	4,897	5,876
5	3,391	3,875	4,359	4,844	5,813
6	3,347	3,826	4,304	4,782	5,739
7	3,299	3,770	4,242	4,713	5,656
8	3,247	3,710	4,174	4,638	5,566
9	3,190	3,646	4,102	4,558	5,469
10	3,132	3,579	4,026	4,474	5,368
11	3,070	3,509	3,948	4,386	5,264
12	3,008	3,438	3,867	4,297	5,156
13	2,944	3,365	3,785	4,206	5,047
14	2,880	3,291	3,703	4,114	4,937
15	2,815	3,217	3,620	4,022	4,826
16	2,751	3,144	3,537	3,930	4,716
17	2,687	3,071	3,455	3,838	4,606
18	2,624	2,998	3,373	3,748	4,498
19	2,561	2,927	3,293	3,659	4,391
20	2,500	2,857	3,214	3,571	4,286
21	2,440	2,788	3,137	3,486	4,183
22	2,381	2,721	3,061	3,402	4,082
23	2,324	2,656	2,988	3,320	3,983
24	2,268	2,592	2,916	3,240	3,888
25	2,213	2,529	2,846	3,162	3,794
26	2,160	2,469	2,777	3,086	3,703
27	2,109	2,410	2,711	3,013	3,615
28	2,059	2,353	2,647	2,941	3,529
29	2,010	2,298	2,585	2,872	3,446
30	1,963	2,244	2,524	2,805	3,366
32	1,874	2,142	2,409	2,677	3,212
34	1,790	2,046	2,301	2,557	3,068
36	1,711	1,956	2,200	2,445	2,934
38	1,638	1,872	2,106	2,340	2,808
40	1,569	1,793	2,017	2,241	2,690
42	1,505	1,719	1,934	2,149	2,579
44	1,444	1,650	1,857	2,063	2,476
46	1,388	1,586	1,784	1,982	2,379
48	1,335	1,525	1,716	1,907	2,288
50	1,285	1,468	1,652	1,835	2,203

strength s_c is 3,500 pounds per square inch, white pine being the most common wood of this strength.

Class *B* includes all woods whose ultimate crushing strength is 4,000 pounds per square inch, such as Northern or short-leaf yellow pine, red pine, Norway pine, spruce and Eastern pine, hemlock, cypress, cedar, California redwood, and California spruce.

Class *C* represents white oak, its strength being 4,500 pounds per square inch.

Class *D* includes those woods whose strength per square inch is 5,000 pounds, such as Southern long-leaf or Georgia yellow pine, Canadian (Ottawa) white pine, and Canadian (Ontario) red pine.

Class *E* includes such woods as Douglas, Oregon, and Washington yellow pine, or pine whose strength per square inch is 6,000 pounds.

To solve the example in Art. **20** by means of Table II, find the given value of g in the table, then take the number opposite it, in the column headed Class *A*. The value of g being 24, the corresponding value of u is found to be 2,268, which is the result of the preceding calculations. The rest of the calculation for finding the safe load is the same as that in Art. **20.**

22. Actual tests of posts made up of three pieces bolted and keyed together show that, by the combination, their ultimate strength per square inch is not materially increased over the strength of the three separate pieces. In other words, the strength per square inch of three 4″ × 12″ joists bolted together, making a 12″ × 12″ built-up column, is not more than three times that of a column consisting of a 4″ × 12″ joist, and the strength per square inch of the former is not as great as that of a solid 12″ × 12″ column.

EXAMPLE.—What is the difference in strength between three 4″ × 12″ spruce joists bolted and keyed together as a column and a 12″ × 12″ timber, both being 16 feet long?

SOLUTION.—In the first case, $g = \frac{16 \times 12}{4} = 48$, and by Table II, $u = 1{,}525$, the ultimate strength per square inch for this ratio of l to d.

Multiplying u by the area of the three joists, we have $1{,}525 \times 144 = 219{,}600$ lb., the ultimate breaking load of the three pieces bolted together.

For the solid timber, $g = \frac{16 \times 12}{12} = 16$, and by Table II, $u = 3{,}144$. Multiplying this value of u by the area, we get $3{,}144 \times 144 = 452{,}736$ lb., making a difference of $452{,}736 - 219{,}600 = 233{,}136$ lb.

The 12" × 12" timber is therefore more than twice as strong as the three pieces bolted together, although the latter contain just as much timber. Ans.

METHOD OF SECURING POSTS IN BUILDINGS

23. Column Bases.—Posts are usually supported by a separate cast-iron or structural steel **base** set in Portland cement. The practice of anchoring the base to the foundation

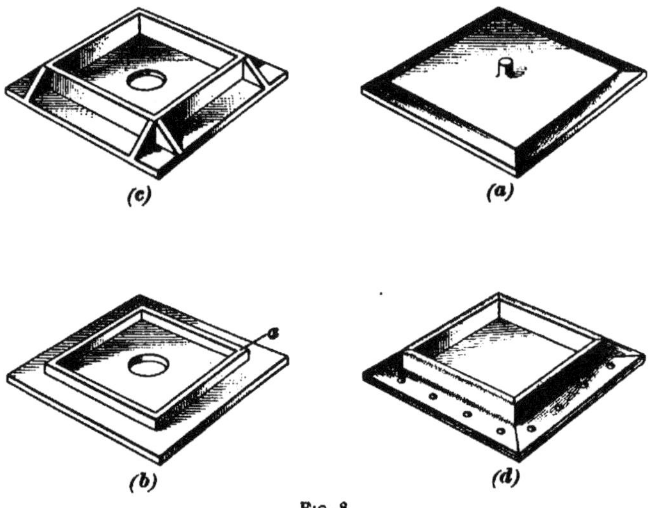

Fig. 8

is generally given up in modern work, except in buildings where the former is likely to be displaced by carts, trucks, etc.

Fig. 8 (*a*) shows the simplest form of cast-iron base that can be used. The short pin at the center is for the purpose of centering the column and holding it in place. In (*b*), the column is held in position by a projecting flange *a* cast on the base; this is usually made a tight fit for the post, which is slipped into it. These designs are only adaptable to cases in which light loads are supported by the posts, as their construction is weak. When a column is heavily loaded, such a base as shown in (*c*) or (*d*) is adopted. The one shown in (*c*) is of cast iron and is strengthened by webs and stiffeners. The base is made somewhat smaller than the column, which is cut away at the end and slipped into it. That shown in (*d*) is made of four angles riveted to the base plate, and is known as the *Van Dorn base*. This has the disadvantage of being more expensive than the others, being a patented shape.

24. Column Caps.—The principal styles of **column caps** are shown in Fig. 9. The cap shown in (*a*) has the advantage of cheapness, but is poor on account of the absence of a socket or guide of any sort to hold the upper column. It may, however, have a pin, such as is shown in Fig. 8 (*a*), cast on it. The cap shown in (*b*) has the advantage that a bolt may be put through the lower end of the upper post *a* at *b* to hold both upper and lower posts in line, in case the beams *c*, *c* should be burned away or broken. The two bolts at the sides go through the beams and hold them in place, thus strengthening the building laterally.

In (*c*) and (*d*) are shown two forms of cast-iron caps in which both the upper and lower columns are held in place by through bolts, which make a rigid connection independent of the beams. The first, (*c*), has a cast-iron ridge *a* at each end across the bed of the cap, which is let into a slot cut in the beam, and assists in holding the beam in place. In the other, (*d*), pins are cast on the bed of the cap. The special advantage of these designs is that the beam can burn away and fall without danger of tearing out the column base.

The cap shown in (*e*) is recommended by the Boston Fire Insurance Company, and has pins on top and bottom that fit into holes bored in the ends of the posts, as in Fig. 8 (*a*). This cap possesses the advantage that the load from the beams may be brought directly over the post. The girders are usually made of two joists, placed one on either side of

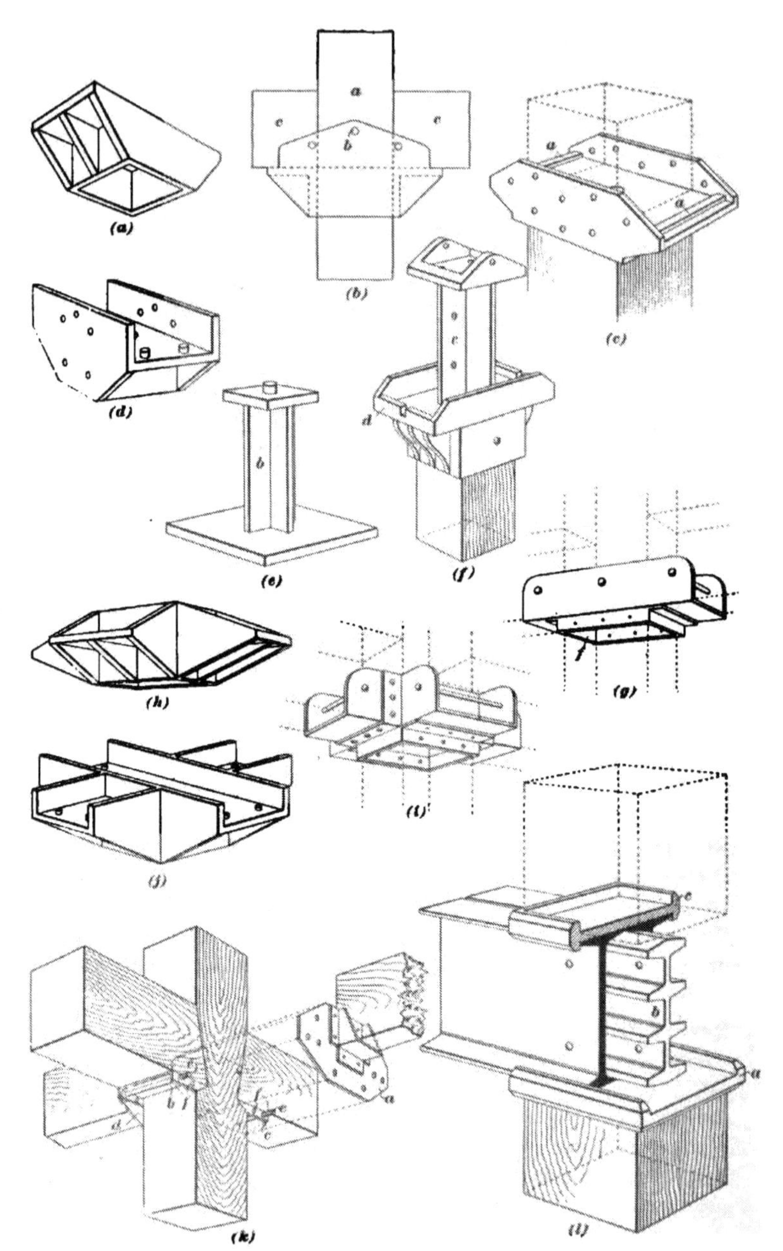

FIG. 9

the vertical web *b* and bolted to it, while these cross-webs transmit the load from column to column.

The column cap shown in (*f*) is modeled after the *Goetz-Mitchell style* shown in (*c*) and (*d*), and embodies the advantages of that form of cap. It is especially designed for flitch-plate girders, the flitch plate being kept back at the end enough to allow the two beams to slip in beside the vertical web *c*. Provisions may be made for bolting the side beams to the vertical web *c*, making a very rigid connection. In this case the rib *d* might be omitted, although by its use additional rigidity is secured, as the liability of the web *c* to shear opposite the bolt holes is overcome by the rib, which tends to prevent any lateral movement. The rib *d* is notched at the center to allow the flitch plate to pass through, while the wooden beams are notched to fit over the rib *d*. A flitch-plate girder will, in case of fire, burn on the outside and form a charcoal covering for the flitch plate, thus protecting it from fire, and the flitch plate being protected, will in turn hold up the beam.

The style of column cap shown in (*g*) is made up of wrought-iron plates and is secured to the lower column by means of spikes driven through holes in the flange *f* at the base. In (*i*) is shown the same cap with brackets for supporting four beams instead of two. In (*h*) and (*j*) are shown two forms of cast-iron caps to be used where four beams are supported by the same column.

The form shown in (*k*) has many advantages. Here the side plate *a* is removed from its true position, as indicated by the dotted lines, in order to show the shape of the post and beams at the joint. The side plates *a* and *d* are secured to the base plates *b* and *c* by means of long and short bolts *e* and *f*, respectively, while long bolts extend through the column, thus tying the side plates securely to it. Like those in views (*g*) and (*i*), this cap is made up of plates and angles, but has the advantage that it can be easily taken apart, and is the only cap that may be used with a continuous column, as by its use columns may be extended through two or even three stories.

All these caps are designed for buildings where wooden girders are used. Sometimes, however, it is desirable to use wooden columns and steel girders on which the wooden beams are supported. This requires a slight variation in the caps, and although a number of these designs may be changed sufficiently to accommodate steel beams, the problem may be solved by using such a cap as is shown in (*l*). This cap consists of a cast-iron plate *a* made with a recess into which the top of the lower column is inserted. The upper side of the cap is designed with a lip on either side, between which the two steel beams forming the box girder are placed, thus keeping them from sliding off the column. The steel beams are connected by a special form of separator *b*, which extends through to the next pair of beams, thus tying the two pairs firmly together. On this joint rests a cap *c*, similar to *a*, placed in the reverse position and extending over the sides of the beams, which supports the upper column. Since the ends of the beams form a portion of the column, the separator *b* should be a snug fit so as to relieve the beams of compression and transmit the stress from the upper to the lower flanges.

The caps shown in (*c*), (*d*), (*e*), (*h*), and (*j*) are known as *Goetz caps*; those shown in (*g*) and (*i*) are the *Van Dorn patents*; and (*k*) is known as the *duplex cap*. These styles of caps are patented and cannot be used without paying a royalty to the patentees.

25. It is a common practice, in using wooden columns, to bore a 1-inch or a 1¼-inch hole the full length of the column through the center, in order to ventilate the interior and prevent dry rot. For the same reason a hole is often left through the cap or base in order to give ventilation to the ends of the column. Such construction is shown in the caps in Fig. 9 (*c*) and (*d*), and in the base, Fig. 8 (*b*). A further advantage of this is that no water can collect and stand at the base of the column.

CAST-IRON COLUMNS

26. Cast-iron columns are most frequently used in buildings of moderate height, but have been used, in some cases, in buildings of sixteen, and even more, stories. The best practice has, during the last few years, so uniformly declared in favor of steel columns that the employment of cast iron is now generally confined to buildings of ordinary height, say four or five stories, or to special cases, where advantages are to be gained in the use, for instance, of a number of ornamental cast-iron columns.

The uncertain strength of cast iron has compelled the adoption of a very low unit stress per square inch; in other words, a very high factor of safety. The uniform strength of structural steel is, on the other hand, so well understood that cast iron, for columns, and especially girders, is falling into disuse. Considerations of economy may, however, in some cases, still justify its employment. But, though cast-iron columns are cheaper per pound and perhaps easier to erect than steel, the declining price of structural steel is removing this advantage of cast iron.

27. Objections to Use of Cast-Iron Columns.—One disadvantage in the use of cast-iron columns is that when fracture occurs, it comes without warning. In high buildings, erected entirely upon cast-iron columns, the danger from wind pressure is very much increased on account of lack of stiffness in the joints of the connections at the several floors. In fact, buildings have been blown 10 inches out of plumb, owing to this lack of rigidity in the connections.

One of the greatest objections to the use of cast-iron columns is that they are liable to be broken by sudden contraction, due to water being played on them in case of fire.

This objection, while held by many engineers, cannot be said to agree with practice, for there are many cases of buildings destroyed by fire in which the beams and girders have been bent and twisted, while the columns have stood the heat remarkably well. The strength of the columns that have stood the test is very uncertain, and every precaution should be taken to test them after a fire.

FORMULAS FOR CAST-IRON COLUMNS

28. The uncertain strength of cast iron gives rise to a number of formulas and rules for calculating the strength of columns made of this material. The only method of testing these formulas is to compare their results with those of actual tests. Unfortunately, these tests are very few in number, but they serve as the basis of calculation, and are undoubtedly more reliable than the theoretical conditions. The formulas for cast-iron columns are usually such that they may be applied to long and short columns, with the exception of the straight-line formulas, whose results for long columns are too high.

GORDON'S FORMULA

29. The strength of a cast-iron column with square ends may be calculated by the following rule:

Rule.—*To find the ultimate strength per square inch of sectional area of a cast-iron column with square ends, divide the ultimate compressive strength per square inch of the material composing the column by 1 plus the quotient obtained by dividing the square of the length of the column, in inches, by 3,600 times the square of the radius of gyration of the section of the column.*

This rule is expressed by the formula

$$u = \frac{S_c}{1 + \frac{l^2}{3{,}600\, r^2}} \qquad (15)$$

in which u = ultimate strength per square inch of column section;

s_c = ultimate compressive strength per square inch of material composing column (for cast iron s_c may be taken as 80,000);

l = length of column, in inches;

r^2 = square of least radius of gyration.

EXAMPLE.—Find the proper working load for a 10-inch, square, cast-iron column, 20 feet long, using a factor of safety of 6, the thickness of the metal being 1 inch.

SOLUTION.—The ultimate compressive strength s_c of cast iron, per square inch, is 80,000 lb., the length $l = 20 \times 12 = 240$ in.; and r^2 for a hollow, square, rectangular column is $\frac{a^2 + a_1^2}{12}$, in which a is the length, in inches, of one side measured outside and a_1, the length of same side measured inside of column. Therefore, $r^2 = \frac{10^2 + 8^2}{12} = 13.6$.

Substituting these values in formula **15**, we have

$$u = \frac{s_c}{1 + \frac{l^2}{3,600\, r^2}} = \frac{80,000}{1 + \frac{240^2}{3,600 \times 13.6}} = \frac{80,000}{2.18} = 36,706 \text{ lb.},$$

the breaking strength of the column in pounds per square inch of section. With a factor of safety of 6, the safe bearing value of the column is $36,706 \div 6 = 6,118$ lb. per sq. in. The net area of the section of the column is $10^2 - 8^2 = 100 - 64 = 36$ sq. in. The entire load that it will support with safety is therefore $36 \times 6,118 = 220,248$ lb. Ans.

30. Formula **15** may be replaced by formula **16.** Though the latter is not as general in its application, it is more convenient in form and may be applied to the usual sections of cast-iron columns.

$$u = \frac{80,000}{1 + a\frac{l^2}{d^2}} \qquad (16)$$

in which u = ultimate strength per square inch of sectional area;

d = least outside dimension of the column, in inches;

l = length of column, in inches;

a = value taken from Table III.

This may be expressed by the following rule:

Rule.—*To find the ultimate strength per square inch of sectional area of a cast-iron column, divide 80,000 by 1 plus the product of a constant and the quotient obtained by dividing the square of the length, in inches, by the square of the least dimension of the column, in inches.*

TABLE III

VALUE OF a FOR DIFFERENT BEARINGS

Cylindrical Columns			Rectangular Columns		
Square Bearing	Pin and Square Bearing	Pin Bearing	Square Bearing	Pin and Square Bearing	Pin Bearing
$\frac{1}{800}$	$\frac{3}{1,600}$	$\frac{1}{400}$	$\frac{3}{3,200}$	$\frac{9}{6,400}$	$\frac{3}{1,600}$

In Table III, the values given in the second column are for columns in square bearing on one end and pin bearing on the other end.

EXAMPLE.—Required, the ultimate strength, per square inch of section, of a square-ended, round, hollow column 15 feet 10 inches long, whose outside diameter is 15 inches, the average thickness of the metal being 1 inch.

SOLUTION.—According to Table III, the value of a for a cylindrical column with square bearing is $\frac{1}{800}$. The value of l in this case is equal to $15 \times 12 + 10$, or 190, and d equals 15.

Substituting these values in formula **16**, we have

$$u = \frac{80,000}{1 + \frac{190^2}{800 \times 15^2}} = 66,666 \text{ lb. Ans.}$$

31. Comparison of Building Laws.—The value of the constant a varies according to the several building laws of New York, Chicago, and Boston, as shown in Table IV, which gives the formulas for obtaining the safe unit strength u_s of columns. These formulas embody their own factors of safety, which vary in the different cities, as mentioned in Art. **34.**

TABLE IV

FORMULAS FOR FINDING THE VALUE u_s FOR COLUMNS

Cylindrical Columns			Rectangular Columns		
New York Old Law	Chicago	Boston	New York Old Law	Chicago	Boston
$\frac{16,000}{1+\frac{l^2}{400d^2}}$	$\frac{10,000}{1+\frac{l^2}{600d^2}}$	$\frac{10,000}{1+\frac{l^2}{800d^2}}$	$\frac{16,000}{1+\frac{l^2}{500d^2}}$	$\frac{10,000}{1+\frac{l^2}{800d^2}}$	$\frac{10,000}{1+\frac{l^2}{1,066d^2}}$

STRAIGHT-LINE FORMULAS

32. It has been demonstrated lately that the results of the formula of the New York building laws, as given in Table IV, were entirely too high. In the revised copy of these laws, this formula has been discarded and the one following is recommended, which embodies its own factor of safety. The results given, however, are rather large, as will be seen in the comparison later.

Rule.—*The safe load per square inch of a cast-iron column may be obtained by subtracting thirty times the ratio of the length, in inches, to the radius of gyration, in inches, from 11,300.*

This rule is expressed by the formula

$$u_s = 11,300 - 30\frac{l}{r} \qquad (17)$$

in which u_s = safe strength, in pounds, per square inch of sectional area;
l = length of column, in inches;
r = radius of gyration, in inches.

EXAMPLE.—What will be the allowable load per square inch on the column in the example in Art. **30**, according to the new formula of the New York building laws?

SOLUTION.—The radius of gyration equals $\frac{\sqrt{15^2 + 13^2}}{4} = 4.9623$;

hence, $u_s = 11,300 - \frac{190 \times 30}{4.96} = 10,151$ lb. Ans.

EULER'S AND THE PARABOLIC FORMULAS

33. The following examples show the application of these formulas to cast-iron columns:

For pin-bearing columns, when $\frac{l}{r}$ is less than 70,

$$u = 60{,}000 - \frac{25}{4}\left(\frac{l}{r}\right)^2 \qquad (18)$$

For pin-bearing columns, when $\frac{l}{r}$ is greater than 70,

$$u = \frac{144{,}000{,}000}{\left(\frac{l}{r}\right)^2} \qquad (19)$$

For square-bearing columns, when $\frac{l}{r}$ is less than 120,

$$u = 60{,}000 - \frac{9}{4}\left(\frac{l}{r}\right)^2 \qquad (20)$$

For square-bearing columns, when $\frac{l}{r}$ is more than 120,

$$u = \frac{400{,}000{,}000}{\left(\frac{l}{r}\right)^2} \qquad (21)$$

in which u = ultimate strength, in pounds per square inch, of sectional area;
l = length of column, in inches;
r = radius of gyration, in inches.

EXAMPLE.—Apply this formula to the column specified in the example in Art. **30**.

SOLUTION.—The radius of gyration found is 4.96 and the length is 190 in.; then, since

$$\frac{9}{4} \times \left(\frac{190}{4.96}\right)^2 = \frac{9}{4} \times \frac{36{,}100}{24.6016} = 3{,}301.6$$

$$u = 60{,}000 - 3{,}301.6 = 56{,}698 \text{ lb. Ans.}$$

EXAMPLES FOR PRACTICE

1. Compare the results of the parabolic and the second form of Gordon's formula, for the ultimate unit strength of a cast-iron column whose length is 10 feet and whose outside diameter is

10 inches, with a uniform thickness of 1 inch, and with flat or square-bearing ends.

Ans. { By parabolic formula, 56,839 lb.
By Gordon's formula, 67,796 lb.

2. Compare the results of the same formulas for pin-bearing ends.

Ans. { By parabolic formula, 51,220 lb.
By Gordon's formula, 58,823 lb.

---

34. Factor of Safety.—In cast-iron columns a generous factor of safety should be used. The old New York building laws required a factor of safety of 5, while the new laws embody their own factor of safety. With formula **15**, a factor of safety of 6 or 7 may be used, but in Boston and Chicago 8 is required.

35. For convenience of comparison, a list is given below of the results of the above formulas for the column mentioned in the example in Art. **30**, by which the student may judge how near these come to the actual results.

	Unit Breaking Load. Pounds per Square Inch	Unit Allowable Load. Pounds per Square Inch
By formula **15**	56,858 ÷ 7	8,123
By Boston building laws . .	66,664 ÷ 8	8,333
Chicago building laws . . .	63,136 ÷ 8	7,892
New York building laws (old)	57,100 ÷ 5	11,420
New York building laws (new)		10,151
By parabolic formula . . .	56,698	
By test, average results . .	30,000	

This shows that, while in Boston a safety factor of 8 is allowed, the actual factor of safety is hardly 4, considering the average result of the tests, and in New York the actual factor of safety is only 3. Great care should, therefore, be taken in the use of cast-iron columns, as the results of the preceding formulas and tests show.

DESIGN OF CAST-IRON COLUMNS

36. General Remarks.—In designing a column, to insure a good casting, no part should be made less than $\frac{3}{4}$ inch thick. It is considered poor designing to make one part or section considerably thicker than an adjacent section, on account of the liability of the thinner parts to cool more quickly than the thick parts and break away from them. Even if they do not break apart in casting, the metal, especially if in tension, is very liable to crack after the column is in place. Therefore, when it is found necessary to thicken a portion of a column, the increase in thickness should be gradual.

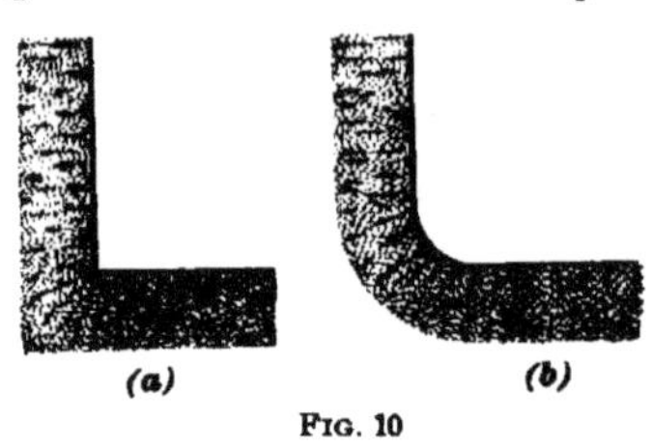

Fig. 10

Cast iron, in cooling in the mold, forms a fiber that runs perpendicular to the surface. When a corner is made as shown in Fig. 10 (*a*), the metal is very liable to crack off; for this reason the corner should have a generous curve connecting the surfaces, as in Fig. 10 (*b*). This curve is called a *fillet* and should have a radius of at least $\frac{1}{4}$ or $\frac{3}{8}$ inch.

37. In casting columns, it is necessary to use a *core* or cylinder made to the inside diameter of the column and composed of sand, flour, and water. After being molded in a

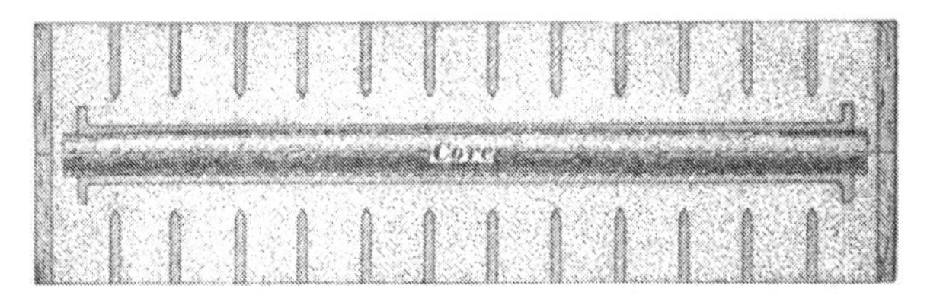

Fig. 11

core box, it is taken out and baked in an oven prepared for that purpose. The core, being relatively light, tends to float on the surface of the molten iron, and if the column is cast on its side, means must be taken to prevent the core from rising and approaching too close to the top side of the column, as shown in Fig. 11. The dotted lines represent the true position of the core before the iron was poured in. The full lines show the position of the core bent by its buoyancy, making the top side of the column thinner than the bottom. The dirt and air caught in the mold float on the molten iron and form blowholes and sand spots on the top side of the column. To avoid these blowholes a small wire rod is forced through the sand mold at intervals, making air vents by which the gases may escape.

There are two ways of avoiding the displacement of the core: (1) If the column is cast in a horizontal position, the core must be held firmly in its position. (2) The best method, however, of casting a cylinder is by standing it on end and admitting the molten metal from the bottom. In this way the core is kept in a central position, with respect to the column, and any dirt or sand that may have been displaced is floated up on top of the metal, thus making a sounder casting. This method, while very useful in casting pipe and engine cylinders, cannot, however, be used in casting columns of any considerable length because the mold will not stand the great pressure due to the head of metal.

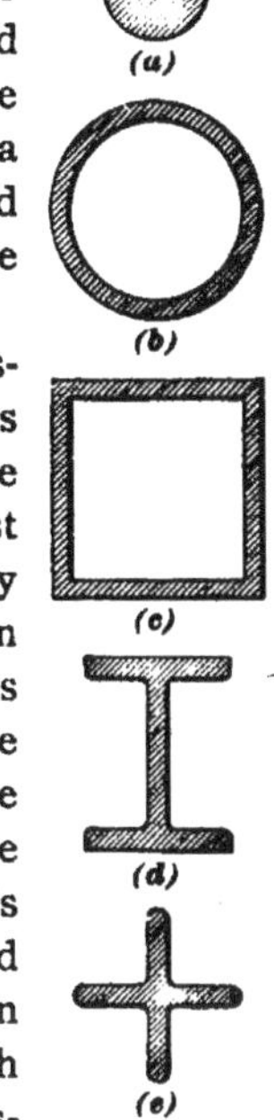

Fig. 12

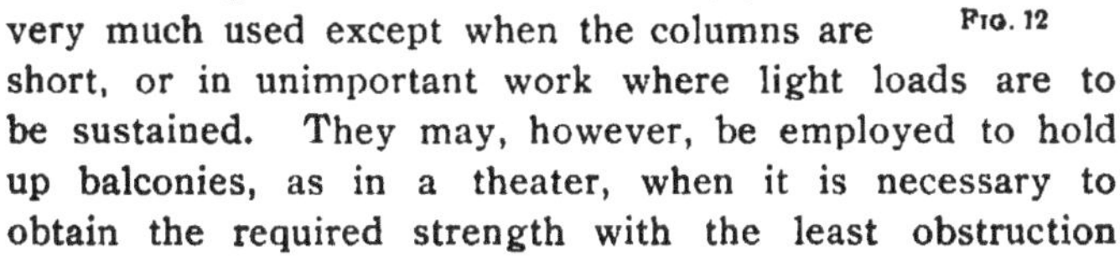

38. Sections of Cast-Iron Columns. The usual sections of cast-iron columns are shown in Fig. 12. The solid column (*a*) is not very much used except when the columns are short, or in unimportant work where light loads are to be sustained. They may, however, be employed to hold up balconies, as in a theater, when it is necessary to obtain the required strength with the least obstruction to view. It is much more economical to use a hollow cylindrical column (*b*), because the same amount of material as in a solid column, when spread into a hollow cylinder, will sustain a much greater load. In (*c*) is shown a rectangular or box column; the sides may be equal or unequal. When unequal, the value of d, or length of a side, is always taken as the shortest side and the value of r, or radius of gyration, is figured for the smallest value of d. This section is almost as efficient for the same amount of material as the one shown in (*b*), with the advantage of being better adapted for forming beam connections. In (*d*) is shown the **H** section, which is very useful, especially in walls where it may be easily built into the masonry and in this way be thoroughly protected from fire. It has the advantage that all surfaces are exposed and open to inspection, so that bad spots may be easily detected and the nature of the casting known. All surfaces may be painted and protected from rust. Also, on account of its shape, it is much more easily cast, no core being required. It has, however, the disadvantage of being an uneconomical section. In (*e*) is shown a form that has the advantage of being easily cast, but it is not an economical shape. This section has been used in old practice for struts in framed structures, such as roof trusses and bridges. In (*f*) is shown a column section that is sometimes used in the construction of store fronts, and, as a cast-iron pilaster, is frequently a structural member in a building. It is not economical, but has the advantage that it is readily built into the walls and thus entirely or partially concealed.

EXAMPLE.—What is the difference in strength of two cast-iron columns, each 16 feet long, having an area of cross-section equal to 28.28 square inches, the one to be a hollow cylindrical column of 10 inches outside diameter and the other to be solid.

SOLUTION.—A solid cylindrical column whose area of cross-section is 28.28 sq. in. will have a diameter of $\sqrt{\frac{28.28}{.7854}} = 6$ in. The square of the radius of gyration for the solid column will then be $\frac{6^2}{16} = 2\frac{1}{4}$, and substituting in formula **15**,

$$u = \frac{80{,}000}{1 + \frac{192^2}{3{,}600 \times 2\frac{1}{4}}} = 14{,}414 \text{ lb.}$$

This multiplied by 28.28 gives $14{,}414 \times 28.28 = 407{,}627.92$ lb. for the strength of the solid column.

A cylinder whose outside diameter is 10 in. and whose area is 28.28 sq. in. would have an inside diameter of

$$\sqrt{d^2 - \frac{28.28}{.7854}} = 8 \text{ in.,}$$

since
$$\frac{\pi d^2}{4} - \frac{\pi d_1^2}{4} = \frac{\pi}{4}(d^2 - d_1^2) = 28.28;$$

and
$$d_1 = \sqrt{d^2 - 28.28 \times \frac{4}{\pi}} = \sqrt{100 - 36} = 8$$

The square of the radius of gyration $r^2 = \frac{d^2 + d_1^2}{16} = 100 + 64 \div 16 = 10\frac{1}{4}$ in. Substituting in formula **15** again, we have

$$u = \frac{80{,}000}{1 + \frac{192^2}{3{,}600 \times 10\frac{1}{4}}} = 40{,}020 \text{ lb.}$$

This result multiplied by 28.28 gives $40{,}020 \times 28.28 = 1{,}131{,}765.6$ lb. Therefore, the hollow cylindrical column is nearly three times as strong as the solid column, in this case, although no more metal has been used. Ans.

39. Column Connections.—Solid cylindrical cast-iron columns are usually fastened together by the simple flange base shown in Fig. 13 (*a*). The flange need only be broad enough to allow bolting together, the minimum bolt circle being expressed by the formula

$$d + o + \tfrac{1}{2} \text{ inch} \qquad \textbf{(22)}$$

in which d = diameter of column;
o = outside diameter of nut.

The flange may be used around hollow cast-iron columns, whether rectangular or circular, as shown in (*b*). In order to reduce the machine work and get a bearing that will be more apt to be true, a design such as is shown in (*c*) has been used. This, however, is not as stiff a connection as that shown in (*b*), the bearing of the base being smaller. Sometimes, to secure a better alinement, a flange is cast on the lower column projecting into the lower end of the upper column. The two connecting flanges are faced and bolted together as shown in (*d*), ordinary bolts being used to hold the two columns together. The connection for **H** columns is shown in (*e*). Here the web, forming the bearing of the

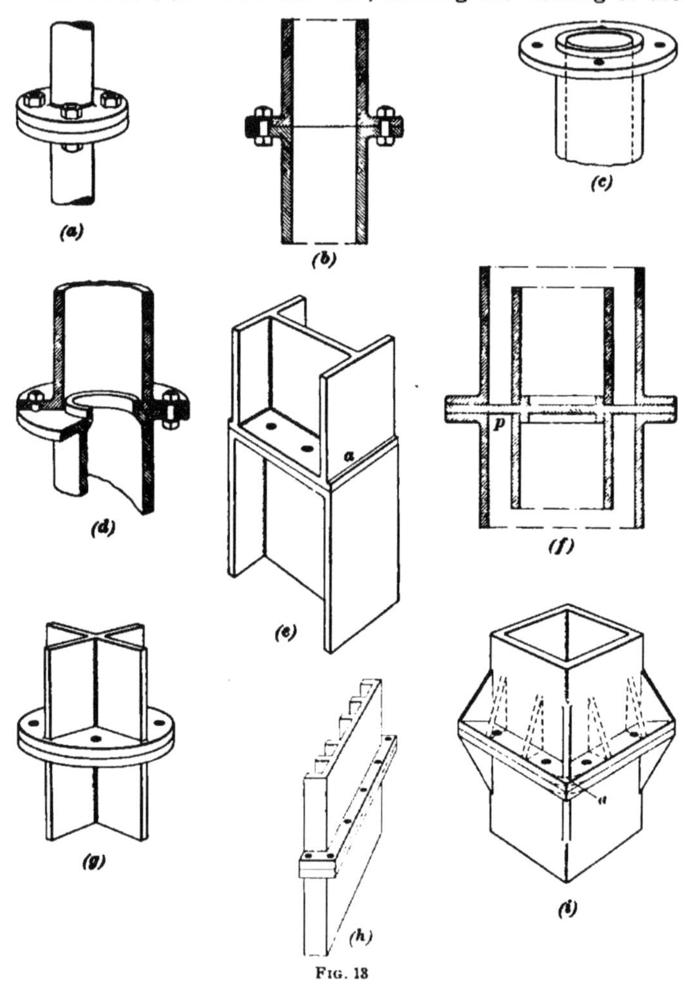

FIG. 18

column, is cast inside the vertical flanges. This is a point in favor of this style of column, because, having no projecting

flanges, the column is easily built into masonry walls. The upper column section is usually reduced in the manner shown in the figure by casting a small lip *a* on the base of the upper column.

The purpose of using two shells, as in (*f*), is for fire-protection, the outer one protecting the inner one from the effect of flames. In this case a plate, such as is shown at *p*, is used, bringing the columns into alinement and adjusting them to their proper position. The objection to the use of this plate is the extra number of joints, there being twice as many as in the ordinary column. Joints are very objectionable features in a column, on account of the difficulty of bringing them to a good bearing. In (*g*) is shown the general form of connection for the **X** shape; this shape is very little used because its form is not economical.

In all the above styles, except (*d*), the bolts should fit the holes tightly, this necessity in the case of (*d*) being obviated on account of the projecting flange or lip of the lower column into the upper column. Some constructors permit the use of tin shims in order to center the bolt, but this is objectionable because it does not hold the columns firmly in line. The best method is to have the holes machine-drilled and use machine bolts that fit the holes accurately. The holes in the bases must then be drilled to a templet to insure their alinement. In order that the bolts may hold the flanges more firmly together, the latter are usually *spot-faced;* that is, the metal around the hole is cut off until the nut has a full and even bearing. (*h*) shows a flat cast-iron pilaster column that it is sometimes necessary to use against masonry walls and in positions where room cannot be provided for a larger column. In calculating the strength of such a column the least radius of gyration should be used, even though the column is securely tied to the wall. In (*i*) is shown a connection frequently used for square columns, though it is probably more usual to put the ribs in, as shown dotted. There is little value in placing the ribs on the diagonal in this instance, from the fact that the corner could be well rounded as at *a*, when the necessity for the corner ribs would not exist.

The reduction of section at the different floors forms an important consideration in selecting the section of a column. The cylindrical column is generally reduced by changing the diameter. Where an abrupt change is to be made and the load carried by the upper column is considerable, great care should be taken to make the flanges at the connection strong enough to stand the shear so that the columns may not telescope.

40. In erecting columns, it frequently happens that the bases are not square to the axis of the column. This throws the column out of a vertical line, which, when brought to its proper position, causes a bearing only on one edge. In order to bring the columns to a fair bearing, sheets of paper, called *shims*, are sometimes used to fill in. If the ends have been left rough, the custom is to use lead or copper sheets between the bases. These two metals are so soft that any slight inequality in the surface of the base sinks into the soft metal, giving the column a full, square bearing.

41. Column Bases.—The conditions that arise in designing a column base to set on the capstone are slightly different from those met with in designing a base at a column connection. The former usually stand directly on a capstone whose unit crushing strength is much less than that of the column. The base must therefore be increased in area so as not to produce a greater pressure per square inch than is safe. Column bases are usually set on a lead or copper plate or in Portland cement. When lead or copper is used the crushing strength of the stone should be considered, as the compressive strength of these metals is greater than that of the stone. When, however, cement is used, the allowable compressive strength per square inch is not more than from 200 to 300 pounds. The Portland cement may be used neat, that is, without any sand or gravel, or it may be used in the proportion of one part cement to one or two parts sand. The strongest mixture is that of one part cement and one part of clean, sharp sand.

Such a design as is shown in Fig. 14 (*a*) is suitable for a column that rests directly on the capstone. The ribs

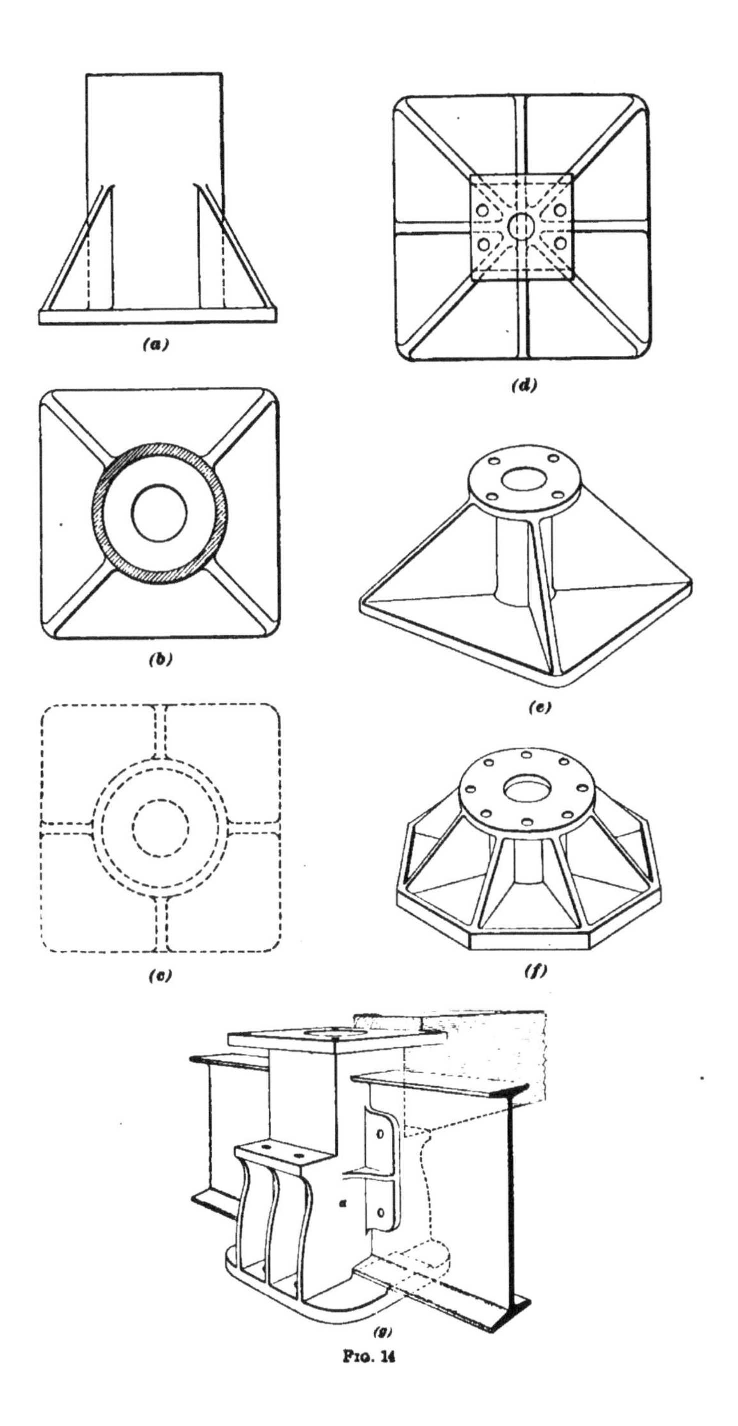

FIG. 14

should be so placed as to protect the corners, as at (*b*) and not as at (*c*), because the corners are the more liable to break off from the load of the column. A much better plan is to rest the column on a separate base, such as is shown in (*e*). This base is suitable for a cylindrical column holding only a comparatively light load. In this case the column has the same flange connection at the base that it has at the top. The advantages of this method are that the bases may be built heavier and may be set so as to give a perfectly level bearing on which to start the column. They are less awkward to handle than a column that has its base cast with it, and they are indispensable where a column is to sustain a heavy load, in order to distribute the pressure evenly over the capstone. Here, as in (*a*), the ribs run to the corners. In (*d*) is shown a base commonly used in connection with the **H** form of column. This base is designed to hold a heavier load than the one shown in (*e*). In (*f*) is a form of base suitable for the heaviest loads to which a cast-iron column is subjected. It is octagonal, having ribs to each angle of the octagon. The lower flange is further stiffened by a vertical rib running around the edge of the base.

A type of base that is sometimes used with advantage for supporting tiers of cast-iron columns is shown in Fig. 14 (*g*). This base is supported on a pier that extends through the basement floor, and the casting *a* provides a support for the first-floor columns and a connection for the first-floor beams.

In these bases, no holding-down bolts are used, as the cement is considered sufficient to stand all shocks, and when the weight of the building comes on the columns, the liability of displacement is still further reduced.

DESIGN OF CAST-IRON BRACKETS

42. The beam connections form an important feature in the selection of a column section. Some column sections are especially adaptable to this service, for example, the rectangular and the **H** section; while the circular columns, especially the smaller ones, are not so easily designed for bracket connections. The usual rule to be followed is that the brackets should be as near the center of the column as possible so as to carry the load to the center line of support. This is done in order to reduce the eccentric loading as much as possible. A bracket should be strong enough to support its load with safety.

The usual form of bracket is shown in Fig. 15 (*a*). The angle made by the outer edge of the vertical web of the bracket should be 30° with the axis of the column. It is sometimes desired to add to the architectural effect by breaking this line into a molding. The cyma reversa shown in (*b*) is well adapted to this treatment, as it gives plenty of metal at the top of the bracket where it is most needed. The cyma recta (*c*), although often used, is not so well adapted to the design of a bracket because the edge is weak, and if a beam were to rest on this edge alone, it is probable that it would break off as shown in (*d*). To avoid this, the top edge may be slightly beveled so that the beam cannot rest on it. The section in (*f*) designates the manner in which the thickness of the shell adjacent to the bracket is frequently increased. This figure, it will be noticed, is somewhat similar to the section in (*d*). A very good design for a strong bracket for an architectural column is shown in (*g*). Here the bed is held up by four vertical webs. In (*e*) is shown a method of supporting a wooden post on a cast-iron column, the post being slipped into the projecting lip *d*. A vertical lug is usually provided to which the beams are bolted. It may be made as shown at *f* in (*e*), or, where a box girder is to be supported, the lug may be designed to slip in between the two webs of the girder, as at *l* in (*h*) and (*i*).

Necessity frequently arises for a cast-iron column that has some pretensions to architectural finish. In Fig. 16 (*a*) is shown a cast-iron column with molded brackets and an architectural cap. The main girders in the building are designated at *a*, and are composed of steel I beams. The secondary girders *b* are of 10″ × 12″ yellow pine. A bracket is provided at *c* for bolting the steel girder *a* securely in

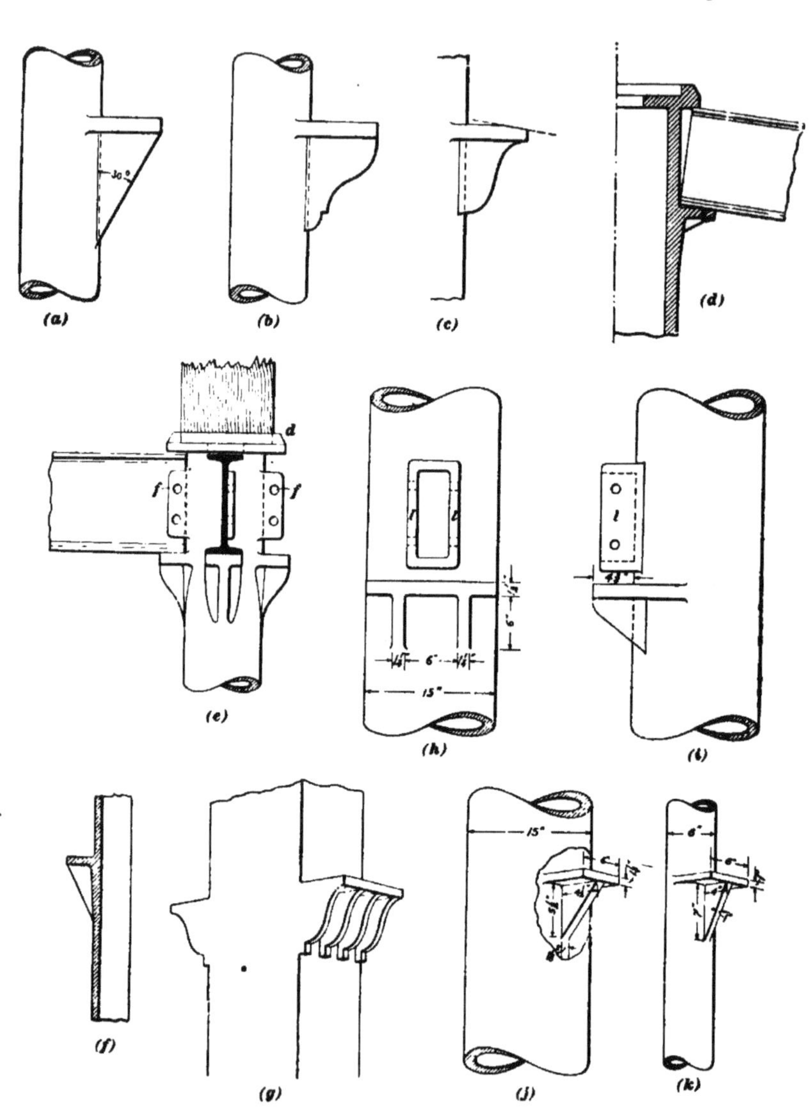

FIG. 15

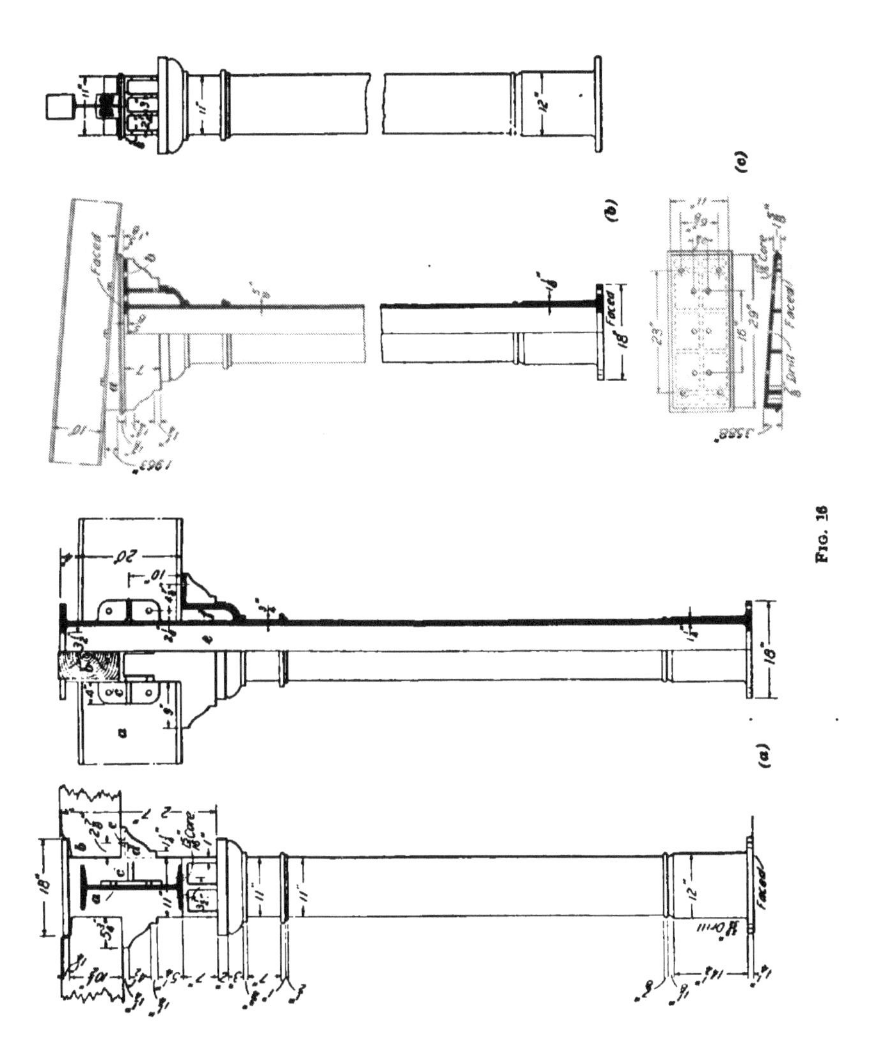

Fig. 16

position, while the secondary wooden beams *b* are held in place by lagscrews passed through the bracket *d* at *e*. The same type of column for an upper story, supporting an inclined rafter, is shown in Fig. 16 (*b*). The block *a* is separate from the column and is secured to it with bolts extending through the bracket at *b*.

43. Many concerns have their own standard designs for column connections and brackets. These are usually embodied in tables that give the required dimensions. Table V gives the standard dimensions of brackets on cast-iron columns for I-beam connections. The face of the shelf should have a pitch away from the column of $\frac{1}{8}$ inch to the foot to allow for the deflection of the beam. As the holes in the column are cored, it will be necessary to have the beams drilled in the field in order to insure alinement.

In this table the values given in the columns marked *A*, *B*, *C*, etc. are the various dimensions for brackets, these dimensions being represented by corresponding letters in the figures accompanying the table. Thus, for a 12-inch I-beam connection, the distance from the bottom of the beam flange to the center of the outside bolt of the vertical lug should be 3 inches, the pitch of the bolts, 3 inches, the projection of the bracket beyond the column, $4\frac{1}{2}$ inches, the depth of the vertical leg of the bracket, $7\frac{3}{4}$ inches, etc.

44. Strength of Cast-Iron Brackets.—In Fig. 15 (*h*), (*i*), (*j*), and (*k*) are shown brackets that were tested to their breaking point. The brackets in almost every case tore out the body of the column, as shown in (*j*). The minimum strength of the bracket shown in (*h*) and (*i*) with a distributed load was about 160,000 pounds, and with a concentrated load at the edge of the bracket, was about 95,000 pounds. The somewhat surprising result was found, in testing brackets of the dimensions of those shown in (*j*) and (*k*), that on different sized columns their strength was practically the same. It will be noticed that these brackets are practically of the same dimensions, with the exception of the height of the web, which is $5\frac{1}{2}$ inches on the 15-inch

TABLE V

DIMENSIONS, IN INCHES, OF STANDARD CONNECTIONS TO CAST-IRON COLUMNS

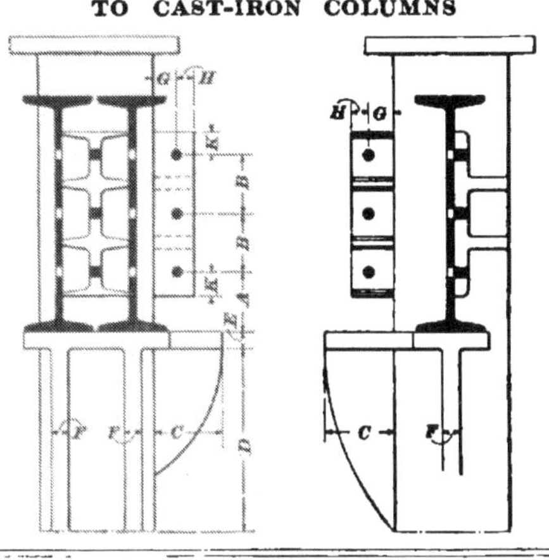

Depth of Beam	A	B	C	D	E	F	G	H	K	Thickness of Lugs	Holes Cored for ¾-Inch Bolts
20	5	5	6	10½	1½	1½	2	1½	2	1	
18	4	5	6	10½	1½	1½	2	1½	2	1	
15	4	3½	5½	9½	1½	1¼	2	1½	1¾	1	
12	3	3	4½	7¾	1¼	1¼	2	1½	1½	1	

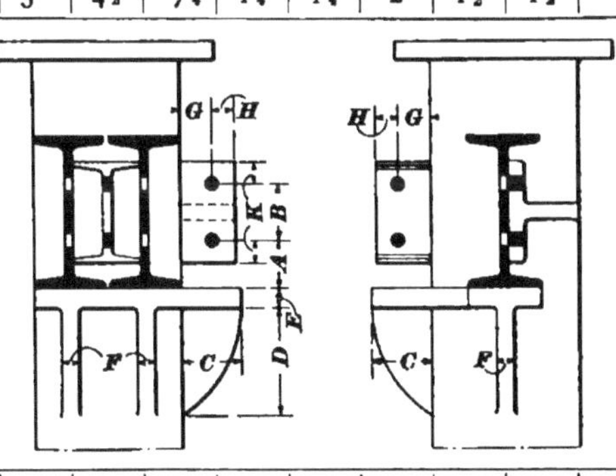

Depth of Beam	A	B	C	D	E	F	G	H	K	Thickness of Lugs	Holes Cored for ¾-Inch Bolts
10	3¼	3½	4	7	1¼	1	2	1½	1½	1	
9	3	3	4	7	1	1	2	1½	1½	1	
8	2½	3	4	7	1	1	2	1½	1½	¾	
7	2¼	2½	4	7	1	1	2	1½	1½	¾	

column and 7 inches on the 6-inch column. When tested to destruction, the breaking strength with a distributed load for the former varied from 122,800 to 142,000 pounds, and for a concentrated load at the edge of the bracket, from 69,800 to 72,900 pounds. The breaking strength of the latter for a distributed load varied from 131,000 to 133,700 pounds, and for a concentrated load at the edge of the bracket was 49,140 pounds. The last test quoted may be discarded, since the manner of failure was by breaking off the lip of the bracket instead of tearing out, as in the other cases. If this result be neglected, it will be seen that the two brackets were practically of the same strength.

If the bracket shown in (*j*) were figured for shear, according to the usual method, the result would be $6 \times 1\frac{1}{4} + 5\frac{1}{2} \times 1\frac{1}{8} = 13\frac{11}{16}$ square inches. This multiplied by the shearing strength of cast iron, 18,000 pounds, gives $18,000 \times 13\frac{11}{16} = 246,375$ pounds. It is easily seen that the calculated shearing strength of the bracket is greatly in excess of the actual strength. If the allowable shearing strength of cast iron is taken at 3,000 pounds per square inch, which is the value required by the New York building laws, we find the safe bearing strength of the bracket to be $13\frac{11}{16} \times 3,000 = 41,062$ pounds. This shows that, while the theoretical factor of safety is about 6, the actual factor of safety is little more than $1\frac{2}{3}$ with a concentrated load at the edge, and about 3 with a distributed load. The following conclusions were made from the results of these tests:

1. A bracket on the usual cylindrical cast-iron column will usually fail by tearing out a portion of the body of the column; the stress that causes this failure is unknown, but it is probably a transverse stress similar to that which exists in a beam.

2. By deductions from tests, figuring the total area of the breaking surface of a bracket, the breaking strength amounts to from 5,700 to 18,000 pounds per square inch with a distributed load, and from 3,600 to 8,400 pounds with a concentrated load at the edge of the bracket.

3. A bracket projecting from 4 to 5 inches from the body of the column will carry only half as much when the load is

concentrated at the edge of the bracket as it would if the load were distributèd over the bracket.

4. The bracket should be made so as to bring the load as close to the column as possible, and thus reduce the bending stress.

45. The following rules have been deduced from tests made and are probably the most reliable for calculating the strength of brackets. A bracket may fail in one or more of the three following ways: (1) by direct shear at the column; (2) by breaking off due to a transverse stress at the column; (3) by tearing out a portion of the column, leaving an approximately elliptical hole.

The First Method.—Under most conditions a bracket will not fail by this method, as most brackets are loaded so as to produce transverse stress rather than shearing. It is well, however, to calculate the strength of the bracket for this stress, assuming the load to be placed close to the column and calculating the shearing strength at 18,000 pounds per square inch of cross-section.

The Second Method.—This method of failure is quite probable, especially with brackets on columns of small diameters. The strength may be figured by applying the formula

$$W = \frac{31{,}000 \times I}{10\,c\,l} \qquad (23)$$

in which W = safe load on the bracket;

I = moment of inertia of section of bracket made by a plane passing through bracket and tangent to outside of column;

c = distance from neutral axis to extreme fiber;

l = distance of center of gravity of load from plane of I.

The Third Method.—The bracket is apt to fail by this method in most cases, especially with columns whose diameters are greater than 6 or 8 inches. It is practically impossible to obtain a method by which the strength of the bracket may be calculated for this kind of failure, as the failure occurs partially from transverse and partially from shearing strains.

If the section modulus of the actual break in the brackets tested could be obtained, it might be possible to find a modulus of rupture that could be utilized in calculating the breaking strength by this method. This section modulus evidently depends on the diameter and thickness of the column, the width of the ledge of the bracket, and the height of the vertical web. If it could be calculated for this form of break, it would probably be found to be less than the section modulus obtained by the second method in columns where the diameter is greater than 6 inches, and greater than the section modulus found by the second method in columns whose diameters are less than 6 inches.

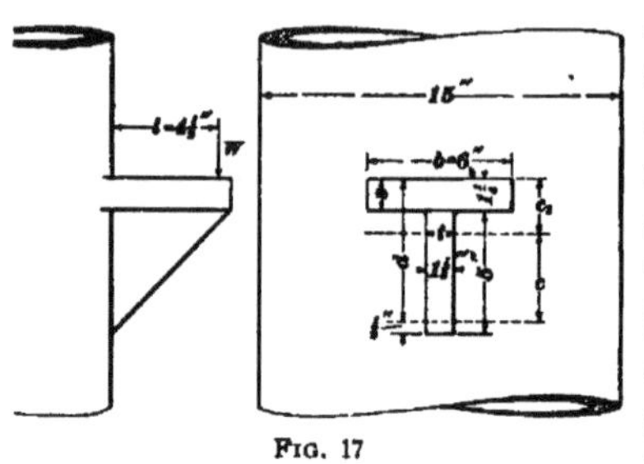

FIG. 17

EXAMPLE.—What will be the safe load of the bracket shown in Fig. 17, if the load is concentrated ½ inch from the outer edge, using a factor of safety of 4?

SOLUTION.—The area resisting the load by shear is $6 \times 1\frac{1}{4} + 4\frac{1}{2} \times 1\frac{1}{8} = 12.5625$ sq. in.; allowing 18,000 lb. per sq. in. shearing, we have 226,125 lb. This divided by 4 gives 56,531 lb., safe load. In measuring the height of the vertical web, ½ inch has been allowed for breaking at the foot of the bracket.

The area at the assumed plane of fracture being 12.56 sq. in., from tables giving the properties of various sections, the distance of the neutral axis from the back of flange is

$$\frac{d^2 t + s^2(b - t)}{2A} = \frac{(5\frac{3}{4}^2 \times 1\frac{1}{8}) + 1\frac{1}{4}^2(6 - 1\frac{1}{8})}{25.12}$$

in which $c_1 = 1.783$ in., and $c = 5.750 - 1.783 = 3.967$ in.

From tables giving the properties of sections, the moment of inertia of the section is $I = \frac{t c^3 + b c_1^3 - (b - t)(c_1 - s)^3}{3}$

$$= \frac{1\frac{1}{8} \times 3.967^3 + 6 \times 1.783^3 - (6 - 1\frac{1}{8})(1.783 - 1\frac{1}{4})^3}{3} = 34.502$$

Then the safe load, by formula **23**, is $W' \times \text{area} = \frac{31{,}000 \times 34.502}{10 \times 3.967 \times 4.5} \times 12.5625 = 75{,}262$ lb.; hence, the safe load is the smaller of these results, or 56,531 lb. Ans.

EXAMPLE FOR PRACTICE

What would be the shearing strength of the bracket shown in Fig. 15 (*k*), allowing ½ inch for breaking at the foot of the bracket?
Ans. 225,000 lb.

---

46. Beam Connections.—Bolt holes for connecting the beams to the columns should always be drilled either in the column or in the steel beams after the latter are in place, because if the holes were cored in the casting and the holes in the beam punched at the mill, it would very likely be found, in the course of erection at the building, that the beams were supported entirely by the shear of the bolts and not by the bracket. The bolts should always fit as tightly as possible, and in the best work both holes should be drilled in the field, the bolts used being machine fit, for on these beam connections depend the rigidity of the building.

47. Inspection.—In examining castings used in building construction, to ascertain their quality and soundness several points are to be considered. The edges should be struck with a light hammer. If the blow makes a slight impression, the iron is probably of good quality, providing it be uniform throughout. If fragments fly off and no sensible indentation be made, the iron is hard and brittle. Air bubbles and blowholes should be searched for by tapping the surface of the casting all over with a hammer. Bubbles, or flaws, filled in with sand from the mold, or purposely stopped with loam, cause a dullness in the sound, leading to their detection. The metal of a casting should be free from bubbles, core nails, or flaws of any kind. The exterior surface should be smooth and clean and the edges of the casting sharp and perfect. An uneven or wavy surface indicates unequal shrinkage, caused by want of uniformity in the texture of the iron.

The surface of a fracture, examined before becoming rusty, should present a fine-grained texture, of a uniform bluish-gray color and high metallic luster.

EXAMPLE.—A long-leaf, yellow-pine, square column 20 feet long is required to sustain a load of 100,000 pounds; provided a factor of safety of 5 is used, what must be the size of the column, according to the parabolic formula for wood columns?

SOLUTION.—In order to solve this problem it is first necessary to assume the size of the column and then determine whether the assumed size meets the requirements of the conditions imposed. In this case a 12″ × 12″ column is assumed. Substituting values in formula **13**, $u_f = 4{,}000 - .8\left(\frac{l}{d}\right)^2$, it is found that $u_f = 4{,}000 - .8\left(\frac{240}{12}\right)^2 = 3{,}680$ lb. per sq. in., which, multiplied by the sectional area of the column, or 144 sq. in., equals 529,920 lb. Using a factor of safety of 5, the safe strength of this column is 529,920 ÷ 5 = 105,984 lb., which is greater than the load to be sustained, and consequently the assumed size 12″ × 12″ is satisfactory and should be used. Ans.

EXAMPLES FOR PRACTICE

1. According to the straight-line formula, what will be the allowable load on an 8″ × 10″ spruce column, 12 feet long, using a factor of safety of 6, the ultimate compressive strength of spruce being 4,000 pounds per square inch? Ans. 43,733 lb.

2. It is required that a short, round, yellow-pine column shall carry 173,000 pounds. What must be the diameter of the column, the safe unit compressive stress of the material being 1,000 pounds? Ans. 15 in.

3. What will be the breaking load of a cast-iron column 20 feet long, 12 inches in diameter outside, made of 1-inch metal, according to Gordon's formula (**15**)? Ans. 1,349,248 lb.

4. The thickness of the metal in a cast-iron column is ¾ inch; if a factor of safety of 6 is used, and the length of the column is 18 feet, what must be its outside diameter to support a load of 133,000 pounds, using formula **15**? Ans. 10.5 in.

5. Find by Gordon's formula (**15**), the ultimate crushing strength of a 10-inch square, outside measurement, cast-iron column; the thickness of the metal is 1 inch, and the length of the column is 20 feet. Ans. 1,327,176 lb.

COLUMNS AND STRUTS

(PART 2)

STEEL COLUMNS

INTRODUCTION

1. It is to the adoption of structural steel shapes in building construction that the modern high building owes its existence. Before steel construction was employed, such loads as 700 or 800 tons on one column were considered visionary, for they could not with safety be sustained by any cast-iron or wood column of practical dimensions. But now loads of 1,000 tons, or more, are often supported on a single column, and the structural steel column is superseding the cast-iron column even in buildings of moderate height.

Although the loads in buildings are generally quiescent and the liability to sudden shock is more remote than in bridges, the columns seldom receive their loads as favorably as in bridges, because in most cases they are subjected to considerable eccentricity in loading; that is, the loads on one side of a column are heavier than those on the other, thus tending to produce bending stress and materially decreasing the ultimate compressive strength of the column.

CLASSIFICATION OF STEEL COLUMNS

2. As in the case of cast-iron columns, the method of securing the ends exercises an important influence on the resistance of steel columns to bending, and consequently on their ability to resist compressive stresses. The classification adopted for structural steel columns is as shown in Fig. 1, where (*a*) is a column with **hinged** or **pinned ends**, (*b*) a column with **flat ends**, and (*c*) a column with **fixed ends.**

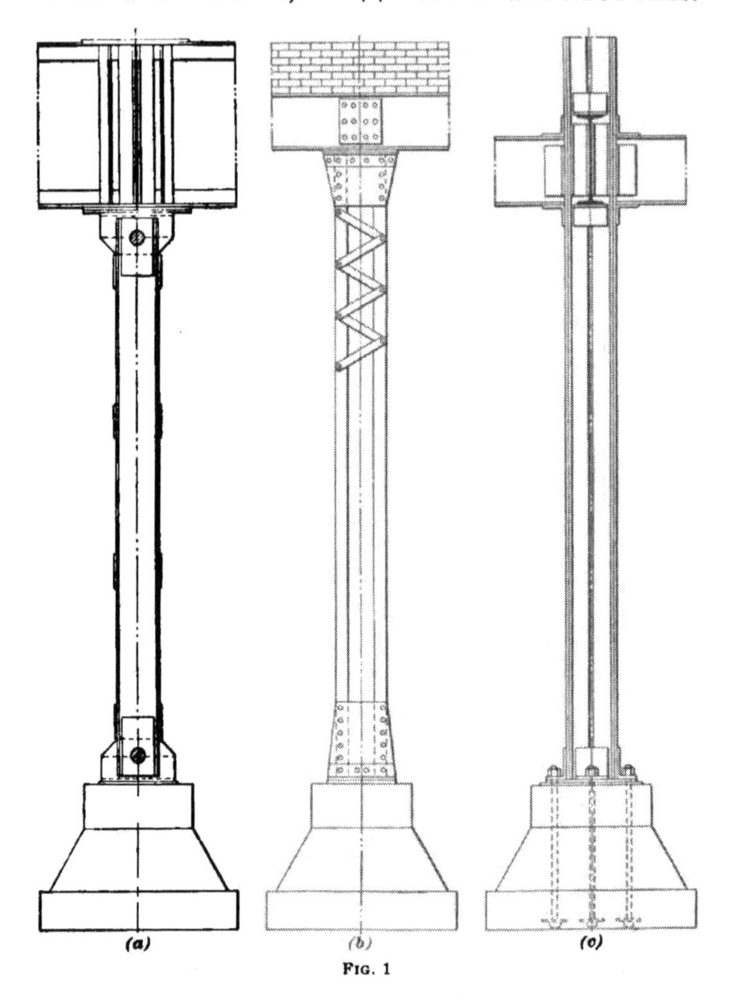

FIG. 1

Hinged-end columns are struts that have only central points or lines of contact, such as are provided by pins passed through reenforced holes in the ends of the member, or, as in unusual instances, resting in sockets in the form of semi-cylindrical depressions in cast-iron bearing blocks. The center of the end joints in a pin-connected member should lie on the central axis of the column.

Flat-end columns are struts that have flat ends normal to the central axis of the strut, but not rigidly secured at the base and top.

Fixed-end columns are rigidly secured at the end to the contiguous parts of the structure, or by anchor bolts to heavy masonry at the base and to portions of the steel frame at the top. In order to realize the conditions of fixed ends the column must be so secured that the attachment would not be severed if the member were subjected to the ultimate load.

FORMS OF STEEL COLUMNS

3. Conditions That Affect the Choice of a Type of Column.—There are at present numerous forms in which rolled-steel shapes are combined to make up columns for structural purposes. The type of column to be used in a building is sometimes prescribed by the owner, or the design of the building is furnished by the architect, thus leaving the engineer little latitude in his choice of design. There are, however, a number of conditions demanded by considerations, partly practical and partly theoretical, that should be carefully studied and compared before selecting the type of column for a particular purpose. In many cases it will be found that these considerations impose conditions that conflict with each other; and in order to make such a compromise as will meet this difficulty in the most satisfactory manner, there is demanded of the engineer a most careful exercise of judgment guided by practical experience. Some of the most important points to be considered in choosing the type of column to be used, in any case, are the following:

1. The cost and availability of the material.
2. The amount of labor required and the facility with which it may be performed in both shop and field.
3. The distribution of material in the column so as to give the maximum strength with the least weight.

4. The facility with which connections may be made between the column and the members which it supports.

5. The application of the connections in such a manner that they will transfer the compressive stresses directly to the axis of the column.

6. The facility with which the thickness of the metal in the different parts composing the column can be reduced in order to meet the reduced loads of the upper floors. It is not desirable to make the columns supporting the upper floors of the building very small, since the beams and girders supporting the upper floors are usually of the same dimensions as those for the lower floors, and consequently require connections as heavy and secure; it is almost impossible to make such connections to small columns, consequently, in order to reduce the weight of the column for the lighter load that it will carry, it is better to reduce the thickness of the material used and keep the section the same.

7. The facility with which fireproofing may be attached to the section. Columns of circular sections may be fireproofed more compactly than rectangular columns.

4. In regard to the cost and availability of the material, such shapes should be selected as are easily rolled and can be placed on the market at a reasonable price. **I** beams, channels, angles, and **Z** bars, together with plates, are the most common commercial sections; these are manufactured at nearly every structural steel mill, and may be obtained promptly in large orders at any locality where a skeleton-constructed building is likely to be erected. Patented sections do not fulfil this condition, and should therefore be avoided in most cases. The consideration of prompt delivery is an important one, and greatly influences the cost and facility with which the modern building is erected.

The consideration of the facility with which the labor in both shop and field can be performed is one that should receive careful attention. In the shop the complexity of the column section, and the number of pieces of which it is composed, greatly influence the element of labor. If there are numerous small pieces, such as brackets or splice plates, each of which requires cutting, bending, and fitting together, with frequent handling, the cost of labor may be proportionately very great. The number of rivets also greatly influences the cost of a column; not only should there be as few rivets as is consistent with strength, but the construction should permit the rivets to be readily driven by machine, so as to avoid hand riveting, which is slow and expensive and now generally admitted to be inferior to machine work. These facts are also true of the labor in the field; the connections should be as simple as possible, and with as few rivets as is consistent with the strength required; they should be easy of access, so that the work on them may be executed conveniently and rapidly.

5. In connection with the question of the distribution of the material in the column so as to develop the maximum amount of strength with the least weight of material, it is

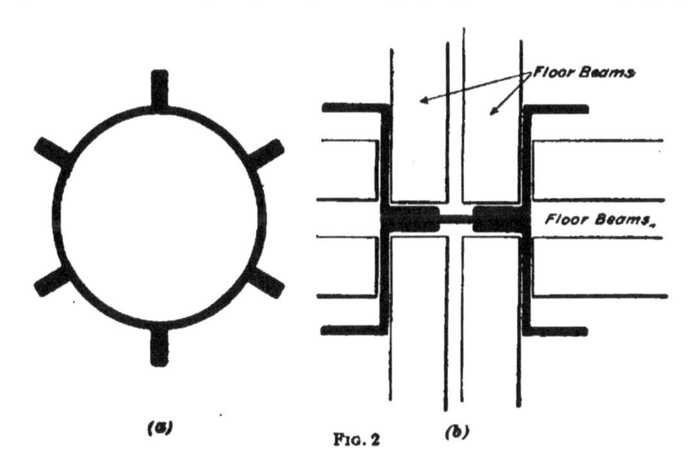

FIG. 2

necessary to consider the facility with which economical connections may be made between the columns and floor-beams, and the directness with which the connections transfer the stresses to the central axis of the column. The relation between these conditions may be illustrated and analyzed by comparing the sections shown in Fig. 2, where (*a*) represents a section of a Phœnix column, while (*b*) is a section of a **Z**-bar column. The apparent advantage of the Phœnix column is that the material composing it is placed where it will be the most efficient, thus fulfilling the third condition of the above list in a satisfactory manner; the radius of gyration of the column is also practically the same on all of its diameters, and the metal is placed at the greatest possible distance from its center.

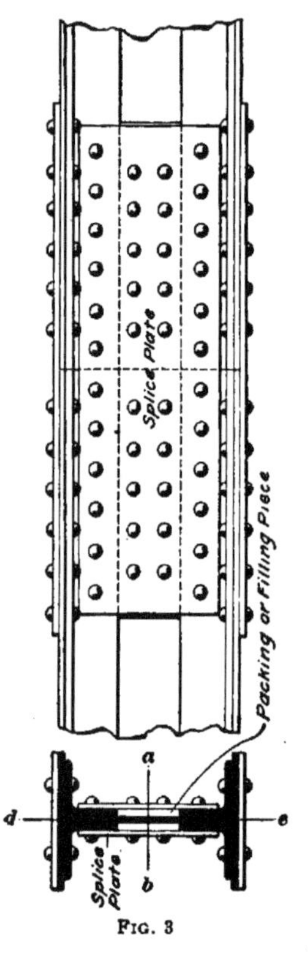

Fig. 3

On the other hand, by referring to the section of the **Z**-bar column, it is seen that a considerable portion of the material is concentrated on the axis, a condition that gives it a relatively small radius of gyration and demands the use of a greater weight of material for a given load. This column, however, offers greater advantages for the proper connection of the floorbeams than does the other, and, owing to its open construction, the beams transmit their loads almost directly to the central axis of the column, thus avoiding the disadvantages of eccentric loading.

Thus it is seen that the section having the best theoretical distribution of material is not always the best to use, on account of these several practical considerations; in fact, the section shown in Fig. 3 is probably as much used for structural steel columns as any other, although it is not an economical section, for the reason that its radius of gyration on the axis *de* is very small in proportion to the weight of material used, and, since the least radius of gyration is always used in calculating the strength of a structural column, all the material that goes to form the greater radius of gyration adds nothing to the theoretical strength.

The great advantage of this section is that it is composed of the cheapest rolled sections, which are put together with a minimum amount of labor. It is also one of the best forms for attaching the beams and girders, and, as will be seen by a study of the details of the splice shown in the figure, for making connections between two adjacent columns.

6. Sections of Columns Frequently Used in Skeleton Building Construction.—The following list shows the general forms of the principal types of steel-column sections now in use:

Larimer column (patented), 1 row of rivets.

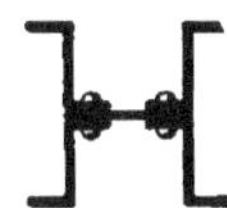

Z-bar column without cover-plates, 2 rows of rivets.

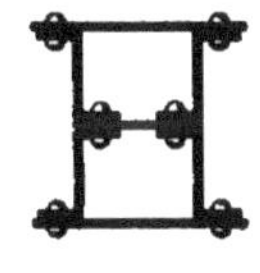

Z-bar column with one cover-plate on each side, 6 rows of rivets.

Z-bar column with two cover-plates on each side, 6 rows of rivets.

Z-bar column, additional section obtained by the use of angles and plates, 8 rows of rivets.

Z-bar column, rectangular section, 6 rows of rivets.

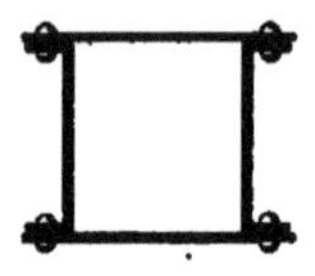

Channel column with plates or lattice, 4 rows of rivets.

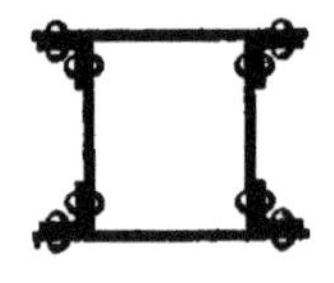

Box column of plates and angles, 8 rows of rivets.

Plate and angle column without cover-plate, 2 rows of rivets.

Plate and angle column with one cover-plate on each side, 6 rows of rivets.

Latticed angle column, 8 rows of rivets.

A column much used by the Pennsylvania Railroad, composed of angles and plates, 10 rows of rivets.

Keystone octagonal column (patented), 4 rows of rivets.

Four-section Phœnix column (patent expired), 4 rows of rivets.

Eight-section Phœnix column (patent expired), 8 rows of rivets.

Grey column (patented), 4 rows of rivets.

FORMULAS FOR STEEL COLUMNS

COLUMNS WITH CONCENTRIC LOADS

7. There are many forms of built-up columns, though several standard sections are preferred by practicing engineers. It becomes necessary, however, to find some general rule or formula by which the strength of columns of different sections may be calculated. The strength of a steel column depends on its length, the ultimate strength of the material, and the radius of gyration of the section.

There are four formulas that consider these factors, namely—*Gordon's* or *Rankin's*, *Euler's*, the *straight-line*, and the *parabolic*. These formulas give the strength of the column per square inch of section, and the actual strength of the column is determined by multiplying the value obtained from the formula by the area of the column section.

8. Gordon's Formula.—The strength of steel columns may be found by the following rule:

Rule.—*To find the ultimate strength per square inch of sectional area of a steel column with fixed ends, divide the ultimate strength of the material composing the column by 1 plus the quotient of the square of the length, in inches, divided by the product of the constant 36,000 and the square of the least radius of gyration of the section.*

For columns with one end fixed and the other end pin-connected use 24,000 for the constant, and for columns with both ends pin-connected use 18,000.

This rule may be expressed by the formulas given in Table I.

TABLE I

GORDON'S FORMULA FOR u

Material	Fixed Ends	Square or Pin-Connected and Fixed Ends	Pin-Connected on Both Ends
Wrought iron	$u = \frac{40{,}000}{1 + \frac{l^2}{40{,}000 r^2}}$	$u = \frac{40{,}000}{1 + \frac{l^2}{30{,}000 r^2}}$	$u = \frac{40{,}000}{1 + \frac{l^2}{20{,}000 r^2}}$
Soft steel	$u = \frac{45{,}000}{1 + \frac{l^2}{36{,}000 r^2}}$	$u = \frac{45{,}000}{1 + \frac{l^2}{24{,}000 r^2}}$	$u = \frac{45{,}000}{1 + \frac{l^2}{18{,}000 r^2}}$
Medium steel	$u = \frac{50{,}000}{1 + \frac{l^2}{36{,}000 r^2}}$	$u = \frac{50{,}000}{1 + \frac{l^2}{24{,}000 r^2}}$	$u = \frac{50{,}000}{1 + \frac{l^2}{18{,}000 r^2}}$

in which u = ultimate strength of column, in pounds per square inch of section;
l = length of columns, in inches;
r = least radius of gyration of section.

It will be noticed from Table I that the same formula applies to columns with square ends as to those that are fixed on one end and pin-connected on the other.

9. Straight-Line Formula.—The following formulas are used in calculating the ultimate unit strength of columns whose lengths are between 50 and 150 radii of gyration:

MEDIUM STEEL

Fixed ends . . . . $60{,}000 - 210\frac{l}{r}$ (1)

Square ends . . . . . . $60{,}000 - 230\frac{l}{r}$ (2)

Pin-connected ends . . $60{,}000 - 260\frac{l}{r}$ (3)

SOFT STEEL

Fixed ends . . . . $54{,}000 - 185\frac{l}{r}$ (4)

Square ends . . . . . . $54{,}000 - 200\frac{l}{r}$ (5)

Pin-connected ends . . $54{,}000 - 225\frac{l}{r}$ (6)

These formulas are derived in practically the same manner as the straight-line formulas used in Fig. 7, *Columns and Struts*, Part 1, the difference being due to the fact that a higher value is used in the above for the ultimate strength of the material.

10. Euler's and the Parabolic Formulas.—As in the case of cast-iron columns, the parabolic formula covers most practical cases. However, because steel is so much stronger, columns whose lengths are great in comparison with their radii of gyration are sometimes used, in which cases Euler's formula may be applied. The limiting value of $\frac{l}{r}$ for wrought iron and steel in these formulas is given in the following table:

WROUGHT-IRON COLUMNS

Pin-connected ends

$$\begin{cases} \text{When } \frac{l}{r} < 170,\ u = 34{,}000 - .67\left(\frac{l}{r}\right)^2 & (7) \\ \text{When } \frac{l}{r} > 170,\ u = \dfrac{432{,}000{,}000}{\left(\frac{l}{r}\right)^2} & (8) \end{cases}$$

Flat ends .

$$\begin{cases} \text{When } \frac{l}{r} < 210,\ u = 34{,}000 - .43\left(\frac{l}{r}\right)^2 & (9) \\ \text{When } \frac{l}{r} > 210,\ u = \dfrac{675{,}000{,}000}{\left(\frac{l}{r}\right)^2} & (10) \end{cases}$$

MILD-STEEL COLUMNS

Pin-connected ends

$$\begin{cases} \text{When } \frac{l}{r} < 150,\ u = 42{,}000 - .97\left(\frac{l}{r}\right)^2 & (11) \\ \text{When } \frac{l}{r} > 150,\ u = \dfrac{456{,}000{,}000}{\left(\frac{l}{r}\right)^2} & (12) \end{cases}$$

Flat ends .

$$\begin{cases} \text{When } \frac{l}{r} < 190,\ u = 42{,}000 - .62\left(\frac{l}{r}\right)^2 & (13) \\ \text{When } \frac{l}{r} > 190,\ u = \dfrac{712{,}000{,}000}{\left(\frac{l}{r}\right)^2} & (14) \end{cases}$$

in which u = ultimate strength of column, in pounds per square inch;

l = length of column, in inches;

r = least radius of gyration, in inches.

In the above formulas, the first in each division is used when the value of $\frac{l}{r}$ is less than the given number and the second when $\frac{l}{r}$ is greater than the given number.

11. Great care is required in the proper classification of a style of column, and it sometimes requires considerable experience to judge in which class a column should be placed. Naturally there would be six divisions, according to the method of securing the ends, as follows: (1) Fixed; (2) fixed and square; (3) square; (4) fixed and pin-connected; (5) pin-connected; and (6) square and pin-connected; but on account of the similarity of the effect of stress on certain classes, the same formula applies in more than one case, and it is therefore only necessary to have three formulas for the three following classes: (1) Fixed, or square and fixed; (2) square, or pin-connected and fixed; (3) pin-connected, or square and pin-connected, as given in Table I. The following examples will assist the student in obtaining a clear understanding of this subject.

12. Columns With Hinged Ends.—To this class belong the compression members of a pin-connected truss,

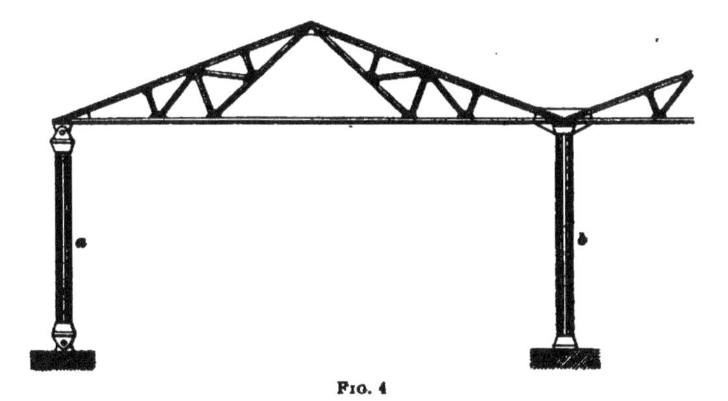

FIG. 4

or the swinging ends of a truss, such as is shown at *a*, Fig. 4. Here the column rests on a pin that permits it to swing to and fro as the truss expands and contracts from changes of temperature.

EXAMPLE.—Using a factor of safety of 4, compute the safe strength of a strut of a pin-connected truss, the length of the strut being 20 feet from center to center of pins, and its section made up of four 6″ × 6″ × $\frac{3}{8}$″ angles, connected as shown in Fig. 5.

SOLUTION.—Since the relation of the section to each of the axes *a b* and *d e* is the same, its moment of inertia and radius of gyration is the same about either axis, and it is therefore necessary to compute these factors for only one of these axes. From a table giving the properties of angles the area of the section of a 6″ × 6″ × ⅝″ angle is 4.36 sq. in., and the distance of its center of gravity from the back of either leg is 1.64 in., nearly. Also from this table, it is found that the moment of inertia of one of the angles, with respect to an axis through its center of gravity parallel with the given axis, is 15.39.

From Fig. 5, the distance of the axis through the center of gravity of the angle from the axis $d\,e$ of the section is 2.14 in.; therefore, substituting the known values in formula **1** in *Properties of Sections*, the moment of inertia of a single angle with respect to the axis $d\,e$, is $I' = I + a\,x^2 = 15.39 + 4.36 \times 2.14^2 = 35.35$. The moment of inertia of the whole section is, therefore, $I_s = \Sigma I' = 4 \times 35.35 = 141.40$.

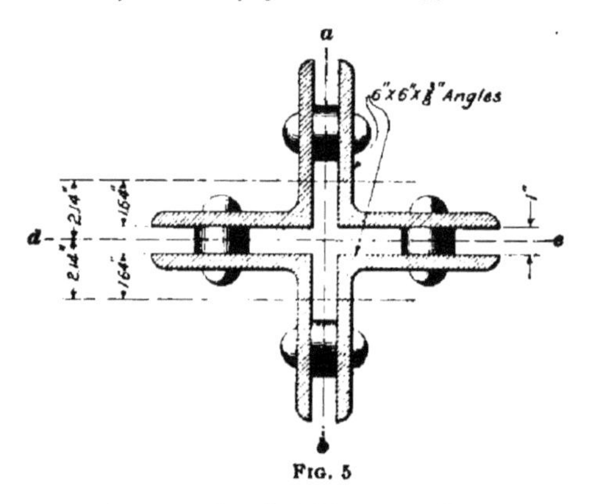

FIG. 5

The total area of the section is 4.36 × 4 = 17.44 sq. in.; therefore, the square of its radius of gyration, from formula **6** of *Properties of Sections*, is $r^2 = \frac{I}{A} = \frac{141.4}{17.44} = 8.1$, nearly.

According to the statement of the problem the length of the strut is 20 ft., or 240 in. Substituting in Gordon's formula for medium steel and pin-connected ends, we find the ultimate strength per square inch of section of the strut to be $u = \dfrac{50{,}000}{1 + \dfrac{240^2}{18{,}000 \times 8.1}} = 35{,}842$ lb.

The safe bearing strength of the strut is therefore

$$\frac{35{,}842 \times 17.44}{4} = 156{,}271 \text{ lb. Ans.}$$

13. Columns With Flat Ends or Fixed at One End and Pin-Connected at the Other.—To this class belong columns whose ends are flat and not held firmly, or where there is a single column extending from cap to foundation as in Fig. 1 (*b*). This class also includes such posts as that shown at *a*, Fig. 6, the lower end being pin-connected and

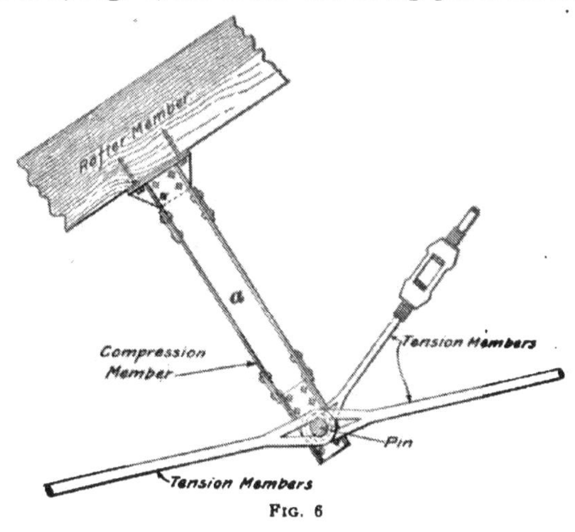

Fig. 6

the upper end securely fastened so that it will assume the shape given in Fig. 4 (*e*), in *Columns and Struts*, Part 1.

Example.—The moment of inertia, with respect to the axis $Y\,Y$ of the medium steel Z-bar column shown in Fig. 7, is 287.92, and, with respect to the axis $X\,X$, 337.17. The total area of the section is 21.36 square inches; what is the safe bearing strength of a column 20 feet in length having flat ends, if a factor of safety of 3 is used?

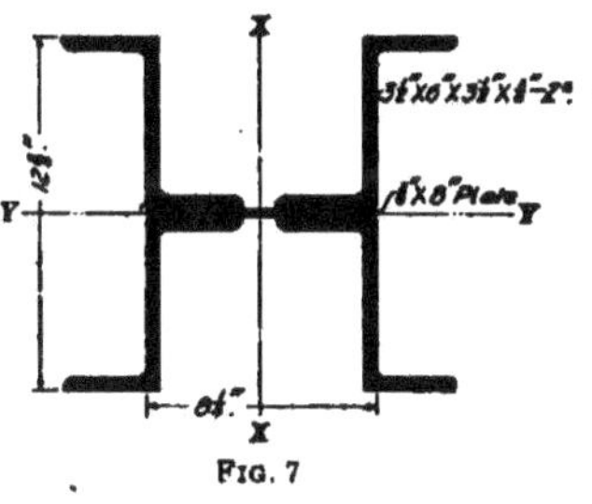

Fig. 7

Solution.—Using the least moment of inertia, which is that with respect to the axis $Y\,Y$, the square of the least radius of gyration is found to be $r^2 = \frac{287.92}{21.36} = 13.48$, nearly; therefore, the ultimate strength per square inch of section is $u = \frac{50,000}{1 + \frac{240^2}{24,000 \times 13.48}} = 42,445$ lb. The safe bearing strength is therefore

$$\frac{42,445 \times 21.36}{3} = 302,208 \text{ lb. Ans.}$$

14. Columns With Fixed Ends.—This type appears in buildings where the column extends through two or more stories or is rigidly secured to the structure so as to restrain its bending out of line.

EXAMPLE.—A section of the compression member of a large structural-steel roof truss is shown in Fig. 8. The ends of the member are firmly riveted to the adjacent members of the truss, and the length of the member is 15 feet; what is its safe bearing strength with a factor of safety of 4?

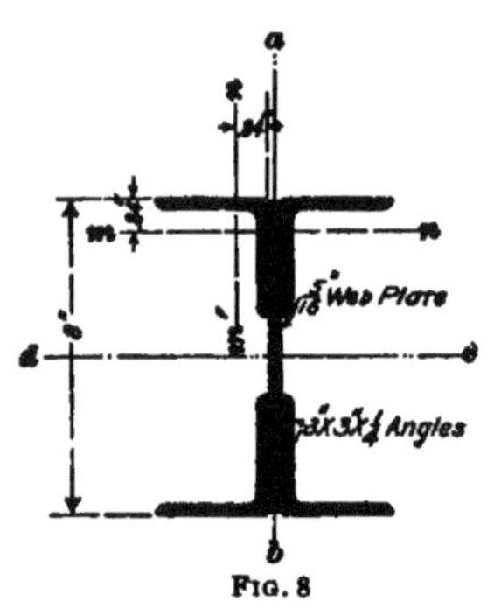

FIG. 8

SOLUTION.—From one of the tables giving the properties of angles, in *Properties of Sections*, the area of a 3″ × 3″ × ¼″ angle is found to be 1.44 sq. in., and the distance from its center of gravity to the back of the flange is .84 in.; also from the same table, the moment of inertia of the angle, with respect to an axis through its center of gravity, is found to be 1.24. In accordance with the dimensions given in the figure, the distance from the axis $d e$ to the centers of gravity of the angles is $4 - .84 = 3.16$ in. The moment of inertia of one of the angles, with respect to the axis $d e$, is, therefore, $I' = 1.24 + 1.44 \times 3.16^2 = 15.62$. The moment of inertia of the web-plate, with respect to $d e$, is $I' = \frac{\frac{5}{16} \times 8^3}{12} = 13.33$. The moment of inertia of the whole section, with respect to $d e$, is $I_s = 4 \times 15.62 + 13.33 = 75.81$. The total area of the section is $4 \times 1.44 + 8 \times \frac{5}{16} = 8.26$ sq. in.; the square of its radius of gyration, with respect to $d e$, is, therefore, $r^2 = \frac{75.81}{8.26} = 9.18$.

Referring now to the axis $a b$, the distance from the axis to the center of gravity of one of the angles is $.84 + \frac{5}{32} = .84 + .156 = .996$, say 1 in. The moment of inertia of the angle, with respect to $a b$, is, therefore, $I' = 1.24 + 1.44 \times 1^2 = 2.68$. The moment of inertia of the web-plate, with respect to $a b$, is $\frac{8 \times \frac{5}{16}^3}{12} = .02$.

Taking the sum of the moments of inertia of the several parts, the moment of inertia of the whole section, with respect to $a b$, is found to be $4 \times 2.68 + .02 = 10.74$; dividing this by the area of the section, the square of the radius of gyration, with respect to this axis, is $r^2 = \frac{10.74}{8.26} = 1.3$, nearly, which, being much less than the value of r^2

with reference to the axis de, is the value to use in computing the strength of the member.

Substituting, in Gordon's formula for medium steel and fixed ends, the ultimate strength per square inch of section is

$$u = \frac{50,000}{1 + \frac{180^2}{36,000 \times 1.3}} = 29,550 \text{ lb.}$$

and the safe bearing strength of the member is

$$29,550 \times 8.26 \div 4 = 61,021 \text{ lb. Ans.}$$

COLUMNS WITH ECCENTRIC LOADS

15. In modern buildings, nearly all columns are subjected to more or less eccentric loading; in fact, it would be difficult to find a column loaded with a perfectly concentric load. However, it is good practice to disregard the eccentric loading generally encountered in ordinary building construction, since the preceding formulas make due allowance for the slight eccentricity caused by the lack of symmetry in the arrangement of the brackets and the application of the loads with regard to the central axis of the column. For example, all the beam connections may be on one side of the column, as shown in Fig. 9, or

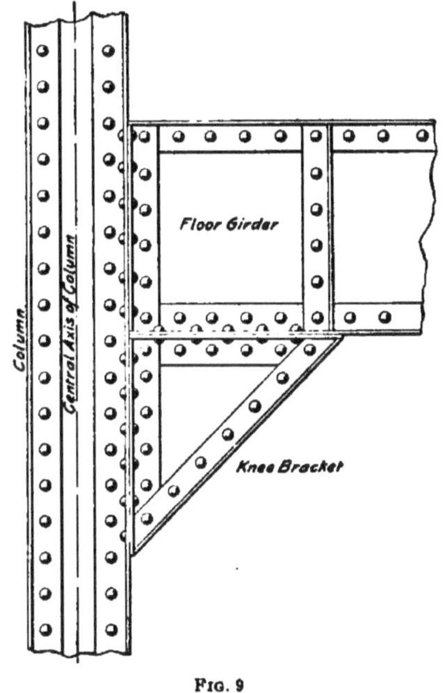

FIG. 9

the beams attached to one side of the column may be more heavily loaded than those attached to the other side. Some of these conditions are usually unavoidable and tend to produce the undesirable effect of eccentric loading.

Should the eccentricity of the load be considerable and liable to produce dangerous transverse or bending stresses, it would materially diminish the ability of the column to withstand direct compressive stresses. The bending or transverse stresses should, in such a case, be calculated and an additional amount of material should be added to the section of the column in order to resist them.

16. The nature of the material in cast-iron columns forbids their use in cases where an eccentric load of any considerable amount is liable to occur. It is very important that any eccentric loading on a cast-iron column should be reduced as much as possible and that the brackets should be so designed that their load is carried as near the center line of the column as possible.

Timber columns are often subjected to an eccentric load due to the attachment of a jib crane to them. They must be calculated for this extra stress according to the rules given hereafter and the maximun stress due to the combined direct and bending stresses should not be more than 25 per cent. above the safe unit stress of the column section if subjected to the same load concentrically placed.

Steel columns are the best adapted to use in cases where eccentric loads exist because their section may be increased with little difficulty to take care of the excessive stress. In the subsequent formulas a sufficient factor of safety has been used to take care of any eccentric load of ordinary magnitude. It is for this reason that the allowable unit stress is increased 25 per cent. over the calculated stress for a concentric load.

17. Strength of Columns With Eccentric Loads. The method, considered in this Section, of proportioning columns for eccentric loads, is pursued through the following steps:

1. The allowable unit stress is determined for the column section when subjected to the entire load without regard to the eccentricity of the loading.

2. The allowable unit stress for the combined compression and bending stresses is found by taking one and one-quarter times the allowable unit stress for concentric loading, as determined in the first step of the process.

3. The unit stress to which the column is subjected by the entire load that it supports is obtained by dividing the total load on the column, in pounds, by the sectional area of the column, in inches. The fiber stress due to the eccentric load is found by dividing the product of the eccentrically placed load and its distance from the neutral axis of the section by the section modulus of the section. By adding these two stresses, the maximum unit stress due to the direct compressive stress combined with the unit stress created by the eccentric load is obtained.

4. When the maximum unit stress due to the combined effect of the direct and eccentric loads is determined, and it is found to exceed the allowable unit stress due to the combined compression and bending, as determined in the second step of the process, the column must be redesigned.

5. In order to determine the section that it is necessary to use for the revised design in case the assumed column section is not sufficient, formula **16** is employed. In this formula, W_e is the equivalent central load that will produce the same stress on the column section as the combined central and eccentric loads. For convenience in determining the approximate value of C, used in formula **16**, Table II is given. If the judgment of the designer dictates that the loads require a light column, the smaller values given in this table are used; while if it is indicated by the conditions of the problem that a heavier column is needed, the greater values are employed. When the equivalent central load, or W_e, has thus been found, the required column may be selected from the tables given at the end of this Section and the column designed accordingly. For instance, if a latticed channel column 20 feet long is to support a load of 105 tons, it is found from Table XI under the column headed 20 Ft., that two 12-inch, 30-pound channels will be required for the column.

18. Formulas for Eccentric Loads.—The several steps described in the method of providing for eccentric loads on columns, as explained in the previous article, may be expressed in the following formulas; the principal formula for determining W_q is obtained as explained in the evolution given here.

Let W_c = central, or axial, load;
W_e = eccentric load;
W = total load on column = $W_c + W_e$;
W_q = equivalent central load;
k = eccentricity of load $W_e = \frac{v}{c}$;
A = area of cross-section of column, in square inches;
r = radius of gyration in direction of bending;
s_{ca} = allowable stress per square inch for direct compression on columns;
s_d = allowable stress per square inch for combined compression and bending = $1.25 s_{ca}$;
s_e = unit stress due to eccentric load;
s_v = direct unit stress per square inch of section due to vertical loads;
c = distance from line passing through center of gravity of column section to extreme edge of section, in inches;
v = distance from eccentric load to line passing through center of gravity of column section, in inches;
$C = \left(\frac{c}{r}\right)^2$.

The bending moment M then equals $W_e v$; the resisting moment M_1 is equal to $\frac{s_e I}{c} = s_e S$, in which I is the moment of inertia, with respect to the neutral axis, and S, the section modulus, is equal to $\frac{I}{c}$. Then, as $M = M_1$,

$$W_e v = s_e S, \text{ or } s_e = \frac{W_e v}{S} \qquad (15)$$

TABLE II

APPROXIMATE VALUES OF C FOR VARIOUS COLUMNS

	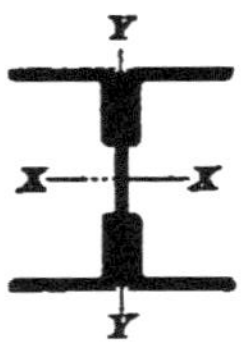			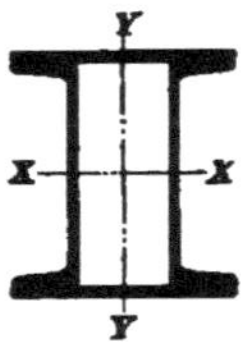	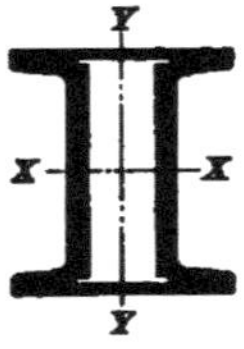	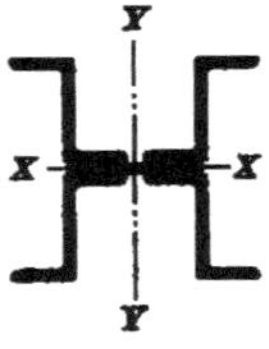	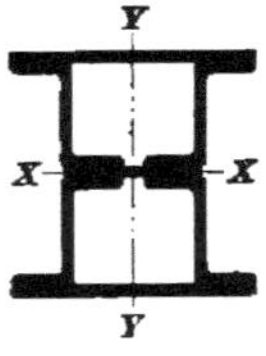	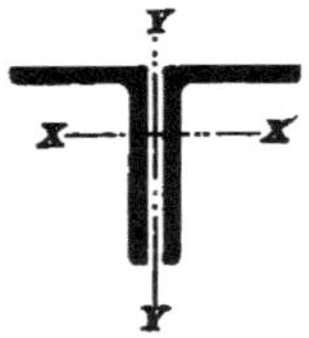
Axis XX	1.60 to 1.75	1.55	1.70 to 2.00	1.40 to 1.60	1.90	2.85 to 3.00	2.30 to 2.00	C_1 1.00 to 1.25 C_2 4.50
Axis YY	6.25 to 5.50	5.00 to 4.00		2.65 to 2.85	3.25 to 3.00	3.50	3.25 to 3.00	

The direct stress s_v is $\frac{W_c + W_e}{A}$. The total stress then is $s_v + s_e$, and this sum may equal s_d but must not exceed it.

Assuming, therefore, that $s_d = s_v + s_e$, and multiplying this equation by A, we have the greatest allowable stress, $A\,s_d = A\,s_v + A\,s_e$. As $s_v = \frac{W_c + W_e}{A}$ and $s_e = \frac{W_e v}{S}$, these values may be substituted for s_v and s_e, when the equation will become $A\,s_d = W_c + W_e + \frac{A\,W_e v}{S}$; $W_c + W_e = W$; therefore, $A\,s_d = W + \frac{A\,W_e v}{S}$. Substituting $\frac{I}{c}$ for S, we have $A\,s_d = W + \frac{W_e A v}{\frac{I}{c}} = W + \frac{W_e A v c^2}{I c}$ after the numerator and denominator have been multiplied by c^2.

As $I = A r^2$, $\frac{1}{r^2} = \frac{A}{I}$; also, $k = \frac{v}{c}$, as previously stated. Arranging the equation in the following form, $A\,s_d = W + W_e \times \frac{A}{I} \times \frac{v}{c} \times c^2$, and substituting $\frac{1}{r^2}$ for $\frac{A}{I}$ and k for $\frac{v}{c}$, we have $A\,s_d = W + W_e \times \frac{c^2}{r^2} \times k = W + W_e C k$, after substituting C for $\frac{c^2}{r^2}$. A, the required area of the column, then equals $\frac{W + W_e C k}{s_d}$, and as $s_d = 1.25 s_{ca}$, $A = \frac{W + W_e C k}{1.25 s_{ca}}$ $= \frac{4(W + W_e C k)}{5 s_{ca}}$. After multiplying by s_{ca}, the equivalent central load $W_q = A\,s_{ca} = \frac{4 s_{ca}(W + W_e C k)}{5 s_{ca}}$, and finally

$$W_q = \tfrac{4}{5}(W + W_e C k) \qquad \textbf{(16)}$$

Table II gives the values of C for various types of columns. These values are only approximate, and considerable judgment must be used in selecting them. They are given to aid the student in finding any required size of column by the formulas previously given. In the case of the two angles placed back to back, C_1 is the value obtained by considering c as the distance from the axis $X X$ to the back of the angles, while C_s is determined by taking c as the distance from the axis $X X$ to the other extremity.

19. In order that a better understanding may be had of the several operations in the method of providing for the eccentric loading on a column, the following example is given:

EXAMPLE.—The 22-foot column shown in Fig. 10, made of four 6-inch Z bars and one 8″ × ⅜″ web-plate, is loaded with 190 tons, 50 tons of which is applied on a bracket at a distance of $8\frac{3}{8}$ inches from the center line of the column. (*a*) What will be the maximum allowable fiber stress of the column? (*b*) What will be the maximum fiber stress produced by the loads on the column? (*c*) How should the column be designed to withstand this load?

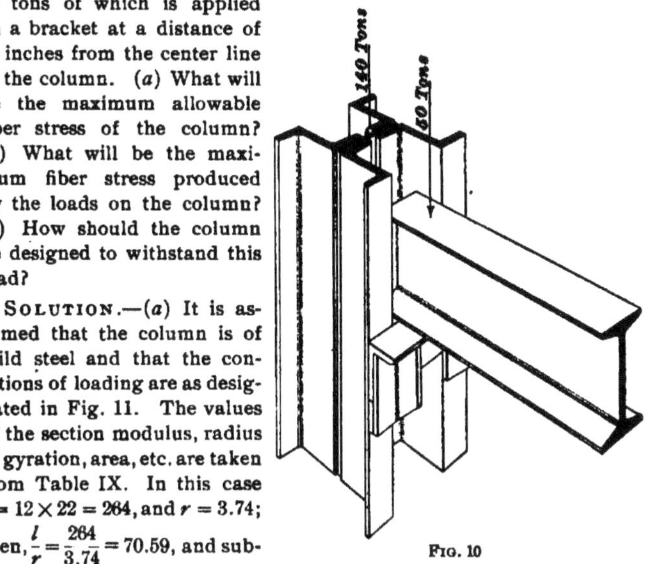

FIG. 10

SOLUTION.—(*a*) It is assumed that the column is of mild steel and that the conditions of loading are as designated in Fig. 11. The values of the section modulus, radius of gyration, area, etc. are taken from Table IX. In this case $l = 12 \times 22 = 264$, and $r = 3.74$; then, $\frac{l}{r} = \frac{264}{3.74} = 70.59$, and substituting in formula **22**, which is used in the tables, $s_{ca} = 15{,}000 - 57 \times 70.59 = 10{,}976$ lb., and the maximum safe fiber stress s_d under eccentric loading will be

$$s_{ca} \times 1.25 = 10{,}976 \times 1.25 = 13{,}720 \text{ lb. Ans.}$$

(*b*) By formula **15**, in which the load given in tons is changed to pounds, the fiber stress due to the eccentric load is $\frac{50 \times 2{,}000 \times 8\frac{3}{8}}{76.8} = 10{,}905$ lb.; the direct stress is $s_v = \frac{W_c + W_e}{A} = \frac{190 \times 2{,}000}{34.8} = 10{,}920$ lb. The maximum fiber stress produced by the loads is

$$10{,}905 + 10{,}920 = 21{,}825 \text{ lb. Ans.}$$

(*c*) This load is much more than the column will safely sustain. Table II shows that the approximate value of C around the axis XX for **Z**-bar columns varies from 2.85 to 3; as the column required will probably be large, 3 may be assumed as the value of C. Also, on account of the increased size of the column, the value of v is assumed as $8\frac{1}{2}$ instead of $8\frac{3}{8}$, and c is increased from $6\frac{3}{8}$ to $6\frac{1}{2}$. In the substitution in the following formula the load is in tons, so that the result is likewise in tons. Then,

$$W_q = \frac{4}{5}\left(190 + \frac{50 \times 3 \times 8\frac{1}{2}}{6\frac{1}{2}}\right) = 308.92 \text{ T.}$$

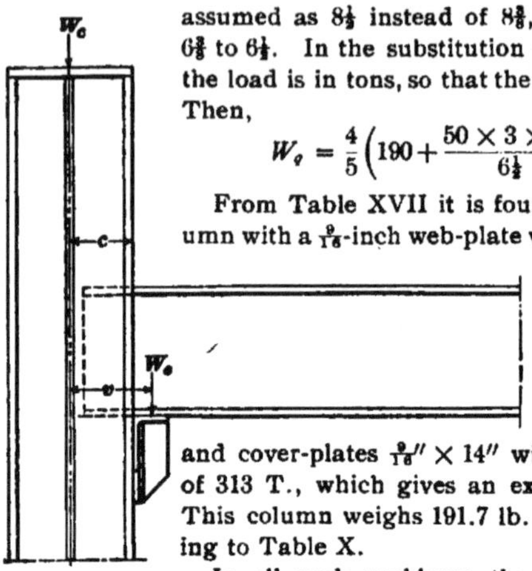

From Table XVII it is found that a 6-in. **Z**-bar column with a $\frac{9}{16}$-inch web-plate will not be strong enough, so that a column with cover-plates will have to be used. Referring to Table XVIII, it is found that a 6-in. **Z**-bar column with a web-plate $8'' \times \frac{3}{4}''$ and cover-plates $\frac{9}{16}'' \times 14''$ will have a safe resistance of 313 T., which gives an excess of strength of 4 T. This column weighs 191.7 lb. per ft. of length, according to Table X.

Fig. 11

In all such problems, the weight is an important factor and the best designer will select the lightest weight of column consistent with the required strength. If the design is changed to a plate and angle column, the value of C, as found in Table II, is assumed as 1.55, and substituting in the formula above,

$$W_q = \frac{4}{5}\left(190 + 50 \times 1.55 \times \frac{8\frac{3}{4}}{6\frac{3}{4}}\right) = 232.2 \text{ T.}$$

Assuming a plate and angle column with a web-plate of $12'' \times \frac{3}{4}''$ and four angles $6'' \times 4'' \times \frac{1}{2}''$ with $\frac{11}{16}$ in. cover-plates, the required strength is found to be 234 T., which is amply sufficient to support the load. As this column weighs only 157.5 lb. per ft. against 191.7 lb. for the **Z**-bar column, a saving of 34 lb. per ft., or 748 lb. for each column, is obtained.

20. Columns With Several Eccentric Loads.—When a column is loaded with a number of eccentric loads, the maximum fiber stress is equal to the algebraic sum of the stresses due to the loads on the column.

EXAMPLE.—An 8-inch channel column composed of two 8-inch, 17-pound channels and two $1'' \times 10''$ cover-plates is loaded with 80 tons, which is distributed on the brackets a, a and b, b, as shown in Fig. 12; what will be the maximum stress due to these loads, and at what part of the column will it occur?

SOLUTION.—Since the maximum fiber stress is equal to the algebraic sum of the stresses due to the loads on the column, it is necessary to determine both the direct stress and the stresses due to eccentric loads.

According to the third step in the method for providing for eccentric loads, as described in Art. **17** the stress due to each eccentric load is equal to its moment about the central axis of the column divided by the section modulus of the column with respect to the same axis. If the moment of a load on one side of the axis XX is considered as being positive, the moment of the load on the other side of the same axis is negative. The same is true regarding the loads on each side of the axis YY. From Table VII, the section modulus for such a column about the axis XX is 98.2 and about the axis YY is 53.4.

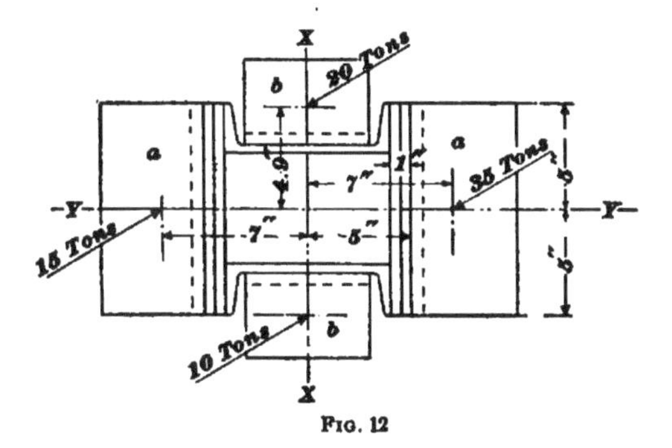

FIG. 12

Consequently, the algebraic sum of the stresses due to the four eccentric loads is $\frac{35 \times 7}{98.2} - \frac{15 \times 7}{98.2} + \frac{20 \times 4.9}{53.4} - \frac{10 \times 4.9}{53.4} = 2.344$ T.

The unit stress due to the direct load is equal to the sum of the loads divided by the area of the column section. In this case, the sum of the loads $= 20 + 35 + 10 + 15 = 80$ T., and the area of the column section found from Table VII is 30 sq. in., so that the direct unit stress is $\frac{80}{30} = 2.666$, which, added to the previous sum, equals $2.666 + 2.344 = 5.01$ T. per sq. in., or 10,020 lb. This stress would occur at the corner between the loads of 20 T. and 35 T. Ans.

21. Columns Under Bending Stress.—Columns are sometimes subjected to a direct bending stress; such cases are met with in columns supporting open sheds, especially where the truss is braced against the column, or the column supports a jib crane. In such cases the bending stress should be calculated and the column proportioned accordingly.

EXAMPLE.—Given a column on which a 5-ton jib crane has been attached, such as is shown in Fig. 13 (*a*). If the column is made up of four 6″ × ¾″ Z bars, one 8″ × ¾″ web-plate, and two 14″ × ⅝″ cover-plates, and is 22¼ feet high to the center of bearing *d*, will it be safe?

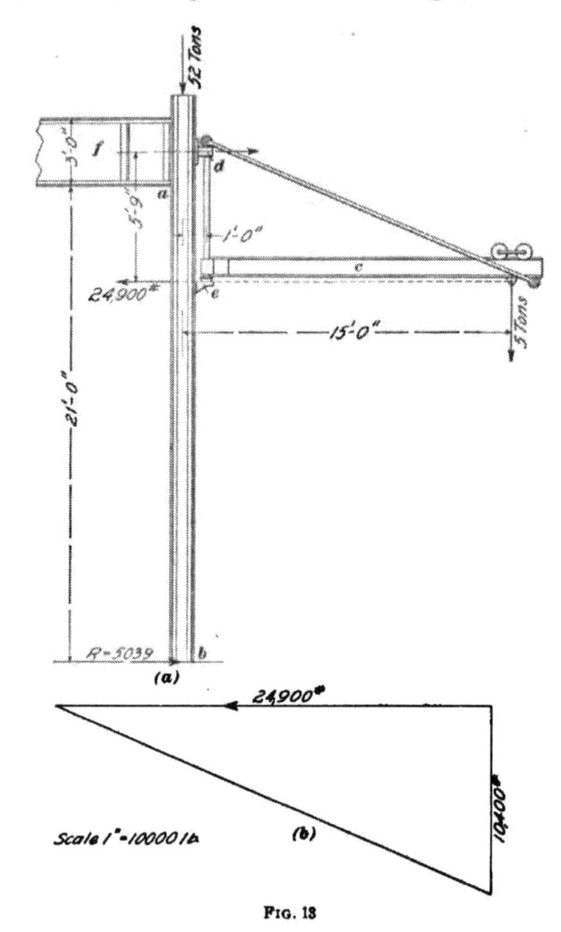

FIG. 13

SOLUTION.—From Table X, we find that the area equals 58 sq. in. The radius of gyration about axis XX is 4.82; the radius of gyration about axis YY is 3.86; the section modulus, with respect to axis XX, is 192.2; the section modulus, with respect to axis YY, is 123.5. In this case, c is 7 in. in both directions. The ultimate strength per square inch, by Gordon's formula, Table I, for medium steel, pin-connected and fixed ends is $u = \dfrac{50,000}{1 + \dfrac{(22.5 \times 12)^2}{24,000 \times 3.86^2}} = 41,535$ lb.

Using a factor of safety of 4, the safe load per square inch is $\dfrac{41,535}{4} = 10,384$ lb., and for eccentric loads is $10,384 \times 1.25 = 12,980$ lb.

In this case, the crane itself is assumed to weigh 800 lb., half of which is concentrated at each end of the I beam c. Resolving the load into its components, and considering the forces as passing through the centers of the bearings d, e, on the column, it is found, by the triangle of forces at (b), that there is a horizontal thrust of 24,900 lb. due to the load, 10,000 + 400 = 10,400 lb.

Considering the column as a beam supported at a and b, and that the center line of bearing d is coincident with the center line of the beam f, then since R : 24,900 = 51 : 252, the reaction R at the base of the column, due to this thrust is 5,039 lb. The maximum stress on the column is:

Stress due to direct load on column,

$$
\begin{aligned}
52 \times 2,000 &= 104000 \text{ lb.}\\
5 \times 2,000 &= 10000 \text{ lb.}\\
& 800 \text{ lb.}\\
\hline
& 114800 \text{ lb.}
\end{aligned}
$$

The stress per square inch of section $= \dfrac{114,800}{58} = 1,979$ lb.

Stress due to bending,

$$
\begin{aligned}
10,800 \times 12 &= 129600 \text{ in.-lb.}\\
5,039 \times 201 &= 1012839 \text{ in.-lb.}\\
\hline
& 1142439 \text{ in.-lb.}
\end{aligned}
$$

The stress per square inch due to bending is then $\dfrac{1,142,439}{S} = \dfrac{1,142,439}{192.2} = 5,944$ lb., and the total stress is 1,979 + 5,944, or 7,923 lb.

Stress when the crane is at right angles to position shown in figure:

$$
\begin{aligned}
\text{Direct stress} &= 1979 \text{ lb.}\\
\text{Bending stress}\\
(10,800 \times 12) \div 192.2 &= 675 \text{ lb.}\\
(5,039 \times 201) \div 123.5 &= 8201 \text{ lb.}\\
\text{Total stress} &= 10855 \text{ lb.}
\end{aligned}
$$

In this case, the bending stress in the column is around the axis YY, when, according to Table X, the section modulus S equals 123.5 instead of 192.2, used above.

The column will, therefore, be amply safe in both directions. Ans.

22. Factors of Safety.—The ability of a column to resist the transverse or bending stresses due to eccentric loading decreases as its length increases, and this ability is less for columns with round or hinged ends than for those with flat or fixed ends. For these reasons it is good practice to assume a minimum factor of safety for the shortest columns or struts and increase the factor with the increase in length, making the rate of increase greater for round or hinged than for flat or fixed ends.

TABLE III

FACTORS OF SAFETY

$\frac{l}{r}$	Fixed and Flat Ends	Hinged Ends	$\frac{l}{r}$	Fixed and Flat Ends	Hinged Ends	$\frac{l}{r}$	Fixed and Flat Ends	Hinged Ends
20	3.2	3.30	110	4.1	4.65	200	5.0	6.00
30	3.3	3.45	120	4.2	4.80	210	5.1	6.15
40	3.4	3.60	130	4.3	4.95	220	5.2	6.30
50	3.5	3.75	140	4.4	5.10	230	5.3	6.45
60	3.6	3.90	150	4.5	5.25	240	5.4	6.60
70	3.7	4.05	160	4.6	5.40	250	5.5	6.75
80	3.8	4.20	170	4.7	5.55	260	5.6	6.90
90	3.9	4.35	180	4.8	5.70	270	5.7	7.05
100	4.0	4.50	190	4.9	5.85	280	5.8	7.20

Assuming a minimum factor of safety of 3 for very short struts, the factors prescribed by good practice for longer struts with flat or fixed ends are found by the formula

$$F = 3 + .01\frac{l}{r} \qquad (17)$$

and for struts with hinged ends,

$$F = 3 + .015\frac{l}{r} \qquad (18)$$

in which F = required factor of safety;

l = length of strut, in inches;

r = least radius of gyration of section.

EXAMPLE.—What is the least factor of safety that should be used with: (*a*) a column with flat ends, and (*b*) a column with hinged ends, the length of the column being 20 feet, and its least radius of gyration, 2.5?

SOLUTION.—(*a*) Applying formula **17**, we have

$$F = 3 + .01 \times \frac{240}{2.5} = 3.96, \text{ say } 4. \quad \text{Ans.}$$

(*b*) Applying formula **18**,

$$F = 3 + .015 \times \frac{240}{2.5} = 4.44. \quad \text{Ans.}$$

Table III gives values for the factor of safety obtained by substituting different values of $\frac{l}{r}$ in formulas **17** and **18.**

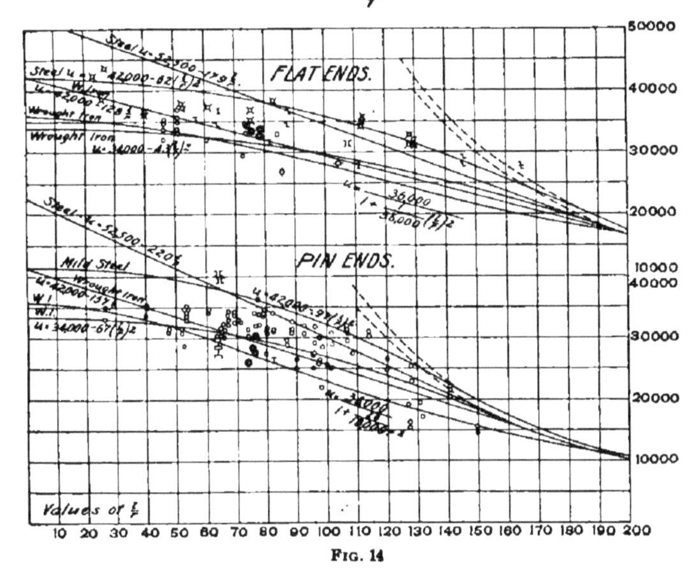

FIG. 14

These values are for columns whose length is as great as 280 times their least radius of gyration; it is not, however, good practice to use a column in which the value of $\frac{l}{r}$ is greater than 150. Another condition to be observed is that the length of the column should not exceed 45 times its least dimension.

23. Results of Tests.—Fig. 14 is a chart giving the results of tests on various forms of structural columns.

The lines showing the results of the parabolic formulas agree with the column tests to a remarkable degree of accuracy. This chart was prepared by Prof. J. B. Johnson, in order to show how closely his parabolic formula agrees with the tests made on columns.

The tests marked with a circle indicate that the test pieces were symmetrical in section with the bearing pin in the center, while those marked with two concentric circles indicate that the column tested was unsymmetrical in section with the bearing pin placed at the center of gravity. The key to the marks used in the chart is given below.

It must not be inferred from this chart that because the Phœnix columns are among the highest in their results per square inch, they are the strongest. It simply means that the columns tested were of steel with flat ends, having a value of $\frac{l}{r}$ varying from 40 to 130.

Wrought iron, Watertown Arsenal	Symmetrical sections, pin in center	○
	Unsymmetrical sections, pin in center of gravity	◎
Phœnix		¤
American		I
Wrought-iron box		▭
Pennsylvania Railroad solid web		⌶
Z bar		Z
Keystone		⌖
Steel latticed channels		][

EXAMPLES FOR PRACTICE

1. What safe load, according to Gordon's formula, can be supported by a plate and angle column made up of four 3″ × 2½″ × ½″ angles and one 6″ × ½″ web-plate, the length of the column being 16 feet and the ends being pin-connected, using a safety factor of 4? Ans. 78,299 lb.

2. Using the straight-line formula, what is the ultimate strength of a 16-inch, 304.1-pound Z-bar column of medium steel 30 feet in height, with flat ends? Ans. 3,754,800 lb.

3. Calculate the allowable maximum fiber stress under an eccentric load of a 14-inch, 179.5-pound Z-bar column 26 feet long, with fixed ends, using Gordon's formula with a factor of safety of 5, the values to be taken from the tables. Ans. 10,562 lb.

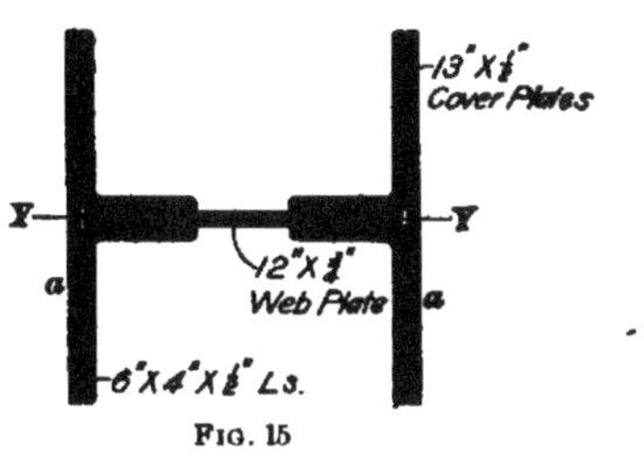

Fig. 15

4. A plate and angle column 28 feet long, fixed at the ends and of the general dimensions shown in Fig. 15, is subjected to a vertical load of 150 tons and a direct horizontal thrust of 6 tons from the knee brace of a truss 5 feet below its top, and parallel to the web-plate of the column. What will be the maximum unit stress in the flange plates *a*, *a*, considering both the transverse and the direct stress? Ans. 10,638 lb.

5. A 10-inch, 15-pound channel column 20 feet high, with square ends, and having ⅜-inch cover-plates, is cross-braced at one-half its height in a plane parallel to the plates of the column. What is its ultimate strength, according to the parabolic formula? Ans. 40,275 lb.

DESIGN OF STEEL COLUMNS

24. Different Column Sections.—Tables giving the properties, dimensions, and safe loads of the common forms of columns are given at the end of this Section, but a few remarks as to some of the forms may be of value at this time. The plate and angle columns are especially adapted for use when cranes or any eccentric loads are to be placed on the column, because they may be made so as to stand the strain without much increase of weight.

The Z bar, channel and plate, and box columns are best adapted to ordinary cases in building, where the load is distributed around the column.

The latticed channel column is used where the load is light. The channels are placed a certain distance apart, as shown in the tables at the end of this Section, and are so arranged that the center of gravity of the channels is, in both cases, the same distance from the center line of the column.

TABLE IV

SIZE OF LATTICE BARS AND STAY-PLATES TO BE USED WITH LATTICED CHANNEL COLUMNS

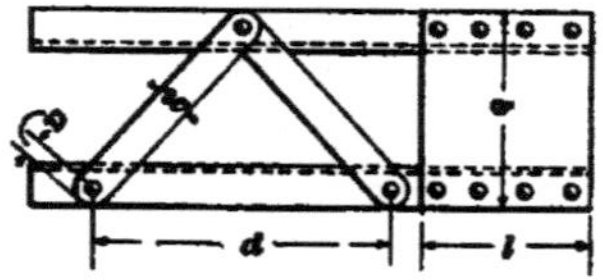

Depth of Channels	Dimensions of Lattice Bars		Weight of Lattice Bars per Foot	Center of Hole to End of Bar a	Distance Center to Center of Rivets d				Size of Stay-Plates at Ends of Columns		Diameter of Rivets
	w	Thickness			Maximum		Minimum		Thickness	l	
Inches	Inches	Inch	Pounds	Inches	Feet	Inches	Feet	Inches	Inch	Inches	Inch
6	$1\frac{1}{2}$	$\frac{1}{4}$	1.28	$1\frac{1}{8}$	0	$11\frac{1}{2}$	0	$6\frac{5}{8}$	$\frac{1}{4}$	$7\frac{1}{2}$	$\frac{5}{8}$
7	$1\frac{3}{4}$	$\frac{1}{4}$	1.49	$1\frac{1}{8}$	1	$1\frac{1}{2}$	0	$7\frac{5}{8}$	$\frac{1}{4}$	10	$\frac{5}{8}$
8	2	$\frac{5}{16}$	2.12	$1\frac{1}{4}$	1	3	0	$8\frac{11}{16}$	$\frac{5}{16}$	9	$\frac{3}{4}$
9	2	$\frac{5}{16}$	2.12	$1\frac{1}{4}$	1	$4\frac{1}{2}$	0	$9\frac{1}{2}$	$\frac{5}{16}$	12	$\frac{3}{4}$
10	2	$\frac{3}{8}$	2.55	$1\frac{1}{4}$	1	$6\frac{1}{2}$	0	$10\frac{11}{16}$	$\frac{3}{8}$	12	$\frac{3}{4}$
12	$2\frac{1}{4}$	$\frac{3}{8}$	2.87	$1\frac{5}{8}$	1	$10\frac{1}{2}$	1	1	$\frac{3}{8}$	15	$\frac{3}{4}$
15	$2\frac{1}{2}$	$\frac{3}{8}$	3.19	$1\frac{1}{2}$	2	$2\frac{1}{2}$	1	$3\frac{5}{16}$	$\frac{3}{8}$	15	$\frac{3}{4}$

25. Spacing of Lattice Bars.—In Table IV are given the necessary dimensions for placing the lattice bars on columns. The value of b for various columns may be found by the following formula. If f is the width of a flange of the channels, then for columns made with the flanges turned out,

$$b = 2f + d \qquad (19)$$

and for columns made with the flanges turned in,

$$b = D \qquad (20)$$

The values of d and D are given in the tables of safe loads for latticed channel columns at the end of this Section, and have nothing to do with the value d in Table IV.

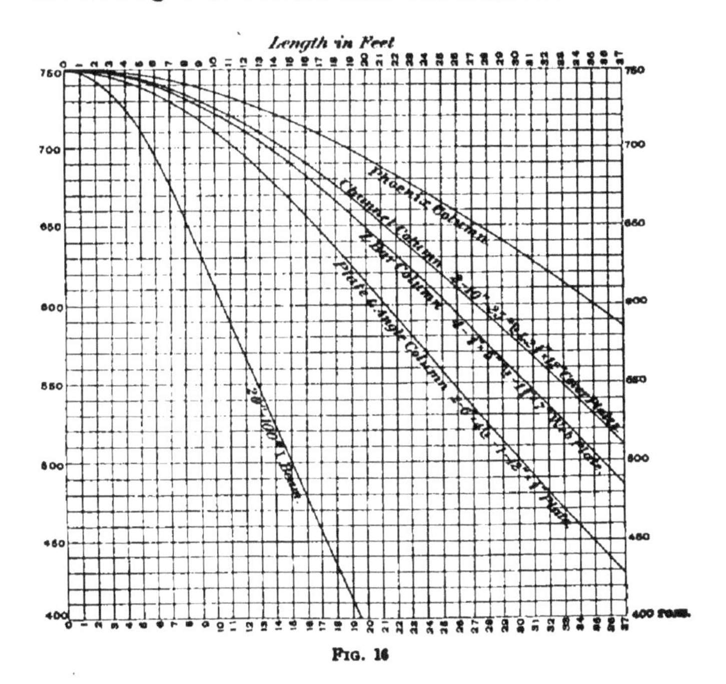

FIG. 16

26. Clearance.—In designing columns with a central web-plate, such as **Z**-bar columns and plate and angle columns, the outside edge of the web-plate should be at least $\frac{1}{8}$ inch back from the faces of the **Z** bars or angles. This **clearance** is allowed in order to provide for any inequality in cutting.

27. Efficiency of Section.—In order to compare the efficiency of the different sections, Fig. 16 has been prepared. This shows the comparative ultimate strength for different lengths of five forms of columns. The columns assumed are those whose greatest diameter is not over 17.4 inches nor less than 15.6 inches, and whose least outside dimension is between 12 and $12\frac{3}{4}$ inches. The area of the section has been reduced to 30 square inches. In other words, the sections are the same weight, about 100 pounds per foot of length, and have about the same limiting extreme dimensions of cross-section.

To ascertain from the diagram the ultimate strength, in tons, of a column of a given section and length, select the corresponding length from either of the horizontal rows of numbers and follow the vertical line belonging to this number until it intersects the curve of the column selected. The value, in tons, of the horizontal line drawn through this point is ascertained from the vertical rows of numbers. For instance, it is desired to find the ultimate strength of a channel column, 20 feet long, of the section indicated in the diagram. On following the 20-foot line, it is seen to have a point of intersection that happens to be common to that of the horizontal line representing 660 tons, which is the ultimate strength of this column.

The strength of a Phœnix column of the same length is seen to be about 693 tons.

The ultimate strength, in tons, has been calculated by Gordon's formula for medium steel columns with fixed ends. This shows the great economy, so far as the weight is concerned, in the use of the box and round shapes, but the advantage of weight is not always the most important consideration, as stated before.

Further, this comparison has been made on the assumption that the load is placed at the center of gravity of the section. It does not take into consideration the problem of eccentric loading, in which case the round column is the least economical. The advantage of the round-column section is realized where floor space is desired and long columns are indispensable.

28. Column Splices and Connections.—The excellence of the detail design of structural steel columns is largely governed by the experience and judgment of the designer, and the care with which he studies the local conditions that will be met with in nearly every new piece of work.

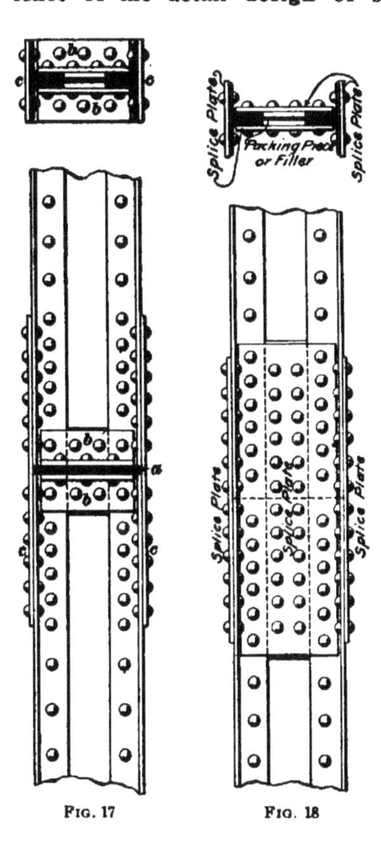

Fig. 17 Fig. 18

As the strength of a building depends almost entirely on the strength of the beam connections to the columns, great care should be taken in their design. Rigid connections for the floorbeams and girders at the several floors, and efficient splice connections between the sections of the columns on the different floors are of the utmost importance, and should be given the most careful attention. The ideal system of column construction would be to make the various columns on the several floors of one continuous set of sections running from foundation to roof. It is evident, however, that this is impossible in high modern-building construction, and therefore the next best thing is to make the spliced connections as rigid as possible.

Columns may be spliced as shown in Fig. 17. The bedplate *a* separates the two columns, and through it, by means of the angles *b b*, the columns are riveted securely together; they are also additionally secured by means of the two splice plates *c*, *c*.

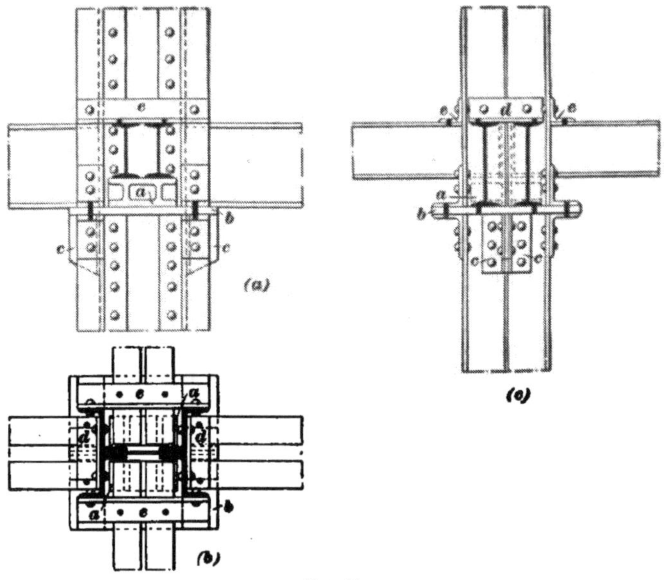

Fig. 19

The abutting ends of structural columns should be milled or planed so as to secure a square and firm bearing and obviate the danger of their being thrown out of line when erected.

Another very good method of connecting two columns is by the use of splice plates on all sides, thus doing away with the bedplates or packing pieces between the ends; this construction is shown in Fig. 18.

In some cases the bedplate extends beyond the column and forms a rest or support for the floorbeams or girders.

This is shown by means of Fig. 19, in which (*a*) is a front elevation, (*b*) a plan view, and (*c*) a side elevation of the joint. It is here advisable to further support the bedplate *b* by a bracket *c* directly under the bearing of the beam. The beams are secured to this bedplate and also to the post by means of angle clips *d*.

If, in the same floor, the beams or girders are of different depths and are at the same level with regard to the top flange, the shallow beams may be supported by introducing cast-iron blocks *a* beneath them, as shown in Fig. 19. In this case angle clips *e* connect the beams and column.

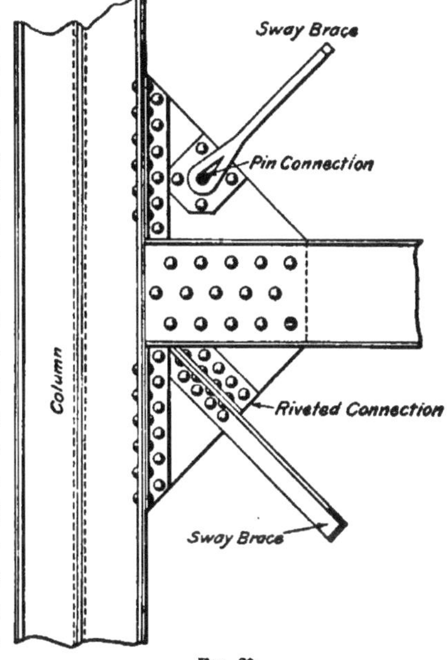

FIG. 20

Where the floor girders are of the built-up plate-girder type, there is no difficulty in securing a rigid connection to the column. The connection to be recommended, where the architectural features of the interior arrangement of the building do not interfere, is shown in Fig. 20. This connection is particularly efficient when the building is high and narrow, and in danger of being acted on by heavy wind pressure.

Other forms of connections, where plate girders are the principal supporting members of a floor system, are shown in Figs. 21 and 22. In Fig. 21, a filler *c* is introduced because the upper column is smaller than the lower, and it becomes necessary to pack between the splice plates. It is, however, preferable to pack equally on both sides of the upper column whenever practicable, as this insures the central axis of the one column being over the central axis of the other.

29. Fig. 23, in which (*a*) is a front elevation and (*b*) a plan view, shows another form of column section made by riveting together four channels and two cover-plates, which makes a very efficient form of section. There are only four rows of rivets, and the column presents four good faces on which to form connections. It will be seen that a single angle *a* has been placed under the 10-inch I beam to support it and take the load. With such a light beam, this is all that is necessary, but where an 18-inch girder is to be connected on the side, additional means must be provided for supporting its load. Two $3'' \times 3'' \times \frac{5}{16}''$ angles *b* have therefore been riveted directly under the angles supporting the beam, in order to transmit the load to the column, this size angle having been selected in order to bring the line of rivets directly under the load. In making up this column, the brackets should be riveted to the outside channel before the channels are riveted together.

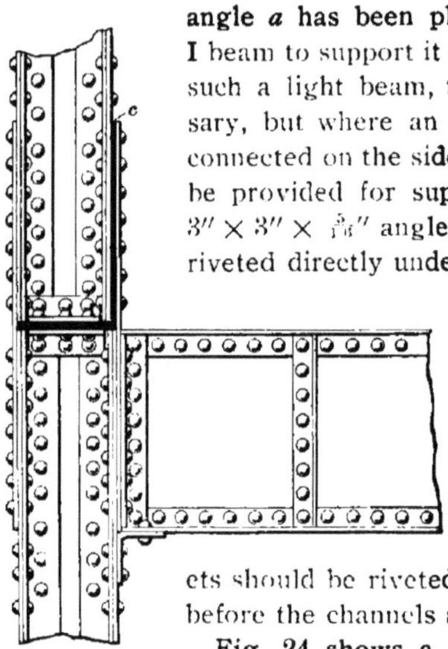

FIG. 21

Fig. 24 shows a connection for a heavy column, designed to support an enormous load, and at first sight the connection seems very light, the splice plate being only ½ inch. The column is made up of four angles with one web-plate and three cover-plates riveted to the angles on two sides of the column. The ends of the columns at such a splice must be planed off very accurately so as to give a full bearing, in order to transmit the load directly to the center of the column below. In such a case, the splice plates are only used as guides to hold the columns in line, and they bear no load. This splice should be as near the floor connection as possible, and not more than 25 inches above the top of the beams.

A splice commonly used in connecting Phœnix columns is illustrated in Fig. 25, which shows an eight-section column in which filler plates have been used. One plate *a*, the form of which is shown in the elevation (*a*), passes directly through the column, tying the parts together and forming a web splice for the floorbeam connections, as shown in the plan view (*b*). At right angles to this plate, two other plates have been placed, which are riveted to the first one by means of angles. These plates greatly stiffen and strengthen the column, but they add inefficient material at the center of the column, which should be figured as part of the load supported.

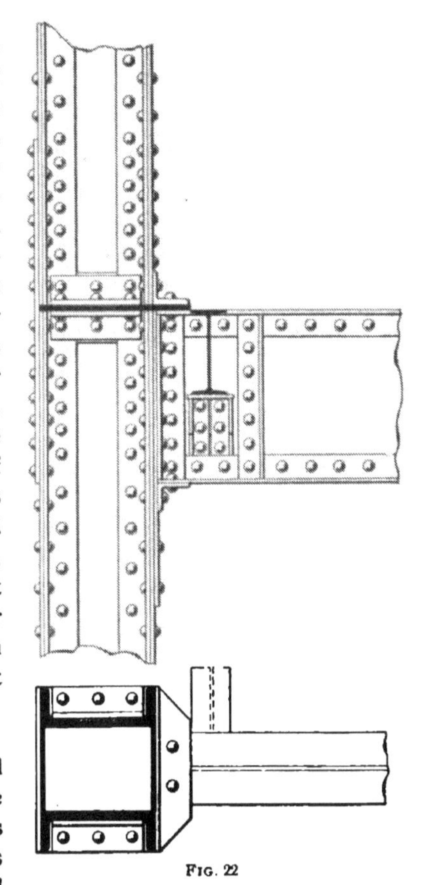

FIG. 22

30. In Figs. 26 and 27 are shown complete detail working drawings of the first-floor columns of a building designed according to the best modern practice. Fig. 26 shows a channel column to which the floorbeams are rigidly secured at a considerable distance below the splice, which has been designed with a bedplate between the two sections. The splice could have been made as well by the use of side-splice plates, similar to those on the other two sides, as previously explained, and would possibly have been more efficient.

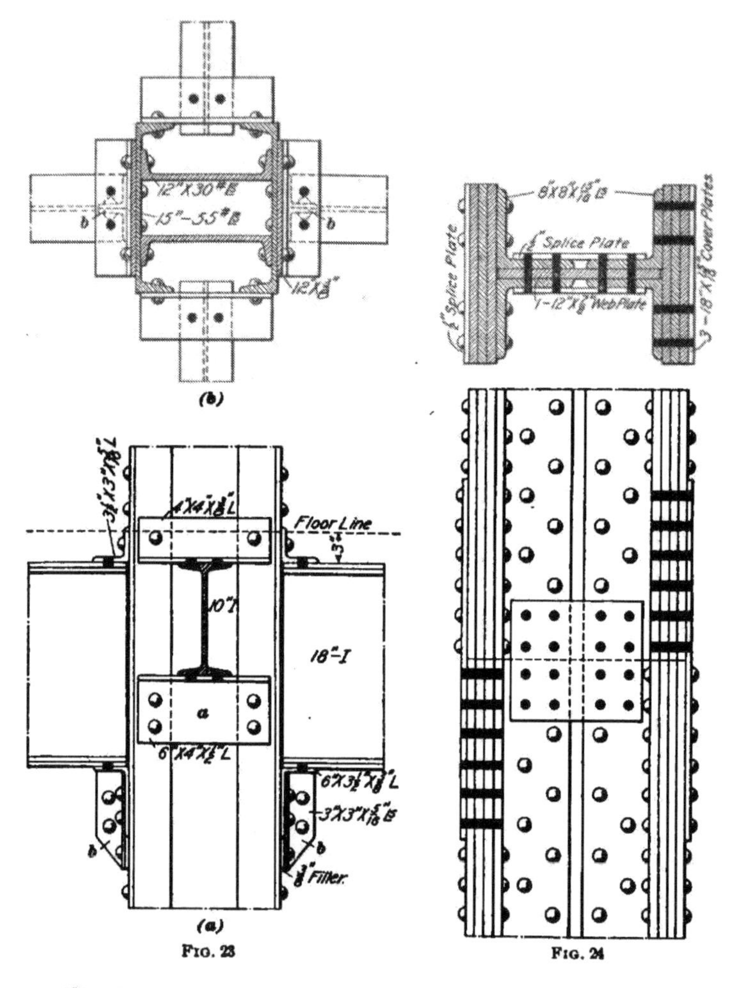

(b)

(a)

FIG. 23

FIG. 24

The base of this column gives very interesting details. It is designed with side plates to distribute the weight over the capstones. The load from the column is partly transmitted to the side plates, which again transmit the load to the 5″ × 5″ angles and thence to the base plate. Part of the load is also transmitted to the base plate through the side angle clips shown in the side view.

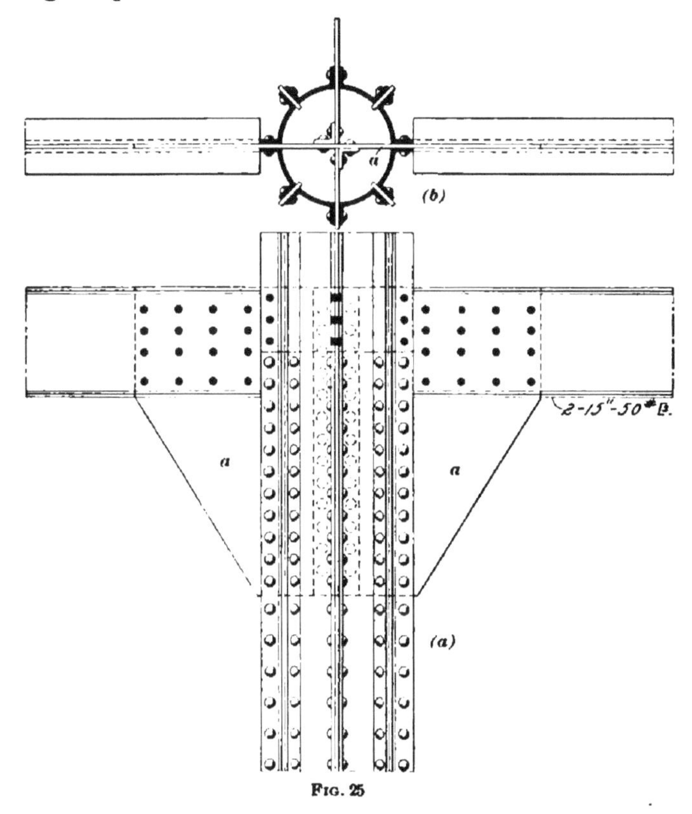

Fig. 25

In Fig. 27 is shown a Z-bar column, the principal points in the detail of which are similar to those in the channel-bar column.

In each of these examples, the shear on the rivets supporting the floorbeam brackets has been calculated to safely sustain the load carried by the floorbeams.

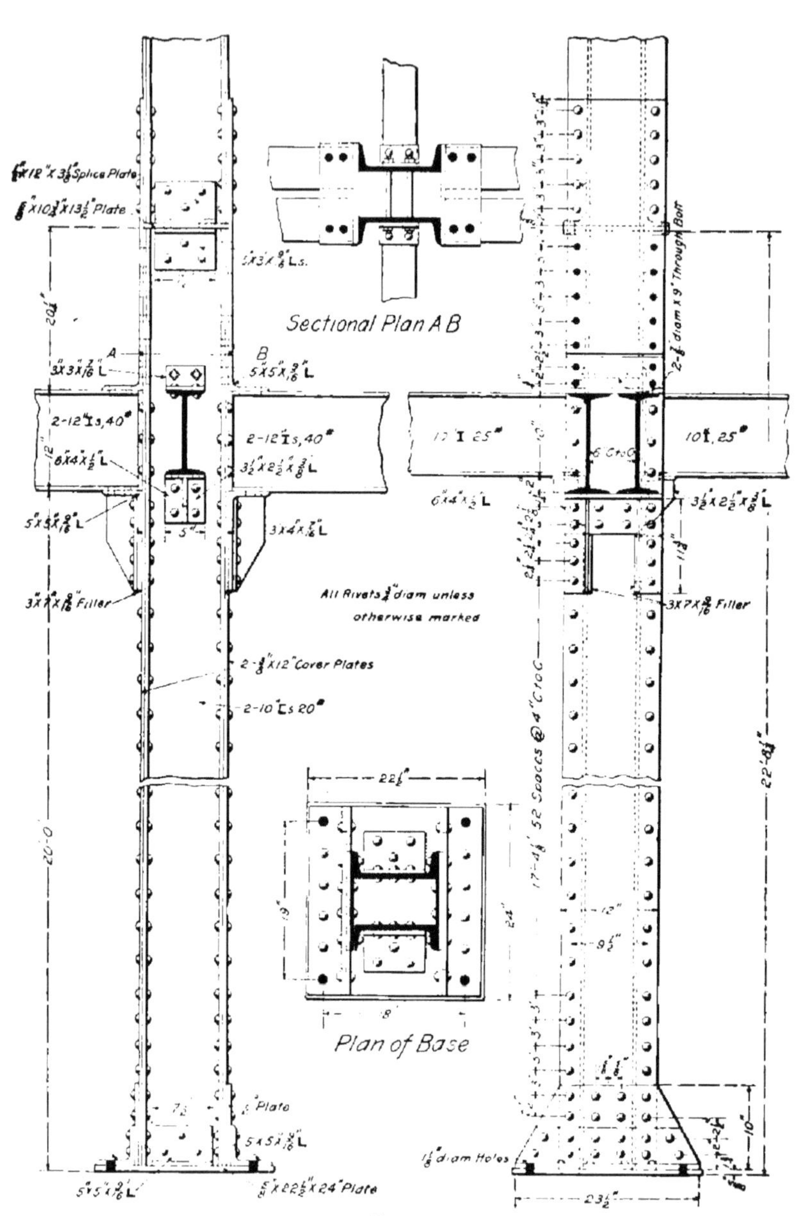
Sectional Plan A B
All Rivets 3/4" diam unless otherwise marked
2-1/2"x12" Cover Plates
2-10" [s 20#
Plan of Base
Through Bolt
Filler
Splice Plate
Plate

FIG. 26

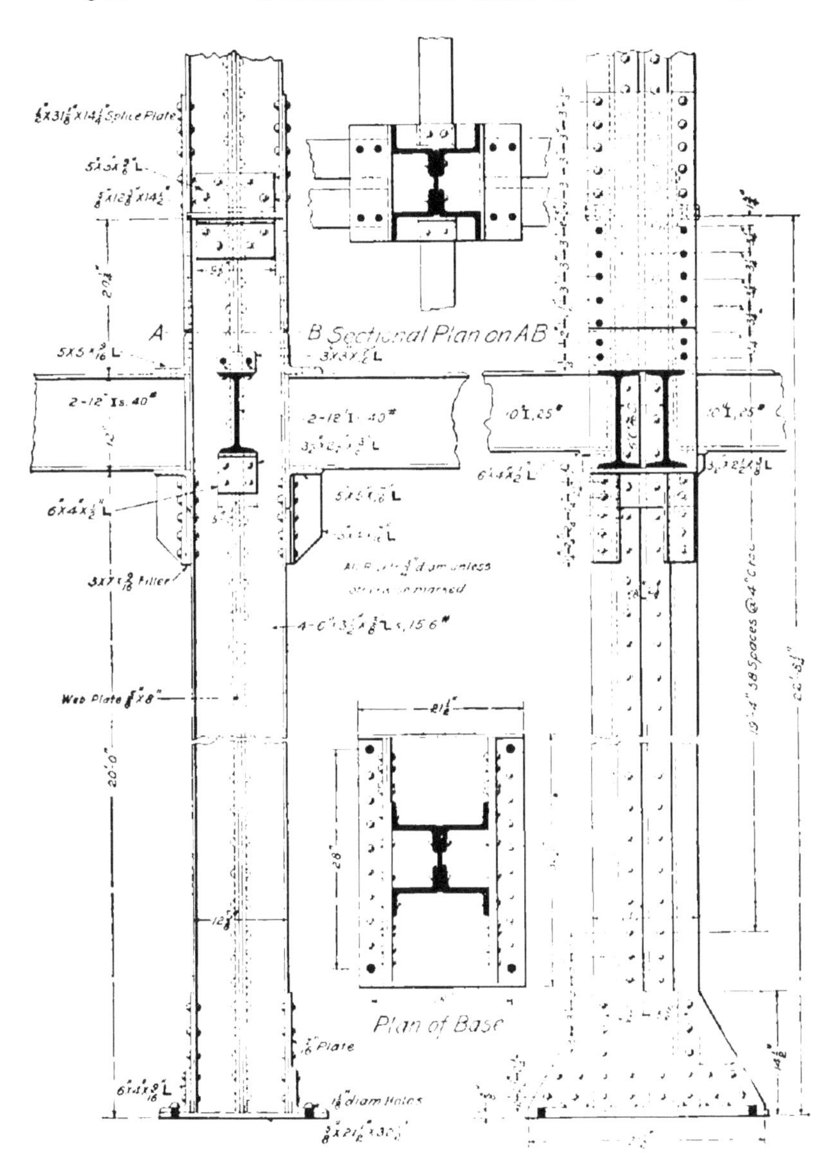

FIG. 27

All the connections and splices to a column should be riveted together with hot rivets; this insures more rigidity against wind pressure than can be obtained by bolted connections.

In Fig. 28 is shown the usual method of forming the base for a Phœnix column. Here the web-plates come into play by helping to distribute the load equally over the bed-

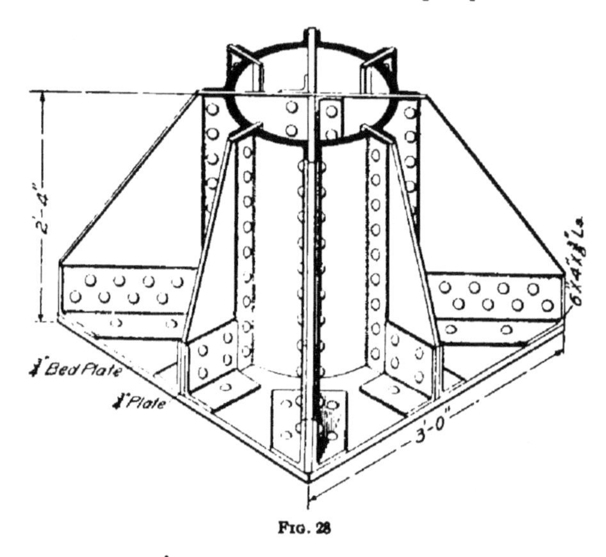

Fig. 28

plate. This base is very strong and is capable of sustaining a load of 100 tons, if set in cement.

31. Assembled Designs.—Fig. 29 gives an example of a column supporting the wall of a building. In the illustration (*a*) is a sectional plan taken along line *A B* of (*c*), (*b*) a side elevation showing the surrounding brickwork in section, and (*c*) a front elevation. Four **Z** bars are riveted to separate plates between which the floor girder passes, forming a cantilever for the support of the wall. The advantages of this design are that the load supported by this girder is transmitted practically to the center of the column, and the walls may be made so as to balance the weight of the floors and partitions. This column section is very efficient for the purpose for which it is intended.

Fig. 30 shows the details of a channel and plate column supporting two 20-inch floor girders *a*, *a*, a 12-inch wall

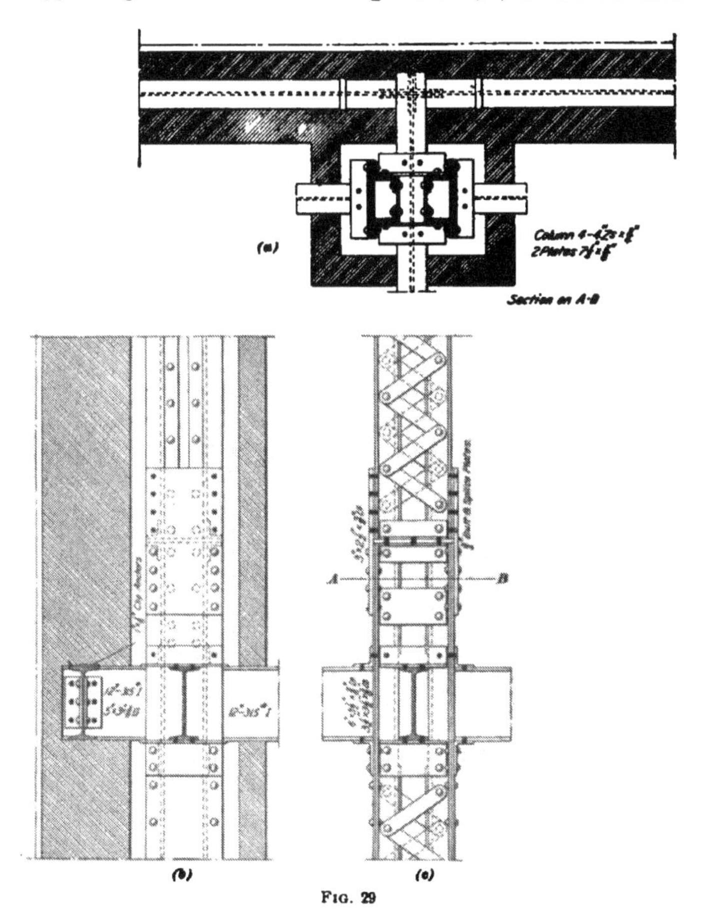

Fig. 29

beam *b*, and a 15-inch floorbeam *c*. An effort has been made to entirely eliminate eccentric loading, and with this object in view a girder has been placed on either side of the column so that the two sides will receive equal loads. The wall beam passes over the projecting ends of the floor girders, the weight of the wall being utilized to balance the floor loads resting on the girders. By this method a very good design can be obtained.

Fig. 31 gives another example of eccentric loading. The design has been made with the object of supporting

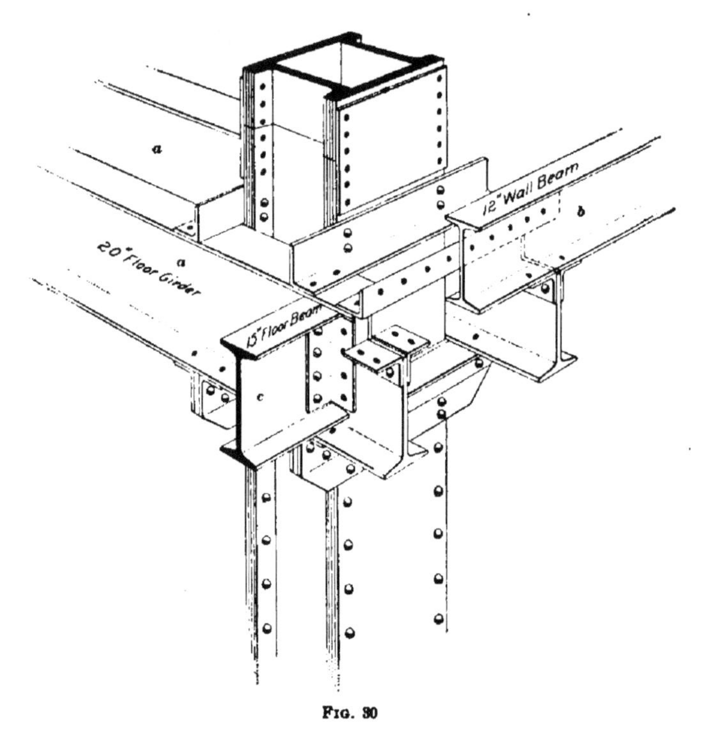

FIG. 30

the large cornice that runs along the face of the building, the upper stories being set back from the line of the lower stories, thereby necessitating the displacement of the upper column.

In the figure, (*a*) is a front elevation, (*b*) a side elevation, (*c*) a section taken along line *A B* in (*a*), and (*d*) a plan

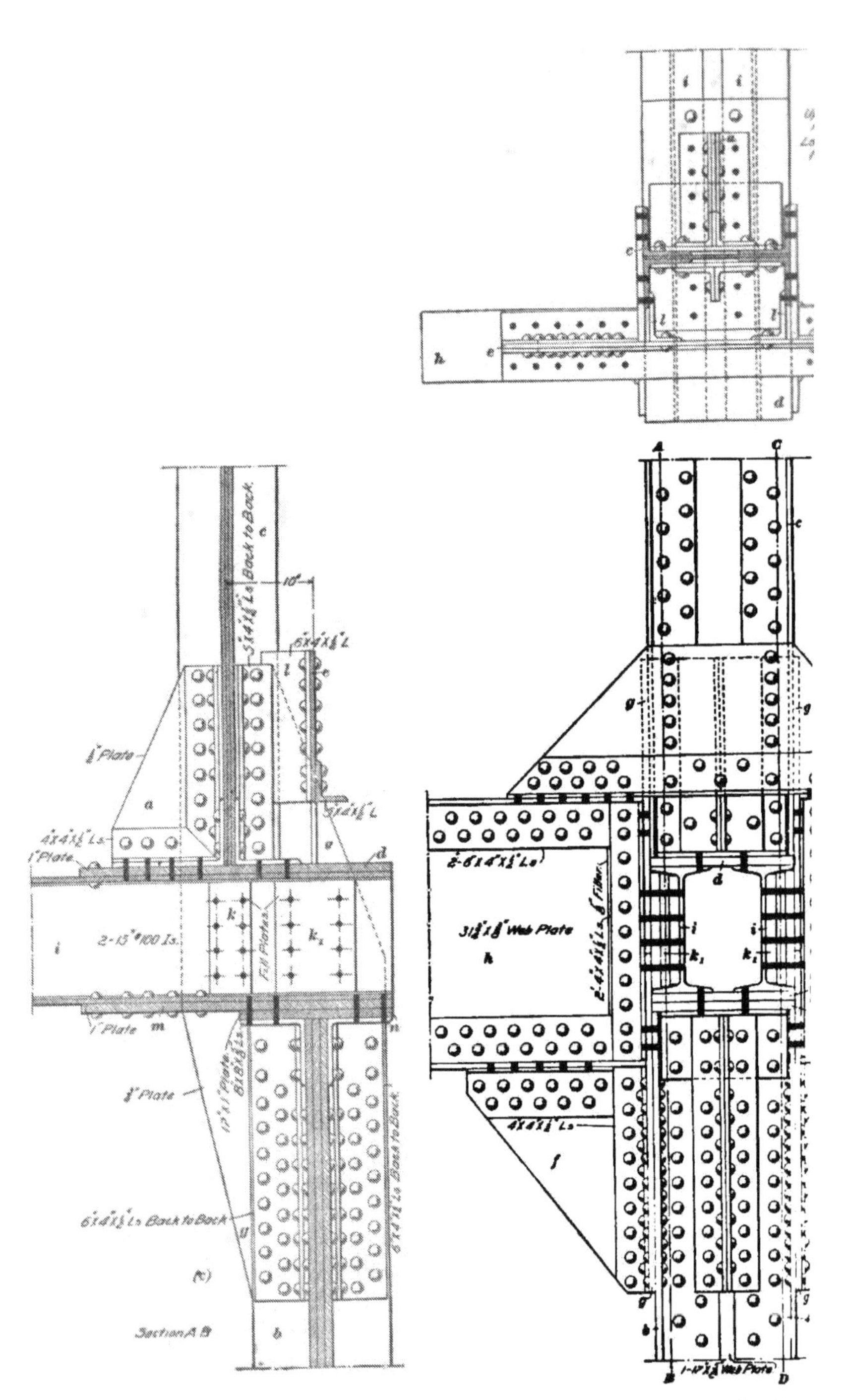

Section A B
2-15" #100 Is.
31 1/2" X 3/8" Web Plate
Fill Plates
4X4X1/2" Ls.

FIG. 31

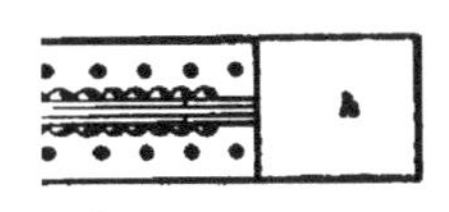
r Column 4-6"X6"X¾" Ls.
"X⅜" Web Plate.
r Column 4-8"X8"X15/16" Ls.
"X⅜" Web Plate..

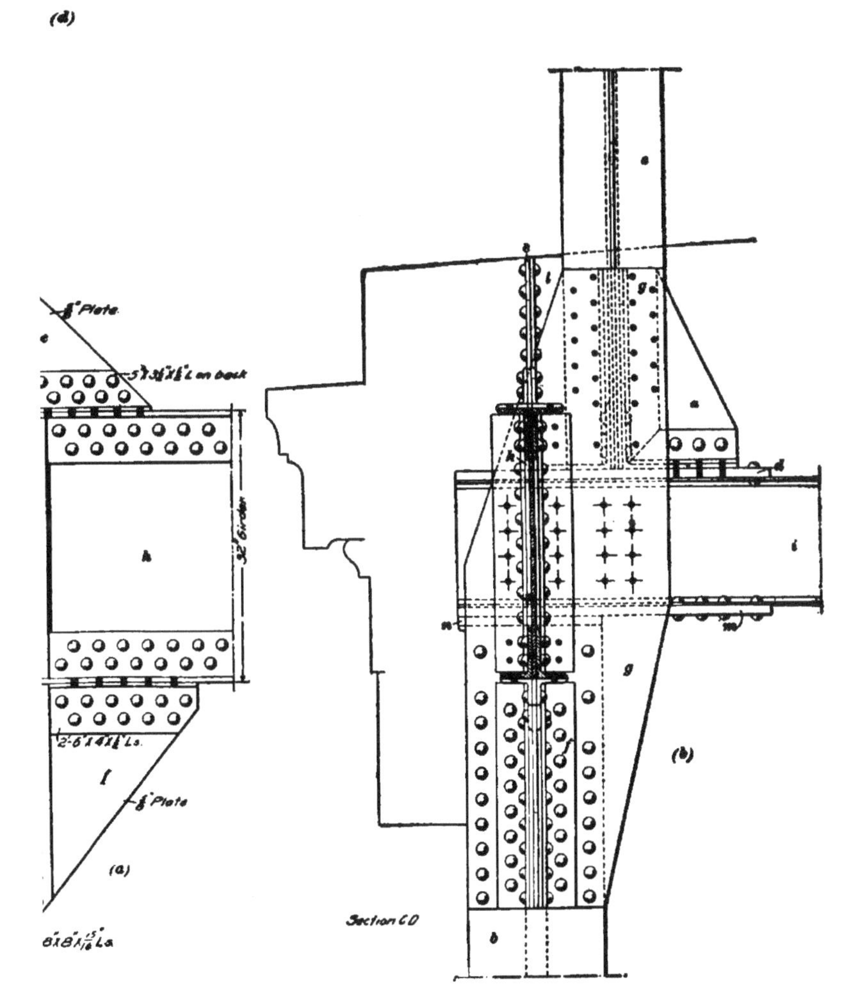
(d)
⅜" Plate
5"X3½"X⅜" L on back
32" Girder
2-6"X4"X⅜" Ls.
⅜" Plate
(a)
6"X8"X15/16" Ls.
Section CD
(b)

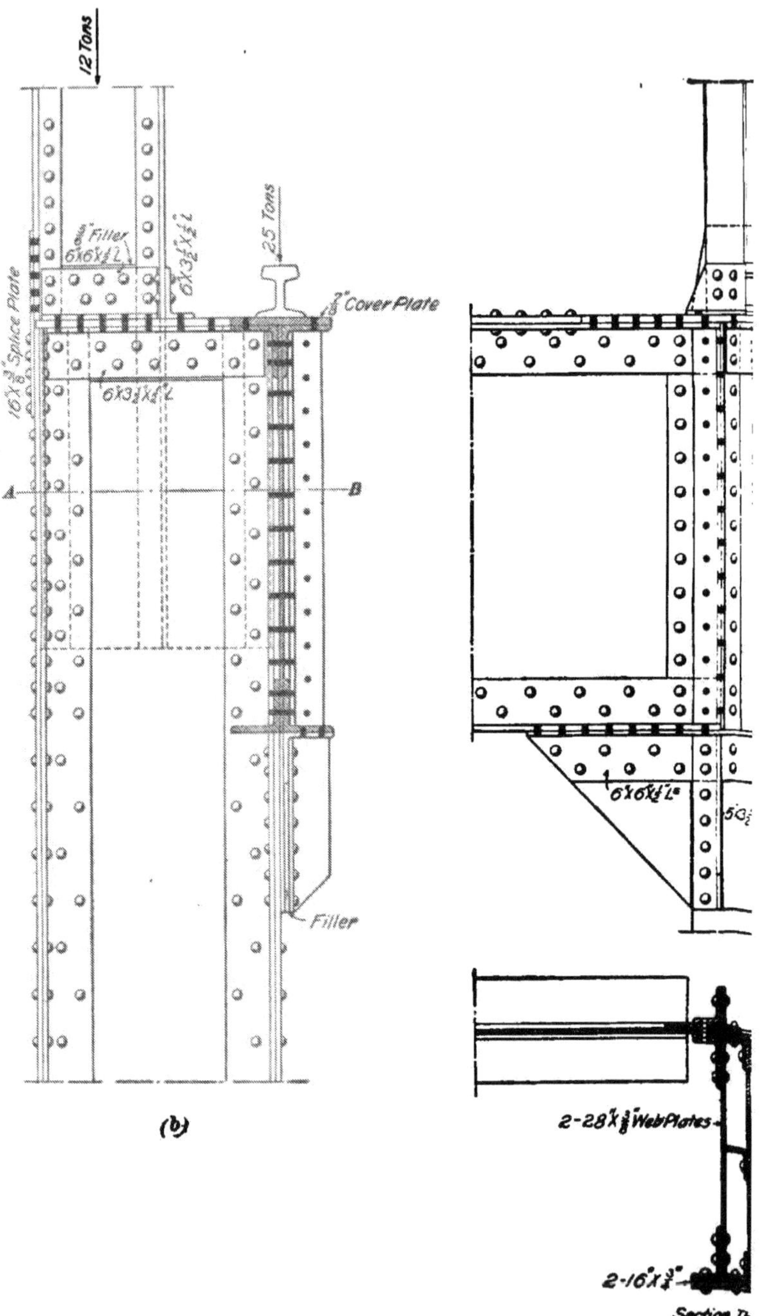
12 Tons
25 Tons
Filler
Cover Plate
Splice Plate
A
B
Filler
2-28"x3/8" Web Plates
2-16"x3/4"
Section Th
Fig. 1

(b)

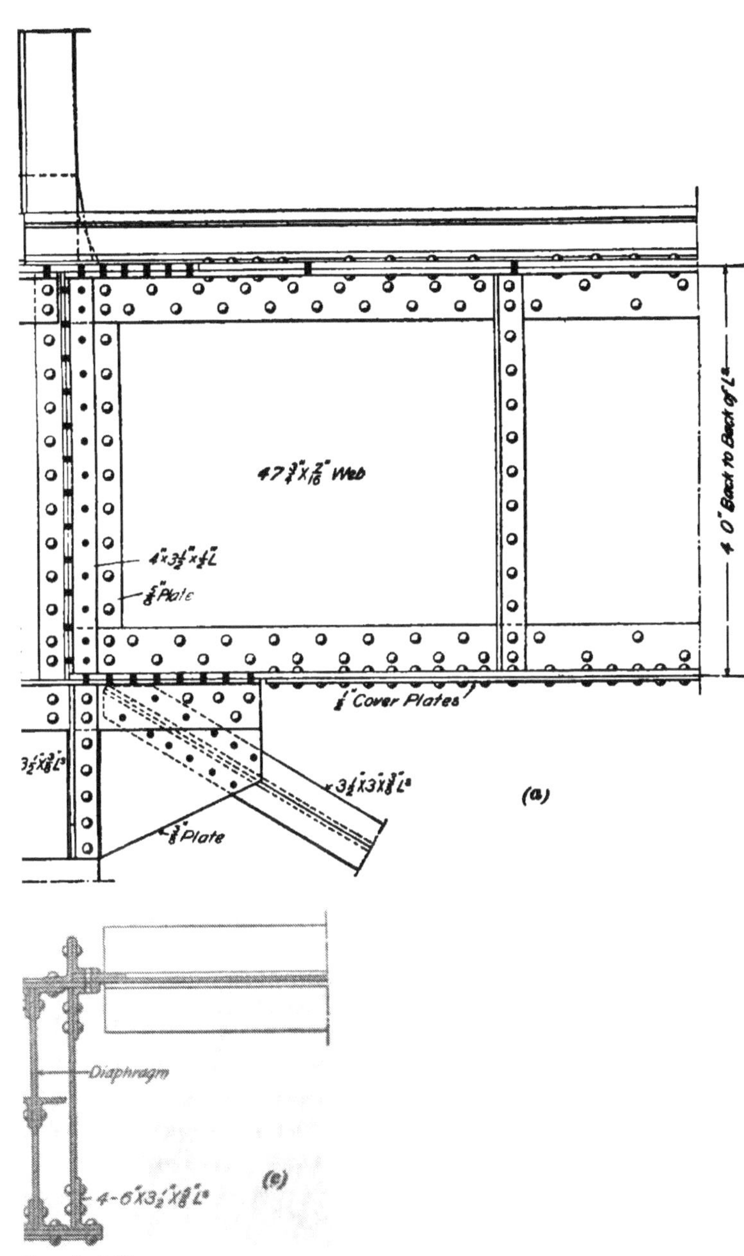

47 3/4" X 7/16" Web
4' 0" Back to Back of L^s
4"x3½"x½" L
⅜" Plate
⅜" Cover Plates
3½"X⅜" L^s
3½"X3"X⅜" L^s
(a)
⅜" Plate
Diaphragm
(c)
4-6"X3½"X⅜" L^s

Through A B

view. The main column *b* is made up of four $8'' \times 8'' \times \frac{15}{16}''$ angles and one $17'' \times \frac{5}{8}''$ web-plate, and supports two 32-inch wall girders *h*, *h*. Each of the latter is made up of four $6'' \times 4'' \times \frac{1}{2}''$ angles, and one $31\frac{3}{4}'' \times \frac{3}{8}''$ web-plate. Two floor girders *i*, *i*, made up of two 15-inch 100-pound **I** beams, run directly over the lower column and support the upper column *c*, which is made up of four $6'' \times 6'' \times \frac{1}{2}''$ angles, with one $17'' \times \frac{5}{8}''$ web-plate. An important part of the structure is the brace *g*, which is of $\frac{3}{4}$-inch plate, located on each side of the column *b* for the purpose of connecting the latter directly with the upper column and also with the girders and the **I** beams. In order to stiffen the **I** beams sufficiently to receive the load of the upper column, four series of filler plates k_1, k_1 have been inserted, which are riveted securely to the webs of the **I** beams, the braces *g*, and the angle stiffeners of the girders *h*.

The load from the upper column is still further distributed over the **I** beams by the use of a 1-inch plate *d* and the vertical web-plate *a*. The two girders are connected by the $\frac{5}{8}$-inch vertical plate *e* and the angles shown, the latter being riveted securely to the upper flanges of the girders. In order to secure the girders *h*, *h* still further against turning over, the $\frac{5}{8}$-inch plates *f* are riveted to the lower flanges of the girders by means of two $6'' \times 4'' \times \frac{1}{2}''$ angles and to the lower column by two $4'' \times 4'' \times \frac{1}{2}''$ angles. To strengthen the connection between the **I** beams and the lower column, two 1-inch plates *m*, *n* have been riveted to the beams and the column connection.

32. Columns for Power Houses and Shops.—Fig. 32 is a detail drawing of a column designed for a power house, in which (*a*) is a front elevation, (*b*) a side elevation, and (*c*) a sectional plan taken on line *A B* in (*b*). The column is one of a series supporting an **I** beam. On top of this beam is a **T** rail over which the side wheels of a 50-ton crane may run. There is a similar beam supported by columns at the other side of the building supporting the other end of the crane. The load to which the column will be subjected from the crane is 25 tons. The column supports also an upper column, which in turn supports the roof and carries about 12 tons. Its section is made up of four $6'' \times 3\frac{1}{2}'' \times \frac{5}{8}''$ angles, two $28'' \times \frac{3}{8}''$ web-plates, and two $16'' \times \frac{3}{4}''$ cover-plates. For the purpose of strengthening the lower column for the load of the upper column, a diaphragm has been inserted in the interior of the lower column directly under the plinth plate. This helps to distribute the load of the upper column more evenly over the whole of the lower column. The diaphragm consists of a vertical plate stiffened by two angles riveted to its sides. Under the plate girder that supports the track for the crane, is a vertical web brace, which acts as a bracket for the crane and also as a connection for the cross-bracing that is inserted between adjacent columns.

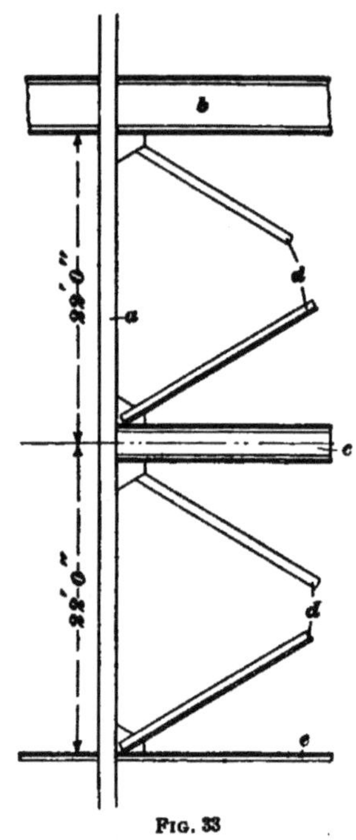

Fig. 33

Fig. 33 shows the connections between the post *a* and one of the adjoining posts. In this figure *b* is the main I beam and *c* a smaller one that connects the post *a* rigidly with adjoining posts on either side; *d* is a cross-brace and *e* an additional longitudinal brace. The columns have a clear height of 44 feet, but by reason of being cross-braced to the adjacent columns, the effective height is only 22 feet.

EXAMPLE.—Given the case shown in Figs 32 and 33; using Gordon's formula: (*a*) what is the maximum fiber stress? (*b*) what load will the column be able to carry, using a factor of safety of 5?

SOLUTION.—(*a*) The moment of inertia of the column about the short axis through the center of the column and parallel with the I beams is:

Moment of inertia of two web-plates $28'' \times \frac{3}{8}'' = 2 \times 686$ . . .	1 3 7 2
Moment of inertia of four $6'' \times 3\frac{1}{2}'' \times \frac{5}{8}''$ angles, $4(20.08 + 5.55 \times 11.87^2)$	3 2 0 8
Moment of inertia of two $16'' \times \frac{3}{4}''$ plates, $2(.5625 + 12 \times 14\frac{3}{8}^2)$	4 9 6 0
Total moment of inertia	9 5 4 0

The area of the column section is as follows:

Two $28'' \times \frac{3}{8}''$ plates	2 1.0 sq. in.
Four $6'' \times 3\frac{1}{2}'' \times \frac{5}{8}''$ angles	2 2.2 sq. in.
Two $16'' \times \frac{3}{4}''$ plates	2 4.0 sq. in.
	6 7.2 sq. in.

The distance from the neutral axis to the extreme fiber is $14\frac{3}{4}$ in.; then the section modulus is $S = \frac{9,540}{14\frac{3}{4}} = 647$.

The maximum fiber stress is $\frac{25 \times 14\frac{3}{4} - 12 \times 7}{647} + \frac{37}{67.2} = .9907$ T. per sq. in. = 1,981.4 lb. per sq. in. Ans.

(*b*) The radius of gyration about the short axis is $r = \sqrt{\frac{9,540}{67.2}}$, or $r^2 = \frac{9,540}{67.2}$.

The moment of inertia of the column about the long axis, at right angles to the I beams is as follows:

Moment of inertia of two web-plates, $2(.123 + 10\frac{1}{2} \times 4\frac{5}{16}^2)$.	3 9 0.8 0
Moment of inertia of four angles, $4(5.08 + 5.55 \times 5.38^2)$ . .	6 6 2.8 8
Moment of inertia of two plates $16'' \times \frac{3}{4}'' = 2 \times 256$	5 1 2.0 0
Total moment of inertia	1 5 6 5.6 8

Using Gordon's formula for medium steel and for flat-ended columns,

$$u = \frac{50,000}{1 + \frac{(44 \times 12)^2}{24,000 \times \frac{9,540}{67.2}}} = \frac{50,000}{1.0818} = 46,219 \text{ lb.}$$

This is the breaking strength calculated with the shorter axis as the base of the calculations. The strength of the column about the longer axis may be determined as follows: Since the column is cross-braced at the middle in this direction, it is assumed to act as a column only half its height. The ultimate strength per square inch, therefore, is

$$u = \frac{50,000}{1 + \frac{(22 \times 12)^2}{24,000 \times \frac{1,565}{67.2}}} = \frac{50,000}{1.124} = 44,484 \text{ lb.}$$

The column therefore is weakest about this axis and its safe load will be

$$\frac{44,484 \times 67.2}{5} = 597,865 \text{ lb. Ans.}$$

33. Tables Giving the Strength and Properties of Steel Columns.—The following tables giving the safe load, in tons, for the usual column sections, will be found convenient in selecting structural steel columns for a determined load. The values in these tables have been calculated from the following formulas:

$$s_{ca} \text{ for medium steel} = \begin{cases} 12{,}000 \text{ for lengths up to 50 radii of gyration} & (21) \\ 15{,}000 - 57\frac{l}{r} \text{ for lengths over 50 radii} & (22) \end{cases}$$

$$s_{ca} \text{ for soft steel} = \begin{cases} 12{,}000 \text{ for lengths up to 30 radii of gyration} & (23) \\ 13{,}500 - 50\frac{l}{r} \text{ for lengths over 30 radii} & (24) \end{cases}$$

These formulas embody their own factor of safety, which varies from 4 to 5. The tables for safe loads on I-beam columns have been calculated for soft steel. The tables of safe loads for plate and angle columns, channel and plate columns, and Z-bar columns have been calculated for medium steel, which is the grade of steel that it is advisable to use for such columns.

TABLE V

PROPERTIES OF STEEL PLATE-AND-ANGLE COLUMNS

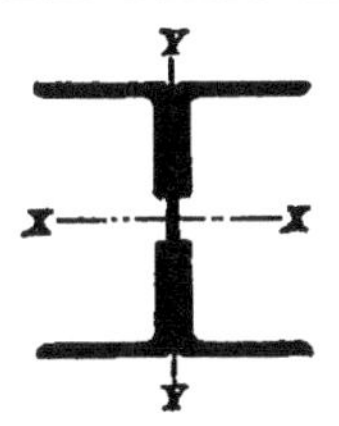

Column Section	Thickness of Plate and Angles. Inch	Area of Section Square Inches	Weight of Column in Pounds per Foot	Axis *X X*			Axis *Y Y*		
				Moment of Inertia	Section Modulus	Radius of Gyration Inches	Moment of Inertia	Section Modulus	Radius of Gyration Inches
One 6-Inch Plate Four 3″ × 2½″ L's	1/4	6.74	22.9	36.3	12.09	2.32	10.4	3.32	1.24
	5/16	8.52	29.0	44.6	14.87	2.29	13.6	4.24	1.26
	7/16	11 71	39.8	59.0	19.68	2.25	21.1	6.42	1.34
	1/2	13.00	44.2	64.6	21.53	2.23	24.7	7.60	1.38
One 7-Inch Plate Four 3½″ × 2½″ L's	1/4	7.51	25.5	58.3	16.65	2.78	16.1	4.43	1.46
	5/16	9.43	32.1	71.9	20.55	2.76	20.8	5.59	1.49
	7/16	12.98	44.1	95.8	27.38	2.72	30.8	8.15	1.54
	1/2	14.50	49.3	105.1	30.02	2.69	36.3	9.69	1.58
One 8-Inch Plate Four 4″ × 3″ L's	5/16	10.86	36.9	107.5	26.88	3.14	30.3	7.30	1.67
	3/8	13.12	44.6	128.5	32.13	3.13	37.4	8.79	1.69
	7/16	14.98	50.9	144.6	36.15	3.11	44.4	10.54	1.72
	1/2	17.24	58.6	163.5	40.88	3.08	53.1	12.29	1.75
	9/16	19.50	66.3	182.9	45.73	3.06	61.9	14.04	1.78
	5/8	20.92	71.1	193.5	48.38	3.04	69.1	16.04	1.82
One 9-Inch Plate Four 4½″ × 3″ L's	5/16	11.81	40.1	154.2	34.26	3.62	42.6	9.15	1.90
	3/8	14.22	48.3	183.5	40.78	3.59	52.9	11.13	1.93
	7/16	16.30	55.5	207.5	46.12	3.57	63.1	13.37	1.97
	1/2	18.74	63.7	235.9	52.44	3.55	75.3	15.64	2.01
	9/16	21.18	72.0	263.0	58.44	3.52	87.9	17.90	2.04
	5/8	22.83	77.6	279.1	62.24	3.50	99.0	20.57	2.08
One 10-Inch Plate Four 5″ × 3″ L's	5/16	12.73	43.3	211.8	42.36	4.08	57.6	11.16	2.13
	3/8	15.35	52.2	252 7	50.54	4.06	71.9	13.68	2.17
	7/16	17.62	59.9	286.4	57.28	4.03	85.9	16.46	2.21
	1/2	20.24	68.8	326.0	65.20	4.01	102.2	19.22	2.25
	9/16	22.35	76.0	355.7	71.14	4.00	118.1	22.36	2.29
	5/8	24.97	84.9	392.3	78.46	3.97	136.6	25.43	2.34
One 12-Inch Plate Four 6″ × 4″ L's	3/8	18.94	64.4	443.6	73.37	4.85	119.6	19.34	2.51
	7/16	22.17	75.4	513.6	85.60	4.81	144.5	23.03	2.55
	1/2	25.44	86.5	584.5	97.42	4.80	171.8	26.96	2.60
	9/16	28.67	97.5	651.0	108.5	4.77	199.7	30.91	2.64
	5/8	30.94	104.9	693.4	115.6	4.75	223.4	35.39	2.69
	11/16	34.17	116.2	760.1	126.8	4.72	255.7	39.88	2.73
	3/4	37.44	127.3	825.3	137.6	4.70	288.7	44.44	2.78

TABLE VI

PROPERTIES OF STEEL PLATE-AND-ANGLE COLUMNS WITH COVER-PLATES

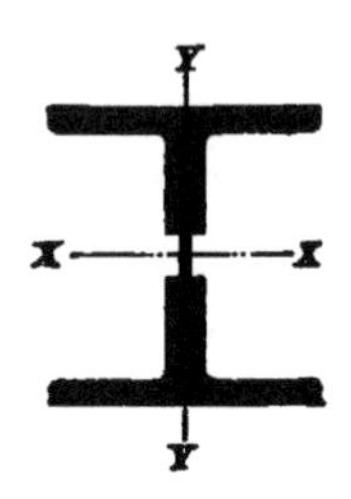

Column Section	Thickness of Cover-Plates Inches	Area of Section Square Inches	Weight of Column in Pounds per Foot	Axis XX			Axis YY		
				Moment of Inertia	Section Modulus	Radius of Gyration Inches	Moment of Inertia	Section Modulus	Radius of Gyration Inches
Four 6″ × 4″ × 9/16″ Angles One 12″ × 9/16″ Web-Plate Two Cover-Plates 13 Inches Wide	1/2	41.44	140.9	1,129	174.0	5.22	366.1	56.32	2.98
	9/16	43.07	146.5	1,199	182.6	5.27	389.0	59.84	3.00
	5/8	44.69	152.0	1,269	192.0	5.33	411.8	63.36	3.04
	11/16	46.32	157.5	1,340	200.3	5.38	434.7	66.88	3.07
	3/4	47.94	163.0	1,415	209.8	5.44	457.6	70.40	3.10
	13/16	49.57	168.5	1,492	219.3	5.49	480.5	73.92	3.12
	7/8	51.19	174.0	1,563	227.2	5.52	503.4	77.44	3.14
	15/16	52.82	179.6	1,642	237.0	5.59	526.2	81.00	3.16
	1	54.44	185.1	1,723	246.0	5.64	549.1	84.48	3.18
	1 1/16	56.07	190.6	1,803	256.1	5.68	572.0	88.00	3.20
	1 1/8	57.69	196.2	1,884	264.9	5.72	594.9	91.52	3.22
	1 3/16	59.32	201.7	1,965	274.3	5.75	617.8	95.04	3.23
	1 1/4	60.94	207.2	2,050	283.2	5.80	640.6	98.56	3.25
	1 5/16	62.57	212.8	2,143	292.7	5.85	663.5	102.1	3.26
	1 3/8	64.19	218.3	2,224	301.8	5.88	686.4	105.6	3.27
	1 7/16	65.82	223.8	2,311	311.2	5.93	709.3	109.1	3.29
	1 1/2	67.44	229.3	2,406	321.3	5.98	732.2	112.6	3.30
Four 6″ × 4″ × 3/4″ Angles One 14″ × 3/4″ Web-Plate Two Cover-Plates 15 Inches Wide	1/2	53.94	183.2	1,981	264.1	6.05	569.5	75.93	3.25
	9/16	55.82	189.8	2,088	276.2	6.12	604.6	80.61	3.30
	5/8	57.69	196.1	2,195	288.3	6.17	639.8	85.30	3.33
	11/16	59.57	202.6	2,304	299.8	6.22	674.9	89.98	3.37
	3/4	61.44	208.8	2,417	312.8	6.28	710.1	94.66	3.40
	13/16	63.32	215.3	2,533	325.3	6.32	745.2	99.36	3.44
	7/8	65.19	221.6	2,645	336.2	6.36	780.4	104.0	3.46
	15/16	67.07	228.1	2,765	349.4	6.41	815.5	108.7	3.49
	1	68.94	234.4	2,885	361.4	6.48	850.7	113.4	3.52
	1 1/16	70.82	240.8	3,004	373.2	6.51	885.9	118.1	3.54
	1 1/8	72.69	247.1	3,131	386.1	6.57	921.0	122.8	3.56
	1 3/16	74.57	253.6	3,251	398.4	6.61	956.2	127.5	3.58
	1 1/4	76.44	259.9	3,383	409.7	6.66	991.3	132.2	3.60
	1 5/16	78.32	266.3	3,510	421.6	6.70	1,026.5	136.9	3.62
	1 3/8	80.19	272.6	3,635	437.3	6.74	1,061.6	141.5	3.64
	1 7/16	82.07	279.1	3,770	448.2	6.79	1,096.8	146 2	3.66
	1 1/2	83.94	285.4	3,903	459.9	6.83	1,132.0	150.9	3.68

TABLE VII

PROPERTIES OF STEEL CHANNEL COLUMNS

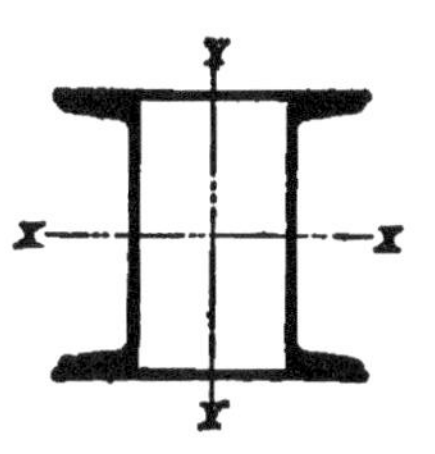

Designation	Weight of Each Channel in Pounds per Foot	Thickness of Cover-Plate Inches	Area of Section Square Inches	Weight of Column in Pounds per Foot	Axis X X			Axis Y Y		
					Moment of Inertia	Section Modulus	Radius of Gyration Inches	Moment of Inertia	Section Modulus	Radius of Gyration Inches
6-Inch Channel Column: Two Channels 6 Inches Deep and Two Cover-Plates 8 Inches Wide	8	1/4	8.70	29.6	64.0	19.9	2.72	46.7	11.7	2.31
	10	1/4	9.88	33.6	68.1	21.0	2.62	50.3	12.6	2.27
	10	5/16	10.88	37.0	78.2	23.6	2.68	56.0	14.0	2.27
	10	3/8	11.88	40.4	90.1	26.6	2.75	61.0	15.3	2.27
	12	3/8	12.96	44.1	98.9	29.2	2.75	71.8	18.0	2.35
	12	7/16	13.96	47.5	110.0	32.0	2.81	77.2	19.3	2.35
	12	1/2	14.96	50.9	122.0	34.9	2.86	82.5	20.6	2.35
	15	1/2	16.72	56.8	127.0	36.4	2.76	86.9	21.7	2.28
	15	9/16	17.72	60.3	138.0	39.3	2.81	92.2	23.1	2.28
	15	5/8	18.72	63.7	152.0	42.1	2.86	97.6	24.4	2.28
	17	5/8	19.70	67.0	161.0	44.4	2.86	111.0	27.8	2.38
	17	11/16	20.70	70.4	174.0	47.2	2.90	116.0	29.1	2.37
	17	3/4	21.70	73.8	188.0	50.2	2.94	122.0	30.4	2.37
	17	13/16	22.70	77.2	203.0	53.1	2.98	127.0	31.8	2.37
	17	7/8	23.70	80.6	217.0	56.0	3.02	132.0	33.1	2.36
	17	15/16	24.70	84.0	233.0	59.0	3.06	138.0	34.4	2.36
	17	1	25.70	87.4	248.0	62.1	3.10	143.0	35.8	2.36
7-Inch Channel Column: Two Channels 7 Inches Deep and Two Cover-Plates 9 Inches Wide	9	1/4	9.72	33.1	97.1	25.9	3.16	71.4	15.8	2.71
	9	5/16	10.85	36.9	113.0	29.7	3.23	79.0	17.6	2.70
	13	5/16	13.23	45.0	129.0	34.1	3.13	100.0	22.3	2.75
	13	3/8	14.35	48.8	146.0	37.8	3.20	108.0	24.0	2.74
	13	7/16	15.48	52.6	163.0	41.6	3.26	115.0	25.7	2.73
	13	1/2	16.60	56.4	181.0	45.4	3.33	123.0	27.4	2.72
	17	1/2	18.95	64.4	191.0	47.8	3.17	133.0	29.6	2.66
	17	9/16	20.08	68.3	209.0	51.5	3.23	141.0	31.4	2.66
	17	5/8	21.20	72.1	228.0	55.3	3.28	149.0	33.1	2.65
	17	11/16	22.33	75.9	247.0	59.1	3.33	156.0	34.7	2.65
	17	3/4	23.45	79.7	267.0	63.0	3.38	163.0	36.4	2.64
	17	13/16	24.58	83.6	288.0	66.8	3.43	171.0	38.1	2.64
	17	7/8	25.70	87.4	309.0	70.7	3.47	179.0	39.8	2.64
	17	15/16	26.83	91.2	331.0	74.7	3.51	187.0	41.5	2.64
	17	1	27.95	95.0	354.0	78.6	3.56	194.0	43.1	2.64

TABLE VII—(*Continued*)

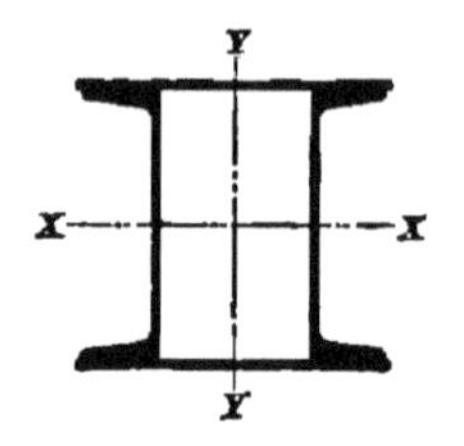

Designation	Weight of Each Channel in Pounds per Foot	Thickness of Cover-Plates Inches	Area of Section Square Inches	Weight of Column in Pounds per Foot	Axis XX Moment of Inertia	Axis XX Section Modulus	Axis XX Radius of Gyration Inches	Axis YY Moment of Inertia	Axis YY Section Modulus	Axis YY Radius of Gyration Inches
8-Inch Channel Column: Two Channels 8 Inches Deep and Two Cover-Plates 10 Inches Wide	10	$\frac{1}{4}$	11.0	37.4	141	33.3	3.58	107	21.5	3.12
	10	$\frac{5}{16}$	12.3	41.7	164	38.1	3.66	118	23.6	3.09
	13	$\frac{5}{16}$	13.9	47.1	179	41.6	3.59	136	27.3	3.14
	13	$\frac{3}{8}$	15.1	51.3	203	46.3	3.66	147	29.3	3.12
	13	$\frac{7}{16}$	16.4	55.6	227	51.2	3.73	157	31.4	3.10
	13	$\frac{1}{2}$	17.6	59.8	252	56.1	3.79	167	33.5	3.08
	17	$\frac{1}{2}$	20.0	67.9	265	58.7	3.64	184	36.8	3.04
	17	$\frac{9}{16}$	21.2	72.1	290	63.8	3.70	194	39.0	3.03
	17	$\frac{5}{8}$	22.5	76.4	317	68.2	3.76	205	40.9	3.02
	17	$\frac{11}{16}$	23.7	80.7	344	73.3	3.81	215	43.0	3.02
	17	$\frac{3}{4}$	25.0	84.9	372	78.2	3.86	225	45.2	3.02
	17	$\frac{13}{16}$	26.2	89.1	400	83.1	3.91	236	47.2	3.00
	17	$\frac{7}{8}$	27.5	93.4	430	88.3	3.96	246	49.3	2.99
	17	$\frac{15}{16}$	28.7	97.6	459	93.1	4.00	257	51.4	2.99
	17	1	30.0	101.9	490	98.2	4.04	267	53.4	2.99
9-Inch Channel Column: Two Channels 9 Inches Deep and Two Cover-Plates 11 Inches Wide	13	$\frac{5}{16}$	14.5	49.2	240	49.8	4.07	167	30.4	3.40
	13	$\frac{3}{8}$	15.9	53.9	272	55.7	4.14	181	32.9	3.38
	16	$\frac{3}{8}$	17.7	60.0	295	60.5	4.09	208	37.8	3.43
	16	$\frac{7}{16}$	19.0	64.7	329	66.7	4.16	222	40.3	3.41
	16	$\frac{1}{2}$	20.4	69.4	364	72.8	4.23	236	42.9	3.41
	21	$\frac{1}{2}$	23.4	79.5	383	76.6	4.05	259	47.0	3.33
	21	$\frac{9}{16}$	24.8	84.2	417	82.5	4.11	273	49.5	3.32
	21	$\frac{5}{8}$	26.1	88.8	453	88.3	4.16	287	52.1	3.31
	21	$\frac{11}{16}$	27.5	93.5	489	94.0	4.21	300	54.6	3.30
	21	$\frac{3}{4}$	28.9	98.2	528	100.0	4.27	314	57.0	3.30
	21	$\frac{13}{16}$	30.3	102.9	566	106.0	4.33	328	59.6	3.29
	21	$\frac{7}{8}$	31.6	107.5	604	113.0	4.38	342	62.2	3.29
	21	$\frac{15}{16}$	33.0	112.2	648	119.0	4.43	356	64.8	3.28
	21	1	34.4	116.9	686	125.0	4.47	370	67.3	3.28
	21	$1\frac{1}{16}$	35.8	121.6	726	131.0	4.50	383	69.6	3.27
	21	$1\frac{1}{8}$	37.1	126.2	771	137.0	4.55	397	72.2	3.27
	21	$1\frac{3}{16}$	38.5	130.9	816	144.0	4.60	411	74.8	3.27
	21	$1\frac{1}{4}$	39.9	135.5	859	149.0	4.64	425	77.3	3.27

TABLE VII—(*Continued*)

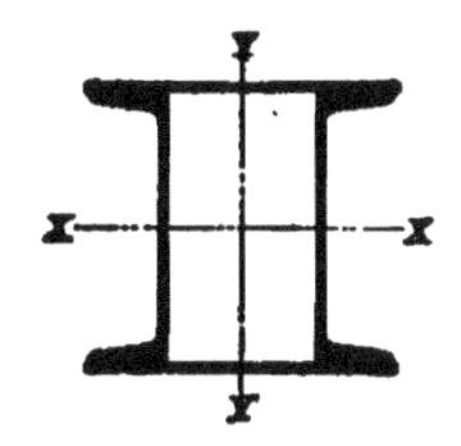

Designation	Weight of Each Channel in Pounds per Foot	Thickness of Cover-Plates Inches	Area of Section Square Inches	Weight of Column in Pounds per Foot	Axis *XX* Moment of Inertia	Axis *XX* Section Modulus	Axis *XX* Radius of Gyration Inches	Axis *YY* Moment of Inertia	Axis *YY* Section Modulus	Axis *YY* Radius of Gyration Inches
10-Inch Channel Column: Two 10-Inch Channels and Two Cover-Plates 12 Inches Wide	15	5/16	16.3	56.8	336	63.2	4.49	227	37.9	3.69
	15	3/8	18.2	61.9	377	70.2	4.55	245	40.9	3.67
	20	3/8	20.8	70.7	412	77.0	4.46	286	47.7	3.71
	20	7/16	22.3	75.8	457	84.0	4.53	304	50.7	3.69
	20	1/2	23.8	80.9	502	91.5	4.60	322	53.7	3.68
	25	1/2	26.7	90.8	526	95.8	4.45	348	58.0	3.61
	25	9/16	28.2	95.9	572	103.0	4.51	366	61.1	3.61
	25	5/8	29.7	101.0	619	110.0	4.56	384	64.0	3.60
	30	5/8	32.6	110.9	643	114.0	4.44	408	68.0	3.54
	30	11/16	34.1	116.0	691	122.0	4.50	426	71.0	3.53
	30	3/4	35.6	121.1	740	129.0	4.56	444	74.0	3.53
	30	13/16	37.1	126.2	790	136.0	4.62	462	77.0	3.53
	30	7/8	38.6	131.3	841	144.0	4.68	480	80.0	3.53
	30	15/16	40.1	136.4	893	150.0	4.73	498	83.0	3.52
	30	1	41.6	141.5	949	158.0	4.78	516	86.0	3.52
	30	1 1/8	44.6	151.7	1,059	172.0	4.87	552	92.0	3.52
	30	1 1/4	47.6	161.9	1,173	188.0	4.97	588	98.0	3.51
	30	1 3/8	50.6	172.1	1,292	203.0	5.05	624	104.0	3.51
	30	1 1/2	53.6	182.3	1,416	217.0	5.14	660	110.0	3.51
12-Inch Channel Column: Two 12-Inch Channels and Two Cover-Plates 14 Inches Wide	20	3/8	22.3	75.8	650	102.0	5.40	429	61.3	4.39
	20	7/16	24.1	81.8	724	112.0	5.48	457	65.3	4.36
	25	7/16	27.1	91.9	760	118.0	5.30	505	72.1	4.31
	25	1/2	28.8	97.9	833	128.0	5.38	534	76.3	4.31
	30	1/2	31.6	107.4	891	137.0	5.32	600	85.7	4.36
	30	9/16	33.4	113.4	964	147.0	5.37	628	89.7	4.34
	30	5/8	35.1	119.3	1,043	157.0	5.45	657	93.9	4.33
	30	11/16	36.9	125.3	1,118	168.0	5.51	686	98.0	4.31
	30	3/4	38.6	131.3	1,198	178.0	5.57	714	102.0	4.30
	35	3/4	41.6	141.5	1,234	183.0	5.44	753	108.0	4.25
	35	13/16	43.4	147.4	1,316	193.0	5.50	782	112.0	4.25
	35	7/8	45.1	153.3	1,396	204.0	5.56	810	116.0	4.24
	35	15/16	46.9	159.3	1,482	214.0	5.63	840	120.0	4.24
	35	1	48.6	165.2	1,565	224.0	5.68	867	124.0	4.22
	35	1 1/8	52.1	177.2	1,742	245.0	5.79	925	132.0	4.21
	35	1 1/4	55.6	189.0	1,922	266.0	5.90	981	140.0	4.21
	35	1 3/8	59.1	200.9	2,105	287.0	5.98	1,039	148.0	4.19
	35	1 1/2	62.6	212.7	2,302	308.0	6.08	1,096	157.0	4.19

TABLE VII—(*Continued*)

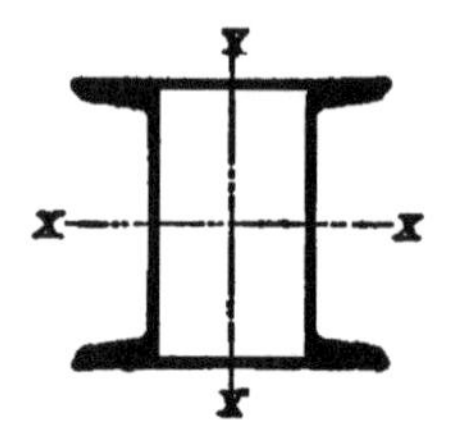

Designation	Weight of Channel	Thickness of Cover-Plates Inches	Area of Section Square Inches	Weight of Column in Pounds per Foot	Axis XX			Axis YY		
					Moment of Inertia	Section Modulus	Radius of Gyration Inches	Moment of Inertia	Section Modulus	Radius of Gyration Inches
15-Inch Channel Column, Light Section: Two 15-Inch Channels and Two Cover-Plates 17 Inches Wide	33	$\frac{1}{2}$	36.4	123.8	1,630	204	6.69	1,084	128	5.46
	33	$\frac{9}{16}$	38.5	131.0	1,767	219	6.77	1,136	134	5.43
	35	$\frac{9}{16}$	39.7	135.2	1,789	222	6.71	1,166	137	5.42
	35	$\frac{5}{8}$	41.9	142.3	1,928	237	6.79	1,217	143	5.39
	40	$\frac{5}{8}$	44.9	152.5	1,983	244	6.65	1,288	152	5.36
	40	$\frac{11}{16}$	47.0	159.7	2,124	259	6.72	1,339	158	5.34
	45	$\frac{11}{16}$	49.8	169.3	2,180	266	6.62	1,405	165	5.31
	45	$\frac{3}{4}$	51.9	176.5	2,324	282	6.69	1,456	171	5.30
	50	$\frac{3}{4}$	54.9	186.7	2,379	288	6.58	1,525	179	5.27
	50	$\frac{13}{16}$	57.0	193.9	2,527	304	6.66	1,576	185	5.26
	50	$\frac{7}{8}$	59.2	201.1	2,673	319	6.72	1,627	191	5.24
	50	$\frac{15}{16}$	61.3	208.4	2,822	335	6.79	1,678	197	5.23
	50	1	63.4	215.6	2,975	350	6.85	1,730	203	5.22
	50	$1\frac{1}{8}$	67.7	230.0	3,288	381	6.97	1,832	216	5.21
	50	$1\frac{1}{4}$	71.9	244.5	3,608	412	7.08	1,934	228	5.19
	50	$1\frac{3}{8}$	76.2	258.9	3,938	444	7.19	2,037	240	5.17
	50	$1\frac{1}{2}$	80.4	273.4	4,278	475	7.30	2,139	252	5.16

TABLE VIII

PROPERTIES OF HEAVY STEEL CHANNEL COLUMNS

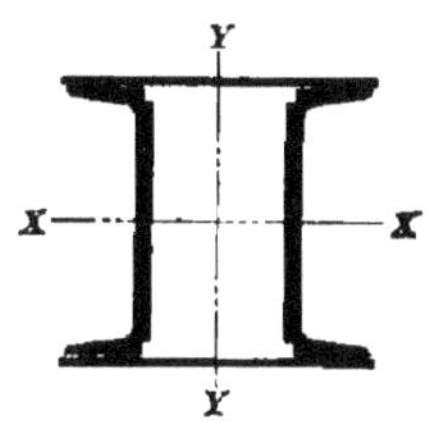

Designation	Column Section	Thickness of Cover-Plates Inches	Area of Section Square Inches	Weight of Column in Pounds per Foot	Axis XX			Axis YY		
					Moment of Inertia	Section Modulus	Radius of Gyration Inches	Moment of Inertia	Section Modulus	Radius of Gyration Inches
10-Inch Extra-Heavy Channel Column	Two 10-Inch 30-Pound Channels Two Cover-Plates 12 Inches Wide and Two Web-Plates 9 In. × 3/4 In.	3/4	49.1	167.0	831	145	4.11	522	87	3.26
		13/16	50.6	172.1	881	152	4.18	540	90	3.27
		7/8	52.1	177.2	932	159	4.23	558	93	3.27
		15/16	53.6	182.3	985	166	4.30	576	96	3.28
		1	55.1	187.4	1,039	173	4.35	594	99	3.28
		1 1/8	58.1	197.6	1,149	188	4.45	630	105	3.29
		1 1/4	61.1	207.8	1,264	203	4.55	666	111	3.30
		1 3/8	64.1	218.0	1,384	218	4.65	702	117	3.31
		1 1/2	67.1	228.2	1,507	233	4.75	738	123	3.31
		1 5/8	70.1	238.4	1,637	247	4.84	774	129	3.32
		1 3/4	73.1	248.6	1,766	263	4.92	810	135	3.33
		1 7/8	76.1	258.8	1,910	279	5.00	846	141	3.33
		2	79.1	269.0	2,057	293	5.10	882	147	3.34
		2 1/8	82.1	279.2	2,208	311	5.19	918	153	3.34
		2 1/4	85.1	289.4	2,361	326	5.27	954	159	3.34
12-Inch Extra-Heavy Channel Column	Two 12-Inch 35-Pound Channels Two Cover-Plates 14 Inches Wide and Two Web-Plates 11 In. × 3/4 In.	3/4	58.1	197.6	1,402	208	4.91	927	132	3.99
		13/16	59.9	203.5	1,478	217	4.97	956	136	3.99
		7/8	61.6	209.4	1,563	228	5.05	985	141	4.00
		15/16	63.4	215.4	1,646	237	5.10	1,013	145	4.00
		1	65.1	221.3	1,729	247	5.15	1,041	149	4.00
		1 1/8	68.6	233.3	1,907	268	5.28	1,099	157	4.00
		1 1/4	72.1	245.1	2,090	288	5.38	1,156	165	4.01
		1 3/8	75.6	257.0	2,272	309	5.49	1,213	173	4.01
		1 1/2	79.1	269.0	2,466	329	5.59	1,271	181	4.01
		1 5/8	82.6	280.8	2,665	349	5.69	1,328	189	4.02
		1 3/4	86.1	292.7	2,876	371	5.78	1,385	198	4.02
		1 7/8	89.6	304.7	3,081	391	5.86	1,442	206	4.02
		2	93.1	316.5	3,313	415	5.97	1,499	214	4.02
		2 1/8	96.6	328.4	3,538	435	6.05	1,557	222	4.02
		2 1/4	100.1	340.4	3,773	458	6.15	1,614	231	4.02

TABLE VIII—(*Continued*)

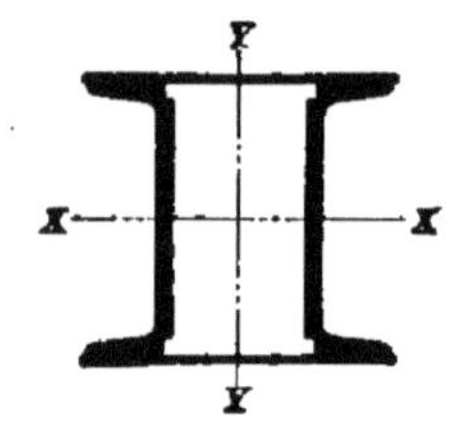

Designation	Column Section	Thickness of Cover-Plates Inches	Area of Section Square Inches	Weight of Column in Pounds per Foot	Axis XX			Axis YY		
					Moment of Inertia	Section Modulus	Radius of Gyration Inches	Moment of Inertia	Section Modulus	Radius of Gyration Inches
15-Inch Extra-Heavy Channel Column	Two 15-Inch 50-Pound Channels Two Cover-Plates 17 Inches Wide and Two Web-Plates 14 In. × $\frac{3}{4}$ In.	$\frac{3}{4}$	75.9	258.1	2,722	330	5.99	1,920	226	5.03
		$\frac{13}{16}$	78.0	265.3	2,870	345	6.06	1,971	232	5.03
		$\frac{7}{8}$	80.2	272.5	3,016	360	6.14	2,022	238	5.02
		$\frac{15}{16}$	82.3	279.8	3,165	375	6.20	2,074	244	5.02
		1	84.4	287.0	3,318	390	6.27	2,125	250	5.02
		$1\frac{1}{8}$	88.7	301.4	3,631	420	6.40	2,227	262	5.01
		$1\frac{1}{4}$	92.9	315.9	3,951	452	6.52	2,330	274	5.01
		$1\frac{3}{8}$	97.2	330.3	4,281	482	6.64	2,432	286	5.00
		$1\frac{1}{2}$	101.4	344.8	4,621	513	6.75	2,534	298	5.00
		$1\frac{5}{8}$	105.7	359.2	4,970	545	6.86	2,637	310	5.00
		$1\frac{3}{4}$	109.9	373.7	5,328	576	6.96	2,739	322	4.99
		$1\frac{7}{8}$	114.2	388.1	5,696	608	7.06	2,841	334	4.99
		2	118.4	402.6	6,075	640	7.16	2,944	346	4.99
		$2\frac{1}{8}$	122.7	417.0	6,463	672	7.26	3,046	358	4.98
		$2\frac{1}{4}$	126.9	431.5	6,864	704	7.36	3,148	370	4.98

TABLE IX

PROPERTIES AND DIMENSIONS OF STANDARD Z-BAR COLUMNS

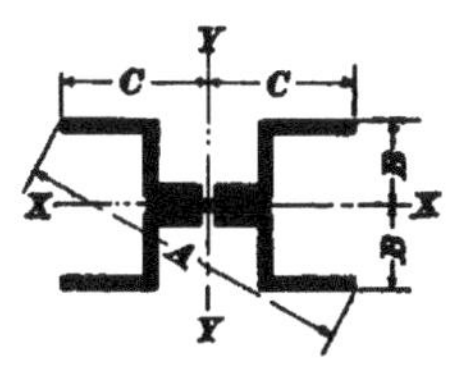

Designation	Section of Column	Thickness of Z Bars and Web-Plate	Area of Section Square Inches	Weight of Column in Pounds per Foot	Axis XX Moment of Inertia	Axis XX Section Modulus	Axis XX Radius of Gyration Inches	Axis YY Moment of Inertia	Axis YY Section Modulus	Axis YY Radius of Gyration Inches	Dimensions A	Dimensions B	Dimensions C
12-Inch Z-Bar Column	Four Z-Bars 6 Inches Deep and One Web-Plate 8 Inches Wide	3/8	21.4	72.7	287	46.5	3.67	337	46.5	3.97	19 1/16	6 3/16	7 1/4
		7/16	25.1	85.2	347	55.2	3.72	391	54.0	3.95	19 3/16	6 9/32	7 1/4
		1/2	28.8	97.8	409	64.1	3.77	445	61.3	3.92	19 5/16	6 3/8	7 1/4
		9/16	31.2	106.2	427	67.9	3.69	469	66.4	3.88	18 7/8	6 9/32	7 1/16
		5/8	34.8	118.5	489	76.8	3.74	518	73.4	3.86	19	6 3/8	7 1/16
		11/16	38.5	130.9	556	85.9	3.79	567	80.0	3.83	19 1/8	6 15/32	7 1/16
		3/4	40.5	137.8	562	88.2	3.72	579	84.2	3.78	18 3/4	6 3/8	6 7/8
		13/16	44.1	149.9	629	97.3	3.77	624	90.7	3.76	18 7/8	6 15/32	6 7/8
		7/8	47.7	162.1	700	106.6	3.82	664	96.5	3.73	19	6 9/16	6 7/8
10-Inch Z-Bar Column	Four Z-Bars 5 Inches Deep and One Web-Plate 7 Inches Wide	5/16	15.8	53.7	149	29.0	3.08	197	30.1	3.54	16 11/16	5 5/32	6 9/16
		3/8	19.0	64.7	186	35.5	3.13	235	35.8	3.52	16 13/16	5 1/4	6 9/16
		7/16	22.3	75.8	225	42.0	3.17	272	42.1	3.50	16 15/16	5 11/32	6 9/16
		1/2	24.5	83.3	236	44.9	3.10	290	45.5	3.44	16 1/2	5 1/4	6 3/8
		9/16	27.7	94.2	275	51.5	3.16	324	50.8	3.42	16 5/8	5 11/32	6 3/8
		5/8	30.9	105.2	318	58.4	3.21	358	56.1	3.40	16 3/4	5 7/16	6 3/8
		11/16	32.7	111.2	320	59.9	3.13	365	59.0	3.34	16 3/8	5 11/32	6 3/16
		3/4	35.8	121.8	363	66.8	3.18	393	63.5	3.32	16 1/2	5 7/16	6 3/16
		13/16	39.0	132.6	411	74.3	3.25	428	69.2	3.30	16 5/8	5 17/32	6 3/16
8-Inch Z-Bar Column	Four Z-Bars 4 Inches Deep and One Web-Plate 6 1/4 Inches Wide	1/4	11.3	38.4	68.7	16.6	2.47	123	20.0	3.31	14 7/8	4 1/8	6 3/16
		5/16	14.2	48.2	89.8	21.3	2.52	152	24.6	3.28	15	4 7/32	6 3/16
		3/8	17.1	58.1	113	26.1	2.57	184	29.8	3.28	15 1/16	4 5/16	6 3/16
		7/16	19.0	64.5	118	28.1	2.49	198	33.1	3.23	14 11/16	4 7/32	6
		1/2	21.9	74.4	142	32.9	2.54	225	37.6	3.21	14 3/4	4 5/16	6
		9/16	24.7	83.9	167	37.8	2.59	252	41.9	3.19	14 7/8	4 13/32	6
		5/8	26.3	89.4	167	38.8	2.52	258	44.3	3.13	14 1/2	4 5/16	5 13/16
		11/16	29.0	98.6	193	43.8	2.58	281	48.4	3.11	14 9/16	4 13/32	5 13/16
		3/4	31.9	108.3	221	49.0	2.63	305	52.4	3.09	14 11/16	4 1/2	5 13/16
6-Inch Z-Bar Column	Four Z-Bars 3 Inches Deep and One Web-Plate 6 Inches Wide	1/4	9.4	31.9	32.3	10.3	1.86	86.7	15.6	3.04	12 3/4	3 1/8	5 9/16
		5/16	11.8	40.1	42.8	13.3	1.91	108	19.3	3.02	12 13/16	3 7/32	5 9/16
		3/8	13.7	46.6	48.0	15.1	1.87	121	22.3	2.97	12 5/8	3 3/16	5 7/16
		7/16	16.1	54.7	59.5	18.1	1.92	140	25.8	2.95	12 11/16	3 9/32	5 7/16
		1/2	17.8	60.5	63.6	19.6	1.89	150	28.3	2.91	12 7/16	3 1/4	5 5/16
		9/16	20.1	68.3	76.0	22.7	1.94	168	31.7	2.89	12 9/16	3 11/32	5 5/16

TABLE X

PROPERTIES AND DIMENSIONS OF STEEL Z-BAR COLUMNS WITH COVER-PLATES

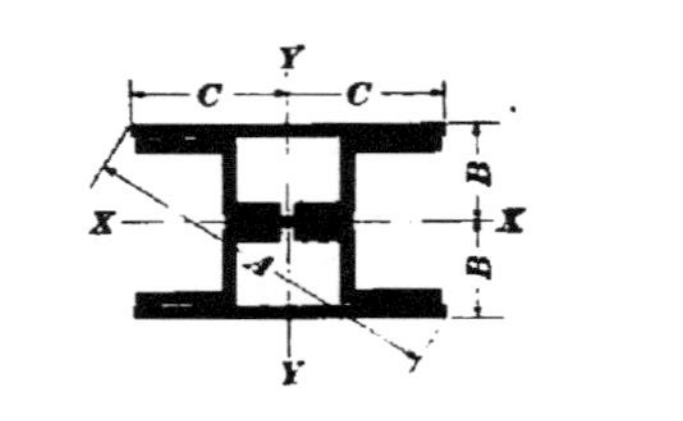

Designation	Section of Column	Thickness of Cover-Plates	Area of Section Square Inches	Weight of Column in Pounds per Foot	Axis XX			Axis YY			Dimensions		
					Moment of Inertia	Section Modulus	Radius of Gyration Inches	Moment of Inertia	Section Modulus	Radius of Gyration Inches	A	B	C
14-Inch Z-Bar Column	Four 6″ × $\frac{3}{4}$″ Z-Bars One 8″ × $\frac{3}{4}$″ Web-Plate Two Cover-Plates 14 Inches Wide	$\frac{3}{8}$	51.0	173.5	1,014	150.0	4.46	750.5	107.2	3.84	$19\frac{7}{16}$	$6\frac{3}{4}$	7
		$\frac{7}{16}$	52.8	179.5	1,094	160.7	4.55	779.2	111.3	3.84	$19\frac{1}{2}$	$6\frac{13}{16}$	7
		$\frac{1}{2}$	54.5	185.3	1,180	171.6	4.65	808.0	115.4	3.85	$19\frac{5}{8}$	$6\frac{7}{8}$	7
		$\frac{9}{16}$	56.3	191.7	1,260	181.6	4.72	836.2	119.5	3.85	$19\frac{3}{4}$	$6\frac{15}{16}$	7
		$\frac{5}{8}$	58.0	197.3	1,344	192.2	4.82	864.7	123.5	3.86	$19\frac{13}{16}$	7	7
		$\frac{11}{16}$	59.8	203.5	1,431	202.7	4.89	893.7	127.7	3.87	$19\frac{7}{8}$	$7\frac{1}{16}$	7
		$\frac{3}{4}$	61.5	209.1	1,511	212.0	4.96	922.0	131.7	3.88	20	$7\frac{1}{8}$	7
		$\frac{13}{16}$	63.3	215.2	1,609	223.9	5.04	951.2	135.9	3.88	$20\frac{1}{16}$	$7\frac{3}{16}$	7
		$\frac{7}{8}$	65.0	220.6	1,701	234.5	5.11	979.5	139.9	3.88	$20\frac{1}{8}$	$7\frac{1}{4}$	7

14-Inch Z-Bar Column. Four 6 1/16″ × 7/8″ Z-Bars, One 10″ × 7/8″ Web-Plate, Two Cover-Plates 16 Inches Wide											
11/16	66.9	227.5	1,618	223.2	4.92	979.3	139.7	3.83	20 1/8	7 1/4	7
3/4	68.7	233.5	1,711	234.0	4.99	1,007.0	143.8	3.84	20 1/4	7 5/16	7
13/16	70.5	239.7	1,805	244.8	5.06	1,035.0	147.9	3.84	20 5/16	7 3/8	7
7/8	72.2	245.2	1,901	255.7	5.13	1,064.0	152.0	3.84	20 7/16	7 7/16	7
15/16	74.0	251.6	1,999	266.5	5.20	1,092.0	156.2	3.84	20 1/2	7 1/2	7
1	75.7	257.3	2,098	277.5	5.26	1,121.0	160.2	3.85	20 5/8	7 9/16	7
1 1/16	77.5	263.3	2,198	288.3	5.32	1,150.0	164.2	3.85	20 11/16	7 5/8	7
1 1/8	79.2	269.4	2,300	299.1	5.39	1,178.0	168.2	3.85	20 13/16	7 11/16	7
1 3/16	81.0	275.4	2,405	310.4	5.45	1,207.0	172.5	3.86	20 7/8	7 3/4	7
1 1/4	82.7	281.2	2,510	321.3	5.51	1,236.0	176.5	3.86	21	7 13/16	7

16-Inch Z-Bar Column. Four 6″ × 7/8″ Z-Bars, One 10″ × 7/8″ Web-Plate, Two Cover-Plates 16 Inches Wide											
1	81.4	276.8	2,298	303.8	5.31	1,726.0	216.2	4.60	22	7 9/16	8
1 1/16	83.4	283.7	2,413	316.5	5.38	1,769.0	221.6	4.60	22 1/8	7 5/8	8
1 1/8	85.4	290.4	2,531	329.5	5.44	1,811.0	226.8	4.60	22 3/16	7 11/16	8
1 3/16	87.4	297.2	2,650	341.9	5.50	1,854.0	232.2	4.60	22 1/4	7 3/4	8
1 1/4	89.4	304.1	2,771	354.4	5.56	1,897.0	237.6	4.60	22 3/8	7 13/16	8
1 5/16	91.4	310.8	2,895	367.6	5.62	1,939.0	242.9	4.60	22 7/16	7 7/8	8
1 3/8	93.4	317.6	3,019	380.4	5.69	1,982.0	248.2	4.60	22 9/16	7 15/16	8
1 7/16	95.4	324.5	3,146	393.3	5.74	2,025.0	253.6	4.60	22 5/8	8	8
1 1/2	97.4	331.2	3,275	406.3	5.80	2,067.0	258.9	4.60	22 11/16	8 1/16	8
1 9/16	99.4	338.0	3,406	419.2	5.86	2,110.0	264.1	4.60	22 13/16	8 1/8	8
1 5/8	101.4	344.9	3,539	432.3	5.91	2,153.0	269.4	4.61	22 7/8	8 3/16	8
1 11/16	103.4	351.7	3,674	445.5	5.96	2,195.0	274.8	4.61	23	8 1/4	8
1 3/4	105.4	358.4	3,811	458.5	6.01	2,238.0	280.1	4.61	23 1/16	8 5/16	8
1 13/16	107.4	365.2	3,951	471.8	6.06	2,280.0	285.4	4.61	23 1/8	8 3/8	8
1 7/8	109.4	372.0	4,092	485.0	6.12	2,323.0	290.8	4.61	23 1/4	8 7/16	8
1 15/16	111.4	378.8	4,235	498.3	6.17	2,366.0	296.2	4.61	23 5/16	8 1/2	8
2	113.4	385.7	4,381	511.7	6.21	2,409.0	301.4	4.61	23 7/16	8 9/16	8
2 1/16	115.4	392.5	4,528	524.9	6.26	2,451.0	306.8	4.61	23 1/2	8 5/8	8
2 1/8	117.4	399.2	4,679	538.6	6.31	2,494.0	312.2	4.61	23 5/8	8 11/16	8
2 3/16	119.4	406.1	4,831	552.1	6.36	2,537.0	317.4	4.61	23 11/16	8 3/4	8
2 1/4	121.4	412.8	4,985	565.3	6.41	2,579.0	322.9	4.61	23 13/16	8 13/16	8

TABLE XI

SAFE LOADS FOR STEEL, LATTICED, CHANNEL COLUMN

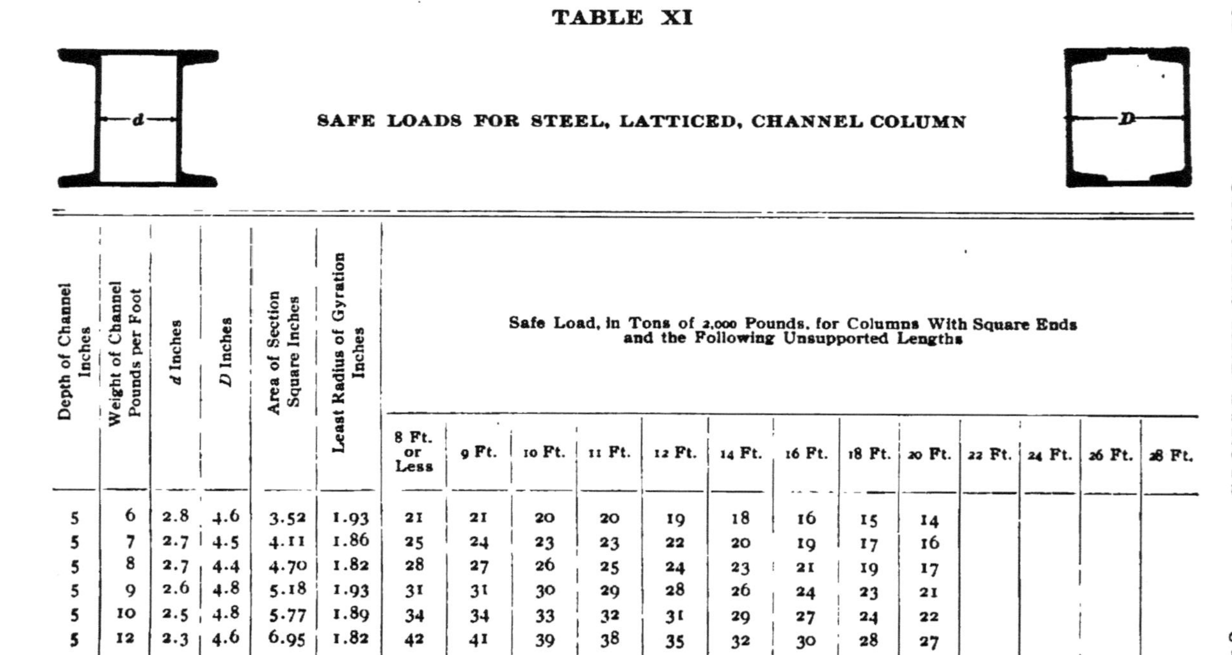

Depth of Channel Inches	Weight of Channel Pounds per Foot	d Inches	D Inches	Area of Section Square Inches	Least Radius of Gyration Inches	Safe Load, in Tons of 2,000 Pounds, for Columns With Square Ends and the Following Unsupported Lengths												
						8 Ft. or Less	9 Ft.	10 Ft.	11 Ft.	12 Ft.	14 Ft.	16 Ft.	18 Ft.	20 Ft.	22 Ft.	24 Ft.	26 Ft.	28 Ft.
5	6	2.8	4.6	3.52	1.93	21	21	20	20	19	18	16	15	14				
5	7	2.7	4.5	4.11	1.86	25	24	23	23	22	20	19	17	16				
5	8	2.7	4.4	4.70	1.82	28	27	26	25	24	23	21	19	17				
5	9	2.6	4.8	5.18	1.93	31	31	30	29	28	26	24	23	21				
5	10	2.5	4.8	5.77	1.89	34	34	33	32	31	29	27	24	22				
5	12	2.3	4.6	6.95	1.82	42	41	39	38	35	32	30	28	27				

6	8	3.5	5.6	4.70	2.33	28	28	28	28	27	26	24	23	22	20	19		
6	9	3.4	5.4	5.29	2.28	32	32	32	31	30	29	27	25	24	22	20		
6	10	3.3	5.3	5.88	2.23	35	35	35	34	33	31	30	28	26	24	23		
6	12	3.2	5.8	6.96	2.32	42	42	42	41	40	38	36	34	31	29	27		
6	13	3.1	5.7	7.55	2.28	45	45	45	44	42	40	37	35	33	31	28		
6	15	2.9	5.5	8.72	2.21	52	52	52	51	49	47	44	41	39	36	33		
6	17	2.8	5.9	9.70	2.28	58	58	58	57	56	53	50	47	44	41	38		
6	18	2.7	5.9	10.3	2.26	62	62	62	60	59	56	53	50	46	43	40		
6	20	2.6	5.8	11.5	2.22	69	69	68	66	65	61	58	54	51	47	44		
7	9	4.3	6.3	5.22	2.70	31	31	31	31	31	30	28	27	26	24	23	22	20
7	10	4.1	6.1	5.81	2.64	35	35	35	35	35	33	32	30	29	27	26	24	23
7	12	4.0	6.0	6.98	2.55	42	42	42	42	41	39	37	35	33	32	30	28	26
7	13	4.0	6.4	7.60	2.69	46	46	46	46	45	43	41	39	37	36	34	32	30
7	15	3.7	6.2	8.78	2.60	53	53	53	53	52	50	47	45	43	40	38	36	33
7	17	3.6	6.1	9.95	2.53	60	60	60	60	58	55	53	50	48	45	42	40	37
						12 Ft. or Less	14 Ft.	16 Ft.	18 Ft.	20 Ft.	22 Ft.	24 Ft.	26 Ft.	28 Ft.	30 Ft.			
8	10	5.0	7.1	6.00	3.08	36	36	35	33	32	31	29	28	26	25			
8	11	4.9	6.9	6.59	3.01	40	39	37	36	34	33	31	30	28	27			
8	12	4.8	6.8	7.18	2.96	43	42	40	39	37	35	34	32	31	29			
8	13	4.9	7.2	7.60	3.07	46	45	43	42	40	38	37	35	34	32			
8	15	4.7	7.0	8.78	2.97	54	52	50	48	46	44	42	39	37	35			
8	17	4.5	6.8	9.97	2.90	60	58	55	53	51	48	46	44	41	39			

TABLE XI—(*Continued*)

Depth of Channel Inches	Weight of Channel Pounds per Foot	d Inches	D Inches	Area of Section Square Inches	Least Radius of Gyration Inches	Safe Load, in Tons of 2,000 Pounds, for Columns With Square Ends and the Following Unsupported Lengths												
						12 Ft. or Less	14 Ft.	16 Ft.	18 Ft.	20 Ft.	22 Ft.	24 Ft.	26 Ft.	28 Ft.	30 Ft.	32 Ft.	36 Ft.	40 Ft.
9	13	5.7	7.9	7.60	3.46	46	46	45	44	42	41	39	38	36	35	33	30	
9	14	5.6	7.8	8.19	3.40	49	49	48	45	43	42	41	39	37	36	35	31	
9	15	5.5	7.7	8.79	3.36	53	53	51	49	47	45	43	42	41	39	37	34	
9	16	5.5	8.1	9.40	3.48	56	56	56	54	52	50	49	47	45	43	41	38	
9	18	5.3	8.0	10.6	3.40	64	64	62	60	58	56	54	52	50	47	45	41	
9	21	5.1	7.7	12.4	3.30	74	74	72	69	67	64	62	59	57	54	51	46	

						18 Ft. or Less	20 Ft.	22 Ft.	24 Ft.	26 Ft.	28 Ft.	30 Ft.	32 Ft.	36 Ft.	40 Ft.	44 Ft.	48 Ft.	52 Ft.
10	15	6.3	8.9	8.80	3.89	53	53	53	53	51	49	48	46	45	43	42	38	35
10	17	6.1	8.7	10.1	3.77	60	60	60	59	57	55	54	52	50	48	46	43	39
10	18	6.0	8.6	10.7	3.73	64	64	64	63	61	59	57	55	53	51	49	45	41
10	20	6.0	8.9	11.8	3.81	71	71	71	69	67	65	63	61	59	57	55	50	46
10	25	5.7	8.6	14.7	3.65	88	88	88	85	82	80	77	74	71	69	66	60	55
10	30	5.4	8.4	17.6	3.53	106	106	105	101	98	94	91	87	84	81	77	70	64
12	20	7.7	10.4	11.8	4.59	71	71	69	67	66	64	62	60	57	53	49	46	
12	23	7.4	10.2	13.6	4.47	82	81	79	77	75	73	71	69	64	60	56	52	
12	25	7.3	10.0	14.8	4.39	89	88	86	83	81	79	76	74	69	65	60	56	
12	27	7.4	10.5	15.8	4.54	95	95	93	90	88	85	83	81	76	71	66	61	
12	30	7.1	10.2	17.6	4.42	106	105	102	100	97	94	91	89	83	78	73	67	
12	33	7.0	10.1	19.4	4.34	116	115	112	109	106	103	100	97	91	85	79	72	
12	35	6.9	10.0	20.6	4.29	123	122	119	116	112	109	105	102	95	89	83	76	
15	33	9.5	12.7	19.4	5.64	116	116	116	116	114	111	109	107	102	98	93	88	84
15	35	9.4	12.5	20.6	5.53	124	124	124	124	121	118	116	113	108	103	98	93	88
15	40	9.1	12.2	23.6	5.40	142	142	142	141	138	135	132	129	123	117	111	105	99
15	45	8.9	12.0	26.4	5.29	158	158	158	157	153	150	147	143	136	130	123	116	109
15	50	8.7	11.8	29.4	5.21	176	176	176	174	170	166	162	159	151	143	135	128	120

TABLE XII

SAFE LOADS FOR STEEL I BEAMS USED AS COLUMNS. BEAMS SUPPORTED AGAINST YIELDING SIDEWISE

Depth of Beam Inches	Weight of Beam per Foot Pounds	Area of Section Square Inches	Radius of Gyration Inches	Safe Load, in Tons of 2,000 Pounds, for Columns With Square Ends and the Following Unsupported Lengths														
				12 Ft. or Less	14 Ft.	16 Ft.	18 Ft.	20 Ft.	22 Ft.	24 Ft.	26 Ft.	28 Ft.	30 Ft.	32 Ft.	34 Ft.	36 Ft.	38 Ft.	40 Ft.
20	90	26.4	7.55	159.0	159.0	159.0	159.0	158.0	156.0	154.0	151.0	149.0	147.0	145.0	143.0	141.0	139.0	137.0
20	80	23.5	7.55	141.0	141.0	141.0	141.0	140.0	138.0	136.0	134.0	132.0	130.0	128.0	127.0	125.0	123.0	121.0
20	75	22.1	7.53	133.0	133.0	133.0	133.0	132.0	130.0	128.0	126.0	125.0	123.0	121.0	119.0	118.0	116.0	114.0
20	65	19.1	7.76	115.0	115.0	115.0	115.0	114.0	113.0	111.0	109.0	108.0	106.0	105.0	104.0	102.0	101.0	99.0
18	80	23.5	6.94	141.0	141.0	141.0	140.0	138.0	136.0	134.0	132.0	130.0	128.0	126.0	124.0	122.0	120.0	118.0
18	70	20.6	6.87	124.0	124.0	124.0	123.0	121.0	119.0	117.0	116.0	114.0	112.0	110.0	108.0	107.0	105.0	103.0
18	65	19.1	6.81	115.0	115.0	115.0	114.0	112.0	110.0	109.0	107.0	105.0	104.0	102.0	100.0	99.0	97.0	95.0
18	60	17.6	6.94	106.0	106.0	106.0	105.0	104.0	102.0	101.0	99.0	97.0	96.0	94.0	93.0	91.0	90.0	88.0
18	55	16.2	7.08	97.0	97.0	97.0	97.0	96.0	94.0	93.0	91.0	90.0	89.0	87.0	86.0	85.0	83.0	82.0
15	80	23.5	5.64	141.0	141.0	139.0	136.0	134.0	131.0	129.0	126.0	124.0	121.0	119.0	116.0	114.0	111.0	109.0
15	75	22.1	5.72	133.0	133.0	131.0	128.0	126.0	124.0	121.0	119.0	117.0	114.0	112.0	110.0	107.0	105.0	103.0
15	60	17.6	6.02	106.0	106.0	105.0	103.0	101.0	100.0	98.0	96.0	94.0	93.0	91.0	89.0	87.0	86.0	84.0
15	50	14.7	6.00	88.0	88.0	87.0	86.0	85.0	83.0	82.0	80.0	79.0	77.0	76.0	74.0	73.0	71.0	70.0
15	42	12.4	5.90	74.0	74.0	74.0	72.0	71.0	70.0	69.0	67.0	66.0	65.0	64.0	62.0	61.0	60.0	58.0

12	65	19.1	4.55	114.0	111.0	109.0	106.0	104.0	101.0	99.0	96.0	94.0	91.0	89.0	86.0	84.0	81.0	78.0
12	55	16.2	4.72	97.0	95.0	93.0	91.0	89.0	87.0	85.0	82.0	80.0	78.0	76.0	74.0	72.0	70.0	68.0
12	50	14.7	4.65	88.0	86.0	84.0	82.0	80.0	78.0	76.0	75.0	73.0	71.0	69.0	67.0	65.0	63.0	61.0
12	40	11.8	4.90	71.0	70.0	68.0	67.0	65.0	64.0	62.0	61.0	59.0	58.0	57.0	55.0	54.0	52.0	51.0
12	35	10.3	4.77	62.0	60.0	59.0	58.0	57.0	55.0	54.0	53.0	51.0	50.0	49.0	47.0	46.0	45.0	44.0
12	31.5	9.3	4.88	56.0	55.0	53.0	52.0	51.0	50.0	49.0	48.0	46.0	45.0	44.0	43.0	42.0	41.0	40.0
				4 Ft. or Less	6 Ft.	8 Ft.	10 Ft.	12 Ft.	14 Ft.	16 Ft.	18 Ft.	20 Ft.	22 Ft.	24 Ft.	26 Ft.	28 Ft.	30 Ft.	32 Ft.
10	40	11.8	3.89	70.8	70.8	70.8	70.6	68.7	66.9	65.1	63.3	61.4	59.6	57.8	56.0	54.2	52.3	50.5
10	35	10.3	4.02	61.8	61.8	61.8	61.8	60.3	58.8	57.2	55.7	54.1	52.6	51.1	49.5	48.0	46.4	44.9
10	30	8.8	3.90	52.8	52.8	52.8	52.6	51.3	49.9	48.6	47.2	45.9	44.5	43.2	41.8	40.5	39.1	37.8
10	25	7.3	4.06	43.8	43.8	43.8	43.8	42.8	41.8	40.7	39.6	38.5	37.4	36.3	35.2	34.2	33.1	32.0
9	33	9.7	3.54	58.2	58.2	58.2	57.2	55.6	53.9	52.3	50.6	49.0	47.4	45.7	44.1	42.4	40.8	39.1
9	27	7.9	3.72	47.4	47.4	47.4	47.0	45.7	44.4	43.1	41.9	40.6	39.3	38.0	36.8	35.5	34.2	32.9
9	25	7.3	3.54	43.8	43.8	43.8	43.1	41.8	40.6	39.4	38.1	36.9	35.7	34.4	33.2	32.0	30.7	29.5
9	21	6.2	3.70	37.2	37.2	37.2	36.8	35.8	34.8	33.8	32.8	31.8	30.8	29.8	28.8	27.7	26.7	25.7
8	27	7.9	3.14	47.4	47.4	47.3	45.8	44.3	42.8	41.3	39.8	38.3	36.7	35.2	33.7	32.2		
8	22	6.4	3.30	38.4	38.4	38.4	37.4	36.2	35.0	33.9	32.7	31.5	30.4	29.2	28.0	26.9		
8	18	5.2	3.30	31.2	31.2	31.2	30.4	29.4	28.5	27.5	26.6	25.6	24.7	23.7	22.8	21.8		
7	20	5.7	2.85	34.2	34.2	33.7	32.5	31.3	30.1	28.9	27.7	26.5	25.3	24.1				
7	15	4.4	2.89	26.4	26.4	26.0	25.1	24.2	23.3	22.4	21.5	20.6	19.7	18.7				
6	15	4.4	2.47	26.4	26.4	25.4	24.3	23.3	22.2	21.1	20.0	19.0	17.9					
6	12	3.6	2.47	21.6	21.6	20.8	19.9	19.0	18.2	17.3	16.4	15.5	14.6					
5	13	3.8	2.06	22.8	22.3	21.2	20.1	19.0	17.9	16.8	15.7							
5	9¾	2.9	2.06	17.4	17.0	16.2	15.3	14.5	13.7	12.8	12.0							
4	10	2.9	1.53	17.3	16.2	15.0	13.9	12.7	11.6									
4	7½	2.2	1.63	13.2	12.4	11.6	10.8	10.0	9.2									
4	6	1.8	1.61	10.8	10.1	9.5	8.8	8.1	7.5									

TABLE XIII

SAFE LOADS FOR STEEL I BEAMS USED AS COLUMNS. BEAMS NOT SUPPORTED AGAINST YIELDING SIDEWISE

Depth of Beam Inches	Weight per Foot-Pounds	Area of Section Square Inches	Radius of Gyration Inches	Safe Load, in Tons of 2,000 Pounds, for Columns With Square Ends and Unsupported for the Length of														
				2 Ft. or Less	3 Ft.	4 Ft.	5 Ft.	6 Ft.	7 Ft.	8 Ft.	9 Ft.	10 Ft.	11 Ft.	12 Ft.	13 Ft.	14 Ft.	15 Ft.	16 Ft.
20	90	26.4	1.27	158.0	158.0	153.0	147.0	141.0	135.0	128.0	122.0	116.0	110.0	103.0	97.1	90.9	84.6	78.4
20	80	23.5	1.19	141.0	141.0	135.0	129.0	123.0	117.0	111.0	105.0	99.0	93.4	87.4	81.5	75.6	69.6	63.7
20	75	22.1	1.13	133.0	132.0	126.0	120.0	114.0	108.0	102.0	96.0	91.0	84.7	78.8	73.0	67.1	61.2	55.4
20	65	19.1	1.16	115.0	114.0	109.0	104.0	99.0	94.0	89.0	84.0	80.0	74.6	69.6	64.7	59.7	54.8	49.8
18	80	23.5	1.30	141.0	141.0	137.0	132.0	126.0	121.0	115.0	110.0	104.0	99.0	93.6	88.1	82.7	77.2	71.8
18	70	20.6	1.18	124.0	123.0	118.0	113.0	108.0	102.0	97.0	92.0	87.0	81.4	76.2	70.9	65.7	60.4	55.2
18	65	19.1	1.12	115.0	114.0	108.0	103.0	98.0	93.0	88.0	83.0	78.0	72.6	67.5	62.4	57.2	52.1	47.0
18	60	17.6	1.13	106.0	105.0	100.0	95.0	91.0	86.0	81.0	77.0	72.0	67.4	62.7	58.0	53.4	48.7	44.0
18	55	16.2	1.16	97.0	97.0	93.0	88.0	84.0	80.0	76.0	72.0	67.0	63.2	59.1	54.9	50.7	46.5	42.3
15	80	23.5	1.25	141.0	141.0	136.0	130.0	125.0	119.0	114.0	108.0	102.0	96.6	91.0	85.3	79.7	74.0	68.4
15	75	22.1	1.25	133.0	133.0	128.0	123.0	117.0	112.0	107.0	101.0	96.1	90.8	85.5	80.2	74.9	69.6	64.3
15	60	17.6	1.29	106.0	106.0	102.0	98.0	94.0	90.2	86.1	82.0	77.9	73.8	69.7	65.6	61.5	57.4	53.3
15	50	14.7	1.20	88.0	88.0	85.0	81.0	77.0	73.5	69.8	66.2	62.5	58.8	55.1	51.5	47.8	44.1	40.4
15	42	12.4	1.08	74.0	73.0	70.0	66.0	63.0	59.6	56.1	52.7	49.2	45.8	42.3	38.9	35.4	32.0	28.5

12	65	19.1	1.23	114.0	114.0	110.0	106.0	101.0	96.4	91.7	87.1	82.4	77.8	73.1	68.5	63.8	59.2	54.5
12	55	16.2	1.25	97.2	97.2	93.8	89.9	86.0	82.1	78.2	74.4	70.5	66.6	62.7	58.8	54.9	51.0	47.1
12	50	14.7	1.15	88.2	87.7	83.9	80.0	76.2	72.4	68.6	64.7	60.9	57.1	53.2	49.4	45.6	41.8	37.9
12	40	11.8	1.20	70.8	70.8	67.9	64.9	62.0	59.0	56.1	53.1	50.2	47.2	44.3	41.3	38.4	35.4	32.5
12	35	10.3	1.01	61.8	60.4	57.3	54.3	51.2	48.2	45.1	42.1	39.0	36.0	32.9	29.9	26.8	23.8	20.7
12	31½	9.3	1.04	55.8	54.7	52.1	49.4	46.7	44.0	41.3	38.6	36.0	33.3	30.6	27.9	25.2	22.5	19.9
				1 Ft.	2 Ft.	3 Ft.	4 Ft.	5 Ft.	6 Ft.	7 Ft.	8 Ft.	9 Ft.	10 Ft.	11 Ft.	12 Ft.	13 Ft.	14 Ft.	15 Ft.
10	40	11.8	1.07	70.8	70.8	69.9	66.5	63.2	59.9	56.6	53.2	49.9	46.6	43.3	40.0	36.6	33.3	30.0
10	35	10.3	1.09	61.8	61.8	61.0	58.2	55.4	52.5	49.7	46.8	44.0	41.2	38.3	35.5	32.7	29.8	27.0
10	30	8.8	.96	52.8	52.8	51.1	48.4	45.6	42.9	40.1	37.4	34.6	31.9	29.1	26.4	23.6	20.9	18.1
10	25	7.3	.99	43.8	43.8	42.6	40.4	38.2	36.0	33.8	31.6	29.3	27.1	24.9	22.7	20.5	18.3	16.1
9	33	9.7	1.03	58.2	58.2	57.0	54.2	51.4	48.6	45.7	42.9	40.1	37.2	34.4	31.6	28.8	25.9	23.1
9	27	7.9	1.07	47.4	47.4	46.7	44.5	42.2	40.0	37.8	35.6	33.4	31.2	30.0	26.7	24.5	22.3	20.1
9	25	7.3	.91	43.8	43.8	42.1	39.7	37.2	34.8	32.4	30.0	27.6	25.2	22.8	20.3	17.9	15.5	13.1
9	21	6.2	.95	37.2	37.2	36.0	34.0	32.1	30.1	28.2	26.2	24.2	22.3	20.3	18.4	16.4	14.4	12.5
8	27	7.9	.93	47.4	47.4	45.7	43.2	40.6	38.1	35.5	33.0	30.4	27.9	25.3	22.8	20.2	17.7	
8	22	6.4	.97	38.4	38.4	37.3	35.3	33.3	31.3	29.3	27.4	25.4	23.4	21.4	19.4	17.5	15.5	
8	78	5.2	.87	31.2	31.2	29.7	27.9	26.1	24.3	22.5	20.8	19.0	17.2	15.4	13.6	11.8	10.0	
7	20	5.7	.92	34.2	34.2	32.9	31.0	29.2	27.3	25.5	23.6	21.7	19.9	18.0	16.2	14.3		
7	15	4.4	.84	26.4	26.4	25.0	23.4	21.8	20.2	18.7	17.1	15.5	14.0	12.4	10.8	9.3		
6	15	4.4	.79	26.4	26.3	24.7	23.0	21.3	19.7	18.0	16.3	14.7	13.0	11.3				
6	12	3.6	.73	21.6	21.3	19.8	18.4	16.9	15.4	13.9	12.4	11.0	9.5	8.0				
5	13	3.8	.72	22.8	22.5	20.9	19.3	17.7	16.2	14.6	13.0	11.4	9.8					
5	9¾	2.9	.67	17.4	17.0	15.7	14.4	13.1	11.8	10.5	9.2	7.9	6.6					
4	10	2.9	.55	17.4	16.4	14.8	13.3	11.7	10.1	8.5								
4	7½	2.2	.56	13.2	12.5	11.3	10.2	9.0	7.8	6.6								
4	6	1.8	.47	10.8	9.8	8.7	7.5	6.3	5.2	4.0								

TABLE XIV

SAFE LOAD FOR STEEL PLATE-AND-ANGLE COLUMNS

Column Section	Thickness of Plate and Angles Inch	Safe Load, in Tons of 2,000 Pounds, for Columns With Square Ends and Unsupported Lengths of												
		6 Ft. or Less	7 Ft.	8 Ft.	9 Ft.	10 Ft.	11 Ft.	12 Ft.	14 Ft.	16 Ft.	18 Ft.	20 Ft.	22 Ft.	
One 6-Inch Plate. Four 3″ × 2½″ L's	$\frac{1}{4}$	39	37	35	34	32	30	28	24	21				
	$\frac{5}{16}$	50	48	45	43	41	39	36	32	27				
	$\frac{7}{16}$	70	67	64	61	58	55	52	46	40				
	$\frac{1}{2}$	78	75	72	69	65	62	59	52	46				
One 7-Inch Plate. Four 3½″ × 2½″ L's	$\frac{1}{4}$	45	44	42	41	39	37	35	32	28	25			
	$\frac{5}{16}$	56	56	54	52	49	47	45	41	36	32			
	$\frac{7}{16}$	78	77	74	71	69	66	63	57	52	46			
	$\frac{1}{2}$	87	87	84	81	77	74	71	65	58	52			
One 8-Inch Plate Four 4″ × 3″ Angles	$\frac{5}{16}$	65	65	64	62	60	57	55	51	46	42	37	33	
	$\frac{3}{8}$	79	79	77	74	72	69	66	61	56	51	45	40	
	$\frac{7}{16}$	90	90	89	86	83	80	77	71	65	59	53	47	
	$\frac{1}{2}$	103	103	102	99	95	92	89	82	76	69	63	56	
	$\frac{9}{16}$	117	117	116	112	109	105	101	94	86	79	71	64	
	$\frac{5}{8}$	126	126	126	122	118	114	110	102	95	87	79	71	
		9 Ft. or Less	10 Ft.	11 Ft.	12 Ft.	14 Ft.	16 Ft.	18 Ft.	20 Ft.	22 Ft.	24 Ft.	26 Ft.	28 Ft.	30 Ft.
One 9-Inch Plate Four 4½″ × 3″ Angles	$\frac{5}{16}$	70	67	65	63	59	55	50	46	42	38			
	$\frac{3}{8}$	84	82	79	76	71	66	61	56	51	46			
	$\frac{7}{16}$	97	94	91	88	83	77	71	66	60	54			
	$\frac{1}{2}$	112	109	106	103	96	90	83	77	71	64			
	$\frac{9}{16}$	127	123	120	116	109	102	95	88	81	74			
	$\frac{5}{8}$	137	133	130	126	118	111	104	96	89	81			
One 10-Inch Plate Four 5″ × 3″ Angles	$\frac{5}{16}$	76	75	73	71	67	63	59	55	51	46	42		
	$\frac{3}{8}$	92	91	89	86	81	77	72	67	62	57	52		
	$\frac{7}{16}$	106	105	102	100	94	89	83	78	72	67	61		
	$\frac{1}{2}$	122	121	118	115	109	103	96	90	84	78	72		
	$\frac{9}{16}$	134	134	131	128	121	114	108	101	94	88	81		
	$\frac{5}{8}$	150	150	147	143	136	129	121	114	106	99	92		
One 12-Inch Plate Four 6″ × 4″ Angles	$\frac{3}{8}$	114	114	114	111	106	101	96	90	85	80	75	70	65
	$\frac{7}{16}$	133	133	133	130	124	118	112	106	100	94	89	83	77
	$\frac{1}{2}$	152	152	152	151	144	138	131	125	118	112	105	98	92
	$\frac{9}{16}$	172	172	172	171	164	156	149	142	134	127	120	112	105
	$\frac{5}{8}$	186	186	186	186	178	170	163	155	147	139	131	123	116
	$\frac{11}{16}$	205	205	205	205	197	188	180	171	163	154	146	138	129
	$\frac{3}{4}$	224	224	224	224	216	208	199	190	181	172	163	154	145

TABLE XV

SAFE LOAD FOR STEEL PLATE-AND-ANGLE COLUMNS WITH COVER-PLATES

Column Section	Thickness of Cover Plate in Inches	Safe Load, in Tons of 2,000 Pounds, for Columns With Square Ends and Unsupported Lengths of												
		12 Ft. or Less	14 Ft.	16 Ft.	18 Ft.	20 Ft.	22 Ft.	24 Ft.	26 Ft.	28 Ft.	30 Ft.	32 Ft.	36 Ft.	40 Ft.
One Plate 12″ × 3/4″. Four 6″ × 4″ × 3/4″ Angles. Two Cover-Plates 13 Inches Wide	1/2	248	244	235	225	216	206	197	187	178	168	159	140	
	9/16	259	255	245	235	225	215	205	196	186	176	166	146	
	5/8	268	265	255	245	235	225	215	205	194	184	174	154	
	11/16	277	276	266	255	245	234	224	213	203	192	182	161	
	3/4	288	286	275	265	254	243	233	222	212	201	190	169	
	13/16	298	296	285	274	263	252	241	231	220	209	198	176	
	7/8	307	306	295	284	273	261	250	239	228	217	205	183	
	15/16	317	316	305	293	282	270	259	247	236	224	213	190	
	1	328	327	315	303	292	280	268	256	244	232	221	197	173
	1 1/16	337	337	325	313	301	289	277	265	253	241	229	205	178
	1 1/8	345	345	334	322	310	297	285	273	261	249	236	212	188
	1 3/16	356	356	345	333	321	308	296	283	270	257	244	219	194
	1 1/4	365	365	354	341	328	316	303	290	277	264	252	226	200
	1 5/16	375	375	364	351	338	325	312	299	286	273	260	234	208
	1 3/8	385	385	375	361	348	334	321	307	294	280	267	240	213
	1 7/16	395	395	384	371	357	342	330	316	302	289	275	248	221
	1 1/2	405	405	394	380	366	353	338	324	310	296	282	254	226

Column Section	Thickness of Cover Plate in Inches	14 Ft. or Less	16 Ft.	18 Ft.	20 Ft.	22 Ft.	24 Ft.	26 Ft.	28 Ft.	30 Ft.	32 Ft.	36 Ft.	40 Ft.	44 Ft.
One Plate 14″ × 3/4″. Four 6″ × 4″ × 3/4″ Angles. Two Cover-Plates 15 Inches Wide	1/2	324	313	302	290	279	268	256	245	234	222	200	177	
	9/16	335	326	314	300	291	280	268	256	245	233	210	187	
	5/8	346	338	326	314	302	291	279	267	255	243	220	196	
	11/16	357	350	338	326	314	302	290	277	265	253	229	205	
	3/4	368	363	351	338	326	313	301	288	276	263	238	213	
	13/16	380	374	361	349	336	324	311	298	286	273	248	223	
	7/8	391	386	373	360	347	334	321	308	296	283	257	231	
	15/16	402	398	385	372	358	345	332	319	306	293	266	240	
	1	413	409	396	382	369	356	342	329	315	302	275	249	222
	1 1/16	425	421	407	394	380	366	353	339	325	312	285	257	230
	1 1/8	436	433	419	405	391	377	363	349	335	322	294	266	238
	1 3/16	447	445	431	417	402	388	374	360	346	331	303	274	246
	1 1/4	458	457	443	428	414	399	385	370	356	341	312	283	254
	1 5/16	470	469	454	439	424	410	395	380	365	350	320	291	261
	1 3/8	481	481	466	451	436	421	405	390	375	360	330	299	269
	1 7/16	492	492	478	463	447	432	416	401	385	370	339	308	277
	1 1/2	503	503	489	473	458	442	426	411	395	380	348	317	286

TABLE XVI

SAFE LOAD FOR STEEL CHANNEL COLUMNS WITH COVER-PLATES

Designation	Weight of Each Channel Pounds per Foot	Thickness of Cover-Plate Inches	Safe Load, in Tons of 2,000 Pounds, for Columns With Square Ends and Unsupported Lengths of											
			10 Ft. or Less	11 Ft.	12 Ft.	13 Ft.	14 Ft.	16 Ft.	18 Ft.	20 Ft.	22 Ft.	24 Ft.	26 Ft.	28 Ft.
6-Inch Channel Column: Two Channels 6 Inches Deep and Two Cover-Plates 8 Inches Wide	8	$\frac{1}{4}$	52	51	49	48	47	44	42	39	37	34	32	29
	10	$\frac{1}{4}$	59	57	55	53	51	50	47	44	41	38	35	32
	10	$\frac{5}{16}$	65	63	62	61	59	55	52	49	45	42	39	36
	10	$\frac{3}{8}$	71	69	67	66	64	61	57	53	50	46	43	39
	12	$\frac{3}{8}$	78	76	74	72	71	67	63	59	55	51	47	43
	12	$\frac{7}{16}$	84	81	79	77	75	71	67	63	58	54	50	46
	12	$\frac{1}{2}$	90	87	84	82	80	76	71	67	62	58	53	49
	15	$\frac{1}{2}$	100	98	95	92	90	85	80	75	70	65	60	55
	15	$\frac{9}{16}$	106	104	101	98	95	89	84	79	73	68	63	58
	15	$\frac{5}{8}$	112	109	106	103	100	95	89	84	78	73	67	62
	17	$\frac{5}{8}$	118	116	113	110	107	102	96	91	85	80	74	69
	17	$\frac{11}{16}$	124	121	119	116	113	107	101	95	90	84	78	72
	17	$\frac{3}{4}$	130	127	124	121	118	112	106	100	94	88	82	76
	17	$\frac{13}{16}$	136	132	128	125	122	116	110	104	97	91	85	79
	17	$\frac{7}{8}$	142	138	134	131	128	122	115	108	102	95	89	82
	17	$\frac{15}{16}$	148	144	141	137	134	127	120	113	106	99	92	85
	17	1	154	150	147	144	140	132	124	117	110	102	95	88

Designation	Weight of Each Channel Pounds per Foot	Thickness of Cover-Plate Inches	11 Ft. or Less	12 Ft.	13 Ft.	14 Ft.	16 Ft.	18 Ft.	20 Ft.	22 Ft.	24 Ft.	26 Ft.	28 Ft.	30 Ft.	32 Ft.
7-Inch Channel Column: Two Channels 7 Inches Deep and Two Cover-Plates 9 Inches Wide	9	$\frac{1}{4}$	58	57	56	54	52	50	47	45	43	40	38	36	34
	9	$\frac{5}{16}$	65	63	61	60	57	55	53	50	47	45	42	39	37
	13	$\frac{5}{16}$	79	77	75	74	71	68	65	62	59	56	53	50	47
	13	$\frac{3}{8}$	86	84	82	80	77	74	71	67	64	60	57	53	50
	13	$\frac{7}{16}$	93	92	90	89	85	81	77	73	69	66	62	58	54
	13	$\frac{1}{2}$	100	98	96	94	90	86	82	78	74	70	66	62	58
	17	$\frac{1}{2}$	114	112	109	107	102	98	93	88	83	79	74	69	64
	17	$\frac{9}{16}$	122	119	116	113	108	103	98	93	88	83	78	73	68
	17	$\frac{5}{8}$	127	124	122	119	114	109	103	98	93	87	82	76	71
	17	$\frac{11}{16}$	134	131	129	126	120	114	109	103	97	92	86	81	75
	17	$\frac{3}{4}$	140	136	133	130	124	118	112	106	100	94	88	82	76
	17	$\frac{13}{16}$	148	145	142	139	132	125	119	113	107	100	94	88	82
	17	$\frac{7}{8}$	154	150	147	144	138	131	125	118	112	105	99	92	86
	17	$\frac{15}{16}$	161	157	154	150	143	137	130	123	116	110	103	96	89
	17	1	168	164	161	157	150	143	136	129	121	114	107	100	93

TABLE XVI—(*Continued*)

Designation	Weight of Each Channel Pounds Per Foot	Thickness of Cover-Plate Inches	Safe Load, in Tons of 2,000 Pounds, for Columns With Square Ends and Unsupported Lengths of 12 Ft. or Less	13 Ft.	14 Ft.	16 Ft.	18 Ft.	20 Ft.	22 Ft.	24 Ft.	26 Ft.	28 Ft.	30 Ft.	32 Ft.	34 Ft.	36 Ft.
8-Inch Channel Column: Two Channels 8 Inches Deep and Two Cover-Plates 10 Inches Wide	10	$\frac{1}{4}$	66	66	66	64	62	60	57	54	52	49	46	44	41	39
	10	$\frac{5}{16}$	74	74	73	70	67	65	62	59	57	54	51	49	46	43
	13	$\frac{5}{16}$	83	83	83	80	77	74	71	68	65	62	59	56	53	50
	13	$\frac{3}{8}$	91	91	90	87	84	80	77	74	70	67	64	60	57	54
	13	$\frac{7}{16}$	98	98	97	94	90	87	83	79	76	72	69	65	62	58
	13	$\frac{1}{2}$	106	106	105	102	98	94	90	86	82	78	74	70	66	62
	17	$\frac{1}{2}$	120	120	118	114	110	105	101	96	92	87	82	78	73	69
	17	$\frac{9}{16}$	127	127	126	120	115	110	106	101	96	91	87	83	78	73
	17	$\frac{5}{8}$	135	135	133	128	123	117	112	107	102	97	92	87	82	77
	17	$\frac{11}{16}$	142	142	140	135	129	123	118	113	108	102	97	92	86	81
	17	$\frac{3}{4}$	150	150	147	141	136	130	125	119	114	108	103	97	92	86
	17	$\frac{13}{16}$	157	157	155	149	143	137	131	125	119	113	107	101	95	89
	17	$\frac{7}{8}$	165	165	162	156	150	144	137	131	125	118	112	106	99	93
	17	$\frac{15}{16}$	172	172	169	163	156	150	143	136	130	123	117	110	104	97
	17	1	180	180	177	171	164	157	150	143	136	129	122	115	108	101

Designation	Weight of Each Channel Pounds Per Foot	Thickness of Cover-Plate Inches	14 Ft. or Less	16 Ft.	18 Ft.	20 Ft.	22 Ft.	24 Ft.	26 Ft.	28 Ft.	30 Ft.	32 Ft.	36 Ft.	40 Ft.
9-Inch Channel Column: Two Channels 9 Inches Deep and Two Cover-Plates 11 Inches Wide	13	$\frac{5}{16}$	87	85	82	79	76	73	70	67	65	62	56	50
	13	$\frac{3}{8}$	95	93	90	87	83	80	77	74	71	67	61	55
	16	$\frac{3}{8}$	106	104	100	97	93	90	86	83	79	76	69	62
	16	$\frac{7}{16}$	114	112	108	105	101	98	94	90	86	82	75	67
	16	$\frac{1}{2}$	122	120	115	111	107	103	99	95	91	87	79	71
	21	$\frac{1}{2}$	140	136	131	126	121	117	112	107	103	98	89	79
	21	$\frac{9}{16}$	149	146	140	135	130	125	120	115	110	104	94	84
	21	$\frac{5}{8}$	157	153	148	142	137	131	126	120	115	109	99	88
	21	$\frac{11}{16}$	165	161	156	150	145	139	133	127	121	116	104	92
	21	$\frac{3}{4}$	173	168	163	157	151	145	139	133	127	121	109	97
	21	$\frac{13}{16}$	182	177	171	164	158	151	145	139	132	126	113	101
	21	$\frac{7}{8}$	190	186	180	173	166	160	153	146	139	133	119	105
	21	$\frac{15}{16}$	198	194	187	180	173	166	159	152	145	138	124	110
	21	1	206	202	195	187	180	173	165	158	151	143	129	114
	21	$1\frac{1}{16}$	215	209	202	194	187	179	172	164	157	149	134	119
	21	$1\frac{1}{8}$	224	218	211	203	195	187	179	171	163	155	139	123
	21	$1\frac{3}{16}$	231	227	218	210	201	193	185	177	169	160	144	128
	21	$1\frac{1}{4}$	239	233	227	218	209	201	192	184	175	167	149	132

TABLE XVI—(*Continued*)

Designation	Weight of Each Channel Pounds per Foot	Thickness of Cover-Plates Inches	Safe Load, in Tons of 2,000 Pounds, for Columns With Square Ends and Unsupported Lengths of												
			15 Ft. or Less	16 Ft.	18 Ft.	20 Ft.	22 Ft.	24 Ft.	26 Ft.	28 Ft.	30 Ft.	32 Ft.	34 Ft.	36 Ft.	40 Ft.
10-Inch Channel Column: Two Channels 10 Inches Deep and Two Cover-Plates 12 Inches Wide	15	$\frac{5}{16}$	100	100	97	94	91	88	85	82	79	76	73	70	64
	15	$\frac{3}{8}$	109	109	106	103	100	97	93	90	86	83	79	75	68
	20	$\frac{3}{8}$	125	125	121	117	113	110	106	102	98	94	90	87	79
	20	$\frac{7}{16}$	135	135	130	126	122	118	114	109	105	101	97	93	85
	20	$\frac{1}{2}$	143	143	139	135	130	126	121	117	113	108	104	99	90
	25	$\frac{1}{2}$	160	160	155	150	145	139	134	129	124	119	114	109	99
	25	$\frac{9}{16}$	169	169	163	157	152	147	141	136	131	126	121	115	104
	25	$\frac{5}{8}$	178	178	172	166	160	155	149	143	138	132	126	121	110
	30	$\frac{5}{8}$	196	193	187	180	174	168	162	156	149	143	137	131	119
	30	$\frac{11}{16}$	205	202	196	189	183	176	169	163	156	150	144	137	124
	30	$\frac{3}{4}$	214	212	206	199	192	185	178	171	164	157	150	143	129
	30	$\frac{13}{16}$	223	221	214	207	200	192	185	178	171	163	156	149	135
	30	$\frac{7}{8}$	232	230	222	215	208	200	192	185	178	170	163	155	140
	30	$\frac{15}{16}$	240	238	230	223	215	207	199	192	184	176	168	161	145
	30	1	250	247	239	231	223	215	207	198	190	182	174	166	150
	30	$1\frac{1}{8}$	268	265	256	247	239	230	222	213	204	196	187	179	162
	30	$1\frac{1}{4}$	286	283	273	264	255	245	236	227	218	208	199	190	171
	30	$1\frac{3}{8}$	304	301	290	280	271	261	251	242	232	222	213	203	184
	30	$1\frac{1}{2}$	322	318	308	297	287	277	266	256	245	235	225	214	193
			18 Ft. or Less	20 Ft.	22 Ft.	24 Ft.	26 Ft.	28 Ft.	30 Ft.	32 Ft.	34 Ft.	36 Ft.	38 Ft.	40 Ft.	42 Ft.
12-Inch Channel Column: Two Channels 12 Inches Deep and Two Cover-Plates 14 Inches Wide	20	$\frac{3}{8}$	134	132	129	125	122	118	115	111	108	104	101	97	94
	20	$\frac{7}{16}$	144	143	140	136	133	129	125	121	117	114	110	106	102
	25	$\frac{7}{16}$	163	160	155	151	147	143	138	134	130	125	121	117	113
	25	$\frac{1}{2}$	173	171	165	160	156	151	147	142	138	133	129	124	120
	30	$\frac{1}{2}$	190	188	183	178	173	168	163	158	153	148	143	138	133
	30	$\frac{9}{16}$	200	197	191	186	181	175	170	165	160	154	149	144	139
	30	$\frac{5}{8}$	211	208	202	197	191	186	181	175	169	164	158	153	147
	30	$\frac{11}{16}$	221	218	212	206	200	194	189	183	177	171	165	159	154
	30	$\frac{3}{4}$	232	228	222	216	210	204	198	192	186	180	174	168	162
	35	$\frac{3}{4}$	250	248	242	235	228	221	214	207	200	193	186	179	172
	35	$\frac{13}{16}$	260	254	247	240	233	227	220	213	206	200	193	186	179
	35	$\frac{7}{8}$	271	265	258	251	243	236	229	222	214	207	200	193	186
	35	$\frac{15}{16}$	281	275	267	260	252	245	238	230	223	215	208	200	193
	35	1	292	286	278	270	262	254	246	238	230	222	214	206	199
	35	$1\frac{1}{8}$	313	306	298	290	281	273	264	256	247	239	231	222	214
	35	$1\frac{1}{4}$	333	326	317	308	299	290	281	272	263	254	245	236	227
	35	$1\frac{3}{8}$	354	346	337	327	318	308	299	289	280	270	261	251	242
	35	$1\frac{1}{2}$	375	366	356	346	336	326	316	306	296	286	276	266	256

TABLE XVI—(*Continued*)

Designation	Weight of Each Channel Pounds per Foot	Thickness of Cover-Plates Inches	Safe Load, in Tons of 2,000 Pounds, for Columns With Square Ends and Unsupported Lengths of												
			22 Ft. or Less	24 Ft.	26 Ft.	28 Ft.	30 Ft.	32 Ft.	34 Ft.	36 Ft.	38 Ft.	40 Ft.	44 Ft.	48 Ft.	52 Ft
15-Inch Channel Column, Light Section: Two 15-Inch Channels and Two Cover-Plates 17 Inches Wide	33	1/2	218	218	214	209	205	200	196	191	186	182	173	164	155
	33	9/16	231	231	226	221	216	211	207	202	197	192	182	173	163
	35	9/16	238	238	233	228	223	218	213	208	203	198	188	178	168
	35	5/8	251	250	245	240	234	229	224	218	213	208	197	186	176
	40	5/8	269	268	262	256	250	245	239	233	228	222	210	199	187
	40	11/16	282	280	274	268	262	256	250	244	238	232	220	208	196
	45	11/16	299	296	290	284	277	271	264	258	252	245	232	219	207
	45	3/4	311	309	302	295	289	282	275	269	262	255	242	229	215
	50	3/4	329	326	319	312	305	298	290	283	276	269	255	240	226
	50	13/16	342	338	331	324	316	309	301	294	287	279	265	250	235
	50	7/8	355	351	343	335	328	320	312	304	297	289	273	258	243
	50	15/16	368	363	355	347	339	331	323	315	307	299	283	267	251
	50	1	380	376	367	359	351	342	334	326	318	309	293	276	260
	50	1 1/8	406	401	392	383	374	365	356	348	339	330	312	294	277
	50	1 1/4	431	426	416	407	397	388	378	369	359	350	331	312	293
	50	1 3/8	457	450	440	430	420	410	400	390	379	369	349	329	308
	50	1 1/2	482	475	464	454	443	432	422	411	400	390	368	347	326
			20 Ft. or Less	22 Ft.	24 Ft.	26 Ft.	28 Ft.	30 Ft.	32 Ft.	34 Ft.	36 Ft.	40 Ft.	44 Ft.	48 Ft.	52 Ft
15-Inch Channel Column, Heavy Section: Two 15-Inch Channels, Two Cover-Plates 17 Inches Wide and Two Web-Plates 14" × 3/4"	50	3/4	455	455	445	435	425	414	404	394	383	363	342	322	301
	50	13/16	468	468	458	447	436	426	415	405	394	373	352	330	309
	50	7/8	481	481	470	459	448	437	426	416	405	383	361	339	318
	50	15/16	494	494	482	471	460	449	438	427	415	393	371	348	326
	50	1	506	506	495	483	472	460	449	438	426	403	380	357	334
	50	1 1/8	532	532	520	508	496	484	471	459	447	423	399	375	351
	50	1 1/4	557	557	544	532	519	506	494	481	468	443	418	392	367
	50	1 3/8	583	583	569	556	543	529	516	503	490	463	437	410	384
	50	1 1/2	608	608	594	580	566	552	539	525	511	483	455	428	400
	50	1 5/8	634	633	619	604	590	575	561	547	532	503	474	445	416
	50	1 3/4	659	658	644	629	614	598	583	568	553	523	493	463	433
	50	1 7/8	685	684	668	653	637	621	606	590	575	543	512	481	449
	50	2	710	709	693	677	661	644	628	612	596	563	531	498	466
	50	2 1/8	736	735	718	701	684	667	651	634	617	583	550	516	482
	50	2 1/4	761	760	743	725	708	690	673	656	638	603	568	534	499

TABLE XVI—(*Continued*)

Safe Load, in Tons of 2,000 Pounds, for Columns With Square Ends and Unsupported Lengths of

Designation	Column Section	Thickness of Cover-Plates Inches	14 Ft. or Less	16 Ft.	18 Ft.	20 Ft.	22 Ft.	24 Ft.	26 Ft.	28 Ft.	30 Ft.	32 Ft.	34 Ft.	36 Ft.	40 Ft.
10-Inch Extra-Heavy Channel Column	Two 10-Inch 30-Pound Channels, Two Cover-Plates 12 Inches Wide and Two Web-Plates 9″ × ¾″	¾	295	286	275	265	255	245	234	224	214	203	193	183	162
		13/16	304	295	284	273	263	252	242	231	221	210	200	189	168
		7/8	313	304	293	282	271	260	249	238	227	216	205	194	172
		15/16	321	312	300	288	277	266	255	244	233	222	211	200	178
		1	331	322	310	299	287	276	264	253	241	230	218	207	184
		1⅛	349	340	329	317	305	291	279	267	255	243	231	218	194
		1¼	367	358	345	333	320	307	295	282	269	257	244	231	206
		1⅜	385	375	361	348	335	322	309	296	283	270	257	243	217
		1½	403	393	381	367	353	339	325	311	297	283	269	255	227
		1⅝	421	410	396	381	367	353	338	324	310	295	281	267	238
		1¾	439	427	412	397	382	367	352	337	322	308	293	278	248
		1⅞	457	445	429	414	398	383	367	352	336	320	305	289	258
		2	475	463	446	430	414	398	382	366	350	334	318	301	269
		2⅛	493	481	465	448	431	415	398	382	365	348	331	314	280
		2¼	511	500	484	466	448	431	413	396	378	361	343	326	291

Designation	Column Section	Thickness of Cover-Plates Inches	16 Ft. or Less	18 Ft.	20 Ft.	22 Ft.	24 Ft.	26 Ft.	28 Ft.	30 Ft.	34 Ft.	38 Ft.	42 Ft.	46 Ft.	50 Ft.
12-Inch Extra-Heavy Channel Column	Two 12-Inch 35-Pound Channels, Two Cover-Plates 14 Inches Wide and Two Web-Plates 11″ × ¾″	¾	349	346	336	326	316	306	296	286	266	246	226	206	186
		13/16	359	357	346	336	326	316	306	296	275	254	233	212	191
		7/8	370	368	356	345	334	324	313	303	283	262	241	220	199
		15/16	381	378	367	357	346	335	324	313	292	270	248	226	204
		1	391	388	377	366	355	344	332	320	298	276	254	232	210
		1⅛	411	409	398	387	375	364	352	339	316	292	268	245	222
		1¼	433	431	419	407	395	382	370	358	333	308	283	258	233
		1⅜	454	452	439	426	413	400	387	374	348	322	296	270	244
		1½	476	473	459	445	432	418	405	391	364	337	310	283	256
		1⅝	496	493	478	464	450	436	422	408	380	352	324	296	268
		1¾	518	515	501	487	472	457	443	428	398	369	339	309	280
		1⅞	538	536	520	504	489	474	458	443	413	382	352	321	291
		2	560	556	541	526	510	494	478	462	430	398	366	334	302
		2⅛	580	577	561	545	529	512	496	479	446	413	380	347	314
		2¼	602	598	581	565	547	530	513	495	461	427	393	359	325

TABLE XVII

SAFE LOAD FOR STEEL Z-BAR COLUMNS

Designation	Thickness of Z-Bars and Web-Plate Inches	Safe Load, in Tons of 2,000 Pounds, for Columns With Square Ends and Unsupported Lengths of												
		8 Ft. or Less	9 Ft.	10 Ft.	11 Ft.	12 Ft.	13 Ft.	14 Ft.	16 Ft.	18 Ft.	20 Ft.	22 Ft.	24 Ft.	
6-Inch Column	$\frac{1}{4}$	56	55	53	52	50	48	46	43	39	36	32	29	
	$\frac{5}{16}$	71	70	68	66	64	61	59	55	51	47	42	38	
	$\frac{3}{8}$	83	79	77	74	72	69	67	62	57	53	48	43	
	$\frac{7}{16}$	97	94	91	88	85	83	80	74	69	63	57	52	
	$\frac{1}{2}$	107	104	100	97	94	91	89	83	75	70	63	57	
	$\frac{9}{16}$	121	118	115	110	107	103	100	93	86	80	73	66	
		11 Ft. or Less	12 Ft.	13 Ft.	14 Ft.	16 Ft.	18 Ft.	20 Ft.	22 Ft.	24 Ft.	26 Ft.	28 Ft.	30 Ft.	
8-Inch Column	$\frac{1}{4}$	67	66	64	63	59	56	53	50	47	44	41	38	
	$\frac{5}{16}$	85	83	81	79	76	72	68	64	60	57	53	49	
	$\frac{3}{8}$	103	101	99	97	92	88	83	78	74	69	65	60	
	$\frac{7}{16}$	114	111	108	106	100	95	90	85	80	74	69	64	
	$\frac{1}{2}$	131	128	125	122	117	111	105	99	93	87	82	76	
	$\frac{9}{16}$	149	146	142	139	133	126	120	113	107	100	94	87	
	$\frac{5}{8}$	158	155	151	148	140	133	126	119	111	104	97	90	
	$\frac{11}{16}$	174	171	166	162	155	147	140	132	125	117	110	102	
	$\frac{3}{4}$	191	189	185	181	172	164	156	148	140	131	123	115	
		13 Ft. or Less	14 Ft.	16 Ft.	18 Ft.	20 Ft.	22 Ft.	24 Ft.	26 Ft.	28 Ft.	30 Ft.	32 Ft.	34 Ft.	36 Ft.
10-Inch Column	$\frac{5}{16}$	95	95	91	88	84	81	77	75	71	66	63	60	56
	$\frac{3}{8}$	114	114	110	106	101	97	93	89	85	81	76	72	68
	$\frac{7}{16}$	134	134	129	124	119	114	109	104	99	94	90	85	80
	$\frac{1}{2}$	147	146	141	135	130	124	119	113	108	102	97	91	86
	$\frac{9}{16}$	166	166	160	154	148	142	135	129	123	117	111	105	99
	$\frac{5}{8}$	185	185	179	172	164	158	151	145	138	131	126	119	113
	$\frac{11}{16}$	196	195	188	181	173	166	159	152	145	138	130	123	116
	$\frac{3}{4}$	215	213	204	198	190	183	174	167	160	152	145	137	130
	$\frac{13}{16}$	234	234	226	218	210	201	193	185	177	169	160	152	144
		16 Ft. or Less	18 Ft.	20 Ft.	22 Ft.	24 Ft.	26 Ft.	28 Ft.	30 Ft.	32 Ft.	34 Ft.	36 Ft.	38 Ft.	40 Ft.
12-Inch Column	$\frac{3}{8}$	128	125	120	116	112	108	104	100	96	92	88	84	80
	$\frac{7}{16}$	150	146	141	137	132	128	123	119	114	109	105	101	96
	$\frac{1}{2}$	173	170	165	160	154	149	144	138	133	128	123	117	112
	$\frac{9}{16}$	187	182	177	171	165	159	153	147	141	135	130	124	118
	$\frac{5}{8}$	209	204	198	191	185	178	172	165	159	152	146	139	133
	$\frac{11}{16}$	231	227	220	214	207	200	193	186	180	172	165	158	151
	$\frac{3}{4}$	243	237	230	222	215	207	200	192	185	177	170	162	155
	$\frac{13}{16}$	265	258	250	242	234	226	248	210	202	194	186	178	170
	$\frac{7}{8}$	286	280	271	262	253	245	236	227	218	209	201	192	183

TABLE XVIII

SAFE LOAD FOR STEEL Z-BAR COLUMNS WITH COVER-PLATES

Section of Column	Thickness of Cover-Plates Inches	Safe Load, in Tons of 2,000 Pounds, for Columns With Square Ends and Unsupported Lengths of												
		16 Ft. or Less	18 Ft.	20 Ft.	22 Ft.	24 Ft.	26 Ft.	28 Ft.	30 Ft.	32 Ft.	34 Ft.	36 Ft.	38 Ft.	40 Ft.
Four Z-Bars 6″ × 3/4″ One Web-Plate 8″ × 3/4″ Two Cover-Plates 14 Inches Wide	3/8	306	300	291	282	273	264	255	246	237	228	219	210	201
	7/16	317	311	301	292	283	273	264	254	245	235	226	216	207
	1/2	327	321	311	302	292	282	273	263	253	244	234	224	215
	9/16	338	332	323	313	303	293	283	273	263	253	243	233	223
	5/8	348	342	332	322	311	301	291	281	270	260	250	240	229
	11/16	358	352	342	331	321	310	300	289	279	268	258	247	237
	3/4	368	362	352	342	331	321	310	300	288	277	266	255	244
	13/16	380	374	363	352	341	329	318	307	296	285	274	262	251
	7/8	389	384	372	361	350	338	327	315	304	293	281	270	258
Four Z-Bars 6 1/8″ × 7/8″ One Web-Plate 8″ × 7/8″ Two Cover-Plates 14 Inches Wide	11/16	401	393	381	370	358	346	334	322	310	298	286	274	262
	3/4	412	404	392	380	368	355	343	331	318	306	294	282	270
	13/16	422	415	402	390	377	365	352	340	328	315	303	290	278
	7/8	433	426	413	400	388	375	362	349	336	324	311	298	285
	15/16	444	436	423	410	397	384	371	357	344	331	318	305	292
	1	454	446	433	419	407	393	380	366	353	339	326	312	299
	1 1/16	465	457	443	430	416	402	388	375	361	348	334	321	307
	1 1/8	475	467	454	440	426	412	398	384	370	356	342	328	314
	1 3/16	486	477	463	449	434	420	406	392	377	363	349	335	321
	1 1/4	496	487	472	457	443	428	414	399	385	370	356	341	327
		20 Ft. or Less	22 Ft.	24 Ft.	26 Ft.	28 Ft.	30 Ft.	32 Ft.	34 Ft.	36 Ft.	38 Ft.	40 Ft.	42 Ft.	44 Ft.
Four Z-Bars 6″ × 1″ One Web-Plate 10″ × 1″ Two Cover-Plates 16 Inches Wide	1	488	476	464	452	440	428	416	404	392	380	368	356	344
	1 1/16	501	488	475	463	450	439	427	415	402	390	378	365	353
	1 1/8	513	501	488	475	463	450	438	425	412	400	387	374	362
	1 3/16	525	513	500	487	474	461	448	435	422	409	396	383	370
	1 1/4	537	523	510	497	484	470	457	444	431	418	404	391	378
	1 5/16	548	536	522	509	495	482	468	455	441	428	414	401	387
	1 3/8	561	548	534	520	506	493	479	465	451	437	424	410	396
	1 7/16	573	559	545	531	517	503	489	475	460	446	432	418	404
	1 1/2	585	571	556	542	527	513	498	484	470	455	441	426	412
	1 9/16	597	583	568	553	539	524	509	494	480	465	450	436	421
	1 5/8	609	594	579	564	549	534	519	505	490	475	460	445	430
	1 11/16	621	607	592	577	561	546	530	515	500	484	469	453	438
	1 3/4	633	618	602	587	571	556	540	525	509	494	478	463	447
	1 13/16	645	631	616	600	584	568	552	536	520	504	488	472	456
	1 7/8	657	643	626	610	593	577	561	544	528	512	496	480	464
	1 15/16	669	654	637	621	603	587	570	554	538	521	505	488	472
	2	681	666	648	631	615	598	581	565	548	531	514	498	481
	2 1/16	693	676	659	642	625	608	591	574	557	540	523	506	489
	2 1/8	705	689	671	654	636	619	601	584	567	549	532	514	497
	2 3/16	717	701	682	664	647	629	612	594	576	559	541	524	506
	2 1/4	729	713	694	676	658	640	622	604	586	568	550	532	514

DETAILS OF CONSTRUCTION

RIVETS AND PINS

INTRODUCTION

1. Rivets and **pins** are the elements by which the different sections and members of a steel structure are connected. Pins are also used in the better class of timber construction, in which case the tension members are usually made of steel or wrought iron.

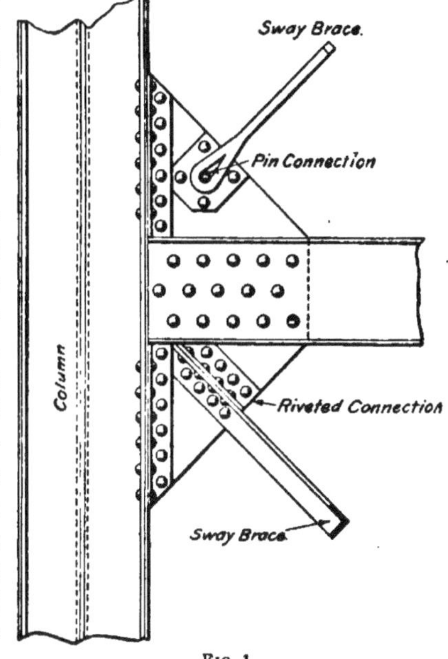

Fig. 1

Where plates or rolled shapes are joined by pins or rivets, as in Fig. 1, there is friction between the parts, which tends to prevent their being pulled apart. This is especially true when rivets are driven close against the plates while hot; in cooling, they contract between the heads and bind the plates tightly together. The friction between plates connected by rivets and pins, however, is an uncertain quantity and should not be considered in calculating the strength of the joint.

RIVETED CONNECTIONS

RIVETS

2. Structural rivets are made either of steel or of wrought iron. Those used in the shop are made of soft steel, but those used in the field are of wrought iron, because this material is less likely to be damaged in heating, and also because considerable heat is lost in passing the rivets from the forge to the riveters, thus reducing them to a temperature that would be too low to work on steel.

3. Sizes.—Rivets used in structural work vary in diameter from $\frac{1}{2}$ inch to 1 inch, being made in $\frac{1}{2}$-, $\frac{5}{8}$-, $\frac{3}{4}$-, $\frac{7}{8}$-, and 1-inch sizes. The $\frac{3}{4}$- and $\frac{7}{8}$-inch rivets are most generally used, the smaller sizes being employed only on unimportant details, or where the clearance will not allow the insertion of larger ones, and those larger than $\frac{7}{8}$ inch are used only where there are very thick plates or very great stresses. Field rivets should never be more than $\frac{7}{8}$ inch.

On account of the difficulty of punching holes of less diameter than the thickness of the plate, the diameter of the rivet should not be less than the thickness of the thickest plate through which the rivet passes. Sections with small flanges necessitate the use of small rivets so as to allow sufficient distance from the edge of the piece and enough space for the tool in driving the rivet.

When the size cannot be determined by the above conditions, such a size should be used as will resist the existing stresses and be economical. It is cheaper to use the same size rivets throughout a piece, and the smaller rivets are more economical as they reduce the weight and the cost of punching and driving. The standard sizes, from the Cambria Steel Company's handbook, to be used in **I** beams, channels, and angles are given in Table I.

4. Diameter of Hole.—The diameter of the rivet hole when drilled should be $\frac{1}{16}$ inch larger than the rivet, as the latter is inserted while hot, and when driven, is upset, thus

TABLE I

MAXIMUM SIZE OF RIVETS IN BEAMS, CHANNELS, AND ANGLES

I Beams						Channels			Angles			
Depth of Beam Inches	Weight per Foot Pounds	Size of Rivet Inches	Depth of Beam Inches	Weight per Foot Pounds	Size of Rivet Inches	Depth of Channel Inches	Weight per Foot Pounds	Size of Rivet Inches	Length of Leg Inches	Size of Rivet Inches	Length of Leg Inches	Size of Rivet Inches
3	5.50	$\frac{3}{8}$	15	42.00	$\frac{3}{4}$	3	4.00	$\frac{3}{8}$	$\frac{3}{4}$	$\frac{1}{4}$	$2\frac{1}{2}$	$\frac{3}{4}$
4	7.50	$\frac{1}{2}$	15	60.00	$\frac{3}{4}$	4	5.25	$\frac{1}{2}$	1	$\frac{3}{8}$	$2\frac{3}{4}$	$\frac{3}{4}$
5	9.75	$\frac{1}{2}$	15	80.00	$\frac{7}{8}$	5	6.50	$\frac{1}{2}$	$1\frac{1}{4}$	$\frac{1}{2}$	3	$\frac{7}{8}$
6	12.25	$\frac{5}{8}$	18	55.00	$\frac{7}{8}$	6	8.00	$\frac{5}{8}$	$1\frac{5}{16}$	$\frac{1}{2}$	$3\frac{1}{2}$	1
7	15.00	$\frac{5}{8}$	20	65.00	1	7	9.75	$\frac{5}{8}$	$1\frac{3}{8}$	$\frac{1}{2}$	4	1
8	17.75	$\frac{3}{4}$	20	80.00	1	8	11.25	$\frac{3}{4}$	$1\frac{1}{2}$	$\frac{1}{2}$	$4\frac{1}{2}$	1
9	21.00	$\frac{3}{4}$	24	80.00	1	9	13.25	$\frac{3}{4}$	$1\frac{3}{4}$	$\frac{5}{8}$	5	1
10	25.00	$\frac{3}{4}$				10	15.00	$\frac{3}{4}$	2	$\frac{5}{8}$	6	1
12	31.50	$\frac{3}{4}$				12	20.50	$\frac{3}{4}$	$2\frac{1}{4}$	$\frac{3}{4}$	7	1
12	40.00	$\frac{3}{4}$				15	33.00	$\frac{3}{4}$	$2\frac{5}{16}$	$\frac{3}{4}$		

completely filling the hole, but when the hole is punched, it is well to allow $\frac{1}{8}$ inch excess of diameter of hole over diameter of rivet. Holes are not drilled except in very important work, or when hard steel is used.

5. Rivet Heads.—The **round head** or **full head,** also called **button head,** is a spherical segment slightly less than a hemisphere.

Countersunk heads are countersunk into the surrounding material so that they are flush on top with the surface of the plate into which they are sunk. These are used when a button head would interfere with the bearing, as at the base of a column where it rests on the bearing block or capstone.

Flattened heads are made by flattening the button heads and are used in extreme cases where a round head would be too high.

6. Dimensions of Heads.—The height and diameter of a rivet head bear the following relations to the diameter of the rivet.

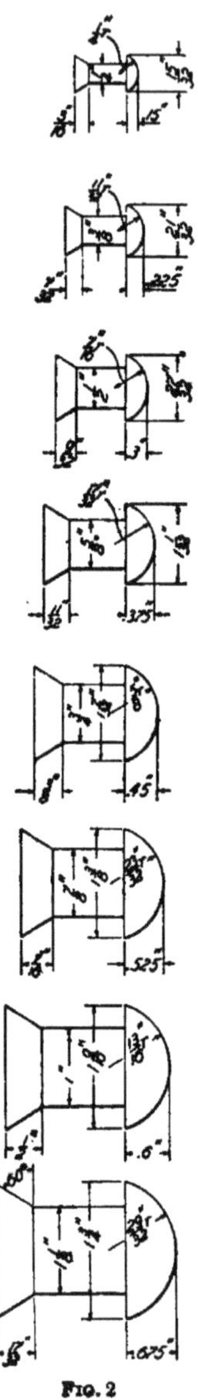

FIG. 2

For button heads the standard height of head is six-tenths the diameter of the rivet and the radius of the head is three-fourths of the diameter of the rivet plus $\frac{1}{16}$ inch. These dimensions are shown in Fig. 2.

For countersunk heads, the diameter of the head is the same as for a button head, and its bevel is 60°; from these data the height, or depth, of the head can be determined. A simple rule is to make the depth equal to one-half of the diameter of the rivet. Countersunk rivets should not be used in plates of less thickness than one-half the diameter of the rivet. Where economy is to be considered, their use should be avoided if possible, as the extra work required in countersinking makes them expensive.

For flattened heads, the height is never less than $\frac{5}{16}$ inch.

7. Clearance.—When one riveted section is close to another, it is necessary to keep them a certain distance apart to allow a clearance for the driven rivets. This clearance is $\frac{5}{8}$ inch for $\frac{3}{4}$-inch rivets and $\frac{3}{4}$ inch for $\frac{7}{8}$-inch rivets. It is always better to provide for a head $\frac{1}{8}$ inch higher than the figured height, to allow for the upsetting of the rivet and discrepancies in the material.

8. Length or Grip.—In order to drive tight rivets, it is necessary to keep them within a certain length or grip. The maximum **grip**, or distance between heads, is generally accepted as not more than four times the diameter of the rivet. In determining the length to be given in ordering rivets, first find the grip, which is considered as the thickness of the parts joined plus $\frac{1}{32}$ inch for each joint between plates; this $\frac{1}{32}$ inch is to allow for unevenness in the surfaces of the plates which might prevent their being in close contact. This grip must be increased in the ratio of the area of the hole to the area of the rivet section, to allow for upsetting, the hole being $\frac{1}{16}$ inch larger in diameter than the rivet. To this result must be added the height required for the head, and the sum is the length of rivet under the formed head. Assume a $\frac{3}{4}$-inch button-head rivet with a grip of $2\frac{1}{8}$ inches. Increasing this grip in the ratio of the area of the hole to the area of the rivet, or $\frac{(\frac{13}{16})^2 \times .7854}{(\frac{3}{4})^2 \times .7854} = \frac{169}{144}$ $\left(\text{since } \frac{.7854}{16^2} \text{ is common to both numerator and denominator}\right)$, gives $2\frac{1}{8} \times \frac{169}{144}$, or about $2\frac{1}{2}$ inches. Adding $1\frac{1}{8}$ inches for the head, the length of rivet required is found to be $2\frac{1}{2} + 1\frac{1}{8} = 3\frac{5}{8}$ inches. Table II gives the length required to form one head, including the increase due to excess of area of hole over area of rivet.

TABLE II

LENGTH OF RIVET SHANK REQUIRED TO FORM ONE RIVET HEAD

Grip Inches	Button Head					Countersunk Head				
	Diameter of Rivet. Inches					Diameter of Rivet. Inches				
	$\frac{1}{2}$	$\frac{5}{8}$	$\frac{3}{4}$	$\frac{7}{8}$	1	$\frac{1}{2}$	$\frac{5}{8}$	$\frac{3}{4}$	$\frac{7}{8}$	1
$\frac{1}{2}$ to $1\frac{3}{8}$	1	$1\frac{1}{4}$	$1\frac{3}{8}$	$1\frac{1}{2}$	$1\frac{5}{8}$	$\frac{5}{8}$	$\frac{3}{4}$	$\frac{3}{4}$	$\frac{7}{8}$	$\frac{7}{8}$
$1\frac{1}{2}$ to $2\frac{7}{8}$	$1\frac{1}{8}$	$1\frac{3}{8}$	$1\frac{1}{2}$	$1\frac{5}{8}$	$1\frac{3}{4}$	$\frac{5}{8}$	$\frac{3}{4}$	$\frac{7}{8}$	$\frac{7}{8}$	1
3 to $4\frac{3}{8}$	$1\frac{1}{4}$	$1\frac{1}{2}$	$1\frac{5}{8}$	$1\frac{3}{4}$	$1\frac{7}{8}$	$\frac{3}{4}$	$\frac{7}{8}$	1	1	$1\frac{1}{8}$
$4\frac{1}{2}$ to $5\frac{1}{2}$	$1\frac{3}{8}$	$1\frac{5}{8}$	$1\frac{3}{4}$	$1\frac{7}{8}$	2				1	$1\frac{1}{8}$

Rivets are ordered in even eighths of an inch. Table III shows the lengths of rivets required for the different grips.

TABLE III

LENGTH OF RIVETS REQUIRED FOR VARIOUS GRIPS, INCLUDING AMOUNT NECESSARY TO FORM ONE HEAD

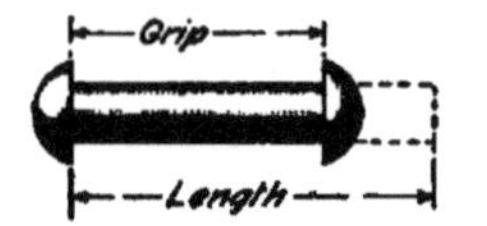

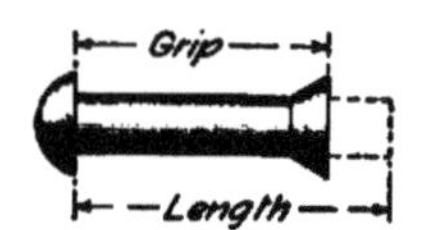

Grip of Rivet Inches	Diameter of Rivet. Inches								Grip of Rivet Inches	Diameter of Rivet. Inches							
	$\frac{1}{4}$	$\frac{3}{8}$	$\frac{1}{2}$	$\frac{5}{8}$	$\frac{3}{4}$	$\frac{7}{8}$	1	$1\frac{1}{8}$		$\frac{1}{4}$	$\frac{3}{8}$	$\frac{1}{2}$	$\frac{5}{8}$	$\frac{3}{4}$	$\frac{7}{8}$	1	$1\frac{1}{8}$
$\frac{1}{2}$	1	$1\frac{1}{4}$	$1\frac{1}{2}$	$1\frac{3}{4}$	$1\frac{7}{8}$	2	$2\frac{1}{8}$	$2\frac{1}{4}$	$3\frac{3}{8}$	$3\frac{7}{8}$	$4\frac{1}{4}$	$4\frac{1}{2}$	$4\frac{7}{8}$	5	$5\frac{1}{8}$	$5\frac{1}{4}$	$5\frac{3}{8}$
$\frac{5}{8}$	$1\frac{1}{8}$	$1\frac{3}{8}$	$1\frac{5}{8}$	$1\frac{7}{8}$	2	$2\frac{1}{8}$	$2\frac{1}{4}$	$2\frac{3}{8}$	$3\frac{1}{2}$	4	$4\frac{3}{8}$	$4\frac{5}{8}$	5	$5\frac{1}{8}$	$5\frac{1}{4}$	$5\frac{3}{8}$	$5\frac{1}{2}$
$\frac{3}{4}$	$1\frac{1}{4}$	$1\frac{1}{2}$	$1\frac{3}{4}$	2	$2\frac{1}{8}$	$2\frac{1}{4}$	$2\frac{3}{8}$	$2\frac{1}{2}$	$3\frac{5}{8}$	$4\frac{1}{8}$	$4\frac{1}{2}$	$4\frac{3}{4}$	$5\frac{1}{8}$	$5\frac{1}{4}$	$5\frac{3}{8}$	$5\frac{1}{2}$	$5\frac{5}{8}$
$\frac{7}{8}$	$1\frac{3}{8}$	$1\frac{5}{8}$	$1\frac{7}{8}$	$2\frac{1}{8}$	$2\frac{1}{4}$	$2\frac{3}{8}$	$2\frac{1}{2}$	$2\frac{5}{8}$	$3\frac{3}{4}$	$4\frac{1}{4}$	$4\frac{5}{8}$	$4\frac{7}{8}$	$5\frac{1}{4}$	$5\frac{3}{8}$	$5\frac{1}{2}$	$5\frac{5}{8}$	$5\frac{3}{4}$
1	$1\frac{1}{2}$	$1\frac{3}{4}$	2	$2\frac{1}{4}$	$2\frac{3}{8}$	$2\frac{1}{2}$	$2\frac{5}{8}$	$2\frac{3}{4}$	$3\frac{7}{8}$	$4\frac{3}{8}$	$4\frac{3}{4}$	5	$5\frac{3}{8}$	$5\frac{1}{2}$	$5\frac{5}{8}$	$5\frac{3}{4}$	$5\frac{7}{8}$
$1\frac{1}{8}$	$1\frac{5}{8}$	$1\frac{7}{8}$	$2\frac{1}{8}$	$2\frac{3}{8}$	$2\frac{1}{2}$	$2\frac{5}{8}$	$2\frac{3}{4}$	$2\frac{7}{8}$	4	$4\frac{1}{2}$	$4\frac{7}{8}$	$5\frac{1}{8}$	$5\frac{1}{2}$	$5\frac{5}{8}$	$5\frac{3}{4}$	$5\frac{7}{8}$	6
$1\frac{1}{4}$	$1\frac{3}{4}$	2	$2\frac{1}{4}$	$2\frac{1}{2}$	$2\frac{5}{8}$	$2\frac{3}{4}$	$2\frac{7}{8}$	3	$4\frac{1}{8}$	$4\frac{5}{8}$	5	$5\frac{1}{4}$	$5\frac{5}{8}$	$5\frac{3}{4}$	$5\frac{7}{8}$	6	$6\frac{1}{8}$
$1\frac{3}{8}$	$1\frac{7}{8}$	$2\frac{1}{8}$	$2\frac{3}{8}$	$2\frac{5}{8}$	$2\frac{7}{8}$	3	3	$3\frac{1}{8}$	$4\frac{1}{4}$	$4\frac{3}{4}$	$5\frac{1}{8}$	$5\frac{1}{2}$	$5\frac{3}{4}$	$5\frac{7}{8}$	6	$6\frac{1}{8}$	$6\frac{1}{4}$
$1\frac{1}{2}$	2	$2\frac{1}{4}$	$2\frac{1}{2}$	$2\frac{3}{4}$	3	$3\frac{1}{8}$	$3\frac{1}{8}$	$3\frac{1}{4}$	$4\frac{3}{8}$	$4\frac{7}{8}$	$5\frac{1}{4}$	$5\frac{5}{8}$	$5\frac{7}{8}$	6	$6\frac{1}{8}$	$6\frac{1}{4}$	$6\frac{3}{8}$
$1\frac{5}{8}$	$2\frac{1}{8}$	$2\frac{3}{8}$	$2\frac{5}{8}$	$2\frac{7}{8}$	$3\frac{1}{8}$	$3\frac{1}{4}$	$3\frac{1}{4}$	$3\frac{1}{2}$	$4\frac{1}{2}$	5	$5\frac{3}{8}$	$5\frac{3}{4}$	6	$6\frac{1}{8}$	$6\frac{1}{4}$	$6\frac{3}{8}$	$6\frac{1}{2}$
$1\frac{3}{4}$	$2\frac{1}{4}$	$2\frac{1}{2}$	$2\frac{3}{4}$	3	$3\frac{1}{4}$	$3\frac{3}{8}$	$3\frac{1}{2}$	$3\frac{5}{8}$	$4\frac{5}{8}$	$5\frac{1}{8}$	$5\frac{1}{2}$	$5\frac{7}{8}$	$6\frac{1}{8}$	$6\frac{1}{4}$	$6\frac{3}{8}$	$6\frac{1}{2}$	$6\frac{5}{8}$
$1\frac{7}{8}$	$2\frac{3}{8}$	$2\frac{5}{8}$	$2\frac{7}{8}$	$3\frac{1}{4}$	$3\frac{3}{8}$	$3\frac{1}{2}$	$3\frac{5}{8}$	$3\frac{3}{4}$	$4\frac{3}{4}$	$5\frac{1}{4}$	$5\frac{5}{8}$	6	$6\frac{1}{4}$	$6\frac{1}{2}$	$6\frac{5}{8}$	$6\frac{3}{4}$	$6\frac{3}{4}$
2	$2\frac{1}{2}$	$2\frac{3}{4}$	$3\frac{1}{8}$	$3\frac{3}{8}$	$3\frac{1}{2}$	$3\frac{5}{8}$	$3\frac{3}{4}$	$3\frac{7}{8}$	$4\frac{7}{8}$	$5\frac{3}{8}$	$5\frac{3}{4}$	$6\frac{1}{8}$	$6\frac{1}{2}$	$6\frac{5}{8}$	$6\frac{3}{4}$	$6\frac{7}{8}$	$6\frac{7}{8}$
$2\frac{1}{8}$	$2\frac{5}{8}$	$2\frac{7}{8}$	$3\frac{1}{4}$	$3\frac{1}{2}$	$3\frac{5}{8}$	$3\frac{3}{4}$	$3\frac{7}{8}$	4	5	$5\frac{1}{2}$	$5\frac{7}{8}$	$6\frac{1}{4}$	$6\frac{5}{8}$	$6\frac{3}{4}$	$6\frac{7}{8}$	7	7
$2\frac{1}{4}$	$2\frac{3}{4}$	3	$3\frac{3}{8}$	$3\frac{5}{8}$	$3\frac{3}{4}$	$3\frac{7}{8}$	4	$4\frac{1}{8}$	$5\frac{1}{8}$	$5\frac{5}{8}$	6	$6\frac{3}{8}$	$6\frac{3}{4}$	$6\frac{7}{8}$	7	$7\frac{1}{8}$	$7\frac{1}{8}$
$2\frac{3}{8}$	$2\frac{7}{8}$	$3\frac{1}{8}$	$3\frac{1}{2}$	$3\frac{3}{4}$	$3\frac{7}{8}$	4	$4\frac{1}{8}$	$4\frac{1}{4}$	$5\frac{1}{4}$	$5\frac{3}{4}$	$6\frac{1}{8}$	$6\frac{1}{2}$	$6\frac{7}{8}$	7	$7\frac{1}{8}$	$7\frac{1}{4}$	$7\frac{1}{4}$
$2\frac{1}{2}$	3	$3\frac{1}{4}$	$3\frac{5}{8}$	$3\frac{7}{8}$	4	$4\frac{1}{8}$	$4\frac{1}{4}$	$4\frac{3}{8}$	$5\frac{3}{8}$	$5\frac{7}{8}$	$6\frac{1}{4}$	$6\frac{5}{8}$	7	$7\frac{1}{8}$	$7\frac{1}{4}$	$7\frac{3}{8}$	$7\frac{3}{8}$
$2\frac{5}{8}$	$3\frac{1}{8}$	$3\frac{1}{2}$	$3\frac{3}{4}$	4	$4\frac{1}{8}$	$4\frac{1}{4}$	$4\frac{3}{8}$	$4\frac{1}{2}$	$5\frac{1}{2}$	6	$6\frac{3}{8}$	$6\frac{3}{4}$	$7\frac{1}{8}$	$7\frac{1}{4}$	$7\frac{3}{8}$	$7\frac{1}{2}$	$7\frac{1}{2}$
$2\frac{3}{4}$	$3\frac{1}{4}$	$3\frac{5}{8}$	$3\frac{7}{8}$	$4\frac{1}{8}$	$4\frac{1}{4}$	$4\frac{3}{8}$	$4\frac{1}{2}$	$4\frac{5}{8}$	$5\frac{5}{8}$	$6\frac{1}{8}$	$6\frac{1}{2}$	$6\frac{7}{8}$	$7\frac{1}{4}$	$7\frac{3}{8}$	$7\frac{1}{2}$	$7\frac{5}{8}$	$7\frac{5}{8}$
$2\frac{7}{8}$	$3\frac{3}{8}$	$3\frac{3}{4}$	4	$4\frac{1}{4}$	$4\frac{3}{8}$	$4\frac{1}{2}$	$4\frac{5}{8}$	$4\frac{3}{4}$	$5\frac{3}{4}$	$6\frac{1}{4}$	$6\frac{3}{4}$	7	$7\frac{3}{8}$	$7\frac{5}{8}$	$7\frac{5}{8}$	$7\frac{3}{4}$	$7\frac{3}{4}$
3	$3\frac{1}{2}$	$3\frac{7}{8}$	$4\frac{1}{8}$	$4\frac{3}{8}$	$4\frac{1}{2}$	$4\frac{5}{8}$	$4\frac{3}{4}$	$4\frac{7}{8}$	$5\frac{7}{8}$	$6\frac{3}{8}$	$6\frac{7}{8}$	$7\frac{1}{8}$	$7\frac{1}{2}$	$7\frac{3}{4}$	$7\frac{3}{4}$	$7\frac{7}{8}$	$7\frac{7}{8}$
$3\frac{1}{8}$	$3\frac{5}{8}$	4	$4\frac{1}{4}$	$4\frac{1}{2}$	$4\frac{3}{4}$	$4\frac{3}{4}$	5	5	6	$6\frac{1}{2}$	7	$7\frac{1}{4}$	$7\frac{5}{8}$	$7\frac{7}{8}$	$7\frac{7}{8}$	8	$8\frac{1}{8}$
$3\frac{1}{4}$	$3\frac{3}{4}$	$4\frac{1}{8}$	$4\frac{3}{8}$	$4\frac{3}{4}$	$4\frac{7}{8}$	5	$5\frac{1}{8}$	$5\frac{1}{4}$									

Amount, in Inches, to Be Subtracted From Above Lengths for Countersunk Heads

	$\frac{1}{8}$	$\frac{1}{4}$	$\frac{1}{2}$	$\frac{1}{2}$	$\frac{5}{8}$	$\frac{3}{4}$	$\frac{7}{8}$	$\frac{7}{8}$		$\frac{1}{8}$	$\frac{1}{4}$	$\frac{1}{2}$	$\frac{1}{2}$	$\frac{5}{8}$	$\frac{3}{4}$	$\frac{7}{8}$	$\frac{7}{8}$

SPACING OF RIVETS

9. Pitch.—The pitch, or distance from center to center of rivets, should be at least three diameters of the rivet. If spaced closer, the material is likely to fracture or become otherwise injured. For members in compression, the pitch in the direction of the stress should not be greater than 6 inches, or sixteen times the thickness of the thinnest outside plate. If possible, the pitch should be kept in even inches, making it either 2, 3, 4, 5, or 6 inches, as the case may require, especially when the rows are long. In girder and bridge work, the maximum pitch of rivets used is 6 inches, although some authorities recommend less. It is usual to space the rivets closer at the joints and foot of a column than in the body; when $\frac{3}{4}$-inch rivets are used, it is customary to space them 3 inches on centers at the joints and bottom, and from $4\frac{1}{2}$ to 6 inches on centers in the remainder of the column; when the rivets are 1 inch, they should be spaced about 4 inches at the joints and 6 inches throughout the length of the column.

In Table IV are given the maximum and minimum pitches of the different sized rivets and the minimum and usual distances from the end of piece.

TABLE IV

RIVET SPACING

Size of Rivet Inch	Minimum Pitch Inches	Maximum Pitch at Ends of Compression Members Inches	Maximum Pitch in Flanges of Chords and Girders Inches	Distance From End of Piece to Center of Rivet Inches	
				Minimum	Usual
$\frac{1}{4}$	$\frac{3}{4}$				
$\frac{3}{8}$	$1\frac{1}{8}$				
$\frac{1}{2}$	$1\frac{1}{2}$				
$\frac{5}{8}$	$1\frac{7}{8}$	$2\frac{1}{2}$	4	$\frac{15}{16}$	$1\frac{1}{4}$
$\frac{3}{4}$	$2\frac{1}{4}$	3	4	$1\frac{1}{8}$	$1\frac{1}{2}$
$\frac{7}{8}$	$2\frac{5}{8}$	$3\frac{1}{2}$	4	$1\frac{5}{16}$	$1\frac{3}{4}$
1	3	4	4	$1\frac{1}{2}$	2

TABLE V

STANDARD SPACING OF RIVETS AND BOLTS THROUGH FLANGES AND CONNECTION ANGLES OF I BEAMS, AND TANGENT DISTANCES BETWEEN FILLETS MEASURED ALONG THE WEB

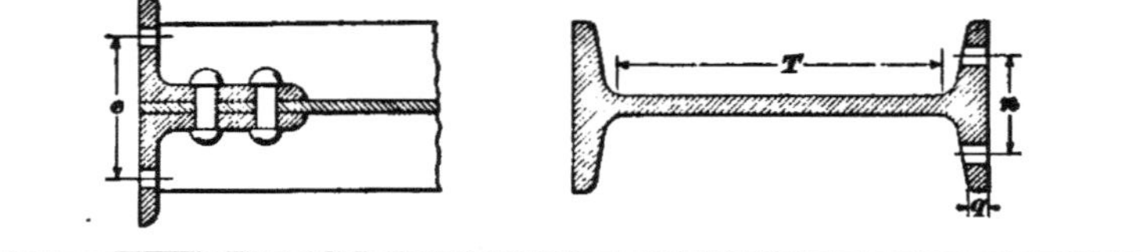

Depth of Beam Inches	Weight Pounds per Foot	n Inches	e Inches	q Inches	T Inches	Depth of Beam Inches	Weight Pounds per Foot	n Inches	e Inches	q Inches	T Inches
3	5.50	$1\frac{7}{16}$	$4\frac{21}{32}$	$\frac{1}{4}$	$1\frac{13}{16}$	12	55.0	3	$5\frac{5}{16}$	$\frac{11}{16}$	$9\frac{5}{16}$
3	6.50	$1\frac{7}{16}$	$4\frac{3}{4}$	$\frac{1}{4}$	$1\frac{13}{16}$	15	42.0	3	$4\frac{29}{32}$	$\frac{5}{8}$	$12\frac{1}{2}$
3	7 50	$1\frac{7}{16}$	$4\frac{7}{8}$	$\frac{1}{4}$	$1\frac{13}{16}$	15	45.0	3	$4\frac{31}{32}$	$\frac{5}{8}$	$12\frac{1}{2}$
4	7 50	$1\frac{1}{2}$	$4\frac{11}{16}$	$\frac{9}{32}$	$2\frac{11}{16}$	15	50.0	3	$5\frac{1}{16}$	$\frac{5}{8}$	$12\frac{1}{2}$
4	8.50	$1\frac{1}{2}$	$4\frac{3}{4}$	$\frac{9}{32}$	$2\frac{11}{16}$	15	55 0	3	$5\frac{5}{32}$	$\frac{5}{8}$	$12\frac{1}{2}$
4	9.50	$1\frac{1}{2}$	$4\frac{27}{32}$	$\frac{5}{16}$	$2\frac{11}{16}$	15	60.0	3	$5\frac{1}{4}$	$\frac{21}{32}$	$12\frac{1}{2}$
4	10.50	$1\frac{1}{2}$	$4\frac{29}{32}$	$\frac{5}{16}$	$2\frac{11}{16}$	15	60.0	$3\frac{1}{4}$	$5\frac{3}{32}$	$\frac{13}{16}$	$11\frac{3}{4}$
5	9.75	$1\frac{3}{4}$	$4\frac{23}{32}$	$\frac{5}{16}$	$3\frac{5}{8}$	15	65 0	$3\frac{1}{4}$	$5\frac{3}{16}$	$\frac{13}{16}$	$11\frac{3}{4}$
5	12.25	$1\frac{3}{4}$	$4\frac{7}{8}$	$\frac{5}{16}$	$3\frac{5}{8}$	15	70.0	$3\frac{1}{4}$	$5\frac{9}{32}$	$\frac{27}{32}$	$11\frac{3}{4}$
5	14.75	$1\frac{3}{4}$	5	$\frac{11}{32}$	$3\frac{5}{8}$	15	75.0	$3\frac{1}{4}$	$5\frac{3}{8}$	$\frac{27}{32}$	$11\frac{3}{4}$
6	12.25	2	$4\frac{23}{32}$	$\frac{11}{32}$	$4\frac{7}{16}$	15	80.0	$3\frac{1}{4}$	$5\frac{15}{32}$	$\frac{27}{32}$	$11\frac{3}{4}$

6	14.75	2	4 27/32	11/32	4 7/16	15	80.0	3 3/4	5 5/16	1 1/32	10 15/16
6	17.25	2	4 31/32	3/8	4 7/16	15	85.0	3 3/4	5 13/32	1 1/32	10 15/16
7	15.00	2 1/4	4 3/4	3/8	5 3/8	15	90.0	3 3/4	5 1/2	1 1/32	10 15/16
7	17.50	2 1/4	4 27/32	3/8	5 3/8	15	95.0	3 3/4	5 19/32	1 1/32	10 15/16
7	20.00	2 1/4	4 31/32	3/8	5 3/8	15	100.0	3 3/4	5 11/16	1 1/16	10 15/16
8	18.00	2 1/4	4 25/32	13/32	6 3/16	18	55.0	3 1/4	4 31/32	11/16	15 3/16
8	20.25	2 1/4	4 27/32	7/16	6 3/16	18	60.0	3 1/4	5 1/16	11/16	15 3/16
8	22.75	2 1/4	4 15/16	7/16	6 3/16	18	65.0	3 1/4	5 1/8	23/32	15 3/16
8	25.25	2 1/4	5 1/32	7/16	6 3/16	18	70.0	3 1/4	5 7/32	23/32	15 3/16
9	21.00	2 1/2	4 25/32	7/16	7 1/16	20	65.0	3 1/2	5	25/32	16 15/16
9	25.00	2 1/2	4 29/32	7/16	7 1/16	20	70.0	3 1/2	5 3/32	25/32	16 15/16
9	30.00	2 1/2	5 1/16	15/32	7 1/16	20	75.0	3 1/2	5 5/32	25/32	16 15/16
9	35.00	2 1/2	5 7/32	15/32	7 1/16	20	80.0	4	5 3/32	29/32	16 7/16
10	25.00	2 5/8	4 13/16	15/32	7 15/16	20	85.0	4	5 5/32	29/32	16 7/16
10	30.00	2 5/8	4 15/16	1/2	7 15/16	20	90.0	4	5 1/4	29/32	16 7/16
10	35.00	2 5/8	5 3/32	1/2	7 15/16	20	95.0	4	5 5/16	29/32	16 7/16
10	40.00	2 5/8	5 1/4	17/32	7 15/16	20	100.0	4	5 3/8	15/16	16 7/16
12	31.50	2 3/4	4 27/32	17/32	9 3/4	24	80.0	4	5	27/32	20 11/16
12	35.00	2 3/4	4 15/16	17/32	9 3/4	24	85.0	4	5 1/16	27/32	20 11/16
12	40.00	2 3/4	5 1/16	9/16	9 3/4	24	90.0	4	5 1/8	7/8	20 11/16
12	40.00	3	4 31/32	21/32	9 5/16	24	95.0	4	5 3/16	7/8	20 11/16
12	45 00	3	5 3/32	21/32	9 5/16	24	100.0	4	5 1/4	7/8	20 11/16
12	50.00	3	5 3/16	21/32	9 5/16						

TABLE VI

STANDARD SPACING OF RIVETS AND BOLTS IN FLANGES AND CONNECTION ANGLES OF CHANNELS, AND TANGENT DISTANCES BETWEEN FILLETS MEASURED ALONG THE WEB

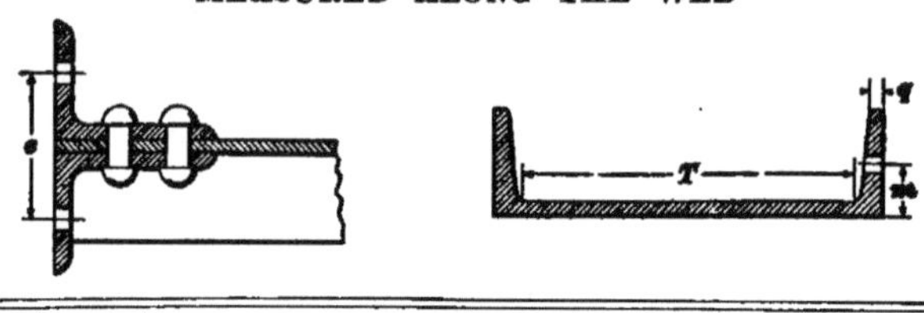

Depth of Channel Inches	Weight Pounds per Foot	m Inches	e Inches	q Inches	T Inches	Depth of Channel Inches	Weight Pounds per Foot	m Inches	e Inches	q Inches	T Inches
3	4.00	15/16	4 21/32	1/4	1 13/16	8	21.25	1 9/16	5 3/32	13/32	6 5/16
3	5.00	15/16	4 3/4	1/4	1 13/16	9	13.25	1 3/8	4 3/4	13/32	7 1/4
3	6.00	15/16	4 7/8	9/32	1 13/16	9	15.00	1 7/16	4 25/32	13/32	7 1/4
4	5.25	1	4 11/16	9/32	2 11/16	9	20.00	1 9/16	4 15/16	3/8	7 1/4
4	6.25	1	4 3/4	9/32	2 11/16	9	25.00	1 3/4	5 1/8	13/32	7 1/4
4	7.25	1	4 27/32	5/16	2 11/16	10	15.00	1 1/2	4 3/4	7/16	8 3/16
5	6.50	1	4 11/16	5/16	3 5/8	10	20.00	1 5/8	4 27/32	7/16	8 3/16
5	9.00	1 1/4	4 27/32	9/32	3 5/8	10	25.00	1 3/4	5 1/32	3/8	8 3/16
5	11.50	1 1/4	4 31/32	5/16	3 5/8	10	30.00	1 15/16	5 3/16	13/32	8 3/16
6	8.00	1 1/16	4 11/16	11/32	4 1/2	10	35.00	2 1/16	5 5/16	7/16	8 3/16
6	10.50	1 3/16	4 13/16	11/32	4 1/2	12	20.50	1 3/4	4 25/32	15/32	9 15/16
6	13.00	1 5/16	4 15/16	11/32	4 1/2	12	25.00	1 7/8	4 7/8	1/2	9 15/16
6	15.50	1 7/16	5 1/16	11/32	4 1/2	12	30.00	2	5	15/32	9 15/16
7	9.75	1 3/16	4 23/32	11/32	5 7/16	12	35.00	2 1/8	5 1/8	1/2	9 15/16
7	12.25	1 5/16	4 13/16	3/8	5 7/16	12	40.00	2 1/4	5 1/4	17/32	9 15/16
7	14.75	1 7/16	4 29/32	3/8	5 7/16	15	33.00	1 7/8	4 29/32	21/32	12 3/8
7	17.25	1 1/2	5 1/32	3/8	5 7/16	15	35.00	1 15/16	4 15/16	21/32	12 3/8
7	19.75	1 5/8	5 1/8	3/8	5 7/16	15	40.00	2	5 1/32	11/16	12 3/8
8	11.25	1 1/4	4 23/32	3/8	6 5/16	15	45.00	2 1/8	5 1/8	5/8	12 3/8
8	13.75	1 5/16	4 13/16	13/32	6 5/16	15	50.00	2 1/4	5 7/32	21/32	12 3/8
8	16.25	1 3/8	4 29/32	3/8	6 5/16	15	55.00	2 5/16	5 5/16	21/32	12 3/8
8	18.75	1 1/2	5	13/32	6 5/16						

10. Distance From End.—In order to avoid the danger of fracturing the material, rivets should not be placed too near the edge or end of the plate. (Owing to the greater strength of wrought iron in the direction of the fiber, rivets in wrought iron may be placed nearer the edge than the end.) According to practice, the distance between the edge of any piece and the center of the rivet hole should not be less than $1\frac{1}{4}$ inches for $\frac{3}{4}$-inch and $\frac{7}{8}$-inch

TABLE VII

STANDARD SPACING OF RIVETS AND BOLTS IN ANGLES, WITH MAXIMUM SIZE OF RIVETS TO BE USED

Length of Leg Inches	m Inches	Diameter of Rivet Inches	Length of Leg Inches	m Inches	Diameter of Rivet Inches	Length of Leg Inches	m Inches	Diameter of Rivet Inches
$\frac{3}{4}$	$\frac{7}{16}$	$\frac{1}{4}$	2	$1\frac{1}{8}$	$\frac{5}{8}$	$3\frac{1}{2}$	Variable, depending on diameter of rivet, thickness of metal, and length of leg.	1
1	$\frac{9}{16}$	$\frac{3}{8}$	$2\frac{1}{4}$	$1\frac{1}{4}$	$\frac{3}{4}$	4		1
$1\frac{1}{4}$	$\frac{11}{16}$	$\frac{1}{2}$	$2\frac{5}{16}$	$1\frac{1}{4}$	$\frac{3}{4}$	$4\frac{1}{2}$		1
$1\frac{5}{16}$	$\frac{3}{4}$	$\frac{1}{2}$	$2\frac{1}{2}$	$1\frac{3}{8}$	$\frac{3}{4}$	5		1
$1\frac{3}{8}$	$\frac{3}{4}$	$\frac{1}{2}$	$2\frac{3}{4}$	$1\frac{1}{2}$	$\frac{3}{4}$	6		1
$1\frac{1}{2}$	$\frac{13}{16}$	$\frac{1}{2}$	3	$1\frac{3}{4}$	$\frac{7}{8}$	7		1
$1\frac{3}{4}$	$\frac{15}{16}$	$\frac{5}{8}$						

rivets, except in bars less than $2\frac{1}{2}$ inches wide; when practicable, it should be at least two diameters of the rivet for all sizes and should not exceed eight times the thickness of the plate. A simple rule is to make the end distance twice the diameter of the rivets and the side distance $\frac{1}{4}$ inch less than this.

11. Clearance.—A certain amount of clearance is required for tools in forming the heads of rivets in shapes having short flanges. For a $\frac{7}{8}$-inch rivet, the distance from

TABLE VIII

STANDARD SPACING OF RIVETS AND BOLTS IN T BARS, WITH MAXIMUM SIZE OF RIVETS TO BE USED

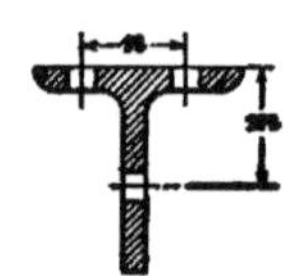

Width of Flange Inches	Depth of Bar Inches	Weight Pounds per Foot	m Inches	Maximum Diameter of Rivet in Stem Inches	n Inches	Maximum Diameter of Rivets in Flange Inches
$3\frac{1}{2}$	$3\frac{1}{2}$	9.30	2	1	$1\frac{3}{4}$	$\frac{5}{8}$
3	3	7.80	$1\frac{3}{4}$	$\frac{7}{8}$	$1\frac{1}{2}$	$\frac{1}{2}$
3	3	6.60	$1\frac{3}{4}$	$\frac{7}{8}$	$1\frac{1}{2}$	$\frac{1}{2}$
$2\frac{1}{2}$	$2\frac{1}{2}$	5.50	$1\frac{3}{8}$	$\frac{3}{4}$	$1\frac{1}{4}$	$\frac{1}{2}$
$2\frac{1}{4}$	$2\frac{1}{4}$	4.90	$1\frac{1}{4}$	$\frac{3}{4}$	$1\frac{1}{8}$	$\frac{1}{2}$
$2\frac{1}{4}$	$2\frac{1}{4}$	4.10	$1\frac{1}{4}$	$\frac{3}{4}$	$1\frac{1}{8}$	$\frac{1}{2}$
2	2	4.30	$1\frac{1}{8}$	$\frac{5}{8}$	1	$\frac{1}{2}$
2	2	3.70	$1\frac{1}{8}$	$\frac{5}{8}$	1	$\frac{1}{2}$
$1\frac{3}{8}$	$1\frac{3}{8}$	1.85	$\frac{3}{4}$	$\frac{1}{2}$	$\frac{5}{8}$	$\frac{1}{4}$
$1\frac{1}{4}$	$1\frac{1}{4}$	1.61	$\frac{11}{16}$	$\frac{1}{2}$	$\frac{5}{8}$	$\frac{1}{4}$
$1\frac{3}{16}$	$1\frac{3}{16}$	1.53	$\frac{5}{8}$	$\frac{1}{2}$	$\frac{5}{8}$	$\frac{1}{4}$
$1\frac{1}{8}$	$1\frac{1}{8}$	1.39	$\frac{9}{16}$	$\frac{1}{2}$	$\frac{5}{8}$	$\frac{1}{4}$
1	1	.89	$\frac{9}{16}$	$\frac{3}{8}$	$\frac{5}{8}$	$\frac{1}{4}$
$3\frac{1}{2}$	4	9.90	$2\frac{1}{4}$	1	$1\frac{3}{4}$	$\frac{5}{8}$
5	3	13.60	$1\frac{3}{4}$	$\frac{7}{8}$	$2\frac{3}{8}$	$\frac{3}{4}$
3	$2\frac{1}{2}$	7.20	$1\frac{3}{8}$	$\frac{3}{4}$	$1\frac{1}{2}$	$\frac{1}{2}$
$2\frac{1}{2}$	$1\frac{1}{4}$	2.90	$\frac{11}{16}$	$\frac{1}{2}$	$1\frac{1}{4}$	$\frac{1}{2}$
$1\frac{1}{4}$	$1\frac{1}{16}$	1.49	$\frac{9}{16}$	$\frac{3}{8}$	$\frac{5}{8}$	$\frac{1}{4}$

the center of the rivet to the back of the angle should not be less than $1\frac{1}{4}$ inches; and for a $\frac{3}{4}$-inch rivet, this distance should not be less than $1\frac{1}{8}$ inches. The perpendicular distance between the center line of a rivet in one leg of an angle and the top of a rivet head on the other leg should not be less than $1\frac{1}{8}$ inches.

Tables V and VI give the standard spacing (adopted by the Cambria Steel Company) of rivets and bolts in flanges and connection angles of **I** beams and channels, and the tangent distances between fillets measured along the web. In Tables VII and VIII are given the standard spacings of rivets and bolts in angles, **T** bars, and **Z** bars, with the maximum size of rivets to be used.

ANALYSIS OF RIVETED JOINTS

12. Every riveted joint is one of two kinds: a *lap joint*, as shown in Fig. 3, or a *butt joint*, as in Fig. 4, either of which may be single-, double-, triple-, or quadruple-riveted. When there is more than one row,

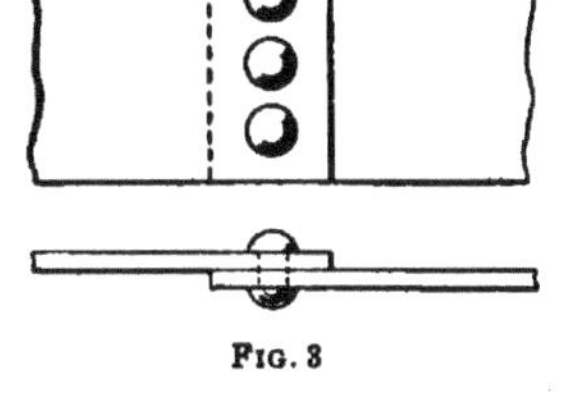

FIG. 3

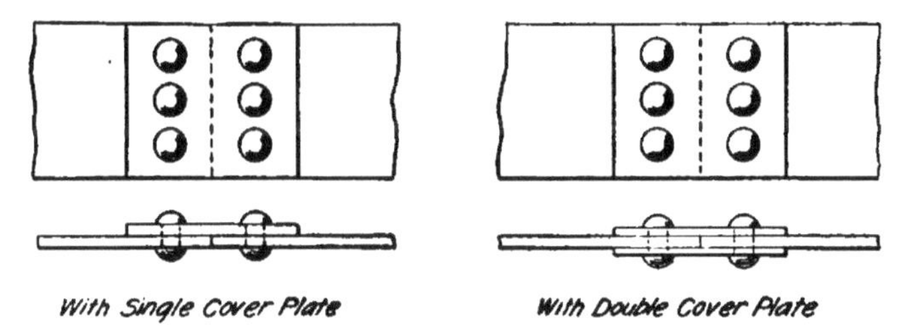

With Single Cover Plate *With Double Cover Plate*

FIG. 4

the riveting may be either *chain* or *zigzag*, usually called *staggered*.

The one most frequently used is the butt joint, it being more symmetrical and more effective. It is always preferable to the lap joint, which is made use of only in unimportant details.

TABLE IX

STANDARD SPACING OF RIVETS AND BOLTS IN Z BARS, WITH MAXIMUM SIZE OF RIVETS TO BE USED

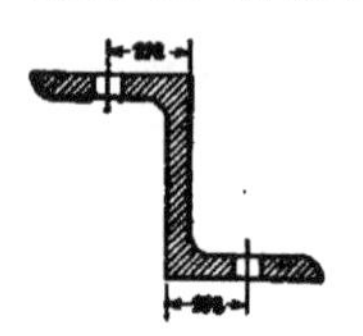

Depth of Bar Inches	Weight Pounds per Foot	m Inches	Maximum Diameter of Rivets Inches
3 to $3\frac{1}{16}$	6.7 to 14.2	$1\frac{5}{8}$	$\frac{3}{4}$
4 to $4\frac{1}{8}$	8.2 to 23.0	$1\frac{3}{4}$	$\frac{7}{8}$
5 to $5\frac{1}{8}$	11.6 to 28.3	$1\frac{7}{8}$	$\frac{7}{8}$
6 to $6\frac{1}{8}$	15.6 to 34.6	2	1

13. Joints in Tension.—A simple lap joint in tension with only one rivet, as shown in Fig. 5, will first be considered. This joint is liable to fail:

1. By one or both bars tearing across, as in Fig. 6.
2. By the rivet being sheared, as in Fig. 7.
3. By the rivet cutting into the bar, or the bar cutting into the rivet, as in Fig. 8, according to which metal is the softer.
4. By the bar being sheared by the rivet, as in Fig. 9.

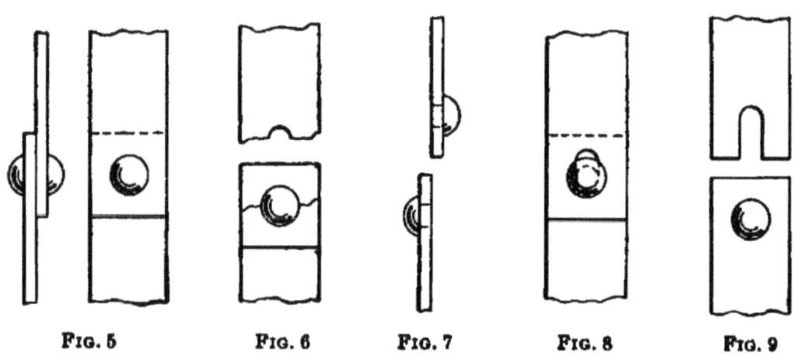

FIG. 5 FIG. 6 FIG. 7 FIG. 8 FIG. 9

In order to secure a good joint, therefore, it is necessary to calculate the stresses coming upon the joint and to arrange the number, size, position, and pitch of rivets and the dimensions of the plates to effectively resist those stresses.

The cause of the first method of failure is the inability of the net section of the bar to resist the stress. To find the net section, deduct the diameter of the rivet hole from the width of the bar and multiply by the thickness. Table X gives the areas to be deducted to obtain the net sections of riveted plates when the thickness of plate and diameter of rivet hole are given. Therefore, to insure strength, the area of this net section multiplied by the safe tensile strength of the steel per square inch must equal the tensile stress on the joint.

14. Bearing Value on Plates.—When the diameter of the rivet is large in proportion to the thickness of the plate, the joint or connection is liable to fail by the crimping of the plate, as shown in Fig. 10. This occurs when the resistance of the plate to crushing or crimping is less than the resistance of the rivet to shear.

FIG. 10

Considering the second method of failure, the strength depends on the bearing value of the plate or rivet, and as this varies with the bearing area of the rivet on the plate, which is assumed to be td, the bearing value B of the plate is expressed by the formula,

$$B = t d r_b \qquad (1)$$

in which t = thickness of plate;
d = diameter of rivet;
r_b = safe bearing resistance per square inch.

15. The condition of the plates in the joint shown in Fig. 11 (*a*) is called *ordinary bearing*, while the plate m connected as shown in (*b*) is said to be in *web bearing*. This distinction is important, because the bearing value of a web-plate is greater than that of an outside plate, the value for

TABLE X

AREAS, IN SQUARE INCHES, TO BE DEDUCTED FROM RIVETED PLATES OR SHAPES TO OBTAIN NET AREAS

Thickness of Plates Inches	Size of Hole. Inches												
	1/4	5/16	3/8	7/16	1/2	9/16	5/8	11/16	3/4	13/16	7/8	15/16	1
1/4	.06	.08	.09	.11	.13	.14	.16	.17	.19	.20	.22	.23	.25
5/16	.08	.10	.12	.14	.16	.18	.20	.21	.23	.25	.27	.29	.31
3/8	.09	.12	.14	.16	.19	.21	.23	.26	.28	.30	.33	.35	.38
7/16	.11	.14	.16	.19	.22	.25	.27	.30	.33	.36	.38	.41	.44
1/2	.13	.16	.19	.22	.25	.28	.31	.34	.38	.41	.44	.47	.50
9/16	.14	.18	.21	.25	.28	.32	.35	.39	.42	.46	.49	.53	.56
5/8	.16	.20	.23	.27	.31	.35	.39	.43	.47	.51	.55	.59	.63
11/16	.17	.21	.26	.30	.34	.39	.43	.47	.52	.56	.60	.64	.69
3/4	.19	.23	.28	.33	.38	.42	.47	.52	.56	.61	.66	.70	.75
13/16	.20	.25	.30	.36	.41	.46	.51	.56	.61	.66	.71	.76	.81
7/8	.22	.27	.33	.38	.44	.49	.55	.60	.66	.71	.77	.82	.88
15/16	.23	.29	.35	.41	.47	.53	.59	.64	.70	.76	.82	.88	.94
1	.25	.31	.38	.44	.50	.56	.63	.69	.75	.81	.88	.94	1.00
1 1/16	.27	.33	.40	.46	.53	.60	.66	.73	.80	.86	.93	1.00	1.06
1 1/8	.28	.35	.42	.49	.56	.63	.70	.77	.84	.91	.98	1.05	1.13
1 3/16	.30	.37	.45	.52	.59	.67	.74	.82	.89	.96	1.04	1.11	1.19
1 1/4	.31	.39	.47	.55	.63	.70	.78	.86	.94	1.02	1.09	1.17	1.25
1 5/16	.33	.41	.49	.57	.66	.74	.82	.90	.98	1.07	1.15	1.23	1.31
1 3/8	.34	.43	.52	.60	.69	.77	.86	.95	1.03	1.12	1.20	1.29	1.38
1 7/16	.36	.45	.54	.63	.72	.81	.90	.99	1.08	1.17	1.26	1.35	1.44
1 1/2	.38	.47	.56	.66	.75	.84	.94	1.03	1.13	1.22	1.31	1.41	1.50
1 9/16	.39	.49	.59	.68	.78	.88	.98	1.07	1.17	1.27	1.37	1.46	1.56
1 5/8	.41	.51	.61	.71	.81	.91	1.02	1.12	1.22	1.32	1.42	1.52	1.63
1 11/16	.42	.53	.63	.74	.84	.95	1 05	1.16	1.27	1.37	1.47	1.58	1.69
1 3/4	.44	.55	.66	.77	.88	.98	1.09	1.20	1.31	1.42	1.53	1.64	1.75
1 13/16	.45	.57	.68	.79	.91	1.02	1.13	1.25	1.36	1.47	1.59	1.70	1.81
1 7/8	.47	.59	.70	.82	.94	1.05	1.17	1.29	1.41	1.52	1.64	1.76	1.88
1 15/16	.48	.61	.73	.85	.97	1.09	1.21	1.33	1.45	1.57	1.70	1.82	1.94
2	.50	.63	.75	.88	1.00	1.13	1.25	1.38	1.50	1.63	1.75	1.88	2.00

Thickness of Plates Inches	Size of Hole. Inches															
	$1\frac{1}{16}$	$1\frac{1}{8}$	$1\frac{3}{16}$	$1\frac{1}{4}$	$1\frac{5}{16}$	$1\frac{3}{8}$	$1\frac{7}{16}$	$1\frac{1}{2}$	$1\frac{9}{16}$	$1\frac{5}{8}$	$1\frac{11}{16}$	$1\frac{3}{4}$	$1\frac{13}{16}$	$1\frac{7}{8}$	$1\frac{15}{16}$	2
$\frac{1}{4}$	.27	.28	.30	.31	.33	.34	.36	.38	.39	.41	.42	.44	.45	.47	.48	.50
$\frac{5}{16}$	.33	.35	.37	.39	.41	.43	.45	.47	.49	.51	.53	.55	.57	.59	.61	.63
$\frac{3}{8}$	.40	.42	.45	.47	.49	.52	.54	.56	.59	.61	.63	.66	.68	.70	.73	.75
$\frac{7}{16}$	.46	.49	.52	.55	.57	.60	.63	.66	.68	.71	.74	.77	.79	.82	.85	.88
$\frac{1}{2}$	.53	.56	.59	.63	.66	.69	.72	.75	.78	.81	.84	.88	.91	.94	.97	1.00
$\frac{9}{16}$	.60	.63	.67	.70	.74	.77	.81	.84	.88	.91	.95	.98	1.02	1.05	1.09	1.13
$\frac{5}{8}$	.66	.70	.74	.78	.82	.86	.90	.94	.98	1.02	1.05	1.09	1.13	1.17	1.21	1.25
$\frac{11}{16}$	.73	.77	.82	.86	.90	.95	.99	1.03	1.07	1.12	1.16	1.20	1.25	1.29	1.33	1.38
$\frac{3}{4}$	.80	.84	.89	.94	.98	1.03	1.08	1.13	1.17	1.22	1.27	1.31	1.36	1.41	1.45	1.50
$\frac{13}{16}$	.86	.91	.96	1.02	1.07	1.12	1.17	1.22	1.27	1.32	1.37	1.42	1.47	1.52	1.57	1.63
$\frac{7}{8}$	.93	.98	1.04	1.09	1.15	1.20	1.26	1.31	1.37	1.42	1.48	1.53	1.59	1.64	1.70	1.75
$\frac{15}{16}$	1.00	1.05	1.11	1.17	1.23	1.29	1.35	1.41	1.46	1.52	1.58	1.64	1.70	1.76	1.82	1.88
1	1.06	1.13	1.19	1.25	1.31	1.38	1.44	1.50	1.56	1.63	1.69	1.75	1.81	1.88	1.94	2.00
$1\frac{1}{16}$	1.13	1.20	1.26	1.33	1.39	1.46	1.53	1.59	1.66	1.73	1.79	1.86	1.93	1.99	2.06	2.13
$1\frac{1}{8}$	1.20	1.27	1.34	1.41	1.48	1.55	1.62	1.69	1.76	1.83	1.90	1.97	2.04	2.11	2.18	2.25
$1\frac{3}{16}$	1.26	1.34	1.41	1.48	1.56	1.63	1.71	1.78	1.86	1.93	2.00	2.08	2.15	2.23	2.30	2.38
$1\frac{1}{4}$	1.33	1.41	1.48	1.56	1.64	1.72	1.80	1.88	1.95	2.03	2.11	2.19	2.27	2.34	2.42	2.50
$1\frac{5}{16}$	1.39	1.48	1.56	1.64	1.72	1.80	1.89	1.97	2.05	2.13	2.21	2.30	2.38	2.46	2.54	2.63
$1\frac{3}{8}$	1.46	1.55	1.63	1.72	1.80	1.89	1.98	2.06	2.15	2.23	2.32	2.41	2.49	2.58	2.66	2.75
$1\frac{7}{16}$	1.53	1.62	1.71	1.80	1.89	1.98	2.07	2.16	2.25	2.34	2.43	2.52	2.61	2.70	2.79	2.88
$1\frac{1}{2}$	1.59	1.69	1.78	1.88	1.97	2.06	2.16	2.25	2.34	2.44	2.53	2.63	2.72	2.81	2.91	3.00
$1\frac{9}{16}$	1.66	1.76	1.86	1.95	2.05	2.15	2.25	2.34	2.44	2.54	2.64	2.73	2.83	2.93	3.03	3.13
$1\frac{5}{8}$	1.73	1.83	1.93	2.03	2.13	2.23	2.34	2.44	2.54	2.64	2.74	2.84	2.95	3.05	3.15	3.25
$1\frac{11}{16}$	1.79	1.90	2.00	2.11	2.21	2.32	2.43	2.53	2.64	2.74	2.85	2.95	3.06	3.16	3.27	3.38
$1\frac{3}{4}$	1.86	1.97	2.08	2.19	2.30	2.41	2.52	2.63	2.73	2.84	2.95	3.06	3.17	3.28	3.39	3.50
$1\frac{13}{16}$	1.93	2.04	2.15	2.27	2.38	2.49	2.61	2.72	2.83	2.95	3.06	3.17	3.29	3.40	3.51	3.63
$1\frac{7}{8}$	1.99	2.11	2.23	2.34	2.46	2.58	2.70	2.81	2.93	3.05	3.16	3.28	3.40	3.52	3.63	3.75
$1\frac{15}{16}$	2.06	2.18	2.30	2.42	2.54	2.66	2.79	2.91	3.03	3.15	3.27	3.39	3.51	3.63	3.75	3.88
2	2.13	2.25	2.38	2.50	2.63	2.75	2.88	3.00	3.13	3.25	3.38	3.50	3.63	3.75	3.88	4.00

web bearing being about one-third greater than that for ordinary bearing. Owing to the support that the material around the hole receives from the rest of the plate, its bearing value is greater than the compressive strength of the material when not so supported. The safe bearing strength of the rivet on the plate, for ordinary bearing, is assumed to be one and one-half times its compressive strength, and for web bearing the safe bearing strength is assumed to be double the compressive strength of the material.

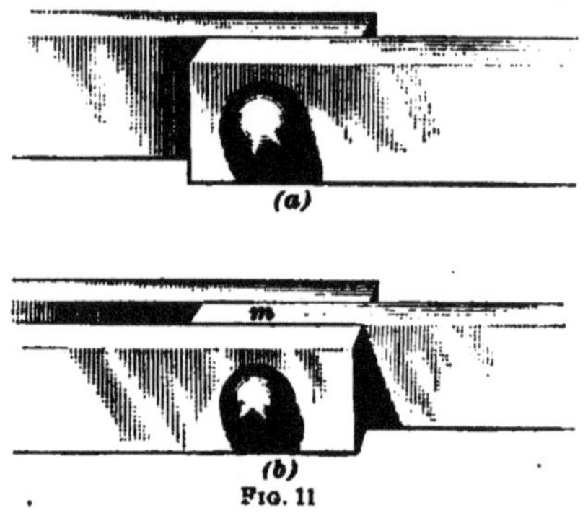

Fig. 11

In deducting the rivet holes, to ascertain the net section of a riveted plate, the diameter of the hole is taken as $\frac{1}{8}$ inch larger than the diameter of the rivet.

As the tensile strength of iron and steel used in the manufacture of rivets, pins, and plates for structural work is more easily determined by tests, and, therefore, better known than either its compressive or shearing strength, it is customary to use this as a basis from which to calculate the bearing value of plates and the shearing strength of rivets and pins.

Good practice assumes that the compressive strength of steel or high-test iron is about thirteen-fifteenths of its tensile strength; that is, if the safe tensile strength of the material per square inch of section is 15,000 pounds, the safe compressive strength may be taken as thirteen-fifteenths of 15,000 = 13,000 pounds.

The bearing strength of the rivet on the plate, for ordinary bearing, is found to be 1.8 times the shearing strength of the rivet, according to the assumptions made in this and the following paragraphs, but it is common in practice to consider the bearing value as twice that for shear, as shown in Table XIII.

16. Shearing Strength of Rivets.—When a rivet shears, the tendency is to cut straight through it across its section, as shown in Fig. 12 (*a*) and (*b*). The resistance of the rivet to shear depends on the area of its cross-section.

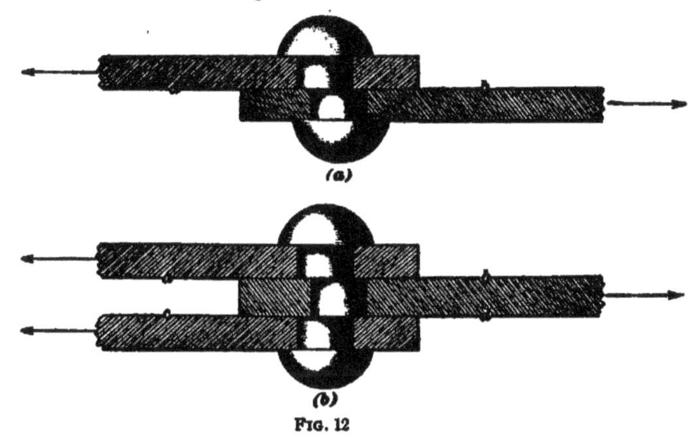

(*a*)

(*b*)

FIG. 12

The shearing strength is considered to be five-sixths of the compressive strength. For example, the tensile strength is 15,000 pounds and the compressive strength 13,000 pounds, the shearing strength becomes five-sixths of 13,000, or 10,833 pounds per square inch of section.

Where there are only two plates connected, as shown in Fig. 12 (*a*), the tendency is to cut the rivet on the single plane *a b*. A rivet in this position is said to be in *single shear*.

At (*b*), the tendency is to cut through the rivet on both the planes *a b* and *c d*; under these conditions the rivet is in *double shear*, and it is evident that, since the rivet will shear across at two places, it will be twice as strong as when the tendency is to shear through only one section.

17. Shearing of Plate.—To prevent the fourth method of failure, the lap should be sufficient to prevent the plate from shearing, as shown in Fig. 9. Since two surfaces have to be sheared, each having an area of $t \times \frac{l}{2}$ (in which l equals the lap), the resistance to shear is $2t \times \frac{l}{2} \times r_s$, or $t\,l\,r_s$, when r_s is the shearing strength of the plate per square inch.

18. A **butt joint with a single cover-plate** may be considered as two lap joints placed close together, its strength depending on the number of rivets on one side of the butt, or half the entire number of rivets. The cover-plate should have at least the same thickness as the plates to be joined.

19. In a **butt joint with two cover-plates,** the thickness of each cover-plate should be at least half the thickness of the plates joined. The only difference in the strength of this joint and that having one cover-plate is that the rivets have twice as much shearing resistance.

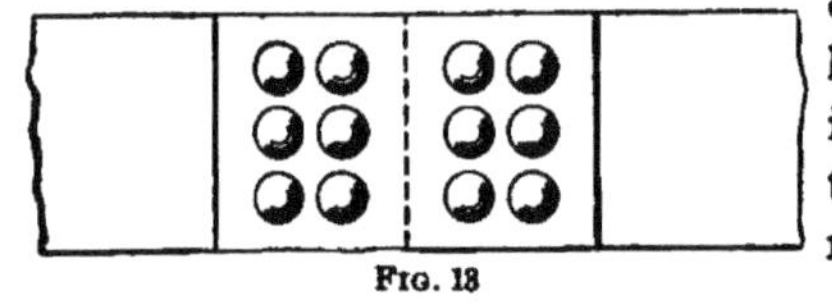

Fig. 13

It must also be remembered that as the plates connected are in web bearing, they have a greater bearing value than the cover-plates, which are in ordinary bearing.

The methods of calculation are the same for double-, triple-, or quadruple-riveted joints, the entire bearing value and shearing value depending on the number of rivets used.

20. Tensile Strength of Plates.—In Fig. 13, the tensile strength of the plate depends on the net section through the outside row of rivets, as it does also in **Fig. 14.** The first method, therefore, weakens the plates more than the second by one rivet hole, but both have the same bearing value and the same shearing strength, since the number of rivets in each case is the same. In Fig. 15, the bearing and shearing values are the same as in the previous cases, but the plates joined are stronger, since they are weakened by only one rivet hole, for,

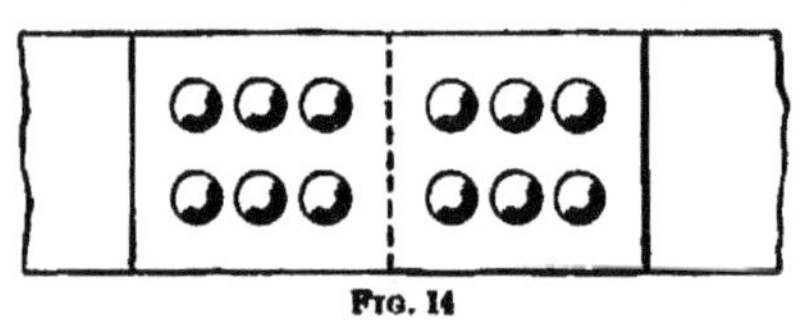

Fig. 14

before the plate can tear across the section A–A, the leading rivet a must fail by shearing or bearing. This counterbalances the weakening of the plate caused by one rivet hole, so that the section A–A is really only weakened by one rivet hole and is as strong as the one through the leading rivet a. It will also be seen that the section through d, e, and f is even stronger than either of the other two, since before it can fail the three rivets a, b, and c must fail.

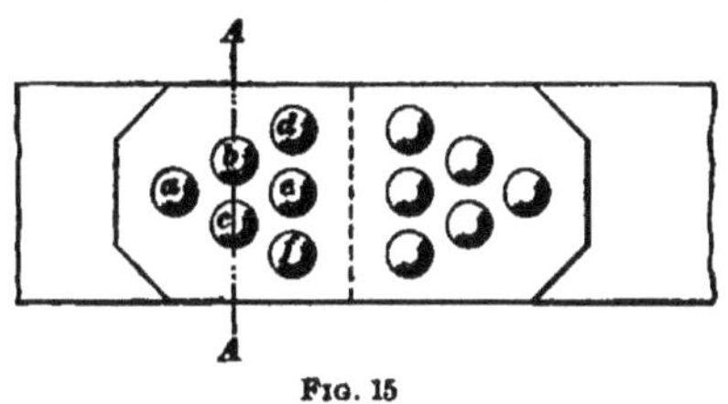

FIG. 15

21. The cover-plates, however, have their section reduced by three rivet holes and must therefore be thicker than half the thickness of the plates joined. Knowing the tensile strength of the plate, its breadth, the diameter of the rivet holes, and the unit tensile strength of the steel or iron, the thickness of the cover-plates to be used in a case like Fig. 15 can be found by the formula,

$$t' = \frac{R_t}{2(b - 3d)r_t} \qquad (2)$$

in which t' = thickness of one cover-plate;
R_t = tensile strength of the plate;
b = breadth of plate;
d = diameter of rivet hole;
r_t = safe unit tensile strength of steel or iron.

22. Rivets in Single Shapes.—The difficulty in using single shapes for tension members is to provide proper connections at the ends. The arrangement of the rivets should be such as to make the net area as great as possible, so that the efficiency of the metal will be maximum. The rivets, therefore, should be placed symmetrically with respect to the axis of the piece, and the cross-section should be reduced as little as possible, as in Fig. 16 (a), where the cross-section of the plate is reduced by only one or two rivet holes.

Where the rivets are arranged as shown in Fig. 16 (*b*), a section at right angles to the axis of the member cuts two and in most cases three rivet holes, and consequently this design is not as efficient as the one shown in (*a*).

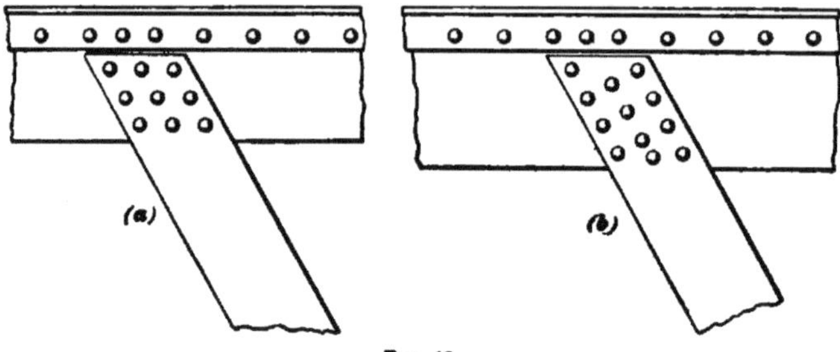

Fig. 16

Angle irons should be attached through both legs to develop the greatest strength and the rivets should be arranged symmetrically, as in plates. Fig. 17 shows the detail as it is usually made.

23. Rivets in Compound Sections.—Sections composed of two or more simple shapes riveted together are used in cases where the member requires special stiffness. If the end attachment is made by means of a pin, the built member is usually reenforced by pin plates, which reduce the unit pressure on the pin.

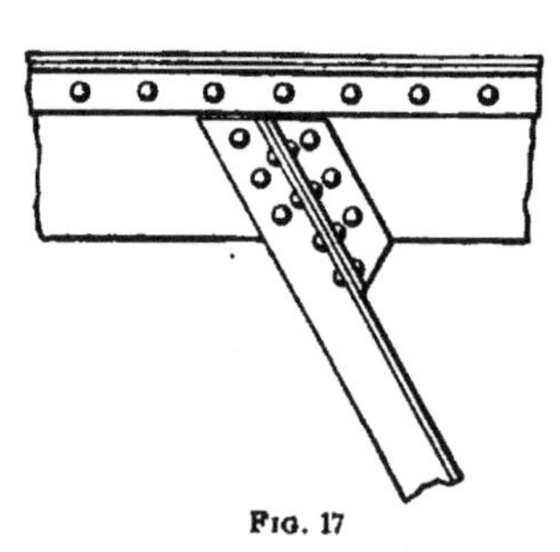

Fig. 17

In designing pin plates in important members, they are often made strong enough to realize the utmost working capacity of the member and not merely for the total stress to which the member will be subjected.

24. Joints in Compression.—If it were possible to fit a joint perfectly, the compressive stress would be transmitted across the joint entirely by the plates and the only

TABLE XI

SHEARING VALUE OF RIVETS

Shearing value = area of rivet × allowable shearing stress per square inch

Diameter of Rivet Inch	Area in Square Inches	Single Shear at 6,000 Pounds per Square Inch	Double Shear at 12,000 Pounds per Square Inch	Single Shear at 6,750 Pounds per Square Inch	Double Shear at 13,500 Pounds per Square Inch	Single Shear at 7,500 Pounds per Square Inch	Double Shear at 15,000 Pounds per Square Inch	Single Shear at 9,000 Pounds per Square Inch	Double Shear at 18,000 Pounds per Square Inch	Single Shear at 10,000 Pounds per Square Inch	Double Shear at 20,000 Pounds per Square Inch
$\frac{3}{8}$	.1105	663	1,325	746	1,491	828	1,657	994	1,989	1,105	2,209
$\frac{1}{2}$	.1964	1,178	2,356	1,325	2,651	1,473	2,945	1,767	3,535	1,964	3,927
$\frac{5}{8}$	.3068	1,841	3,682	2,071	4,142	2,301	4,602	2,761	5,522	3,068	6,136
$\frac{3}{4}$	.4418	2,651	5,301	2,982	5,964	3,313	6,627	3,976	7,952	4,418	8,836
$\frac{7}{8}$	.6013	3,608	7,216	4,059	8,118	4,510	9,020	5,411	10,823	6,013	12,026
1	.7854	4,712	9,425	5,301	10,603	5,891	11,781	7,068	14,137	7,854	15,708

TABLE XII

BEARING VALUE, PER SQUARE INCH, OF RIVETED PLATES

Bearing value = diameter of rivet × thickness of plate × allowable bearing stress per square inch

Diameter of Rivet Inch	Thickness of Plate. Inch												
	$\frac{1}{4}$	$\frac{5}{16}$	$\frac{3}{8}$	$\frac{7}{16}$	$\frac{1}{2}$	$\frac{9}{16}$	$\frac{5}{8}$	$\frac{11}{16}$	$\frac{3}{4}$	$\frac{13}{16}$	$\frac{7}{8}$	$\frac{15}{16}$	1
	Bearing Value of Plates at 12,000 Pounds per Square Inch												
$\frac{3}{8}$	1,125	1,406	1,688										
$\frac{1}{2}$	1,500	1,875	2,250	2,625	3,000								
$\frac{5}{8}$	1,875	2,344	2,813	3,281	3,750	4,219	4,688						
$\frac{3}{4}$	2,250	2,813	3,375	3,938	4,500	5,063	5,625	6,188	6,750				
$\frac{7}{8}$	2,625	3,281	3,938	4,594	5,250	5,906	6,563	7,219	7,875	8,531	9,188	9,844	
1	3,000	3,750	4,500	5,250	6,000	6,750	7,500	8,250	9,000	9,750	10,500	11,250	12,000
	Bearing Value of Plates at 13,500 Pounds per Square Inch												
$\frac{3}{8}$	1,266	1,582	1,898										
$\frac{1}{2}$	1,688	2,109	2,531	2,953	3,375								
$\frac{5}{8}$	2,109	2,637	3,164	3,691	4,219	4,746	5,273						
$\frac{3}{4}$	2,531	3,164	3,797	4,430	5,063	5,695	6,328	6,961	7,594				
$\frac{7}{8}$	2,953	3,691	4,430	5,168	5,906	6,645	7,383	8,121	8,859	9,598	10,336	11,074	
1	3,375	4,219	5,063	5,906	6,750	7,594	8,438	9,281	10,125	10,969	11,813	12,656	13,500

Bearing Value of Plates at 15,000 Pounds per Square Inch

3/8	1,406	1,758	2,109										
1/2	1,875	2,344	2,813	3,281	3,750								
5/8	2,344	2,930	3,516	4,102	4,688	5,273	5,859						
3/4	2,813	3,516	4,219	4,922	5,625	6,328	7,031	7,734	8,438				
7/8	3,281	4,102	4,922	5,742	6,563	7,383	8,203	9,023	9,844	10,664	11,484	12,305	
1	3,750	4,688	5,625	6,563	7,500	8,438	9,375	10,313	11,250	12,188	13,125	14,063	15,000

Bearing Value of Plates at 18,000 Pounds per Square Inch

3/8	1,688	2,109	2,531										
1/2	2,250	2,913	3,375	3,938	4,500								
5/8	2,813	3,516	4,219	4,922	5,625	6,328	7,031						
3/4	3,375	4,219	5,063	5,906	6,750	7,593	8,438	9,281	10,125				
7/8	3,938	4,922	5,906	6,891	7,875	8,859	9,844	10,828	11,813	12,797	13,781	14,766	
1	4,500	5,625	6,750	7,875	9,000	10,125	11,250	12,375	13,500	14,625	15,750	16,875	18,000

Bearing Value of Plates at 20,000 Pounds per Square Inch

3/8	1,875	2,344	2,813										
1/2	2,500	3,125	3,750	4,375	5,000								
5/8	3,125	3,906	4,688	5,469	6,250	7,031	7,813						
3/4	3,750	4,688	5,625	6,563	7,500	8,438	9,375	10,313	11,250				
7/8	4,375	5,469	6,563	7,656	8,750	9,844	10,938	12,031	13,125	14,219	15,313	16,406	
1	5,000	6,250	7,500	8,750	10,000	11,250	12,500	13,750	15,000	16,250	17,500	18,750	20,000

use of the rivets would be to hold the plates together. But as a perfect fit cannot be insured, the entire stress is assumed to be transmitted by the rivets, and the calculations are the same as for rivets in tension joints. The only

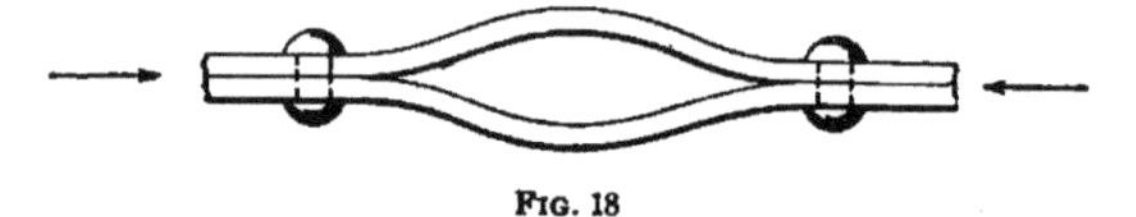

FIG. 18

difference is that there is no deduction made from the cross-section for rivet holes, as the rivet completely fills up the hole and transmits the stress.

It is necessary sometimes to have the longitudinal pitch less than in tension joints to avoid the danger of buckling the plates between the rivets, as in Fig. 18.

25. Tables of Bearing Values of Riveted Plates and Shearing Values of Rivets.—In order to avoid the necessity of calculating the shearing value of the rivets and the bearing value of the riveted plates and rolled sections, Tables XI and XII have been prepared. It will be noticed that the areas and shearing values for both double and single shear are given in Table XI for rivets from $\frac{3}{8}$ inch to 1 inch in diameter, using unit allowable shearing stresses of 6,000, 6,750, 7,500, and 9,000 pounds for single shear, and double these amounts for double shear.

Table XII gives the bearing values of riveted plates for different thicknesses of plate and unit allowable bearing stresses of 12,000, 13,500, 15,000, and 18,000 pounds.

The usual allowable strains, in pounds per square inch, on riveted work are given in Table XIII.

26. In designing a riveted joint, it is necessary to calculate the number of rivets required to resist the shear and the number required to provide sufficient bearing on the plate. The greater value must be adopted as the number of rivets to be used. By dividing the total tensile stress in the bar by the values taken from Tables XI and XII, the number of rivets required for bearing and the number for shearing are obtained.

In Table XII, the bearing values between the lower and upper zigzag heavy lines in each group are greater than single shear and less than double shear for the corresponding dimensions, so that in case of single shear, the single-shearing value governs, and in case of double shear, the bearing value governs the design.

The bearing values above and to the right of the upper zigzag heavy lines are greater than double shear for the corresponding dimensions, so that in these cases the shearing values govern the design.

The bearing values below and to the left of the lower zigzag heavy lines are less than single shear, and in these cases the bearing values govern the design.

TABLE XIII

ALLOWABLE STRAINS ON RIVETED WORK

Rivets	Shearing Pounds per Square Inch	Bearing Pounds per Square Inch
Iron rivets in railroad bridges	6,000	12,000
Iron rivets in highway bridges and buildings	7,500	15,000
Steel rivets in railroad bridges	7,500	15,000
Steel rivets in highway bridges and buildings	9,000	18,000

EXAMPLE 1.—Two pieces of structural steel are joined by rivets, as shown in Fig. 19. If the tensile strength of the steel is 60,000 pounds per square inch, and a factor of safety of 4 is used, what is the safe strength of this joint?

SOLUTION.—The safe tensile strength of the steel is $60,000 \div 4 = 15,000$ lb. per sq. in. The width of the pieces connected is $2\frac{1}{2}$ in., from which is to be deducted 1 in. for the rivet hole, leaving a net width of $1\frac{1}{2}$ in., which, multiplied by the thickness of the plate, gives a net area of $1\frac{1}{2} \times \frac{3}{8} = .5625$ sq. in. Then, $.5625 \times 15,000 = 8,437$ lb., the strength of the plate.

To determine whether the strength of the rivets is equal to the net section of the plate: Taking the compressive value of the plate as thirteen-fifteenths of 15,000 lb., or 13,000 lb. per sq. in., and the rivets being in ordinary bearing, the safe bearing strength is $13,000 \times 1\frac{1}{2} = 19,500$ lb. per sq. in. of bearing area. The bearing area is $\frac{7}{8} \times \frac{3}{8} = .328$ sq. in.; therefore, the safe bearing strength for one rivet is $19,500 \times .328 = 6,396$ lb., and for the two it is $2 \times 6,396 = 12,792$ lb.

The shearing strength of the steel is five-sixths of $13,000 = 10,833$ lb. per sq. in. The area of a $\frac{7}{8}$-in. rivet is .601 sq. in., which, multiplied by 10,833, gives 6,510 lb., the shearing strength of one rivet. The total resistance to shear of the rivets in the joint is, therefore, $6,510 \times 2 = 13,020$ lb.

The safe resistance of the three elements entering into the strength of the joint is, therefore, as follows: Resistance of net section of the plate = 8,437 lb.; bearing value of the plate = 12,792 lb.; shearing strength of the two rivets = 13,020 lb.; from which it is easily seen that the strength of the joint is that of the net section of the plate, 8,437 lb. Ans.

Since the bearing value of the plate and the shearing strength of the rivets are considerably in excess of this amount, it appears that the rivets are large for the joint, and it is probable that $\frac{3}{4}$-in. rivets would give better results.

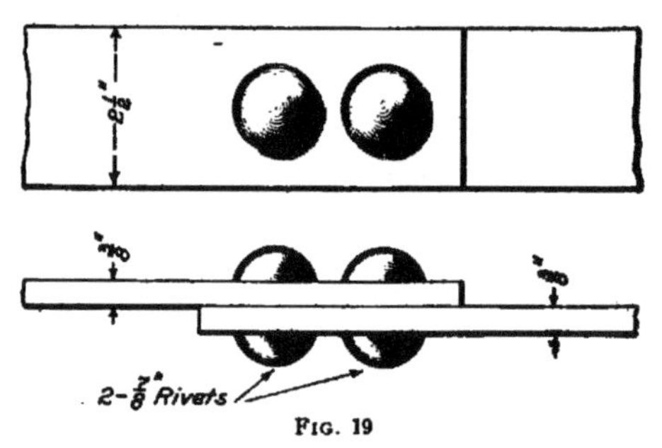

Fig. 19

Example 2.—Determine the safe strength of the riveted joint shown in Fig. 20, in which the plates and rivets each have a safe tensile strength of 16,000 pounds per square inch.

Solution.—The safe tensile strength of the material being 16,000 lb., the safe compressive strength is thirteen-fifteenths of 16,000 lb., or 13,867 lb., and the shearing strength of the rivets is five-sixths of 13,867, or 11,556 lb. per sq. in. of section. The area of the section of a $\frac{7}{8}$-in. rivet is .601 sq. in.; therefore, the total shearing strength of the three rivets, each of which is in double shear, is $2 \times .601 \times 11,556 \times 3 = 41,670$ lb.

The two outside plates are in ordinary bearing, and their bearing value is $1\frac{1}{2} \times 13,867 = 20,800$ lb. per sq. in. There are three rivet holes in each plate, each with a bearing area of $\frac{3}{8} \times \frac{7}{8} = .328$ sq. in.; the total bearing strength of the two plates is, therefore, $20,800 \times .328 \times 3 \times 2 = 40,934$ lb.

The bearing value of the central or web-bearing plate at one rivet hole is $2 \times 13,867 = 27,734$ lb. per sq. in., and the bearing area is $\frac{3}{4} \times \frac{7}{8} = .656$ sq. in.; the total bearing strength for the three rivets is, therefore, $27,734 \times .656 \times 3 = 54,580$ lb.

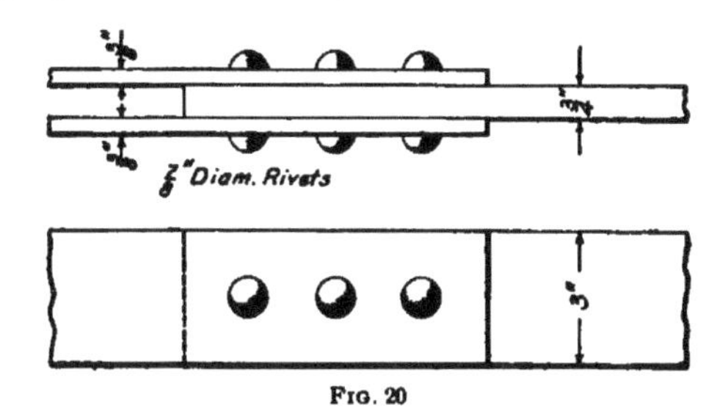

Fig. 20

The safe tensile strength of the central plate is equal to its net section multiplied by 16,000, the safe unit tensile strength of the material. The net width of the plate is $3 - 1 = 2$ in., and its net area, $2 \times \frac{3}{4} = 1.5$ sq. in.; therefore, the safe strength is $1.5 \times 16,000 = 24,000$ lb. The strength of the two outside plates, calculated in the same manner, is found to be 24,000 lb. also; therefore, it is evident that, since the strength of the net section of the plate is much less than either the strength of the rivets or the bearing value of the plates, it determines the strength of the joint, which is 24,000 lb. Ans.

Example 3.—The lower chord of a roof truss is made up of two angles, as shown in Fig. 21. A butt joint in this member is reenforced by the vertical splice plates *a*, *b*, and *d*, and a horizontal plate *c*. Calculate the number of ¾-inch rivets to be used, the full safe strength of the material being realized, using a unit tensile stress of 15,000 pounds, for angles and plates.

Solution.—The sectional area of a $4'' \times 3\frac{1}{2}'' \times \frac{3}{8}''$ angle is 2.67 sq. in. The cross-section of the two angles is $2.67 \times 2 = 5.34$ sq. in. If the rivets are staggered in the two legs of the angles, their section will be reduced by two rivet holes, or $2 \times \frac{7}{8} \times \frac{3}{8} = .65625$ sq. in. This leaves a net sectional area of 4.68375 sq. in. Since the unit tensile stress is 15,000 lb., the total stress in the joint is $4.68375 \times 15,000 = 70,256$ lb., which stress must be taken up by the four plates *a*, *b*, *c*, and *d*.

First determine the number of rivets in the horizontal plate *c*. The net section of this plate $= (7\frac{1}{4} \times \frac{3}{8}) - (2 \times \frac{7}{8} \times \frac{3}{8}) = 2.0625$ sq. in., and its resistance is $2.0625 \times 15,000 = 30,937$ lb. Using the usual unit shearing stress of iron rivets in buildings, which is 7,500 lb., the value of a $\frac{3}{4}$-in. rivet in single shear will be found, from Table XI, to be 3,313. Therefore, the number of rivets required in the plate c is $30,937 \div 3,313 = 9$; but to make the joint symmetrical, ten rivets must be used, five being placed on each side, at a pitch of 4 in., the end distance being $1\frac{3}{8}$ in., and the distance from center of rivet to the butt joint $1\frac{1}{8}$ in., as in Fig. 21.

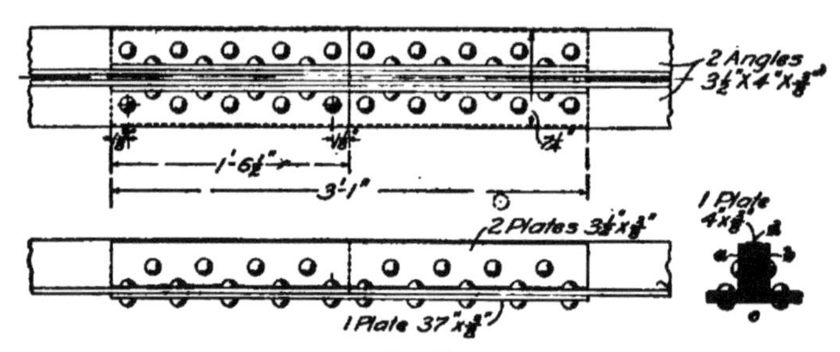

FIG. 21

Next ascertain whether the resistance of plates a, b, and d is sufficient to resist the remaining stress, not taken up by c. This equals $70,256 - 30,937 = 39,319$ lb. The net section of plate $d = (4 \times \frac{3}{8}) - (\frac{7}{8} \times \frac{3}{8}) = 1.172$ sq. in. The net section of a and $b = 2 \times [(3.5 \times \frac{3}{8}) - (\frac{7}{8} \times \frac{3}{8})] = 1.968$ sq. in. The total net section of these three plates is, therefore, $1.968 + 1.172 = 3.14$ sq. in., and their resistance equals $3.14 \times 15,000 = 47,100$ lb., which more than equals the stress 39,319 lb.

Now calculate the number of rivets through the vertical flanges and plates. The bearing values of these vertical members, taken from Table XII, are as follows:

Three splice plates	d in web bearing $4,219 \times 1\frac{1}{3} =$	5,625 lb.
	a and b in ordinary bearing	8,438 lb.
	Sum	14,063 lb.
Two angle legs in web bearing		11,250 lb.
Rivets	in double shear in one plate d	6,627 lb.
	in single shear in two plates a and b	6,627 lb.
	Sum	13,254 lb.
Rivet in double shear in two angle legs		13,254 lb.

From the above, it will be seen that the value 11,250 is the least and therefore it governs the design. The number of rivets through these members is $47,100 \div 11,250 = 4$, which are staggered with respect to those in the horizontal members.

EXAMPLE 4.—Fig. 22 represents the lower end of the main strut in a Fink roof truss, which is connected by means of a pin to the tension members. Each channel is reenforced by a pin plate, as shown. Determine the number of ¾-inch rivets to be used in each pin plate, the allowable unit shearing stress being 7,500 pounds.

The detailing is to be done for the actual stress to which the member is subjected, the total compressive stress in the member being 48,000 pounds.

SOLUTION.—The distribution of stresses between the pin plate and channel is in direct proportion to their respective bearings on the pin. Hence, the amount of stress in one pin plate $= \frac{3 \times .5625}{3 \times (.5625 + .32)} \times \frac{48,000}{2} = 15,288$ lb., the thickness of the channel web being .32 in.

The shearing value of a ¾-in. rivet in single shear is found, from Table XI, to be 3,313 and the bearing value from Table XII of a ¾-in. rivet in ordinary bearing in a $\frac{5}{16}$-in., or .3125-in. plate, which is nearly the thickness of the web, is 3,516. From this it will be seen that the shearing value is less than the bearing value and therefore governs the design. The number of rivets, then, is 15,288 ÷ 3,313 = 4.6, or 5; hence, in order to make the joint symmetrical, six rivets in three rows will be used, two rows above the pin and one below, as shown in Fig. 22. Ans.

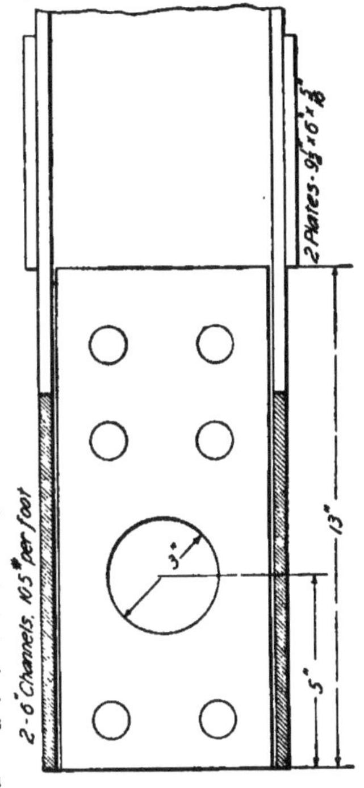

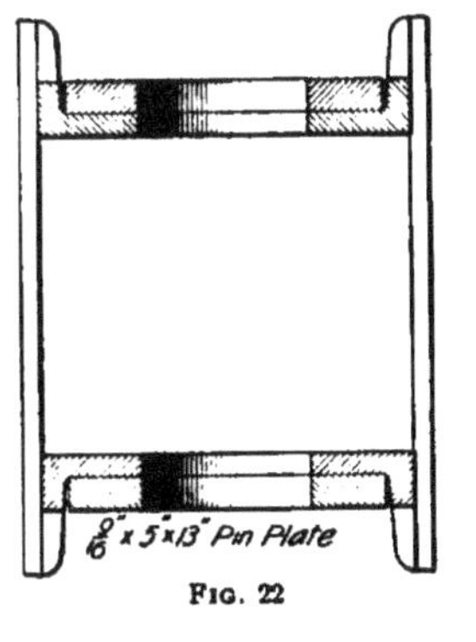

FIG. 22

EXAMPLES FOR PRACTICE

1. What are the safe strengths of ⅞-, ¾-, and ⅝-inch rivets, in double shear, and also in single shear,

assuming that the safe tensile strength of the material used in their manufacture is 15,000 pounds per square inch of section?

	Diameter of Rivet Inch	Double Shear Pounds	Single Shear Pounds
Ans.	$\frac{7}{8}$	13,028	6,514
	$\frac{3}{4}$	9,572	4,786
	$\frac{5}{8}$	6,648	3,324

2. What pulling force will two pieces of $\frac{3}{8}'' \times 2\frac{1}{2}''$ bar safely resist, provided that they are connected at the ends by two $\frac{3}{4}$-inch rivets, as shown in Fig. 23? The safe tensile strength of the material in rivets and bars is 15,000 pounds. Ans. 9,140 lb.

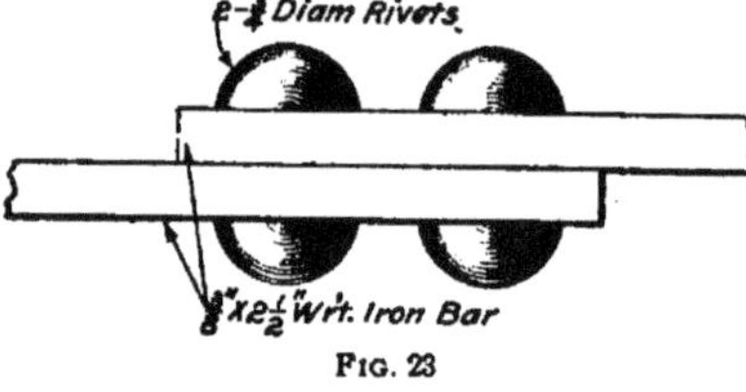

Fig. 23

3. What is the safe resisting moment of a pin 5 inches in diameter, if the safe strength of the material is 20,000 pounds per square inch? Ans. 245,400 in.-lb.

4. It is necessary to construct the connection of a tension member as shown in Fig. 24. What is the safe load that this member will

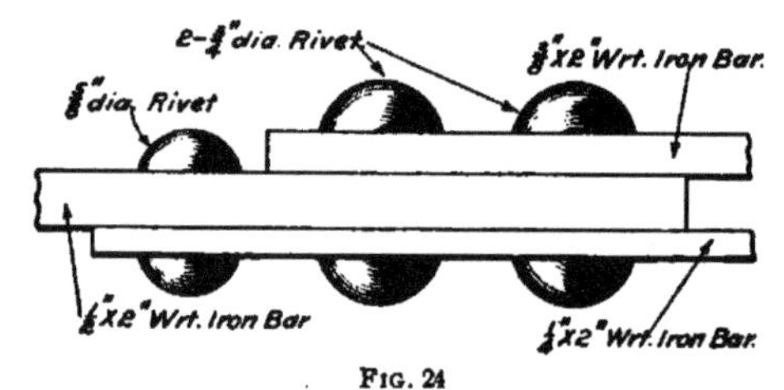

Fig. 24

carry, if the safe tensile strength of the material in both the rivets and bars is 18,000 pounds per square inch? Ans. 11,250 lb.

PIN CONNECTIONS

COMPARISON WITH RIVETED CONNECTIONS

27. While riveted connections make a stiffer structure and one that is less liable to accident from the failure of a single member, pin connections can be more satisfactorily designed since they are almost free from secondary stresses. The riveted system generally requires more material than the pin system and the latter requires more skilled workmanship. The pin system is the cheaper for a long span but for a short span it is sometimes more expensive. Good results can be obtained by either system, and the item of cost will usually be the controlling factor.

METHODS OF FAILURE

28. A pin is subject to strain in three ways: by shearing, by bending, and by crushing of the plates. Frequently, the first is not considered, as the size of pin required to resist the bending and bearing stresses generally exceeds that required for the shear.

The usual stresses allowed for pins, in pounds per square inch, are given in Table XIV.

TABLE XIV

ALLOWABLE STRESSES ON PINS

Kind of Pin	Shearing Pounds per Square Inch	Bearing Pounds per Square Inch	Bending Pounds per Square Inch
Iron pins, railroad bridges	7,500	12,000	15,000
Iron pins, highway bridges and buildings	9,000	15,000	18,000
Steel pins, railroad bridges	9,000	15,000	18,000
Steel pins, highway bridges and buildings	11,250	18,000	22,500

CALCULATIONS FOR RESISTANCE

29. Shearing.—As the shearing strength of a pin depends on its cross-section, the method of calculating this value is the same as for rivets, and the maximum shear at any section on the pin is equal to the square root of the sum of the squares of the horizontal and vertical shears at that point.

The shearing values of pins of different diameters are given in Table XV, the allowable stresses given in the previous article being used.

TABLE XV

SHEARING VALUES OF PINS

Shearing value = area of pin × shearing stress per square inch

Diameter of Pin Inches	Area of Pin Square Inches	Shearing Value at 1,000 Pounds per Square Inch Pounds	Shearing Value at 7,500 Pounds per Square Inch Pounds	Shearing Value at 9,000 Pounds per Square Inch Pounds	Shearing Value at 11,250 Pounds per Square Inch Pounds
$1\frac{7}{16}$	1.62	1,620	12,150	14,600	18,200
$1\frac{11}{16}$	2.24	2,240	16,800	20,200	25,200
$1\frac{15}{16}$	2.95	2,950	22,100	26,550	33,200
$2\frac{3}{16}$	3.76	3,760	28,200	33,800	42,300
$2\frac{7}{16}$	4.67	4,670	35,000	42,000	52,500
$2\frac{11}{16}$	5.67	5,670	42,500	51,000	63,800
$2\frac{15}{16}$	6.78	6,780	50,850	61,000	76,300
$3\frac{3}{16}$	7.98	7,980	59,850	71,800	89,800
$3\frac{7}{16}$	9.28	9,280	69,600	83,500	104,400
$3\frac{11}{16}$	10.68	10,680	80,100	96,100	120,150
$3\frac{15}{16}$	12.18	12,180	91,350	109,600	137,000
$4\frac{3}{8}$	15.03	15,030	112,700	135,300	169,100
$4\frac{5}{8}$	16.80	16,800	126,000	151,200	189,000
$4\frac{7}{8}$	18.66	18,660	140,000	167,900	209,900
$5\frac{3}{8}$	22.69	22,690	170,150	204,200	255,200
$5\frac{5}{8}$	24.85	24,850	187,000	224,000	280,000
$5\frac{7}{8}$	27.11	27,110	203,000	244,000	305,000
6	28.27	28,270	212,000	254,000	318,000
$6\frac{1}{4}$	30.68	30,680	230,000	276,000	345,000
$6\frac{1}{2}$	33.18	33,180	249,000	299,000	373,000
$6\frac{3}{4}$	35.79	35,790	268,000	322,000	403,000
7	38.48	38,480	289,000	346,000	433,000
$7\frac{1}{2}$	44.18	44,180	331,000	398,000	497,000
8	50.27	50,270	377,000	452,000	568,000
$8\frac{1}{2}$	56.75	56,750	426,000	511,000	638,000
9	63.62	63,620	477,000	573,000	716,000
$9\frac{1}{2}$	70.88	70,880	532,000	638,000	797,000
10	78.54	78,540	589,000	707,000	884,000
11	95.03	95,030	712,000	855,000	1,069,000
12	113.10	113,100	848,000	1,018,000	1,272,000

30. Bending.—Pins, when used to connect the several members of a structure, besides being subjected to shearing in the same manner as rivets, may be required to resist heavy bending stresses; they may then be regarded as solid cylindrical beams and calculated to resist the greatest bending moment that may come on them.

Take, for example, the pin in Fig. 25, which connects the three tension bars *a*, *a*, and *c*; the pull of the two bars *a*, *a*,

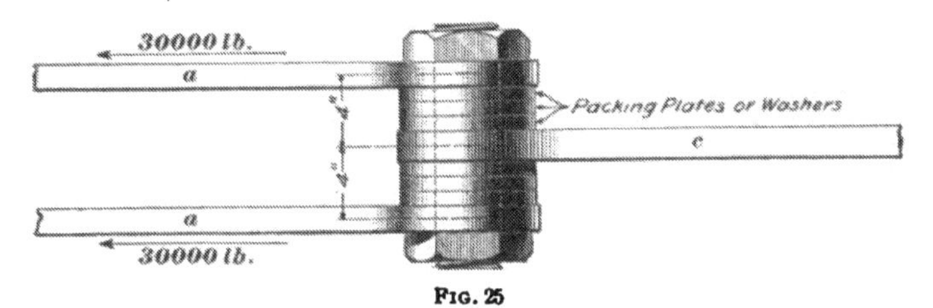

FIG. 25

both acting in the same direction, is transmitted to the bar *c* by means of the pin. The stress on each bar *a* is 30,000 pounds, consequently the stress on the bar *c* must be 60,000 pounds. Assuming a maximum unit stress of 15,000 pounds per square inch, it is desired to find what diameter of pin is required to resist the bending moment produced by the stresses exerted on it.

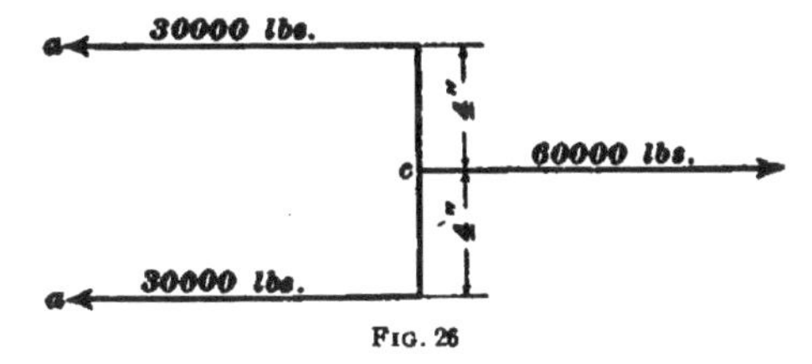

FIG. 26

In calculating the bending moment on a pin, the forces acting on it through the several members are considered as being applied at the center of the bearings. In Fig. 25, the distance between the centers of the bearings of the members is 4 inches, and by referring to the diagram, Fig. 26, it is seen that the greatest bending moment, which is at *c*, is equal to $30,000 \times 4 = 120,000$ inch-pounds.

Having found the greatest bending moment on the pin, it is necessary to determine its diameter, in order that its resisting moment may equal the moment of the bending stresses.

In Table I in *Properties of Sections*, the section modulus of a circular section is found to be

$$S = \frac{\pi d^3}{32} = .098 d^3$$

the allowable unit stress on the material is $s = 15{,}000$ pounds per square inch, and the bending moment is $M = 120{,}000$ inch-pounds; substituting these values in the formula $M = sS$, gives $120{,}000 = 15{,}000 \times .098 d^3$, from which

$$d^3 = \frac{120{,}000}{15{,}000 \times .098} = 81.63$$

The diameter of the pin is, therefore, $d = \sqrt[3]{81.63} = 4.338 = 4\frac{3}{8}$ inches, nearly.

31. Table of Maximum Bending Moments on Pins. To avoid the necessity of calculating the resisting moment of pins, Table XVI, Maximum Bending Moments on Pins, will be found convenient. This table gives the resisting moments of pins from $1\frac{7}{16}$ to 12 inches in diameter, calculated for allowable fiber stresses of 1,000, 15,000, 18,000, and 22,500 pounds per square inch.

The values for a stress of 1,000 pounds per square inch are given so that the values for other fiber stresses may be determined. For instance, it is desired to find the bending moment on a $4\frac{3}{8}$-inch pin for a fiber stress of 20,000 pounds per square inch. The bending moment for a fiber stress of 18,000 pounds per square inch is 147,960 and for 1,000 pounds it is 8,220; then the bending moment, allowing a fiber stress of 20,000 pounds per square inch, is $147{,}960 + 8{,}220 \times 2 = 164{,}400$ inch-pounds.

32. Resultant Moment of Several Stresses.—When a pin is used at a joint at which several members, extending in different directions, meet, as shown in Fig. 27, it is necessary to combine the stresses so as to find the resultant that

TABLE XVI

MAXIMUM BENDING MOMENTS ON PINS

Bending moment = diameter³ × .0982 × allowable fiber stress per square inch

Diameter of Pin Inches	Area of Pin Square Inches	Moments, in Inch-Pounds, for Fiber Stresses of				Diameter of Pin Inches
		1,000 Pounds per Square Inch	15,000 Pounds per Square Inch	18,000 Pounds per Square Inch	22,500 Pounds per Square Inch	
1 7/16	1.62	291	4,370	5,240	6,550	1 7/16
1 11/16	2.24	471	7,070	8,480	10,610	1 11/16
1 15/16	2.95	713	10,700	12,840	16,050	1 15/16
2 3/16	3.76	1,026	15,400	18,480	23,110	2 3/16
2 7/16	4.67	1,420	21,310	25,580	31,970	2 7/16
2 11/16	5.67	1,904	28,570	34,280	42,850	2 11/16
2 15/16	6.78	2,487	37,310	44,770	55,960	2 15/16
3 3/16	7.98	3,178	47,670	57,200	71,500	3 3/16
3 7/16	9.28	3,986	59,790	71,750	89,680	3 7/16
3 11/16	10.68	4,920	73,810	88,570	110,710	3 11/16
3 15/16	12.18	5,990	89,860	107,830	134,790	3 15/16
4 3/8	15.03	8,220	123,300	147,960	185,000	4 3/8
4 5/8	16.80	9,713	145,700	174,800	218,500	4 5/8
4 7/8	18.66	11,373	170,600	204,700	255,900	4 7/8
5 3/8	22.69	15,246	228,700	274,400	343,000	5 3/8
5 5/8	24.85	17,473	262,100	314,500	393,100	5 5/8
5 7/8	27.11	19,906	298,600	358,300	447,900	5 7/8
6	28.27	21,206	318,100	381,700	477,100	6
6 1/4	30.68	23,966	359,500	431,400	539,300	6 1/4
6 1/2	33.18	26,960	404,400	485,300	606,600	6 1/2
6 3/4	35.79	30,193	452,900	543,500	679,400	6 3/4
7	38.48	33,673	505,100	606,100	757,700	7
7 1/2	44.18	41,420	621,300	745,500	931,900	7 1/2
8	50.27	50,266	754,000	904,800	1,131,000	8
8 1/2	56.75	60,293	904,400	1,085,200	1,356,600	8 1/2
9	63.62	71,566	1,073,500	1,288,200	1,610,300	9
9 1/2	70.88	84,173	1,262,600	1,515,100	1,893,900	9 1/2
10	78.54	98,173	1,472,600	1,767,100	2,208,900	10
11	95.03	130,673	1,960,100	2,352,100	2,940,100	11
12	113.10	169,646	2,544,700	3,053,600	3,817,000	12

gives the greatest bending moment. This is conveniently done by first resolving the stresses on each member into vertical and horizontal components, and calculating the bending moments produced in each of these directions by all the corresponding components. The maximum bending moment is then given by the resultant of these two bending moments.

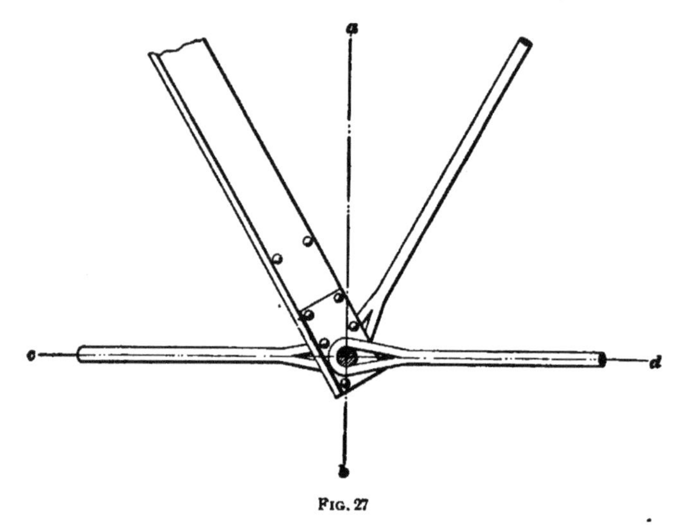

FIG. 27

The details of the method for finding the maximum bending stress on a pin, will be made clear by a study of the following illustrative example:

Fig. 28 shows one of the lower joints of a roof truss. At this joint there are four sets of members, two of which act in a horizontal, while one acts in a vertical direction; since they already act in the directions of the required components, these forces need not be resolved. However, there is one inclined member in which there is a compressive stress of 40,000 pounds, which stress must be resolved into its vertical and horizontal components. Draw the line *a b* parallel to the strut and of such a length as to represent the magnitude of the stress. From *a*, draw the horizontal line *a c* intersecting the vertical line at the point *c*. The direction of the forces around the triangle is shown by the arrows. On measuring the line *a c*, the horizontal component of the stress in *a b* is found to be 20,000 pounds, while the vertical component of the stress is found to be 34,650 pounds.

Having determined these components, a diagram showing all the horizontal stresses acting on the pin and tending to

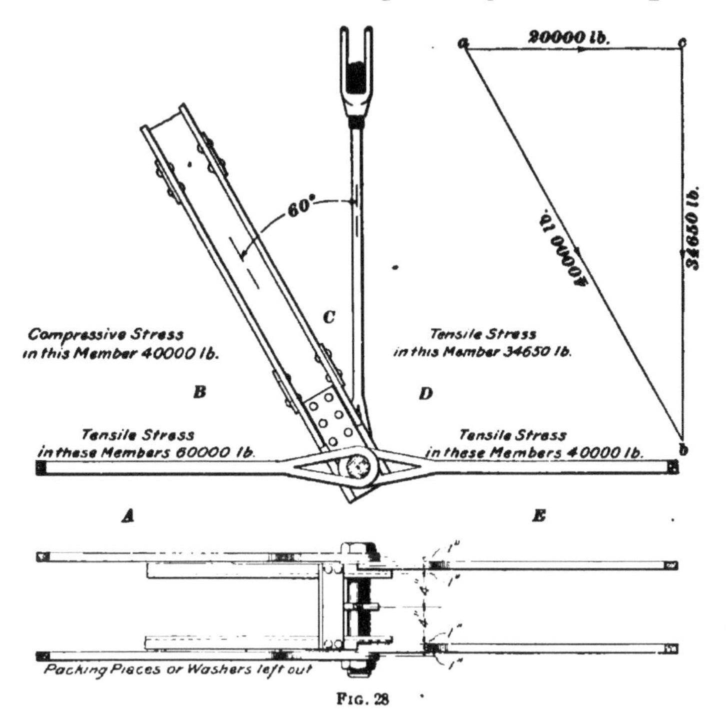

Fig. 28

bend it, and also another showing all the vertical forces, should be drawn as illustrated in Fig. 29 (*a*) and (*b*), the distance from center to center of the members being taken from the detail plan of the joint, Fig. 28. It must always be remembered that, in accordance with the principles of equilibrium, the sum of the resultants of the forces acting on the pin in any one direction must equal the sum of all the resultants acting in the opposite direction; otherwise, the pin will move in the direction of the greater sum, and the structure will fail. Thus, from Fig. 28, it is seen that the vertical component of the stress in BC acts in an opposite direction to the stress in the member CD, while the horizontal component acts in opposition to the stress in the member AB and in the same direction as the stress in DE. This makes the algebraic sum of all the components in either the horizontal or vertical direction equal to zero, and fulfils the condition of equilibrium.

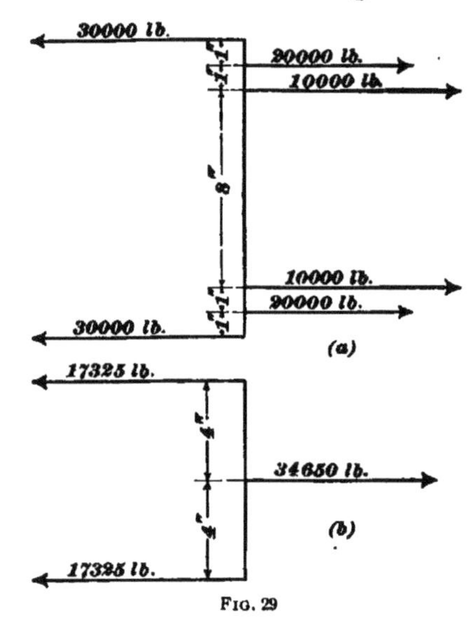

Fig. 29

The resultant of the vertical and horizontal bending moments may also be calculated by the rule given for finding the length of the hypotenuse of a right-triangle; for example, in this case the lengths of the sides are represented by the horizontal bending moment of 40,000 inch-pounds, and the vertical bending moment of 69,300 inch-pounds; the resultant bending moment is, therefore, $\sqrt{40,000^2 + 69,300^2} = 80,015$ inch-pounds.

In order to determine the required size of pin for this joint, assume a safe fiber stress of 15,000 pounds per square inch; then, by referring to Table XVI, it is seen that a pin $3\frac{15}{16}$ inches in diameter, under a fiber stress of 15,000 pounds per square inch, has a resisting moment of 89,860 inch-pounds, which is slightly greater than the value required by the conditions.

The pin must also be proportioned for the bearing value, as explained in Art. **34**.

33. The Graphic Method of Obtaining Bending Moment.—This method will be explained by applying it to the pin-connected joint shown in Fig. 28. A diagrammatic representation of the pin with the loads upon it is shown in isometric perspective in Fig. 30 (*a*), *a b* being laid off at an angle of 45° with the horizontal. The distances between

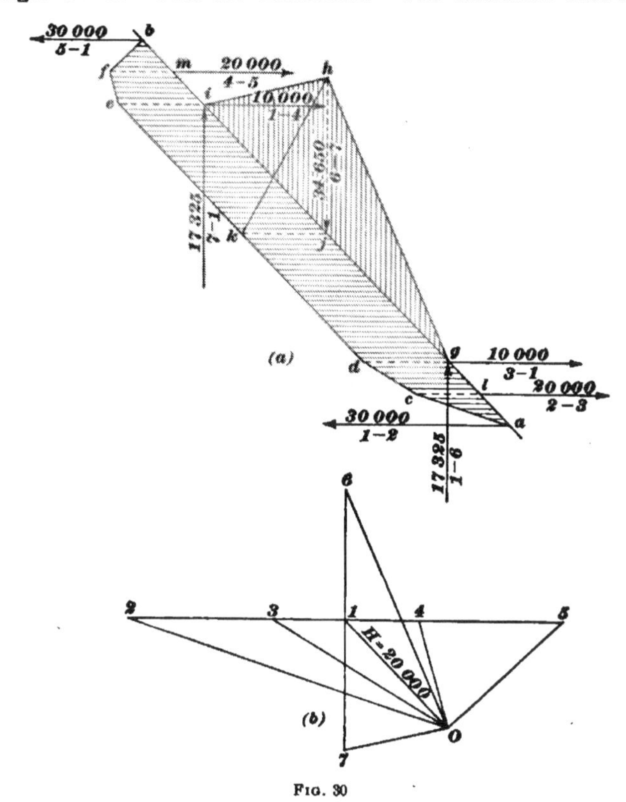

Fig. 30

loads are laid off on *a b* to the scale ¼ inch = 1 inch. In (*b*), the horizontal load line is laid off from *1* to *2*, *2* to *3*, *3* to *1*, *1* to *4*, *4* to *5*, and *5* to *1*, the scale being $\frac{1}{80}$ inch = 800 pounds. The vertical load line is laid off from *1* to *6*, *6* to *7*, and from *7* to *6*. Take a line *H*, located in a horizontal plane and perpendicular to both the vertical and horizontal load lines; this figure is also shown in isometric perspective, the line *H* being drawn at an angle of 45° with the horizontal line *2–5*. Lay off, to scale, any amount, in this case 20,000 pounds, on the line *H* and connect the point *0* thus located with the points *2*, *3*, *4*, *5*, *6*, and *7* on the load lines. Then *6–1–7–0* represents the vertical force polygon and *2–3–1–4–5–0*, the horizontal force polygon.

From *a*, the point of application of the first load, lay off *a c* in (*a*) parallel to *2–0* in (*b*) until it intersects the continuation of the second load line at the point *c*; from *c* draw *c d* parallel to *3–0* until it intersects the line of action of the 10,000-pound load, at the point *d*. Since the bending moment due to the horizontal forces remains the same between the lines of action of the two 10,000-pound loads, the width of the polygon representing the bending moment is the same between these two loads, consequently *d e* is laid off parallel to *g i*. *e f* and *f b* are drawn parallel to *4–0* and *5–0*, respectively. The horizontal bending moment at any point may be found by scaling the width of the polygon *a c d e f b* on a horizontal line at the point in question, and multiplying by *H*. It may be readily observed that the greatest horizontal bending moment is at any point between *g* and *i*.

The polygon for the vertical bending moment is obtained in a similar manner, *i h* and *h g* being drawn parallel to *7–0* and *6–0*, respectively. The maximum vertical bending moment occurs at the point *j*, or the center of the pin, and it is evident that the greatest resultant moment is at this point. It is equal to the resultant of *h j* and *k j*, or *h k*, multiplied by *H*. Scaling *h k*, it is found to equal 4 inches, and 20,000 × 4 = 80,000 inch-pounds. This result is practically correct, for the amount obtained by computation, in Art. **32**, was 80,015 inch-pounds.

34. Bearing Value.—The bearing or crushing value of a pin depends on its bearing area on the bars or plates composing the joint. This area is considered as the product of the diameter of the pin and the thickness of the plate.

TABLE XVII

BEARING VALUES OF PIN PLATES FOR 1 INCH THICKNESS OF PLATE

Bearing value = diameter of pin × thickness of plate × stress per square inch

Diameter of Pin Inches	Bearing Value at 1,000 Pounds per Square Inch Pounds	Bearing Value at 12,000 Pounds per Square Inch Pounds	Bearing Value at 15,000 Pounds per Square Inch Pounds	Bearing Value at 18,000 Pounds per Square Inch Pounds	Diameter of Pin Inches
$1\frac{7}{16}$	1,437.5	17,200	21,600	25,900	$1\frac{7}{16}$
$1\frac{11}{16}$	1,687.5	20,200	25,300	30,400	$1\frac{11}{16}$
$1\frac{15}{16}$	1,937.5	23,200	29,100	34,900	$1\frac{15}{16}$
$2\frac{3}{16}$	2,187.5	26,200	32,800	39,400	$2\frac{3}{16}$
$2\frac{7}{16}$	2,437.5	29,200	36,600	43,900	$2\frac{7}{16}$
$2\frac{11}{16}$	2,687.5	32,200	40,300	48,400	$2\frac{11}{16}$
$2\frac{15}{16}$	2,937.5	35,200	44,100	52,900	$2\frac{15}{16}$
$3\frac{3}{16}$	3,187.5	38,200	47,800	57,400	$3\frac{3}{16}$
$3\frac{7}{16}$	3,437.5	41,200	51,600	61,900	$3\frac{7}{16}$
$3\frac{11}{16}$	3,687.5	44,200	55,300	66,400	$3\frac{11}{16}$
$3\frac{15}{16}$	3,937.5	47,200	59,100	70,900	$3\frac{15}{16}$
$4\frac{3}{8}$	4,375.0	52,500	65,600	78,750	$4\frac{3}{8}$
$4\frac{5}{8}$	4,625.0	55,500	69,400	83,300	$4\frac{5}{8}$
$4\frac{7}{8}$	4,875.0	58,500	73,100	87,750	$4\frac{7}{8}$
$5\frac{3}{8}$	5,375.0	64,500	80,600	96,750	$5\frac{3}{8}$
$5\frac{5}{8}$	5,625.0	67,500	84,400	101,250	$5\frac{5}{8}$
$5\frac{7}{8}$	5,875.0	70,500	88,100	105,750	$5\frac{7}{8}$
6	6,000.0	72,000	90,000	108,000	6
$6\frac{1}{4}$	6,250.0	75,000	93,800	112,500	$6\frac{1}{4}$
$6\frac{1}{2}$	6,500.0	78,000	97,500	117,000	$6\frac{1}{2}$
$6\frac{3}{4}$	6,750.0	81,000	101,300	121,500	$6\frac{3}{4}$
7	7,000.0	84,000	105,000	126,000	7
$7\frac{1}{2}$	7,500.0	90,000	112,500	135,000	$7\frac{1}{2}$
8	8,000.0	96,000	120,000	144,000	8
$8\frac{1}{2}$	8,500.0	102,000	127,500	153,000	$8\frac{1}{2}$
9	9,000.0	108,000	135,000	162,000	9
10	10,000.0	120,000	150,000	180,000	10
11	11,000.0	132,000	165,000	198,000	11
12	12,000.0	144,000	180,000	216,000	12

Table XVII gives the bearing values of pins on a plate 1 inch thick, using allowable bearing values of 1,000, 12,000, 15,000, and 18,000 pounds per square inch. The use of the column giving bearing values at 1,000 pounds per square inch is similar to that of the corresponding column in Table XVI, as previously explained. The application of this table may be illustrated by the following example. In Fig. 28, assume the thickness of the horizontal bars resisting 60,000 pounds to be $1\frac{1}{4}$ inches and those having a stress of 40,000 pounds to be $\frac{3}{4}$ inch. The bearing value for a 1-inch plate will be one and one-third times as great as that for a $\frac{3}{4}$-inch plate; hence, $(40{,}000 \div 2) \times 1\frac{1}{3} = 26{,}667$ pounds represents the bearing value for a 1-inch plate corresponding to 20,000 pounds for a $\frac{3}{4}$-inch plate. From Table XVII, it is seen that a $1\frac{15}{16}$-inch pin, having a safe unit fiber stress of 15,000 pounds, is sufficient to provide the required bearing, and as a $3\frac{15}{16}$-inch pin is to be used, it will give ample bearing value. For a $1\frac{1}{4}$-inch plate, the bearing value is one and one-fourth times as much as for a 1-inch plate, and therefore, the value corresponding to the bearing that would be required for a 1-inch plate is $(60{,}000 \div 2) \div 1\frac{1}{4} = 24{,}000$ pounds. Assuming a fiber stress of 15,000 pounds, the size of pin required to provide the bearing is $1\frac{11}{16}$ inches. In the same manner, the size of pin required to give the bearing value for each plate may be determined.

35. As an example in the design of a pin, the joint shown in Fig. 31 will be considered. It is first necessary to determine the horizontal and vertical components of the oblique forces. These components may be obtained graphically or by computation, but in this case the latter method will be followed.

The member *b* in the frame diagram of the joint shown in Fig. 32 (*a*), is at an angle of 15° with the horizontal. The vertical component of this force will be equal to the product of the sine of 15° and the stress in the member, while the horizontal component will equal the stress multiplied by

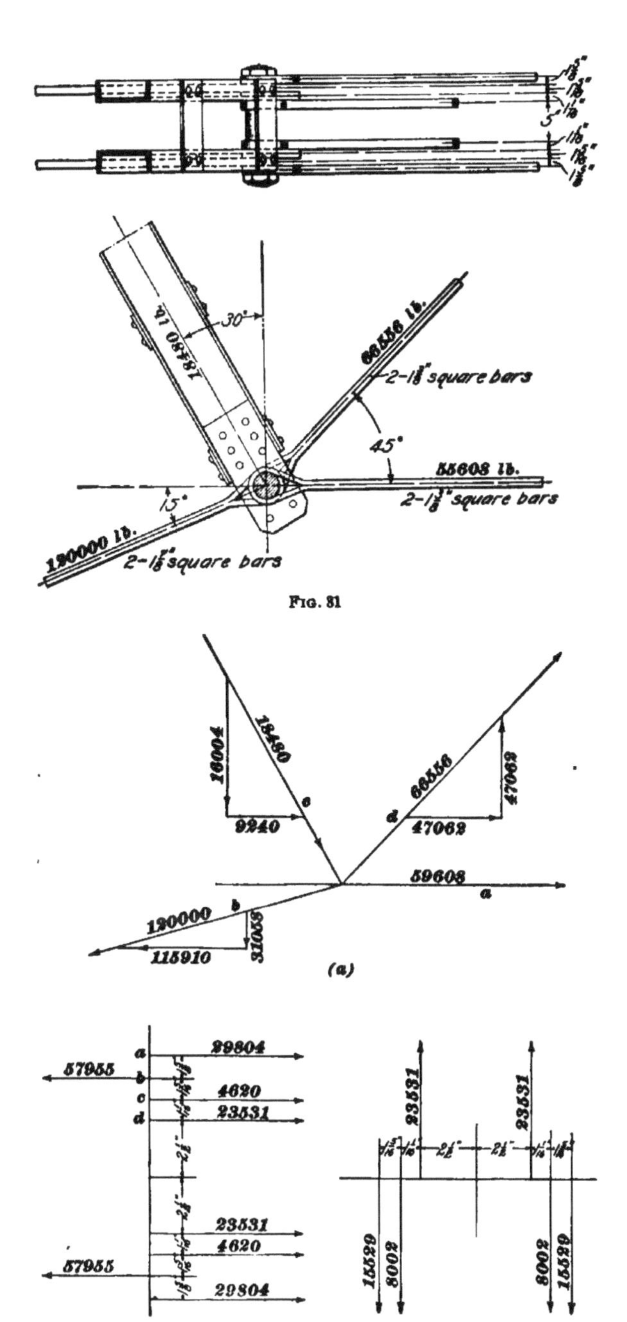

18480 lb.
30°
66556 lb.
2-1⅞" square bars
45°
59608 lb.
2-1$\frac{3}{16}$" square bars
15°
120000 lb.
2-1⅞" square bars
16004
18480
9240
c
d
66556
47062
47062
59608
a
b
120000
115910
31058
(a)
29804
57955
a
b
c
d
4620
23531
23531
4620
57955
29804
(b)
23531
23531
15529
8002
8002
15529
(c)

Fig. 31

Fig. 32

the cosine of 15°. From a table of natural sines and cosines, the following values are obtained:

$$\sin 15° = .258819$$
$$\cos 15° = .965926$$

Then the vertical component of the force is 120,000 × .258819 = 31,058 pounds, and the horizontal component equals 120,000 × .965926 = 115,910 pounds. The values for the other angles, obtained from the table, are:

$$\sin 30° = .500000$$
$$\cos 30° = .866025$$
$$\sin 45° = .707107$$
$$\cos 45° = .707107$$

The components of the oblique force of 18,480 pounds are:

Vertical component = 18,480 × .866025 = 16,004 pounds
Horizontal component = 18,480 × .500000 = 9,240 pounds

For the oblique force of 66,556 pounds the horizontal and vertical components are equal, their amount being 66,556 × .707107 = 47,062 pounds.

The action of the horizontal forces and components on the pin is illustrated in Fig. 32 (*b*), and from this diagram the horizontal bending moment may be computed. The bending moment about the point *b* due to the horizontal forces is equal to $29,804 \times 1\frac{5}{8} = 48,432$ inch-pounds; about the point *c*, the bending moment is $(29,804 \times 2\frac{15}{16}) - (57,955 \times 1\frac{5}{16}) = 11,483$ inch-pounds; and about the point *d*, the horizontal bending moment is $(57,955 \times 2\frac{3}{8}) - [(29,804 \times 4) + (4,620 \times 1\frac{1}{16})] = 13,518$ inch-pounds.

There is no vertical bending moment about the points *a* and *b*, as will be observed from the diagram at (*c*), but about the points *c* and *d* it is as follows:

About *c*, $15,529 \times 1\frac{5}{16} = 20,382$ inch-pounds
About *d*, $(15,529 \times 2\frac{3}{8}) + (8,002 \times 1\frac{1}{16}) = 45,383$ inch-pounds

It is readily determined by observation that the maximum bending moment will be the horizontal moment of

48,432 inch-pounds at the point *b*, or the resultant of the horizontal and vertical moments at the point *d*. This resultant equals $\sqrt{13,518^2 + 45,383^2} = 47,353$ inch-pounds; consequently, this is the maximum bending moment and it occurs at the point *d*. The size of pin required to resist this bending moment may be determined by the method explained in Art. **30**, or from Table XVI. Considering a safe fiber stress of 22,500 pounds per square inch, it is found from the table that a $2\frac{15}{16}$-inch pin must be used to supply the required resisting moment. It is readily observed that the shear between the points *a* and *b* is equal to 29,804 pounds; between *b* and *c* it is the resultant of the horizontal shear of $57,955 - 29,804 = 28,151$ pounds and the vertical shear of 15,529 pounds. Between *c* and *d* the vertical and horizontal shears are each equal to 23,531 pounds and their resultant, or $\sqrt{2 \times 23,531^2} = 33,278$ pounds, the maximum shear on the pin. From Table XV, it is found that a $2\frac{3}{16}$-inch pin having a safe unit shearing strength of 11,250 pounds will be sufficient to resist the shear.

It is next necessary to consider the bearing value of the pin on the different members of the joint. In the member carrying 120,000 pounds, there are two $1\frac{7}{8}$-inch bars, each having 60,000 pounds stress. The bearing value of such a bar will be equivalent to that of a 1-inch bar having a stress of $60,000 \div 1\frac{7}{8} = 32,000$ pounds. Therefore, referring to Table XVII, it is seen that a $1\frac{15}{16}$-inch pin having a safe unit bearing value of 18,000 pounds will supply the required resistance. The size of pin necessary to give the bearing value required by the other members of the joint may be computed, but it will be found that a $1\frac{15}{16}$ inch pin is sufficient in each case. From these calculations for bending moment, shear, and bearing value, it is found that a $2\frac{15}{16}$-inch pin will be required for this joint.

DETAILS OF DESIGN

36. Position of Pins.—Pins are not always placed on the center lines of compression members or those inclined to the horizontal less than 45°, through which they pass, but at a

TABLE XVIII

STANDARD PINS AND NUTS

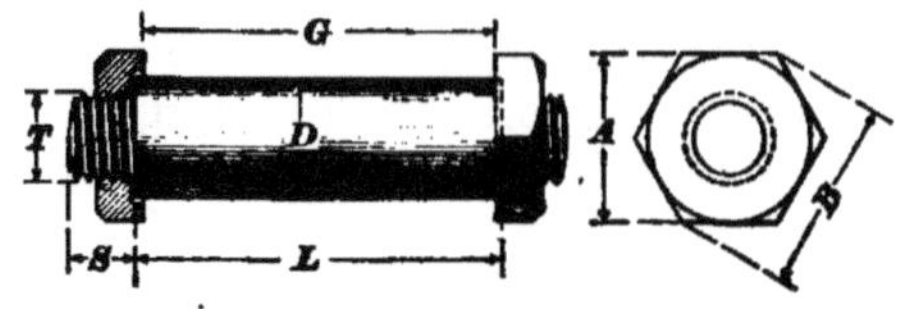

D Diameter of Pin Inches	*T* Diameter of Thread Inches	*S* Length of Thread Inches	*A* Short Diameter of Nut Inches	*B* Long Diameter of Nut Inches	Weight of One Nut Pounds
1 3/16	1	1 1/2	1 3/4	2	
1 7/16	1	1 1/2	1 3/4	2	
1 11/16	1 1/2	1 1/2	3 1/4	3 3/4	1.5
1 15/16	1 1/2	1 1/2	3 1/4	3 3/4	1.5
2 3/16	1 1/2	1 1/2	3 1/4	3 3/4	1.5
2 7/16	1 3/4	1 1/2	3 1/4	3 3/4	1.5
2 11/16	2	1 1/2	3 3/4	4 1/4	2.5
2 15/16	2 1/4	1 1/2	4 1/2	5 1/4	3.0
3 3/16	2 1/2	1 1/2	4 1/2	5 1/4	2.8
3 7/16	2 1/2	1 1/2	4 1/2	5 1/4	2.8
3 11/16	2 3/4	1 1/2	4 3/4	5 1/2	3.0
3 15/16	3	1 1/2	4 3/4	5 1/2	3.0
4 3/8	3 1/2	1 1/2	5 1/2	6 1/4	3.8
4 5/8	3 1/2	1 1/2	5 1/2	6 1/4	3.8
4 7/8	4	1 1/2	6	7	6.7
5 3/8	4	2	6	7	6.7
5 7/8	4	2	7	8	9.1
7	5	2 1/4	8	9 1/4	12.0
8	6	2 1/4	10 1/2	12	22.8
9	7	2 1/4	10 1/2	12	18.8

TABLE XIX

STANDARD SCREW THREADS AND NUTS

Diameter of Rod or Bolt	Threads per Inch	Diameter at Root of Thread	Area of Rod or Bolt at Root of Thread	Short Diameter of Nuts	Long Diameter, Hexagon Nuts	Long Diameter, Square Nuts	Thickness of Nuts
1/4	20	.185	.026	1/2	37/64	7/10	1/4
5/16	18	.240	.045	19/32	11/16	10/12	5/16
3/8	16	.294	.067	11/16	51/64	63/64	3/8
7/16	14	.344	.092	25/32	7/8	1 7/64	7/16
1/2	13	.400	.125	7/8	1	1 15/64	1/2
9/16	12	.454	.161	31/32	1 1/8	1 23/64	9/16
5/8	11	.507	.201	1 1/16	1 7/32	1 1/2	5/8
3/4	10	.620	.301	1 1/4	1 7/16	1 49/64	3/4
7/8	9	.731	.419	1 7/16	1 21/32	2 1/32	7/8
1	8	.837	.550	1 5/8	1 7/8	2 19/64	1
1 1/8	7	.940	.693	1 13/16	2 3/32	2 9/16	1 1/8
1 1/4	7	1.065	.890	2	2 5/16	2 53/64	1 1/4
1 3/8	6	1.160	1.056	2 3/16	2 17/32	3 3/32	1 3/8
1 1/2	6	1.284	1.294	2 3/8	2 3/4	3 23/64	1 1/2
1 5/8	5 1/2	1.389	1.515	2 9/16	2 31/32	3 5/8	1 5/8
1 3/4	5	1.491	1.746	2 3/4	3 3/16	3 57/64	1 3/4
1 7/8	5	1.616	2.051	2 15/16	3 13/32	4 5/32	1 7/8
2	4 1/2	1.712	2.301	3 1/8	3 5/8	4 27/64	2
2 1/4	4 1/2	1.962	3.023	3 1/2	4 1/64	4 61/64	2 1/4
2 1/2	4	2.176	3.718	3 7/8	4 1/2	5 31/64	2 1/2
2 3/4	4	2.426	4.622	4 1/4	4 29/32	6	2 3/4
3	3 1/2	2.629	5.428	4 5/8	5 3/8	6 17/32	3
3 1/4	3 1/2	2.879	6.509	5	5 13/16	7 1/16	3 1/4
3 1/2	3 1/4	3.100	7.547	5 3/8	6 7/64	7 39/64	3 1/2
3 3/4	3	3.318	8.641	5 3/4	6 21/32	8 1/8	3 3/4
4	3	3.567	9.993	6 1/8	7 3/32	8 41/64	4

distance below the center, so that the stress acting along the neutral axis will produce a moment that will counteract, or neutralize, the moment due to the weight of the member itself.

The distance from the edge of the pinhole to the end of the plate must be great enough to prevent splitting along the line of the least net section.

37. Pin Plates.—Pin plates should be used at all pinholes in built-up members and should extend at least 6 inches within the member, to provide for at least two transverse rows of rivets at that place. The net section through the pinhole should be 40 per cent. more than the net section through the body of the member, and the net section outside of the pinhole along the center line of stress should equal the net section of the body.

38. Packing of Joints.—The members of a pin-connected joint should be packed so as to produce the least bending moment on the pin and should be arranged symmetrically on both sides of the center line. Sufficient clearance should be allowed for inaccuracies in manufacture. Interior vacant spaces must be filled with steel fillers if their omission will permit the members to move on the pin. All bars must be, as nearly as possible, parallel to the central plane of the truss.

39. Standard Sizes of Pins.—Table XVIII gives standard sizes and dimensions of pins and nuts. These sizes have been fixed on by the several steel companies in the United States and are standard in this country.

40. Standard Screw Threads and Nuts.—Table XIX gives the proportions for United States standard screw threads and nuts.

EXAMPLES FOR PRACTICE

1. One of the tension members in a structure is connected as shown in Fig. 33. The tension bars are made of structural steel with a safe tensile strength of 15,000 pounds per square inch. (*a*) What is the bearing value of the bar *c*? (*b*) What is the bearing value of the two bars *a*?

Ans. { (*a*) 104,000 lb.
(*b*) 97,500 lb.

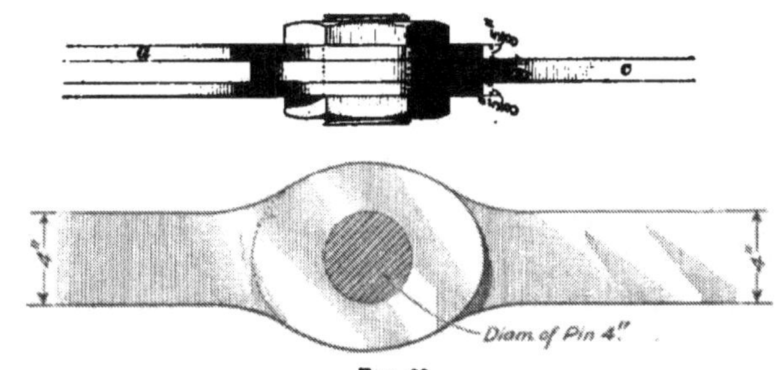

FIG. 33

FIG. 34

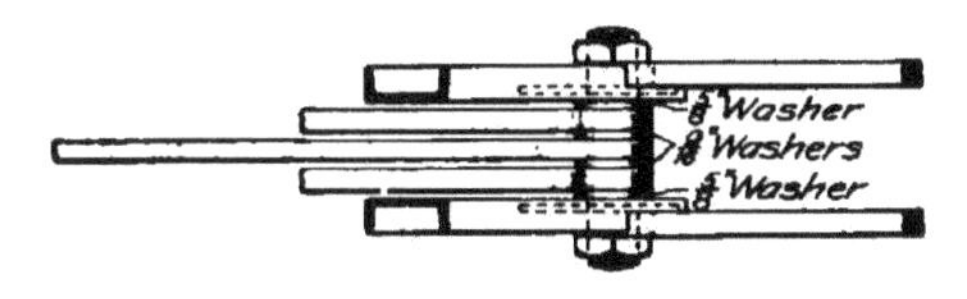

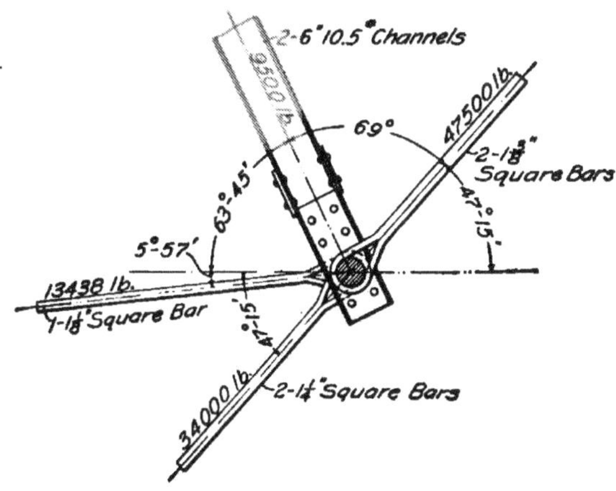

FIG. 35

2. In Fig. 34 is shown a pin connection, the pull on the tension bar *a* being 140,000 pounds. If the safe shearing strength of the material in the pin is 10,000 pounds per square inch, and the safe fiber stress in bending is 15,000 pounds per square inch: (*a*) what size of pin will be required to resist the shear? (*b*) what size will be required to resist the bending?

Ans. { (*a*) $3\frac{3}{16}$ in. in diameter
(*b*) $4\frac{3}{8}$ in. in diameter

3. (*a*) What is the maximum resultant bending moment on the pin used in the connection shown in Fig. 35? (*b*) What size pin is required for this connection when the highest unit strength values given in the tables are assumed?

Ans. { (*a*) 71,480 in.-lb.
(*b*) $3\frac{3}{16}$-in. pin

GRAPHICAL ANALYSIS OF STRESSES

(PART 1)

INTRODUCTION

1. The **graphical analysis of stresses** is the study, by means of diagrams, of the stability or equilibrium of structures and the relation between the external forces and the stresses created in the members of the frame. It is founded on the principle that any force may be designated, both as to direction and intensity, by a straight line by letting the direction of the line be identical with that of the force and adjusting its length according to an arbitrary unit adopted for the forces under consideration. The advantage existing in the use of graphical statics for the solution of stresses in framed structures is that when the principles are correctly applied no important error can exist, and though the stresses determined by this means will not be extremely accurate, they cannot be radically wrong. The approximations obtained by the diagrams give results as nearly accurate as the practical design of any member in the structure can be.

2. Definitions.—Before studying this subject the meaning of several terms should be thoroughly understood. Other terms, which require an extensive explanation, will be defined when first introduced.

A **force**, in graphical statics, is understood to be a weight or pressure applied in a certain direction at a particular point. It may be either an external or internal, or a concurrent or non-concurrent, or even a coplanar, or non-coplanar force.

An **external force** on any structure represents either a weight or a reaction; for instance, the vertical dead or snow loads on a roof truss are external forces, as are also the reactions created by the resistance of the piers or abutments beneath the ends of the roof trusses.

An **internal force** is the stress created in any member of a framed structure. Sometimes it is denominated as a strain, but this is incorrect for strain is now generally understood to mean the distortion, or amount of change of form created in a piece of material by a stress or the force that must exist before the change of form is accomplished. In a roof truss, the internal forces are the compression in the rafter members and struts, and the tension in the tie-rods.

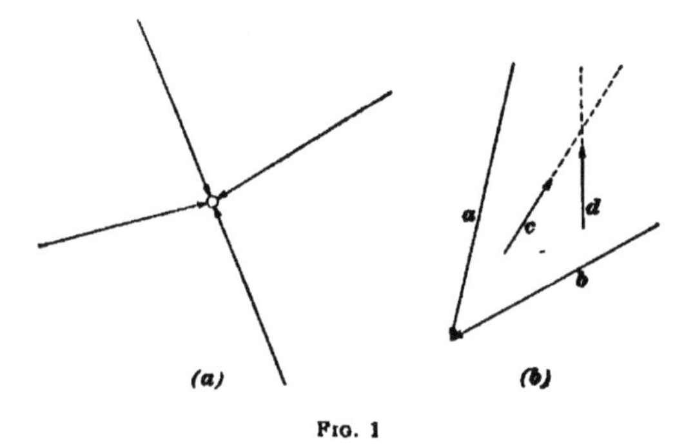

FIG. 1

Concurrent forces are those forçes that act or meet at a single point, as in Fig. 1 (*a*).

Non-concurrent forces comprise any system of forces that do not meet at a single point, as in Fig. 1 (*b*). While *a* and *b* meet at a certain point, and *c* and *d* can be extended until they meet, the four forces cannot act at one point.

Coplanar and **non-coplanar forces** are the forces that act in the same plane, and in different planes, respectively. Fig. 2 (*a*) illustrates the first force and (*b*) the second.

Equilibrium exists in any structure when there is no tendency for the structure to move; or if it is moving uniformly, there is no tendency for the rate or direction of the motion to change. In graphical statics, which term is derived from the study of the equilibrium of structures, when a body is said to be in equilibrium, it is inferred that there is no tendency for the body or structural frame to change from a state of rest to one of motion; nor from a state of motion to one of rest. A body may be in equilibrium against motion either in a lateral direction or in a rotary direction.

Equilibrium of translation is a term that has been applied to a body in equilibrium against lateral motion.

Rotary equilibrium is applied to a body that is in equilibrium about a particular point; that is, the condition that exists in a body that has no rotary tendency about a fixed point.

Complete equilibrium can only be considered to exist when there is no tendency for a body toward translatory or

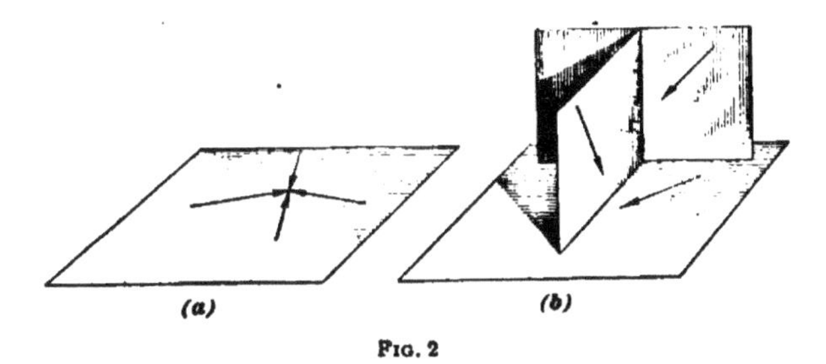

FIG. 2

rotary motion; that is, when both the conditions of translatory equilibrium and rotary equilibrium are fulfilled.

A couple is the term given to two equal and opposing forces not coincident with respect to their line of action; or, in the terms of the definitions, two equal and opposing non-concurrent forces form a couple.

The **moment** of a force about any point not located in its line of action is its tendency to rotate about that point; the amount of the moment is obtained by multiplying the intensity, or amount, of the force by its lever arm, or the perpendicular distance from the line of action of the force to the point around which it tends to rotate, this point being termed the center of moments. The moment of any force is never expressed in either units of weight or units of length, but by a combination of the two, as inch-pounds, foot-pounds, and foot-tons.

The **resultant** of any system of forces is any force that, by its direction and amount, will equal in its effect on a body or structure the effect of all the forces of the system. Any force that will destroy the action of all the forces of the system is equal and opposed to the resultant; such a force might be termed **resultant reaction.** Where the resultant of a system of forces equals zero, the system is in translatory equilibrium, while if the moment of the resultant about any point equals zero, the system is in rotary equilibrium. Where any system of forces is resolved into a couple, that is, two equal and opposing non-concurrent forces, the system cannot be in complete equilibrium and no single force will replace the two non-concurrent forces.

FUNDAMENTAL PRINCIPLES OF GRAPHICAL ANALYSIS

EFFECTS OF A FORCE

3. The **effect of a force** on a body may be compared with another force when the three following conditions are fulfilled in regard to both forces:

1. The point of application, or point at which the force acts on the body, must be known.

2. The direction of the force, or the straight line along which the force tends to move the point of application, must be known.

3. The magnitude, or value, of the force, when compared with a given standard, must be known.

The unit of magnitude of forces will be taken as 1 pound, throughout this Course, and all forces spoken of as a certain number of pounds.

4. The fundamental principles of the relations between force and motion, which were first stated by Sir Isaac Newton and are called "Newton's Three Laws of Motion," are as follows:

Law I.—*All bodies continue in a state of rest, or of uniform motion in a straight line, unless acted on by some external force that compels a change.*

Law II.—*A force acting on a body in motion or at rest, produces the same effect whether it acts alone or with other forces.*

Law III.—*To every action there is always opposed an equal and contrary reaction.*

From the first law of motion, it is inferred that a body once set in motion by any force, no matter how small, will move forever in a straight line, and always with the same velocity, unless acted on by some force that compels a change.

The deduction from the second law is that, if two or more forces act on a body, their final effect on that body will be in proportion to their magnitude and to the directions in which they act. Thus, if the wind is blowing due west, with a velocity of 50 miles per hour, and a ball is thrown due north, with the same velocity, the wind will carry the ball just as far west as the force of the throw carried it north, and the combined effect will be to cause it to move northwest. The amount of departure from due north will be proportional to the force of the wind, and independent of the velocity due to the force of the throw.

A——B

FIG. 3

The third law states that action and reaction are equal and opposite. A man cannot lift himself by his boot straps, for the reason that he presses downward with the same force that he pulls upward; the downward reaction equals the upward action, and is opposite to it.

5. A force may be represented by a line; thus, in Fig. 3, let A be the point of application of the force, let the length of the line $A\ B$ represent its magnitude, and let the arrowhead indicate the direction in which the force acts, then the line $A\ B$ fulfils the three required conditions and shows the point of application, the direction, and the intensity of the force.

THE COMPOSITION OF FORCES

6. Parallelogram of Forces.—When two forces act on a body at the same time, but at different angles, their final result may be obtained as follows:

In Fig. 4, let *A* be the common point of application of two forces, and let *A B* and *A C* represent the magnitude and direction of the forces. The final effect of the movement due to these two forces will be the same whether they act singly or together. For instance, let the line *A B* represent the distance that the force *A B* would cause the body to move; similarly, let *A C* represent the distance that the force *A C* would cause the body to move, when both forces were acting separately. The force *A B*, acting alone, would carry the body to *B*; if the force *A C* were now to act on the body, it would carry it along the line *B D*, parallel to *A C*. to a point *D*, at a distance from *B* equal to *A C*. Join *C* and *D*, then *C D* is parallel to *A B*, and *A B D C* is a parallelogram. Draw the diagonal *A D*. According to the second law of motion the body will stop at *D* whether the forces act separately or together, but if they act together, the path of the body will be along *A D*, the diagonal of the parallelogram. Moreover, the length of the line *A D* represents the magnitude of a force, which acting at *A* in the direction *A D*, would cause the body to move from *A* to *D*; in other words, *A D* measured to the same scale as *A B* and *A C*, represents the magnitude and direction of the combined effect of the two forces *A B* and *A C*, and is called the resultant. Suppose that the scale used was 50 pounds to the inch, then, if $A\ B = 50$ pounds, and $A\ C = 62\frac{1}{2}$ pounds, the length of *A B* would be $\frac{50}{50} = 1$ inch, and the length of *A C* would be $\frac{62.5}{50} = 1\frac{1}{4}$ inches. If *A D*, or the resultant, measures $1\frac{3}{4}$ inches, its magnitude would be $1\frac{3}{4} \times 50 = 87\frac{1}{2}$ pounds. Therefore, a

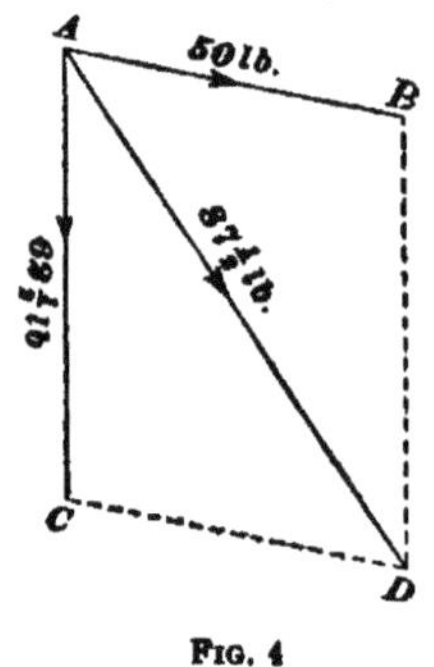

FIG. 4

force of $87\frac{1}{2}$ pounds, acting on a body at A, in the direction $A\ D$, will produce the same result as the combined effects of a force of 50 pounds acting in the direction $A\ B$, and a force of $62\frac{1}{2}$ pounds acting in the direction $A\ C$.

7. This method of finding the resulting action of two forces acting on a body at a common point, is correct for forces of any direction and magnitude. Hence, to find the resultant of two forces when their common point of application, their direction, and magnitudes are known:

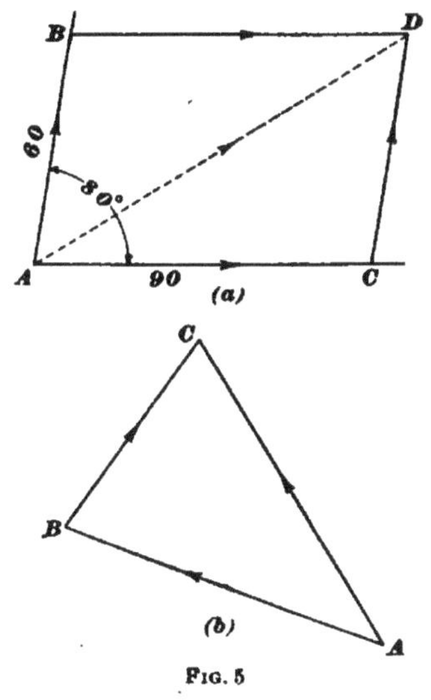

Fig. 5

Rule.—*Through an assumed point, draw two lines parallel with the direction of the two forces. With any scale, measure from the point of intersection, in the direction of the forces, distances corresponding to the magnitudes of the respective forces, and from the points thus obtained complete the parallelogram. Draw the diagonal of the parallelogram from the point of intersection of the two forces; this diagonal will be the resultant, and its direction will be away from the point of intersection of the two forces. Its magnitude must be measured with the same scale that was used to lay off the two forces.*

This method is called the **graphical method of the parallelogram of forces.**

Example.—If two forces act on a body at a common point, both acting away from the body, and the angle between them is 80°, what is the value of the resultant, the magnitude of the two forces being 60 and 90 pounds, respectively.

Solution.—Draw two indefinite lines having an angle of 80° between them. With any convenient scale, say 10 lb. to the inch, measure off $A\ B = 60 \div 10 = 6$ in., and $A\ C = 90 \div 10 = 9$ in., as shown in Fig. 5 (*a*). Through B draw $B\ D$ parallel to $A\ C$, and through C

draw CD parallel to AB. Then draw AD, which will be the resultant; its direction is toward the point D, as shown by the arrow.

Measuring AD, its length is found to be 11.7 in. Hence, 11.7 × 10 = 117 lb. Ans.

8. Triangle of Forces.—The above example might also have been solved by the method called the **triangle of forces**, which is as follows: In Fig. 5 (*b*), suppose that the two forces AB and BC act separately, first from A to B, and then from B to C, in the direction of the arrows. Connect A and C; then AC is the resultant of the forces AB and BC. It will also be noticed in following the direction of the forces around the triangle, that the direction of the resultant AC is opposite to that of AB and BC. Hence, to find the resultant of two forces acting on a body at a common point, by the method of triangle of forces:

Rule.—*Draw the lines of action of the two forces as if each force acted separately, the lengths of the lines being proportional to the magnitude of the forces. Join the extremities of the two lines by a straight line, which will be the resultant; its direction will be opposite to that of the two forces.*

When the resultant is spoken of as being opposite in direction to the other forces around the polygon, it is meant that, starting from the point where the drawing of the polygon was commenced, and tracing each line in succession, the pencil will have the same general direction around the polygon as if passing around a circle, from left to right or from right to left, but the closing line or resultant must have an opposite direction; that is, the two arrowheads, the one on the resultant and the other on the last side, must point toward the intersection of the resultant and the last side.

9. Resultant of Several Forces.—When three or more forces act on a body at a given point, their resultant may be found by the following rule:

Rule.—*Find the resultant of any two forces; treat this resultant as a single force, and combine it with a third force to find a second resultant. Combine this second resultant with a fourth force, to find a third resultant, etc. After all the forces have* *been thus combined, the last resultant will be the resultant of all of the forces, both in magnitude and direction.*

EXAMPLE.—Find the resultant of all the forces acting on the point O in Fig. 6, the length of the lines being proportional to the magnitude of the forces.

SOLUTION.—Draw $O E$ parallel and equal to $A O$, and $E F$ parallel and equal to $B O$, then $O F$ is the resultant of these two forces, and its direction is from O to F, opposed to $O E$ and $E F$. Treat $O F$ as if $O E$ and $E F$ did not exist, and draw $F G$ parallel and equal to $O C$; $O G$ will be the resultant of $O F$ and $F G$; but $O F$ is the resultant of $O E$ and $E F$; hence, $O G$ is the resultant of $O E$, $E F$, and $F G$, and therefore of $A O$, $B O$, and $C O$. Likewise draw $G L$ parallel

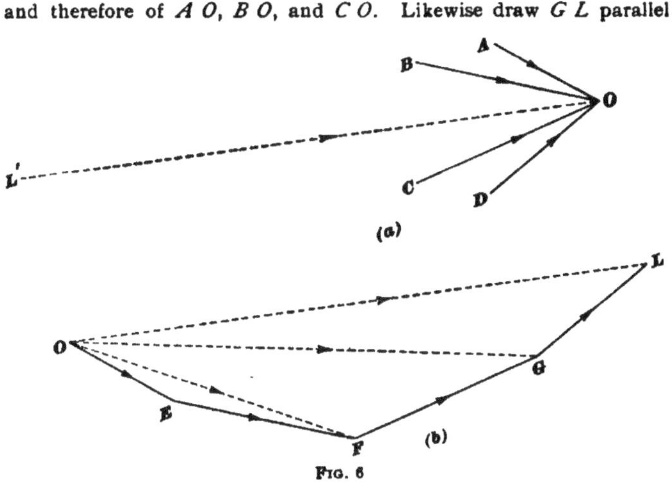

FIG. 6

and equal to $D O$. Join O and L, and $O L$ will be the resultant of all the forces $A O$, $B O$, $C O$, and $D O$ (both in magnitude and direction), acting at the point O. If $L' O$ were drawn parallel and equal to $O L$, and having the same direction, it would represent the effect produced on the body by the combined action of the forces $A O$, $B O$, $C O$, and $D O$.

10. In Fig. 6, it will be noticed that $O E$, $E F$, $F G$, $G L$, and $L O$ are sides of a polygon $O E F G L$, in which $O L$, the resultant, is the closing side, and that its direction is opposed to that of all the other sides. This fact is made use of in what is called the **method of the polygon of forces.** To find the resultant of several forces acting on a body at the same point by this method:

Rule.—*Through any point, draw a line parallel to one of the forces, and having the same direction and magnitude. At the end of this line, draw another line, parallel to, and having the same direction and magnitude as a second force; at the end of the second line, draw a line parallel and equal in magnitude and direction to a third force. Thus continue until lines have been drawn parallel and equal in magnitude and direction to all of the forces.*

The straight line joining the free ends of the first and last lines will be the closing sides of the polygon; mark it opposite in direction to that of the other forces around the polygon, and it will be the resultant of all the forces.

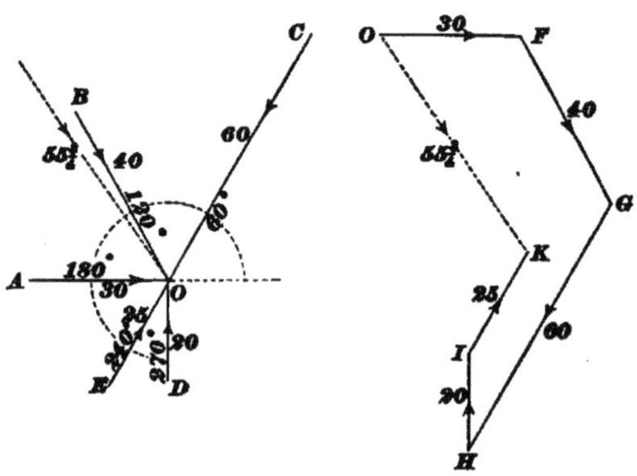

Fig. 7

Example.—If five forces act on a body at angles of 60°, 120°, 180°, 240°, and 270°, toward the same point, and their respective magnitudes are 60, 40, 30, 25, and 20 pounds, find the magnitude and direction of their resultant by the method of polygon of forces.*

Solution.—From a common point O, Fig. 7, draw the lines of action of the forces, making the given angles with a horizontal line through O, and mark them as acting toward O, by means of arrowheads, as shown. Choose some convenient scale, such that the whole figure may be drawn in a space of the required size on the drawing. Select any one of the forces, as $A\ O$, and draw $O\ F$ parallel to it, and equal in length

*All the angles in the figure are measured from a horizontal line in a direction opposite to the movement of the hands of a watch, from 1° up to 360°.

to 30 lb. on the scale. It must also act in the same direction as *A O*. At *F*, draw *F G* parallel to *B O*, and equal to 40 lb. In a similar manner, draw *G H*, *H I*, and *I K* parallel to *C O*, *D O*, and *E O*, and equal to 60, 20, and 25 lb., respectively. Join *O* and *K* by *O K*, and *O K* will be the resultant of the combined action of the five forces; its direction is opposite to that of the other forces around the polygon *O F G H I K*, and its magnitude = 55¾ lb. Ans.

11. If the resultant *O K*, Fig. 7, were to act alone on the body in the direction shown by the arrowhead with a force of 55¾ pounds, it would produce exactly the same effect as

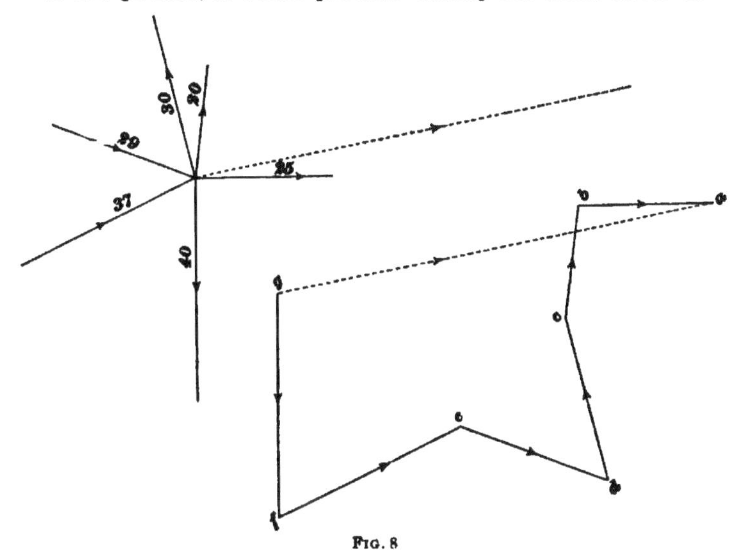

FIG. 8

the combined action of the five forces. If *O F*, *F G*, *G H*, *H I*, and *I K* represent the distances and directions that the forces would move the body, if acting separately, *O K* is the direction and distance of movement of the body when all the forces act together. It is evident, therefore, that any number of forces acting on a body at the same point, or having their lines of action pass through the same point, can be replaced by a single force (resultant), whose line of action shall pass through that point.

Heretofore, it has been assumed that the forces acted on a

single point on the surface of the body, but it will make no difference where they act, so long as the lines of action of all the forces intersect at a single point either within or without the body, only so that the resultant can be drawn through the point of intersection. If two forces act on a body in the same straight line and in the same direction, their resultant is the sum of the two forces; but if they act in opposite directions, their resultant is the difference of the two forces, and its direction is the same as that of the greater force. If they are equal and opposite, the resultant is zero, or one force just balances the other.

EXAMPLE.—Find the resultant of the forces whose lines of action pass through a single point, as shown in Fig. 8.

SOLUTION.—Take any convenient point *g*, and draw a line *g f*, parallel to one of the forces, say the one marked *40*, making it equal in length to 40 lb. on the scale, and indicate its direction by the arrowhead. Take some other force—the one marked *37* will be convenient; the line *fe* represents this force. From the point *e* draw a line parallel to some other force, say the one marked *29*, and make it equal in magnitude and direction to it. So continue with the other forces, taking care that the general direction around the polygon is not changed. The last force drawn in the figure is *a b*, representing the force marked *25*. Join the points *a* and *g*; then, *ag* is the resultant of all the forces shown in the figure. Its direction is from *g* to *a*, opposed to the general direction of the others around the polygon. It does not matter in what order the different forces are taken, the resultant will be the same in magnitude and direction, if the work is done correctly.

These various methods of finding the resultant of several forces are all grouped under one head: The Composition of Forces.

THE RESOLUTION OF FORCES

12. Since two forces can be combined to form a single resultant force, a single force may also be treated as if it were the resultant of two or more forces, whose action on a body will be the same as that of the single force. Thus, in Fig. 9, the force *O A* may be resolved into two forces, *O B* and *B A*, whose directions are opposed to *O A*. If the force *O A* acts on a body, moving or at rest on a horizontal plane, and the resolved force *O B* is vertical, and *B A* horizontal, *O B*, measured to the same scale as *O A*, is the magnitude of that part of *O A* that pushes the body downwards, while *B A* is the magnitude of that part of the force *O A*, which is exerted in pushing the body in a horizontal direction. *O B* and *B A* are called the **components** of the force *O A*, and when these components are vertical and horizontal, as in the present case, they are called the *vertical component* and the *horizontal component* of the force *O A*. These components may be drawn in any direction and the angle at their intersection is not necessarily a right angle.

13. It frequently happens that the position, magnitude, and direction of a certain force are known, and that it is desired to know the effect of the force in some direction other than that in which it acts. Thus, in Fig. 9, assume that *O A* represents, to some scale, the magnitude, direction, and line of action of a force acting on a body at *A*, and that it is desired to know what effect *O A* produces in the direction *B A*, which may be any direction. To find the value of the component of *O A* that acts in the direction *B A*, it is necessary to employ the following rule:

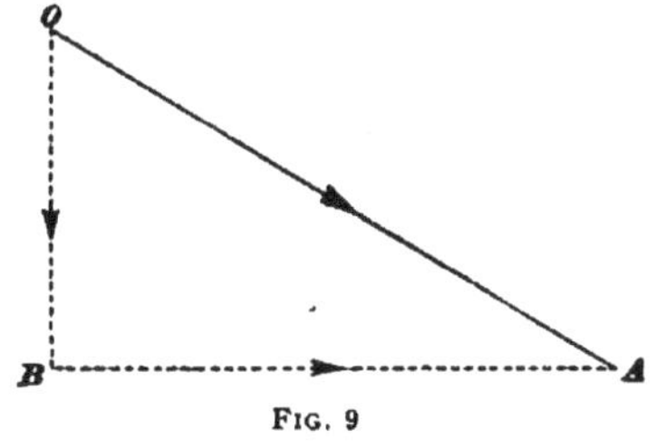

FIG. 9

Rule.—*From one extremity of the line representing the given force, draw a line parallel to the direction in which it is desired that the component shall act; from the other extremity of the given force, draw a line perpendicular to the component first drawn, and intersecting it. The length of the component, measured from the point of intersection to the intersection of the component with the given force, will be the magnitude of the effect produced by the given force in the required direction.*

Thus, suppose that *O A*, Fig. 9, represents a force acting on a body resting on a horizontal plane, and that it is desired to know what vertical pressure *O A* produces on the body.

Here the desired direction is vertical; hence, from one extremity, as O, draw $O B$ parallel to the desired direction (vertical in this case), and from the other extremity draw $A B$ perpendicular to $O B$, and intersecting $O B$ at B. Then $O B$, when measured to the same scale as $O A$, will be the value to the vertical pressure produced by $O A$.

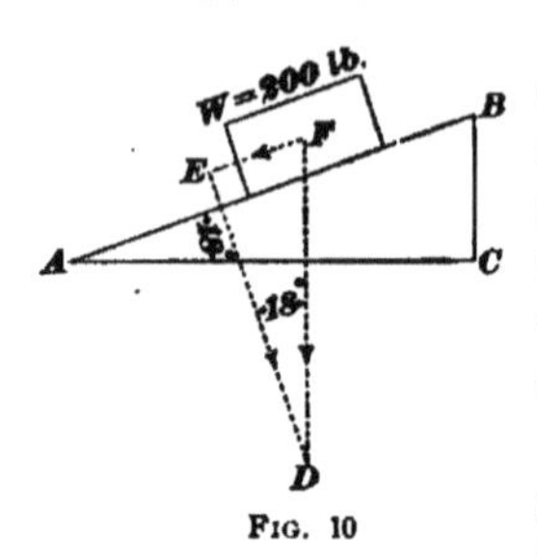

Fig. 10

Example.—If a body weighing 200 pounds rests on an inclined plane whose angle of inclination to the horizontal is 18°, what force does it exert perpendicular to the plane, and what force does it exert parallel to the plane, tending to slide downwards?

Solution.—Let $A B C$, Fig. 10, be the plane, the angle A being equal to 18°, and let W be the weight. Draw a vertical line $F D$ = 200 lb., to represent the magnitude of the weight. Through F, draw $F E$ parallel to $A B$, and through D draw $D E$ perpendicular to $E F$, the two lines intersecting at E. $F D$ is now resolved into two components, one $F E$ tending to pull the weight downwards, and the other $E D$ acting as a perpendicular pressure on the plane.

On measuring $F E$ with the same scale by which the weight $F D$ was laid off, its intensity is found to be about 61.8 lb., and the perpendicular pressure $E D$ on the plane is found to measure 190.2 lb. Ans.

EQUILIBRIUM

14. When a body is at rest, all of the forces that act on it must balance one another; the forces are then said to be in equilibrium. The most important of the forces acting on the body is gravity, which acts on every particle. But a force that must be considered when determining the equilibrium of framed structures is the wind pressure.

A body is in *stable equilibrium* when, if slightly displaced from its position of rest, the forces acting on it tend to return it to that position; for example, a cube, a cone resting on its base, a pendulum, etc.

A body acted on by a system of forces is in *unstable equilibrium* when the application of a small force is sufficient to produce motion; for example, a cone standing on its apex, an egg balanced on end, etc.

Since two kinds of motion may be produced in a body acted upon by external forces, the following conditions must be fulfilled in order that a body be in equilibrium:

1. The resultant of all the forces tending to move the body in any direction must be zero.

2. The resultant of all the forces tending to turn the body about any center must be zero.

But, if either of the two following conditions prevails the body will be in unstable equilibrium or unrest:

1. If the forces acting on a body create or influence motion of the body in the direction of the line of action of the force.

2. If the force acting on the body tends to move or rotate the body around some fixed point, which point is always necessarily outside the line of action of the force.

EQUILIBRIUM AT THE JOINTS OF A STRUCTURE

15. The several forces acting at any joint of a framed structure must theoretically be concurrent, and a single joint can, in consequence, be subjected only to translatory motion. But, where several forces meet at a joint in any stable structure there must be no translatory motion, for if the joint could move in any direction it would fail and cause the destruction of the frame.

In Fig. 11 (*a*) are shown five concurrent forces that act in the directions indicated by the arrows. By drawing the stress diagram for these forces, as shown at (*b*), their resultant is determined by the line *b c*, which force is the combined effect, both in direction and in intensity, of all of the concurrent forces acting at the joint *c*. Since this line *b c* is the resultant of the system of forces, it is evident that if no other force is substituted at the joint, it will move in the oblique direction indicated by the arrow on this line in the stress diagram. In order to produce equilibrium in the joint, a force *c b* in (*a*) acting in opposition to *c a*, and of the same amount, must be applied to the joint *c*, as shown. Consequently, the joint is in unstable equilibrium and can never be in equilibrium of translation until the points *c* and *b* in the stress diagram coincide.

The stress diagram of the five concurrent forces, acting on the joint shown in Fig. 11 (c), is shown in (d). It will be observed that there is no resultant, but that the polygon closes at the point c, the end of the last force $f\,c$ coinciding with the beginning of the first force $a\,c$. It is evident, therefore,

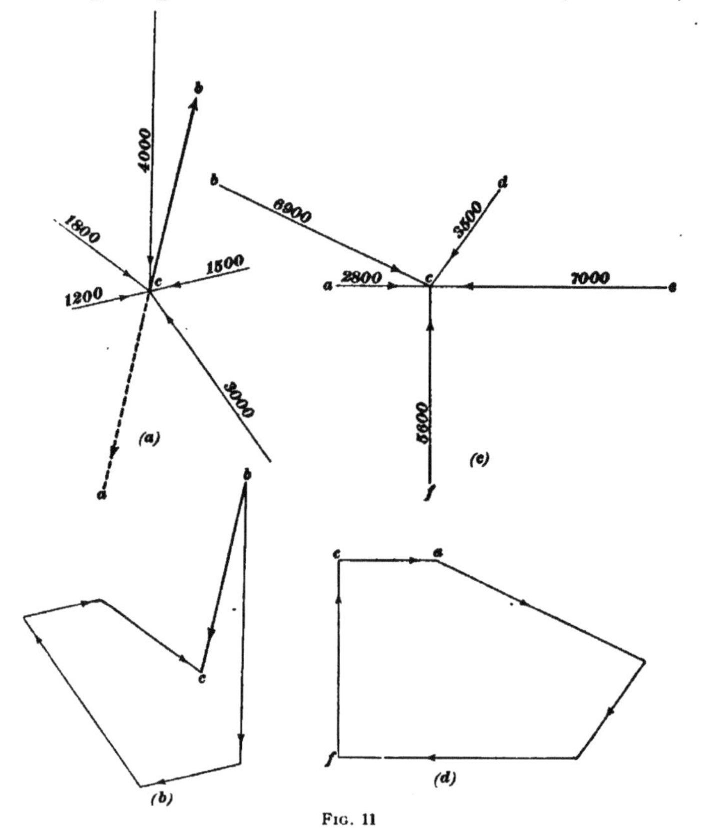

Fig. 11

since the resultant of all the forces about the joint c is zero, that no force need be substituted in the frame diagram to produce equilibrium of translation, and in consequence the joint in the structure will be stable and no failure of the frame, through weakness at this joint, can exist.

16. In order to analyze the conditions that create equilibrium of translation in any system of concurrent forces, resolve each of the forces of the system shown at Fig. 11 (*c*) into vertical and horizontal components, as shown in Fig. 12, by the method explained under The Resolution of Forces. The forces *a c* and *e c*, Fig. 11 (*c*), are horizontal and *f c* is vertical, but the forces *b c* and *d c* are oblique and may be resolved into vertical and horizontal components, as shown by y and y_1, x and x_1, respectively, Fig. 12. By scaling the horizontal components x and x_1, it is found that $x = 2,100$ pounds and $x_1 = 6,300$ pounds, and it will also be observed that the direction of the component x is opposed to the

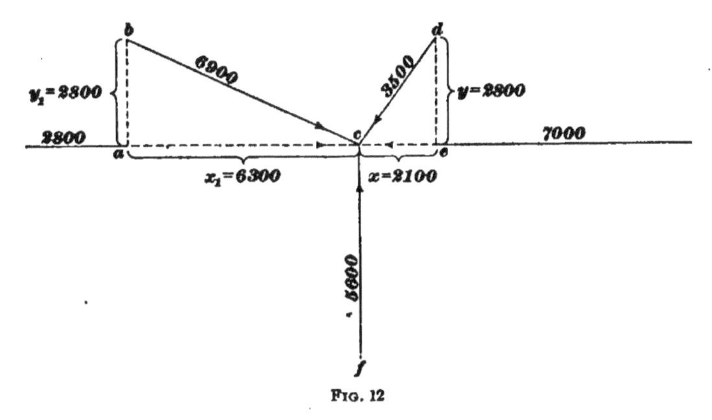

FIG. 12

direction of the force *a c*, and x_1 is opposed to the force *e c*. By studying the diagram, Fig. 12, it will be observed that the horizontal forces acting toward the left are exactly equal to the horizontal forces acting toward the right, and that they act and react so as to completely annul all action in a horizontal direction. In other words, the algebraic sum of all of the horizontal forces acting at the joint is zero, or, as it might be stated, $x + ec = x_1 + ac$ and $(x + ec) - (x_1 + ac) = 0$.

Again, the vertical components y_1 and y of the forces *b c* and *d c* both act in opposition to the force *f c* and the algebraic sum of all the vertical forces is equal to zero, for the sum of y and y_1 is equal to 5,600, as may be proved by scaling the diagram. The algebraic sum of all the vertical forces at the joint is consequently equal to zero. From these deductions the condition of equilibrium at a joint may therefore be expressed by the following rule:

Rule.—*In order that any joint in a frame, or any system of concurrent forces, shall be in equilibrium of translation, the algebraic sum of all the horizontal and vertical forces and all the horizontal and vertical components of the oblique forces shall equal zero.*

17. The conclusion that in order to have equilibrium of translation the algebraic sum of the several forces at a joint must equal zero, can be reached in another way. For instance, it has been stated and proved that the resultant of any system of concurrent forces must equal zero and it is evident that when all of the forces about a joint are resolved into their horizontal and vertical components, the final resultant is the hypotenuse of the triangle formed by the resultants of the components. In Fig. 13 (*a*), the several oblique forces have been resolved into their components x, x, x, and y, y, y, and x_1 and y_1 are the algebraic sums of these components, as may be proved by scaling the diagram. By laying off x_1 and y_1, as indicated, and drawing the hypotenuse R, the resultant is obtained. This resultant is therefore $\sqrt{\Sigma x^2 + \Sigma y^2}$, in which x and y equal, respectively, the horizontal and vertical components of each force about the joint. The components x_1 and y_1 may also be obtained directly from the stress diagram, Fig. 13 (*b*), by resolving the resultant R into its vertical and horizontal components.

The direction of the resultant R can be determined by obtaining the tangent of the angle that R makes with the horizontal or vertical components. The tangent of the angle marked z in Fig. 13 (*a*), is equal to $\frac{y_1}{x_1}$. Where, therefore, the vertical and horizontal components of each of the forces are considered, their direction being represented by V and H, respectively,

$$\tan z = \frac{\Sigma y}{\Sigma x}, \text{ or } \frac{\Sigma\, V \text{components}}{\Sigma\, H \text{components}}$$

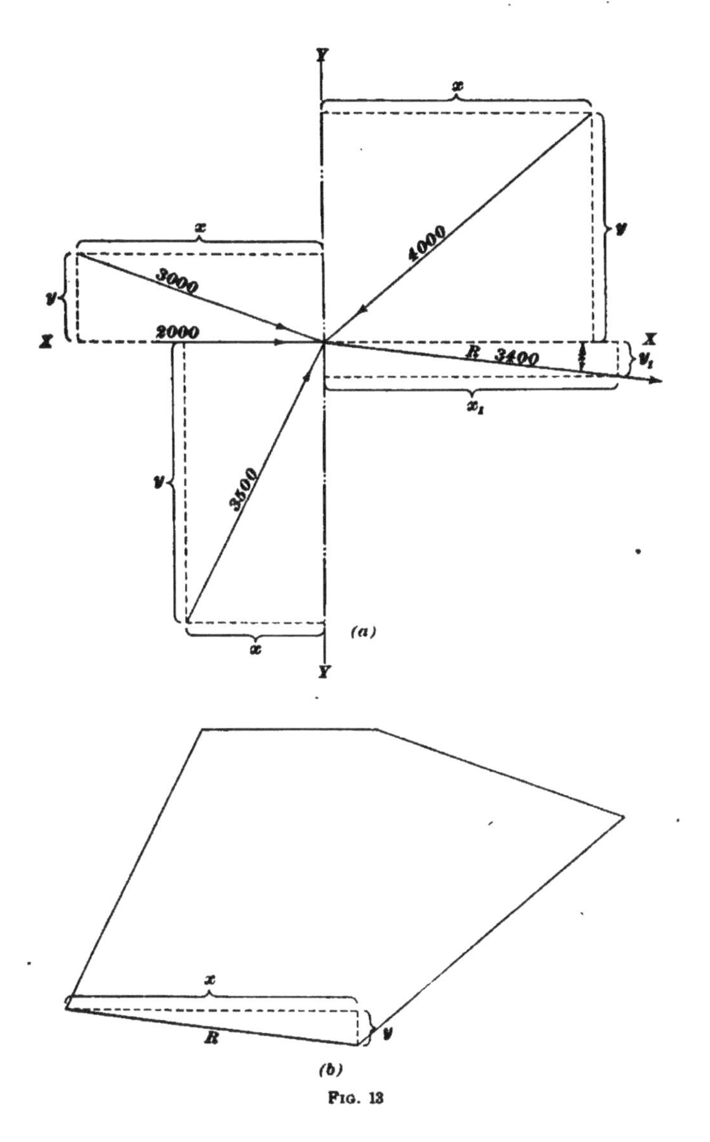

FIG. 13

But, from a previous statement and deduction, the resultant R of any system of concurrent forces in equilibrium must equal zero, so that $R^2 = (\Sigma H \text{ components})^2 + (\Sigma V \text{ components})^2 = 0$. R having an inappreciable amount, consequently, has no direction and it is conclusively shown that for equilibrium of translation the sum of the horizontal and vertical components must equal zero. This condition must therefore exist at every joint in a framed structure in order to secure the equilibrium or stability of the entire structure.

EQUILIBRIUM OF THE ENTIRE FRAME

18. Any framed structure, besides having each joint at which concurrent forces occur in equilibrium, must be in equilibrium with regard to the action of the external forces, which are usually non-concurrent. The external forces on any frame, such as a roof truss, are the loads and their reactions. On a roof truss, or other exposed structure, the external forces consist of the vertical loads due to the weight of the structure and snow, and the horizontal or oblique forces due to the wind pressure. The reactions occur at the abutments or supports of the structure and act in opposition to the loads, to hold the structure in equilibrium. It is sometimes necessary to introduce in the analysis of stresses imaginary or assumed reactions and forces that replace the resistance to bending offered by some member; this is due to the fact that a transverse stress cannot be shown in a diagram in conjunction with direct stresses.

Fig. 14 (*a*) shows an iron bracket resting on the ledge *a*, tied into the wall at *b*, and supporting, at its end, a weight W. As the external, or non-concurrent, forces alone are to be taken into account, the members of the bracket may be disregarded and the structure considered as a solid triangular body held in equilibrium by the four forces W, C, D, and E, as shown in (*b*). The force W acts vertically while C, which is a tensile stress, acts horizontally; D and E may be regarded as reactions. For the frame to be in equilibrium, the sum of the vertical forces must equal zero, as must also the sum of the horizontal forces. W is known, both in direction and intensity, and since E is negative with respect to the force W, their algebraic sum must be $W - E = 0$; therefore, $W = E$. Since the resultant of the horizontal forces must equal zero, $C - D = 0$, or $C = D$, thus proving that the triangular frame is in translatory equilibrium; that is, there is no tendency for the frame to move laterally in any direction.

The conditions of complete equilibrium are not fulfilled unless equilibrium of both translation and rotation exists about any point of the structure. Assume the point c as the center of moments. The force C acts about this point with a lever arm equal to y and the weight W tends to rotate in the opposite direction with a lever arm equal to x. Since the forces E and

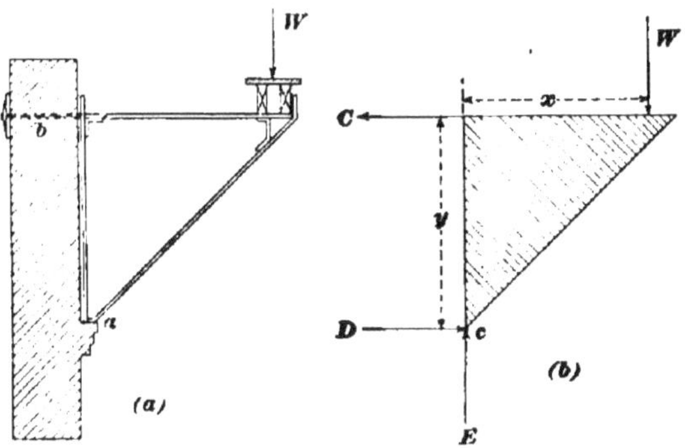

FIG. 14

D intersect at the center of moments c, there is no tendency for them to rotate the structure about this point, for their lever arm is zero; so that $D \times 0 = 0$, as does $E \times 0 = 0$. $W \times x$ must equal $C \times y$ in order to produce rotary equilibrium; therefore, $W \times x - C \times y = 0$. The algebraic sum of the moments of all the forces acting about c and tending to rotate the triangular frame is consequently equal to zero.

THE FORCE AND EQUILIBRIUM POLYGON

19. Since, in order for any structure to be in equilibrium the algebraic sum of the forces acting on the structure must equal zero, when a structure supports a number of parallel loads, the reactions at the ends of the structure, which are coincident with the line of action of the loads, must equal, when added together, the sum of the loads.

If the loads are not all parallel, but exert their forces along different lines of action, and the reactions in consequence do not coincide with the line of action of the several forces, the sum of the reactions does not equal the sum of the loads; but, in order for the structure to be in equilibrium, the sum of the vertical and horizontal components of all the forces must equal the sum of the vertical and horizontal components of the reaction.

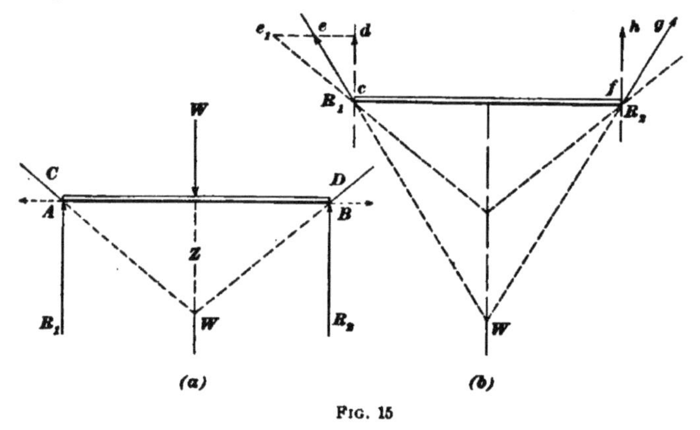

Fig. 15

Assume that in Fig. 15 (a), a beam loaded at the center with the weight W is held in equilibrium by the reactions R_1 and R_2; or, designating them by the usual system of notation, the forces $A Z$ and $B Z$. Suppose, now, that instead of a beam that has a certain transverse strength, which cannot be represented graphically, the load W is supported by a flexible cord attached at R_1 and R_2. This cord might have any length and consequently might drop any distance below the beam, but no matter what this drop may be, the vertical reactions R_1 and R_2 will always be the same, for the sum of R_1 and R_2 equals the sum of the vertical components of all of the forces acting in one direction on the cord, while W, or the amount of the load, must be equal and act in the opposite direction to fulfil the condition that the algebraic sum of the vertical loads shall equal zero. Besides the reactions R_1 and R_2, in order to create equilibrium in the cord supporting the load W, there must be horizontal forces A and B at each end of the cord to prevent their approaching each other. These two forces could as well be supplied by a horizontal strut extending between the ends of the cord, and could be considered as the compressive strength of the beam.

It is evident, therefore, since there are horizontal and vertical forces at each end of the beam, that a single force equal to their resultant could be applied to the ends of the cord in order that equilibrium might be maintained. These oblique resultants C and D of the vertical and horizontal components must, therefore, in order that they may alone create equilibrium, have their line of action coincident with the direction of the cord, as shown. Assume that the cord, instead of occupying the position designated in Fig. 15 (*a*), is considerably longer and that the weight is applied at twice the distance below the beam, as in Fig. 15 (*b*). The directions of the oblique reactions coincident with the present direction of the cord are indicated by the lines ce and fg. But it was stated that the vertical reactions, which are the vertical components of these oblique forces and are represented by the vertical lines cd and fh, Fig. 15 (*b*), must always be of the same amount when the load W is located in the center of the cord; that is, their sum must equal the load. It is evident, then, that if the vertical component at R_1 is laid off from c to d, and de is drawn horizontally from d, de will equal the horizontal component of the oblique force at this abutment. It will also be noticed that when the oblique reaction C, Fig. 15 (*a*), is designated by the dotted line ce_1, Fig. 15 (*b*), de_1 is the horizontal component for the oblique reaction at R_1. As seen in Fig. 15 (*b*), the horizontal reaction of the cord having the lesser drop is much greater than the horizontal reaction of the cord having the greater drop, and it is therefore evident that as the drop of the cord becomes greater, the horizontal component of the oblique force becomes less. Likewise, that the greater the drop of the cord, the less will be the oblique reaction until when the oblique reaction approaches verticality the thrust or horizontal force will become zero while the reaction will equal one-half of the load.

20. Assume that to the beam shown in Fig. 16 (*a*) is applied a load *W* equal to 1,000 pounds. Lay off in the

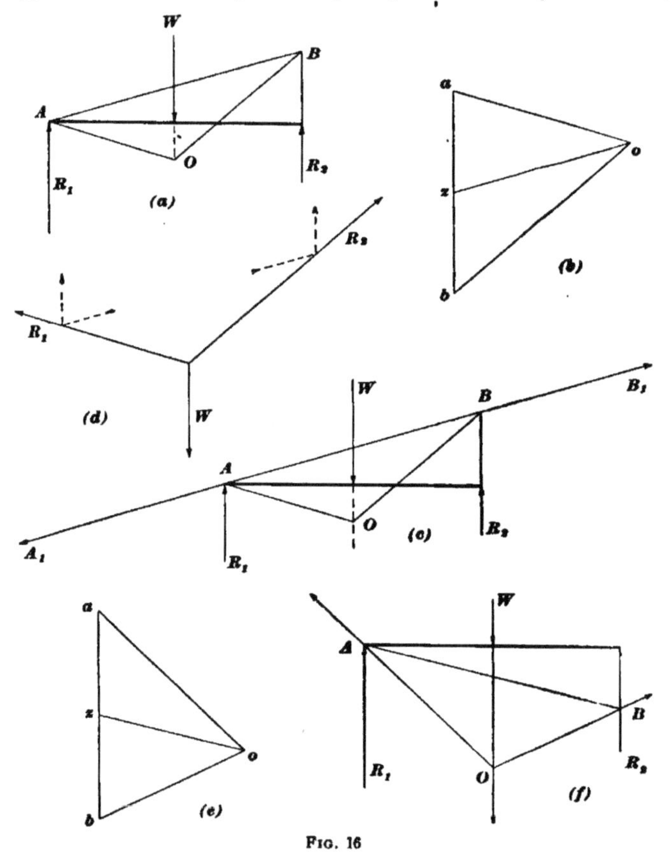

Fig. 16

diagram at (*b*) a line *a b* equal, by scale, to 1,000 pounds. Choose any point, or pole, such as *o*, and draw lines from *a* and *b* to this point. From the point *A* in (*a*), extend a line *A O* parallel with *a o* in (*b*). Extend the line of action of the force W acting on the beam until it intersects this line just drawn at the point O. From the point O thus found, draw a line $O\,B$ parallel with $o\,b$ in (b) and designate the point of intersection of this line and the line of action of the reaction R_2, as B. Connect the points A and B just found in this diagram, and returning to the diagram (b), extend from the pole, or point o, a line coincident in direction with the line $A\,B$ in (a). Where this line intersects the line $a\,b$, designate the point as z. Fig. 16 (b) is known as the *force polygon*, while Fig. 16 (a) is known as the *equilibrium*, or *funicular, polygon*.

On analyzing these diagrams, it is found that the distances $b\,z$ and $z\,a$ are equal and since $a\,b$ represents the amount of the load and $b\,z$ and $z\,a$ are one-half of the load, it is evident that these two forces represent, in amount and direction, the reactions R_1 and R_2. No matter where the pole o had been chosen or assumed, the result would have been the same, for the line $o\,z$ is always drawn parallel with the line $A\,B$ found in the equilibrium polygon. On scaling the lines $a\,o$ and $b\,o$ in Fig. 16 (b), the amount of the oblique forces necessary to create equilibrium, when acting in the direction shown, is found, and $o\,z$ is the thrust exerted or the compression produced in the strut introduced between A and B in (a).

A few explanations at this point may help the student in understanding the meaning of the various lines in force and funicular polygons. When the line $a\,b$, Fig. 16 (b), representing the total load W, has been laid off, and a pole o selected, from which the lines $o\,a$ and $o\,b$ are drawn, the force $a\,b$ has been resolved into the components $a\,o$ and $b\,o$. This construction is based on the triangle of forces, with this difference, that here the resultant $a\,b$ is given and the components found. It is clear that there is a great latitude as to the selection of location of the point o, and therefore of the direction and magnitude of these components.

In a force polygon, any of the components may be resolved into other components, the direction of which may have been given. For instance, in Fig. 16 (b) it is desired to resolve the forces $a\,o$ and $o\,b$ into components, two of which will be vertical and two parallel with the line AB in the funicular polygon. The line oz, Fig. 16 (b), was drawn parallel with AB and constitutes a component that will be common both to ao and ob; the other component for ao will be az and for ob will be bz. The force triangle, or polygon, aob has thus been divided into two smaller triangles, in one of which the force ao has been resolved into the two components az and oz, while the other force ob has been resolved into the components oz and bz. It is evident that the force oz can be resolved into other components and these again into others; in fact, this process may be carried on indefinitely.

In constructing the funicular polygon in Fig. 16 (a), only the total load $W = ab$ and its components ao and ob were given. The known component of R_1, that is, ao, which is also one of the components of W, was drawn through the point A and intersected the line of action of the force W at O. From this point, a line was drawn parallel with the component ob in the force polygon, and intersecting the line of action of the reaction R_2 at B. The lines AO and OB represent the components of the load W, the directions and magnitudes of which were determined by the selection of the point o in Fig. 16 (b). As the lines of action and magnitude of the other components of the reactions R_1 and R_2 must be the same, they must of necessity be located on the line connecting the points A and B. The line oz in the force polygon parallel with AB, will determine the magnitude of the other components of the reactions R_1 and R_2, az being the vertical reaction R_1, bz the other reaction R_2, and oz the components along the line AB.

To illustrate the locations and actions of the various components more clearly, Fig. 16 (c) has been introduced. In it AA_1 and AO represent the components of the reaction R_1, AO and OB, those of the load W, and OB and BB_1, those of the reaction R_2. In general, it is understood that the lines of action of the components AA_1 and BB_1 are located along the line AB and it is therefore unnecessary to show them separately, as has been done in this case.

The directions in which the various components act are

found from the force polygon in the manner described, but it is necessary to bear in mind that when a certain force in a polygon is to establish equilibrium, its direction conforms to that of the other forces, but if it is to serve as a resultant, it must act in the opposite direction.

Considering the structure in (*a*) as a cord suspending a weight with its ends separated by a compression member in the position of *A B*, it is evident that the force polygon (*b*) contains the reaction diagram giving the reactions *a z* and *z b* for each end of the equilibrium polygon. In Fig. 16 (*d*), which represents the weighted cord, the reactions R_1 and R_2 coincident in direction with the lines of action of the cord, are the only forces required at the ends of the cord to preserve equilibrium. These reactions are represented in the force polygon by *a o* and *b o*, and are equal to the tension in the cord. For the reactions, the two components could be substituted as shown in (*d*) by the dotted lines; without the reactions these forces would create equilibrium. The components of the oblique reaction R_1 in (*d*) are represented in the force polygon by *z a* and *o z*, while the corresponding components of R_2 are *b z* and *z o*.

Another pole could have been chosen and the force polygon drawn as at (*e*), in which case the equilibrium polygon would have changed correspondingly, assuming the form shown in (*f*). It will be noticed from (*e*) that the vertical components *b z* and *z a* of the oblique reactions always remain the same length and consequently the same amount when *o z* is drawn parallel with *A B* in the diagram (*f*). The direction of the reactions coincident with the direction of the cords and the oblique components of this reaction are alone changed.

21. Assume the conditions of loading shown in Fig. 17 (*a*). The load *W* of 1,000 pounds is not centrally placed on the beam; consequently, the principle of moments involved in the theory of beams makes the reaction R_1 considerably less than R_2. Lay off in the force polygon, Fig. 17 (*b*), to a scale of $\frac{1}{4}$ inch equals 100 pounds, the line *a b* equal to 1,000 pounds. Intersecting at any pole *o* draw *a o* and *b o*. From the point *A*, in (*a*), extend a line parallel with *a o* and intersecting the line of action of the weight or the force *W* extended, at the point *O*. From *O*, draw a line parallel with *b o*, Fig. 17 (*b*), intersecting the reaction R_2 at *B*. Connect *A* and *B* as before, and from *o* in (*b*) draw a line parallel with *A B* in the equilibrium polygon and intersecting *a b*. By measuring *b z* and *z a*, the reactions R_2 and R_1 will be found to equal 770 and 230 pounds, respectively.

The student should not be guided by the size of these diagrams, as the size of those employed in practice for solving graphical problems should be considerably larger to give

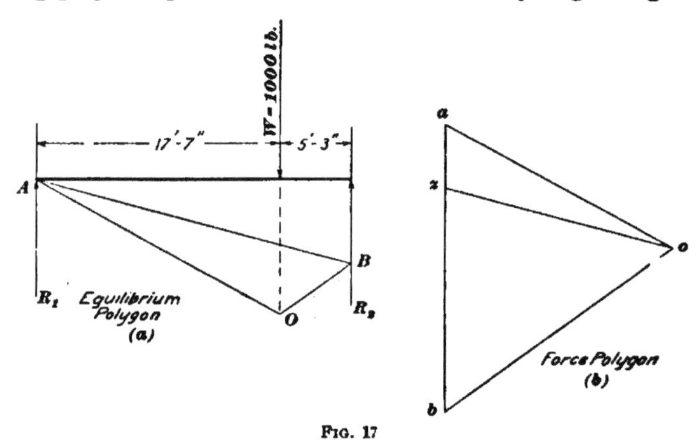

FIG. 17

accurate results. The stress diagram would be laid out to a scale of from 500 pounds to 5,000 pounds to 1 inch, while the scale used for the frame diagram would be from $\frac{1}{16}$ inch to $\frac{1}{4}$ inch to the foot.

22. The application of the force and equilibrium polygons for determining the reactions of non-concurrent forces is not limited to a beam or structure supporting one weight, for they can be used for obtaining the reactions of a simple beam supporting any number of loads acting in any direction. For example, assume that a beam, Fig. 18 (*a*), supports the two loads W_1 and W_2, which are placed at the position fixed by the given dimensions. In order to determine the reactions R_1 and R_2, lay off the force polygon (b) by drawing the load line $a\,c$ and, measuring with some convenient scale, make $a\,b$ and $b\,c$ equal, respectively, to W_1 and W_2. Choose any pole o and connect it with the points a, b, and c by radial lines. Draw the equilibrium polygon by commencing at A and drawing $A\,O$ parallel with $a\,o$ of the force polygon, intersecting the line of action of the load W_1 at O. From this point, draw a line parallel with the line $o\,b$ in the force polygon, intersecting the line of action of the load W_2 at B. Finally, draw from B a line parallel with $o\,c$ in the force polygon and intersecting the reaction R_2 at C. Connect A and C, as explained in the previous cases, and from o in the force polygon draw a line parallel with $A\,C$, intersecting the load line at z; the reactions R_1 and R_2 are found by scaling $z\,a$ and $c\,z$. If, in the equilibrium polygon, $A\,O$ and $C\,B$ are extended until they intersect at the point O_1, a vertical line drawn upwards from O_1 will divide $A\,C$ into two such parts that the ratio of X to X_1 will be inversely proportional to the ratio of R_1 to R_2; the amount of R_1 and R_2 may be found by designating $A\,C$ equal to the sum of the loads W_1 and W_2, when R_1 will have the same ratio to the total load as X_1 to the whole distance $A\,C$, and R_2 will bear the same relation to the total load as X to $A\,C$.

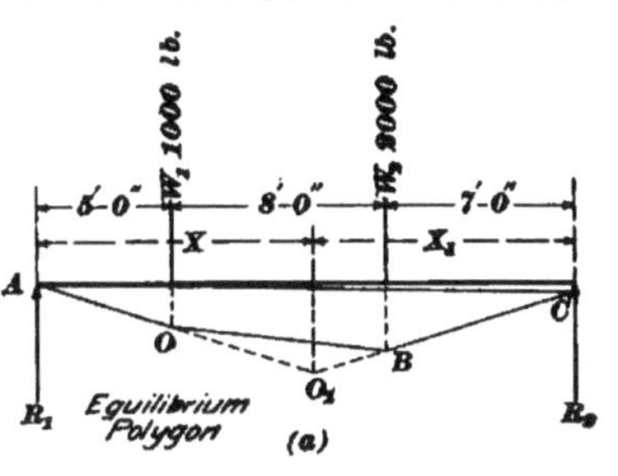

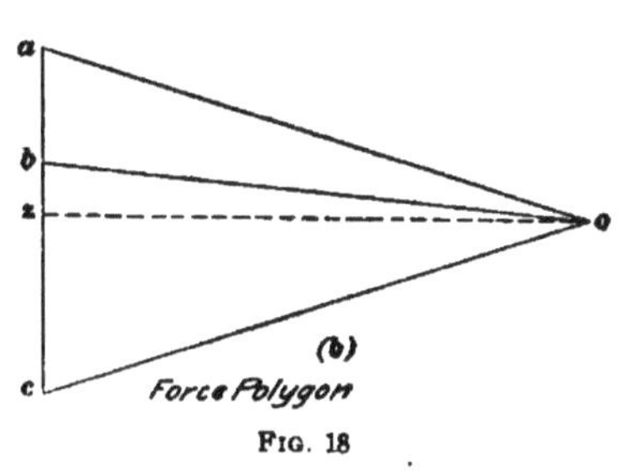

FIG. 18

23. The example shown in Fig. 19 (a) is similar to the one just described with the exception that several loads are applied and there are vertical loads also at each end of the beam. The force polygon, Fig. 19 (*b*), is drawn as in the previous cases, and in drawing the equilibrium polygon the lines of action of the several forces are extended as shown by the dotted lines. In drawing this equilibrium polygon, however, it will be noticed that the line marked *1* is parallel with the second line from the top of the force polygon, or

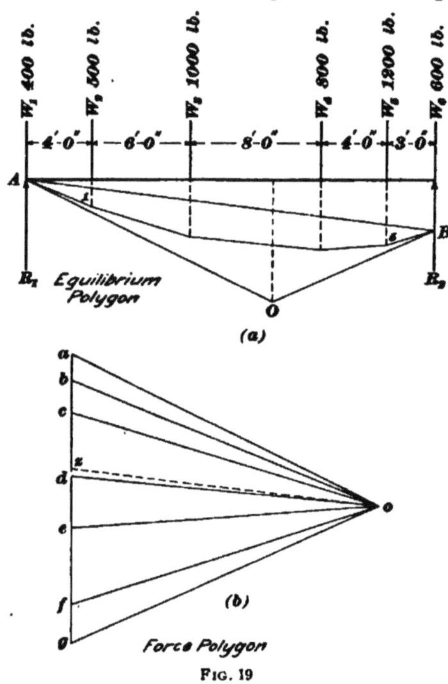

FIG. 19

b o, and that the first line of the force polygon *a o* is represented in the equilibrium polygon by *A O*, while reference to the right-hand end of the figure will show that the line marked *5* in the equilibrium polygon is parallel with the line marked *o* in the force polygon. The reason for this is that the loads W_1 and W_6 are coincident with the reactions R_1 and R_2, respectively. When the load W_1, represented in the force polygon by the force *a b*, is resolved into its two components *o a* and *o b*, these should, as already explained, be laid off on either side of the line of action of the load W_1, one intersecting the reaction R_1 and the other the line of action of load W_2. In this case the forces W_1 and R_1 being coincident, only the component *o b* can be drawn. The same conditions prevail at the load W_6. While the loads W_1 and W_6, on account of their position, produce no stress in the girder, yet they add their share to the total load and will therefore have to be included among the forces in the force polygon, in this manner affecting the position of the point *z* and, as a consequence, the amount of the vertical reactions R_1 and R_2. If the forces *a b*, *b c*, . . . *f g* were combined, they would equal the total load and would be represented by the line *a g*, the components of which would be *o a* and *o g*. If *A O* and *O B* were drawn parallel with these components they would intersect at the point *O*, which would be located on the line of action of a load, the location and magnitude of which would be such that it would have the same effect as the six loads. The point of intersection *O* of the lines *A O* and *B O* is the center of action of the loads; that is, a vertical force equal and opposed to the loads when located at this point will just balance the loads and create rotary equilibrium, or conversely, it is the position at which a weight, equal to the weights on the beam, suspended from the point *O* on a cord fastened at *A* and *B* will produce the same reactions as the weights on the beam. The reactions at the end of the beam, which are found by measuring *z a* and *g z*, are represented in the equilibrium polygon by R_1 and R_2. The construction of the equilibrium polygon as described usually exists where there are two end loads on the beam or structure, and the student should always bear in mind the fact that the equilibrium polygon is contained between the lines of action of the reactions, and that in the equilibrium polygon there exists the same number of lines as radiate from the pole *o* in the force polygon.

24. The solution of the complicated problem shown in Fig. 20 (*a*) is as readily accomplished as in the simpler

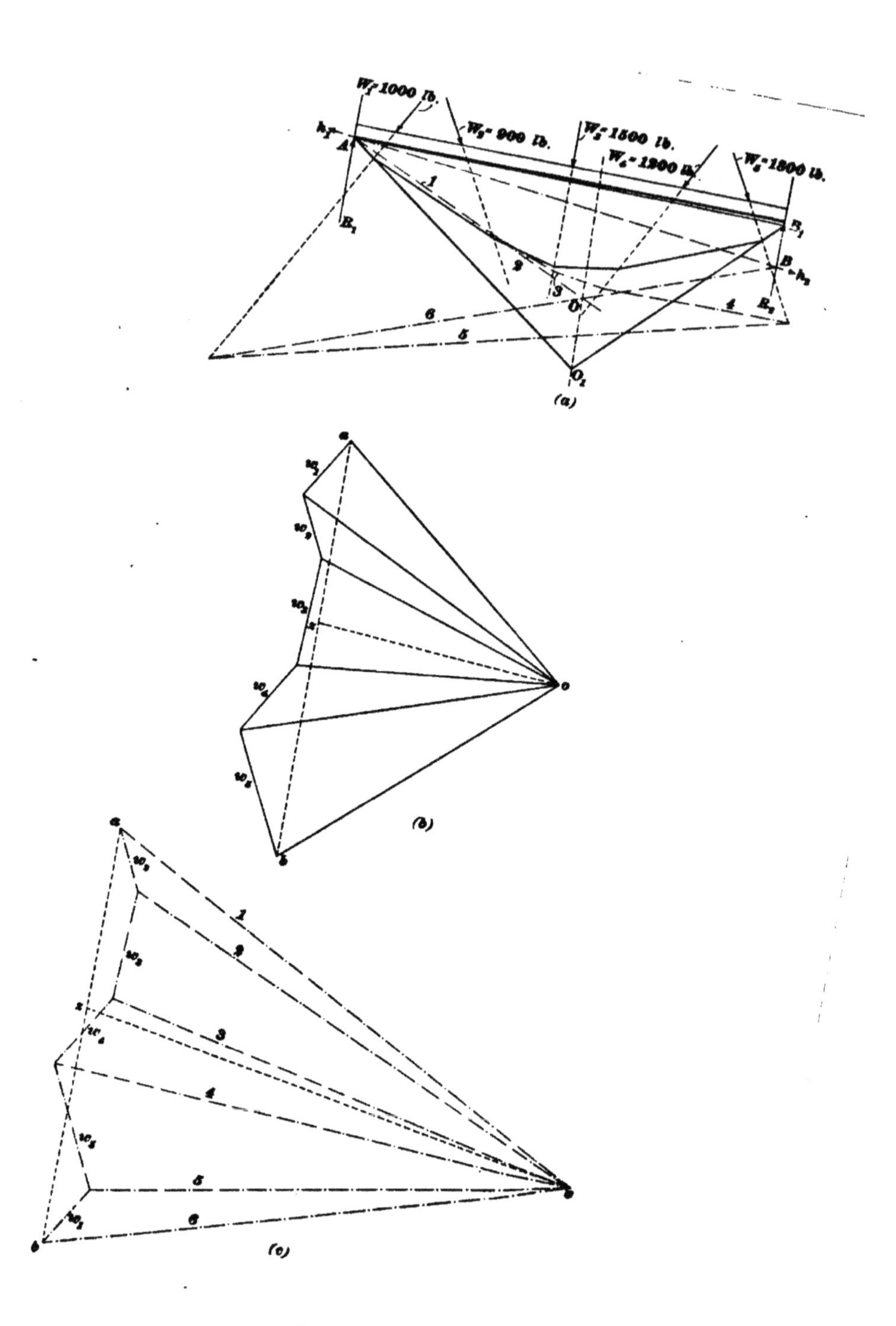

W1=1000 lb.
W2=900 lb.
W3=1500 lb.
W4=1200 lb.
W5=1800 lb.
A
B1
B
R1
R2
O
O1
(a)
(b)
(c)

Fig. 20

cases previously assumed. The several loads need not be laid off in the force polygon in the regular order, as shown in Fig. 20 (*b*); but in whatever order they are laid off in the force polygon, the same sequence must be observed in laying out the equilibrium polygon. Hence, the load line of the force polygon may be obtained by laying off the loads W_2, W_3, W_4, W_5, and W_1, as shown in Fig. 20 (*c*); and by connecting the points at the ends of the load line, as designated by the line *a b*, the direction of the reactions at the ends of the beam is found. Choose any pole *o* and draw radial lines to the end of each force on the load line. The reactions R_1 and R_2 in the equilibrium polygon will then lie in the direction of the line *a b* in either force polygon, and both equilibrium polygons will be included between these two lines.

Both equilibrium polygons are shown in Fig. 20 (*a*); the one corresponding to the force polygon at (*b*) is shown with solid lines, while the one described from the force polygon (*c*) is shown with dot-and-dash lines. In drawing the latter, commence at the point *A* on the line of action of the reaction R_1 and draw the line marked *1* of the equilibrium polygon parallel with the first radial line in the force polygon. This line in the force polygon connects the intersection of the reaction line *a z* and the force or load w_2 with the pole *o* and should, consequently, be drawn in the equilibrium polygon from the line of action of the reaction R_1 to the line of action of W_2. From this intersection, draw line *2* parallel with the line in the force polygon similarly marked; this extends to the line of action of the force W_3, since line *2* in the force polygon connects the intersection of loads w_2 and w_3 to the pole *o*.

Draw from this intersection, line *3* parallel with the ray *3* in the force polygon, intersecting at W_4 extended. Proceed in this manner around the entire equilibrium polygon until the point *B* has been obtained. From *B* in the equilibrium polygon draw a line to the point *A* and draw from *o* in the force polygon a line parallel to *A B* intersecting the line of the reactions at *z*. The required reactions are measured from *b* to *z* and from *z* to *a*, and are equal, by measurement, to R_2 and R_1, respectively.

The system of non-concurrent forces W_1, W_2, W_3, etc. can be held in equilibrium by either one of two systems of reactions: first, by reactions acting in the direction of *B O* and *A O* in the equilibrium polygon and equal in amount to the forces determined by measuring *o b* and *o a* in the force polygon (*c*); second, by the reactions R_1 and R_2 and the thrusts h_1 and h_2, these forces being the components of *a o* and *o b* in (*c*).

It is shown in this solution, by comparing the diagrams (*b*) and (*c*), that the reactions have the same amount and direction no matter in what order the loads are taken, if the same sequence is followed in drawing the equilibrium polygon. To make this clearer, it may be further stated that in drawing the equilibrium polygon, all that need be observed is that, commencing at the point *A*, the lines *1*, *2*, *3*, *4*, etc. of the equilibrium polygon are laid off between the lines of action of the several forces and extended in the direction determined by the ray in the force polygon relating to the particular force in the equilibrium polygon from which it is drawn. Thus, the line *1* in the equilibrium polygon, extending between the lines of action of the reaction R_1 and the load W_1 corresponds with line *1* in the force polygon drawn from the intersection of the line *a z*, which represents the reaction R_1, and the load line w_1. The line *2* in the equilibrium polygon corresponds with ray *2* in the force polygon and is drawn from the point of intersection of line *1* and the line of action of W_1 extended. Likewise, line *3* in the equilibrium polygon coincides in direction with ray *3* in the force polygon and is drawn from the intersection of line *2* with the line of action of W_2 extended.

The reason for this method of constructing the equilibrium polygon from the force polygon, Fig. 20 (*c*), is easily seen when it is remembered what was stated in Art. 23 regarding the components of each of the forces on the load line *a b*. In each of the force triangles into which the force polygon has been divided, the two rays are components of the force that constitutes the third side. Lines drawn in the equilibrium polygon parallel with these components should in each case intersect on the line of action of that load or force that, in the force polygon, is represented by the third side of the triangle in which the components are located.

The lines in the equilibrium polygons, which are drawn through the points O and O_1 in the direction of the reactions, give the positions on the lines AB and AB_1, respectively, at which it would be necessary to apply a force equal to the sum of the reactions and acting in their direction, to balance the loads and to produce both rotary and translatory equilibrium. In this instance, these lines are coincident and pass through both points O and O_1.

The results obtained from the two force polygons will be the same, though probably it is always more convenient to take the forces in the order in which they occur. The student should thoroughly familiarize himself with the principles involved in the application of the equilibrium and force polygons; and that he may understand the usefulness of this system of analysis, he should work out for himself by the graphical method the problems called for in the following examples.

EXAMPLES FOR PRACTICE

1. A main girder, having a span of 40 feet, is subjected to the superimposed loads shown in Fig. 21; determine the amount of the reaction at both ends, by the graphical method. Ans. $\begin{cases} R_1 = 7{,}184 \\ R_2 = 5{,}516 \end{cases}$

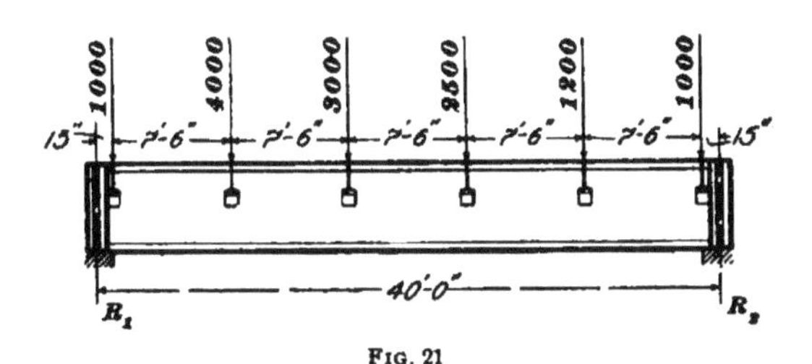

FIG. 21

2. A steel I beam is required to sustain the foot of a slanting column or strut forming one of the supports of a water tank, as

shown in Fig. 22; what will be the direction of the reactions on the walls and their amounts, by the graphical determination, provided the compression in the strut equals 3,000 pounds?

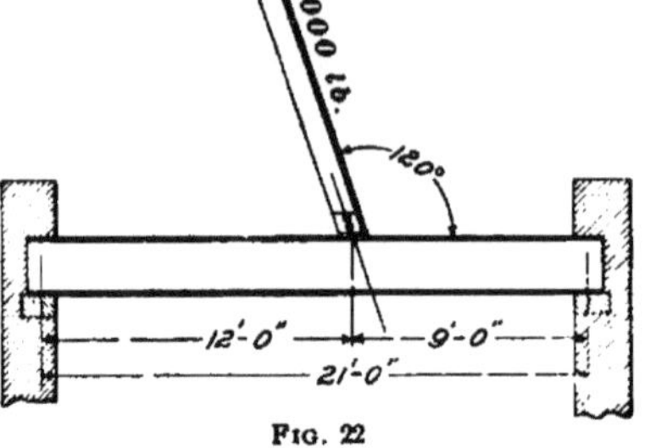

FIG. 22

Ans. { Direction: same as load; Amounts: R_1 = 1,275, R_2 = 1,725 }

3. A pair of light steel channels in a machine shop is subjected to the pull of several belts and is also required to sustain a load from a hoist and traveler, as shown in Fig. 23; what direction and amounts will the graphical method give the reactions at the column supports?

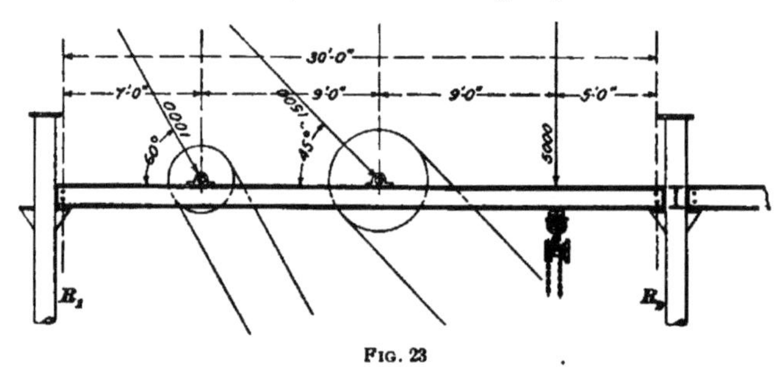

FIG. 23

Ans. { Direction: approximately 13° with vertical axis; Amounts: R_1 = 2,050, R_2 = 5,050 }

APPLICATION OF GRAPHICAL ANALYSIS

SCOPE OF GRAPHICAL SOLUTIONS

25. The study of the composition of forces has shown that the final effect of two forces can be represented by a third force having a different direction and intensity; and also that any oblique force can be resolved into its vertical and horizontal components. Therefore, when the direction and intensity of one of three forces acting at a joint or connection in a framed structure are given, the other forces can be obtained, provided that the directions or amounts of both, or

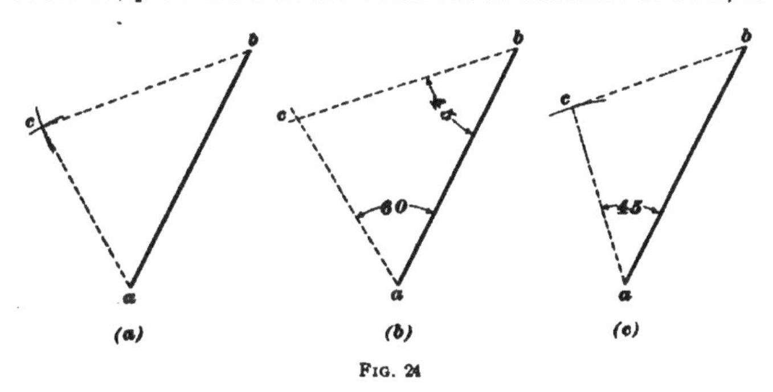

FIG. 24

the direction and amount of one are known; for instance, the forces *a b* in Fig. 24 (*a*), (*b*), and (*c*) are known both as to intensity and direction. It is assumed in the case shown in (*a*) that the amount of each of the two unknown forces is given and that in order to determine these forces it is necessary to know the direction in which they act. To find the unknown directions of the two forces the dividers should be set to scale for one of the forces and an arc struck from *a*, as designated. The dividers being then set for the other force, an arc is struck from b and their intersection at c determines the direction of ac and cb. In (b) the directions of the two unknown forces are known either from the fact that they are parallel with another line already determined, or because the angle that they make with ab is known. Assuming that the angle that the unknown force extending from a makes with the line ab is 60°, the line representing this force may be extended indefinitely, as shown. If the angle of the force drawn from b, with reference to ab, is equal to 45°, it is evident that a line drawn at this angle from b will intersect the other force at c. Then by scaling ac and bc the two unknown forces are completely determined, that is, both their direction and intensity are known. In (c) the direction and intensity of both ab and ac are known; therefore, the point c can be located by scale and bc drawn. Consequently, both the direction and amount of bc may be obtained in this manner.

26. From the foregoing, it has been determined that the following statements are true of any system of three forces:

1. That the unknown direction of two forces can be obtained when their amounts and the amount and direction of the third force are known.
2. That the amount of two unknown forces may be found when their direction and the amount and direction of the third force are known.
3. That the direction and amount of a force may be obtained when the direction and the amounts of the other two forces are known.

In the application of graphical statics, the truth conveyed by the first statement is seldom employed, but the second and third principles form the basis of the science of graphical statics as applied in the solution of stresses in framed structures.

In Fig. 25 (a) is shown a diagrammatical figure that represents a panel point, or joint, on the rafter member of a roof truss. The known stresses about this point are the compression in the lower portion of the rafter member and the vertical force W that represents the load at the panel point applied through the purlin that is secured to the frame at this position. These two known forces are shown by heavy lines, while the unknown stresses exist in the upper portion of the rafter member and the strut, the latter being designated by light lines.

Assume that the weight W is equal to 1,000 pounds and that the stress in the lower portion of the rafter member is 6,000 pounds. The problem is to determine the stress in the upper portion of the rafter member and in the strut. In (*b*), lay off on a line parallel with the rafter member, a distance equal, by scale, to 6,000 pounds. It is known that the direction of the stress in this member is upwards

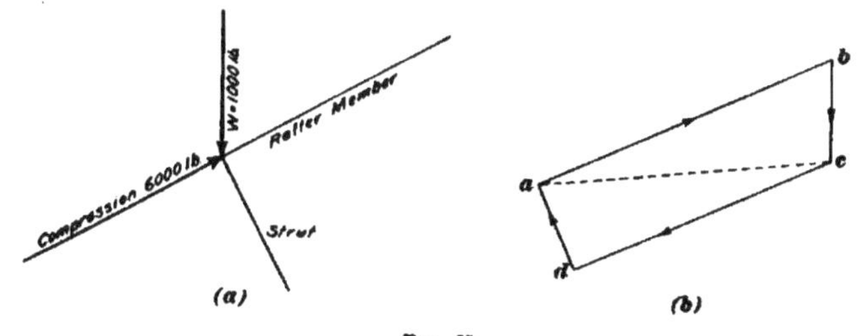

FIG. 25

toward the joint, so that the line just drawn extends from a to b in the direction of the arrow, as shown. From b, the line bc is laid off equal, by the same scale, to the amount of W or 1,000 pounds, and downwards. The resultant of these two forces will extend, as shown by the dotted line, from a to c. Let this resultant, therefore, represent the base line on which the two unknown forces are to be constructed; that is, let it represent the line ab shown in Fig. 24 (*a*), (*b*), and (*c*). The length and amount of this resultant ac is known.

By applying, therefore, the second principle stated above, the amount of the two unknown forces may be found. In order to determine these forces, draw from c a line parallel with the rafter member and from a a line parallel with the strut. Then, by measuring with the same scale to which *a b* and *b c* were laid off, the stress in the strut can be found from *a d*, while the stress in the upper portion of the rafter member is found by measuring *c d*. The polygon of forces around the entire joint extends from *a* to *b*, from *b* to *c*, from *c* to *d*, and from *d* to *a*; since the polygon closes, the joint is in equilibrium of translation, and because the forces are concurrent, the joint has no rotary tendency. Though it was not necessary to draw the resultant *a c*, this was done to show that every force polygon is made up of a series of triangles of forces, and the resultant of any number of forces of the system is obtained by a line connecting the end of any force with the point of commencement of the polygon. For instance, the resultant of *a b* and *b c* is *a c*, while the resultant of *a b*, *b c*, and *c d* is *d a*; in the same manner the resultant of *b c*, *c d*, and *d a* is *a b*.

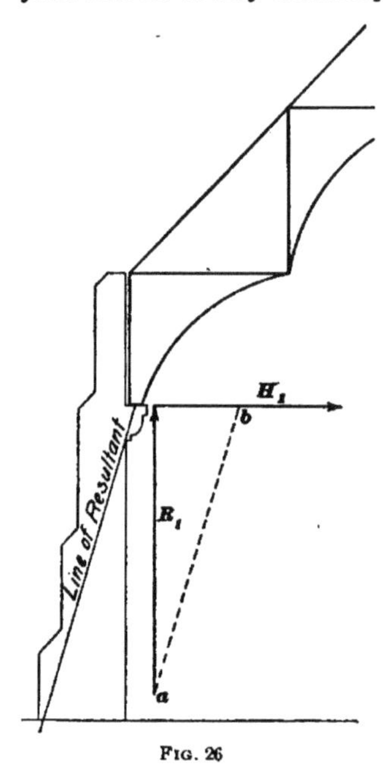

FIG. 26

The application of the third principle is usually employed where it is necessary to determine one of the unknown forces by means of calculation, as will be further explained; it is also used in drawing the reaction diagrams. In explanation of the latter, assume that in order to hold the foot of the roof truss designated diagrammatically in Fig. 26, there necessarily exists a vertical force, as shown by R_1, and a horizontal thrust, as designated by H_1. It is desired to determine the resultant of these two forces, in order to find at what position on the base line this force will intersect, so that it may be decided whether the abutment is in equilibrium. The amount of the thrust H_1 is laid off to scale, as shown. The reaction

R_1 is known and the points *a* and *b* are consequently located. The resultant will therefore extend from *a* to *b*, and when drawn from the center of the foot of the truss, the effect of these two forces on the abutment is determined and exists as shown.

DIAGRAMS

27. In the application of graphical statics to the determination of stresses in roof trusses and framed structures, the direction of all internal and external forces, excepting the reactions, are known, and the solution resolves itself into the determination of the amounts of the forces only. In order, therefore, that the direction of the several external and internal forces that are known may be applied in drawing the diagram for obtaining the stresses, it is necessary that an accurate outline of the truss or framed structure shall be laid out to scale. In this diagram, which is drawn according to the principal dimensions, the exact direction of all the external forces and the internal members, that represent the internal forces, are accurately designated. Such a diagram, which is practically the preliminary sketch or study of the structure, is called the **frame diagram,** from the fact that it represents the framework, or skeleton, of the structure. The diagram that is drawn in order to determine the stresses and in which each line represents graphically, in its relation with the other lines of the diagram, the several forces exerted on the structure, either internally or externally, is called the **stress diagram,** though sometimes erroneously termed *strain diagram.*

The frame diagram for a simple roof truss is shown in Fig. 27 (*a*), while the stress diagram is shown in (*b*) of the same figure. The frame diagram is drawn to a convenient scale of a certain number of inches to the foot, but since lines in the stress diagram represent forces in direction and intensity, the stress diagram is always drawn to a convenient scale of pounds equal to some unit of linear measurement, such as 1 inch; for instance, the frame diagram shown in the figure may be drawn to a scale of $\frac{1}{4}$ inch = 1 foot, while the stress diagram may be laid out to such a scale that every inch in the length of the line represents 1,000 pounds, or if a tenth scale is used each $\frac{1}{10}$ inch equals 100 pounds. The lines in the frame diagram representing the forces are never drawn to scale, as they simply show the direction of the forces. Sometimes, however, it is customary to lay out a stress diagram upon or connected with the frame diagram in order to obtain, possibly, the vertical or horizontal components of an oblique force, or to find the resultant reaction for a system of forces representing the independent reactions due to different external forces acting on the structure; for example, in Fig. 28 is shown, in light lines, a stress diagram drawn in conjunction with the frame diagram. Here it was deemed expedient to determine one force, such as R_1, that would equal in effect the horizontal force H_1, the vertical force V_1,

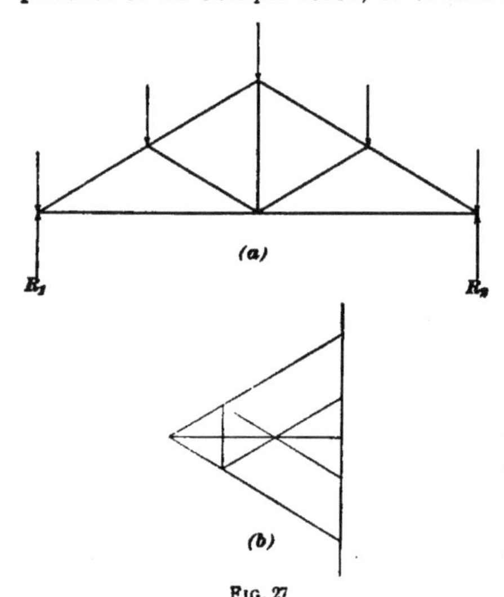

FIG. 27

and the oblique force O_1. The direction and amount of the resultant R_1 was found by laying off, to scale, the amounts of V_1, H_1, and O_1 on their respective coincident or parallel lines, ab, bc, and cd, the resultant R_1 being equal in intensity and direction to da. When this force R_1 is introduced instead of the three forces, the equilibrium of the structure is still maintained.

In drawing the frame diagram, any usual scale, as $\frac{1}{8}$, $\frac{1}{4}$, $\frac{3}{8}$, $\frac{3}{4}$, or $\frac{1}{2}$ inch to the foot, can be used, though probably the $\frac{1}{8}$- and $\frac{1}{4}$-inch scale will be found most convenient. In laying out stress diagrams, the tenth scale can be more readily employed than the usual inch scale divided into 8ths and 16ths.

28. The frame and stress diagrams are often called *reciprocal diagrams*, from the fact that for every system of forces in the frame diagram there exists in the stress diagram a series of lines that are respectively parallel with the lines of the forces in the frame diagram; and for every joint at which a system of concurrent forces exists in the frame diagram, there is a corresponding polygon of forces in the stress diagram. In determining the stresses on any framed structure, it is usually necessary to draw several stress

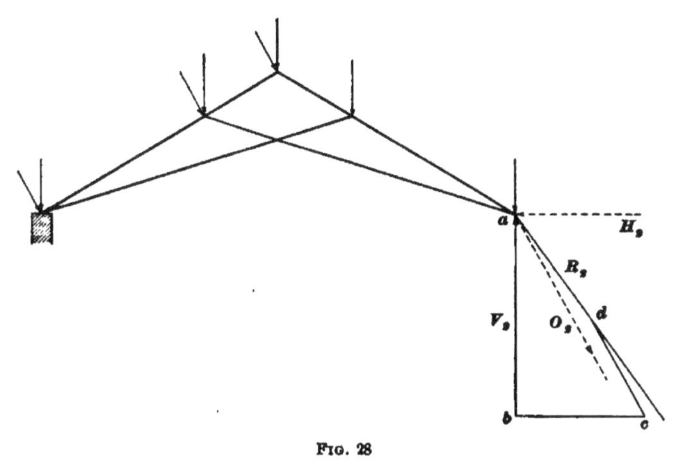

FIG. 28

diagrams; for instance, to find the stresses existing in the truss, it is customary to draw at least a vertical-load diagram and a wind-load diagram, and sometimes it is advisable to draw also a snow-load diagram and even a stress diagram for the wind load on one side of the roof and the snow load on the other. For each stress diagram, the frame diagram remains the same, with the exception that the direction and amounts of the loads at the panel points, as well as the reactions, change. Oftentimes, one outline of the framed structure embodies all of the forces acting on the truss, though where a number of stress diagrams are to be drawn, considerable confusion is usually avoided by drawing separate frame diagrams. In the design of all roof trusses, the wind and vertical loads should be considered separately, though the less conservative engineers do not consider the separate effects of the wind and snow on spans of less than 100 feet.

NOTATION

29. In order that the external forces or the internal stresses exerted in the frame diagram may be intelligently designated in the stress diagram, so that by inspection one may know that a particular line in the stress diagram designates the amount and intensity of the stress in a particular member in the frame diagram, some convenient **notation** must be employed. A notation should be used such that, in analyzing the stresses around a single joint, the polygon of forces may be readily traced and the completion of the polygon of forces will be immediately known when the point from which the polygon was started has been finally reached. The system of lettering, or notation, commonly used in graphical statics, consists in placing a capital letter in every space throughout the frame diagram between the several members and the forces acting externally. The lettering is usually commenced at the left-hand end of the diagram and runs around the figure in the direction traveled by the hands of a clock; the internal spaces are then lettered in alphabetical order, commencing likewise at the left-hand end of the figure. The middle point of the diagram, or that space included between the two principal reactions, is usually denominated by the letter Z.

In Fig. 29 (a) is shown a frame diagram lettered with the notation just described. The first vertical load at the left-hand end of the truss is known as the load, or force, AB; the second load as BC, etc. The three divisions of the tie-member beginning at the left are designated as GZ, IZ, and KZ, while the several portions of the rafter members are known as GB, CH, JD, and EK. The struts are GH and JK, while the tension rods are denominated HI and IJ.

The stress diagram that would be laid out in order to analyze the stresses in the frame diagram at (*a*) is shown at (*b*); the same letters designate the stresses or forces similarly marked in the frame diagram, with the exception that small, or lower-case, letters are used instead of capitals; for instance, *a b* represents to scale the force *A B*; *b c* represents the direction and intensity of the force *B C*; *c d*

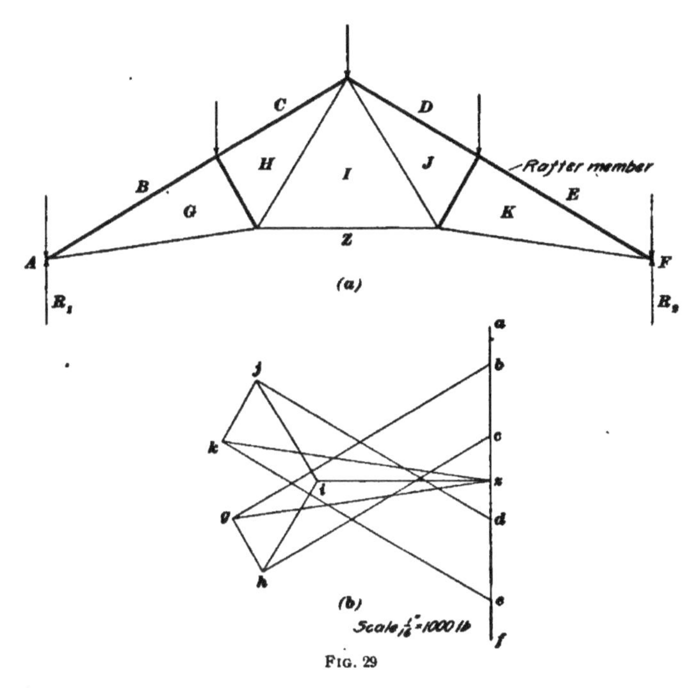

FIG. 29

likewise stands for *C D* in the frame diagram. The reaction R_1 is designated in the stress diagram by *z a*, while *f z* will give, on measuring, the amount of the reaction R_2.

The convenience of this notation consists in the fact that the frame and stress diagrams bear a peculiar relation to each other in that where any system of forces in the frame diagram acts at a single point, there is a corresponding closed polygon represented in the stress diagram. For example, the apex of the truss is a point at which five forces meet—they are the vertical load *CD*, the stress in the right-hand rafter member *DJ*, the stress in the oblique tension member *JI*, and the stresses in the left-hand rafter member and tension member, respectively, designated as *HC* and *IH*. In the stress diagram these forces do not meet at a single point, but form the polygon *cdjih*, the last line *hc* of the polygon making a closed figure. On the other hand, any closed figure in the frame diagram, such as the triangular space including the letter *G*, is designated in the stress diagram by a number of lines meeting at a common point; that is, the space *G* in the frame diagram is enclosed by the members of the frame designated as *BG*, *GZ*, and *GH*, while in the stress diagram the lines that represent the stresses in these several members meet at the point *g*, and are *bg*, *gz*, and *gh*.

It is not absolutely necessary to use this system of lettering, for any letters or numbers may be employed, so long as no two spaces in the truss are marked alike. It is well, however, to have some definite system in engineering work, and the foregoing, being as convenient as any that can be suggested, is universally adopted.

DETERMINATION OF EXTERNAL FORCES

30. Before the stress diagram for any framed structure can be drawn, it is necessary to complete the polygon of external forces, which consists of the loads on the frame and their reactions. This step is one of the most important in graphical solutions; for if this polygon is incorrectly laid out, it will be impossible to close the stress diagram. The loads supported on a frame are always considered as concentrated at the joints of the frame; in designing roof trusses, the purlins are either located at the joints along the rafter members or else the loads are considered as being transmitted to these points by the transverse strength of the member between the panel points. In Fig. 30 is shown a simple roof truss with its loads and reactions. The panel points, at which the loads on the rafter members are considered as being concentrated, are *a*, *b*, *c*, *d*, and *e*. The load at *a* is the portion of the wind pressure or the weight of the roof supported by one-half of the portion of the rafter member between *a* and *b*, while the load on the joint *b* would be the wind pressure, or the weight of the roofing material supported by one-half of the rafter member, or one-half of *a b* and *b c*. The load at the apex of the roof is equal to the weight on one-half that portion of the truss included between *b c* and *c d*.

The principal difficulty in determining the polygon of external forces is to find the amount of the reactions, or the forces opposed to the loads, in order to create equilibrium in the structure. When the frame or truss is symmetrical and the

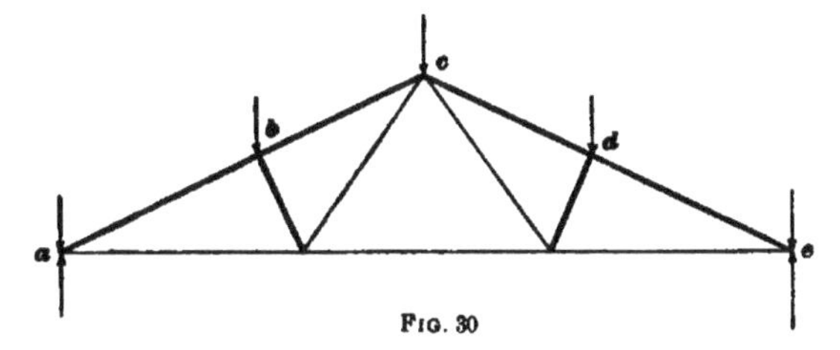

FIG. 30

line of action of the reactions is parallel with the direction of the loads, the reactions are obtained by inspection, for they are each equal to one-half the sum of the loads on the truss or frame. Should the truss or frame, however, be unsymmetrically loaded or should the reactions be other than parallel with the line of action of the loads, their direction must be assumed and their amounts determined by calculation, or both their amounts and direction may usually be found by the graphical method.

DETERMINING THE REACTIONS BY CALCULATION

31. The method of determining, by calculation, the reactions for any framed structure involves the principle of moments. If the frame will always be regarded as a solid section or body acted on by external forces, in this way eliminating the internal members or stresses, little difficulty will be experienced in determining the method of procedure for finding the reactions.

In Fig. 31 (*a*), (*b*), and (*c*) is shown a type of simple roof truss loaded in several different ways. In each instance, to determine the reactions by calculation, it is first necessary to evolve the formula and then, deciding on certain values, make the necessary calculations.

In (*a*) is shown a truss unsymmetrically loaded, that is, corresponding loads each side of the center line are not of the same amount, so that the reactions will not each be equal to one-half of the total load on the truss, though from the fact that the reactions coincide with the direction of the loads, it is known that the sum of the reactions is equal to the sum of the loads.

In order to determine the reaction R_2, the truss is considered as being hinged at the point c and the algebraic sum of the moments of all the external forces about this point should equal zero, or the sum of the moments of the loads equals the moment of the reaction R_2, the latter moment being the product of R_2 and the length S. If the sum of the moments due to the loads is divided by the lever arm of R_2, the amount of R_2 will be determined. If the loads on this truss are represented by w_1, w_2, w_3, etc., and their respective lever arms about the point c designated by x_2, x_3, etc., the algebraic sum of the moments about the point c will equal $w_2x_2 + w_3x_3$, etc.; and since the lever arm of R_2 is equal to the span of the truss, or S, the amount of R_2 may be determined by the expression

$$R_2 = \frac{w_2x_2 + w_3x_3 + w_4x_4 + w_5x_5}{S} \qquad (1)$$

As the load w_1 is coincident with the reaction R_1, its moment is zero and is therefore not included in the formula.

For example, assume that the span of the truss is equal to 40 feet and that the distances x_2, x_3, x_4, x_5 are equal, respectively, to 10, 20, 30, and 40 feet; w_1 equals 1,000 pounds; w_2 equals 2,000 pounds; w_3 equals 2,500 pounds; w_4 equals 800 pounds; w_5 equals 600 pounds.

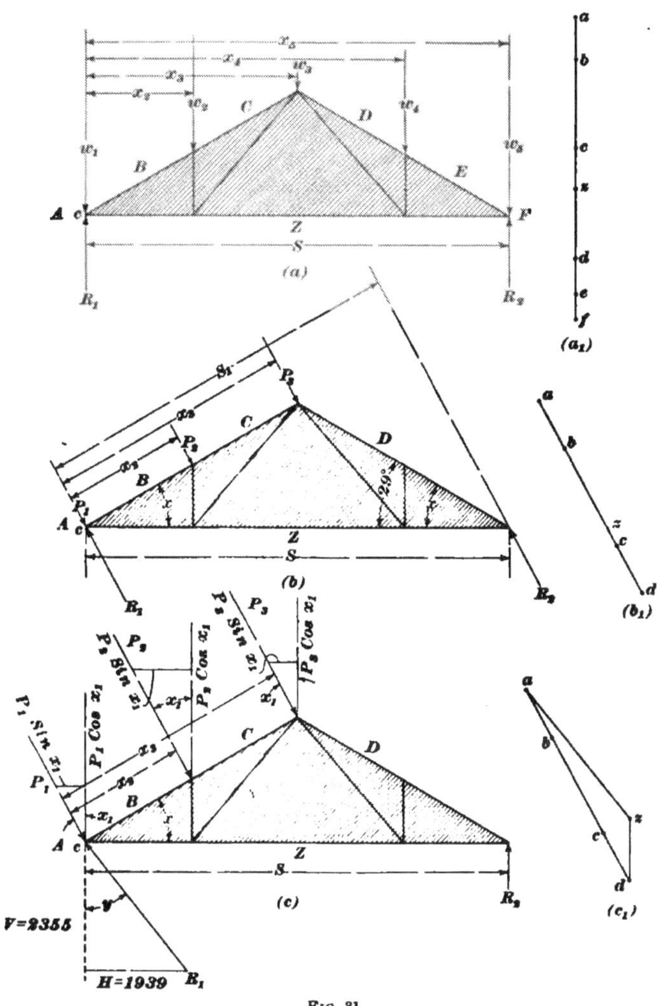

FIG. 31

Since the calculation for moments can best be systematically arranged by tabulation, the sum of the moments of the loads will be expressed as follows:

$w_2 x_2$ = 2,000 lb. × 10 ft. = 20000 ft.-lb.
$w_3 x_3$ = 2,500 lb. × 20 ft. = 50000 ft.-lb.
$w_4 x_4$ = 800 lb. × 30 ft. = 24000 ft.-lb.
$w_5 x_5$ = 600 lb. × 40 ft. = 24000 ft.-lb.

Total moment about c = 118000 ft.-lb.

The moment of R_2 about the point c must equal the sum of the moments of the loads, and hence the reaction R_2 equals 118,000 ÷ 40, or 2,950 pounds. The sum of the reactions must equal the sum of the loads, and since the sum of the loads equals 6,900, the amount of R_1 will equal 6,900 − 2,950, or 3,950 pounds.

The polygon of external forces, as designated at (a_1), may now be laid out, and because all the external forces on the truss are coincident in direction, the force polygon will extend in a straight line. This polygon will be laid out from a to b, from b to c, from c to d, from d to e, from e to f, and back from f, locating the point z, by scale, at a distance from f equal to the reaction R_2, or 2,950 pounds, as determined by the calculation. If the polygon of the external forces has been accurately laid out, the distance $z\,a$ will, on scaling, be found to equal the reaction R_1, or, as determined, 3,950 pounds.

32. In Fig. 31 (b) is shown the frame diagram of a roof truss sustaining on the left-hand rafter member, at the several panel points, the wind loads P_1, P_2, and P_3. It is considered that the roof truss is securely fixed to a wall or other support at both ends, and under such conditions the reactions coincide with the direction of the normal wind loads. Since this is similar to the preceding case, the sum of the reactions is equal to the sum of the loads, but the reactions, by inspection, are evidently not equal to each other.

The amount of R_2 may be determined by considering the moments of the external forces with the point c as the center of moments. The lever arms of the loads P_2, P_3 are not, as before, horizontal, but being at right angles with the line of action of the wind pressure, which is normal to the slope, they are parallel with the rafter member or slope. The reaction R_2 does not act, in this instance, with a leverage equal to the span of the truss, but exerts its force through a lever arm at right angles to its action, as represented by S_1, and this distance may be obtained by extending the line of action of R_2 and scaling, or it may be calculated. The distance S_1 is always equal to the span of the truss divided by the secant of the angle x; or expressed algebraically, where S = the span of the truss, $S_1 = \frac{S}{\sec x}$. In case a secant table is not at hand, S_1 can also be found by means of the cosine of the angle x, as follows: $S_1 = S \cos x$.

The equation for determining the reaction R_2 of the roof truss shown in (*b*) may then be expressed by the formula

$$R_2 = \frac{P_2 x_2 + P_3 x_3}{S_1}$$

which is the same as

$$R_2 = \frac{(P_2 x_2 + P_3 x_3) \sec x}{S}$$

or

$$R_2 = \frac{P_2 x_2 + P_3 x_3}{S \cos x} \qquad (2)$$

To apply this information it will be assumed that it is desirable to obtain the amount of R_2 when P_2 = 2,000 and P_3 = 1,000 pounds, the distances x_2 and x_3 being equal, respectively, to 11.43 and 22.86 feet. The angle x is, approximately, 29°; then S_1 equals the secant of 29°, or 1.14335 divided into the span, or 40 feet, which gives 34.98 feet. When this value has been obtained, the calculation for the reaction R_2 will be as follows:

Positive moments of wind pressure about the point *c* are:

$P_2 x_2$ = 2,000 lb. × 11.43 ft. = 22860 ft.-lb.
$P_3 x_3$ = 1,000 lb. × 22.86 ft. = 22860 ft.-lb.

Total moment due to wind pressure = 45720 ft.-lb.

This moment, in order that the frame may be held in equilibrium, must be equal to the moment of R_2 about the point *c*, and by dividing the amount just obtained by S_1, R_2 is found to be equal to 45,720 ÷ 34.98, or 1,307 pounds. The amount of R_1 is found by deducting the amount of R_2 from the sum of the loads; therefore, if P_1 equals 1,000 pounds, R_1 equals 4,000 − 1,307, or 2,693 pounds.

The stress diagram, or polygon of external forces, for the frame shown in (*b*) is designated in (b_1). The load line in this instance is oblique, since it must coincide with the direction of the external forces, and when the forces *a b*, *b c*, and *c d* have been laid off and the amount of the calculated reaction R_2 measured from *d*, thus locating the point *z*, the length measured from *z* to *a*, by scale, will check the calculations, if it equals R_1. Having in this manner located the point *z*, the polygon of external forces will extend from *a* to *b*, from *b* to *c*, from *c* to *d*, from *d* back to *z*, and from *z* to the starting point, thus completing the figure.

33. The frame shown in Fig. 31 (*c*) is loaded with several oblique panel loads caused by the wind pressure; the moments of these loads about the point *c* are determined in the same manner as were the moments of the loads shown in (*b*). In this instance, however, the truss is considered as being supported on a roller bearing at the right-hand end in order to allow lateral play for the contraction and expansion of the metallic frame. This bearing is regarded as frictionless, so that there is no horizontal resistance whatever, in consequence of which condition the reaction R_2 under a roller bearing is always considered as vertical in direction. Hence, the problem resolves itself into the determination of the amount of this reaction R_2. The sum of the moments due to the several wind loads on the rafter members is the same as for the frame in (*b*) and is equal to $P_1 x_1 + P_2 x_2$. The leverage of R_2, since it acts vertically about the point *c*, is equal to the span of the truss, so that the amount of the reaction R_2 is determined from the formula

$$R_2 = \frac{P_1 x_1 + P_2 x_2}{S} \qquad (3)$$

For example, assume that P_1 and P_2 equal, respectively, 2,000 and 1,000 pounds, while x_1 and x_2 equal 11.43 and

22.86 feet; the span, as before, being equal to 40 feet, then

$$R_2 = \frac{2{,}000 \times 11.43 + 1{,}000 \times 22.86}{40} = 1{,}143 \text{ pounds}$$

The other reaction, or R_1, is not vertical, because the truss is fixed at the end, and in consequence the actual sum of the reactions is not equal to the sum of the loads.

It is known, however, that for any frame to be in equilibrium, the algebraic sum of the vertical and horizontal components of all the forces acting on that frame must equal zero. The reaction R_2 is vertical, and consequently has no horizontal component. It is evident, therefore, that the reaction R_1 must have a horizontal component equal to the horizontal components of all the loads, in order that the algebraic sum of the horizontal components of all the loads acting on the truss may be equal to zero. Each of the wind loads P_1, P_2, and P_3 may be regarded as one side of a triangle in which its components constitute the remaining sides. Knowing that the angle x_1 is equal to the angle x, the value of either of the components may be found by means of one of the trigonometric functions. Thus, the horizontal component is equal to $P \sin x_1$, and the sum of the horizontal components of the three oblique forces on the rafter member will equal $(P_1 + P_2 + P_3) \sin x_1$. The sine of 29° is equal to .48481, and the sum of the loads P_1, P_2, and P_3 is equal to 4,000 pounds, so that the sum of the horizontal components of the oblique forces, or H, equals 4,000 $\times$ .48481, or 1,939 pounds.

Having in this manner obtained the horizontal component that R_1 must possess in order to equal the horizontal components of the loads, it is necessary in a similar manner to determine the vertical component of R_1. Since R_2 acts vertically, its amount must be subtracted from the sum of the vertical components of the oblique forces P_1, P_2, and P_3. Trigonometrically, the sum of the vertical components of the pressures equals $(P_1 + P_2 + P_3) \cos x_1$, or substituting the values, the vertical component $V = 4{,}000 \times .87462$, or 3,498 pounds. From this must be deducted the entire amount of R_2, and when this is done there remains for a vertical component of R_1 an amount equal to $3{,}498 - 1{,}143 = 2{,}355$ pounds.

It is now evident that since both the vertical and horizontal components of R_1 have been obtained, the amount and direction of R_1 may be found.

The amount of R_1 may be found by considering the vertical and horizontal components as the two sides of a right triangle and calculating the hypotenuse by the rule of Pythagoras, that is, by taking the square root of the sum of the squares of the vertical and horizontal components. Employing this method, $R_1 = \sqrt{2{,}355^2 + 1{,}939^2}$, or 3,050. The direction of this reaction is still to be determined. The angle y that it makes with the vertical may be found by either of the following trigonometric functions: secant $y = \frac{R_1}{V} = \frac{3{,}050}{2{,}355}$, or cosine $y = \frac{V}{R_1} = \frac{2{,}355}{3{,}050}$. In the absence of a secant table, the latter formula is used. It will be found from the above that secant y equals 1.295, and cosine y equals .7721, either of which corresponds to an angle of about 39° 30′; therefore, the reaction R_1 will be of the intensity calculated, or 3,050 pounds, and will extend in a direction to the right of the vertical at an angle of 39° 30′.

The calculations for R_1 need not have been made, for both its intensity and direction could be obtained graphically in laying out the polygon of external forces. Fig. 31 (c_1) shows the polygon of external forces for the frame shown in (c). The load line extends from a to b, from b to c, from c to d, and then, since the reaction R_2 is the vertical distance from d to the point z, the distance from d to z is made equal to the amount of the reaction R_2, or 1,143 pounds. When the point z has been located, the amount of the reaction R_1 and its direction may be determined by a line connecting the points z and a. If this line in the particular example that has been considered measures, to scale, 3,050 pounds and extends in a direction of 39° 30′ with the vertical, it will coincide with the values obtained by the calculation and will check the diagram. The polygon of external forces in this case will then extend from *a* to *b*, from *b* to *c*, from *c* to *d*, from *d* vertically to the point *z*, and from *z* back to the starting point of the polygon or *a*.

GRAPHICAL METHOD OF OBTAINING THE REACTIONS

34. The reaction of a system of forces acting on a framed structure may be entirely determined by the graphical method. It is assumed that the roof truss shown in Fig. 32 (*a*) is loaded with the vertical loads *A B*, *B C*, etc. at each panel point and that it is desired to obtain the

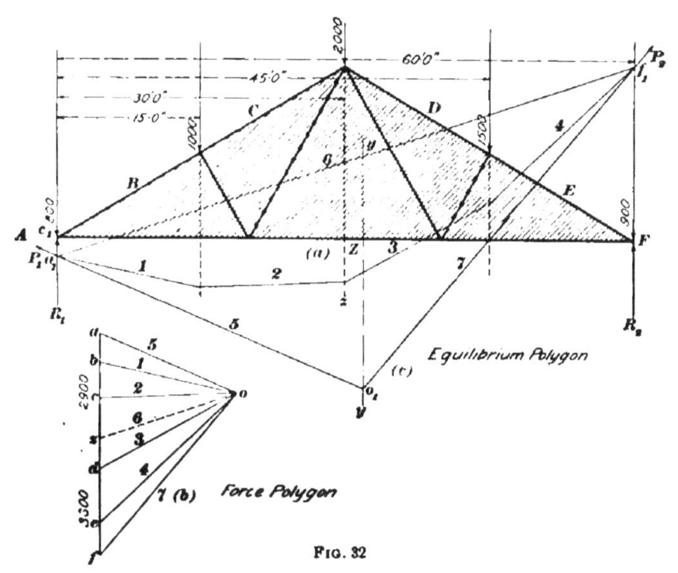

FIG. 32

amount of the reactions R_1 and R_2. If the loads had been symmetrical in amount on each side of the truss, these reactions would have been equal and would have corresponded in amount with one-half the load on the truss. It will be noted, however, that the right-hand half of the truss is more heavily loaded than the left-hand portion, and therefore the reactions will not be equal. Before the stress diagram can be drawn, the unknown external forces R_1 and R_2 must be obtained in order that the location of the point z in the stress diagram may be determined. These reactions can be calculated by the principles of moments explained in Art. **31**, or they can be determined by the graphical method. This method consists in drawing the force polygon (*b*), and from this constructing the equilibrium, or funicular, polygon (*c*).

In laying out the force polygon, lay off the load line designated from a to f. While it is more systematic to lay off the loads on this line in the order in which they occur around the truss, it is not necessary that this procedure should be followed, for these loads may be laid off in any convenient succession. Each portion, however, of the load line must truly represent, to scale, the amount of the force that it designates and its direction. Any point or pole o is selected and lines are drawn from it to the several divisions on the load line. The figure will then appear as shown, with the exception that the line oz has not as yet been determined, for the direction of this line is what is required in order to find the amount of the reactions R_1 and R_2. Having completed the force polygon (*b*), proceed with the equilibrium polygon (*c*). In drawing this polygon, first extend the line of action of all the forces acting on the frame, as shown by the dotted lines, and commence the diagram at the left-hand end by drawing the line *1* parallel with bo in the force polygon and from the intersection of the line *1* with the line of action of the force BC draw line *2* parallel with co in the force polygon; and where this line intersects the line of action of CD draw the line *3* parallel with do. Likewise, from the intersection of *3* with the line of action of DE, draw *4* parallel with eo in the force polygon. This latter line will intersect the line of action of R_2, and from this intersection f_1 draw $f_1 o_1$ in the same direction as of in the force polygon, while from a_1, the left-hand end of the equilibrium polygon, extend the line $a_1 o_1$ parallel with ao in the force polygon. Connect the points f_1 and a_1 in the equilibrium polygon and draw from o in the force polygon a line parallel with $f_1 a_1$. The intersection of this line oz, or *6* in the force polygon, will divide the vertical load line into

fz and za, which, to the scale to which the force polygon was laid out, represent the reactions R_2 and R_1, respectively.

In the equilibrium polygon, the lines *1*, *2*, *3*, and *4* represent the form that a cord or string would assume if it supported the loads BC, CD, and DE at the several points where it changes direction, and its ends were attached at a_1 and f_1. In order that the system might be in equilibrium, the cord would require either the oblique reactions marked P_1 and P_2, or a compression member exerting a compressive stress as represented by the line $a_1 f_1$, and the vertical reactions R_1 and R_2. It is evident from this that P_1 and P_2, or their equivalents, the forces $a_1 o_1$ and $f_1 o_1$ in the equilibrium polygon, are the resultants of two sets of components R_1 and $f_1 a_1$, and R_2 and $a_1 f_1$, respectively. A force designated by two letters should be read in the direction in which the force acts; that is, the letter toward which the arrow representing the direction of the force points should be read last. In this instance, $a_1 f_1$ represents a compressive stress and when speaking with reference to the point a_1, the force is designated as $f_1 a_1$, while if the point f_1 is being considered, the force should be read $a_1 f_1$. In the force polygon, ao corresponds with the force P_1 in the equilibrium polygon and of, with P_2. The components of ao in the force polygon are az and zo, while the components of the force fo are fz and zo. By measuring fz to the scale to which the load line was laid out, it will be found to equal, approximately, 3,300 pounds, while za, measured with the same scale, equals 2,900 pounds.

To prove, or check, the diagrams, the moments of the several external forces may be calculated about the point of application of either R_1 or R_2. Considering the loads on the roof truss as giving positive moments about c_1, their sum is obtained by the following calculation:

Moment of AB = 800 lb. × 0 ft. = 0 0 0 0 0 ft.-lb.
Moment of BC = 1,000 lb. × 15 ft. = 1 5 0 0 0 ft.-lb.
Moment of CD = 2,000 lb. × 30 ft. = 6 0 0 0 0 ft.-lb.
Moment of DE = 1,500 lb. × 45 ft. = 6 7 5 0 0 ft.-lb.
Moment of EF = 900 lb. × 60 ft. = 5 4 0 0 0 ft.-lb.

The sum of moments = 1 9 6 5 0 0 ft.-lb.

The leverage of R_2 about c_1 is equal to 60 feet, and consequently the value of R_2 is found by dividing the sum of the positive moments by this distance; by the calculation it equals 3,275 pounds, while the value of R_1 is the difference between this amount and the sum of the loads, or 2,925 pounds, which corresponds, approximately, with the values obtained by the diagram, thus proving the correctness of the graphical solution.

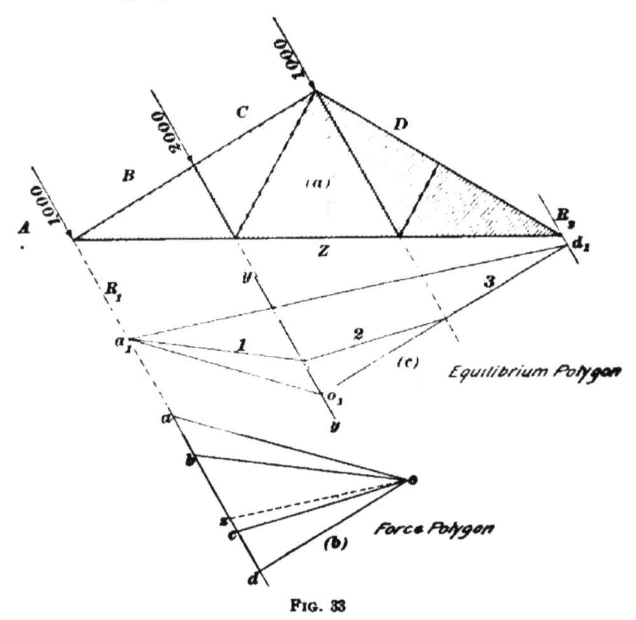

FIG. 33

35. In Fig. 33 (a) is shown a similar roof truss of the same dimensions, but supporting a wind load only on its left-hand portion. These wind stresses act parallel with each other and normal to the slope of the roof, so that if the truss is fixed at the ends, the reactions R_1 and R_2 will coincide in direction with these forces.

To determine the amount of the wind reactions R_1 and R_2, construct the force polygon (b). Lay off the load line from a to d and represent to the proper scale the forces $a\,b$, $b\,c$, and *c d*, which correspond with the wind loads *A B*, *B C*, and *C D*. Connect any point or pole *o* with the several divisions on the load line, thus obtaining the oblique lines *a o*, *b o*, *c o*, and *d o*. Commence the equilibrium polygon (*c*) at any point on the left-hand wind reaction, as a_1, and draw line *1* parallel with *b o*, line *2* parallel with *c o*, and line *3* parallel with *d o*, this last line intersecting the reaction R_2 at the point d_1; $a_1 o_1$ is drawn parallel with *a o* and line *3* is extended until it intersects $a_1 o_1$ at the point o_1. By connecting the points d_1 and a_1 in the equilibrium polygon and drawing in the force polygon, from *o*, a line parallel with the line just described, the point *z* is located and the reaction R_2 is found by measuring *d z*, while the amount of R_1 is known by scaling *z a*.

It will be noticed in drawing this diagram that the lines *1*, *2*, and *3* were drawn between the lines of action of the forces *A B*, *B C*, and *C D*, and that the force *3* coincides with the reaction component $d_1 o_1$. The reason that the last force *2* does not close on the line of action of the reaction R_2 is because there is no force on the right-hand portion of the truss. If there had been a force corresponding with *A B*, the reaction component $d_1 o_1$ would have been distinct from the force *3* and the diagram to the right would have been similar to the portion at the left. On being measured, *d z* and *z a* are found to equal, respectively, 1,350 and 2,650 pounds.

36. In Fig. 34 is shown the method just described applied to a roof truss supporting both the vertical dead and snow loads, and the oblique loads due to the wind on the left-hand portion of the truss. The truss is considered fixed at the ends, but neither the amount nor the direction of the reactions R_1 and R_2 is known, so that these must be determined by the graphical method. As in the previous cases, the force polygon is laid out as shown at (*b*). Commence this diagram by laying off the load line, which extends from *a* to *k*. In designating the several forces making up this line, they were for convenience taken in sequence, that is, in the order in which they occur around the truss. For instance, in the force polygon *a b* represents *A B* in the frame diagram; *b c*, the oblique wind force *B C*; *c d*, the vertical load *C D*; *d e*, the oblique wind load *D E*; etc. Since the forces in the load line are not coincident with regard to direction, their resultant will not lie along the load line, but

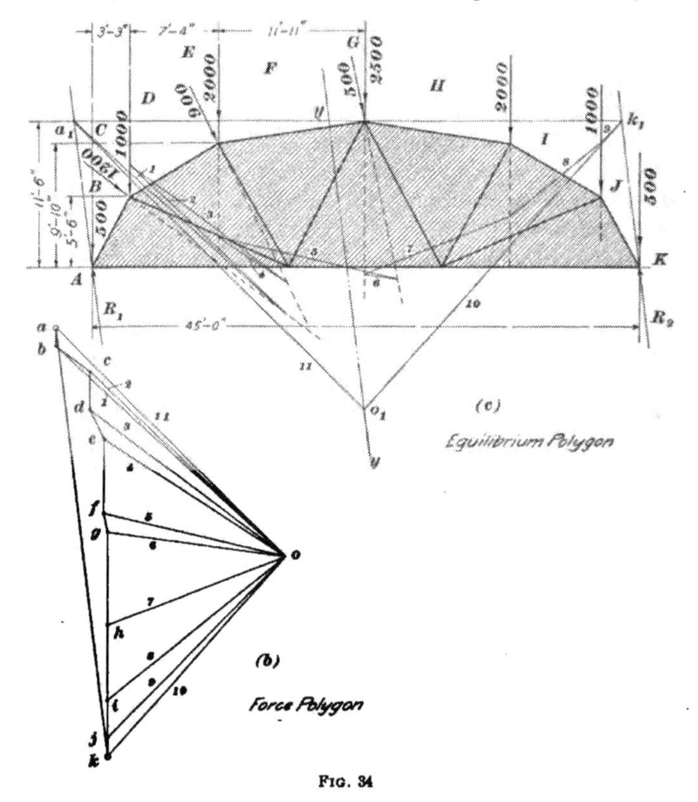

FIG. 34

will extend from *a* to *k*. This line will represent, by scale, the sum of the reactions R_1 and R_2 and its direction coincides with the direction of the reactions. It is then in order to determine what percentage of the length of this line will represent R_1 and what percentage will represent R_2; to do this, the equilibrium polygon (*c*) must be drawn.

Before commencing this polygon, connect some pole or point o, in the force polygon, with each point on the load line, as shown by the oblique lines $a\,o$, $b\,o$, $c\,o$, $d\,o$, $e\,o$, etc.; also, extend the line of action of each of the forces acting on the truss, as shown by the dotted lines. Start the equilibrium polygon from any point a_1 on the line of action of R_1 by drawing line *1* parallel with $b\,o$ in the force polygon. This line should be drawn until it intersects the line of action of $B\,C$. The line *2* is drawn from this intersection in a direction parallel with $c\,o$ in the force polygon and of such a length that it intersects the line of action of the force $C\,D$. From this intersection, line *3* is drawn parallel with $d\,o$ in the force polygon until it intersects $D\,E$ extended. From this intersection, line *4* is drawn parallel with $e\,o$, and where this line intersects the force $E\,F$ extended, line *5* is drawn parallel with $f\,o$ intersecting $F\,G$, and from this point line *6* is drawn parallel with $g\,o$ intersecting the force $G\,H$ extended. Line *7* is drawn from this intersection parallel with $o\,h$, and from the intersection of line *7* with the extended force $H\,I$, line *8* is drawn parallel with $i\,o$, intersecting $I\,J$, and finally, line *9* is drawn in a direction parallel with $j\,o$, intersecting the line of action of the reaction R_2 at the point k_1. From this point, line *10* is drawn parallel with $k\,o$ in the force polygon and intersecting at o_1 line *11* drawn from a_1 parallel with $a\,o$ in the force polygon.

37. The reactions for the several frames described in connection with Figs. 32, 33, and 34 can be found from the equilibrium polygon. In each case a line $y\,y$ drawn through the point o_1 and parallel with the reactions will divide the cord connecting the points on the two reactions into two such parts as to inversely represent the amount of the reactions; for instance, in the equilibrium polygon in Fig. 32, the line $y\,y$ divides the line $a_1\,f_1$ into two such parts that if the entire line $a_1\,f_1$ is considered as representing the sum of the reactions R_1 and R_2, $y\,f_1$ will equal R_1 and $a_1\,y$ will equal R_2. Again, in the equilibrium polygon, Fig. 33, the line $y\,y$ drawn through o_1 parallel with the load line divides the line $a_1\,d_1$

into two such parts that the portion to the right equals the amount of the reaction R_1, while the portion to the left equals the amount of the reaction R_2, considering that the entire length of the line represents the sum of the

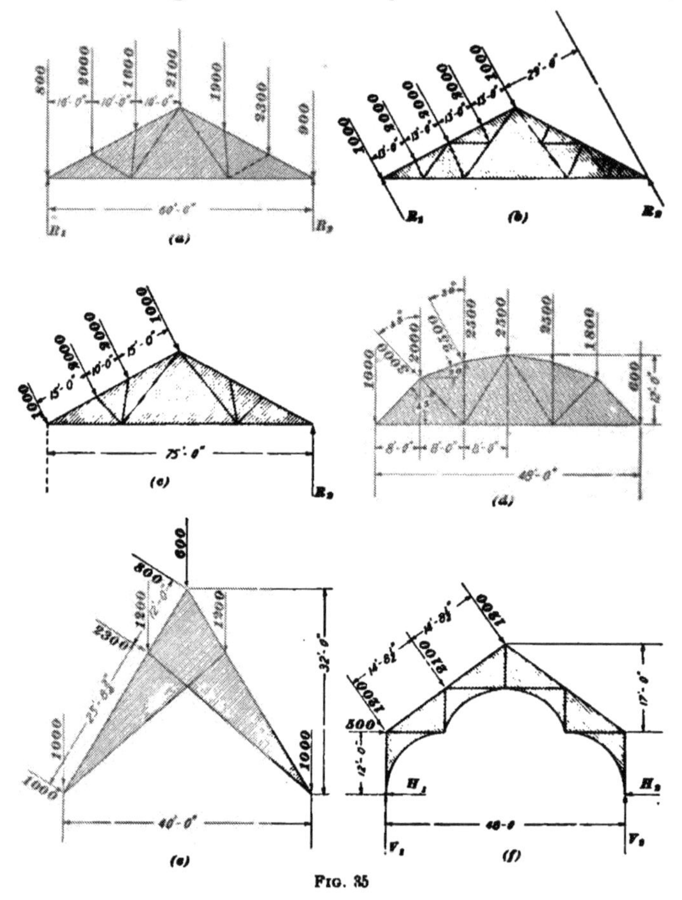

Fig. 35

reactions. Again, in Fig. 34, the line yy coincides with the direction of the reactions and divides $a_1\,k_1$ into such proportion of parts that $y\,k_1 : a_1\,k_1 = R_1 : R_1 + R_2$, and $a_1\,y : a_1\,k_1 = R_2 : R_1 + R_2$.

EXAMPLES FOR PRACTICE

1. Determine, by calculation, the amount of the reactions R_1 and R_2 for the frame shown in Fig. 35 (*a*).

Ans. $\begin{cases} R_1 = 5,600 \\ R_2 = 6,000 \end{cases}$

2. Calculate the amount of the reactions R_1 and R_2 of the frame shown in Fig. 35 (*b*).

Ans. $\begin{cases} R_1 = 5,432 \\ R_2 = 2,568 \end{cases}$

3. Find the amount of R_2, by calculation, for the frame in Fig. 35 (*c*), and determine the amount and direction of R_1 by drawing the polygon of external forces. The right-hand end of the truss is on rollers, and the left-hand end is fixed.

Ans. $\begin{cases} R_1 = 4,566 \\ R_2 = 1,600 \end{cases}$

4. Determine, by the graphical method, the direction and the amount of the reactions for a frame loaded as shown in Fig. 35 (*d*).

Ans. $\begin{cases} R_1 = 8,000 \\ R_2 = 9,500 \\ \text{Direction: } 11^\circ \text{ with vertical} \end{cases}$

5. Employ the graphical method in determining the reactions for the scissor truss, shown in Fig. 35 (*e*), when loaded with vertical loads and oblique forces as designated.

Ans. $\begin{cases} R_1 = 2,950 \\ R_2 = 5,050 \\ \text{Direction: } 26^\circ \text{ with vertical} \end{cases}$

6. In Fig. 35 (*f*) is shown a hammer-beam roof truss. Find the vertical and horizontal components of the resultant reactions at both ends, either graphically or by calculation.

Ans. $\begin{cases} H_1 = 1,275 \\ H_2 = 1,875 \\ V_1 = 1,525 \\ V_2 = 2,150 \end{cases}$

DETERMINATION OF INTERNAL STRESSES

METHOD BY SOLUTION OF JOINTS

38. Maxwell's Method.—The method generally adopted for determining the internal stresses of a framed structure is known as **Maxwell's method.** After the reactions have been found, either analytically or graphically, the force polygon for each joint in the frame is found. The assembled polygons of forces, which are laid out for each joint, form the stress diagram, which it is usual to commence at the left-hand end of the frame and work around consecutive joints. It is not always possible, however, to follow this order, for often the joint in sequence contains more than two unknown stresses, in which case it becomes necessary to analyze some other joint in the frame around which there are only two unknown forces. In working around the frame, it is usual to begin at the left-hand end of the truss and read off and lay out the stresses around the joint in the same direction as the movement of the hands of a watch. It would be possible to work around each joint in the opposite direction, that is, in a direction opposite to the movement of the hands of a watch. But whatever may be the direction of the first joint analyzed it must be followed in analyzing the others.

In order that this important method, which will be used extensively in the application of graphical statics in analyzing the internal stresses of a framed structure may be thoroughly understood, the example illustrated in Fig. 36 will be assumed. In this figure is shown the frame and vertical-load stress diagrams for a quadrilateral truss. The first procedure in the analysis of any frame by the graphical method is to determine all the external forces. In this instance the loads are given, and since the truss is symmetrically loaded, the reactions are each equal to one-half of the loads on the truss, so that R_1 and R_2 are each equal in amount to 2,100 pounds, and are designated, respectively, as AZ and FZ, when the notation explained in Art. **29** is employed. The polygon of external forces, since all of these forces are coincident in direction, is contained within a straight line, which is represented in the stress diagram from a to f. This load line is determined by laying off the force ab equal in amount to AB in the frame diagram; likewise, bc equal to BC; and following around the truss in this manner until the point f has been located. When f has been determined, measure upwards on the load line an amount equal to R_2, or 2,100 pounds to scale, in this manner locating the point z, so that fz represents, by scale, the amount of the reaction R_2. The distance from z to a should then measure, by scale, an amount equal to R_1. The polygon of external forces will now be a closed figure and will extend, beginning at the left-hand foot of the truss and taking the forces in consecutive order toward the right, from the point a in the stress diagram to b,

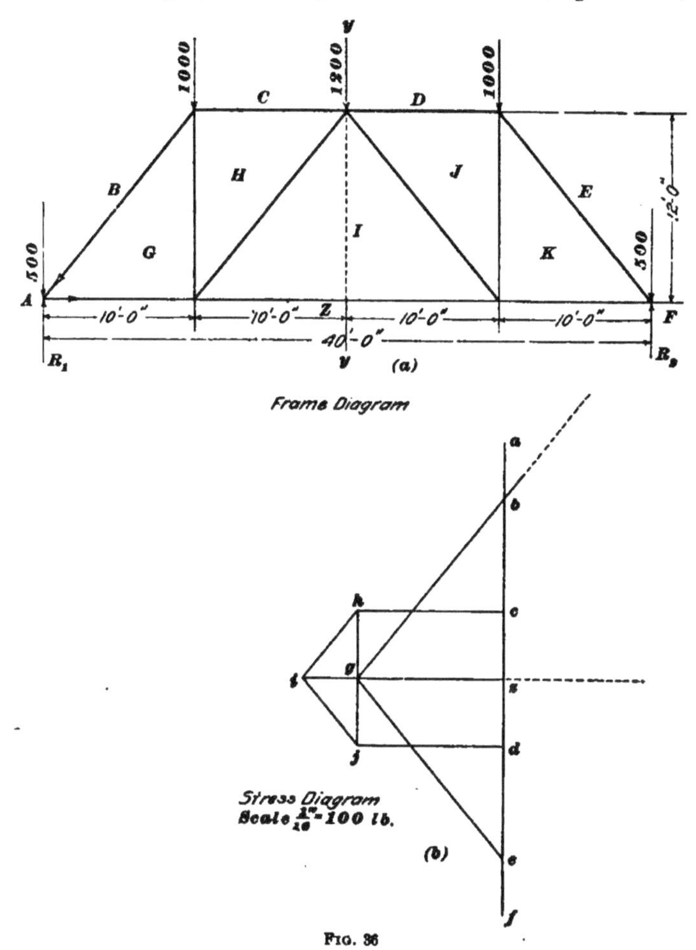

FIG. 36

from b to c, from c to d, from d to e, from e to f, and from f to z upwards, corresponding with the direction of R_2, and from z upwards to a, likewise corresponding in direction to R_1, the last line thus closing the polygon. It is important to remember that before the stress diagram can be attempted, the polygon of external forces must be completed. If this polygon is incorrectly drawn or does not close, the diagram cannot be correctly laid out.

Having determined the polygon of external forces, the joint *A B G Z* in the frame diagram can be analyzed. At this joint in the frame there are only too unknown stresses, *B G* and *G Z*. The known stresses, *A B* and *Z A*, are already laid out in the stress diagram and are correspondingly lettered. To find the stresses in the members *B G* and *G Z*, draw an indefinite line from the point *b* in the stress diagram parallel with *B G* in the frame diagram, and from *z* draw a line parallel with *G Z*. The point of intersection of these two lines will be the point *g* and the polygon for the system of forces meeting at the extreme left-hand joint of the truss will read, in the stress diagram, from *a* to *b*, from *b* to *g*, from *g* to *z*, and from *z* back to *a*, the point of commencement.

39. Determination of the Kind of Stress.—In drawing the stress diagram the question naturally arises as to the direction in which these lines should be drawn, since a line might be drawn from *b* upwards, as shown by the dotted line, being still parallel with *B G* in the frame diagram; and in like manner the line *z g* might be drawn in a direction from *z* toward the right instead of toward the left. It is quite evident, however, that if the lines were drawn as suggested and shown dotted, they would not intersect and the polygon for the system of forces around the left-hand joint of the frame would not close. It is clear, therefore, that the lines that represent the stresses in the members *B G* and *G Z* can be drawn only in one direction and the direction in which they are read determines the kind of stress that exists in the member. For instance, in drawing the polygon for the stresses around the joint *A B G Z*, the force *a b* in the stress diagram is downwards. The stress from *b* to *g* is likewise downwards and may be designated by an arrow on *B G* in the frame diagram; *g z* is to the right, likewise shown by the arrow on this member in the frame diagram, the final line of the polygon being upwards from *z* to *a* and thus representing the amount of the reaction R_1.

As the stresses are read, their direction should be marked with arrows on the frame diagram; this shows the direction of the stress and designates whether it is compression or tension. Forces acting away from a joint are always tensile stresses; those acting toward a joint are always compressive stresses.

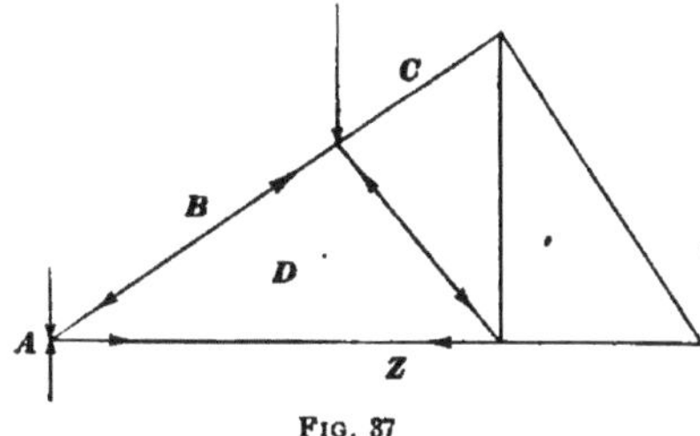

Fig. 37

If there is tension at one end of a member, it is evident there must be an equal amount of tension at the other end; if there is compression at one end of a member, there is an equal amount at the other.

An easy way to remember whether the arrows designate compression or tension by their direction, as shown on the members in a frame diagram, may be seen by referring to Figs. 37 and 38. The member *D Z*, Fig. 37, is a tension member; the arrows point away from the joints and toward each other, and resemble the form of an elastic material, stretched, as shown at *A*, Fig. 38; while in the compression member *B D*, the arrows act toward the joints, or away from each other, and resemble the form that a plastic material assumes on being compressed, as shown at *B*, Fig. 38.

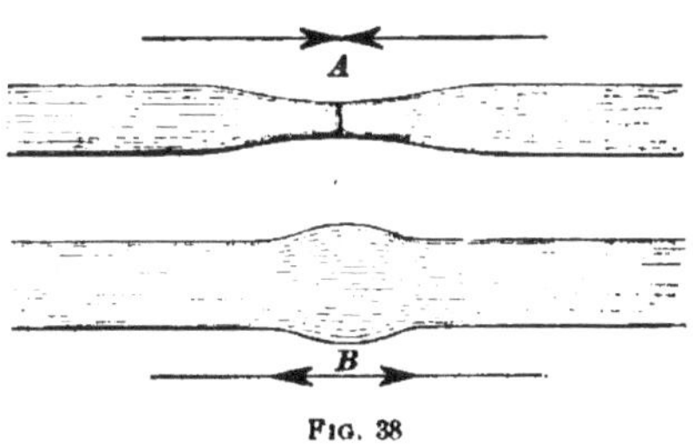

Fig. 38

40. Proceeding with the solution of the frame diagram shown in Fig. 36, the joint *B C H G* may next be analyzed. Around this joint the only unknown forces are the stresses in the members *C H* and *H G*. To find the amount of these stresses, draw a line in the stress diagram from the point *c* parallel with *CH*, and from *g* draw a line upwards parallel with *GH*; the point of intersection will be *h*. The polygon of forces about the joint under consideration will then extend in the stress diagram from *b* to *c*, from *c* to *h*, from *h* to *g*, from *g* back to the starting point *b*. The joint *CDJIH* cannot as yet be analyzed, for there are three unknown forces represented by the members *DJ*, *JI*, and *IH*, but the joint marked *GHIZ* can be analyzed, for the only unknown stresses exist in the members *HI* and *IZ*. To analyze this joint, draw from the point *h*, already determined, an indefinite line parallel with *HI* in the frame diagram, and from *z* a line parallel with *IZ*, intersecting this first line at *i*; thus, the polygon of forces around the joint *GHIZ* is complete and extends in the stress diagram from *g* to *h*, from *h* to *i*, from *i* to *z*, and from *z* back to *g*. In analyzing this last joint, the stress in the member *h i* was determined, so that there now exist only two unknown forces at the joint *CDJIH*. By analyzing this joint, the polygon of external forces will be found to extend in the stress diagram from *c* to *d*, from *d* to *j*, from *j* to *i*, from *i* to *h*, and from *h* back to the point *c*, and it will be found that the stress diagram will begin to repeat on the lower side of the line *z i*. If the frame diagram is further analyzed, it will be found that the stress diagram is balanced on the line *z i* and is symmetrical about this axis, showing that the truss is symmetrically loaded. Since this is the case, only one-half of the stress diagram need be drawn, for the stresses in the members on each side of the center line of the frame diagram marked *yy* are the same, owing to the fact that the truss is symmetrically loaded.

The stresses in the several members may now be determined by measuring their respective lines in the stress diagram, and a tabulation of these stresses will give a convenient table for designing the several members to sustain tensile and compressive stresses.

GRAPHICAL STATICS BY THE METHOD OF SECTIONS

41. Another method of drawing the stress diagrams for any framed structure is by the **method of sections,** and while it differs somewhat from the method just explained, the results are the same. In general, this method consists of passing a plane through any section of the frame in such a manner that it will intersect only two members in which the forces are unknown, these two unknown stresses being obtained by completing the polygon of forces necessary to produce equilibrium in that portion of the frame lying to one side or the other of the plane. Since the method is more readily explained by an application of its principles to an example, the frame diagram of a quadrangular truss, shown in Fig. 39, will be analyzed. It is necessary to lay off the load line in the same manner as in the analysis of the joints of the frame by considering the forces about the joints. The load line for the frame diagram, Fig. 39, will then be a vertical line extending from *a* to *b*, *b* to *c*, etc., from *f* back to *z*, giving the reaction R_2, and from *z* to the point *a*, the length of the latter line being equal to the reaction R_1. Commencing at the left-hand joint of the frame diagram, pass a plane through the frame, cutting the several members along the line *a b*. The portion of the frame to the left of this plane is held in equilibrium by the reaction R_1, the force *A B*, and the stresses in the members *B G* and *G Z*, the only two unknown forces about the joint consisting of the latter. These four forces form a system that acts on the portion of the truss to the left of the line *a b*, as is shown in (*a*). The direction of the unknown forces being known, they may be drawn in the stress diagram as shown, and their point of intersection will be *g*. Another plane is now passed through the frame, cutting the members along the line *c d* and the left-hand portion of the truss is held in equilibrium by the reaction R_1, the loads *A B* and *B C*, and the stresses *C H*, *H G*, *G Z*. The portion of the frame held in equilibrium by these forces to the left of the line *c d* is shown in (*b*). The only two unknown forces acting on the left-hand portion of the frame are CH and HG, for all the other forces have been shown in the stress diagram. To find these forces, then, and close the polygon for this portion of the frame, all

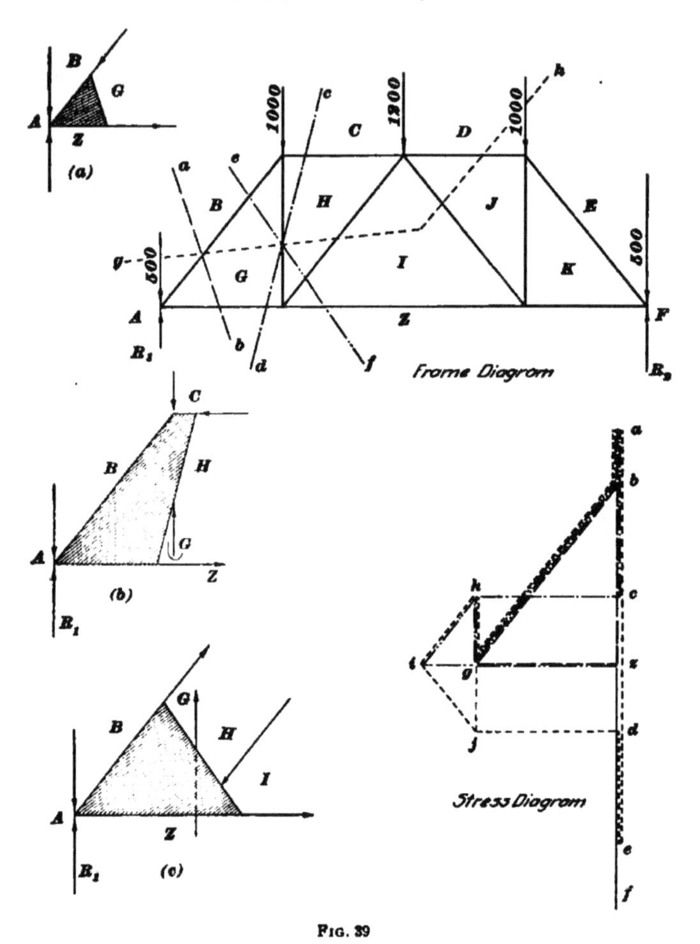

FIG. 39

that it is necessary to do is to draw the lines ch and hg in the stress diagram, thus locating the point h. Again, assume that a plane is passed through the frame along the line ef;

the left-hand portion of the frame will be held in equilibrium by the forces shown in (*c*), which are the reaction R_1, the load *A B*, and the stresses in the members *B G*, *G H*, *H I*, and *I Z*. The two unknown stresses are *H I* and *I Z*, and their amounts are determined when the lines *h i* and *i z* have been drawn in the stress diagram, and the polygon of forces for the left-hand portion of the truss has been completed. A separate line has been used in the stress diagram to represent the polygon of forces in each case, in order that the solution of each section may be clear.

This system can be continued throughout the truss, and it will be found on comparison of the frame diagram with the stress diagram, that any plane, such as *g h*, will cut the truss in such a manner that each portion will be in equilibrium; and that when the stress diagram has been entirely completed the polygon will close for either portion of the frame diagram cut by the plane. This system of graphical statics is preferred by many, but the other is generally practiced.

GRAPHICAL ANALYSIS OF STRESSES

(PART 2)

DIAGRAMS FOR SIMPLE FRAMES

1. The principles stated in *Graphical Analysis of Stresses*, Part 1, may be used to analyze the stresses in any framed structure and will be applied in this Section principally to the determination of stresses in the members of roof trusses. An effort will be made to include in the applications all the principal types of trusses, and when special features of the solution are introduced they will be explained so that they may be used advantageously elsewhere than in the particular type under discussion.

2. In Fig. 1 is shown a simple frame, in which the compression member *B D* and the tension members *A D* and *D C* form a triangular structure that supports the downward pull of 1,000 pounds; the triangular frame is supported, in turn, by the reactions *A B*, *B C*. The problem consists in drawing the stress diagram to determine the stress in the various members.

With a scale of 40 pounds to $\frac{1}{10}$ inch, draw the vertical line *c a*, in the stress diagram, equal to 1,000 pounds. This line represents the force *C A* in the frame diagram. Working around the joint *A D C* in the direction of the dotted arrow, the first member encountered is *A D*; hence, from *a* in the stress diagram, draw a line parallel to *A D* in the frame diagram. *D C* being the next member met, from *c* in the stress diagram draw a line parallel to it. The point of intersection of these two lines is *d*. This done, go around the joint again, to see that none of the members have been omitted, and also to get the direction in which the stresses act. Starting at *c* in the stress diagram, and going around

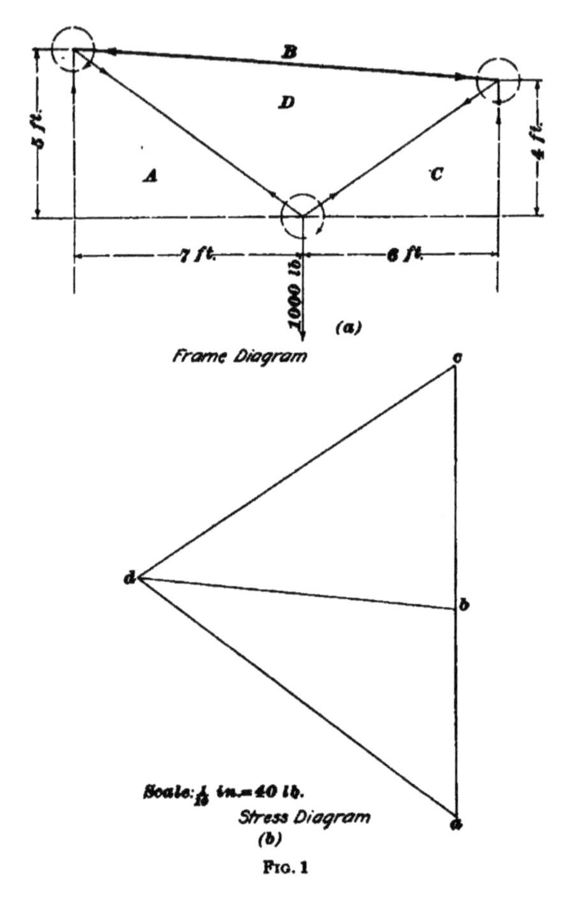

Fig. 1

the joint *C A D*, the polygon of forces is from *c* to *a* to *d*, and from *d* back to *c*, thus arriving at the point from which the start was made. The next joint in the frame diagram is *A B D*. The point *b* on the line *c a* is not known, but may be determined by calculating the reactions AB and BC by the principles of moments. The load of 1,000 pounds is located on the frame 6 feet from the reaction BC. The moment about CB is $1,000 \times 6 = 6,000$ foot-pounds, while the reaction at AB equals $6,000 \div 13 = 461$ pounds. Knowing that the force AB is 461 pounds and that it acts upwards, the point b can easily be located by measuring from a on the line ac in the stress diagram; then the line bd may be drawn and if found parallel to the member BD in the frame diagram, the stress diagram is correct. In this case, however, it is not necessary to calculate the reactions AB and BC, for the point d having been already determined and it being known that the line db must be parallel to DB, all that is needed is to draw a line from d parallel to DB, and the point where it cuts the line ca is b. Having drawn the line db, go around the joints ABD and CDB, marking the direction of the stress by the arrowheads, as shown in the frame diagram. Around the joint ABD the polygon of forces is from a to b, from b to d, and from d back to a. Working around the joint CDB, the polygon of forces is from c to d, from d to b, and from b back to c. This completes the stress diagram; the amount of the stress in the several members of the frame diagram is found by measuring the corresponding lines in the stress diagrams.

3. Diagram for a Small Roof Truss.—Fig. 2 is the frame diagram for a small roof truss, in which the two rafter members EB and CE are connected at their foot by the tension member EZ; the loads and their reactions are as shown in the frame diagram. The problem is to determine the stresses in the several members composing the truss.

Draw the vertical line ad, shown in the stress diagram, Fig. 2. Lay off to any scale, say, 400 pounds to $\frac{1}{16}$ inch, the loads ab, bc, and cd. From the point d, the reaction dz acts upwards, with a force of 8,000 pounds, which determines the point z. Going around the joint $ABEZ$, the reaction ZA acts upwards and AB downwards. Then, from the point b in the stress diagram draw the line be parallel to BE in the frame diagram, and from *z* draw the line *e z* parallel to the member *E Z* in the frame diagram. The point of

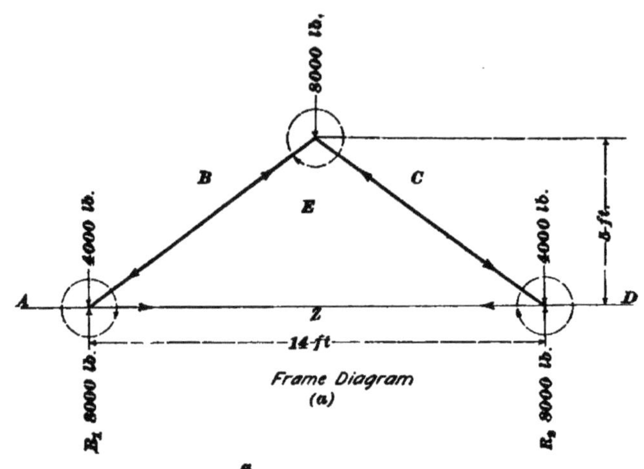

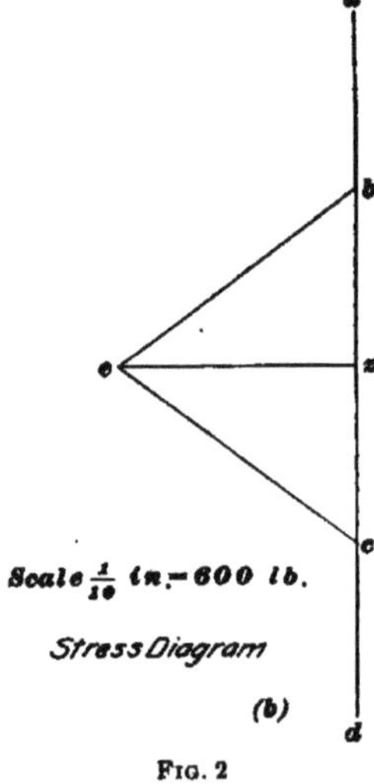

Fig. 2

intersection will be *e*. Having gone thus far, again go around the joint, to get the direction of the stresses in the members and to see whether the polygon of forces is correctly drawn. Go, for instance, from *z* to *a* upwards; *a* to *b* downwards; then from *b* to *e*, and from *e* back to *z*, the starting point. The next joint in the frame diagram is *B C E*. The force *b c* being already determined, from point *c* draw the line *c e* parallel to *C E* in the frame diagram; this line will pass through the point *e* if the diagram has been drawn correctly. The polygon of forces at *E B C* is from *b* to *c* downwards, then from *c* to *e*, and back from *e* to *b*, the starting point. The next joint in the frame diagram is *E C D Z*, and the polygon of forces in the stress diagram is from *e* to *c*, already drawn, from *c* to *d*, *d* to *z*, and back to *e*.

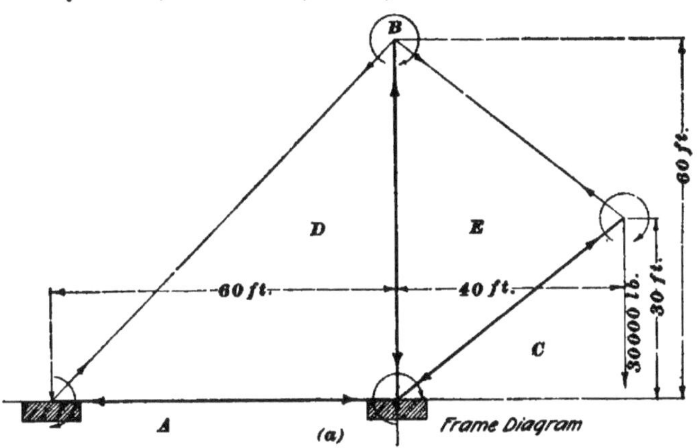

(a)

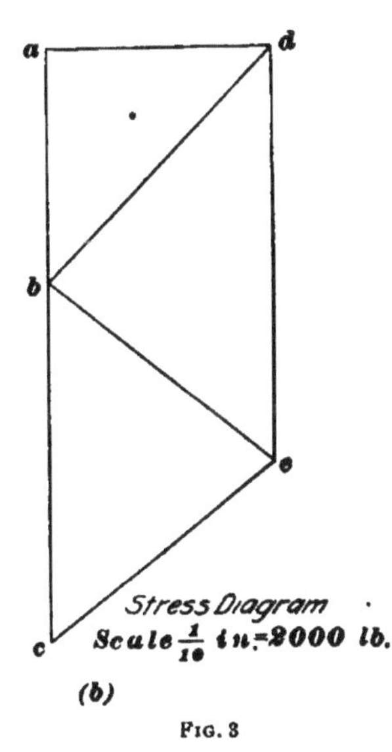

(b)

Fig. 3

The stress diagram being completed, all that remains is to measure the various lines, in the stress diagram, that represent the corresponding members in the frame diagram. Thus, *e b* measures $\frac{17\frac{1}{2}}{10}$ inches, the scale being 400 pounds to $\frac{1}{10}$ inch; the stress in this member is 7,000 pounds; the line *e z* measures about $\frac{14\frac{1}{4}}{10}$ inches, and the stress in the member *e z* is 5,700 pounds. Thus, the stress in any member may be determined.

4. Diagram for a Jib Crane. A jib crane, proportioned as in Fig. 3, has a load of 30,000 pounds suspended at the end of the jib; what are the stresses in the guy ropes, and in the different members of the jib, and what are the reactions *C A* and *D A*?

In the stress diagram draw the vertical line *bc* equal to 30,000 pounds, and from the point *c*, draw the line *ce* parallel to *CE* in the frame diagram. Then from *b* draw the line *eb* parallel to *EB* in the frame diagram. Going around the joint to check the polygon of forces, it is found to be from *c* to *e*, from *e* to *b*, and from *b* back to *c*. The next joint encountered is *EDB*. Hence, from *e*, draw *ed* upwards parallel to *ED*, and from *b* draw *db* parallel to *DB*, the point where these two lines intersect being *d*. The polygon of forces about the joint *EDB* is from *b* to *e* to *d*, and from *d* back to *b*, the starting point. Next, in the joint *CADE*, continue *bc* upwards indefinitely; then from *d* draw *ad* parallel to *AD* in the frame diagram; where the lines just drawn intersect will be the point *a*; *de* and *ec* have already been drawn. The remaining joint to work around is *ABD*. The stress diagram shows that the forces around this joint have already been determined, and the stress diagram is complete.

The stresses in the members may be determined by measuring the lines corresponding to them in the stress diagram, with the scale to which the diagram has been drawn.

5. Roof Truss With a 40-Foot Span.—Fig. 4 (*a*) shows the frame diagram for a roof truss having a 40-foot span. The loads are as shown, the compression members being indicated by heavy lines, and the tension members by light lines; required, to draw the stress diagram for this truss.

First draw the vertical line *af*, as shown in the stress diagram at (*b*); mark the point *a* and lay off on this vertical line, to any scale, using, in this case, 1,500 pounds to $\frac{1}{10}$ inch, the loads *ab*, *bc*, *cd*, *de*, and *ef*, corresponding to the loads *AB*, *BC*, *CD*, *DE*, and *EF* in the frame diagram. The truss being symmetrically loaded, the loads are the same, in amount, on both sides of the center line; the reactions R_1 and R_2 are, therefore, each equal to one-half the load, in this case, 16,500 pounds. Hence, *za* may be laid off on the vertical line, and as R_1 equals R_2, *za* must equal *fz*; consequently, *z* is located centrally between *a* and *f*, or between *c* and *d*. The point *z* having been determined, proceed with the diagram by going around the joint *A B G Z*. Draw *b g* in the stress diagram parallel to *B G* in the frame diagram;

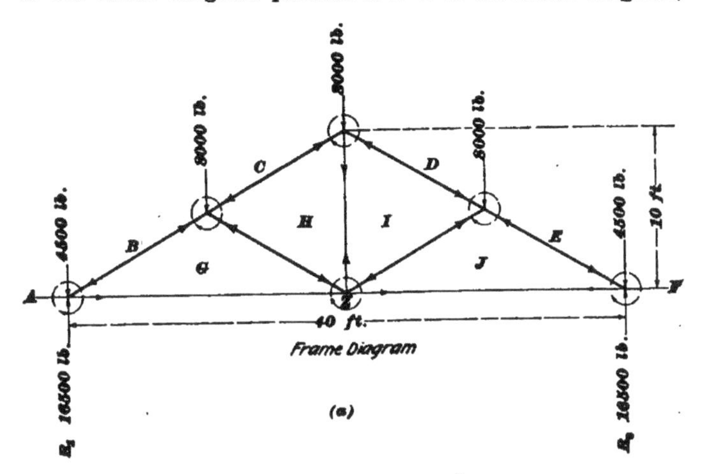

Frame Diagram

(a)

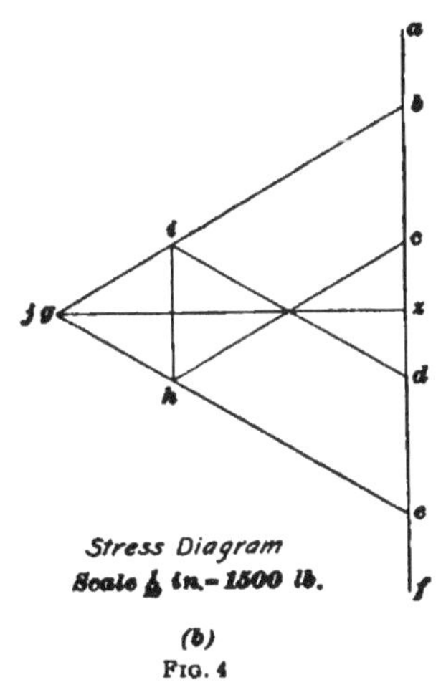

Stress Diagram
Scale 1/8 in.-1500 lb.

(b)
FIG. 4

then from *z* draw *g z* parallel to *G Z*, the point where the two lines intersect being *g*. The next joint is *B C H G*; as *b c* in the stress diagram is already known, draw *c h* parallel to *C H* and *h g* parallel to *H G*. Then the polygon of forces around this joint will be from *b* to *c* to *h* to *g*, and from *g* back to *b*. It is now expedient to analyze the joint *C D I H*; in the stress diagram, *h c* and *c d* have already been obtained; then, from *d*, draw *d i* parallel to *D I*, and from the point *h*, already known, draw *i h* parallel to *I H*. The polygon of forces around this joint will be from *c* to *d* to *i* to *h* to *c*, the starting point; arrowheads on the members in the frame diagram always mark the direction in which the stresses act in the stress diagram. Of the forces around the joint *GHIJZ*, *zg*, *gh*, and *hi* have been obtained. Then from the point *i*, *ij* is drawn parallel to *IJ*; and from *z*, *jz*, parallel to *JZ*; the point *j* is found to fall on the point *g*, while the polygon of forces around this joint is from *z* to *g* to *h* to *i* to *j*, and from *j* back to *z*, the starting point. Of the stresses in the members around the joint *IDEJ*, *ji*, *id*, and *de* are known; then from *e* draw *ej* parallel to *EJ*. The polygon of forces around this joint will then be from *i* to *d* to *e* to *j*, and from *j* back to *i*.

The only remaining joint is *JEFZ*. The stress diagram shows that the stresses in these members have been determined, while the polygon of forces around this joint is from *j* to *e* to *f* to *z*, and back from *z* to *j*.

The stress diagram being completed, the amount of the stresses may be determined by measuring the various lines with the scale to which the diagram has been drawn.

HOWE ROOF TRUSS

6. In Fig. 5 is shown a frame suitable for timber construction with iron tension members known as a **Howe truss** when constructed as in the figure, and as a **Pratt truss** when the vertical members are made to resist compression and the oblique members, tension. In this problem it is desired to draw the stress diagrams for the dead load and the wind load; also, to design and properly proportion the roof truss to resist the stresses that the various members may be required to sustain.

Draw the frame diagram for the vertical load shown in Fig. 5, and mark the dead load coming on the different panel points, or joints, in the truss. The truss being symmetrically loaded, the reactions R_1 and R_2 are each equal to half the load on the truss.

Draw the stress diagram for the dead load to the scale of, say, 800 pounds to $\frac{1}{10}$ inch. Draw the vertical load line *aj*, and determine the point *z*. Then draw the stress diagram by the methods previously given. Only one-half the diagram need be drawn, as the stresses obtained on one side of the center of the truss apply to the other side; for instance, *FQ* is the same as *PE*. Having completed half the stress diagram for the dead load, redraw the frame diagram for the wind load. The direction and amount of the wind pressure at the several panel points, or joints, of the truss are shown in the frame diagram.

As both ends of the truss are secured against sliding, the reactions act in a parallel direction to the wind pressure. If the left-hand side of the truss be secured and the right-hand side placed on rollers, to allow for expansion, as is sometimes the case with iron or structural steel trusses, the right-hand reaction, instead of being parallel to the direction of the wind, will be vertical. This makes considerable difference in the stress diagram, as will be explained subsequently.

To determine the amount of the reactions R_1 and R_2, let c be the center around which the moment of R_2 is taken; then the perpendicular distance between the line of action of R_2 and the point c will be 71.22 feet. Extend the left-hand rafter until it cuts the line of action of the force R_2 at the point y'. Regard this extension and the rafter as a beam, and calculate the amount of the reactions R_1 and R_2 by the principle of moments.

The moments about the point c are as follows:

2,800 × 11.18 =	3 1 3 0 4	foot-pounds
2,800 × 22.36 =	6 2 6 0 8	foot-pounds
2,800 × 33.54 =	9 3 9 1 2	foot-pounds
1,400 × 44.72 =	6 2 6 0 8	foot-pounds
Total,	2 5 0 4 3 2	foot-pounds

and 250,432 ÷ 71.22 = 3,516 pounds, the reaction R_2. Having found R_2, find R_1 by subtracting R_2 from the sum of the loads.

The sum of the loads is: 1,400 + 2,800 + 2,800 + 2,800 + 1,400 = 11,200 pounds. Then, 11,200 − 3,516 = 7,684 pounds, the reaction R_1.

7. Next, lay out the wind stress diagram by drawing the load line *a f* parallel to the direction of the wind in the frame diagram for the wind load. Lay off to scale (in this case

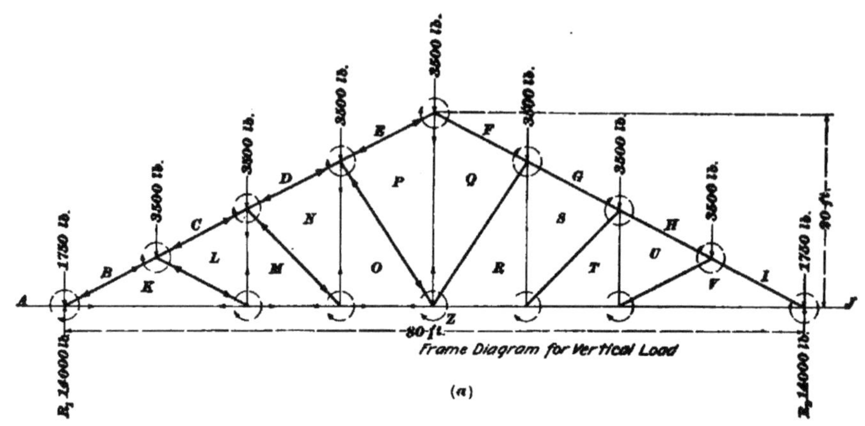

Frame Diagram for Vertical Load

(a)

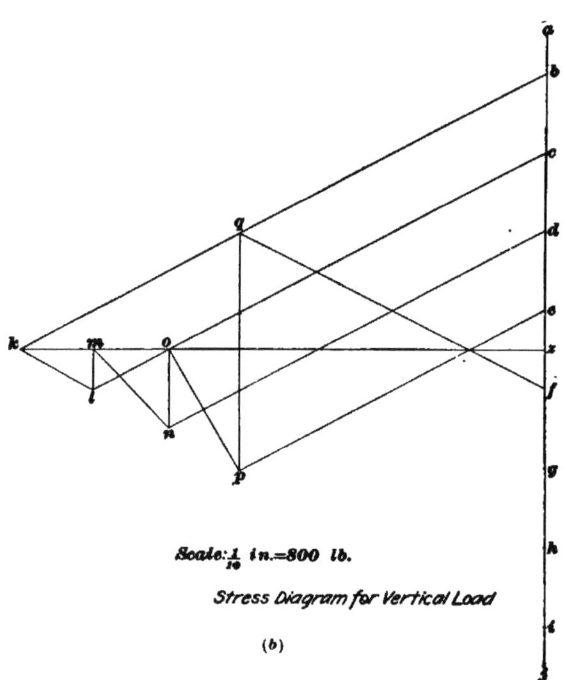

Stress Diagram for Vertical Load

(b)

Fig. 5

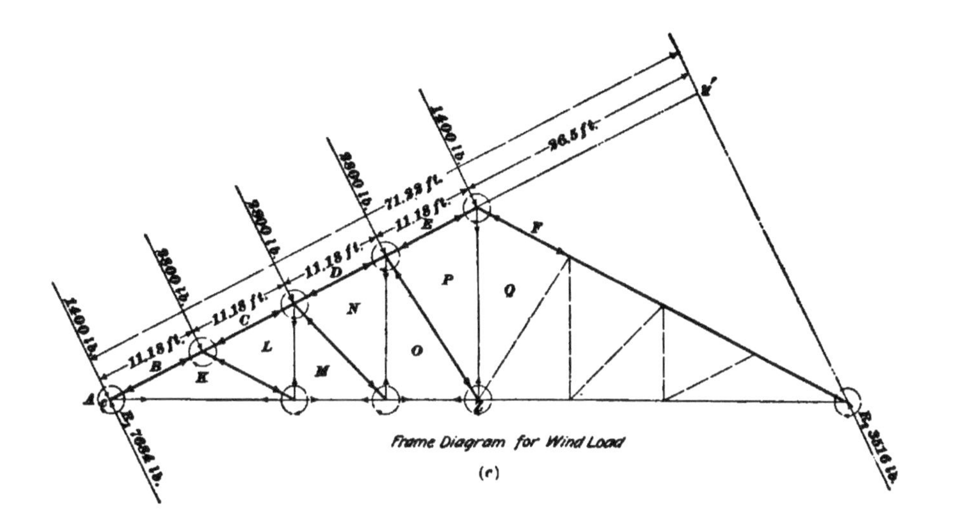

Frame Diagram for Wind Load

(c)

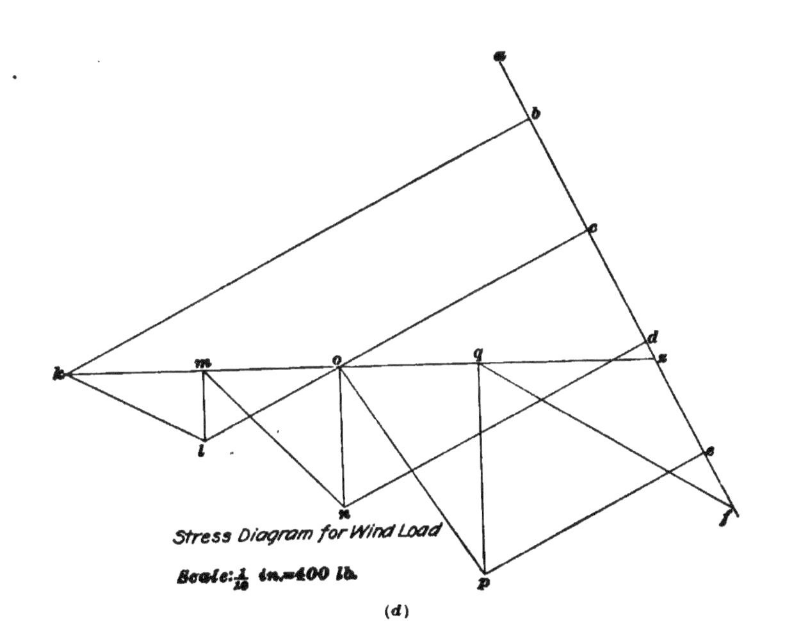

Stress Diagram for Wind Load

Scale: $\frac{1}{16}$ in. = 400 lb.

(d)

Fig. 5

400 pounds to $\frac{1}{10}$ inch) the forces *a b*, *b c*, *c d*, *d e*, and *e f* equal to *A B*, *B C*, *C D*, *D E* and *E F*, in the frame diagram. Then from *a*, lay off *a z* equal to the reaction *Z A*, or R_1. If the other forces or loads have been laid off accurately, *f z* should, on measurement, be found equal to the right-hand reaction R_2.

The first joint to analyze is *A B K Z*. Start at *b* and draw the line *b k* parallel to *B K* in the frame diagram; then from *z* draw *z k* parallel to *K Z*; where the two lines intersect is the point *k*. The polygon of forces around this joint is from *a* to *b* to *k* to *z*, and from *z* back to the starting point *a*. The next joint to analyze is *B C L K*. From the point *c* draw the line *c l* parallel to *C L* in the frame diagram and from *k* draw the line *k l* parallel to the member *L K*, the point of intersection being *l*. The polygon of forces around this joint is from *b* to *c* to *l* to *k*, and from *k* back to *b*.

To analyze the joint *K L M Z*: *k l* being already known, the next member is *L M*; so from the point *l* draw the line *l m* parallel to *L M* in the frame diagram. As the next member around this joint is *M Z*, to which *m z* in the stress diagram is parallel, the point *m* is located where the line *l m* intersects the line *m z*; this completes the joint, the polygon of forces around it being from *k* to *l* to *m* to *z*, and from *z* back to *k*.

To determine the stresses in the members around the joint *C D N M L*, draw, from the point *d*, the line *d n*, parallel to the member *D N* in the frame diagram; then, from the point *m* draw *m n* parallel to *N M*. The polygon of forces around this joint is from *c* to *d* to *n* to *m* to *l*, and from *l* back to *c*.

To analyze the forces around the joint *M N O Z*, draw from *n* the line *n o* upwards, parallel to *N O* in the frame diagram; as the next member *O Z* is horizontal, the point *o* must be at the intersection *n o* and *o z*. This completes the joint, and the polygon of forces around it is from *m* to *n* to *o* to *z*, and from *z* back to *m*, the starting point.

Now analyze the joint *D E P O N*. From *e* draw the line *e p* parallel to *E P* in the frame diagram, and from *o* draw the line *o p* parallel to *P O*. The intersection of these lines determines the point *p*, and the polygon of forces around this joint is from d to e to p to o to n, and from n back to d, the starting point.

The analysis of the joint $OPQZ$ is made by drawing the vertical line pq from the point p; the point where pq intersects bk is q. Then the polygon of forces around this joint is from o to p to q to z, and from z back to o. The members shown in dotted lines do not sustain any stresses from the pressure of the wind when it strikes the left-hand side of the truss.

TABLE I

Member	Dead-Load Diagram	Wind-Load Diagram	Total of Both
BK	27,000 +	12,000 +	39,000 +
CL	23,500 +	10,000 +	33,500 +
DN	19,500 +	7,600 +	27,100 +
EP	16,000 +	5,680 +	21,680 +
KZ	24,000 −	13,300 −	37,300 −
MZ	21,000 −	10,000 −	31,000 −
OZ	17,500 −	7,000 −	24,500 −
LK	4,000 +	3,500 +	7,500 +
NM	5,000 +	4,400 +	9,400 +
PO	6,500 +	5,400 +	11,900 +
ML	1,600 −	1,500 −	3,100 −
ON	3,500 −	3,000 −	6,500 −
QP	10,600 −	4,500 −	15,100 −
FQ	16,000 +	6,600 +	22,600 +

The final joint to go around, completing the stress diagram, is $EFQP$. The only unknown force around this joint is the stress in the member FQ. A line drawn from f in the stress diagram, parallel with the member FQ, should pass through the point q; if it does not, the diagram has been inaccurately drawn. This is always a test of the accuracy of the stress diagram, and if the last line in this diagram does not close on the proper point, when drawn parallel to the member it represents, the stress diagram should be redrawn, to determine whether the loads and reactions have been laid out correctly, and whether any of the joints or members in the structure have been omitted.

8. The two diagrams being completed, measure the different lines and obtain the stresses in the various members, as given in Table I. It must be noticed, however, that while the wind, acting on one side of a truss, does not create stresses in those members shown dotted on the opposite side of the frame, these members should be proportioned like the others, because the wind is just as likely to blow on this side of the roof and reverse the conditions.

The values given in this table represent the stresses, in round numbers, on the various members in the truss, as obtained from the dead-load and wind-stress diagrams. The + sign after a value, signifies that the stress is compressive, while the − sign shows that it is a tensile stress.

SCISSORS ROOF TRUSS

9. The type of roof truss shown in Fig. 6 (*a*) is known as the **scissors truss**; it is used for spans of from 20 to 35 feet, and is generally constructed of wood. An inspection of the frame diagram in (*a*) shows that the space *H* is a quadrangle, so that underloading the frame as constructed is liable to cause distortion and make it impossible for the vertical loads and the reactions R_1 and R_2 to keep it in equilibrium. If it is not clear that the space *H* is liable to distortion from the vertical loads and their reaction, a rough bristol-board model, hinged at the joints, should be made and notice taken of the change of form created by any vertical load applied to the truss.

To prove that such a condition exists and that it is necessary to introduce some member in order to prevent the distortion of the truss, the diagram shown in (*b*) is made. The load line is represented by the line *a f*, and the several vertical loads on the rafter member of the truss are marked off, by scale, from *a* to *b* to *c* to *d* to *e* to *f*. Since the truss is symmetrically loaded and the reactions R_1 and R_2

are equal, the point *z* will be located midway between the points *a* and *f*, or it will divide the space *c d* into halves.

Around the joint *A B G Z* there are but two unknowns;

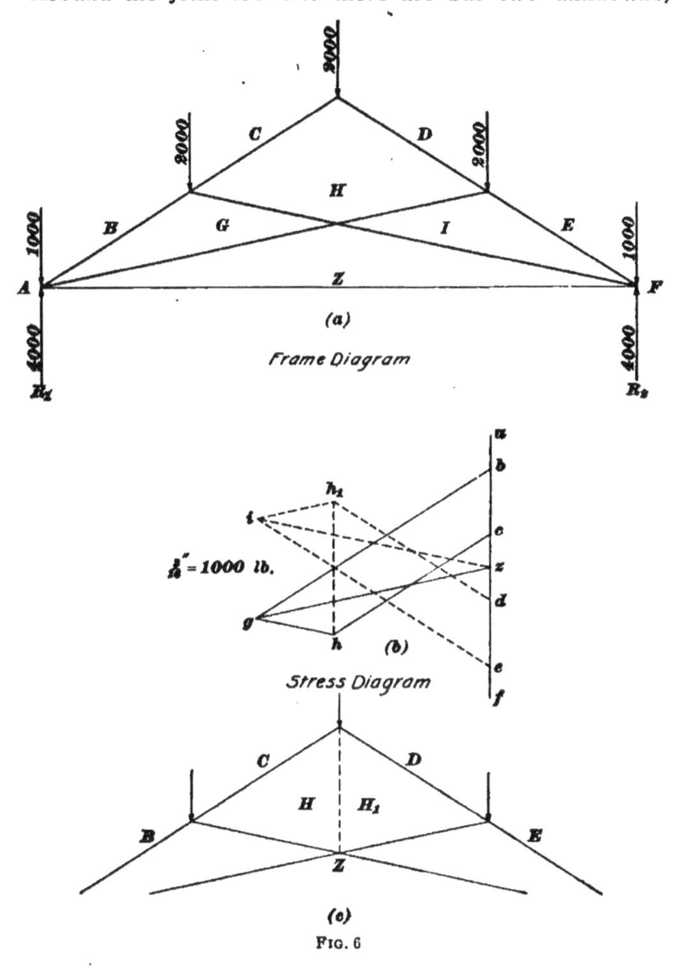

FIG. 6

therefore, the stresses in the members may be solved by drawing from *b*, in the stress diagram, a line parallel with *B G*, and from *z* a line parallel with *G Z* in the frame diagram. The point *g* will be located at the intersection of these lines, and the polygon of forces around this joint will extend from *a* to *b* to *g* to *z*, and from *z* back to *a*, the starting point.

The stresses in the members of the joint *BCHG* are represented in the stress diagram when the lines *ch* and *gh* are drawn parallel, respectively, to their members in the frame diagram.

In analyzing the next joint *CDH*, a difficulty is encountered, as the point *h* is already located, and it is impossible to draw in the stress diagram a line from *d* to *h* and parallel with that member in the frame diagram, but a line can be drawn from *d* to such a point as h_1, which will be parallel with the right-hand rafter member of the truss. In fact, when the point h_1 has been found, the other half of the diagram may be completed as shown dotted, for it is exactly similar to the stress diagram for the left-hand portion of the truss. Even though the diagrams can be drawn in this way, the polygon of forces is incomplete around this joint, for in tracing its stresses, there is no closed polygon and equilibrium is not maintained. If, however, the points *h* and h_1 in the stress diagram are connected, the polygon of forces about the apex joint of the truss will be complete. This line of stress $h h_1$ is represented in the frame diagram of the truss by introducing the vertical member HH_1, as shown in (*c*). The truss could be held in equilibrium by inserting a member between the joints *BCHG* and *DEIH*, or by placing a tie between the feet of the roof truss at R_1 and R_2. It is not unusual for carpenters to construct this truss as shown in (*a*). This is radically wrong, for, unless the walls will withstand a side thrust, the stability of the truss depends on the transverse strength of the rafter members at the joints *BCHG* and *DEIH*.

10. The scissors truss shown in Fig. 7 (*a*) can be analyzed without the insertion of another member, provided there is sufficient stability at the feet of the truss to resist an oblique thrust. That this thrust is fixed in direction and amount may

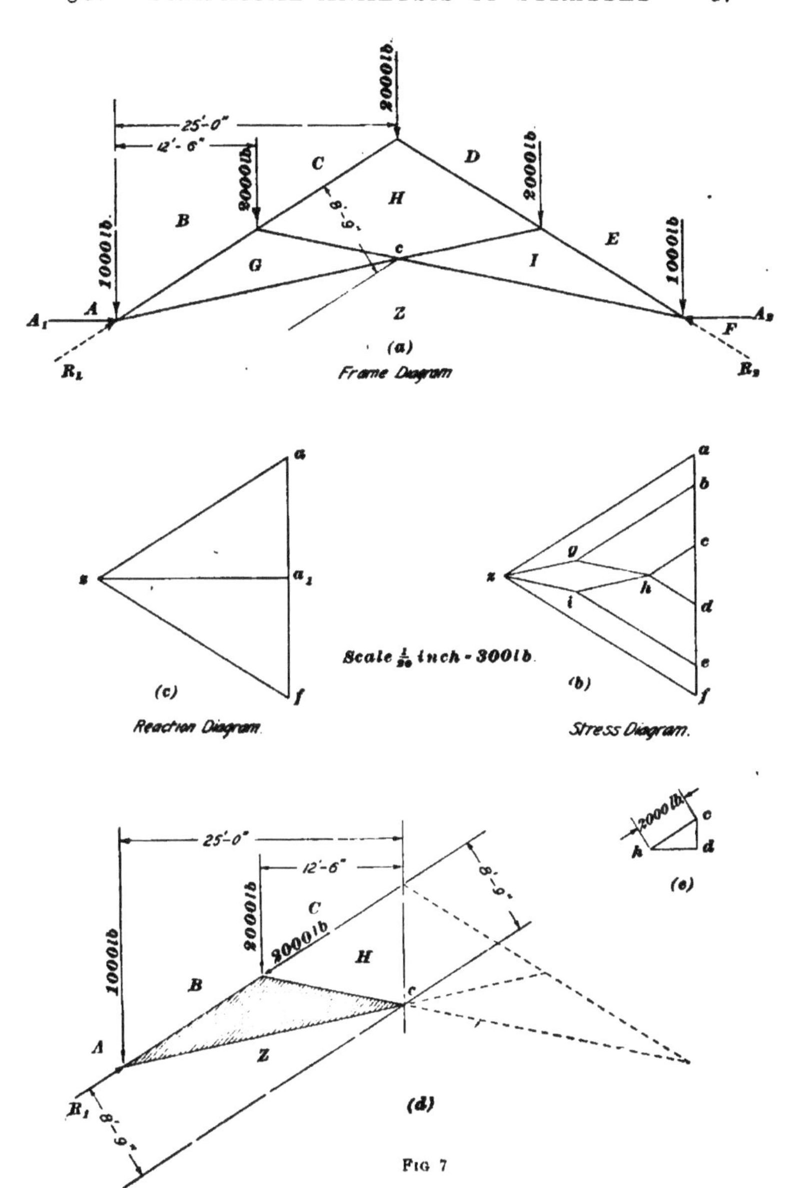

25'-0"
12'-6"
2000 lb.
2000 lb.
2000 lb.
1000 lb.
1000 lb.
8'-9"
A
B
C
D
E
F
G
H
I
Z
c
A_1
A_2
R_1
R_2
(a)
Frame Diagram
Scale $\frac{1}{20}$ inch = 300 lb.
(c)
Reaction Diagram
(b)
Stress Diagram.
2000 lb
(e)
(d)

FIG 7

be determined by drawing the stress diagram shown at (*b*). Commence the analysis of the truss, instead of at the lower right- or left-hand joint, at the apex or the joint CDH. Lay off the load line, as designated from a to f, and draw from the points c and d lines parallel with the members CH and DH, respectively, in the frame diagram, thus locating the point h.

The joints $BCHG$ and $DEIH$ can be analyzed by drawing the lines in the stress diagram designated by bg, ei, gh, and ih. Having the points g and i, the stress lines gz and iz can be found by drawing, from the points g and i, lines parallel with the members ZG and IZ, respectively. Having determined the point z, by connecting the points a and f with z, the reactions za and fz are obtained, both in direction and in amount; and it is evident that since the diagram closes and the polygons of both the external and internal forces are complete, the truss is in equilibrium when R_1 and R_2 are inclined, as shown. The inclined reactions R_1 and R_2 are the oblique resultants of the vertical reactions of 4,000 pounds shown in Fig. 6 (*a*) and horizontal thrusts introduced at A_1 and A_2, as designated in Fig. 7 (*a*).

The reaction diagram is shown in Fig. 7 (*c*). Here aa_1 is equal to 4,000 pounds, or the vertical reaction. The horizontal thrust at each abutment is measured by the length of the line a_1z and the oblique resultants of these two forces, which are R_1 and R_2 in the frame diagram in (*a*), are shown from a to z and from f to z. These lines, it will be noticed, coincide with the same lines in the stress diagram in (*b*). The resultants, or the oblique reactions R_1 and R_2, are shown dotted, in the frame diagram (*a*), at each abutment.

11. While the fact that the diagram (*b*) closes is sufficient proof of its correctness and the accuracy with which it has been drawn, it is interesting to check the reactions R_1 and R_2, by calculating the moments of the external forces about some point, such as c.

Regard the left-hand portion of the truss as being held in rotatory equilibrium about the point c by the forces designated in the diagram (*d*). Here the member CH of the frame diagram (*a*) is replaced by a force *C H*, the amount of which may be found, assuming that the stress diagram has not as yet been drawn, by laying off in the vertical direction of the force *C D*, from the point *c* shown in (*e*), a line equal in length to one-half of *C D*, or 1,000 pounds, and by intersecting a horizontal line from *d* by a line drawn parallel with *C H* in the frame diagram. On measuring the length of *c h* in (*e*), the amount of the stress, or force, *C H*, designated in diagram (*d*), is found.

The forces *A B*, *B C*, and *C H* in this figure, acting through lever arms about the point *c*, are opposed by the moment of the reaction R_1 or *Z A* about the same point. The sum of the moments of the forces *A B*, *B C*, and *C H* is found from the following:

Moment of *A B* = 1,000 × 25	=	2 5 0 0 0 foot-pounds
Moment of *B C* = 2,000 × 12.5	=	2 5 0 0 0 foot-pounds
Moment of *C H* = 2,000 × 8.75	=	1 7 5 0 0 foot-pounds
Sum of positive moments	=	6 7 5 0 0 foot-pounds

The leverage of R_1 is equal to 8.75 feet, so that its amount equals 67,500 ÷ 8.75, or 7,714 pounds. By scaling the lines *a z* or *f z* in either the stress diagram (*b*) or in the reaction diagram (*c*) it will be found to correspond with the result just found.

12. Dead-Load Diagram.—Fig. 8 (*a*) is the frame diagram of a scissors roof truss of which the stress diagrams for the dead load and wind pressures are to be found.

First draw the stress diagram, Fig. 8 (*b*), for the dead load. As the dead loads on the truss are symmetrical, both in amount and location with regard to the center line of the truss, the reactions are the same at either end of the truss, and each is equal in amount to one-half the load, in this case, 7,575 pounds. Before going around the joints, draw the vertical line *a f* in the stress diagram; then starting at the point *a*, lay off to the scale of, say, 400 pounds to every $\frac{1}{16}$ inch, the force *a b* equal to *A B* in the frame diagram; then lay off *b c* equal to *B C*, *c d* equal to *C D*, *d e* equal to *D E*, and *e f* equal to *E F*. Then, as the truss is symmetrically loaded, the

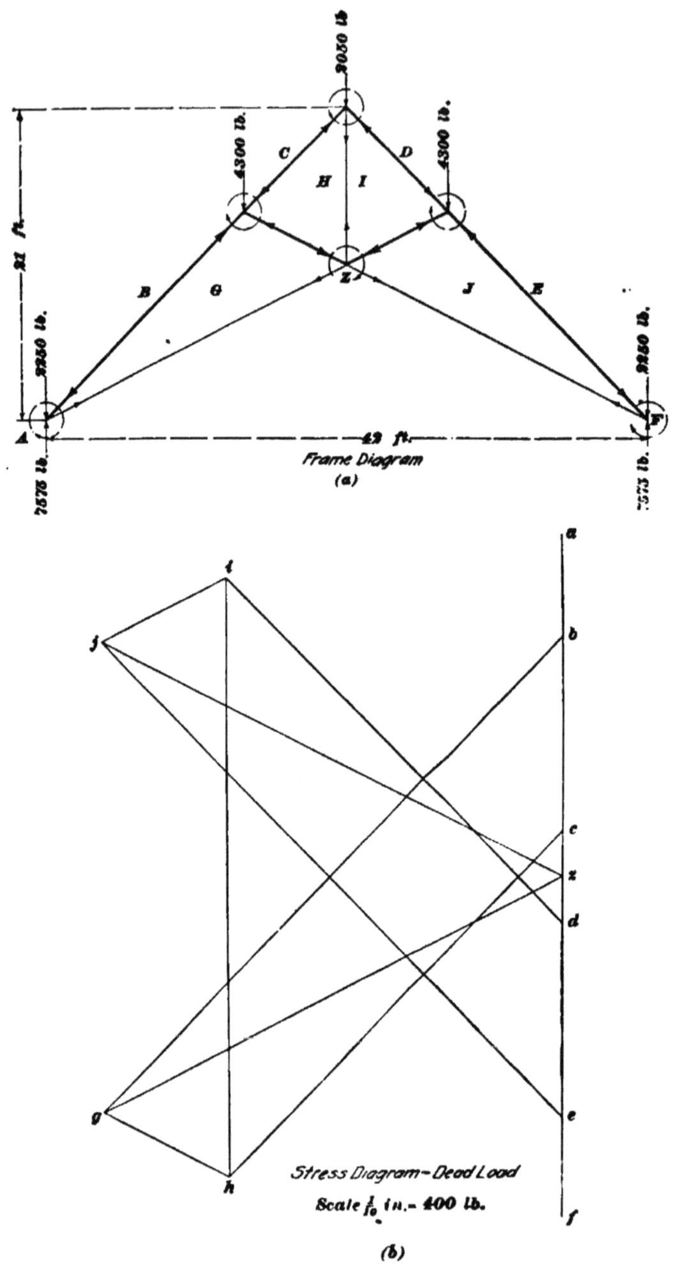

9050 lb
4300 lb.
4300 lb.
C
D
H
I
21 ft.
B
G
Z
J
E
9250 lb.
9250 lb.
A
F
42 ft.
7575 lb.
7575 lb.
Frame Diagram
(a)
a
b
c
z
d
e
f
i
j
g
h
Stress Diagram—Dead Load
Scale $\frac{1}{16}$ in.= 400 lb.
(b)

FIG. 8

point *z* is located midway between the points *a* and *f*. If the truss were not symmetrically loaded, the reactions would have to be calculated by the principle of moments.

Having located the loads and their reactions on the vertical line *af*, obtain the stresses in the members around the joint *ABGZ* from the point *b*, by drawing a line *bg* parallel to *BG* in the frame diagram; then from *z* draw the line *gz* parallel to *GZ*; the intersection of these lines will be the point *g*. The polygon of forces around this joint is from *b* to *g* to *z* to *a*, from *a* back to *b*, the starting point. Bear in mind that the forces in the stress diagram, representing the reactions, must have the same direction as the reactions in the frame diagram. The lines determining the stresses around the joint *BCHG* should next be drawn; *bc* having been determined from *c*, draw a line *ch* parallel to *CH*, and from *g* draw a line *gh* parallel to *HG*, the intersection of these two lines determining the point *h*, while the polygon of forces is from *b* to *c* to *h* to *g*, and from *g* back to *b*.

Now work around the joint *CDIH*; *cd* being already known, from the point *d* draw the line *di*, parallel to *DI*, and from the point *h* draw the line *hi*, parallel to *IH*, the intersection of the two lines being the point *i*. In going around the joint *GHIJZ*, the stresses in the members *zg*, *gh*, *hi*, have already been determined and drawn in the stress diagram. Then, from the point *i* draw the line *ij* parallel to *IJ*, and from *z* draw the line *zj* parallel to *JZ*, and the intersection of these two lines will be the point *j*. The polygon of forces around this joint is from *z* to *g* to *h* to *i* to *j*, and from *j* back to *z*, the starting point.

The next joint to analyze, in going around the truss, is *DEJI*; *ji*, *id*, and *de* being known, the only remaining force to determine is the stress in the member *EJ*. The point *e* being fixed, draw the line *ej* parallel to *EJ* in the frame diagram, and if this line, which completes the diagram, passes through the point *j*, the diagram is correct and accurately drawn. The stresses around the right-hand heel of the truss are all known, the line *ej* just drawn having been the only unknown member at this joint. The polygon of

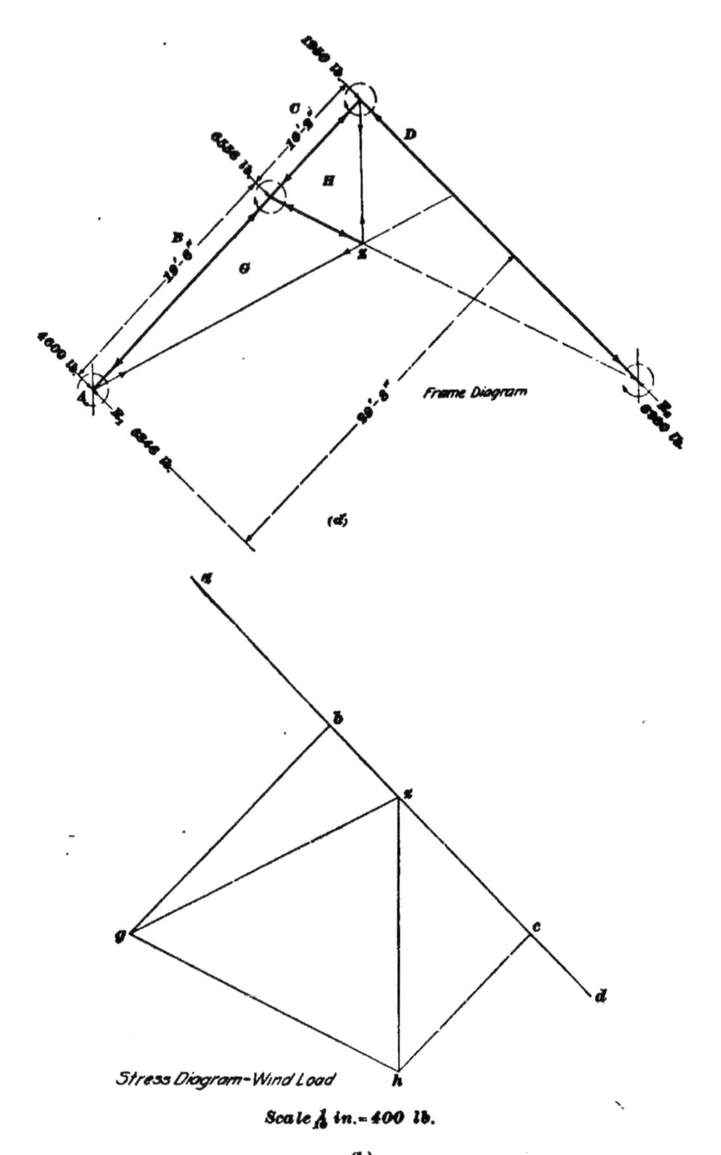
D
H
G
Frame Diagram
(a)
a
b
x
g
c
d
h
Stress Diagram-Wind Load
Scale $\frac{1}{16}$ in.=400 lb.

(b)

Fig. 9

forces around the joint *DEJI* is from *d* to *e* to *j* to *i*, and from *i* to *d*, the starting point. The polygon of forces around the joint *EFZJ* is from *e* to *f* to *z* to *j*, and from *j* back to *e*, completing the stress diagram for the dead or vertical load on the roof truss.

13. The **wind-load diagram,** however, remains to be drawn. Redraw the frame diagram, as shown in Fig. 9. The wind is always considered as acting normally, or at right angles, to the roof, the amount of its pressure at the different joints of the truss being shown on the frame diagram. As the heels of the trusses are fixed, the reactions act in lines parallel with the wind pressure.

To estimate the amount of the reactions R_1 and R_2, consider the left-hand rafter member as a beam, and R_1 and R_2 as the reactions supporting it. The moments due to the wind pressure *BC* and *CD* acting about R_1 are:

At *BC*, 6,556 × 19.5	= 127842	foot-pounds
At *CD*, 1,950 × 29.66	= 57837	foot-pounds
Total,	185679	foot-pounds

The lever arm with which R_2 resists the wind pressure acting at the joints is 29 feet 8 inches, so 185,679 ÷ 29.66 = 6,260 pounds, the amount of the reaction at R_2. The sum of the loads being 4,600 + 6,556 + 1,950 = 13,106 pounds, the reaction at R_1 is 13,106 − 6,260 = 6,846 pounds.

First draw, in the stress diagram, the load line *a d* parallel to the direction of the wind pressure and the reactions. Lay off to the scale to which the stress diagram is drawn—in this case, 400 pounds to $\frac{1}{16}$ inch—the force *a b* equal to *AB* in the frame diagram, then lay off *bc* equal to *BC*, and *c d* equal to *CD*. From *d* lay off the amount of the reaction R_2 or *dz*, which determines the point *z*, and the distance *z a*, according to scale, equals in amount the left-hand reaction R_1. The polygon of external forces is, then, from *a* to *b* to *c* to *d* to *z*, and from *z* back to *a*, the starting point, forming a straight line, as in all cases so far analyzed.

Continue the stress diagram by going around the joint *ABGZ*; from the point *b* draw the line *bg*, and from the point *z* draw the line *zg* parallel to the corresponding members in the frame diagram, the point where the two lines intersect being *g*. The polygon of forces around this point is from *a* to *b* to *g* to *z*, and from *z* back to *a*.

The next joint is *BCHG*; *bc* having already been obtained, from the point *c* draw the line *ch* parallel to *CH*, and from *g* draw the line *gh* parallel to *HG*, *h* being the point where these two lines intersect. Disregard the two members in the frame diagram shown in dotted lines, which do nothing toward sustaining the wind pressure. Now work around the joint *HCDZ*, of which *cd* and *dz* are known; draw, from *z*, the line *zh* parallel with *ZH* in the frame diagram; if this closing line of the diagram passes through the point *h*, the diagram has been drawn accurately. The polygon of forces around the joint *BCHG* is from *b* to *c* to *h* to *g*, and from *g* back to *b*. The polygon of forces around the joint *CDZH* is from *c* to *d* to *z* to *h*, and from *h* back to *c*. The polygon of forces around the joint *ZGH* is from *z* to *g* to *h*, and from *h* to *z*, the starting point.

The stress diagrams for both the dead and the wind load being complete, the stress in each member of the truss is obtained by determining, by scale, the stress due to both the dead and wind loads in each member, adding the two together for the maximum load in the member. To determine, for instance, the stress in the strut *HG*, measure the length of the line *hg* in the stress diagram, Fig. 8, for the dead load, and obtain its amount; then measure the same line *hg* in the stress diagram for the wind load, Fig. 9; find the sum of the two measurements, and determine the maximum stress in the strut *hg* from the assumed scale of the drawing.

It must be remembered that while the wind acting on one side of a truss does not create stresses in all the members on the opposite side, these members should be proportioned in a like manner to the other members, because the wind is quite as likely to blow on this side of the roof and reverse the conditions.

TRUSS WITH LOADS ON LOWER CHORD

14. Stress Diagram.—The type of roof truss shown in Fig. 10 is admirably adapted for spans of 40 and 50 feet, and when built of structural steel makes excellent principals for

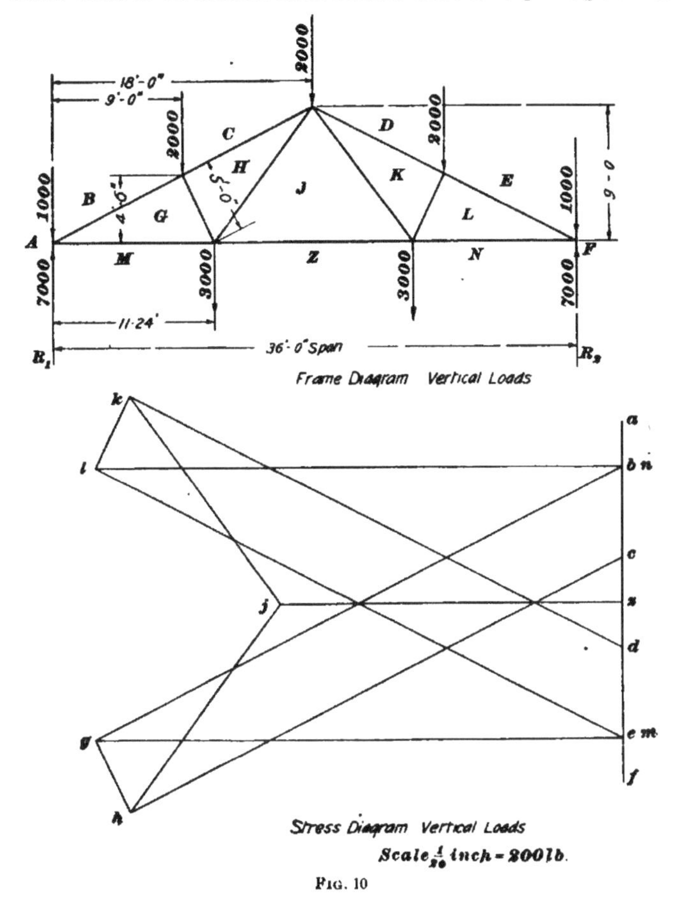

Fig. 10

the support of roofs over boiler houses, shops, etc. In such positions, or where they support a ceiling beneath the lower chord, or tie-member, they are often required to carry considerable weight, which is transmitted to the joints *MZ* and *ZN*.

By introducing these loads suspended from the panel points on the lower chord, no confusion need result in drawing the stress diagram if these loads and the correct reactions are introduced in the load line of the stress diagram, which represents the force polygon of the external forces. The sum of the reactions R_1 and R_2 must equal the sum of the loads on the frame, including the weights suspended from the lower chord, and as the truss is symmetrically loaded the reactions R_1 and R_2 are equal, and each is half the entire load. The system of lettering for the designation of the forces on the frame may be followed in alphabetical order, the letter *Z* being used to designate the space below the lower chord of the truss. The external forces on the frame diagram of Fig. 10 will then read, on the load line, downwards from *a* to *b* to *c* to *d* to *e* to *f*; then upwards from *f* to *n*, which distance is laid off to equal, in this instance, the reaction R_2, or 7,000 pounds. The force polygon will now continue from *n* to *z* downwards, thus designating the suspended weight of 3,000 pounds on the right-hand portion of the lower chord. The force *z m* is now laid off, also, downwards, in order to designate the suspended weight of 3,000 pounds on the left-hand portion of the lower chord. When *m* has been located on the load line, the reaction R_1 is laid off upwards equal to 7,000 pounds and is represented in the stress diagram by *m a*. It will be noticed in laying off the polygon of external forces in the stress diagram that the point *n* lies at the same place as the point *b* and that the point *m* coincides with the point *e*. This duplicate designation of a point is frequently necessary, and especially so when weights are suspended from points on the lower chord.

In this case, the polygon of external forces is actually a single straight line, but, nevertheless, it is a many sided figure, being made up of the several forces around the truss. That it has no appreciable area or shape is due to the fact that all the external forces are vertical and, coinciding with one another, form a straight line.

In drawing the stress diagram for the vertical loads, commence at *b* and draw *bg* of indefinite length, and parallel with the member *B G* in the frame diagram. From *m* draw a line parallel with *M G* in the frame diagram, thus obtaining the point *g* at the intersection of the two lines. The polygon of forces about the joint *A B G M* will then read from *a* to *b* to *g* to *m* and from *m* back to the starting point *a*. From this it will be observed that the stress in *B G* at its lower end acts toward the joint, and, consequently, this member is in compression, while the stress at the left-hand end of the member *G M* is represented in the stress diagram by a line drawn in a direction away from the joint, so that this member is in tension. Having proceeded this far, one may hesitate as to whether to proceed with the joint *B C H G* or *G H J Z M*. An inspection of the latter joint, however, will show that there are a number of members gathered at this point and that so far only two of the forces are known.

It will be observed, however, that *B C H G* contains only two unknown forces, and that when this joint has been solved, *G H J Z M* is likewise solvable. These two joints may then be represented in the stress diagram by the polygons *b c h g* and *g h j z m*. In this manner the stresses in the members *G H*, *H J*, and *C H*, together with *J Z*, are determined and the stresses in all the members of the truss obtained. Since the loads are symmetrical, the stress diagram will be the same on both sides of the line *z j*.

15. If, after determining the stresses in *G M* and *G H*, there had still remained three unknown forces about the joint *G H J Z M*, it would have been necessary to figure the stress in *J Z* by taking the moments about the apex of the truss. This stress will equal the algebraic sum of the moments of the external forces about the apex divided by the perpendicular distance between that point and the member *J Z*, or the rise of the truss. All vertical loads on the left-hand side of the truss tend to turn this half of the truss in a downward direction, while the reaction R_1 acts in opposition to these forces and tends to revolve it in an upward direction. The moments of the several forces *A B*, *B C*, and *M Z* are equal to their amount multiplied by the perpendicular, which, in this case, is the horizontal distance from their line of action to the apex of the truss, while the moment of R_1 about the apex of the truss is equal to its amount multiplied by one-half the span. The several distances entering into the calculation of moments for these forces are given in the figure, and the calculations are as follows.

Negative moments, due to the vertical loads on the truss around the apex, equal:

Moment of *A B*, 1,000 × 18	=	1 8 0 0 0 foot-pounds
Moment of *B C*, 2,000 × 9	=	1 8 0 0 0 foot-pounds
Moment of *M Z*, 3,000 × 6.76	=	2 0 2 8 0 foot-pounds
Sum of negative moments	=	5 6 2 8 0 foot-pounds

Positive moments about the same point equal:

Moment of *M A* = 7,000 × 18 = 126,000 foot-pounds

The algebraic sum of these moments equals 126,000 − 56,280 = 69,720 foot-pounds. It will be observed that the force *C D* does not exert a moment about the apex of the truss, so that, in figuring the negative moments, this force is neglected. The perpendicular distance from the member *J Z* to the apex of the truss is 9 feet, so that the stress in this member is equal to 69,720 ÷ 9 = 7,746 pounds, which is proved to be correct by scaling the line *J Z* in the stress diagram.

The stress in the member *C H* can be checked by figuring the moments about the joint *G H J Z M*. The negative moments about this joint, due to the vertical loads to the left, equal:

Moment of *A B*, 1,000 × 11.24	=	1 1 2 4 0 foot-pounds
Moment of *B C*, 2,000 × 2.24	=	4 4 8 0 foot-pounds
Total of negative moments	=	1 5 7 2 0 foot-pounds

The positive moment about the same point equals 7,000 × 11.24 = 78,680, and the algebraic sum of the moments about the joint is 78,680 − 15,720 = 62,960. The perpendicular distance from *C H* to the joint *G H J Z M*, according to the frame diagram, is 4 feet 6 inches, and the stress in the member *CH* is found by dividing 62,960 by 4.5, which is found to equal 13,991 pounds. This amount is approximately the stress found on scaling the line *c h* in the stress diagram.

16. Wind-Load Diagram.—The frame and stress diagrams of the wind load for the same truss are shown in Fig. 11. It is first necessary to calculate the reactions R_1 and R_2. Since the truss is fixed at both ends, these reactions coincide in direction with the normal wind pressures at each panel point. The amount of R_2 may be calculated by taking the moments about *c* and finding the algebraic sum of all the external forces. If the normal wind loads acting at each panel point on the rafter member are considered as exerting negative moments about *c*, the reaction R_2 exerts a positive moment about the same point, and the moments of the several forces are as given in the following calculation. It must be remembered in these calculations that the moment of any force is obtained by multiplying the amount of the force by the perpendicular distance between its line of action and the point around which it tends to rotate.

Moment of *B C*, 2,400 × 10.06 = 2 4 1 4 4 foot-pounds
Moment of *CD*, 1,200 × 20.12 = 2 4 1 4 4 foot-pounds
Sum of negative moments = 4 8 2 8 8 foot-pounds

Since the amount of R_2 multiplied by its lever arm must equal the sum of the products of each force on the rafter member multiplied by its lever arm, it is evident, since the lever arm of R_2 is known, that R_2 will equal the sum of the negative moments divided by the perpendicular distance between the line of action of R_2 and the point *c*. This distance, from the dimensions given in the frame diagram, is 32.18 feet and the value of R_2 is equal to 48,288 divided by 32.18, or 1,500 pounds. Since the directions of all the external forces coincide, it is evident that the condition of translatory equilibrium is fulfilled when the algebraic sum of the reactions and the loads equals zero, so that if one of the reactions is known the other can be found by deducting the known reaction from the sum of the loads; hence, R_1 equals

4,800 − 1,500, or 3,300 pounds. In this way all the external forces necessary to produce equilibrium in the truss have been determined and the load line and polygon of

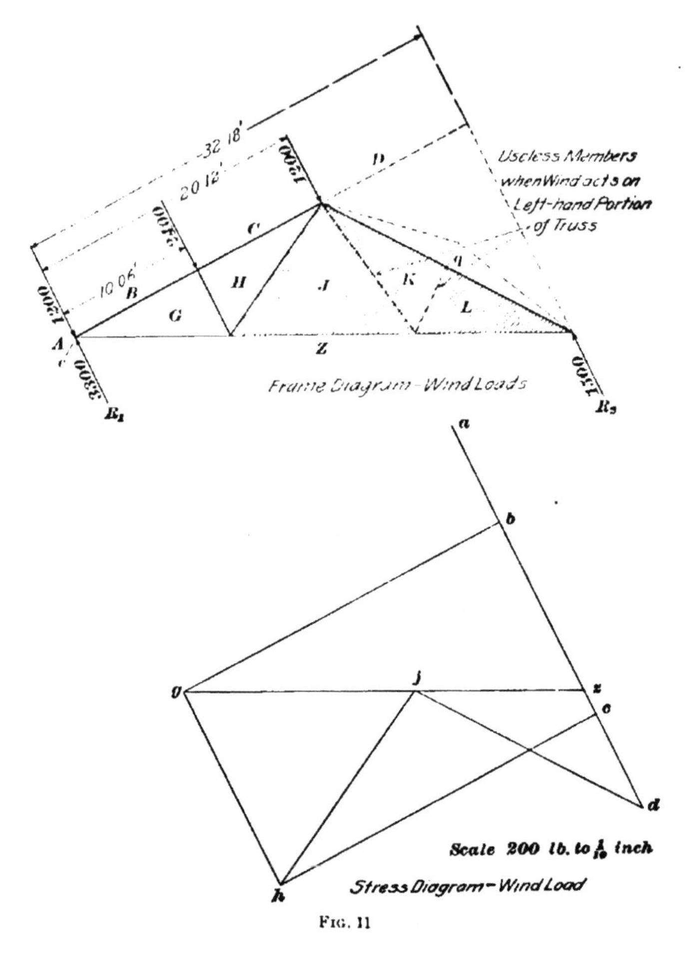

FIG. 11

external forces may be drawn. This force polygon is drawn from *a* to *b* to *c* to *d*. Then the amount of the right-hand reaction, or R_2, is laid off from *d* upwards and the point *z* located. The distance $z a$, according to scale, should equal, in amount, the value of R_1 previously found.

There will be no difficulty encountered in drawing the stress diagram for the wind load except, possibly, in concluding the diagram. It will be observed, however, in working around the last joint *CDJH* that the members *JK* and *KL* will be subjected to no stress, so that they may be disregarded in drawing the diagram. When a stress diagram is closed without certain members of the frame being represented, it shows that no stresses exist in these members under the condition of loading presented. For instance, in the stress diagram of Fig. 11, it is manifestly impossible to represent the stress in the member *JK*, for a line drawn from *d* parallel with the member *DK* intersects the line drawn from *h* parallel with the member *HJ* at a point which, if the stress *HJ* is to be represented, must be the point *j*, and no line can be drawn from *j* parallel with the member *JK* that will give a point *k* on a line representing the member *DK*. Having no point *k*, *k l* cannot be drawn.

Frequently, the useless members in a frame may be determined by inspection. Consider the right-hand rafter member *DK* of the frame diagram shown in Fig. 11. This member extends in a straight line from the apex to the heel of the truss and represents a force acting in the direction of its axis. It is evident that any force, such as *KL*, applied laterally to the force *DK*, will not be opposed by any portion of the force *DK*, but will, if the rafter member is joined at the point *q*, push or deflect the rafter member out of line, as designated by the dotted lines. Evidently, therefore, no force *KL* can exist and the member *KL* is consequently useless with the wind acting on the left-hand slope of the truss.

The forces around the last joint of the frame are the stresses in the members *HC*, *DJ*, *JH*, and the force *CD*. The strut and tie being omitted, the letters *K* and *L* are neglected and the entire space, as sectioned, can be designated as *J*, so that the stress diagram for the wind loads will be closed by working around the joint at the apex of the truss.

If the reactions have been carefully calculated and the stress diagram accurately drawn, the polygon of forces around this last joint of the truss will read, in the stress diagram, from *c* to *d* to *j* to *h*, and from *h* to *c*, the starting point.

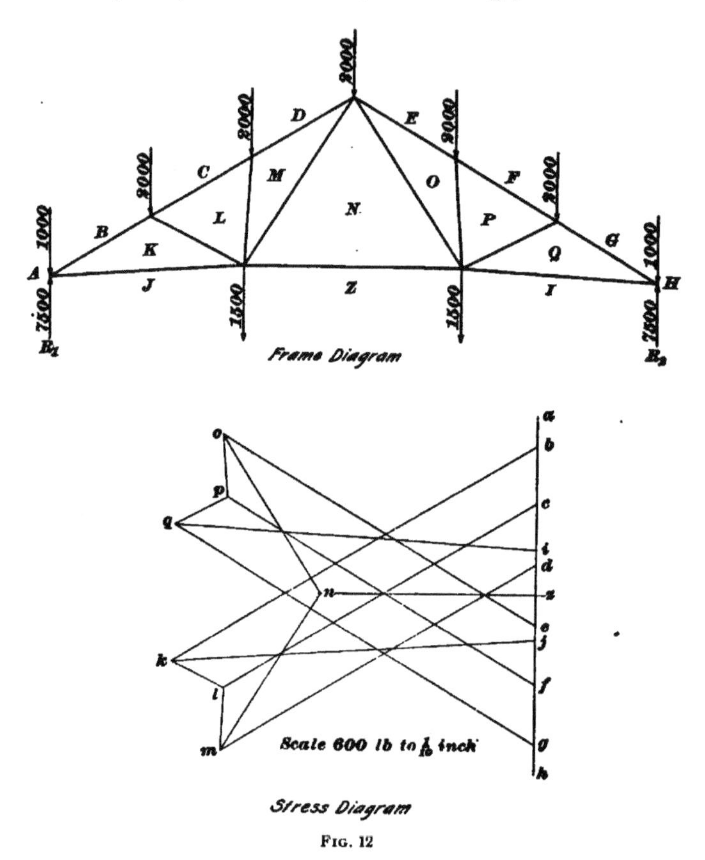

Fig. 12

17. Diagram for Cambered Truss.—Figs. 12 and 13 show the graphical analysis for the stresses of a frame diagram that differs somewhat from the one described. It is an excellent design for roof trusses of small span. The lower chord of this truss is cambered, that is, it does not extend in a straight line, but is raised at the center; besides, the rafter member is supported at two places instead of one. The stress diagrams for this frame should be laid out as an

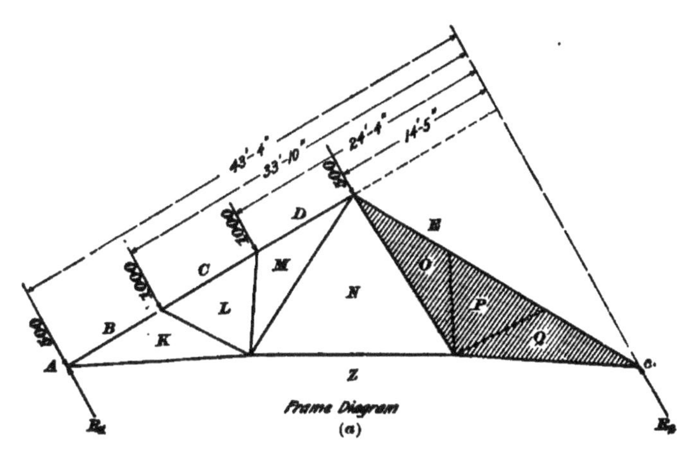

Frame Diagram
(a)

exercise. In the solution, no difficulty will be experienced in laying out the stress diagram, Fig. 12, for the vertical loads on the truss.

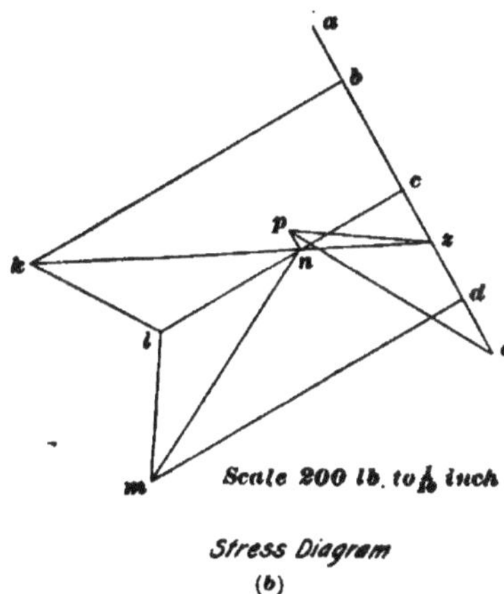

Stress Diagram
(b)
Fig. 13

The stress diagram, Fig. 13, for the wind loads cannot so readily be drawn, from the fact that the reactions R_1, R_2 must be determined by the principle of moments. After the reactions have been determined and the stress diagram drawn, with the exception of the lines representing the stresses around the apex of the truss, the frame diagram should be studied to determine which members in the right-hand portion of the truss are useless when the wind blows from the left-hand side. By inspection, it will be perceived that OP and PQ are of no use in resisting the wind loads, so that the entire space that is section-lined can be designated by the single letter P. When this has been determined, the joint at the apex of the truss can be analyzed by drawing a line from e, in the stress diagram, parallel with the direction of the right-hand rafter member and a line from n parallel with the tension member NP; the point of intersection of these two lines will be P. The analysis of this joint, however, does not complete the diagram, from the fact that the stress has not been determined for the member pz. This stress may be found by connecting the points p and z in the stress diagram, and unless this line coincides in direction with the member PZ, the stress diagram is incorrectly drawn, either on account of inaccuracy or because the amounts of the reactions have been incorrectly determined.

In reading the several stresses in the polygon of forces last determined, it will be found that ZN and NP, as well as PZ, are subjected to tensile stresses. It will be observed, also, on inspection of the stress diagram, that ZK sustains a greater stress than PZ, but as the wind is as likely to blow on one side of the roof as on the other, the member KZ and the corresponding member on the right-hand portion of the truss must be proportioned for the maximum stress. The small stress in the member PZ, owing to this consideration, should be disregarded. Likewise, since the lower portion of the rafter member BK must be proportioned for the maximum stress, the stress determined for EP should be disregarded, from the fact that if BK is proportioned for the maximum stress, the rafter member throughout its length will be capable of sustaining any minor stress. The only necessity, therefore, for finishing the diagram is the desirability of checking the accuracy of the work by having the diagram closed, that is, by determining whether the last line in the diagram connecting two points, otherwise determined, lies in a direction coincident with the final member of the truss that it represents.

QUEEN-POST TRUSS

18. In Fig. 14 is shown a type of truss known as the **queen post,** from the fact that the vertical members, shown dotted, are made of timber, and are so constructed as to resemble posts. These members, *FG* and *GH*, are theoretically useless when the truss is loaded at its upper panel points with vertical loads equal in amount and symmetrical as to location about the center of the truss. Disregarding

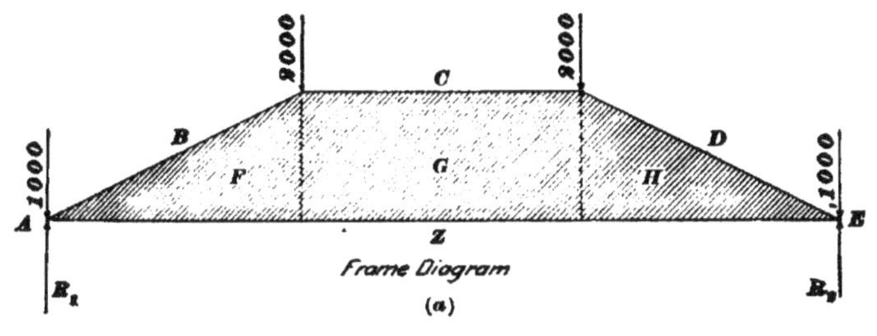

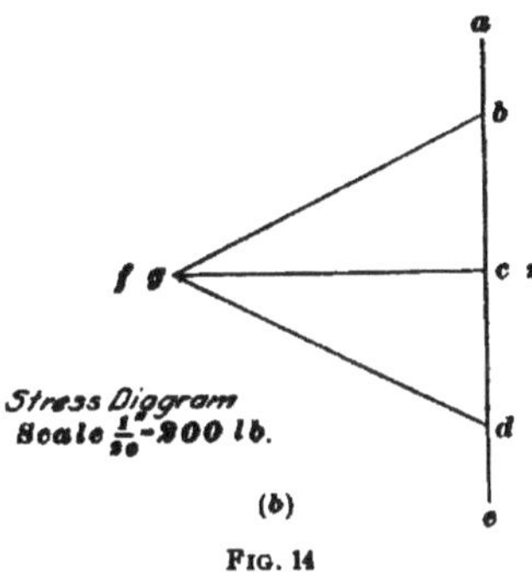

Fig. 14

these members, therefore, and denominating all of the section-lined space as *G*, the stress diagram is drawn by laying out the load line *a e* and constructing the polygon of external forces that is enclosed by *a b*, *b c*, *c d*, *d e*, *e z*, and *z a*. From *b* the line *b f* is drawn parallel with *B F* in the frame diagram, the intersection *f* being found by extending from *z*, which coincides with the point *c* since the truss is symmetrically loaded, a horizontal line parallel with *G Z* in the frame diagram.

In analyzing the next joint *B C G*, it is found that when a line *b g* is drawn parallel with the member *B G*, it will be identical with the line *b f*, already drawn in the stress diagram; *B C* is already known. It will be found that the stress in the member *C G* must lie on the line *z f* in the stress diagram; consequently, it is proved that there can be no force in the member *F G*, for the vertical distance between *cg* and *gz*, in the stress diagram, is zero. The stresses in the members *F G* and *G H* being nothing, the stress diagram can be finished by drawing, from *d* upwards, a line parallel with *D G* in the frame diagram. If the interior space of the frame is not regarded as one space, but is maintained in three portions, such as *F*, *G*, and *H*, the point *h* will be coincident with *f* and *g*.

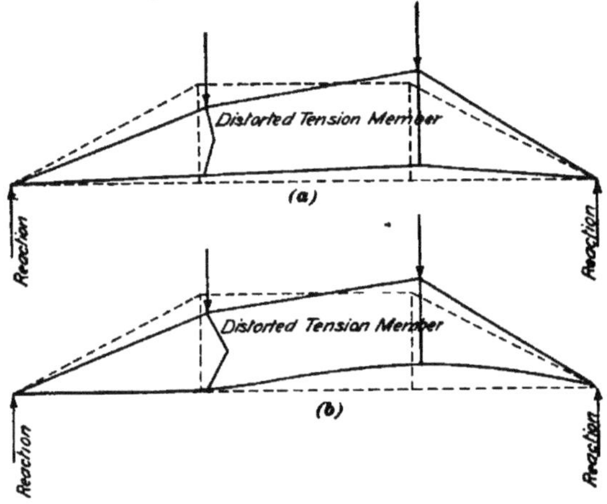

Fig. 15

By inspection of the stress diagram we observe that the stresses in *C G* and *G Z* are equal, though they differ in kind, for *C G* is in compression and *G Z* is in tension. By referring to the frame diagram, it will be observed that the space *G* is a rectangle, and, on consideration, it is evident that a rectangle hinged at the corners, which is the condition assumed in all framed structures, is liable to distortion. This cannot occur when the vertical loads are symmetrically placed, and when corresponding loads on each side of the vertical center line of the truss are equal, for a tendency of one to distort the frame is counteracted by the other, which tends to produce distortion in the opposite direction. If, however, there is a great variation between the vertical loads

BC and *CD*, the frame will be distorted, as shown in Fig. 15 (*a*), provided it is hinged at the joints. In practice,

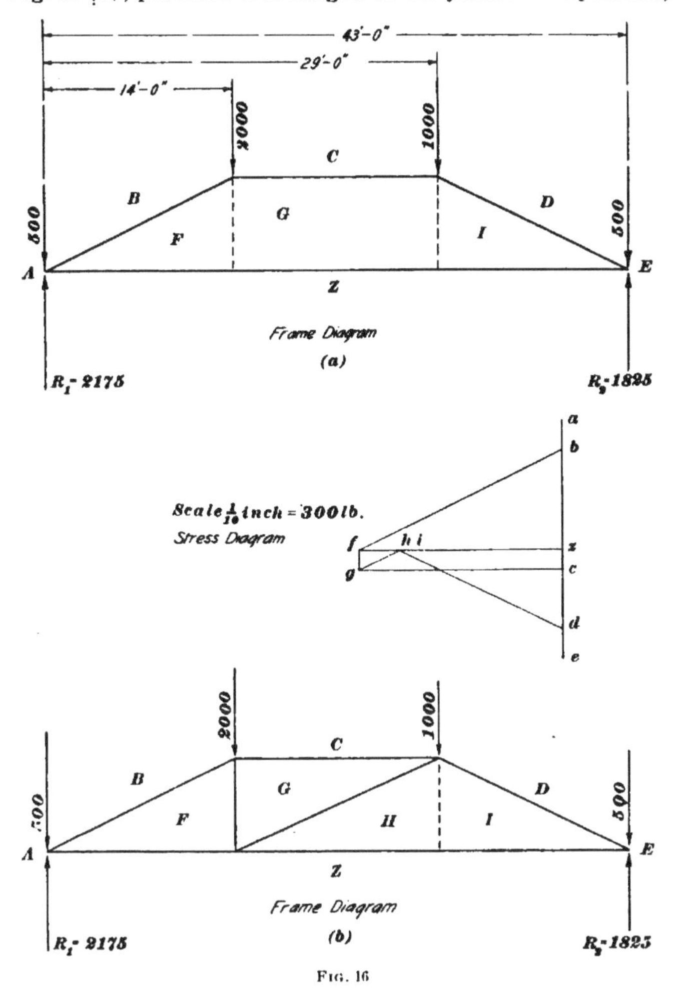

FIG. 16

though the distortion would tend toward that shown in the figure, the lower tie-member, or chord, has considerable transverse strength and will be deflected as shown in (*b*).

That this type of truss under unequal loads is unstable, and that equilibrium cannot be obtained without the introduction of another member, will be demonstrated by the failure of an effort to draw the stress diagram under such a condition of loading. For instance, in laying out the stress diagram for the truss loaded as shown in Fig. 16 (*a*), the load line or polygon of external forces is laid off the same as that for the stress diagram previously completed. No difficulty is experienced in drawing the stresses around the joint *B C G F*, but when the joint *F G Z* is analyzed it is found that the point *g* in the stress diagram cannot be connected with *z* by a line parallel to *G Z* in the frame diagram. So, it is clear that under these conditions the stress diagram cannot be drawn in such a manner as to make a closed polygon, and, in consequence, equilibrium cannot be maintained without the introduction of another force or its equivalent, a member exerting a stress. The question then arises what member must be placed in the frame diagram of the queen-post truss in order to obtain equilibrium? Since a rectangular frame is not a stable form and because a triangular frame is always rigid under any condition of loading, it is reasonable to divide the frame into triangles by a diagonal member, as has been done in Fig. 16 (*b*). It will be found in drawing the stress diagram that there is no place for the member *H I*, for the point *h* coincides with the point *i*, and consequently the stress in the member *H I* is zero. By going over the lines in the stress diagram for the frame shown at (*b*), it will be found that the diagonal member *G H* is in tension. Equilibrium in a truss could have been maintained as well by inserting a diagonal extending between the joints *H I Z* and *B C G F*. If the greatest load were still maintained at *B C*, this diagonal member would be in compression and the member *F G* would be subject to no stress.

19. In the rectangular space subdivided by reversing the diagonal in this manner, the stresses in the vertical members *F G* and *H I* are changed, but the stresses *G C* and *H Z* are not altered. If, however, the space included between the vertical members *F G* and *H I* and the horizontal members *C G* and *H Z* had been any other quadrangle instead of a rectangle, the stresses in the members *C G* and *H Z* would have been changed as well. It is an important point to remember that the reversal of a diagonal dividing a quadrangle, other than a rectangle, would change the amount of stress in each of the sides, or in each of the members enclosing the figure.

20. It was observed in drawing the stress diagram for the queen-post truss, Fig. 14, that there was no stress in the dotted vertical members. It was also found, in drawing the stress diagram for an unsymmetrically loaded queen-post truss, as shown in the frame diagram, Fig. 16 (*b*), that there was no stress in the vertical member *H I*, shown dotted. It is evident, therefore, that these members are useless, that the stress in each case is zero, and that the members are not needed to close the polygon of forces about any joint. In each of these cases it is also clear that the conditions of equilibrium are realized in the frame without the insertion of these members. A careful inspection of the frame diagram in each case reveals the fact that in every case where these members have been proved useless, two of the forces at the joint *G H Z* in Fig. 14 and *H I Z* in Fig. 16 (*b*) act in the same straight line, and that there is only one remaining force at the joint. In order that equilibrium in the joint can be maintained, the algebraic sum of the vertical forces or vertical components acting at a joint must equal zero, and the algebraic sum of the horizontal forces or horizontal components at any joint must likewise equal zero.

Where two opposing forces act in a straight line, the introduction of a third force at any angle to the line of action of the two forces will only tend to disturb the equilibrium and can never be introduced in such a manner that the third force or its components will produce an algebraic sum of zero. This is an important principle in graphical statics and will clearly assist the student in the analysis of intricate framed structures and the determination of the stresses in the members by the graphical method.

As it will be advisable to bear this principle in mind, it has been placed in the following concise form:

Rule.—*Where the lines of action of two opposing forces coincide and there is but one other force acting at the same point, the two opposing forces must be equal, and the third force will destroy the equilibrium of the system, unless it equals zero.*

DIAGRAMS FOR COMPLEX FRAMES

TRUSSES WITH JOINTS ACTED ON BY THREE UNKNOWN FORCES

FINK ROOF TRUSS

21. Methods of Obtaining One of Three Unknown Forces.—In analyzing the stresses in framed structures by the graphical method, joints are often encountered where three of the several stresses acting at the point are unknown. As it is impossible to determine the stress in a third unknown at a particular joint by the usual means employed in the graphical solution, when such a difficulty arises some expedient must be resorted to in order to obtain the amount of one of the unknown forces. The three methods employed to accomplish this purpose are:

1. A study of the symmetry of the frame diagram so as to determine by inspection, if possible, whether one of the unknown members or stresses does not equal some force or stress previously determined.
2. A calculation of one of the unknown forces or the stress in some adjacent member that may assist in determining one of the forces, by the theory of moments.
3. Reversing the direction of a member in the frame diagram. This method, which is known *as the solution by the reversal of the diagonal*, as it involves no calculations, is clearly graphical, surer than the first, and more expedient than the second.

These three methods will be explained in conjunction with the solution of stresses for the Fink, or Polonceau, type of roof truss shown by the frame diagram in Fig. 17.

22. Method by Symmetry.—The Fink, or Polonceau, roof truss is particularly favored by many on account of its symmetrical appearance, and from the fact that it can be

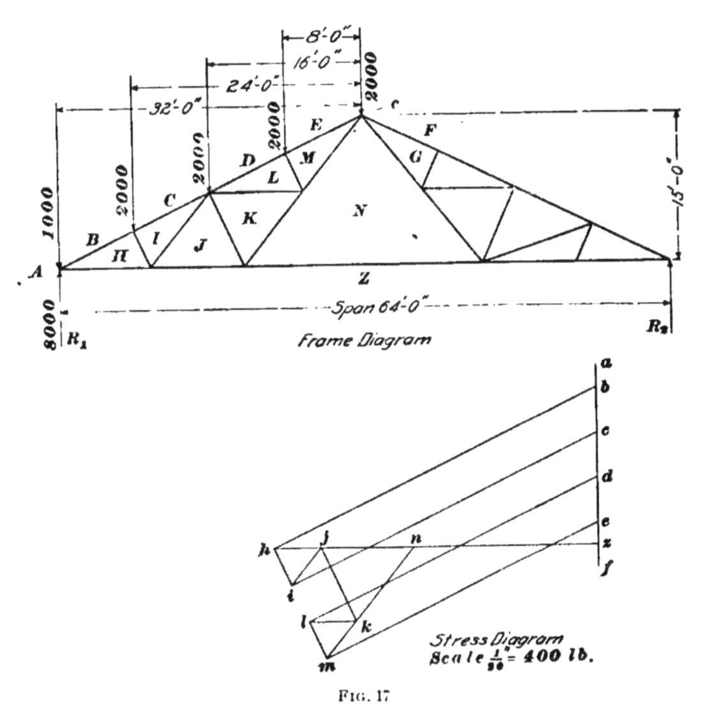

FIG. 17

readily and economically constructed. There is a difficulty, however, in analyzing the stresses, from the fact that when the joint *CDLKJI* is attempted, the operation is immediately halted since the stresses in the members *DL*, *LK*, and *KJ* are unknown. Likewise, on referring to the joint *JKNZ*, the stresses in the members *JK*, *KN*, and *NZ* are unknown. By studying the diagram in an effort to determine by the symmetry of the truss whether any of the unknown stresses can reasonably be supposed to have the same stress as another member—the stress in which can be determined—it is found that *L M* must have the same stress as *H I*; this being true, and *L K* being the reverse of *J I*, necessarily sustains the same stress from the foot of the strut *M L* as *I J* obtains from the foot of *H I*. With this fact in view, commence the diagram by laying out a portion of the load line from *a* to *f*, and since the reactions R_1 and R_2 are equal, on account of the loads being symmetrical, the point *z* will be located midway between *e* and *f*. Draw, from *b*, an indefinite line extending parallel with *B H* in the frame diagram, and, from *z*, a line *h z* extending parallel with *H Z*. The point of intersection will be *h* and the polygon of forces around the left-hand joint of the roof truss will be traced in the stress diagram from *a* to *b* to *h* to *z*, and from *z* upwards to the starting point *a*. The joints *B C I H* and *H I J Z* can now be laid out in the stress diagram and their respective polygons of forces obtained, thus locating the points *i* and *j*.

Now the difficult joints in the truss are encountered, but it has been suggested that *M L* and *I H* carry the same stress. Therefore, it is evident, on inspection of the frame diagram, that *l* and *m* must lie somewhere along the lines drawn from *d* and *e* parallel with *D L* and *E M*. It is also known that a line drawn downwards from *j* parallel with *J K* must contain, somewhere along its length, the point *k*, but on further consideration it is known that *k* and *l* must be connected by a line parallel with *L K* in the frame diagram, and also, which is still more important, that *l k* must equal, in length, *i j*; then by trial, *l k* can be located as shown. The point *m* is also located and the diagram completed by drawing from *m* a line parallel with *M N* in the stress diagram, which line will pass through *k* and intersect a line drawn from *z* parallel with *N Z* in the frame diagram at the point *n*. The polygon of forces around this last joint *J K N Z* reads in the stress diagram from *j* to *k* to *n* to *z*, and from *z* back to *j*. The figure being thus closed, the proof of the method is evident.

Since the truss is symmetrical with regard to a vertical center line and is likewise symmetrically loaded, the other half of the diagram will be the reverse of the one drawn, and the line *z h* in the stress diagram will be an axis of symmetry.

23. Method by Moments.—Again referring to the frame diagram, Fig. 17, it is observed that if the stress in the member *N Z* could be calculated, the joint *J K N Z* would no longer contain three unknown forces and consequently could be solved by the usual graphical method.

The stress in the member *N Z* may be readily calculated by taking moments about the point *c*; that is, by considering the truss as hinged at this point with all the loads on the rafter members tending to revolve the left-hand half of the truss in a direction opposed to the movement of the hands of a watch, while R_1, the left-hand reaction, opposes this movement and tends to rotate the left-hand half of the truss in a direction about *c* agreeing with the movement of the hands of a watch. But the algebraic sum of this system of forces will not equal zero, so that the member *N Z* must be able to resist such a stress as to retain equilibrium of rotation for the left-hand half of the truss. The moment of the stress in the member *N Z* about the point *c* is equal to the product of its amount by the perpendicular distance between its line of action and the point in question, or 15 feet, and the amount of the stress in this member can be obtained by dividing the algebraic sum of the moments of the external forces R_1, *A B*, *B C*, *C D*, *D E*, and *E F* about the point *c* by 15. For instance, the moments of the loads on the rafter members are as follows:

Moment of *A B*	= 1,000 × 32 =	32000 foot-pounds
Moment of *B C*	= 2,000 × 24 =	48000 foot-pounds
Moment of *C D*	= 2,000 × 16 =	32000 foot-pounds
Moment of *D E*	= 2,000 × 8 =	16000 foot-pounds
Moment of *E F*	= 2,000 × 0 =	0 foot-pounds
Sum of moments opposed to R_1	=	128000 foot-pounds

The moment of R_1 about the point *c* = 8,000 × 32 = 256,000 foot-pounds. If the sum of the moments for the loads on the rafter member is regarded as negative and the moment of R_1 is regarded as positive, the algebraic sum of the moments due to the external forces on one-half of the truss about the point *c* will equal 256,000 − 128,000, or 128,000 foot-pounds. This, divided by 15, equals 8,533 pounds, the stress in the member *NZ*.

The amount, 8,533, just obtained, may now be laid off from the point *z* in the stress diagram along the line parallel with the member *NZ* in the frame diagram and the point *n* thus located. Having determined the point *n*, the solution of the stresses around the joint *JKNZ* is readily made. By scaling the length of the line *nz* in the stress diagram, which was previously drawn, it is found that the calculated amount of the stress in the member *NZ* agrees with that obtained by the method of symmetry.

Having obtained the stress in the member *NZ*, by calculation, no difficulty in drawing the stress diagram will be encountered, for by the solution of the joint *JKNZ* the stress in the member *JK* is obtained. Therefore, since *CI* and *IJ* are already known, there remain around the joint *CDLKJI* only the two unknown forces *DL* and *LK*. The diagram can now readily be completed by working around the other joints.

24. Method by Reversal of Diagonal.—This method consists in reversing the direction of a diagonal in the frame diagram, in order to determine one of the unknown forces about the joint *DEML* or *JKNZ*, Fig. 18. The omitted member *LK* is shown light in the frame diagram, and is replaced by the member *KM*. The former member *LM*, shown dotted, if considered in the frame diagram, will produce a joint at *KMN* in unstable equilibrium and consequently the stress in this member must be zero, so that it is disregarded in drawing the stress diagram.

By omitting the useless members *LK* and *LM*, the frame diagram becomes as shown by the heavy lines and the stress diagram may be commenced. The load line and the points *h*, *i*, and *j* are found as usual and the stresses in the joint

CDKJI may be laid out in the stress diagram by drawing a line from *d* parallel with *DK* and an intersecting line from

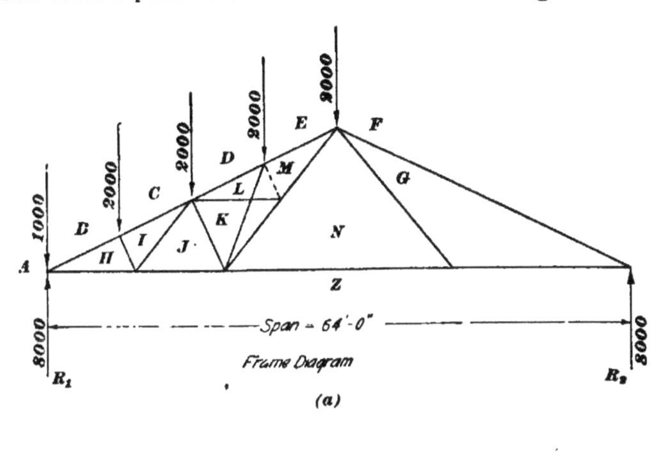

(a)

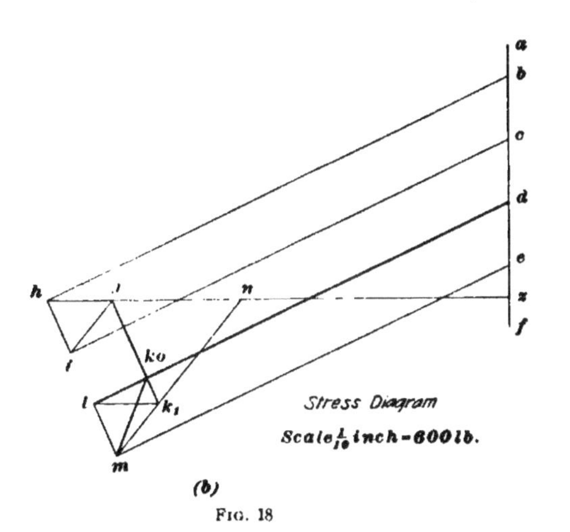

(b)

Fig. 18

the point *j* parallel with *JK*, the point of intersection being designated as k_0.

The forces around the joint *JKMNZ* may be found by first analyzing the joint *DEMK*. This joint is laid out in the stress diagram by drawing a line from e parallel with EM and by describing from k_0, a line parallel with KM in the frame diagram. In this way m is located at the intersection of these two lines. The joint $JKMNZ$ may now be solved by drawing mn parallel with MN in the frame diagram and locating n at the intersection of the line just drawn and the line zh. The points n and m having been obtained, the omitted members LK and LM may be replaced and the reverse diagonal eliminated. The stress diagram is completed for the original truss by drawing from the point m a line ml parallel with ML in the frame diagram and extending the line jk_0 to its new intersection k_1 on the line mn, the diagram being closed by the line lk_1, which is parallel with LK, and the extension of the line drawn from d parallel with DL in the frame diagram. It will be observed that the reversal of the diagonal was resorted to simply in order to obtain the points m and n, and consequently the stress represented by mn was not changed in any way by the reversal of the diagonal LK. The letters k_0 and k_1 are used only provisionally with the reversed diagonal KM. On replacing the diagonal LK, the letter k_0 is dropped and k_1 reverts to the orginal letter k.

QUADRILATERAL ROOF TRUSS

25. An interesting problem that involves the principles just described is offered by the frame diagram shown in Fig. 19 (a), which is the skeleton outline of a roof truss that can conveniently be used for spans of from 70 to 90 feet. The lantern, or skylight, the outline of which is shown dotted, may be of light work erected on the top of the truss, so that its weight need not be considered in proportioning the members in the frame diagram. The truss is shown with the wind acting on the left-hand side. The normal wind loads AB, BC, CD, and DE are calculated from the normal pressure on the roof slope acting on the panel area supported at each joint, while the load EF is the normal load on one-half of the panel and includes the left-hand reaction of the wind load on the lantern, or skylight. The wind load FG is equal to the right-hand wind reaction of the skylight. The truss is considered as being constructed of steel and provided with a roller bearing at the right-hand end to allow for expansion;

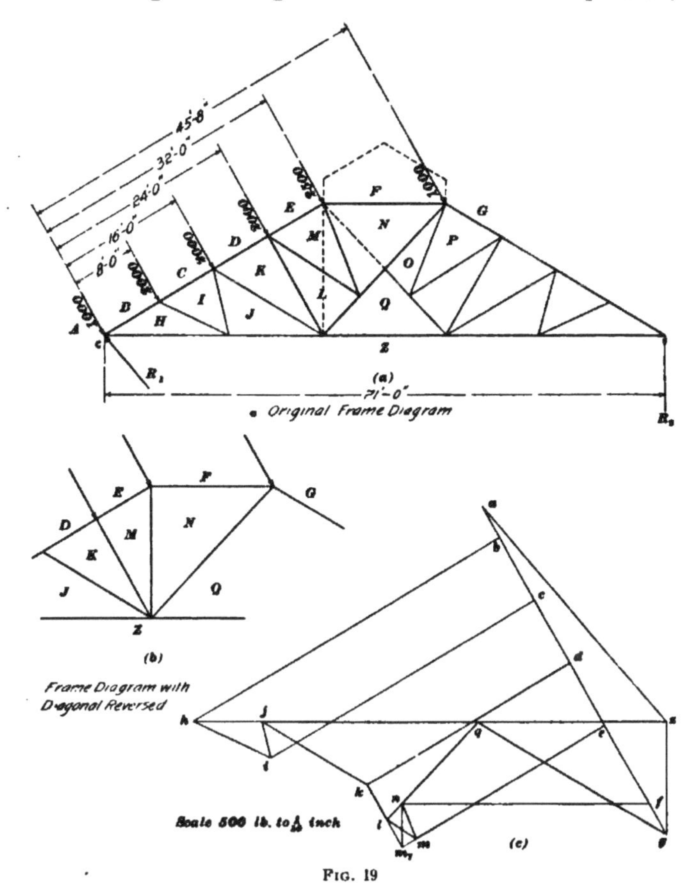

Fig. 19

the other end of the truss is fixed. This condition existing, the direction of R_2 is known to be vertical; as the amount of one of the reactions must be known before the polygon of external forces can be laid out, the amount of R_2 may be calculated by taking moments about the point *c*, that is, the point of application of R_1, and regarding all of the loads as giving positive moments. Their sum is obtained by the following calculation:

Moment of $AB = 1{,}000 \times 0$	=	0 foot-pounds
Moment of $BC = 2{,}000 \times 8$	=	16000 foot-pounds
Moment of $CD = 2{,}000 \times 16$	=	32000 foot-pounds
Moment of $DE = 2{,}000 \times 24$	=	48000 foot-pounds
Moment of $EF = 2{,}500 \times 32$	=	80000 foot-pounds
Moment of $FG = 1{,}000 \times 45.666$	=	45666 foot-pounds
Sum of positive moments	=	221666 foot-pounds

The lever arm of R_2 is equal to the span of the truss or 71 feet, so that the amount of R_2 is equal to $221{,}666 \div 71 = 3{,}122$ foot-pounds. This result having been obtained, construct the polygon of forces shown in the stress diagram of the figure.

26. The force polygon is laid out by marking off, to scale, the forces *ab*, *bc*, *cd*, *de*, *ef*, and *fg*, then laying off upwards, from *g*, the amount of R_2, locating in this way the point *z*. By connecting *z* and *a*, the direction and amount of R_1 is obtained; for the direction of R_1 coincides with the line *az* and its amount is equal to the distance, according to the scale of the diagram, from *z* to *a*. The stresses in the several members around the joint *ABHZ* are obtained by drawing from *b*, in the stress diagram, the line *bh* parallel with *BH*, in the frame diagram, and intersecting this line with a horizontal line from *z* that is parallel with the member *HZ*.

The solution of the joint *BCIH*, at which two unknown forces exist, is readily made by drawing, from *c*, a line parallel with *CI* and extending from *h* a line parallel with the member *HI*, thus locating the point *i*. The stresses around the joints *HIJZ* and *CDKJI* are likewise drawn without any difficulty, but in attempting either the joint *DEMLK* or *JKLQZ*, three unknown forces are encountered. There is no way of eliminating this difficulty by the method of symmetry, for no symmetry of the members exists in this truss. The stress in *QZ* or *FN* can be calculated by taking moments about the joints *FGPON* or *JKLQZ*, but as the problem can be more readily solved by reversing the diagonal *LM*, this method will be employed. It may be determined by inspection that the extension of the member *OQ*, shown dotted, is of no use when the wind acts on the left-hand portion of the structure, though this extension will come into play when the wind load acts on the right-hand portion of the truss, in which case the extension of the member *LQ* will be subjected to no stress, as, likewise, will be the various secondary members of the frame. Redraw and reletter the frame diagram with the diagonal reversed, as shown in (*b*), and proceed with the stress diagram by drawing from the point *k*, which has been previously found, the line $k m_1$ parallel with the member *KM* in the frame diagram (*b*). This line will intersect a line drawn from *e* and parallel with the member *EM* at m_1.

Next analyze the joint *EFNM*, as shown in the frame diagram (*b*), by drawing, from *f*, an indefinite horizontal line parallel with the member *FN* and extending, from m_1, a line upwards parallel with the member *NM*, thus locating the point *n*. The joint *KMNQZJ* in the frame diagram (*b*) can now be analyzed by drawing from *n* a line parallel with *NQ*. Where this line intersects the line drawn from *z* horizontally, or parallel with *QZ*, will be located the point *q* and the diagram for the truss may be closed by a line from *g* to *q*, which, if parallel with the member *GQ*, is proof that in the truss with the diagonal reversed, the diagram is correct and accurately drawn. Since the points *n* and *q* have been located in the stress diagram, the original truss with the diagonal *ML*, as shown in (*a*), may be drawn, and the stresses for the members *LM* and *MN* obtained by drawing from the point *n* a line parallel with *NM* in the original frame diagram. The intersection of this line with the line $e m_1$ will locate the point *m*, and the length of the line *nm* will give the amount of the stress in the member *NM*. If the line *nq* is extended it should pass through the point *l*, when *lq* will indicate the stress in the member *LQ*. By extending a line from *m* parallel with *LM* in the original diagram, until it intersects $k\,m$, at l, the stress in this member may be obtained by measuring the length of the line $l\,m$, and the polygons of forces about the joints with three unknown forces, which prevented the further analysis of the truss, are completed and the stress diagram for the original truss is shown by the heavy lines. The auxiliary lines that it was necessary to draw when the diagonal was reversed are also shown in the stress diagram.

It is an interesting example and will give considerable practice for one to lay out the stress diagram for the wind on the right-hand side of the truss, considering the roller bearing still at R_2. Likewise, lay out the stress diagram for the vertical loads, and if both diagrams close one may be reasonably sure that they have been correctly drawn.

FINK TRUSS WITH CAMBERED TIE-MEMBER

27. The Polonceau, or Fink, truss is frequently built with the lower chord of the truss cambered (raised at the center), as shown in view (*a*), Figs. 20 and 21. Cambering the lower chord in this manner gives greater headroom under the truss at the center, and somewhat improves its appearance; but it increases the stresses on all the members except *KL*, *ML*, and *ON*, and the corresponding members on the other side of the truss.

28. Diagram for Vertical Loads.—Fig. 20 (*a*) is a frame diagram showing the vertical loads on a Fink truss, the span of which is 80 feet, while the lower chord, or tie, is cambered from the horizontal 28 inches. The stress diagram for the vertical loads may be drawn, as shown in Fig. 20 (*b*), by first drawing the vertical load line *a f*, and laying off, to some convenient scale, the loads *a b*, *b c*, *c d*, *d e*, and *e f* designated in the frame diagram by *A B*, *B C*, *C D*, *D E*, and *E F*. Since the truss is symmetrically loaded, only one-half of the stress diagram need be drawn, and consequently the loads only as far as *e f* need be laid off on the vertical load line. The reactions R_1 and R_2 are each equal to one-half the load, or 18,000 pounds; hence, the point *z* is located midway between e and f, and $z\,a$ represents the reaction R_1, 18,000 pounds.

The stresses around the joint $A\,B\,K\,Z$ may be drawn in

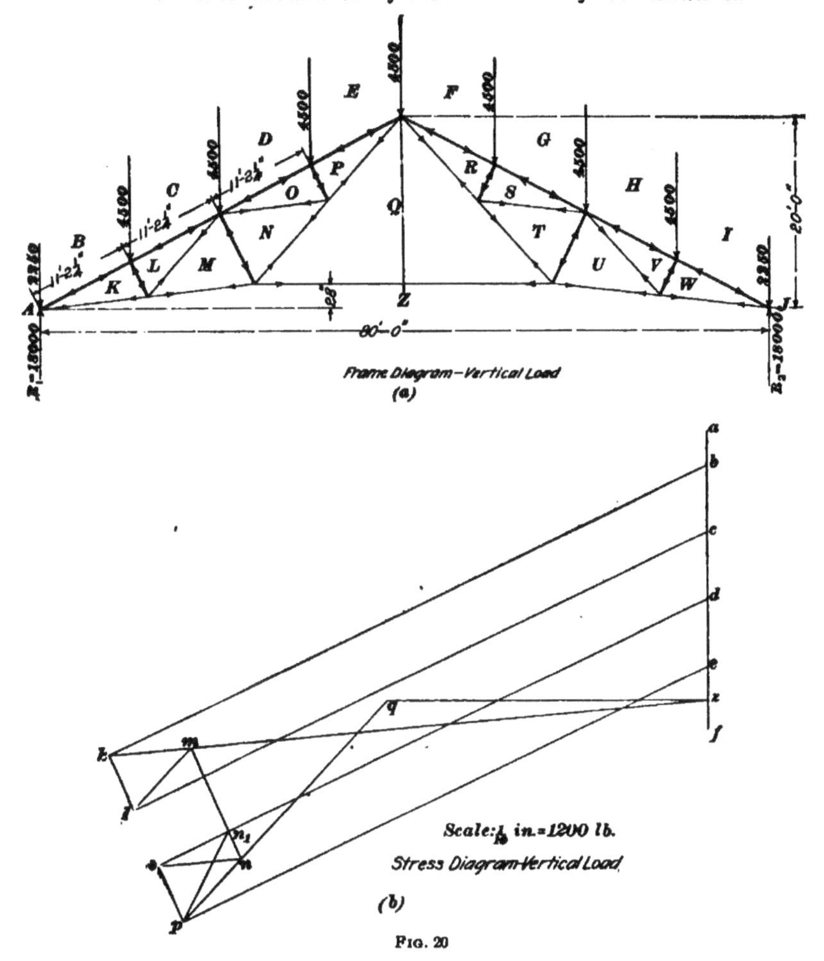

FIG. 20

the stress diagram by commencing at b and drawing $b\,k$ parallel with $B\,K$, and from z, a line parallel with $K\,Z$, intersecting the first line at k. The polygon of forces around this joint will then be from *a* to *b* to *k* to *z*, and from *z* back to *a*, the starting point. From the direction of these forces, the correct direction of the arrowheads in the frame diagram may be marked, and from their direction the kind of stress on the member is observed.

The next joint in the truss to be analyzed is *B C L K*. In the stress diagram begin at *c* and draw *c l* parallel with *CL* in the frame diagram, then from *k*, draw a line parallel with *L K* in the frame diagram, until it intersects the line drawn from *c* at the point *l*. The polygon of forces around the joint is from *b* to *c* to *l* to *k*, and from *k* back to *b*, the starting point.

Around the joint *K L M Z*, the stresses are obtained by drawing from *l* a line parallel with *L M* in the frame diagram, until it intersects the line *z k* at the point *m*. The polygon of forces around this joint is from *k* to *l* to *m* to *z*, and from *z* back to *k*, the starting point.

Difficulty will be encountered on attempting to analyze either the joint *C D O N M L* or *M N Q Z*, for at each of these joints there are three unknown forces or stresses. The difficulty may, however, be overcome by applying either of the three methods given in Art. 21. For instance, by reversing the diagonal *N O* and leaving the member *O P* out of consideration, a provisional point n_1 is located on a line $d n_1$ drawn parallel with *D O*. By means of n_1, the point *p* is located on the line *e p*, after which the diagonal *N O* may be replaced and the truss considered in its original form.

The remaining joints, when taken in their usual order, offer no difficulty, and the other half of the diagram need not be drawn unless it is desired to check the half just completed.

The polygon of forces that is traced in going around the joint *C D O N M L* affords a good illustration of the rule, that the forces that meet at a joint must make a closed polygon in the stress diagram.

One of the peculiarities of the stress diagram, Fig. 20 (*b*), and one that is worthy of note, as it will materially assist in drawing the diagram, is that the triangles *l k m* and *p o n* are equal and are similar to the larger one whose base is *m n*.

29. The **wind-stress diagram** may now be drawn. First the frame diagram is redrawn, and on it, as shown in Fig. 21 (*a*), are designated the wind loads acting at the several joints of the truss, in a direction normal to the slope of the roof.

The reactions R_1 and R_2 may be calculated by the principle of moments. Since the truss is securely fastened at both ends, neither end being free to move in a lateral direction, these reactions will be parallel with the action of the wind on the roof, that is, normal to the slope. The reaction R_2 may first be obtained by extending its line of direction until it intersects the extension of the left-hand rafter member at the point a'. Then by taking the center of moments at the left-hand reaction R_1, the magnitude of the reaction R_2 may be computed.

For convenience, reduce to feet and decimals the distance from each panel point to the point of rotation R_1; the moments about this point will then be as follows:

$5{,}625 \times 11.188 = 62932.50$ foot-pounds
$5{,}625 \times 22.375 = 125859.38$ foot-pounds
$5{,}625 \times 33.563 = 188791.88$ foot-pounds
$2{,}813 \times 44.750 = 125881.75$ foot-pounds

Total, 503465.51 foot-pounds

The distance of the center of moments of the line of action of the reaction R_2 is $44.75 + 27 = 71.75$ feet, and $503{,}465.51 \div 71.75 = 7{,}016$, the magnitude of the reaction R_2, in pounds.

Since the sum of the wind loads is 22,501 pounds, which is equal to the sum of the reactions, the reaction due to the wind at R_1 is $22{,}501 - 7{,}016 = 15{,}485$ pounds.

To construct the wind-stress diagram, Fig. 21 (*b*), draw the load line from *a* to *f* parallel to the reactions and direction of the wind loads at the several joints. Then lay off on the load line the loads *a b*, *b c*, *c d*, *d e*, and *e f*, which are, respectively, equal to the corresponding loads *A B*, *B C*, *C D*, *D E*, and *E F* in the frame diagram. Having located the point *f*, the magnitude of the reaction R_2, represented by *f z*, may be laid off upwards (the direction in which it acts), and the point z thus located; then on scaling za, it should be found

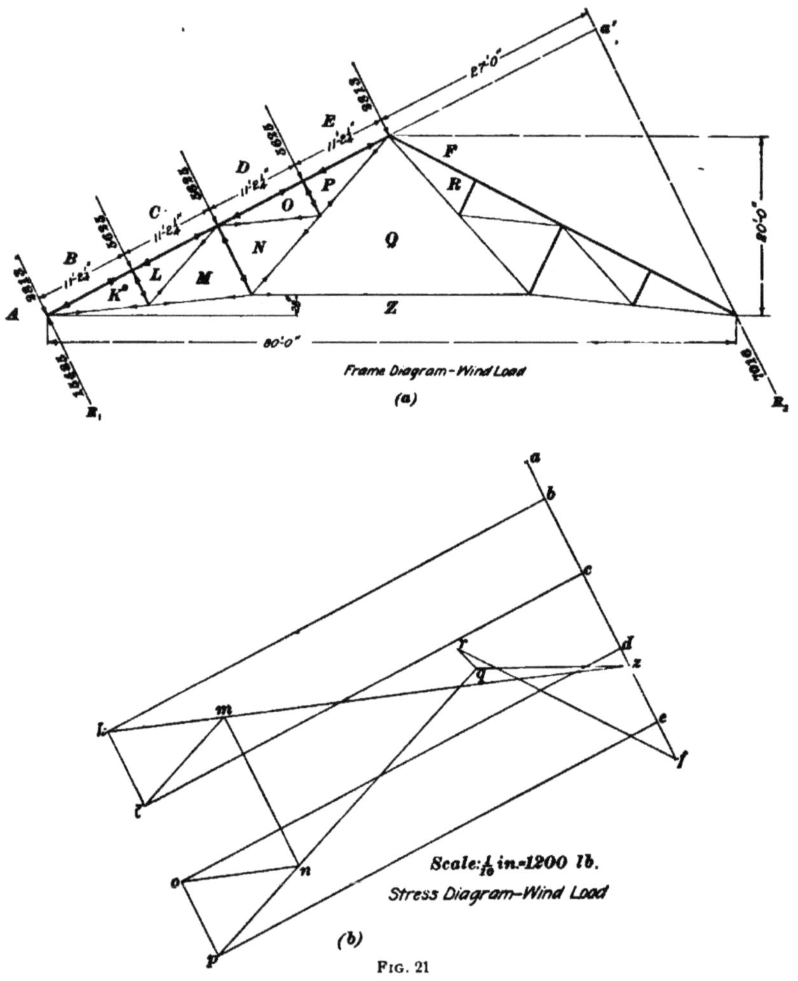

Fig. 21

equal to 15,485 pounds, the reaction R_1. The polygon of external forces will then be: from a to b to c to d to e to f to z, and from z to a, the starting point.

The joint *CDONML* may be solved in this stress diagram in the same manner as it was solved in the vertical-load stress diagram. The analysis of stresses around the last joint *EFRQP* in the frame diagram; is interesting, from the fact that the stress *qr* closes the diagram, and, if the diagram is correctly drawn, this line must be parallel to the member *QR*.

TABLE II

Member	Stress Due to Vertical or Dead Load	Stress Due to Wind Load	Total Stress
BK	+ 46,000	+ 34,500	+ 80,500
CL	+ 44,000	+ 34,500	+ 78,500
DO	+ 42,000	+ 34,500	+ 76,500
EP	+ 40,000	+ 34,500	+ 74,500
KL	+ 4,000	+ 5,500	+ 9,500
MN	+ 8,000	+ 11,250	+ 19,250
OP	+ 4,000	+ 5,500	+ 9,500
ZK	− 41,500	− 36,750	− 78,250
ZM	− 35,500	− 28,500	− 64,000
ZQ	− 21,000	− 11,000	− 32,000
QN	− 15,500	− 18,500	− 34,000
QP	− 21,500	− 26,000	− 47,500
FR		+ 17,000	
RQ		− 2,000	
NO	− 5,750	− 8,000	− 13,750
LM	− 5,750	− 8,000	− 13,750

Having drawn both the wind- and the vertical-load stress diagrams, the stresses in the several members in the truss may be obtained by scaling, and their magnitudes may be tabulated as in Table II.

In Table II, compression is indicated by the plus sign and tension by the minus sign.

WIND AND SNOW LOADS ON OPPOSITE SIDES

CRESCENT TRUSS OF SMALL SPAN

30. In the solutions for determining the stresses in roof trusses, two stress diagrams have been considered: the one for the vertical load on the roof truss and the other for the wind load taken normal to the roof slope, and a combination of the stresses in the several members determined by these diagrams has been assumed to give the maximum stresses in the members throughout the truss. The assumption that the maximum stress is attained by combining the results of the full vertical load and the wind-load diagrams is usually on the side of safety, though there are occasions when the combination of these two diagrams does not give the maximum stress in the members and the conditions as expressed by such a method can seldom exist. In the design of a large roof truss, therefore, where economy is a consideration, it is best to draw several diagrams and combine the stresses found by several combinations of these diagrams; the actual conditions as they will occur under varied wind and snow loads may then be approached. It is usual, therefore, to draw, first, the dead load exclusive of the snow load; second, a diagram of the truss including the entire snow load; third, a diagram of the wind load acting on the roof truss from both the right and the left, the different directions giving different stress diagrams if the truss is supported on rollers at one end; fourth, stress diagrams that will represent the snow load on the leeward side of the truss and the wind load acting on the windward side; that is, stress diagrams that will give the stresses in the several members of the roof truss, when one side of the truss is covered with snow and the wind blows with its maximum velocity on the other side.

This latter condition is the one that will be especially considered in this problem; that such a condition may be approached is reasonable from the fact that when the wind is blowing with its maximum velocity of 80 or 100 miles an hour,

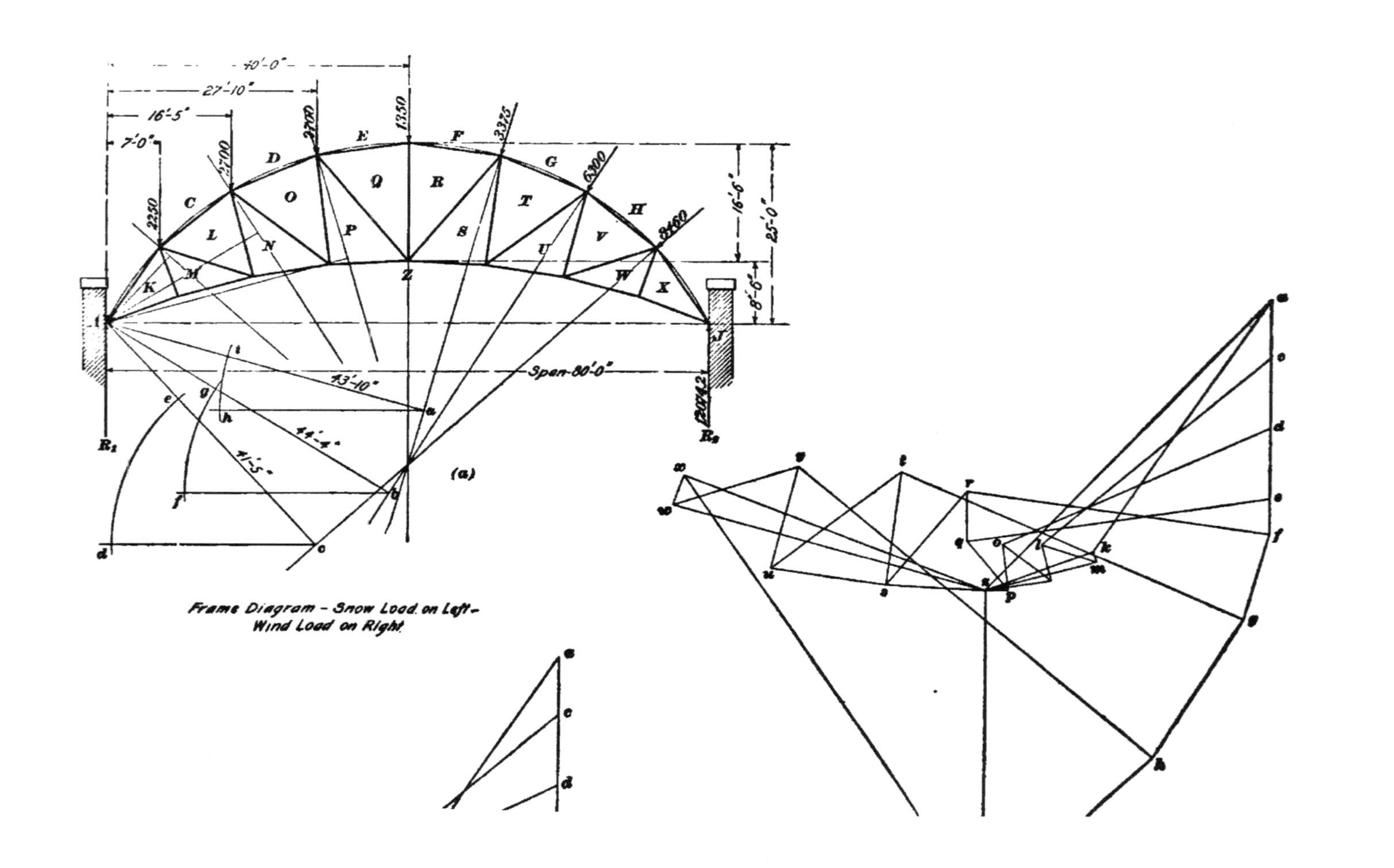

Frame Diagram - Snow Load on Left-
Wind Load on Right.

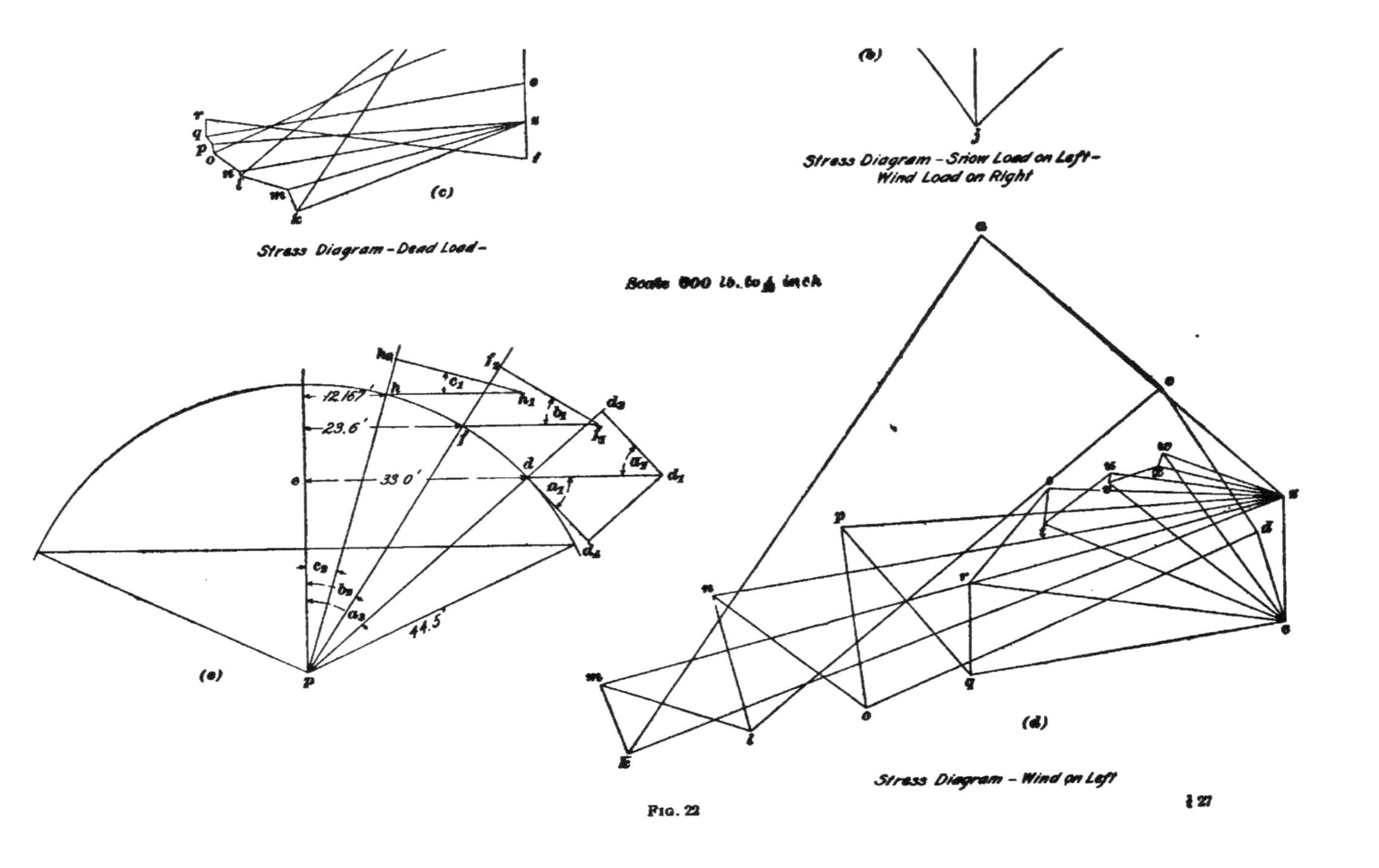
(c)
Stress Diagram - Dead Load -
(b)
Stress Diagram - Snow Load on Left -
Wind Load on Right
Scale 800 lb. to 1/16 inch
(e)
12.167'
23.6'
33.0'
44.5'
(d)
Stress Diagram - Wind on Left

Fig. 22

no snow can possibly be supported on one side of the roof truss, though a full snow load may exist on the other side.

In Fig. 22 (a) is shown the frame diagram of a **crescent-shaped roof truss** with the web members arranged similar to a Howe truss. The lower chord is cambered 8 feet 6 inches to comply with the conditions imposed by the architectural treatment of the interior of the building. The span of the truss is 80 feet, its total rise from the *springing line*, or level of the foot of the truss, is 25 feet, and the radii of the top and bottom chords are 44 feet 6 inches and 96 feet 5 inches, respectively. On the left-hand portion of the truss at each panel point are designated the snow loads, while on the right-hand half is shown the resultant wind pressure at the several panel points. There being no load or wind pressure at the two ends of the truss, the letters B and I are omitted and the first and last loads are termed AC and HJ, respectively.

In determining the snow loads on the left-hand portion of the truss, it might be considered that the slope of the upper chord, and consequently of the roof in the panel AK, is so steep as to preclude the lodgment of any snow on it. In this particular instance, however, since the walls of the building terminate in a parapet, which would be likely to cause the lodgment of considerable snow depending for its support on a portion of the chord AK, it was considered advisable to assume the panel load AC as made up of a snow load equal to 8 pounds per square foot on one-half of the chord AK, and of 12 pounds per square foot on one-half of the chord CL. Since the principals, or roof trusses, are 18 feet from center to center and the distance from panel point to panel point is 12 feet 6 inches, the vertical loads due to the snow at the several panel points may be determined by the following calculation:

Part of AC =	$6.25 \times 18 \times 8$ =	900	
Part of AC =	$6.25 \times 18 \times 12$ =	1,350	2,250
CD =	$12.50 \times 18 \times 12$ =		2,700
DE =	$12.50 \times 18 \times 12$ =		2,700
EF =	$6.25 \times 18 \times 12$ =		1,350

In calculating the amount of the wind resultant at the several points on the right-hand side of the truss, it is advisable to calculate the pressure normal to the curve of the roof slope or the upper chord; that is, perpendicular to tangents at these points, or in the direction of the radius drawn through its point of application. These normal pressures can be calculated by Hutton's formula, or found from a table giving the values for roof slopes of different pitches. It was in this manner that the wind loads were determined in laying out the stress diagram in Fig. 22 (*a*). The above formula and table are given in *Loads in Structures*.

31. Determining Wind Pressure on Truss With Curved or Broken Upper Chord.—A method for determining the wind pressures at the several panel points on a truss with a curved or broken upper chord, which is usually sufficiently accurate for all practical purposes, is shown in Fig. 22 (*e*). It is based on the following rule:

Rule I.—*The wind pressure normal to an inclined surface varies, approximately, as the sine of the angle that this surface makes with the horizon.*

The roof slope, in this instance, being a curved surface, a tangent drawn through the point where the wind resultant is supposed to act will represent the incline. Referring to Fig. 22 (*e*), the wind resultant $d_1 d$ is acting at the point d, through which a tangent $d d_4$ is drawn. According to the above rule the wind pressure $d d_3$, normal to the roof, will vary with the sine of the angle a_1. By drawing line $d_1 d_4$ parallel to $d d_3$, a right triangle $d d_1 d_4$ is produced, in which the hypotenuse $d d_1$ and the angle a_1 are given, or, if not, may be calculated as shown below. The side $d_1 d_4$ may then be found by means of the function $d_1 d_4 = \sin a_1 \times d d_1$. Assuming that the side $d d_1$ represents the wind resultant of 40 pounds per square foot and that the angle a_1 is 48°, the side $d_1 d_4 = \sin 48° \times 40 = .743 \times 40 = 29.7$ pounds. By drawing $d_1 d_3$ parallel to the tangent $d d_4$, a triangle equal to $d d_4 d_1$ will be formed, in which the angle a_2 is equal to a_1, and the side $d d_3$ is equal to $d_1 d_4$. The normal wind pressure $d d_3$ will, therefore, be proportional to the sine of the angle a_2. For obvious reasons, the triangle $e\,d\,p$ is similar to either of these triangles, the angle a_3 being equal to a_1 and a_2. Therefore, if these angles are not known they may be found by means of the function:

$$\sin a_3 = \frac{d\,e}{d\,p} = \frac{33}{44.5} = .7415 = 47° \ 52', \text{ or nearly } 48°$$

the distance $d\,e$ being found from the drawing and the radius calculated, if not given, by means of the rise of 25 feet and the total length of 80 feet.

Other wind resultants are shown at the points f and h, and for the reason already given, it is obvious that the angles b_3 and c_3 are equal to b_1 and c_1, respectively. If these angles are not known they may be found by the following rule:

Rule II.—*The sine of an angle is the quotient of the opposite side divided by the hypotenuse.*

The hypotenuse in this case is equal to the radius of the arc, and the side opposite the angle is one-half of the chord of the arc enclosed by twice the angle.

The problem may be solved graphically in the following manner: Let the line $d\,d_1$ represent the pressure per square foot of the wind resultant, drawn to any convenient scale. From its outer end, draw a line $d_1\,d_2$ perpendicular to the extended radius $p\,d$; then the length of the line $d\,d_2$ will give the normal wind pressure per square foot, measured to the same scale. Or the line $d\,d_1$ may represent the total panel pressure at this place, based on a pressure of 40 pounds per square foot, when the line $d\,d_2$ will give the reduced pressure normal to the panel, or to the tangent representing two adjoining panels. In a similar manner, the normal pressures are found at the points f and h by laying off the lines $f\,f_1$ and $h\,h_1$ equal, in length, to $d\,d_1$ and drawing the lines $f_1\,f_2$ and $h_1\,h_2$ normal to the extended radii $p\,f$ and $p\,h$, respectively, when the lines $f_2\,f$ and $h_2\,h$ will represent the normal wind pressures at these points.

The normal pressure may also be found by means of the following proportion: $d\,p : d\,e =$ the wind resultant $d\,d_1$: to the normal pressure $d_2\,d$. Replacing $d_2\,d$ by the letter x and giving the other terms their respective values: $dp = 44.5$,

$de = 33$, $dd_1 = 40$, we have $44.5 : 33 = 40 : x$, therefore, $x = \frac{33 \times 40}{44.5} = 29.7$ pounds per square foot.

Following the same method $f_1 f = \frac{23.6 \times 40}{44.5} = 21.2$ pounds and $h_1 h = \frac{12.167 \times 40}{44.5} = 10.936$ pounds.

It will be observed that no resultant wind pressure was introduced in the frame diagram at the point EF or at the right-hand springing point. This is reasonable from the fact that the crown of the truss is so near a flat surface that the wind pressure will be almost imperceptible on the upper half of the panel member FR. Similarly, the parapet wall will so protect the lower half of the panel member JX that it would be useless to consider the wind pressure against this portion of the roof surface.

It is now assumed that all the external forces on this roof truss have been determined except the reactions. It is assumed that the right-hand end of the truss is on rollers and that consequently the reaction R_2 is vertical. Its amount, however, is as yet unknown, and the direction and amount of the reaction R_1 can be ascertained only by completing the polygon of external forces. In order to do this, the amount of the reaction R_2 must be ascertained by taking the moments of both the vertical and wind panel loads about the point of application of the reaction R_1 or around the left-hand abutment of the truss. The moments of these forces being the products obtained by multiplying the intensity of the force by its lever arm or the perpendicular distance from its point of action to the point of rotation or center of moments, their sum may be ascertained by the following notation:

Moment of AC	= 2,250 × 7	=	15750 foot-pounds
Moment of CD	= 2,700 × 16.4167	=	44325 foot-pounds
Moment of DE	= 2,700 × 27.8333	=	75150 foot-pounds
Moment of EF	= 1,350 × 40	=	54000 foot-pounds
Moment of FG	= 3,375 × 43.8333	=	147937 foot-pounds
Moment of GH	= 6,300 × 44.3333	=	279300 foot-pounds
Moment of HJ	= 8,438 × 41.4167	=	349474 foot-pounds
			965936 foot-pounds

The sum of the moments about the left-hand abutment equals 965,936 foot-pounds. The lever arm through which the reaction R_2 acts around the center of moments at the left-hand abutment is equal to the span, or 80 feet, and as the moment of R_2 about this point must equal the sum of the moments of the vertical and wind loads at the panel points, the value of R_2 is found by dividing 965,936 by 80, which gives 12,074.2 pounds. The stress diagram for the combined snow and wind load may now be commenced by laying out the polygon of external forces, in which way the amount and direction of the reaction R_1 will be determined.

32. In Fig. 22 (*b*) is shown the stress diagram for the truss loaded as shown in (*a*). This diagram is drawn to a scale of 600 pounds to $\frac{1}{16}$ inch. Lay off the vertical loads *AC*, *CD*, *DE*, and *EF* on the load line in the stress diagram, to scale, as shown at *ac*, *cd*, *de*, and *ef*, respectively. From *f*, in the stress diagram, lay off the oblique lines *fg*, *gh*, and *hj* representing the resultant wind pressures at the several panel points at the right-hand side of the truss. From *j*, extend the vertical line upwards and lay off the distance *jz* equal to the amount of reaction R_2, or 12,074.2 pounds. The point *z* having been thus located, draw a line connecting it with the point *a* and by so doing determine, by the direction and the length of the line *za*, the amount and direction of the reaction R_1. The polygon of external forces is now completed and includes, as in every case of graphical statics, both the loads and their several reactions. In going around the polygon of forces, following the direction from the left to the right in the frame diagram, the forces are *ac*, *cd*, *de*, *ef*, *fg*, *gh*, *hj*, and from *j* to *z*, and from *z* back to *a*, the starting point. This closes the polygon of external forces and the stress diagram may be completed by working around each point, conveniently commencing at the left-hand end of the truss.

Start the diagram by working around the joint *AKZ*, drawing from the point *a*, in the stress diagram, the line *ak* parallel with *AK*, in the frame diagram, and extending from *z* the line *kz* parallel with the member *KZ*, the intersection of the two lines being the point *k*. It will be necessary now to go around the joint *KMZ*, for around the joint *ACLMK* there are more than two unknowns. Therefore, in the stress diagram (*b*), draw from the point *k* an indefinite line *km* parallel with *KM*, in the frame diagram, and from *z* extend a line in the direction of *zm* parallel with *MZ*, in this way locating the point *m* at the intersection of *km* and *zm*. The analysis of the stresses around the joint *ACLMK* is now clear and they may be determined by drawing from *c* an indefinite line parallel with *CL*, and from *m* a line parallel with *ML*, upwards and to the left, intersecting the line drawn from *c* at the point *l*. The polygon of forces around this latter joint will then be from *a* to *c* to *l* to *m* to *k* and from *k* back to the starting point *a*. Each joint may, in turn, be analyzed in this manner and the diagram completed as shown.

The diagram for the dead load, exclusive of the snow load, is shown in Fig. 22 (*c*), the dead loads on the panel points being, in this instance, practically the same as the snow loads. In Fig. 22 (*d*) is designated the diagram for the wind load on the left-hand slope of the truss, neglecting the snow load, and a similar diagram for the wind load on the right-hand portion of the truss, neglecting the snow load, might be drawn.

CRESCENT TRUSS OF LARGE SPAN

33. The **crescent roof truss of large span,** shown in Fig. 23 (*a*), is an excellent truss for exposition buildings, car barns, and similar structures. Its elements are the top and bottom chords and the web members. While the web members are arranged entirely for utility they do not present an unpleasing appearance. The lower member of this truss is usually cambered, as shown on the drawing. Both the top and bottom can be arranged on segments of circles or the panel points may be located by means of ordinates from a top or bottom horizontal line, as shown in the figure. It is proposed, in analyzing this truss, that four diagrams

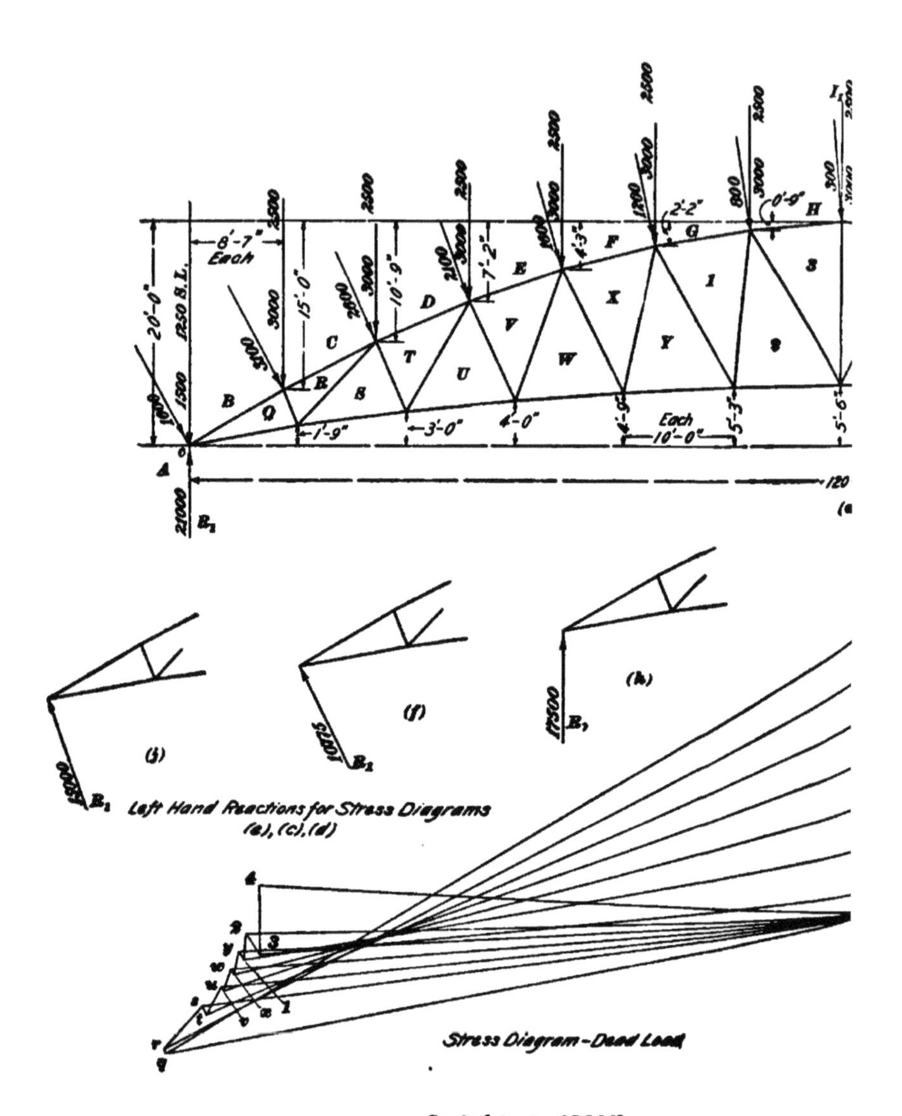

Scale $\frac{1}{10}$ inch = 1200 lb.

(b)

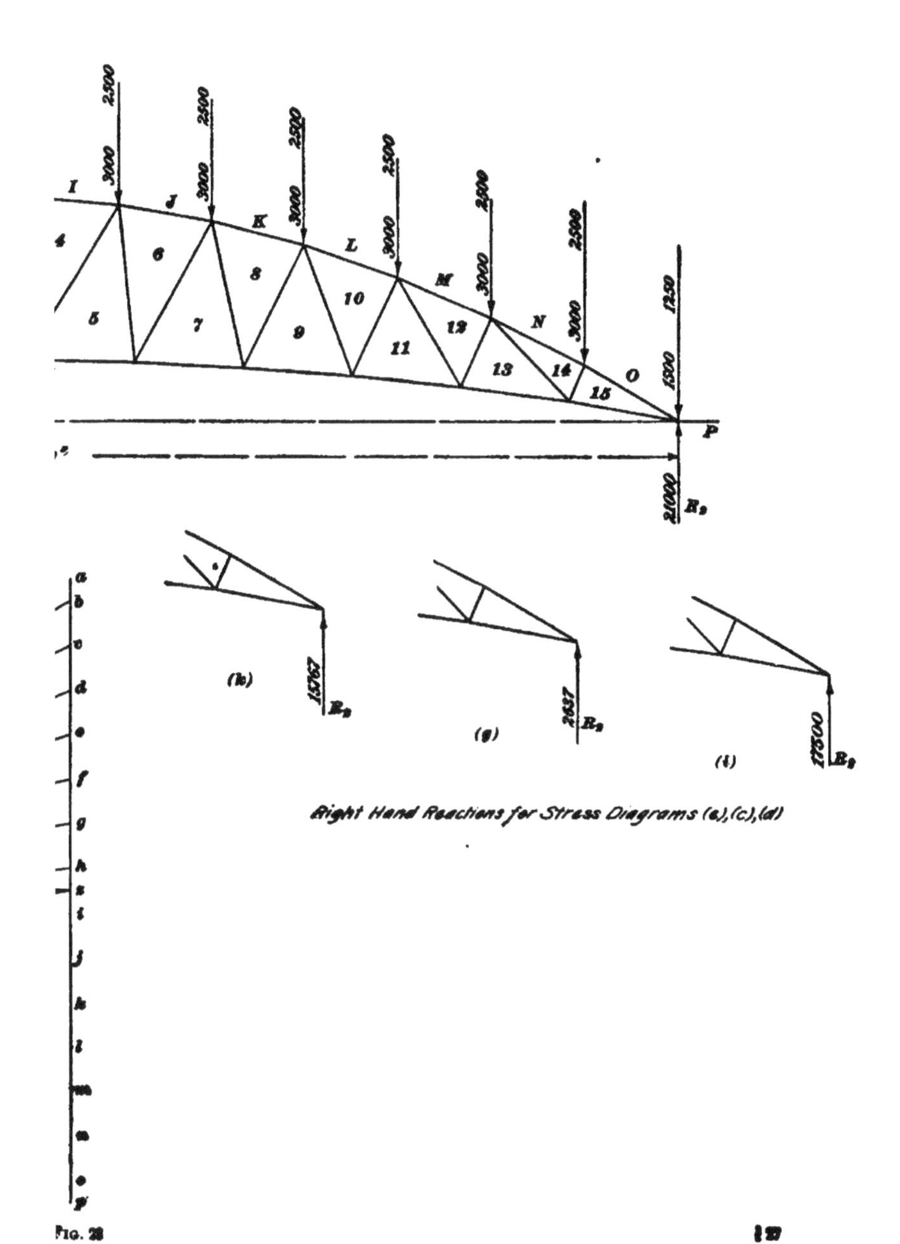

(h)
15767
R_2
(g)
2637
R_2
(i)
17500
R_2
21000
R_2
Right Hand Reactions for Stress Diagrams (e),(c),(d)

FIG. 28

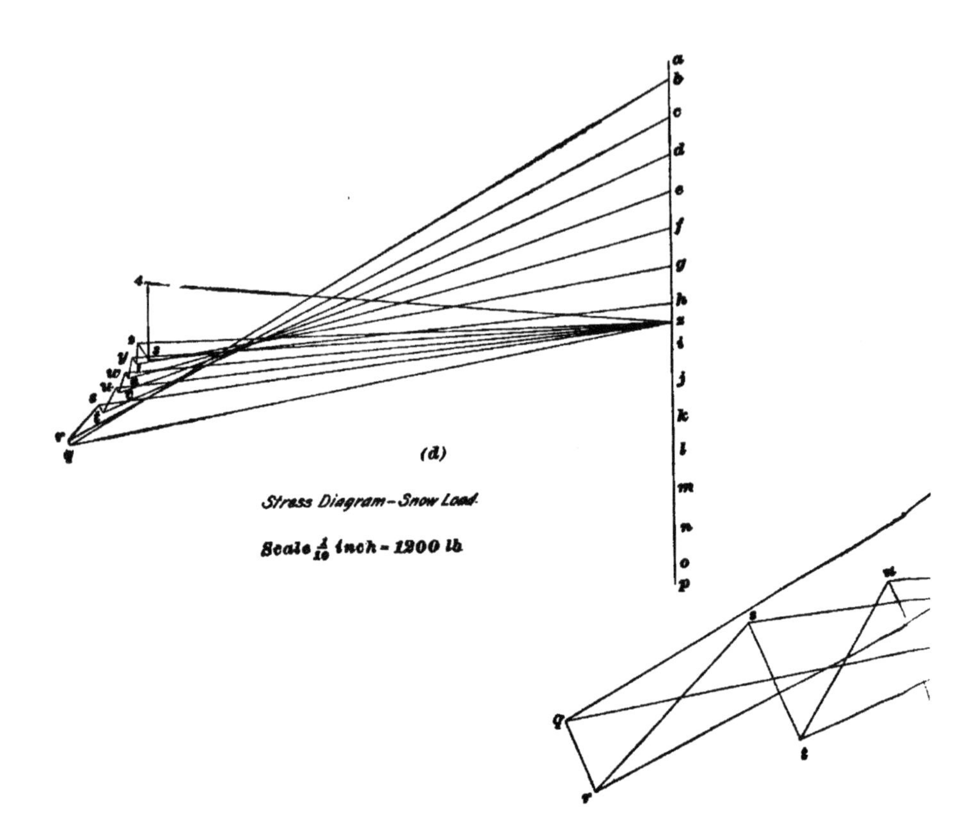
a
b
c
d
e
f
g
h
i
j
k
l
m
n
o
p
(d)
Stress Diagram–Snow Load.
Scale 1/10 inch = 1200 lb
q
r
s
t
u

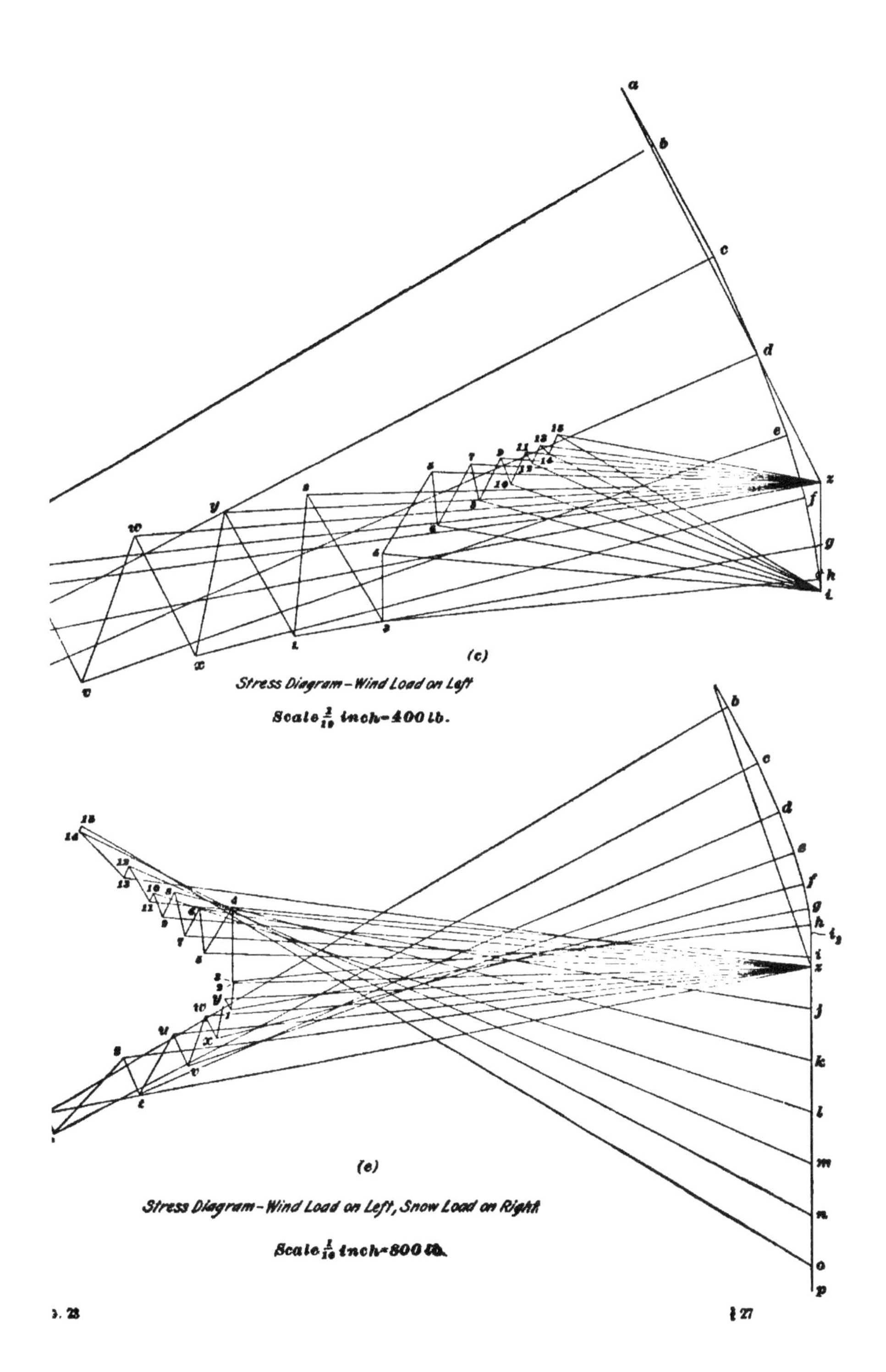
a
b
c
d
e
f
g
h
i
z
w
y
x
v
(c)
Stress Diagram—Wind Load on Left
Scale $\frac{1}{16}$ inch = 400 lb.
b
c
d
e
f
g
h
i_1
i
z
j
k
l
m
n
o
p
s
t
u
v
w
x
y
(e)
Stress Diagram—Wind Load on Left, Snow Load on Right
Scale $\frac{1}{16}$ inch = 800 lb.

be drawn: First, the dead-load diagram (*b*); second, the wind-load diagram (*c*); third, the snow-load diagram (*d*); fourth, a stress diagram (*e*), considering the truss as being subjected to a snow load on one side and the maximum wind load on the other.

In calculating the panel loads for the frame diagram necessary to draw the dead-load diagram shown in Fig. 23 (*b*), the weight of the principal, or roof truss, and the roof covering was considered; and in the calculation for all of the loads at the several panel points, it was assumed that the roof trusses were placed 16 feet from center to center. The dead-load diagram offers no difficulty, and as the truss is symmetrically loaded, only one-half the diagram need be drawn. The loads and reactions lie in vertical lines, so that the load line is a single straight line extending from *a* to *p* and the polygon of external forces for the dead loads is drawn from *a* to *b* to *c* to *p*, back to *z*, then to the starting point *a*. It will be found by working around the truss that at any joint there will not exist more than two unknown forces.

In drawing the wind-stress diagram (*c*), it is assumed that the right-hand end of the truss is on rollers and consequently that the reaction is vertical at this point. Since the direction of R_1, or the right-hand reaction, is known, its amount can be readily calculated by taking moments about the point *c*, or the left-hand reaction. The lever arms may be obtained by scaling the lengths of lines from the center of moments *c* perpendicular to the lines of action of the wind at each panel point. The calculation is as follows:

Moment of AB	= 1,600 × 0	=	0	foot-pounds
Moment of BC	= 3,100 × 9.8	=	30380	foot-pounds
Moment of CD	= 2,600 × 19.4	=	50440	foot-pounds
Moment of DE	= 2,100 × 28.6	=	60060	foot-pounds
Moment of EF	= 1,600 × 37.3	=	59680	foot-pounds
Moment of FG	= 1,200 × 45.5	=	54600	foot-pounds
Moment of GH	= 800 × 53.5	=	42800	foot-pounds
Moment of HI	= 300 × 61.5	=	18450	foot-pounds
	Sum of moments	=	316410	foot-pounds

The leverage of the right-hand reaction, since it acts vertically, is equal to the span of the truss, or 120 feet; and 316,410 ÷ 120 = 2,637 pounds, the amount of the reaction R_2. In laying out the stress diagram, draw the load line $a\,b$, $b\,c$, $c\,d$, $d\,e$, $e\,f$, $f\,g$, $g\,h$, and $h\,i$, being careful to designate each force in its true length and direction, as each force acts in a different direction. When the point i, which is the termination of the last wind load, has been located, the reaction R_2 can be drawn vertically and its amount, 2,637 pounds, laid off to scale, thus locating the point z. By connecting z and a the amount and direction of the left-hand reaction, which is oblique, since the end of the truss is fixed, are determined. The diagram may then be completed, when it will have the appearance shown. The reactions corresponding to this stress diagram are shown to the left and right of the truss in Fig. 23 (f) and (g).

Moment of AB	=	1,600 ×	0	=	0 ft.-lb.
Moment of BC	=	3,100 ×	9.8	=	30380 ft.-lb.
Moment of CD	=	2,600 ×	19.4	=	50440 ft.-lb.
Moment of DE	=	2,100 ×	28.6	=	60060 ft.-lb.
Moment of EF	=	1,600 ×	37.3	=	59680 ft.-lb.
Moment of FG	=	1,200 ×	45.5	=	54600 ft.-lb.
Moment of GH	=	800 ×	53.5	=	42800 ft.-lb.
Moment of HI_1	=	300 ×	61.5	=	18450 ft.-lb.
Moment of I_1I	=	1,250 ×	60.0	=	75000 ft.-lb.
Moment of IJ	=	2,500 ×	68.583	=	171458 ft.-lb.
Moment of JK	=	2,500 ×	77.1667	=	192917 ft.-lb.
Moment of KL	=	2,500 ×	85.75	=	214375 ft.-lb.
Moment of LM	=	2,500 ×	94.333	=	235833 ft.-lb.
Moment of MN	=	2,500 ×	102.9167	=	257292 ft.-lb.
Moment of NO	=	2,500 ×	111.5	=	278750 ft.-lb.
Moment of OP	=	1,250 ×	120.0	=	150000 ft.-lb.
			Sum of moments	=	1892035 ft.-lb.

Dividing this sum by the span gives 1,892,035 ÷ 120, or 15,767 pounds, as the right-hand vertical reaction, which is shown in Fig. 23 (k). The load line may now be laid out in the stress diagram as shown.

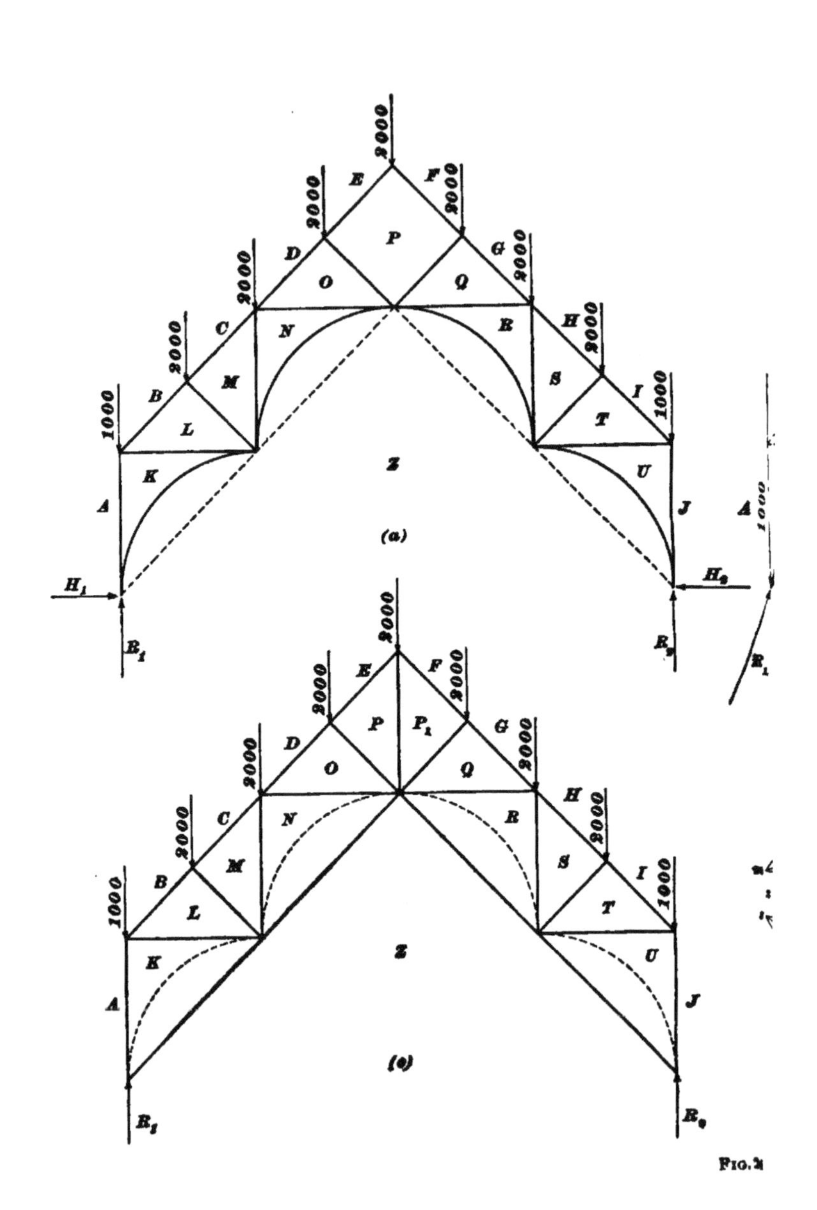

FIG. 2

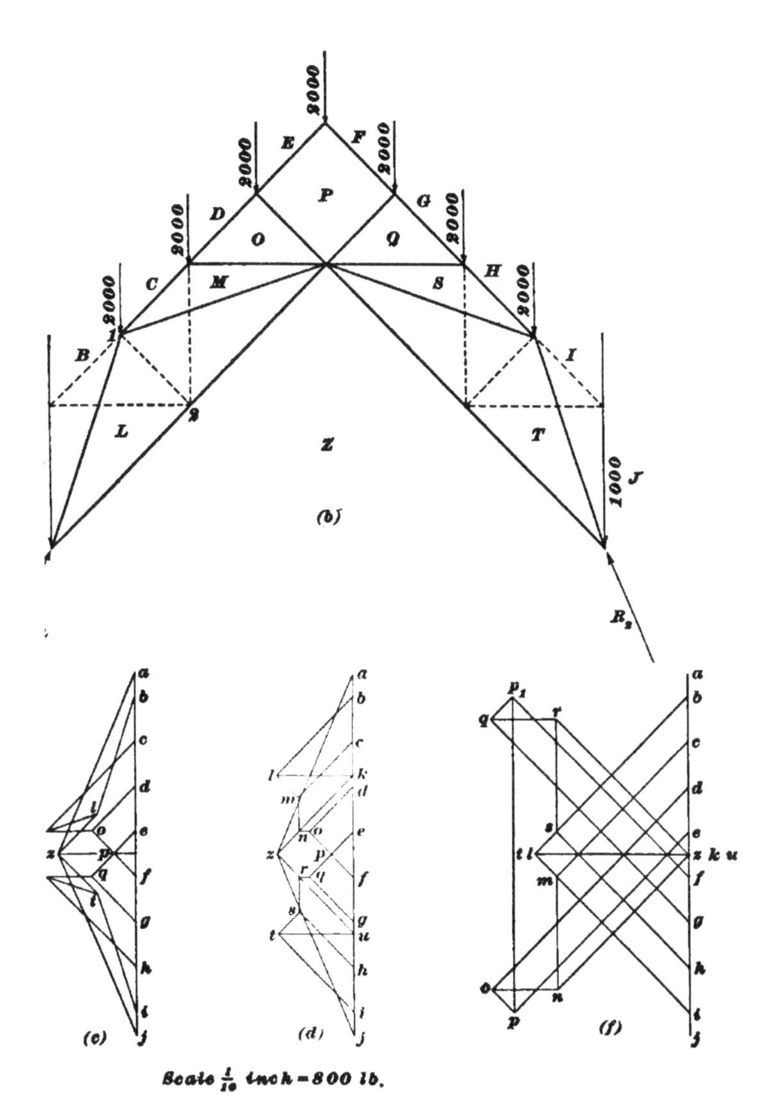
2000
E
F
2000
2000
P
D
G
2000
O
Q
2000
2000
C
M
S
H
2000
1
B
I
2
L
Z
T
1000
J
(b)
R2
(c)
(d)
(f)
Scale 1/16 inch = 800 lb.

The stress diagram for the snow load, which is assumed to be 2,500 pounds on each panel, is shown in Fig. 23 (*d*). As will be observed, this diagram is similar to the dead-load diagram, except that the stresses in the different members are slightly reduced. The corresponding reactions are shown in Fig. 23 (*h*) and (*i*).

In Fig. 23 (*e*), the stress diagram is given for a wind load on the left-hand side of the truss and a snow load on the right. The snow load on the panel point *H I* will be only half of 2,500 pounds in this case, since the snow on one side of the truss only is considered. The right-hand end of the truss is assumed to be on rollers and the reaction R_2 is obtained by taking the moments about the support at the other end of the truss, or the point *c*. The calculations are as given on the preceding page.

The forces *a b*, *b c*, *c d*, *d e*, *e f*, *f g*, *g h*, and $h\,i_1$ should be carefully laid off in the proper directions; the remaining forces $i_1 i$, *i j*, *j k*, *k l*, *l m*, *m n*, *n o*, and *o p* are in a vertical direction. The reaction R_2 is now laid off vertically from *p* to *z* and a line drawn connecting *z* and *a*; this line represents the direction and magnitude of the left-hand reaction, which may be drawn in the frame diagram as shown in Fig. 23 (*j*). The reactions R_1 and R_2, shown on the frame diagram (*a*), are for the dead load, exclusive of the snow load. No difficulty will be encountered in drawing this stress diagram (*e*), as there are not more than two unknown forces at any joint.

ROOF TRUSSES WITH HORIZONTAL OR OBLIQUE THRUSTS

HAMMER-BEAM TRUSS

34. The **hammer-beam roof truss**, shown in Fig. 24 (*a*), is adapted to spans of from 40 to 50 feet and is much used in timber and composite construction for church roofs. It is especially applicable to the usual Gothic treatment of such buildings, for it is ornamental in form, and the several spaces into which it is divided by the members readily lend themselves to Gothic details. This type of truss cannot be economically used unless the walls on which it rests are so buttressed or braced that they can provide a horizontal thrust, for in a truss lacking a vertical member subdividing the space *P*, Fig. 24 (*a*), it is imperative that this thrust should be provided for. If the walls are capable of resisting a vertical reaction only, the apex of the truss will tend to rise and the distortion shown in Fig. 25 will take place; that is, providing the stiffness of the joints, or the resistance of the rafter members to transverse stress at *a*, *a* is not

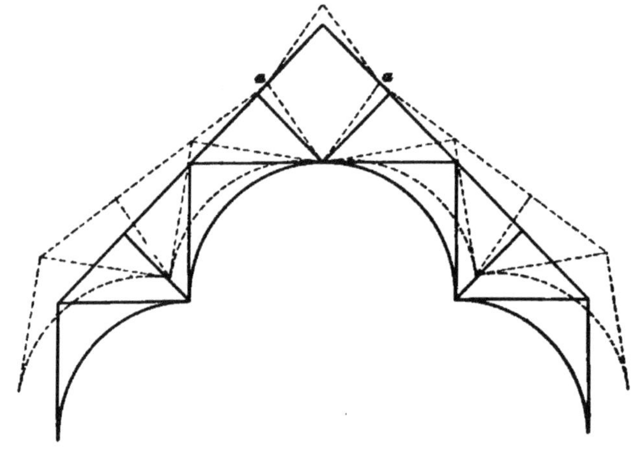

FIG. 25

sufficient. In view of these facts, this truss is usually adapted for the roofs of buildings, such as shown in the cross-section in Fig. 26. Here the truss spans the aisle of the church and rests on the arcades *a*, *a* at either side. The cloisters are covered by a sloped roof *b*, *b*, and their walls are usually buttressed, so that the roofs *b*, *b* provide the thrust necessary to hold the main roof truss in equilibrium. The construction is pleasing, for it is not unusual to provide a clearstory between the straight portions *c*, *c*, of the main truss, the lights and shadows obtained by this arrangement being particularly adaptable to the architectural treatment of a Gothic edifice.

While this horizontal thrust can be calculated by the principle of moments, it is more readily determined by changing the position of some of the members in the truss on the principle of the reversal of the diagonal. It is evident that if only a vertical reaction is supplied, as at *A Z*, Fig. 24 (*a*), on the analysis of the joint there will be three unknowns consisting of *A K*, *K Z*, and the horizontal thrust H_1. The diagonal *L K* can be reversed and will then assume the position shown by the member *B L* in Fig. 24 (*b*). Likewise,

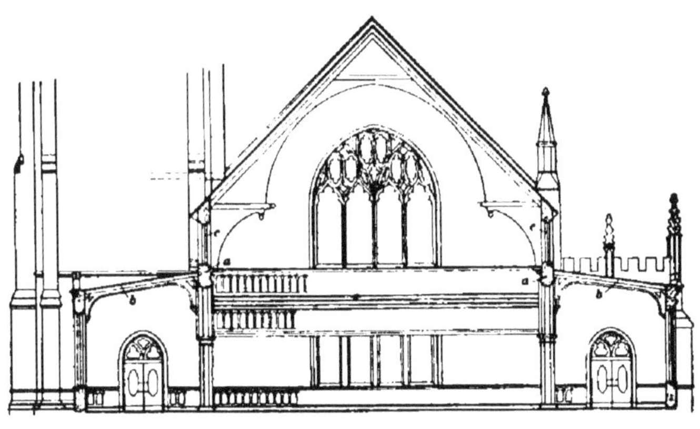

FIG. 26

M N may be reversed and assume the position of *M L* in (*b*). In changing the diagonals in this manner, it leaves the strut *L M* in Fig. 24 (*a*) acting on two members represented by the dotted lines *K Z* and *N Z*, coincident in their line of action, and it is evident that there will be no stress in the strut *1–2* shown dotted in (*b*), for, by the principles explained in Art. **20**, it was shown that where the lines of action of two opposing forces coincide, the introduction of a third force will destroy the equilibrium. This member, marked *L M* in (*a*) and *1–2* in (*b*), may therefore be omitted from the new diagram.

According to the same reasoning *A B*, in Fig. 24 (*a*), is opposed by a direct equivalent compressive stress in *A K*, whose line of action coincides with the line of action of *A B*. When the member *L K* is reversed, as shown in (*b*), the introduction of *B L* in (*a*) tends to destroy the equilibrium and consequently its stress is zero. The new diagram is then completed, as shown, by the solid lines in Fig. 24 (*b*), and there remains to be determined the amounts and directions of the reactions R_1 and R_2 in order that they may be introduced in the old diagram (*a*).

The reciprocal stress diagram for the frame diagram shown in Fig. 24 (*b*) is shown in Fig. 24 (*c*). In laying out this diagram the load line *a j* is laid out as usual, but since there are three unknowns at the lower joint of the truss, the diagram must be commenced at another panel point. An inspection shows that the apex joint *E F P* contains only two unknown forces, and that when *E P* is found the joint at *D E P O* contains only two unknowns. Therefore, it is best to commence the diagram at the apex of the truss. Draw, in the stress diagram, lines *c p* and *f p* parallel, respectively, with the members they represent in the frame diagram, thus locating the point *p*. Then draw the stresses for the joints *D E P O* and *C D O M*; *O M* having been obtained in this manner, there will exist only two unknowns at the joint *B C M L*, and this joint may be analyzed in the stress diagram. Having obtained the stress in the member *B L*, there exist only two unknowns at the left-hand foot of the truss. A line, therefore, drawn from *l* parallel with the member *L Z* will intersect at a point *z* with a horizontal line dividing the vertical load line centrally, for the truss is symmetrically loaded. The point *z* having thus been obtained, by connecting *z* and *a* the reaction R_1 is found in direction and amount. By completing the other half of the diagram the reaction R_2 may be obtained. It will be found, however, that under a symmetrical vertical load and where the proper oblique reactions R_1 and R_2 have been supplied, the vertical component is equal to the sum of one-half the loads on the roof truss, while the horizontal thrust may be obtained from the stress diagram by measuring the horizontal line from the point z to its intersection with the vertical load line.

When the reactions have been obtained they may be placed in the stress diagram for the original truss, where they serve to locate the point z and then k by the line $z\,k$, after which no difficulty will be found in drawing this diagram, as shown in Fig. 24 (*d*).

The dotted lines shown in the frame diagram (*a*) represent members that must be substituted for the curved members, and the analysis of the substituted straight members will give the direct stress acting at the ends of the curved members. These curved members must be proportioned to withstand a bending moment equal to the direct stress in the assumed dotted member multiplied by the perpendicular distance from the line of action of the dotted member to the farthest point of the curve.

35. While it was found necessary to provide horizontal or oblique thrusts at the abutments in order to preserve equilibrium in the hammer-beam truss shown in Fig. 24 (*a*), if this truss had been provided with a vertical member dividing the upper space P, it would have been found that the truss was stable when vertical reactions alone were considered at the abutments. The frame diagram drawn with the vertical member in the upper space is shown in Fig. 24 (*e*); and the stress diagram can easily be drawn for the truss with vertical reactions alone. Before it is drawn, however, inspection will show that since all the loads and the reactions are vertical and the member $A\,K$ is coincident with the line of action of the reaction R_1, there will be no stress in the member $K\,Z$. The truth of this assumption is substantiated by the principle stated in Art. **20.**

This stress diagram, drawn to the same scale as the others, is shown in Fig. 24 (*f*). An inspection will show that the stresses in the several members have been greatly influenced, and that the truss designed with supports that offer only vertical resistance is not economical; besides, the slightest movement that might take place in the connections,

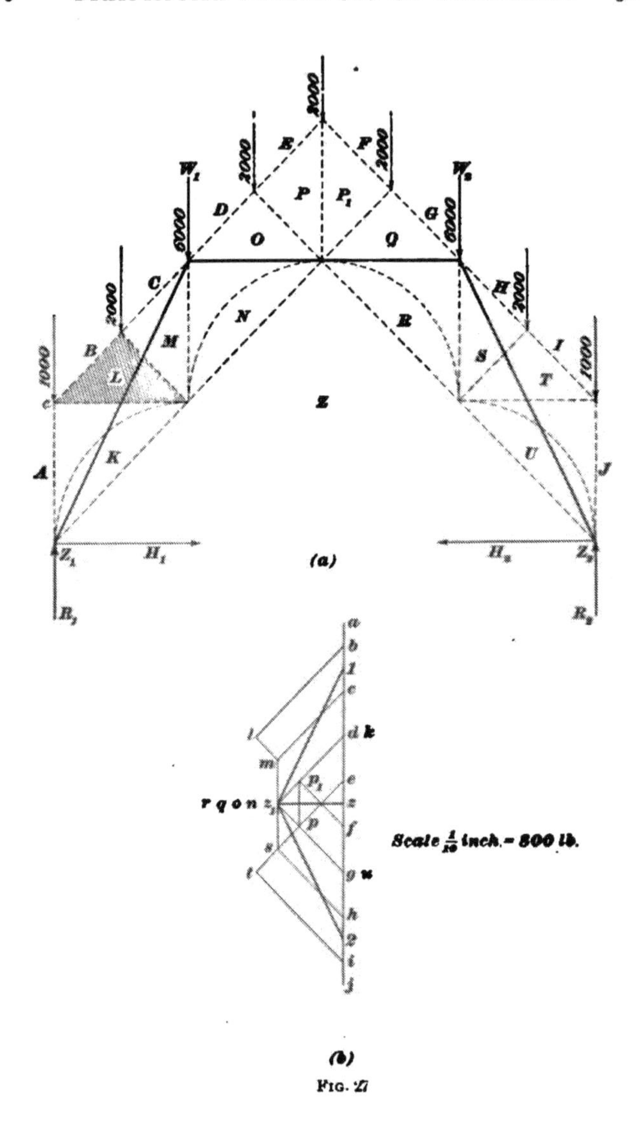

FIG. 27

which is likely to occur when the truss is constructed of timber, will produce a sagging, which distortion will add still more to the stresses already so greatly augmented.

36. These conditions decide that it is best, for economical reasons, to provide horizontal thrusts H_1 and H_2, shown at the foot of the truss in Fig. 27 (*a*). Before the stress diagram can be drawn for the truss, it is necessary to determine what these thrusts will be in order to produce equilibrium in the truss, when the reactions R_1 and R_2 are assumed as each being equal to one-half of the load on the truss. The framed structure furnished by the outline of the truss may be substituted by the trapezoidal frame shown by the heavy lines in Fig. 27 (*a*). This frame can be held in equilibrium by the loads W_1 and W_2, the reactions R_1 and R_2, and the thrusts H_1 and H_2. The loads W_1 and W_2 can each be taken as equal to the sum of the loads *CD* and *DE*, and one-half *EF*, as well as one-half *BC* in Fig. 24 (*a*). It is reasonable to assume these loads as acting at W_1 and W_2 for the upper portion of the hammer-beam truss, that is, that portion of the truss above the horizontal member extending between the two rafter members, is practically a truss of small dimensions superimposed on the other and supported at the joints *CD* and *GH*. It is also a correct assumption to make, that one-half of the forces *BC* and *HI* are conveyed down the dotted oblique members *LM* and *ST* to the vertical members *MN* and *SR*, from the fact that the shaded triangle is isosceles and the load *BC* is located at the apex, so that three-eighths of the load on the truss is concentrated at each point *CD* and *GH*. Commence the stress diagram, Fig. 27 (*b*), by laying off the load line, as usual, from *a* to *j*, marking on it the loads around the truss in the order in which they occur, as from *a* to *b* to *c* to *d*, etc. Since the truss is symmetrically loaded, the stress diagram will be symmetrical and the horizontal thrusts H_1 and H_2 will lie in a line $z_1 z$ bisecting the load line. Draw, therefore, from the points *1* and *2*, midway between *bc* and *hi*, lines parallel with *LK* and *TU* in (*a*). The evident reason for drawing these lines from the points *1* and *2* instead of from *b* and *i* is that in determining W_1 and W_2 only one-half of the loads *B C* and *H I* were considered. From the point z_1, where these lines intersect, the horizontal line $z_1 z$ is drawn. These several lines, together with the portion of the load line included between *1* and *2*, now form the polygon of external forces for the diagrammatical frame represented by the heavy lines and supporting W_1 and W_2 at its points of connection. The line $1 z_1$ represents the amount of the oblique reaction from the load W_1, or the stress in the member *L M*; $z_1 z$ is the amount of the horizontal thrust H_1, while $z\,1$ is the reaction for the frame represented by the heavy lines. The reaction R_1 of the original frame shown by the dotted lines is measured by the distance $z_1 a$ and the compressive stress, or thrust, in the horizontal member represented by the heavy line in the frame diagram is equal to the horizontal thrust at the foot of the oblique member, for the principle of equilibrium must be observed and the sum of the horizontal forces must equal zero. Likewise, the algebraic sum of the vertical forces must equal zero, in consequence of which W_1 and W_2 must equal the vertical reactions at the foot of the frame. From the fact that one-half the load *B C* can be considered as transmitted to the left-hand end of the shaded triangle, the load *A B* is really increased by this amount, and the reaction R_1 of the original frame, shown dotted, is equal to W_1 plus the load *A B* and one-half of the load *B C*. The question might arise as to the correctness of taking $z_1 z$ as representing the entire horizontal thrust of the original frame, but reference to (*a*) shows that by considering one-half of *B C* as transmitted to a joint in a vertical line with *C D* and the other half to the joint *A B L*, that the moments about the point *c* have not been changed. When the point z_1 has been determined, the stresses for the members in the original truss can be laid out without difficulty, except when the joint *C D O N M* is to be analyzed. Here it will be found that the member *O N* has no stress, for the points *o*, *n*, *q*, and *r* coincide. This will undoubtedly be confusing from the fact that it was stated that the horizontal member shown by the heavy line in the frame diagram was subject to a compressive stress equal to H_1, but this stress is equalized when the small framed truss, superimposed on the

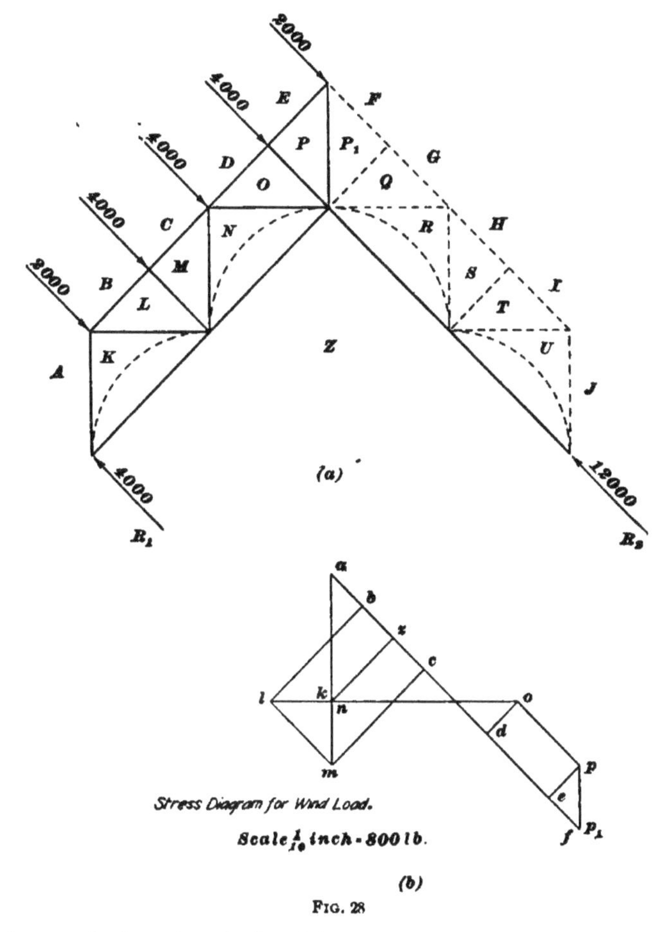

Fig. 28

larger, is introduced, for this truss creates a side thrust at the points of support of W_1 and W_2 equal to the horizontal compression in the member. It is, however, in an opposite direction and completely neutralizes this stress. It is also found, in drawing the stress diagram, that there is no stress in the members *NZ* and *RZ*. There cannot be stresses in these members from vertical and symmetrical loads, but stresses will be created when the truss is subjected to an oblique wind pressure or when the truss is unsymmetrically loaded. The members *NZ* and *RZ* could not, however, be well omitted even for a vertical load, for they must be used to stiffen the truss and to provide against any slight variation that might exist between the loads on opposite sides of the truss.

The frame and stress diagrams for the wind load on this truss are shown in Fig. 28 (*a*) and (*b*). Such a truss is always constructed with ends rigidly fixed, so that the reactions from an oblique wind pressure will coincide with the direction of the forces acting on the rafter members. The amount of these reactions may be determined by the graphical method or they may be calculated by the principle of moments. By reference to Fig. 28, it will be noticed that most of the members in the frame diagram are shown dotted. These members are useless when the wind acts on the left-hand side and when the reactions act in a direction coinciding with the lines of action of the oblique wind pressure on the slope.

ROOF TRUSSES WITH MEMBERS SUBJECTED TO TRANSVERSE STRESS

THE "A" TYPE OF TRUSS

37. In Fig. 29 is shown the **A type of roof truss**, which is particularly adapted for buildings where Gothic architectural treatment is desired; it is most frequently used for high narrow buildings, such as office buildings, as an assembly hall is often required on the top floor. In order that the ceiling may be raised and a vaulted, or arched, effect obtained, it is necessary to raise the tie-member of the truss some distance above the plate line at the top of the wall. Since a member like the one shown dotted at *a* would interfere with the cove, or vaulted, effect of the ceiling, such a member is not permissible; the member *b* must be made of heavy rolled shapes or of a compound riveted section in order that it may not only offer sufficient resistance to the direct stresses, but that it may, as well, provide the necessary transverse strength. That this member is subjected to a great transverse stress and tends to turn about the point *c* when subjected to the upward tendency of the reaction R_1, is evident on inspection of the frame. If the foot of the member *b* and the one on the opposite side were secured to the top of a heavy masonry wall capable of resisting a horizontal thrust, these members would sustain only direct stresses, and the stresses throughout the truss would be reduced to a minimum.

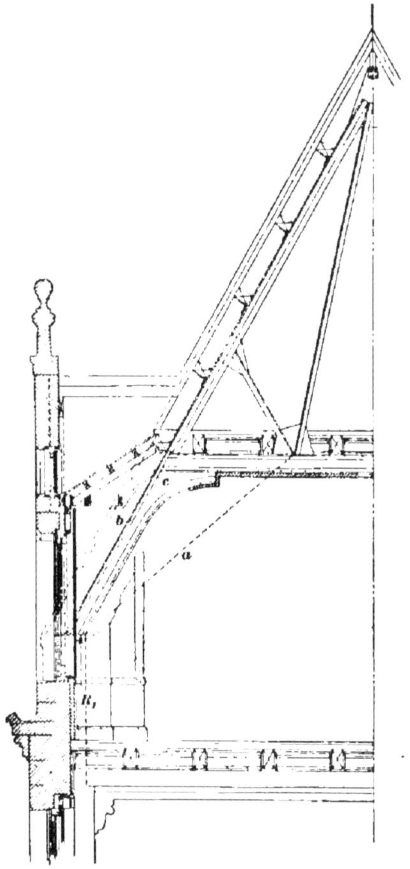

FIG. 29

38. In the frame diagram shown in Fig. 30 (*a*), H_1 and H_2 are readily determined by taking moments about the joints $B\,C_1\,C\,I\,Z$ and $M\,F\,F_1\,G\,Z$. For instance, the positive moment about the left-hand joint is $1{,}000 \times 6 = 6{,}000$ foot-pounds. The negative moment about the same point is

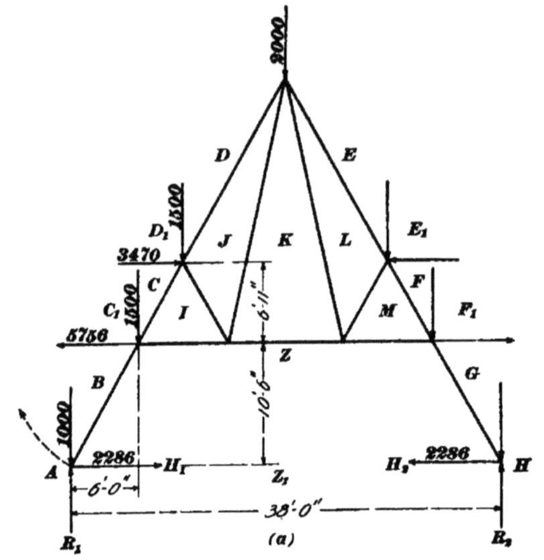

(a)

Frame Diagram-Vertical Load

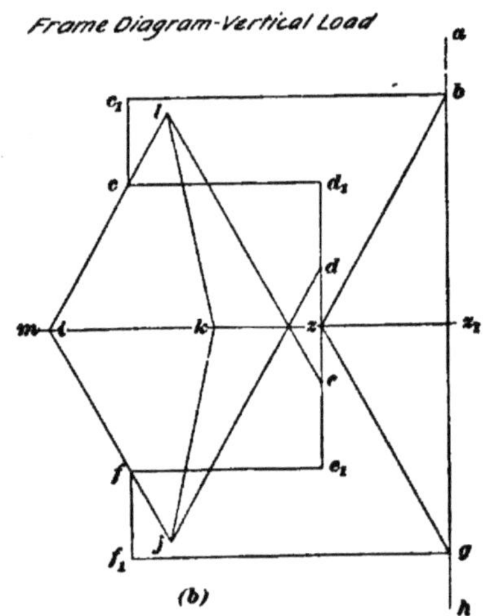

(b)

Stress Diagram-Vertical Load for Horizontal Thrust Supplied by Transverse Resistance of Leg

Scale 400 lb. to $\frac{1}{16}$ inch

FIG. 30

equal to the reaction R_1, or $5{,}000 \times 6 = 30{,}000$ foot-pounds. The difference between the positive and negative moments about the joint will then be $30{,}000 - 6{,}000 = 24{,}000$ foot-pounds. The leverage of the thrust H_1 is, according to the dimensions on the frame diagram, 10 feet 6 inches, so that this thrust will amount to $24{,}000 \div 10.5 = 2{,}286$ pounds.

By substituting the horizontal forces H_1 and H_2, not only will the frame diagram be in equilibrium, but the members BZ and ZG will have no tendency to spread, and there will be no transverse stresses about the joints BC_1CIZ and MFF_1GZ, the two legs of the truss being subjected only to direct stress. These two necessary horizontal thrusts are introduced in the frame diagram so that the stress diagram, Fig. 30 (*b*), may be drawn without difficulty. Trusses of this character are often erected on structural steel columns, or they may be placed on the light upper walls of a building that could not offer the necessary horizontal resistance. When this construction is encountered, the members BZ and GZ must resist the tendency to spread or to revolve about their junction with the frame, by their transverse stress. But this character of stress cannot be represented in the graphical analysis of a framed structure in connection with direct stresses. Therefore, this transverse stress must be replaced by direct stresses without destroying the equilibrium of the frame or any of the joints.

39. Referring to Fig. 30 (*a*), it is evident that the member BZ will tend to revolve about the joint BC_1CIZ, in the direction indicated by the dotted arrow. This rotary tendency may be resisted in two ways—either by the transverse strength of the member BZ or by the horizontal force H_1.

The latter method must be adopted in order to analyze the frame by the method of graphical statics; by the introduction of the imaginary forces H_1 and H_2, there is a tendency to destroy the rotary equilibrium about the joints BC_1CIZ and MFF_1GZ. This tendency to rotate must be overcome by the introduction of the forces CD_1 and FE_1, the amounts of which are readily calculated by taking moments about the joints BC_1CIZ and MFF_1GZ; for instance, the moment of H_1 about the joint BC_1CIZ is equal to 2,286 multiplied by the perpendicular distance from its line of action to the point of rotation, or 10.5 feet, which equals about 24,000 foot-pounds. The lever arm with which the horizontal force CD_1 tends to resist this rotary tendency about the joint BC_1CIZ is equal to 6 feet 11 inches, so that the amount of the force CD_1 is found by dividing 24,000 by 6.9167, or approximately 3,470 pounds. This force tends to produce rotation around the joint BC_1CIZ in the opposite direction to that created by the thrust H_1; consequently, they act in the same direction, and in order to produce translatory equilibrium a horizontal reaction equal to their sum must be introduced at BC_1CIZ, which will be equal, in amount, to 2,286 + 3,470 = 5,756. Since the truss is symmetrically loaded, similar forces must necessarily be introduced at the joints LEE_1FM and FF_1GZM, in order to create equilibrium in that half of the truss subjected to the horizontal thrust H_2. The several forces having been obtained in this manner, the frame diagram with its external forces will be as shown in Fig. 30 (*a*).

40. Commence the stress diagram by laying out the load line, which extends from a to b to c_1 to c to d_1 to d to e to e_1, horizontally to f, vertically to f_1, horizontally to g, then to h and from h to z_1 in a vertical direction corresponding in amount with the reaction R_2, to z, back to z_1, and from z_1 back to a, the starting point, this last line representing the amount of the reaction R_1. The polygon of external forces closes in this manner, and all the lines in the load line represent truly, in amount and direction, the forces that they represent in the frame diagram. When the load line has been completed, commence with the joint $ABZZ_1$ and draw, in the stress diagram, a line from b parallel with BZ in the frame diagram; where this line intersects a horizontal line drawn from z_1, which represents the thrusts H_1 and H_2, will be located the point z. The forces about the lower left-hand foot of the truss will then be represented in the stress diagram by the polygon extending from a to b to z to z_1 and from z_1 back to a.

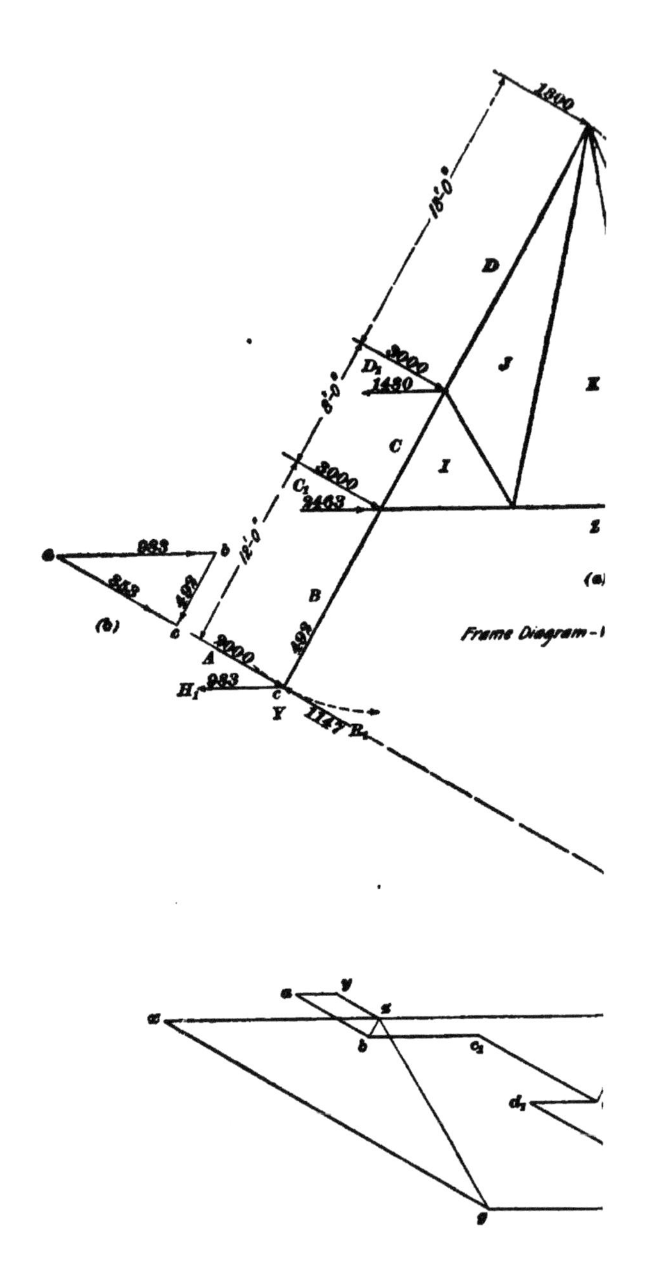
1800
15'-0"
D
3000
D1
1480
8'-0"
J
K
C
I
3000
C1
2463
983
b
852
492
12'-0"
(b)
c
B
(a)
Frame Diagram
3000
A
492
983
H1
c
Y
1147
y
a
x
z
b
c1
d1
g

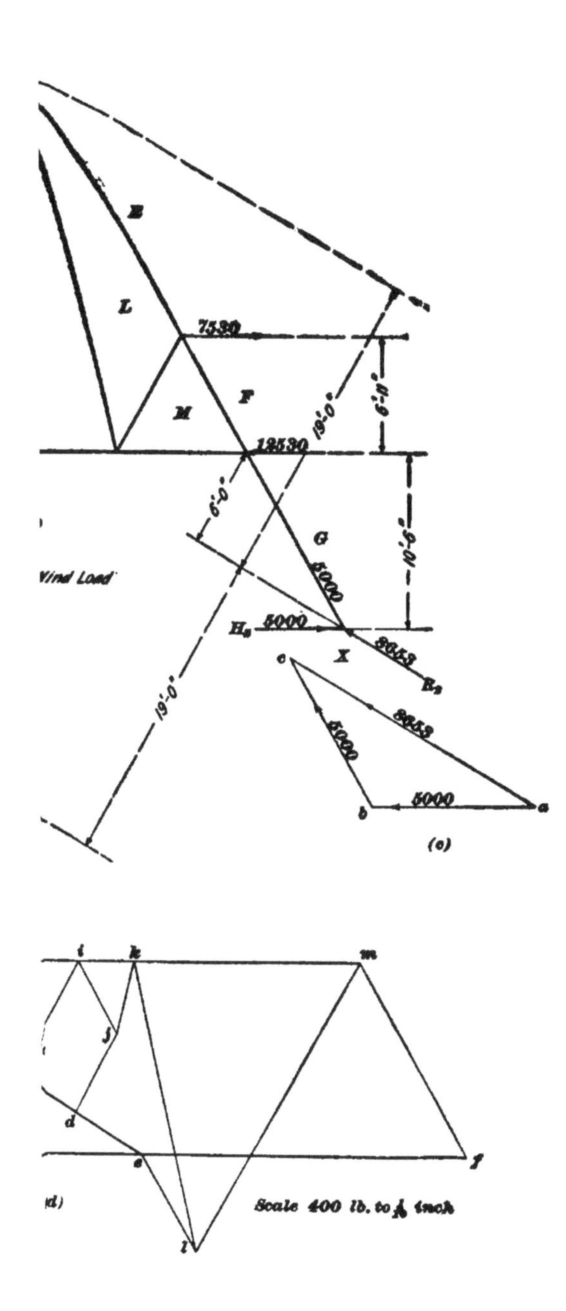

(c)

(d) Scale 400 lb. to $\frac{1}{8}$ inch

In the next joint, $BC_1 CIZ$, three of the forces are known and the polygon of forces is completed by drawing from c a line parallel with ci indefinitely, and then extending the horizontal line $z_1 z$ until it intersects the first line; the point of intersection will be i. The polygon of forces about this joint will read in the stress diagram from b to c_1 to c to i to z, and from z back to the starting point b.

The joint $IJKZ$ cannot as yet be analyzed, from the fact that there are more than two unknown forces in the system surrounding it, so that it is necessary to draw the polygon of forces for the joint $CD_1 DJI$. To analyze this joint, draw, from the point d, a line, indefinite in length, in the direction of the member DJ; and, from i, draw a line parallel with the member IJ in the frame diagram, intersecting the first line drawn at j. The polygon of forces about this joint will then be completed and $IJKZ$ may be analyzed, thus completing one-half of the stress diagram. As the loads are symmetrically placed on the truss and the truss is symmetrical about the center line, the stress diagram for the entire truss will be balanced on the horizontal line $z_1 i$. Since the stress in the member MZ is the same as in the member IZ, m and i will fall at the same point.

41. The problem of determining the stresses in roof trusses of this character becomes somewhat more complicated when the wind stress diagram is considered. The principle of equalizing the bending moment by the introduction of certain forces in the frame diagram is exactly the same, however. If the frame diagram for the wind loads on the left-hand portion of the truss is as shown in Fig. 31 (*a*), the external loads acting on the truss are AB, $C_1 C$, $D_1 D$, and DE. These wind loads produce parallel reactions at R_1 and R_2, the amounts of which may be determined by taking moments about either foot of the truss. Taking moments about the joint $ABZY$, or the point c, and considering the loads as exerting a positive moment, the calculation for obtaining their sum will be as follows, the lever arms being obtained by scaling the drawing:

Moment of AB	$= 2,000 \times 0 =$	0	foot-pounds
Moment of C_1C	$= 3,000 \times 12 =$	36000	foot-pounds
Moment of D_1D	$= 3,000 \times 20 =$	60000	foot-pounds
Moment of DE	$= 1,800 \times 38 =$	68400	foot-pounds
	Sum of moments $=$	164400	foot-pounds

The leverage of the reaction R_2 acting parallel with the direction of the loads is 19 feet, so that its amount is $164,400 \div 19 = 8,653$ pounds. The amount of the reaction R_1 is equal to the difference between the sum of the loads and the amount of the reaction R_2, or 1,147 pounds. Both of these reactions, though establishing equilibrium in the frame, produce, or tend to produce, rotation about the joints BC_1CIZ and MFF_1GZ, so that rotary equilibrium is not maintained at these points.

In actual practice, these bending moments are resisted by the transverse strength of the members BZ and GZ at these points. For analytical purposes, however, it is necessary to substitute some counteracting force against this rotation at the foot of each leg. The amount and direction of these forces are readily determined by drawing the reaction diagrams, Fig. 31 (*b*) and (*c*). Before drawing the reaction diagram (*b*), an inspection of the joint $ABZY$ reveals the fact that there is a force of 2,000 pounds coincident and acting in the same straight line with the reaction of 1,147 pounds, but in the opposite direction. Under such a condition equilibrium of the joint cannot be maintained, for the force of 2,000 pounds will more than counteract the smaller force of 1,147 pounds. It is evident, then, that there is a tendency for the point c to move in a direction shown by the dotted arrows with a force equal to 853 pounds. This, then, is the force that actually tends to create bending about the joint BC_1CIZ and may be resolved into its components ab and bc, shown in the reaction diagram (*b*).

In order to analyze this force of 853 pounds, draw the line ac in the direction of R_1 and lay off, in the diagram (*b*), the amount of 853 pounds to scale, thus locating the points a and c. From c draw a line parallel with the member BZ and from a, a horizontal line; thus it is found that the components $a\,b$ and $b\,c$, acting in the directions shown, will be equal in effect to the single force $a\,c$. Of the two components just found, $b\,c$ is the stress in the member $B\,Z$, while the horizontal force $a\,b$ must be substituted in the frame diagram, as shown. The system of forces about the foot of the left-hand leg of the truss, which maintains equilibrium at this point, will now consist of $A\,B$, $B\,Z$, $Z\,Y$, and $Y\,A$.

The reaction R_2 must, in a similar manner, be analyzed into its components, which are opposite and equal in effect to its action on the foot of the right-hand leg of the truss. These components are shown in direction and amount in the reaction diagram (c), in which the stress in the member $G\,Z$ is represented by the line $c\,b$ and the horizontal thrust, or force, that it is necessary to introduce in the diagram in order to maintain equilibrium, is represented by the line $b\,a$. In this manner the system of forces necessary to create equilibrium at the lower right-hand joint of the truss is completed and consists of the forces $Z\,G$, $G\,X$, and $X\,Z$. The external forces necessary to create equilibrium in the frame and at each joint have now been introduced in the frame diagram, and the stress diagram shown in Fig. 31 (d) may be drawn. The load line is shown by the heavy line extending from the point a along the lines $a\,b$, $b\,c_1$, $c_1\,c$, $c\,d_1$, $d_1\,d$, $d\,e$, $e\,f$, $f\,g$, $g\,x$, $x\,z$, $z\,y$, and $y\,a$. In this manner the polygon of forces is closed on the starting point. The stress diagram may be commenced and worked out from either end of the diagram and offers little difficulty, giving the figure designated.

In designing the truss, the stresses obtained in the vertical-load and wind-load diagrams may be combined and in this manner the maximum stress in any member is determined. Likewise, the maximum bending moment at the joints $B\,C_1\,C\,I\,Z$ and $M\,F\,F_1\,G\,Z$ must be determined and the leg of the truss proportioned for the maximum moment. In this instance the maximum bending moment at $M\,F\,F_1\,G\,Z$ occurs under the wind load and is equal to 8,653 pounds multiplied by 6, or 51,918 foot-pounds. Practically the same result

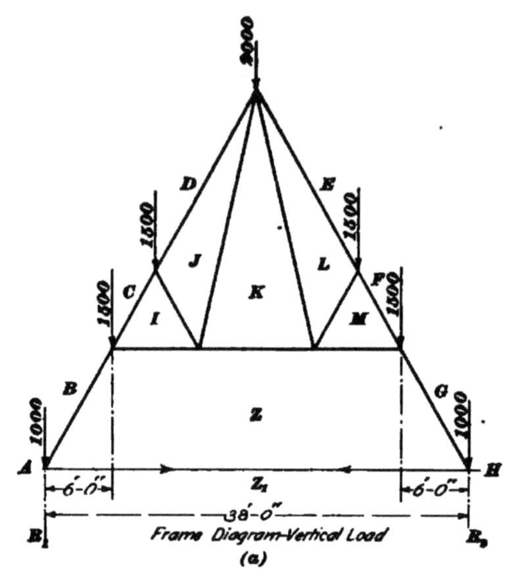
2000
1500
1500
1500
1500
1000
1000
D
E
J
L
C
F
K
I
M
B
G
Z
A
H
6'-0"
6'-0"
38'-0"
Frame Diagram-Vertical Load
(a)

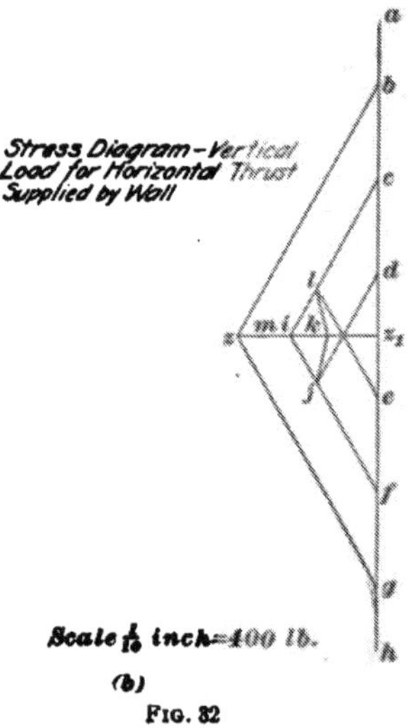
Stress Diagram-Vertical
Load for Horizontal Thrust
Supplied by Wall
a
b
c
d
e
f
g
h
Scale 1/16 inch=400 lb.
(b)

FIG. 82

would have been obtained by multiplying the horizontal force H_2 by its lever arm, which gives $5,000 \times 10.5 = 52,500$ foot-pounds.

Carefully study the difference between the stresses created in the several members of the frame diagram when the horizontal forces H_1 and H_2 in Fig. 30 (*a*) are provided for by the transverse resistance of the legs at the joints $B\,C_1\,C\,I\,Z$ and $M\,F\,F_1\,G\,Z$, and when the necessary horizontal thrusts are supplied by the lateral resistance of a wall, or other stable abutment, as shown in Fig. 32. On comparing the stress diagrams in Figs. 30 and 32, it will be noticed that the lines $b\,z$ are equal, showing that the direct stress in the members $B\,Z$ and $Z\,G$ is not altered by the introduction of the imaginary horizontal forces $Z_1\,Z$, $B\,C_1$, and $C\,D_1$ in Fig. 30 (*a*). Likewise, $c\,i$ remains the same, showing that the stresses in $C\,I$ and $F\,M$ are not altered; but the stresses in $D\,J$ and $E\,L$ are greatly increased, as well as those in $J\,K$ and $K\,L$. $I\,J$ and $L\,M$ also have a great additional stress when the truss is held in equilibrium by the transverse resistance of $B\,Z$ and $Z\,G$ instead of by the horizontal thrusts $Z\,Z_1$, in Fig. 32. The horizontal tie-member, represented by $I\,Z$, $K\,Z$ and $M\,Z$, is also subjected to a great increase in stress.

TRUSSES SUPPORTED BY KNEE-BRACED COLUMNS

42. The type of truss shown in Fig. 33 is most frequently used in the construction of train sheds, rolling mills, and similar buildings. It is usually of a great span, from 80 to 100 feet, and is supported at either end on steel columns. If the connection of the column to the roof truss at *a* and *b* were hinged and no member were introduced between the column and the roof truss, it is evident that the frame would be, theoretically, in equilibrium under a vertical load, but that when a wind pressure acted on the roof, either normal to the roof slope or horizontally against the vertical surfaces of the building, the columns or posts would tend to turn about their lower ends and that when the stiffness of the joints *a* and *b* was overcome, the destruction of the structure would result.

In order to provide against this tendency to lateral motion across the building, due to the wind pressure on the roof and side, knee braces *c* and *d* must be introduced between the column and the truss. These knee braces act in unison and brace the entire structure laterally from the fact that when the wind blows against the left-hand portion of the truss, the member *c* will be in tension, for the angle made by the column and the horizontal member of the truss at the left-hand end will tend to distort under the wind load and form an obtuse angle; while the member *d* will be in compression, for the angle

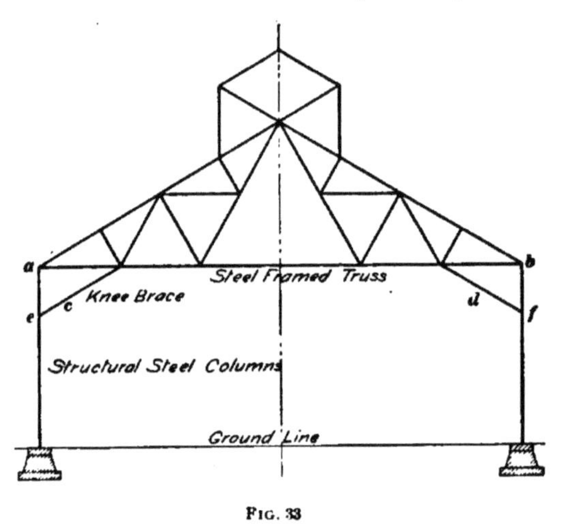

Fig. 33

formed by the column and the horizontal member of the truss at the right-hand end will tend to form an acute angle. If the joints *e* and *f* made by the connection of the knee braces with the columns were considered as being hinged, the structure would tend to fail at these points, as described in connection with the joint at the top of the column. It is therefore necessary to provide, in the column, a sufficient resistance to transverse stress to take up the pull and the thrust of the knee braces *c* and *d*. But bending moments and transverse stresses cannot be represented in the stress diagrams for obtaining direct stresses, and therefore some

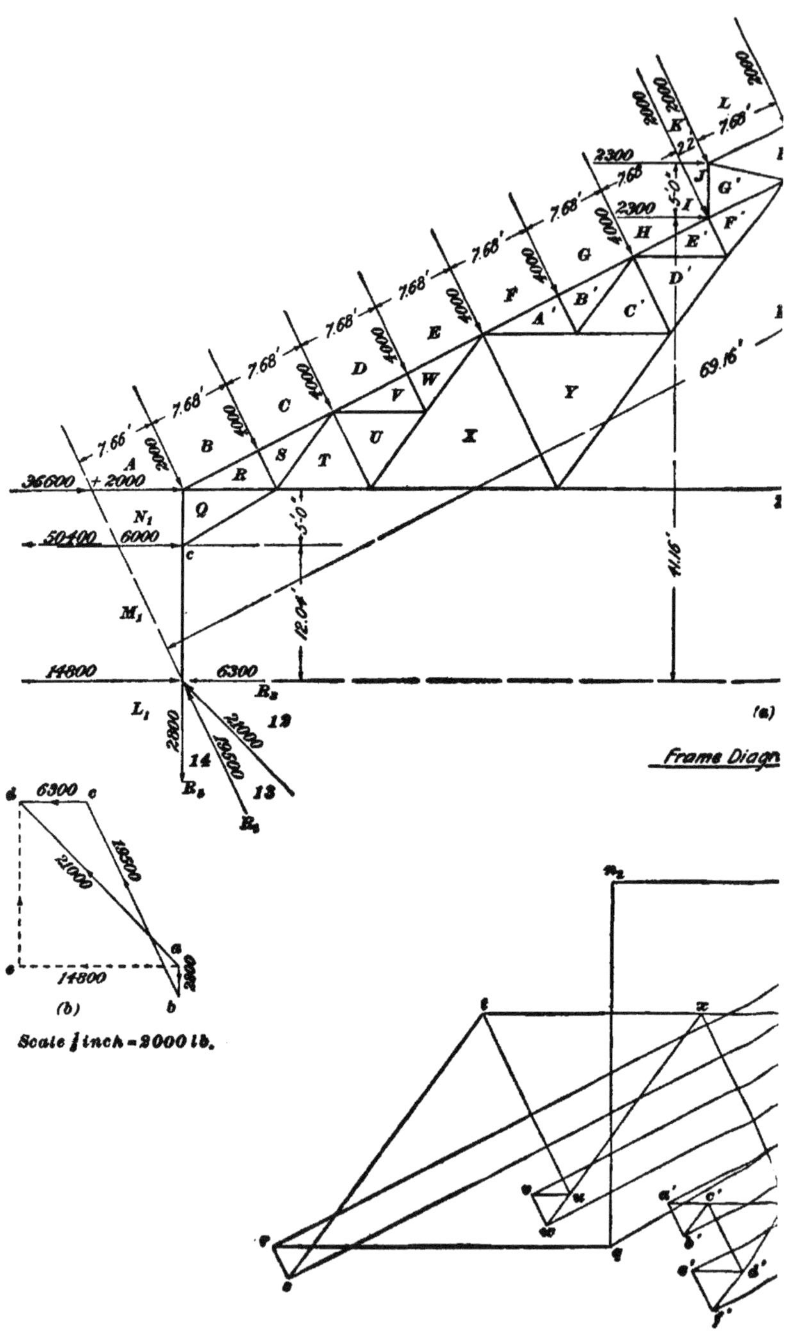

Frame Diagr

Stress L

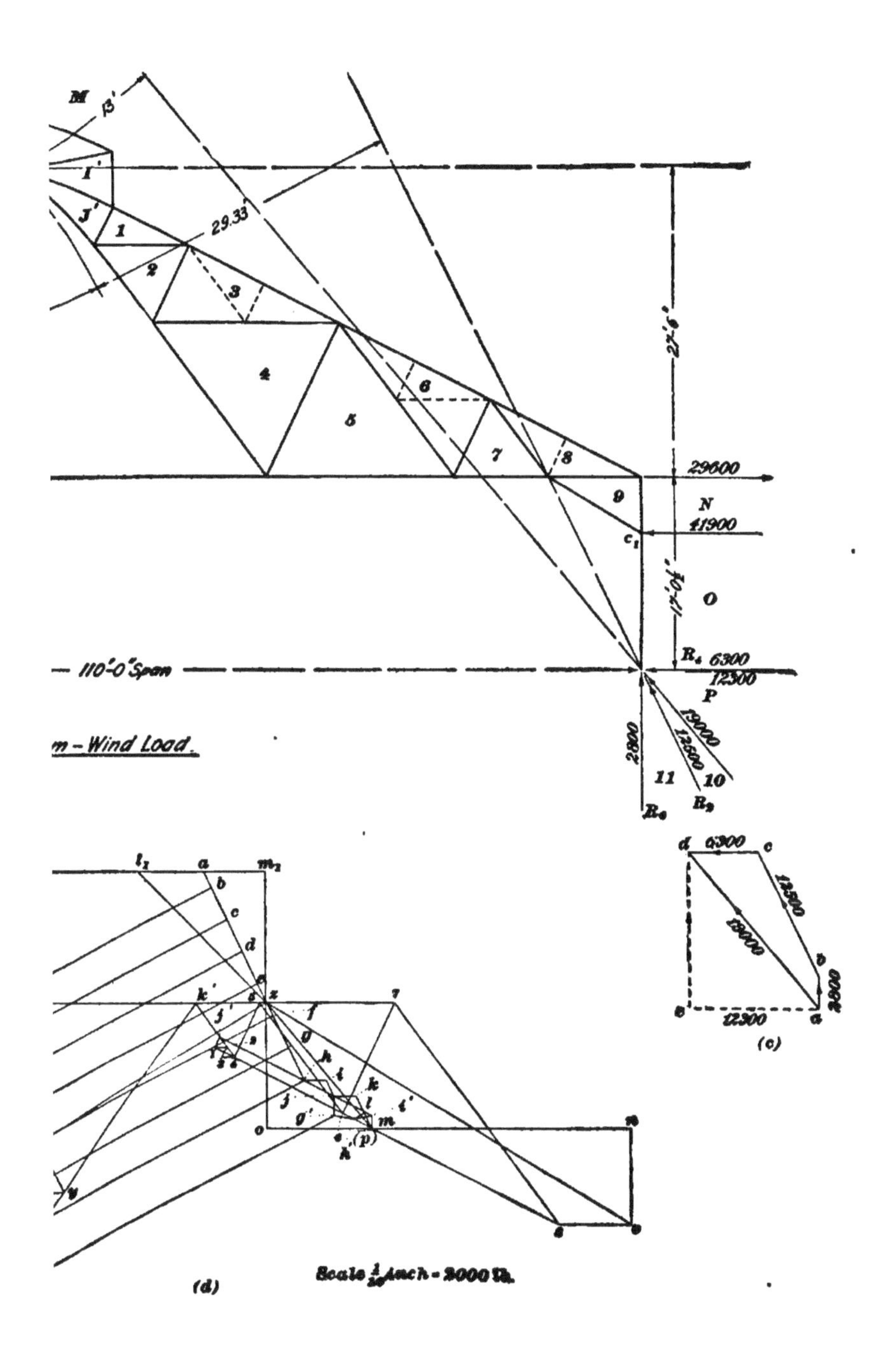

M
13'
29.33'
27'-6"
29600
N
41900
17'-0½"
O
R_4 6300
12300
P
110'-0" Span
m - Wind Load.
2800
13000
12500
11
10
R_4
R_3
d 6300 c
12500
13000
b
2800
e 12300 a
(c)
(d)
Scale $\frac{1}{16}$ inch = 3000 lb.

means must be introduced by which these transverse stresses are eliminated for the purpose of drawing the stress diagram. In eliminating these tranverse stresses, direct stresses must be substituted, as described in Arts. **38** and **39**, so that the frame will be retained in equilibrium and the stress in the members throughout the truss will not be altered.

43. In the frame diagram shown in Fig. 34 (*a*), it will be observed that the wind pressure acts normally to the slope of the oblique rafter member, and horizontally against the vertical surfaces. The oblique forces are AB, BC, CD, DE, EF, FG, GH, IJ, KL, LM, while the horizontal forces are $M_1 N_1$, $N_1 A$, HI, and JK. In order that the frame may be held in equilibrium, reactions must be provided at the foot of the columns, not only to resist the oblique forces but also to resist the horizontal forces acting on the truss. It is convenient, however, to determine these reactions separately and then to draw reaction diagrams for each end of the truss and thus obtain the resultant reactions. In order to determine the reaction acting at the foot of each column coincident with the line of action of the oblique forces and of such an amount as to balance these forces, take the moments of all the oblique forces about the foot of the left-hand column; the calculation will be as follows:

Moment of AB =	2,000 × 7.66 =	15320 foot-pounds
Moment of BC =	4,000 × 15.34 =	61360 foot-pounds
Moment of CD =	4,000 × 23.02 =	92080 foot-pounds
Moment of DE =	4,000 × 30.70 =	122800 foot-pounds
Moment of EF =	4,000 × 38.38 =	153520 foot-pounds
Moment of FG =	4,000 × 46.06 =	184240 foot-pounds
Moment of GH =	4,000 × 53.74 =	214960 foot-pounds
Moment of IJ =	2,000 × 61.42 =	122840 foot-pounds
Moment of KL =	2,000 × 63.62 =	127240 foot-pounds
Moment of LM =	2,000 × 71.30 =	142600 foot-pounds
Sum of positive moments =		1236960 foot-pounds

The reaction R, acts on the left-hand column foot with a lever arm equal to 98.49 feet, so that this reaction will equal, in amount, 1,236,960 ÷ 98.49, or, practically, 12,500 foot-pounds.

The sum of the oblique forces acting on the truss is equal to 32,000 pounds, and consequently R_2 equals 32,000 − 12,500 = 19,500 pounds. The horizontal forces acting on the truss tend not only to push the truss along laterally but to overturn the entire frame; that is, the forces M_1N_1, N_1A, HI, and JK tend to produce a rotary motion about the foot of the right-hand column that must be resisted by a pull in the left-hand column of the frame and a corresponding compression in the right-hand column.

The reactions R_3 and R_4 are horizontal and opposed to the direction of the horizontal wind pressure, but it is assumed that half of the tendency of the structure to move laterally is taken up at the foot of each column; consequently, R_3 and R_4 are equal in amount to one-half of the horizontal forces acting on the frame. The sum of these forces being 12,600 pounds, the horizontal reactions will equal 6,300 pounds. The vertical reactions R_5 and R_6, necessary to overcome the rotary tendency of the horizontal forces M_1N_1, N_1A, etc., and to produce equilibrium of rotation, may be found by taking moments about the foot of the left-hand column. The calculations for these reactions are as follows:

Moment of M_1N_1 = 6,000 × 12.04 = 72240 foot-pounds
Moment of N_1A = 2,000 × 17.04 = 34080 foot-pounds
Moment of HI = 2,300 × 41.16 = 94668 foot-pounds
Moment of JK = 2,300 × 46.16 = 106168 foot-pounds
Sum of positive moments = 307156 foot-pounds

The lever arm of R_6 is equal to the span of the truss, or the distance from center to center of the upright columns or posts. This distance is equal to 110 feet, so that the amount of R_6 is found by dividing 307,156 by 110, which gives the reaction as equal to, in round numbers, 2,800 pounds. The reaction R_5 must necessarily equal this amount in order that the algebraic sum of the vertical forces may be equal to zero.

There exist now at each joint three forces. At the left-hand end of the truss, there are R_1, R_3, and R_5, while at the right-hand end of the truss there are created R_2, R_4, and R_6.

In each system of forces the vertical and horizontal reactions are equal, but the amounts of the oblique reactions are different. In order to obtain the resultant of these three reactions, which may be substituted in their stead, it is necessary to draw the reaction diagrams shown in Fig. 34 (*b*) and (*c*). In the diagram (*b*), *a b* is laid off to scale, equal to 2,800 pounds, and represents, in direction and amount, the reaction R_1. From *b*, a line is drawn parallel with the reaction R_1 and is made of such a length that it will equal to scale 19,500 pounds, thus locating the point *c*. From the point *c*, a horizontal line drawn to *d* truly represents the horizontal force, or reaction, R_1 equal to 6,300 pounds. By connecting *a* and *d*, the amount and direction of the resultant of this system of forces are obtained, and the amount, on measuring, is found equal to 21,000 pounds. In consequence of the substitution of this resultant at the foot of the left-hand truss, the other forces of the system may be discarded, for this reaction is equal to them in effect on the truss. In a similar manner, the resultant reaction at the right-hand end of the truss may be obtained and will be found to equal, approximately, 19,000 pounds.

These oblique reactions can be resolved into their vertical and horizontal components, which are shown by the dotted lines in each of the reaction diagrams. The vertical components designated in each diagram by *d e* are the amounts of direct compressive stress in the columns; that is, by measuring *d e* in the diagram (*b*), the stress in the column $M_1 Z$ is obtained, while the length of the line *d e*, in the stress diagram (*c*), gives the amount of the stress in the column *Z O*. The horizontal component, however, in each case tends to bend the column or to turn the columns about the points *c* and c_1. Before the stress diagram can be drawn, this horizontal component of the reactions must be equalized and the imaginary forces $L_1 M_1$ and *P O* equal, respectively, to 14,800 and 12,300 pounds introduced in a direction opposite to that of the horizontal components *a e*. But these forces destroy the equilibrium of the frame diagram, so that forces must be introduced that will maintain both rotary and translatory equilibrium. Considering each column hinged at the point of connection with the knee brace and extending in a continuous piece from its foot to its connection with the truss, it is evident that the forces $L_1 M_1$ and OP will have a moment about the points c and c_1 equal to their amount multiplied by the length of the columns from the foot to the points c and c_1.

Forces applied as at $N_1 A$ and MN must resist this moment created by $L_1 M_1$ and OP, and since they have a lever arm equal to the length of the portion of the column $N_1 Q$ and $9 N$, or 5 feet, the amounts of the forces necessary to be substituted at the connection of the top of the columns with the truss may be found by dividing the moments of $M_1 L_1$ and OP about c and c_1 by the length $N_1 Q$ and $9 N$. The calculation for the forces $N_1 A$ and NM is therefore as follows:

Moment of $L_1 M_1$ about c equals $14{,}800 \times 12.04 = 178{,}192$ foot-pounds; and $178{,}192 \div 5 =$, in round numbers, 35,600 pounds, the amount of $N_1 A$. The moment of PO about c_1 equals $12{,}300 \times 12.04 = 148{,}092$ foot-pounds; and 148,092 $\div 5 =$, in round numbers, 29,600 pounds, which equals MN. These two forces introduced on each side of the truss as $N_1 A$ and MN maintain rotary equilibrium, but they destroy equilibrium of translation and, in consequence, must be opposed by an opposite and equal force so that it becomes necessary to introduce the two forces $M_1 N_1$ and ON. The left-hand force $M_1 N_1$ is equal to the sum of $L_1 M_1$ and $N_1 A$, while the right-hand force NO is equal to the sum of PO and MN. By the introduction of these last forces, the entire structure is in both rotary and translatory equilibrium and all the external forces acting on the truss are shown in the frame diagram.

44. The stress diagram shown in Fig. 34 (*d*) may now be commenced by first marking out the load line, which extends from *a* to *b* to *c* to *d* to *e* to *f* to *g* to *h*, horizontally to *i*, obliquely to *j*, horizontally to *k*, and obliquely to *l* to *m*; *m n* may then be laid off horizontally, for its amount and direction are known; no may be measured from n and the point o thus located. The point p will coincide with the point m, for the length of the line on is equal to the sum of nm and po, and since om has been laid off, the distance from m to o must equal the force op. An oblique line can be drawn from m extending indefinitely, but parallel with the resultant reaction represented by the force in the frame diagram, Fig. 34 (c), marked dc. In order to locate the point z, however, the line $a n_1$ must be laid off parallel with and equal to the force $N_1 A$ in the frame diagram. Then, from n_1, the point m_1 may be located by measuring off, to scale, the amount of the force $M_1 N_1$. Having located m_1, the point l_1 is readily located by measuring from m_1 the amount of the force $L_1 M_1$. When l_1 has been located, a line drawn from l_1 parallel with the force, or resultant reaction, $L_1 Z$, in the frame diagram, may be drawn, and where it intersects the oblique line $m z$, representing the right-hand resultant reaction, will be located the point z. The points o, m_1, and z should lie in a vertical line; $m_1 z$ is the amount of compression in the left-hand column, and oz is the amount of compression in the right-hand column, these two compressive stresses being equal, respectively, to the vertical components of the resultant reactions shown by the vertical dotted lines in the reaction diagrams, Fig. 34 (b) and (c).

The polygon of external forces has now been completed and extends from a to b to c to d . . . to l to m to n to o to p, obliquely to z, thence along the line coincident in direction with the line of action of the left-hand reaction to l_1; from l_1 to m_1 to n_1, back to a, the starting point. When the polygon of forces thus closes in a satisfactory manner, the stress diagram for the stresses in the members of the structure may be laid out. There are but two unknowns in the joint $M_1 N_1 Q Z$, so that no difficulty is encountered in analyzing this joint. Likewise, the joint $A B R Q N_1$ is readily solved and the stresses around the joints $B C S R$ and $Q R S T Z$ can be laid out without trouble. The next three joints, however, have three unknown forces, so that the method of symmetry must be used. By observation it can readily be seen that the member *VW* in the frame diagram is subject to the same stress as the member *RS*, both sustaining 4,000 pounds. If *VW* is known, therefore, the joint *DEWV* can readily be solved, for in the stress diagram the point *v* must be at such a distance from the line *tu*, the direction of which is known, that the line *vu*, when drawn in parallel with *VU* in the frame diagram, will intersect the line drawn from *t*, parallel to *TU*, at a point *u* that will be central, or midway, between the two lines *dv* and *ew*. By careful inspection in this manner, it is possible to work around the entire truss, and the finished diagram will be as shown. However, it will be found that the members shown dotted in the right-hand half of the truss are useless when the wind acts on the left-hand half of the truss. These members may therefore be neglected and the stresses obtained in the members marked *3-4*, *4-5*, *5-6*, *6-7*, *7-8*, and *8-9*. The last line of the stress diagram will be *9z*, and the diagram should close on the point *z* when *9z* is drawn parallel with the right-hand knee brace of the frame diagram, or the member *9z*.

The analysis of this truss is the same as that shown in Fig. 35. This illustration shows not only the wind-load diagram, but also the dead-load diagram. The framework of the truss is much simpler than that in the example described, though the latter is an excellent example of the application of the principles just described.

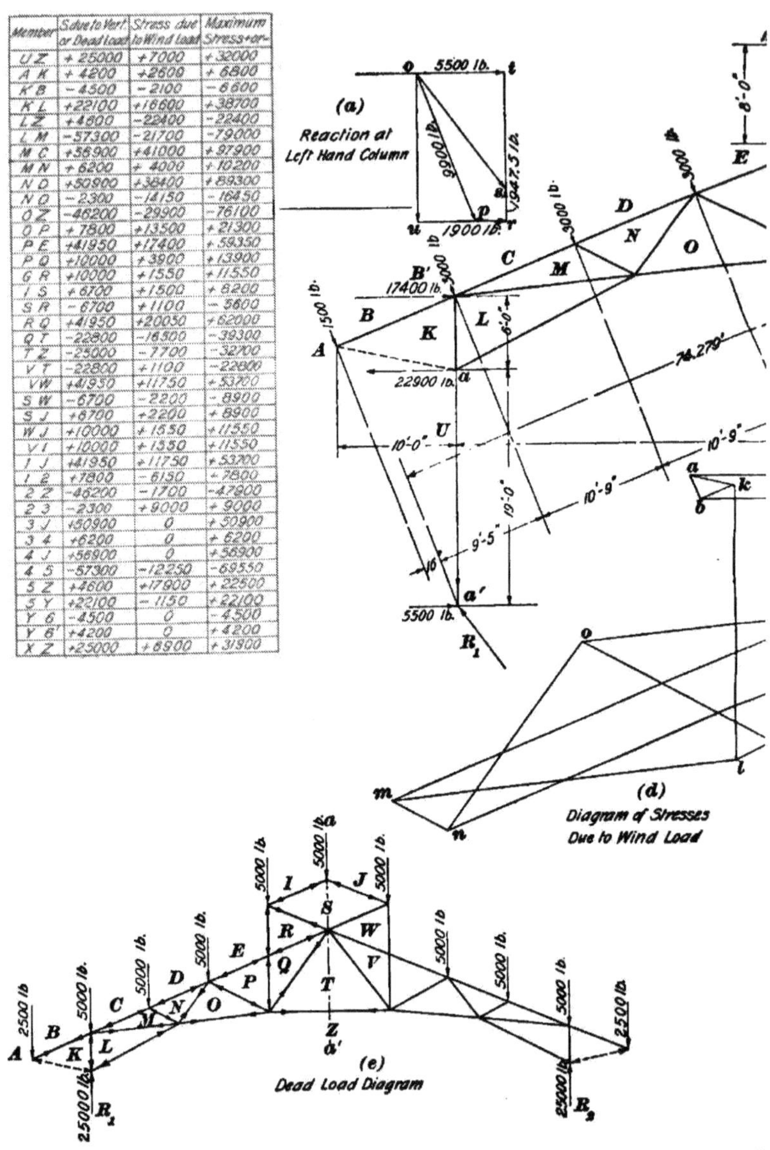

Member	S due to Vert. or Dead Load	Stress due to Wind Load	Maximum Stress + or −
U Z	+ 25000	+7000	+ 32000
A K	+ 4200	+2600	+ 6800
K B	− 4500	− 2100	− 6600
K L	+22100	+16600	+38700
L Z	+4600	−22400	−22400
L M	−57300	−21700	−79000
M C	+56900	+41000	+97900
M N	+ 6200	+ 4000	+ 10200
N O	+50900	+38400	+89300
N O	− 2300	−14150	− 16450
O Z	−46200	−29900	−76100
O P	+ 7800	+13500	+ 21300
P E	+41950	+17400	+ 59350
P Q	+10000	+ 3900	+13900
G R	+10000	+1550	+11550
I S	+ 6700	+1500	+ 8200
S R	− 6700	+1100	− 5600
R Q	+41950	+20050	+62000
Q T	−22800	−16500	−39300
T Z	−25000	− 7700	−32700
V T	−22800	+1100	−22800
V W	+41950	+11750	+ 53700
S W	− 6700	− 2200	− 8900
S J	+6700	+2200	+ 8900
W J	+10000	+ 1550	+11550
V I	+10000	+ 1550	+11550
I J	+41950	+11750	+53700
I 2	+7800	− 6150	+ 7800
2 Z	−46200	− 1700	−47900
2 3	−2300	+ 9000	+ 9000
3 J	+50900	0	+ 50900
3 4	+6200	0	+ 6200
4 J	+56900	0	+56900
4 5	−57300	−12250	−69550
5 Z	+4600	+17900	+ 22500
5 Y	+22100	− 1150	+22100
Y 6	−4500	0	− 4500
Y 6'	+4200	0	+ 4200
X Z	+25000	+ 6900	+ 31900

(a) Reaction at Left Hand Column

(d) Diagram of Stresses Due to Wind Load

(e) Dead Load Diagram

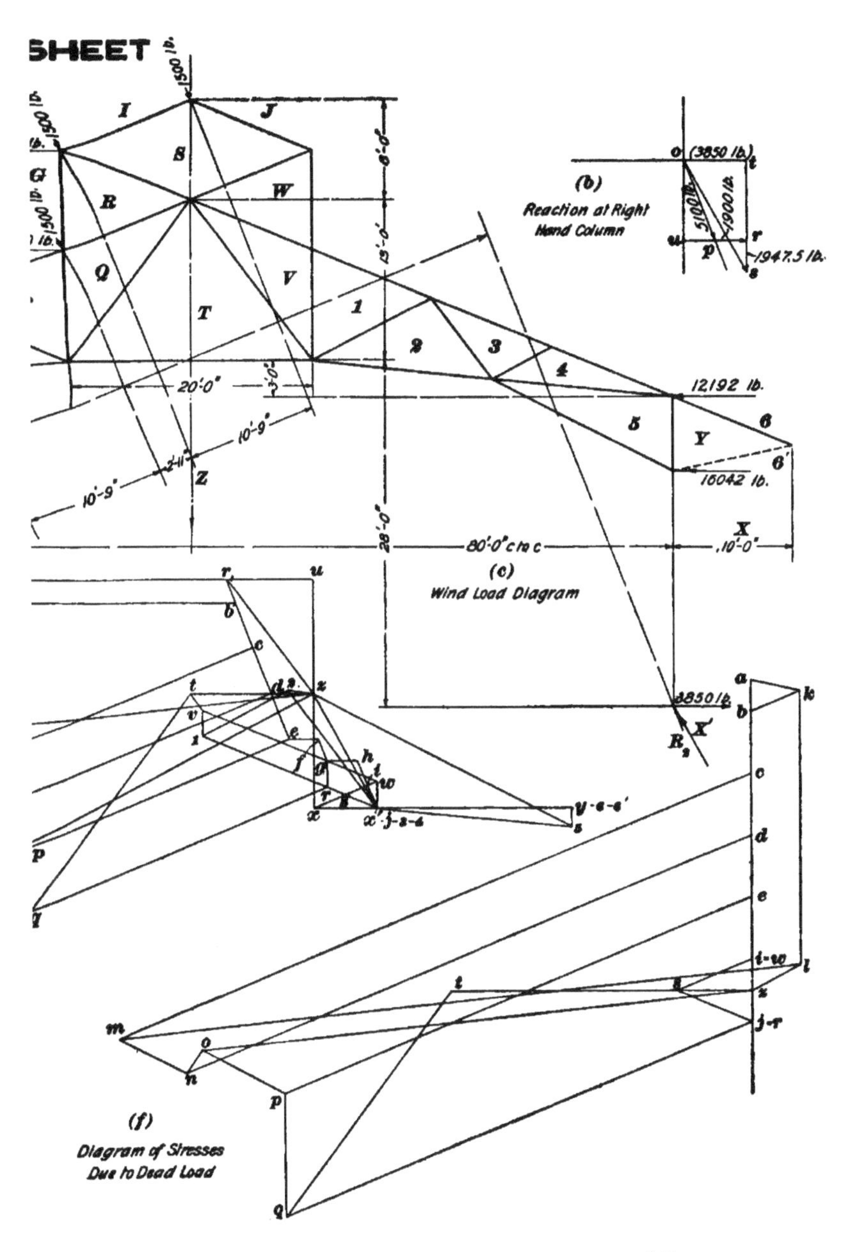
SHEET
1500 lb.
12192 lb.
16042 lb.
3850 lb.
20'-0"
10'-9"
10'-9"
2'-11"
3'-0"
8'-0"
13'-0"
28'-0"
80'-0" c to c
10'-0"
(b)
Reaction at Right
Hand Column
(3850 lb.)
5100 lb.
1900 lb.
1947.5 lb.
(c)
Wind Load Diagram
(f)
Diagram of Stresses
Due to Dead Load

INDEX

NOTE.—All items in this index refer first to the section and then to the page of the section. Thus, "Aluminum 8 9" means that aluminum will be found on page 9 of section 8.

N

O

P

Q

U

V

W

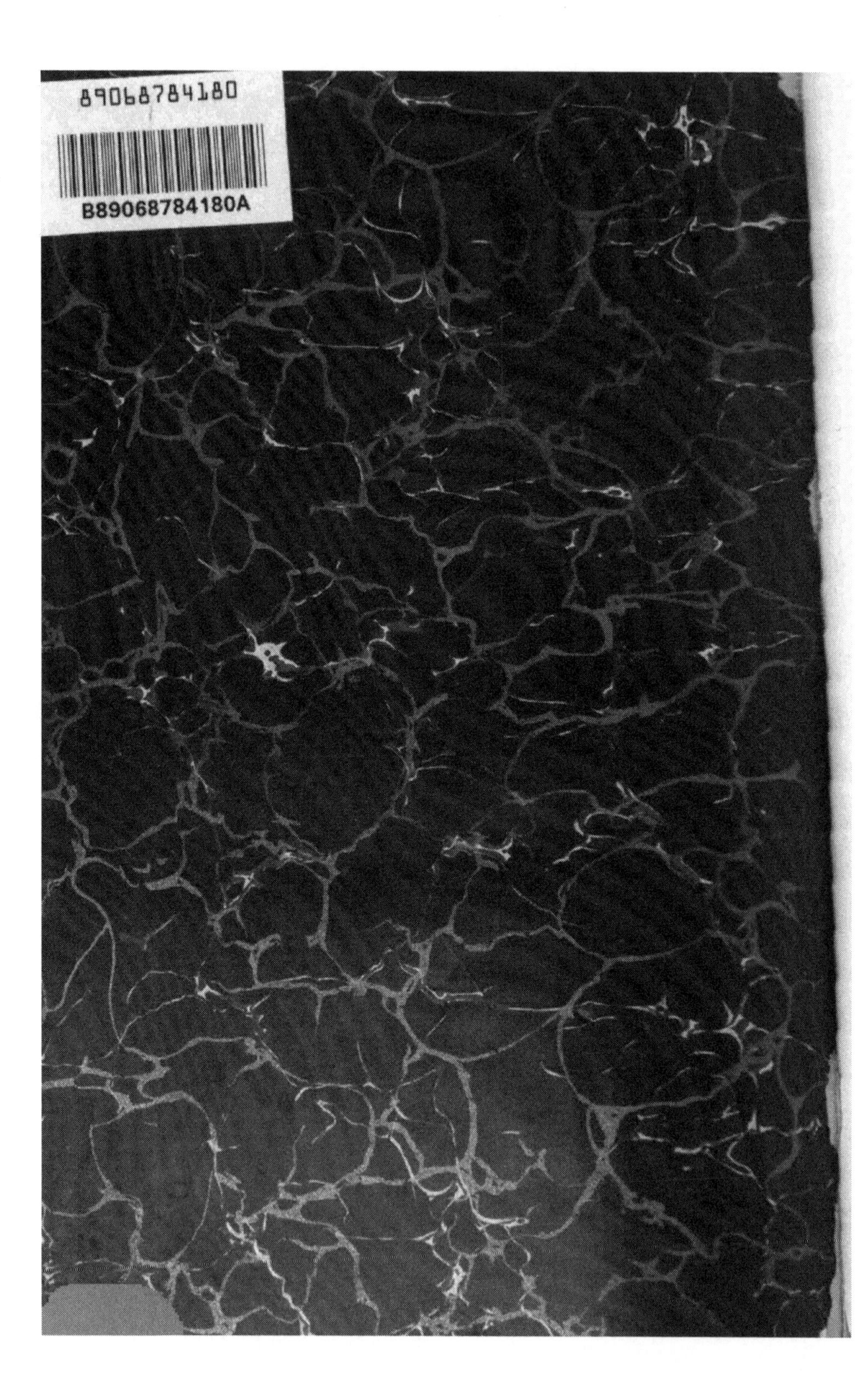